# McGRAW-HILL CONCISE ENCYCLOPEDIA OF CHEMISTRY

McGraw-Hill

New York Chicago San Francisco Lisbon London Madrid Mexico City
Milan New Delhi San Juan Seoul Singapore Sydney Toronto

The McGraw·Hill Companies



**Library of Congress Cataloging in Publication Data**

McGraw-Hill concise encyclopedia of chemistry.
p. cm.
Includes index
ISBN 0-07-143953-6
1. Chemistry—Encyclopedias. I. Title: Concise encyclopedia of chemistry.

QD4.M34 2004
540.3—dc22 2004049906



2 3 4 5 6 7 8 9 0 DOC/DOC 0 9 8 7 6 5

This book was printed on acid-free paper.

It was set in Helvetica Black and Souvenir by TechBooks, Fairfax, Virginia.

The book was printed and bound by RR Donnelley, The Lakeside Press.

# CONTENTS

# EDITORIAL STAFF

**Mark D. Licker,** Publisher

**Elizabeth Geller,** Managing Editor

**Jonathan Weil,** Senior Staff Editor

**David Blumel,** Editor

**Alyssa Rappaport,** Editor

**Charles Wagner,** Manager, Digital Content

**Renee Taylor,** Editorial Assistant

# EDITING, DESIGN, AND PRODUCTION STAFF

**Roger Kasunic,** Vice President—Editing, Design, and Production

**Joe Faulk,** Editing Manager

**Frank Kotowski, Jr.,** Senior Editing Supervisor

**Ron Lane,** Art Director

**Vincent Piazza,** Assistant Art Director

**Thomas G. Kowalczyk,** Production Manager

**Pamela A. Pelton,** Senior Production Supervisor

# CONSULTING EDITORS

# PREFACE

For more than four decades, the *McGraw-Hill Encyclopedia of Science & Technology* has been an indispensable scientific reference work for a broad range of readers, from students to professionals and interested general readers. Found in many thousands of libraries around the world, its 20 volumes authoritatively cover every major field of science. However, the needs of many readers will also be served by a concise work covering a specific scientific or technical discipline in a handy, portable format. For this reason, the editors of the *Encyclopedia* have produced this series of paperback editions, each devoted to a major field of science or engineering.

The articles in this *McGraw-Hill Concise Encyclopedia of Chemistry* cover all the principal topics of this field. Each one is a condensed version of the parent article that retains its authoritativeness and clarity of presentation, providing the reader with essential knowledge in chemistry without extensive detail. The authors are international experts, including Nobel Prize winners. The initials of the authors are at the end of the articles; their full names and affiliations are listed in the back of the book.

The reader will find over 700 alphabetically arranged entries, many illustrated with diagrams or images. Most include cross references to other articles for background reading or further study. Dual measurement units (U.S. Customary and International System) are used throughout. The Appendix includes useful information complementing the articles. Finally, the Index provides quick access to specific information in the articles.

This concise reference will fill the need for accurate, current scientific and technical information in a convenient, economical format. It can serve as the starting point for research by anyone seriously interested in science, even professionals seeking information outside their own specialty. It should prove to be a much used and trusted addition to the reader's bookshelf.

**MARK D. LICKER**
Publisher

# ORGANIZATION OF THE ENCYCLOPEDIA

**Alphabetization.** The more than 700 article titles are sequenced on a word-by-word basis, not letter by letter. Hyphenated words are treated as separate words. In occasional inverted article titles, the comma provides a full stop. The index is alphabetized on the same principles. Readers can turn directly to the pages for much of their research. Examples of sequencing are:

**Acid and base**
**Acid anhydride**
**Acid-base indicator**
**Acid halide**
**Alkali metals**
**Alkaline-earth metals**
**Alkylation (petroleum)**

**Cross references.** Virtually every article has cross references set in CAPITALS AND SMALL CAPITALS. These references offer the user the option of turning to other articles in the volume for related information.

**Measurement units.** Since some readers prefer the U.S. Customary System while others require the International System of Units (SI), measurements in the Encyclopedia are given in dual units.

**Contributors.** The authorship of each article is specified at its conclusion, in the form of the contributor's initials for brevity. The contributor's full name and affiliation may be found in the "Contributors" section at the back of the volume.

**Appendix.** Every user should explore the variety of succinct information supplied by the Appendix, which includes conversion factors, measurement tables, fundamental constants, and a biographical listing of scientists. Users wishing to go beyond the scope of this Encyclopedia will find recommended books and journals listed in the "Bibliographies" section; the titles are grouped by subject area.

**Index.** The 3500-entry index offers the reader the time-saving convenience of being able to quickly locate specific information in the text, rather than approaching the Encyclopedia via article titles only. This elaborate breakdown of the volume's contents assures both the general reader and the professional of efficient use of the *McGraw-Hill Concise Encyclopedia of Chemistry*.

**Absorption** Either the taking up of matter in bulk by other matter, as in the dissolving of a gas by a liquid; or the taking up of energy from radiation by the medium through which the radiation is passing. In the first case, an absorption coefficient is defined as the amount of gas dissolved at standard conditions by a unit volume of the solvent. Absorption in this sense is a volume effect: The absorbed substance permeates the whole of the absorber. In absorption of the second type, attenuation is produced which in many cases follows Lambert's law and adds to the effects of scattering if the latter is present.

Absorption of electromagnetic radiation can occur in several ways. For example, microwaves in a waveguide lose energy to the walls of the guide. For nonperfect conductors, the wave penetrates the guide surface and energy in the wave is transferred to the atoms of the guide. Light is absorbed by atoms of the medium through which it passes, and in some cases this absorption is quite distinctive. Selected frequencies from a heterochromatic source are strongly absorbed, as in the absorption spectrum of the Sun. Electromagnetic radiation can be absorbed by the photoelectric effect, where the light quantum is absorbed and an electron of the absorbing atom is ejected, and also by Compton scattering. Electron-positron pairs may be created by the absorption of a photon of sufficiently high energy. Photons can be absorbed by photoproduction of nuclear and subnuclear particles, analogous to the photoelectric effect.

Sound waves are absorbed at suitable frequencies by particles suspended in the air (wavelength of the order of the particle size), where the sound energy is transformed into vibrational energy of the absorbing particles.

Absorption of energy from a beam of particles can occur by the ionization process, where an electron in the medium through which the beam passes is removed by the beam particles. The finite range of protons and alpha particles in matter is a result of this process. In the case of low-energy electrons, scattering is as important as ionization, so that range is a less well-defined concept. Particles themselves may be absorbed from a beam. For example, in a nuclear reaction an incident particle $X$ is absorbed into nucleus $Y$, and the result may be that another particle $Z$, or a photon, or particle $X$ with changed energy comes out. Low-energy positrons are quickly absorbed by annihilating with electrons in matter to yield two gamma rays. [M.H.H.]

In the chemical process industries and in related areas such as petroleum refining and fuels purification, absorption usually means gas absorption. This is a unit operation in which a gas (or vapor) mixture is contacted with a liquid solvent selected to preferentially absorb one, or in some cases more than one, component from the mixture. The purpose is either to recover a desired component from a gas mixture or to rid the mixture of an impurity. In the latter case, the operation is often referred to as scrubbing.

When the operation is employed in reverse, that is, when a gas is utilized to extract a component from a liquid mixture, it is referred to as gas desorption, stripping, or sparging.

In gas absorption, either no further changes occur to the gaseous component once it is absorbed in the liquid solvent, or the absorbed component (solute) will become involved in a chemical reaction with the solvent in the liquid phase. In the former case, the operation is referred to as physical gas absorption, and in the latter case as gas absorption with chemical reaction. *See* GAS ABSORPTION OPERATIONS; UNIT OPERATIONS.

[W.F.F.]

**Accelerator mass spectrometry** The use of a combination of mass spectrometers and an accelerator to measure the natural abundances of very rare radioactive isotopes. These abundances are frequently lower than parts per trillion. The most important applications of accelerator mass spectrometry are in archeological and geophysical studies, as, for example, in radiocarbon dating by the counting of the rare carbon-14 (radiocarbon; $^{14}C$) isotope. *See* MASS SPECTROSCOPE.

The advantage of counting the radioactive atoms themselves rather than their decay products is well illustrated by radiocarbon dating, which requires the measurement of the number of $^{14}C$ atoms in a sample. The long half-life of 5730 years for $^{14}C$ implies that only 15 beta-particle emissions per minute are observed from 1 g of contemporary carbon. However, an accelerator mass spectrometer can be used to count the $^{14}C$ atoms at over 15 per second from a milligram sample of carbon. Consequently, accelerator mass spectrometry can be used to date samples that are a thousand times smaller than those that are dated by using the beta-particle counting method, and the procedure is carried out about 120 times faster.

For the study of many rare radioactive atoms, accelerator mass spectrometry also has the important advantage that there can be no background except for contamination with the species being studied. For example, significant interference with the beta-particle counting of radiocarbon from cosmic rays and natural radioactivity occurs for

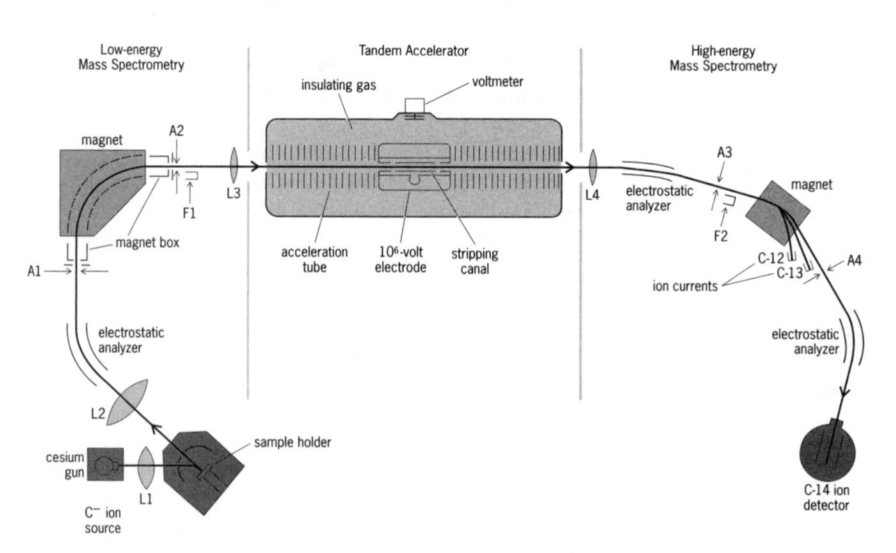


**Simplified diagram of an accelerator mass spectrometer used for radiocarbon dating. The equipment is divided into three sections. Electric lenses L1–L4 are used to focus the ion beams; apertures A1–A4 and charge collection cups F1 and F2 are used for setting up the equipment.**

carbon samples about 25,000 years old. In contrast, accelerator mass spectrometer measurements are affected only by the natural contamination of the sample which becomes serious for samples about 50,000 years old.

**Apparatus.** The success of accelerator mass spectrometry results from the use of more than one stage of mass spectrometry and at least two stages of ion acceleration. The illustration shows the layout of an ideal accelerator mass spectrometer for radiocarbon studies, divided for convenience into three stages.

The first part of the accelerator mass spectrometer is very similar to a conventional mass spectrometer. In the second stage, a tandem accelerator first accelerates negative ions to the central high-voltage electrode, converts them into positive ions by several successive collisions with gas molecules in a region of higher gas pressure, known as a stripping canal, and then further accelerates the multiply charged positive ions through the same voltage difference back to ground potential. In the third stage, the accelerated ions are analyzed further by the high-energy mass spectrometer.

**Distinguishing features.** The features that clearly distinguish accelerator mass spectrometry from conventional mass spectrometry are the elimination of molecular ions and isobars from the mass spectrometry.

A tandem accelerator provides a convenient way of completely eliminating molecular ions from the mass spectrometry because ions of a few megaelectronvolts can lose several electrons on passing through the region of higher gas pressure in the stripping canal. Molecules with more than two electrons missing have not been observed, so that accelerator mass spectrometry utilizing charge $-3$ ions is free of molecular interferences.

The use of a negative-ion source, which is necessary for tandem acceleration, can also ensure the complete separation of atoms of nearly identical mass (isobars). In the case of radiocarbon analysis, the abundant stable $^{14}$N ions and the very rare radioactive $^{14}$C ions are separated completely because the negative ion of nitrogen is unstable whereas the negative ion of carbon is stable. In other cases, it is possible to count the ions without any background in the ion detectors because of their high energy. In many cases, it is also possible to identify the ion. *See* MASS SPECTROMETRY.

[A.E.L.]

**Acetal** A geminal diether ($R_1 = H$). Ketals, considered a subclass of acetals, are also geminal diethers ($R_1 = C$, aliphatic or aromatic). Acetals are (1) independent structural units or a part of certain biological and commercial polymers, (2) blocking or protecting groups for complex molecules undergoing selective synthetic transformations, and (3) entry compounds for independent organic chemical reactions. *See* POLYACETAL.

Acetals are easily prepared by the reaction of aldehydes with excess alcohol, under acid-catalyzed conditions. This is usually a two-step process (see reaction below) in

$$\underset{\text{Aldehyde or ketone}}{R-\underset{\substack{|\\R_1}}{C}=O} \underset{H_2O}{\overset{R_2OH}{\rightleftharpoons}} \underset{\text{Hemiacetal or hemiketal}}{R-\underset{\substack{|\\R_1}}{\overset{\substack{OH\\|}}{C}}-OR_2} \underset{H_2O}{\overset{R_2OH}{\rightleftharpoons}} \underset{\text{Acetal or ketal}}{R-\underset{\substack{|\\R_1}}{\overset{\substack{OR_2\\|}}{C}}-OR_2}$$

which an aldehyde is treated with an alcohol to yield a less stable hemiacetal, which then reacts with additional alcohol to give the acetal. Protonic or Lewis acids are effective catalysts for acetal formation; dehydrating agents, such as calcium chloride and molecular sieves, can also be used for molecules, such as sugars, where acids may cause problems. Less common acetal preparations are Grignard reagent condensation with orthoformates and mercuric-catalyzed additions of alcohols to acetylenes. *See* ALDEHYDE; KETONE; ORGANIC SYNTHESIS. [C.F.B.]

**Acetic acid** A colorless, pungent liquid, $CH_3COOH$, melting at 16.7°C and boiling at 118.0°C. Acetic acid is the sour principle in vinegar. Concentrated acid is called glacial acetic acid because of its readiness to crystallize at cool temperatures.

Acetic acid is manufactured by three main routes: butane liquid-phase catalytic oxidation in acetic acid solvent, palladium–copper salt–catalyzed oxidation of ethylene in aqueous solution, and methanol carbonylation in the presence of rhodium catalyst. Large quantities of acetic acid are recovered in the manufacture of cellulose acetate and polyvinyl alcohol. Some acetic acid is produced in the oxidation of higher olefins, aromatic hydrocarbons, ketones, and alcohols. *See* OXIDATION PROCESS; WOOD CHEMICALS.

Pure acetic acid is completely miscible with water, ethanol, diethyl ether, and carbon tetrachloride, but is not soluble in carbon disulfide. In a water solution, acetic acid is a typical weakly ionized acid ($Ka = 1.8 \times 10^{-5}$). Acetic acid neutralizes many oxides and hydroxides, and decomposes carbonates to furnish acetate salts, which are used in textile dyeing and finishing, as pigments, and as pesticides; examples are verdigris, white lead, and paris green. [F.W.]

**Acetone** A chemical compound, $CH_3COCH_3$. A colorless liquid with an ethereal odor, it is the first member of the homologous series of aliphatic ketones. Its physical properties include boiling point 56.2°C (133.2°F), melting point −94.8°C (−138.6°F), and specific gravity 0.791.

Acetone is used as a solvent for cellulose ethers, cellulose acetate, cellulose nitrate, and other cellulose esters. Cellulose acetate is spun from acetone solution. Lacquers, based on cellulose esters, are used in solution in mixed solvents including acetone. [D.A.S.]

**Acetylcholine** A naturally occurring quaternary ammonium cation ester, with the formula $CH_3(O)COC_2H_4N(CH)_3^+$, that plays a prominent role in nervous system function. The great importance of acetylcholine derives from its role as a neurotransmitter for cholinergic neurons, which innervate many tissues, including smooth muscle and skeletal muscle, the heart, ganglia, and glands. The effect of stimulating a cholinergic nerve, for example, the contraction of skeletal muscle or the slowing of the heartbeat, results from the release of acetylcholine from the nerve endings.

Acetylcholine is synthesized at axon endings from acetyl coenzyme A and choline by the enzyme choline acetyltransferase, and is stored at each ending in hundreds of thousands of membrane-enclosed synaptic vesicles. When a nerve impulse reaches an axon ending, voltage-gated calcium channels in the axonal membrane open and calcium, which is extremely low inside the cell, enters the nerve ending. The increase in calcium-ion concentration causes hundreds of synaptic vesicles to fuse with the cell membrane and expel acetylcholine into the synaptic cleft (exocytosis). The acetylcholine released at a neuromuscular junction binds reversibly to acetylcholine receptors in the muscle end-plate membrane, a postsynaptic membrane that is separated from the nerve ending by a very short distance. The receptor is a cation channel which opens when two acetylcholine molecules are bound, allowing a sodium current to enter the muscle cell and depolarize the membrane. The resulting impulse indirectly causes the muscle to contract.

Acetylcholine must be rapidly removed from a synapse in order to restore it to its resting state. This is accomplished in part by diffusion but mainly by the enzyme acetylcholinesterase, which hydrolyzes acetylcholine.

Acetylcholinesterase is a very fast enzyme: one enzyme molecule can hydrolyze 10,000 molecules of acetylcholine in 1 s. Any substance that efficiently inhibits acetylcholinesterase will be extremely toxic. [I.B.W.]

**Acetylene** An organic compound with the formula $C_2H_2$ or HC≡CH. The first member of the alkynes, acetylene is a gas with a narrow liquid range; the triple point is −81°C (−114°F). The heat of formation ($\Delta H_f^\circ$) is +227 kilojoules/mole, and acetylene is the most endothermic compound per carbon of any hydrocarbon. The compound is thus extremely energy-rich and can decompose with explosive force. At one time acetylene was a basic compound for much chemical manufacturing. It is highly reactive and is a versatile source of other reactive compounds.

The availability of acetylene does not depend on petroleum liquids, since it can be prepared by hydrolysis of calcium carbide ($CaC_2$), obtained from lime (CaO), and charcoal or coke (C). In modern practice, methane ($CH_4$) is passed through a zone heated to 1500°C (2732°F) for a few milliseconds, and acetylene is then separated from the hydrogen in the effluent gas.

The main use of acetylene is in the manufacture of compounds derived from butyne-1,4-diol. The latter is obtained by condensation of acetylene with two moles of formaldehyde and is converted to butyrolactone, tetrahydrofuran, and pyrrolidone. Two additional products, vinyl fluoride and vinyl ether, are also based on acetylene.

Because of the very high heat of formation and combustion, an acetylene-oxygen mixture provides a very high temperature flame for welding and metal cutting. For this purpose acetylene is shipped as a solution in acetone, loaded under pressure in cylinders that contain a noncombustible spongy packing. *See* ALKYNE. [J.A.Mo.]

**Acid and base** Two interrelated classes of chemical compounds, the precise definitions of which have varied considerably with the development of chemistry. Some of these controversies are still unresolved.

Acids initially were defined only by their common properties as substances which had a sour taste, dissolved many metals, and reacted with alkalies (or bases) to form salts. For a time it was believed that a common constituent of all acids was the element oxygen, but gradually it became clear that, if there were an essential element, it was hydrogen, not oxygen. This concept of an acid proved to be satisfactory for about 50 years.

Bases initially were defined as those substances which reacted with acids to form salts (they were the "base" of the salt). The alkalies, soda and potash, were the best-known bases, but it soon became clear that there were other bases, notably ammonia and the amines.

When the concept of ionization of chemical compounds in water solution became established, acids were defined as substances which ionized in aqueous solution to give hydrogen ions, $H^+$, and bases were substances which reacted to give hydroxide ions, $OH^-$. These definitions are sometimes known as the Arrhenius-Ostwald theory of acids and bases. Their use makes it possible to discuss acid and base equilibria and also the strengths of individual acids and bases.

A powerful and wide-ranging protonic theory of acids and bases was introduced by J. N. Brønsted in 1923 and was rapidly accepted. Somewhat similar ideas were advanced almost simultaneously by T. M. Lowry and the new theory is occasionally called the Brønsted-Lowry theory. The Brønsted definitions of acids and bases are: An

**Conjugate acid-base pairs**

| | Acids | | Bases | |
|---|---|---|---|---|
| Strong acids ↑ | $H_2SO_4$ | ——— | $HSO_4^-$ | Weak bases ↑ |
| | HCl | ——— | $Cl^-$ | |
| | $H_3O^+$ | ——— | $H_2O$ | |
| | $HSO_4^-$ | ——— | $SO_4^{2-}$ | |
| | $HF_{(aq)}$ | ——— | $F^-$ | |
| | $CH_3COOH$ | ——— | $CH_3COO^-$ | |
| | $NH_4^+$ | ——— | $NH_3$ | |
| | $NCO_3^-$ | ——— | $CO_3^{2-}$ | |
| ↓ Weak acids | $H_2O$ | ——— | $OH^-$ | ↓ Strong bases |
| | $C_2H_5OH$ | ——— | $C_2H_5O^-$ | |

acid is a species which can act as a source of protons; a base is a species which can accept protons. Compared to the water (Arrhenius) theory, this represents only a slight change in the definition of an acid but a considerable extension of the term base. In addition to hydroxide ion, the bases now include a wide variety of uncharged species, such as ammonia and the amines, as well as numerous charged species, such as the anions of weak acids. In fact, every acid can generate a base by loss of a proton. Acids and bases which are related in this way are known as conjugate acid-base pairs, and the table lists examples.

As the table shows, strengths of acids and bases are not independent. A very strong Brønsted acid implies a very weak conjugate base and vice versa. A qualitative ordering of acid strength or base strength permits a rough prediction of the extent to which an acid-base reaction will go. The rule is that a strong acid and a strong base will react extensively with each other, whereas a weak acid and a weak base will react together only very slightly.

Studies of catalysis have played a large role in the acceptance of a set of quite different definitions of acids and bases, those due to G. N. Lewis: An acid is a substance which can accept an electron pair from a base; a base is a substance which can donate an electron pair. Bases under the Lewis definition are very similar to those defined by Brønsted, but the Lewis definition for acids is very much broader.

Another comprehensive theory was proposed by M. Usanovich in 1939 and is sometimes known as the positive-negative theory. Acids are defined as substances which form salts with bases, give up cations, and add themselves to anions and free electrons. Bases are similarly defined as substances which give up anions or electrons and add themselves to cations. So far, this theory has had little acceptance, quite possibly because the definitions are too broad to be very useful. *See* BASE (CHEMISTRY); BUFFERS (CHEMISTRY); HYDROGEN ION. [F.A.L.; R.H.Bo.]

**Acid anhydride** One of an important class of reactive organic compounds derived from acids via formal intermolecular dehydration.

Anhydrides of straight-chain acids containing from 2 to 12 carbon atoms are liquids with boiling points higher than those of the parent acids. They are relatively insoluble in cold water and are soluble in alcohol, ether, and other common organic solvents. The lower members are pungent, corrosive, and weakly lacrimatory. Anhydrides from acids with more than 12 carbon atoms and cyclic anhydrides from dicarboxylic acids are crystalline solids.

Because the direct intermolecular removal of water from organic acids is not practicable, anhydrides must be prepared by means of indirect processes. A general method involves interaction of an acid salt with an acid chloride.

Acetic anhydride, the most important aliphatic anhydride, is manufactured by air oxidation of acetaldehyde, using as catalysts the acetates of copper and cobalt, shown in the reaction below.

$$CH_3\overset{\overset{O}{\|}}{C}{-}H + O_2 \xrightarrow{\text{catalyst}} CH_3\overset{\overset{O}{\|}}{C}OOH + CH_3CHO \longrightarrow (CH_3CO)_2O + H_2O$$

Cyclic anhydrides are obtained by warming succinic or glutaric acids, either alone, with acetic anhydride, or with acetyl chloride. Under these conditions, adipic acid first forms linear, polymeric anhydride mixtures, from which the monomer is obtained by slow, high-vacuum distillation. Cyclic anhydrides are also formed by simple heat treatment of cis-unsaturated dicarboxylic acids, for example, maleic and glutaconic acids; and of aromatic 1,2-dicarboxylic acids, for example, phthalic acid. Commercially, however, both phthalic (structure **1**) and maleic (**2**) anhydrides are primary products

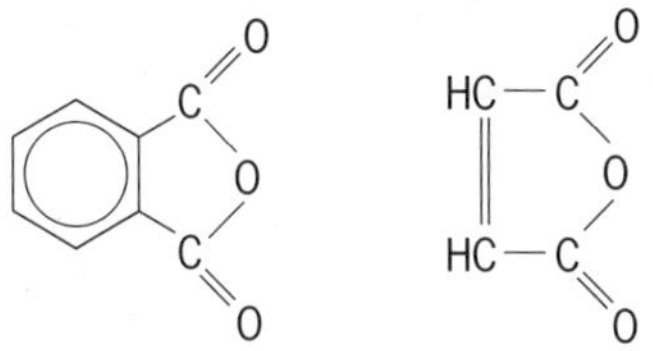

(1) (2)

of manufacture,being formed by vapor-phase, catalytic (vanadium pentoxide), air oxidation of naphthalene and benzene, respectively; at the reaction temperature, the anhydrides form directly.

Anhydrides are used in the preparation of esters. Ethyl acetate and butyl acetate (from butyl alcohol and acetic anhydride) are excellent solvents for cellulose nitrate lacquers. Acetates of high-molecular-weight alcohols are used as plasticizers for plastics and resins. Cellulose and acetic anhydride give cellulose acetate, used in acetate rayon and photographic film. The reaction of anhydrides with sodium peroxide forms peroxides (acetyl peroxide is violently explosive), used as catalysts for polymerization reactions and for addition of alkyl halides to alkenes. In Friedel-Crafts reactions, anhydrides react with aromatic compounds, forming ketones such as acetophenone.

Maleic anhydride reacts with many dienes to give hydroaromatics of various complexities (Diels-Alder reaction). Maleic anhydride is used commercially in the manufacture of alkyd resins from polyhydric alcohols. Soil conditioners are produced by basic hydrolysis of the copolymer of maleic anhydride with vinyl acetate.

Phthalic anhydride and alcohols form esters (phthalates) used as plasticizers for plastics and resins. Condensed with phenols and sulfuric acid, phthalic anhydride yields phthaleins, such as phenolphthalein; with *m*-dihydroxybenzenes under the same conditions, xanthene dyes form, for example, fluorescein. Phthalic anhydride is used in manufacturing glyptal resins (from the anhydride and glycerol) and in manufacturing anthraquinone. Heating phthalic anhydride with ammonia gives phthalimide, used in Gabriel's synthesis of primary amines, amino acids, and anthranilic acid (*o*-aminobenzoic acid). With alkaline hydrogen peroxide, phthalic anhydride yields monoperoxyphthalic acid, used along with benzoyl peroxide as polymerization catalysts, and as bleaching agents for oils, fats, and other edibles.

Anhydrides react with water to form the parent acid, with alcohols to give esters, and with ammonia to yield amides; and with primary or secondary amines, they furnish

*N*-substituted and *N,N*-disubstituted amides, respectively. *See* ACID HALIDE; ACYLATION; CARBOXYLIC ACID; DIELS-ALDER REACTION; ESTER; FRIEDEL-CRAFTS REACTION. [P.E.F.]

**Acid-base indicator** A substance that reveals, through characteristic color changes, the degree of acidity or basicity of solutions. Indicators are weak organic acids or bases which exist in more than one structural form (tautomers) of which at least one form is colored. Intense color is desirable so that very little indicator is needed; the indicator itself will thus not affect the acidity of the solution.

Acid-base indicators are commonly employed to mark the end of an acid-base titration or to measure the existing pH of a solution. Care must be used to compare colors only within the indicator range. A color comparator may also be used, employing standard color filters instead of buffer solutions.

The indicator range is the pH interval of color change of the indicator. In this range there is competition between indicator and added base for the available protons; the color change, for example, yellow to red, is gradual rather than instantaneous. Observers may, therefore, differ in selecting the precise point of change.

**Common-acid base indicators**

| Common name | pH range | Color change (acid to base) | pK |
|---|---|---|---|
| Methyl violet | 0–2, 5–6 | Yellow to blue violet to violet | |
| Metacresol purple | 1.2–2.8, 7.3–9.0 | Red to yellow to purple | 1.5 |
| Thymol blue | 1.2–2.8, 8.0–9.6 | Red to yellow to blue | 1.7 |
| Tropeoline 00 (Orange IV) | 1.4–3.0 | Red to yellow | |
| Bromphenol blue | 3.0–4.6 | Yellow to blue | 4.1 |
| Methyl orange | 2.8–4.0 | Orange to yellow | 3.4 |
| Bromcresol green | 3.8–5.4 | Yellow to blue | 4.9 |
| Methyl red | 4.2–6.3 | Red to yellow | 5.0 |
| Chlorphenol red | 5.0–6.8 | Yellow to red | 6.2 |
| Bromcresol purple | 5.2–6.8 | Yellow to purple | 6.4 |
| Bromthymol blue | 6.0–7.6 | Yellow to blue | 7.3 |
| Phenol red | 6.8–8.4 | Yellow to red | 8.0 |
| Cresol red | 2.0–3.0, 7.2–8.8 | Orange to amber to red | 8.3 |
| Orthocresol-phthalein | 8.2–9.8 | Colorless to red | |
| Phenolphthalein | 8.4–10.0 | Colorless to pink | 9.7 |
| Thymolphthalein | 10.0–11.0 | Colorless to red | 9.9 |
| Alizarin yellow GG | 10.0–12.0 | Yellow to lilac | |
| Malachite green | 11.4–13.0 | Green to colorless | |

The table lists many of the common indicators, their ranges of pH and color change, and pK values. *See* ACID AND BASE; HYDROGEN ION; TITRATION. [A.L.H.]

**Acid halide** One of a large group of organic substances possessing the halocarbonyl group,

$$\mathrm{R{-}\overset{\overset{\displaystyle O}{\|}}{C}{-}X}$$

in which X stands for fluorine, chlorine, bromine, or iodine. The terms acyl and aroyl halides refer to aliphatic or aromatic derivatives, respectively.

The great inherent reactivity of acid halides precludes their free existence in nature; all are made by synthetic processes. In general, acid halides have low melting and boiling points and show little tendency toward molecular association. With the exception of the formyl halides (which do not exist), the lower members are pungent, corrosive, lacrimatory liquids that fume in moist air. The higher members are low-melting solids.

Acid chlorides are prepared by replacement of carboxylic hydroxyl of organic acids by treatment with phosphorus trichloride, phosphorus pentachloride, or thionyl chloride.

Although acid bromides may be prepared by these methods, acid iodides are best prepared from the acid chloride treatment with either $CaI_2$ or HI, and acid fluorides from the acid chloride by interaction with HF or antimony fluoride.

The reactivity of acid halides centers upon the halocarbonyl group, resulting in substitution of the halogen by appropriate structures. Thus with substances containing active hydrogen atoms (for example, water, primary and secondary alcohols, ammonia, and primary and secondary amines), hydrogen chloride is formed together with acids, esters, amides, and *N*-substituted amides, respectively. [P.E.F.]

**Acrylonitrile** An explosive, poisonous, flammable liquid, boiling at 171°F (77.3°C), partly soluble in water. It may be regarded as vinyl cyanide, and its systematic name is 2-propenonitrile. Acrylonitrile is prepared by ammoxidation of propylene over various sorts of catalysts, chiefly metallic oxides.

Most of the acrylonitrile produced is consumed in the manufacture of acrylic and modacrylic fibers. Substantial quantities are used in acrylonitrile-butadiene-styrene (ABS) resins, in nitrile elastomers, and in the synthesis of adiponitrile by electrodimerization. Smaller amounts of acrylonitrile are used in cyanoethylation reactions, in the synthesis of drugs, dyestuffs, and pesticides, and as co-monomers with vinyl acetate, vinylpyridine, and similar monomers.

Acrylonitrile undergoes spontaneous polymerization, often with explosive force. It polymerizes violently in the presence of suitable alkaline substances. *See* NITRILE; POLYMERIZATION. [F.W.]

**Actinide elements** The series of elements beginning with actinium (atomic number 89) and including thorium, protactinium, uranium, and the transuranium elements through the element lawrencium (atomic number 103). These elements have a strong chemical resemblance to the lanthanide, or rare-earth, elements of atomic numbers 57 to 71. Their atomic numbers, names, and chemical symbols are: 89, actinium (Ac), the prototype element, sometimes not included as an actual member of the actinide series; 90, thorium (Th); 91, protactinium (Pa); 92, uranium (U); 93, neptunium (Np); 94, plutonium (Pu); 95, americium (Am); 96, curium (Cm); 97, berkelium (Bk); 98, californium (Cf); 99, einsteinium (Es); 100, fermium (Fm); 101, mendelevium (Md); 102, nobelium (No); 103, lawrencium (Lr). Except for thorium and uranium, the actinide elements are not present in nature in appreciable quantities. The transuranium elements were discovered and investigated as a result of their synthesis in nuclear reactions. All are radioactive and except for thorium and uranium, weighable amounts must be handled with special precautions.

Most actinide elements have the following in common: trivalent cations which form complex ions and organic chelates; soluble sulfates, nitrates, halides, perchlorates, and sulfides; and acid-insoluble fluorides and oxalates. *See* ACTINIUM; LAWRENCIUM; PERIODIC TABLE; PROTACTINIUM; THORIUM; TRANSURANIUM ELEMENTS; URANIUM. [G.T.S.]

**Actinium** A chemical element, Ac, atomic number 89, and atomic weight 227.0. Actinium was discovered by A. Debierne in 1899. Milligram quantities of the element are available by irradiation of radium in a nuclear reactor. Actinium-227 is a beta-emitting element whose half-life is 22 years. Six other radioisotopes with half-lives ranging from 10 days to less than 1 minute have been identified.

The relationship of actinium to the element lanthanum, the prototype rare earth, is striking. In every case, the actnium compound can be prepared by the method used to form the corresponding lanthanum compound with which it is isomorphous in the solid, anhydrous state. *See* ACTINIDE ELEMENTS; LANTHANUM; PERIODIC TABLE. [S.F.]

**Activated carbon** A powdered, granular, or pelleted form of amorphous carbon characterized by very large surface area per unit volume because of an enormous number of fine pores. Activated carbon is capable of collecting gases, liquids, or dissolved substances on the surface of its pores.

Adsorption on activated carbon is selective, favoring nonpolar over polar substances. Compared with other commercial adsorbents, activated carbon has a broad spectrum of adsorptive activity, excellent physical and chemical stability, and ease of production from readily available, frequently waste materials. *See* ADSORPTION.

Almost any carbonaceous raw material can be used for the manufacture of activated carbon. Wood, peat, and lignite are commonly used for the decolorizing materials. Bone char made by calcining bones is used in large quantity for sugar refining. Nut shells (particularly coconut), coal, petroleum coke, and other residues in either granular, briqueted, or pelleted form are used for adsorbent products.

Activation is the process of treating the carbon to open an enormous number of pores in the 1.2- to 20-nanometer-diameter range (gas-adsorbent carbon) or up to 100-nm-diameter range (decolorizing carbons). After activation, the carbon has the large surface area (500–1500 $m^2/g$) responsible for the adsorption phenomena. Carbons that have not been subjected previously to high temperatures are easiest to activate. Selective oxidation of the base carbon with steam, carbon dioxide, flue gas, or air is one method of developing the pore structure. Other methods require the mixing of chemicals, such as metal chlorides (particularly zinc chloride) or sulfides or phosphates, potassium sulfide, potassium thiocyanate, or phosphoric acid, with the carbonaceous matter, followed by calcining and washing the residue. *See* CARBON; CHARCOAL. [H.B.A.]

**Activation analysis** A technique in which a neutron, charged particle, or gamma photon is captured by a stable nuclide to produce a different, radioactive nuclide which is then measured. The technique is specific, highly sensitive, and applicable to almost every element in the periodic table.

In neutron activation analysis (NAA), the most widely used form of activation analysis, the sample to be analyzed is placed in a nuclear reactor where it is exposed to a flux of thermal neutrons. Some of these neutrons are captured by isotopes of elements in the sample; this results in the formation of a nuclide with the same atomic number, but with one more mass unit of weight. A prompt gamma ray is immediately emitted by the new nuclide.

Measurement of the induced radioactivities is the key to activation analysis. This is usually obtained from the gamma-ray spectra of the induced radionuclides. Gamma rays from radioactive isotopes have unique, discrete energies, and a device that converts such rays into electronic signals that can be amplified and displayed as a function of energy is a gamma-ray spectrometer. It consists of a detector [germanium doped with lithium, GeLi, or sodium iodide doped with thallium, NaI(Tl)] and associated electronics.

Activation analysis can also be performed with charged particles (protons or $He^{3+}$ ions, for example), but because fluxes of such particles are usually lower than reactor neutron fluxes and cross sections are much smaller, charged-particle methods are usually reserved for special samples. Charged particles penetrate only a short distance into samples, which is another disadvantage. A variant called proton-induced x-ray emission (PIXE) has been highly successful in analyzing air particulates on filters.

Activation analysis has been applied to a variety of samples. It is particularly useful for small (1 mg or less) samples, and one irradiation can provide information on 30 or more elements. Samples such as small amounts of pollutants, fly ash, very pure experimental alloys, and biological tissue have been successfully studied by neutron activation analysis. Of particular interest has been its use in forensic studies; paint, glass, tape, and other specimens of physical evidence have been assayed for legal purposes. In addition, the method has been used for authentication of art objects and paintings where only a small sample is available. *See* FORENSIC CHEMISTRY; NUCLEAR REACTION; PARTICLE DETECTOR; RADIOISOTOPE. [W.S.L.]

**Activity (thermodynamics)** The activity of a substance is a thermodynamic property that is related to the chemical potential of that substance. Activities are closely related to measures of concentration, such as partial pressures and mole fractions. The conditions that hold in chemical reaction equilibrium and in phase equilibrium can be expressed in terms of activities of the species involved.

The activity $a_i$ of chemical species $i$ in a phase (a homogeneous portion) of a thermodynamic system is defined by Eq. (1), where $\mu_i$ is the chemical potential of $i$ in that

$$a_i \equiv \exp[(\mu_i - \mu_i^\circ)/RT] \quad (1)$$

phase, $\mu_i^\circ$ is the chemical potential of $i$ in its standard state, $R$ is the gas constant, and $T$ is the absolute temperature. Equation (1) shows that the activity $a_i$ depends on the choice of standard state for species $i$. Since $\mu_i$ is an intensive quantity that depends on the temperature, pressure, and composition of the phase, $a_i$ is an intensive function of these variables. $a_i$ is dimensionless. When $\mu_i = \mu_i^\circ$, then $a_i = 1$. The degree of departure of $a_i$ from 1 measures the degree of departure of the chemical potential from its standard-state value. The chemical potential is defined by $\mu_i \equiv \partial G/\partial n_i$, where $G$ is the Gibbs energy of the phase, $n_i$ is the number of moles of substance $i$ in the phase, and the partial derivative is taken at constant temperature, pressure, and amounts of all substances except $i$. *See* CHEMICAL THERMODYNAMICS; FREE ENERGY.

From Eq. (1), it follows that $\mu_i$ is given by Eq. (2). The chemical potentials $\mu_i$ are key

$$\mu_i = \mu_i^\circ + RT \ln a_i \quad (2)$$

thermodynamic quantities of a phase, since all thermodynamic properties of the phase can be found if the chemical potentials in the phase are known as functions of temperature, pressure, and composition. Activities are more convenient to work with than chemical potentials, because the chemical potential of a substance in a phase goes to minus infinity as the amount of that substance in the phase goes to zero; also, one can determine only a value of $\mu_i$ relative to its value in some other state, whereas one can determine the actual value of each $a_i$. [I.N.L.]

**Acylation** A process in which a hydrogen atom in an organic compound is replaced by an acyl group (R—CO, where R = an organic group). The reaction involves substitution by a nucleophile (electron donor) at the electrophilic carbonyl group (C=O) of a carboxylic acid derivative. The substitution usually proceeds by an addition-elimination sequence. Two common reagents, with the general formula RCOX, that bring about acylation are acid halides (X = Cl, Br) and anhydrides (X = OCOR). There are also other acylating reagents. The carboxylic acid (X = OH) itself can function as an acylating agent when it is protonated by a strong acid catalyst as in the direct esterification of an alcohol. Typical nucleophiles in the acylation reaction are alcohols (ROH) or phenols (ArOH), both of which give rise to esters, and ammonia or amines ($RNH_2$), which give amides. *See* ACID ANHYDRIDE; ACID HALIDE; AMIDE; AMINE; ELECTROPHILIC AND NUCLEOPHILIC REAGENTS. [J.A.Mo.]

**Adenosine triphosphate (ATP)** A coenzyme and one of the most important compounds in the metabolism of all organisms, since it serves as a coupling agent between different enzymatic reactions. Adenosine triphosphate is adenosine diphosphate (ADP) with an additional phosphate group attached through a pyrophosphate linkage to the terminal phosphate group (see illustration). ATP is a powerful donor of phosphate groups to suitable acceptors because of the pyrophosphate nature of the bonds between its three phosphate radicals. For instance, in the phosphorylation of glucose, which is an essential reaction in carbohydrate metabolism, the enzyme hexokinase catalyzes the transfer of the terminal phosphate group.

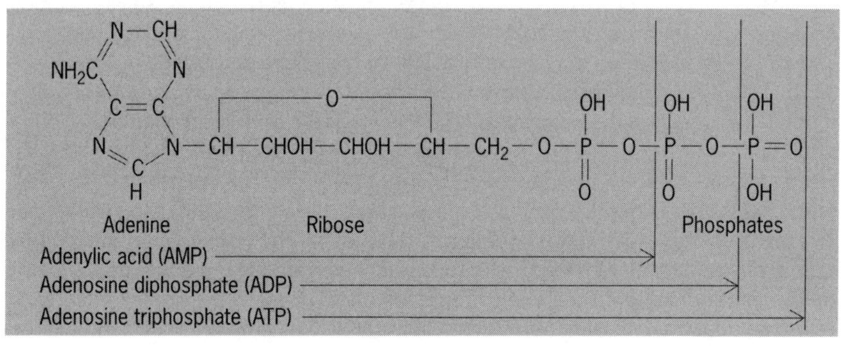


**Structure of adenylic acid and phosphate derivatives ADP and ATP.**

ATP serves as the immediate source of energy for the mechanical work performed by muscle. In its presence, the muscle protein actomyosin contracts with the formation of adenosine diphosphate and inorganic phosphate. ATP is also involved in the activation of amino acids, a necessary step in the synthesis of protein.

In metabolism, ATP is generated from adenosine diphosphate and inorganic phosphate mainly as a consequence of energy-yielding oxidation-reduction reactions. In respiration, ATP is generated during the transport of electrons from the substrate to oxygen via the cytochrome system. In photosynthetic organisms, ATP is generated as a result of photochemical reactions. *See* CYTOCHROME.

By virtue of its energy-rich pyrophosphate bonds, ATP serves as a link between sources of energy available to a living system and the chemical and mechanical work which is associated with growth, reproduction, and maintenance of living substance. For this reason, it has been referred to as the storehouse of energy of living systems. Because ATP, ADP, and adenylic acid are constantly interconverted through participation in various metabolic processes, they act as coenzymes for the coupled reactions in which they function. *See* BIOCHEMISTRY; COENZYME. [M.D.]

**Adhesive** A material capable of fastening two other materials together by means of surface attachment. The terms glue, mucilage, mastic, and cement are synonymous with adhesive. In a generic sense, the word adhesive implies any material capable of fastening by surface attachment, and thus will include inorganic materials such as portland cement and solders. In a practical sense, however, adhesive implies the broad set of materials composed of organic compounds, mainly polymeric, which can be used to fasten two materials together. The materials being fastened together by the adhesive are the adherends, and an adhesive joint or adhesive bond is the resulting assembly. Adhesion is the physical attraction of the surface of one material for the surface of another.

The phenomenon of adhesion has been described by many theories. The most widely accepted and investigated is the wettability-adsorption theory. This theory states that for maximum adhesion the adhesive must come into intimate contact with the surface of the adherend. That is, the adhesive must completely wet the adherend. This wetting is considered to be maximized when the intermolecular forces are the same forces as are normally considered in intermolecular interactions such as the van der Waals, dipole-dipole, dipole-induced dipole, and electrostatic interactions. Of these, the van der Waals force is considered the most important. The formation of chemical bonds at the interface is not considered to be of primary importance for achieving maximum wetting, but in many cases it is considered important in achieving durable adhesive bonds. *See* ADSORPTION; CHEMICAL BONDING; INTERMOLECULAR FORCES.

The greatest growth in the development and use of organic compound-based adhesives came with the application of synthetically derived organic polymers. Broadly, these materials can be divided into two types: thermoplastics and thermosets. Thermoplastic adhesives become soft or liquid upon heating and are also soluble. Thermoset adhesives cure upon heating and then become solid and insoluble. Those adhesives which cure under ambient conditions by appropriate choice of chemistry are also considered thermosets. *See* POLYMER.

Pressure-sensitive adhesives are mostly thermoplastic in nature and exhibit an important property known as tack. That is, pressure-sensitive adhesives exhibit a measurable adhesive strength with only a mild applied pressure. Pressure-sensitive adhesives are derived from elastomeric materials, such as polybutadiene or polyisoprene.

Structural adhesives are, in general, thermosets and have the property of fastening adherends that are structural materials (such as metals and wood) for long periods of time even when the adhesive joint is under load. Phenolic-based structural adhesives were among the first structural adhesives to be developed and used. The most widely used structural adhesives are based upon epoxy resins. An important property for a structural adhesive is resistance to fracture (toughness). Thermoplastics, because they are not cured, can deform under load and exhibit resistance to fracture. As a class, thermosets are quite brittle, and thermoset adhesives are modified by elastomers to increase their resistance to fracture.

Hot-melt adhesives are used for the manufacture of corrugated paper, in packaging, in bookbinding, and in shoe manufacture. Pressure-sensitive adhesives are most widely used in the form of coatings on tapes, such as electrical tape and surgical tape. Structural adhesives are applied in the form of liquids, pastes, or 100% adhesive films. Epoxy liquids and pastes are very widely used adhesive materials, having application in many assembly operations ranging from general industrial to automotive to aerospace vehicle construction. Solid-film structural adhesives are used widely in aircraft construction. Acrylic adhesives are used in thread-locking operations and in small-assembly operations such as electronics manufacture which require rapid cure times. The largest-volume use of adhesives is in plywood and other timber products manufacture. Adhesives for wood bonding range from the natural products (such as blood or casein) to the very durable phenolic-based adhesives. *See* PHENOLIC RESIN.

[A.V.P.]

**Adsorption** A process in which atoms or molecules move from a bulk phase (that is, solid, liquid, or gas) onto a solid or liquid surface. An example is purification by adsorption where impurities are filtered from liquids or gases by their adsorption onto the surface of a high-surface-area solid such as activated charcoal. Other examples include the segregation of surfactant molecules to the surface of a liquid, the bonding of reactant molecules to the solid surface of a heterogeneous catalyst, and the migration of ions to the surface of a charged electrode.

Adsorption is to be distinguished from absorption, a process in which atoms or molecules move into the bulk of a porous material, such as the absorption of water by a sponge. Sorption is a more general term that includes both adsorption and absorption. Desorption refers to the reverse of adsorption, and is a process in which molecules adsorbed on a surface are transferred back into a bulk phase. The term adsorption is most often used in the context of solid surfaces in contact with liquids and gases. Molecules that have been adsorbed onto solid surfaces are referred to generically as adsorbates, and the surface to which they are adsorbed as the substrate or adsorbent. *See* ABSORPTION.

At the molecular level, adsorption is due to attractive interactions between a surface and the species being adsorbed. The magnitude of these interactions covers approximately two orders of magnitude (8–800 kilojoules/mole), similar to the range of interactions found between atoms and molecules in bulk phases. Traditionally, adsorption is classified according to the magnitude of the adsorption forces. Weak interactions ($<40$ kJ/mol) analogous to those between molecules in liquids give rise to what is called physical adsorption or physisorption. Strong interactions ($>40$ kJ/mol) similar to those found between atoms within a molecule (for example, covalent bonds) give rise to chemical adsorption or chemisorption. In physisorption the adsorbed molecule remains intact, but in chemisorption the molecule can be broken into fragments on the surface, in which case the process is called dissociative chemisorption.

The extent of adsorption depends on physical parameters such as temperature, pressure, and concentration in the bulk phase, and the surface area of the adsorbent, as well as on chemical parameters such as the elemental nature of the adsorbate and the adsorbent. Low temperatures, high pressures, high surface areas, and highly reactive adsorbates or adsorbents generally favor adsorption.

Adsorption is directly applied in processes such as filtration and detergent action.

Adsorption also plays an important role in processes such as heterogeneous catalysis, electrochemistry, adhesion, lubrication, and molecular recognition. In heterogeneous catalysis, gas or solution-phase molecules adsorb onto the catalyst surface, and reactions in the adsorbed monolayer lead to products which are desorbed from the surface. In electrochemistry, molecules adsorbed to the surface of an electrode donate or accept electrons from the electrode as part of oxidation or reduction reactions. In adhesion and lubrication, the chemical and mechanical properties of adsorbed monolayers play a role in determining how solid surfaces behave when in contact with one another. In biological systems, the adsorption of atoms and molecules onto the surface of a cell membrane is the first step in molecular recognition. *See* ELECTROCHEMISTRY; HETEROGENEOUS CATALYSIS; MOLECULAR RECOGNITION. [B.E.Be.]

**Adsorption operations** Processes for separation of gases based on the adsorption effect. When a pure gas or a gas mixture is contacted with a solid surface, some of the gas molecules are concentrated at the surface due to gas-solid attractive forces, in a phenomenon known as adsorption. The gas is called the adsorbate and the solid is called the adsorbent.

Adsorption can be either physical or chemical. Physisorption resembles the condensation of gases to liquids, and it may be mono- or multilayered on the surface. Chemisorption is characterized by the formation of a chemical bond between the adsorbate and the adsorbent.

If one component of a gas mixture is strongly adsorbed relative to the others, a surface phase rich in the strongly adsorbed species is created. This effect forms the basis of separation of gas mixtures by gas adsorption operations. Gas adsorption has become a fast-growing unit operation for the chemical and petrochemical industries, and it is being applied to solve many different kinds of gas separation and purification problems of practical importance. *See* ADSORPTION; UNIT OPERATIONS.

Most separations and purifications of gas mixtures are done in packed columns (that is, columns filled with solid adsorbent particles). Desorption of adsorbates from a column is usually accomplished by heating the column with a hot, weakly adsorbed gas; lowering the column pressure; purging the column with a weakly adsorbed gas; or combinations of these methods.

Microporous adsorbents like zeolites, activated carbons, silica gels, and aluminas are commonly employed in industrial gas separations. These solids exhibit a wide spectrum of pore structures, surface polarity, and chemistry which makes them specifically selective for separation of many different gas mixtures. Separation is normally based on the equilibrium selectivity. However, zeolites and carbon molecular sieves can also separate gases based on molecular shape and size factors which influence the rate of adsorption. *See* ACTIVATED CARBON; MOLECULAR SIEVE; ZEOLITE.

The most frequent industrial applications of gas adsorption have been the drying of gases, solvent vapor recovery, and removal of impurities or pollutants. The adsorbates in these cases are present in dilute quantities. These separations use a thermal-swing adsorption process whereby the adsorption is carried out at a near-ambient temperature followed by thermal regeneration using a portion of the cleaned gas or steam. The adsorbent is then cooled and reused.

Pressure-swing adsorption processes are used for separating bulk gas mixtures. The adsorption is carried out at an elevated pressure level to give a product stream enriched in the more weakly adsorbed component. After the column is saturated with the strongly adsorbed component, it is regenerated by depressurization and purging with a portion of the product gas. The cycle is repeated after raising the pressure to the adsorption level. Key examples of pressure-swing adsorption processes are the production of enriched oxygen and nitrogen from air; production of ultrapure hydrogen from various hydrogen-containing streams such as steam-methane reformer off-gas; and separation of normal from branched-chain paraffins. Both thermal-swing and pressure-swing adsorption processes use multiple adsorbent columns to maintain continuity, so that when one column is undergoing adsorption the others are in various stages of regeneration modes. The thermal-swing adsorption processes typically use long cycle times (hours) in contrast to the rapid cycle times (minutes) for pressure-swing adsorption processes. [A.L.M.; S.Sir.]

**Aerosol** A suspension of small particles in a gas. The particles may be solid or liquid or a mixture of both. Aerosols are formed by the conversion of gases to particles, the disintegration of liquids or solids, or the resuspension of powdered material. Aerosol formation from a gas results in much finer particles than disintegration processes (except when condensation takes place directly on existing large particles). Dust, smoke, fume, haze, and mist are common terms for aerosols. Dust usually refers to solid particles produced by disintegration, while smoke and fume particles are generally smaller and formed from the gas phase. Mists are composed of liquid droplets. These special terms are helpful but are difficult to define exactly.

Aerosol particles range in size from molecular clusters on the order of 1 nanometer to 100 micrometers. The stable clusters formed by homogeneous nucleation and the smallest solid particles that compose agglomerates have a significant fraction of their molecules in the surface layer.

Aerosols are important in the atmospheric sciences and air pollution; inhalation therapy and industrial hygiene; manufacture of pigments, fillers, and metal powders; and fabrication of optical fibers. Atmospheric aerosols influence climate directly and indirectly. They directly affect radiation transfer on global and regional scales. Indirect effects result from their role as cloud condensation nuclei in changing droplet size distributions that affect the optical properties of clouds and precipitation. There is evidence that the stratospheric aerosol is significant in ozone destruction.

The atmospheric aerosol consists of material emitted directly from sources (primary component) and material formed by gas-to-particle conversion in the atmosphere (secondary component). The secondary component is usually the result of chemical reactions which take place in either the gas or aerosol phases. Contributions to the atmospheric aerosol come from both natural and anthropogenic sources. The effects of the atmospheric aerosol are largely determined by the size and chemical composition of the individual particles and their morphology (shape or fractal character). For many applications, the aerosol can be characterized sufficiently by measuring the particle size distribution function and the average distribution of chemical components with respect to particle size. The chemical composition of the atmospheric aerosol can be used to resolve its sources, natural or anthropogenic, by a method based on chemical signatures. Particle-to-particle variations in chemical composition and particle structural characteristics can also be measured; they probably affect the biochemical behavior and nucleating properties of aerosols.

Aerosol optical properties depend on particle size distribution and refractive index, and the wavelength of the light. These are determining factors in atmospheric visibility and the radiation balance.

Effects of the atmospheric aerosol on human health have led to the establishment of ambient air-quality standards by the United States and other industrialized nations. Adverse health effects have stimulated many controlled studies of aerosol inhalation by humans and animals. There is much uncertainty concerning the chemical components of the atmospheric aerosol that produce adverse health effects detected in epidemiological studies.

Aerosols containing pharmaceutical agents have long been used in the treatment of lung diseases such as asthma. Current efforts are directed toward systemic delivery of drugs, such as aerosolized insulin, which are transported across the alveolar walls into the blood.

Aerosol processes are used routinely in the manufacture of fine particles. Aerosol reaction engineering refers to the design of such processes, with the goal of relating product properties to the properties of the aerosol precursors and the process conditions. The most important large-scale commercial systems are flame reactors for production of pigments and powdered materials such as titania and fumed silica. Optical fibers are fabricated by an aerosol process in which a combustion-generated silica fume is deposited on the inside walls of a quartz tube a few centimeters in diameter, along with suitable dopant aerosols to control refractive index. Pyrolysis reactors are used in carbon black manufacture. Micrometer-size iron and nickel powders are produced industrially by the thermal decomposition of their carbonyls. Large pilot-scale aerosol reactors are operated using high-energy electron beams to irradiate flue gases from fossil fuel combustion. The goal is to convert sulfur oxides and nitrogen oxides to ammonium sulfate and nitrate that can be sold as a fertilizer.

Atmospheric aerosols and aerosols emitted from industrial sources are normally composed of mixtures of chemical compounds. Each chemical species is distributed with respect to particle size in a way that depends on its source and past history; hence, different substances tend to accumulate in different particle size ranges. This effect has been observed for emissions from pulverized coal combustion and municipal waste incinerators, and it undoubtedly occurs in emissions from other sources. Chemical segregation with respect to size has important implications for the effects of aerosols on public health and the environment, because particle transport and deposition depend strongly on particle size. [S.K.F.]

**Air separation** Separation of atmospheric air into its primary constituents. Nitrogen, oxygen, and argon are the primary constituents of air. Small quantities of neon, helium, krypton, and xenon are present at constant concentrations and can be separated as products. Varying quantities of water, carbon dioxide, hydrocarbons, hydrogen, carbon monoxide, and trace environmental impurities (sulfur and nitrogen oxides, chlorine) are present depending upon location and climate. Typical quantities are shown in the table. These impurities are removed during air separation to maximize efficiency and avoid hazardous operation. *See* AIR; ARGON; HELIUM; KRYPTON; NEON; XENON.

**Composition of dry air**

| Component | Percent by volume | Component | Parts per million by volume |
|---|---|---|---|
| Nitrogen | 78.084 | Carbon dioxide | 350–400 |
| Oxygen | 20.946 | Neon | 18.2 |
| Argon | 0.934 | Helium | 5.2 |
| | | Krypton | 1.1 |
| | | Xenon | 0.09 |
| | | Methane | 1–15 |
| | | Acetylene | 0–0.5 |
| | | Other hydrocarbons | 0–5 |

Three different technologies are used for the separation of air: cryogenic distillation, ambient temperature adsorption, and membrane separations. The latter two have evolved to full commercial status. Membrane technology is economical for the production of nitrogen and oxygen-enriched air (up to about 40% oxygen) at small scale. Adsorption technology produces nitrogen and medium-purity oxygen (85–95% oxygen) at flow rates up to 100 tons/day. The cryogenic process can generate oxygen or nitrogen at flows of 2500 tons/day from a single plant and make the full range of products.

Air separation is a major industry. Nitrogen and oxygen rank second and third in the scale of production of commodity chemicals; and air is the primary source of argon, neon, krypton, and xenon. Oxygen is used for steel, chemicals manufacture, and waste processing. Important uses are in integrated gasification combined cycle production of electricity, waste water treatment, and oxygen-enriched combustion. Nitrogen provides inert atmospheres for fuel, steel, and chemical processing and for the production of semiconductors. [R.M.Th.]

**Alcohol** A member of a class of organic compounds composed of carbon, hydrogen, and oxygen. They can be considered as hydroxyl derivatives of hydrocarbons produced by the replacement of one or more hydrogens by one or more hydroxyl (S—OH) groups.

**Classification.** Alcohols may be mono-, di-, tri-, or polyhydric, depending upon the number of hydroxyl groups they possess. They are classified as primary ($RCH_2OH$), secondary ($R_2CHOH$), or tertiary ($R_3COH$), depending on the number of hydrogen atoms attached to the carbon atom bearing the hydroxyl group. Alcohols can also be characterized by the molecular configuration of the hydrocarbon portion (aliphatic, cyclic, heterocyclic, or unsaturated). There are two systems in use for alcohol nomenclature, the common naming system and the IUPAC (International Union of Pure and Applied Chemistry) naming system. The common name is sometimes associated with the natural source of the alcohol or with the hydrocarbon portion (for example, methyl alcohol, ethyl alcohol). The IUPAC method is a systematic procedure with agreed-upon rules. The name of the alcohol is derived from the parent hydrocarbon which corresponds to the longest carbon chain in the alcohol. The final "e" in the hydrocarbon name is dropped and replaced with "ol"; and a number before the name indicates the position of the hydroxyl. Examples of these two systems are given in the table.

Oxidation of primary alcohols produces aldehydes (RCHO) and carboxylic acids ($RCO_2H$); oxidation of secondary alcohols yields ketones (RCOR′). Dehydration of alcohols produces alkenes and ethers (ROR). Reaction of alcohols with carboxylic acids results in the formation of esters (ROCOR′), a reaction of great industrial importance. The hydroxyl group of an alcohol is readily replaced by halogens or pseudohalogens. *See* ALDEHYDE; ALKENE; CARBOXYLIC ACID; ESTER; ETHER; KETONE.

**Uses.** Industrially, the monohydric aliphatic alcohols are classified according to their uses as the lower alcohols (1–5 carbon atoms), the plasticizer-range alcohols (6–11 carbon atoms), and the detergent-range alcohols (12 or more carbon atoms). The lower alcohols are employed as solvents, extractants, and antifreezes. Esters of the lower alcohols are employed extensively as solvents for lacquers, paints, varnishes, inks, and adhesives. The plasticizer-range alcohols find their primary use in the form of esters as plasticizers and also as lubricants in high-speed applications such as jet engines.

**Alcohols and their formulas**

| Name | | |
|---|---|---|
| Common | IUPAC | Formula |
| Methyl alcohol | Methanol | $CH_3OH$ |
| Ethyl alcohol | Ethanol | $CH_3CH_2OH$ |
| *n*-Propyl alcohol | 1-Propanol | $CH_3CH_2CH_2OH$ |
| Isopropyl alcohol | 2-Propanol | $(CH_3)_2CHO$ |
| *n*-Butyl alcohol | 1-Butanol | $CH_3(CH_2)_2CH_2OH$ |
| *sec*-Butyl alcohol | 2-Butanol | $CH_3CH_2CHOHCH_3$ |
| *tert*-Butyl alcohol | 2-Methyl-2-propanol | $(CH_3)_3COH$ |
| Isobutyl alcohol | 2-Methyl-1-propanol | $(CH_3)_2CHCH_2OH$ |
| *n*-Amyl alcohol | 1-Pentanol | $CH_3(CH_2)_3CH_2OH$ |
| *n*-Hexyl alcohol | 1-Hexanol | $CH_3(CH_2)_4CH_2OH$ |
| Allyl alcohol | 2-Propen-1-ol | $CH_2{=}CHCH_2OH$ |
| Crotyl alcohol | 2-Buten-1-ol | $CH_3CH{=}CHCH_2OH$ |
| Ethylene glycol | 1,2-Ethanediol | $HOCH_2CH_2OH$ |
| Propylene glycol | 1,2-Propanediol | $CH_3CHOHCH_2OH$ |
| Trimethylene glycol | 1,3-Propanediol | $HOCH_2CH_3CH_2OH$ |
| Glycerol | 1,2,3-Propanetriol | $CH_2OHCHOHCH_2OH$ |

The detergent-range alcohols are used in the form of sulfate and ethoxysulfate esters in detergents and surfactants.

Alcohols are derived either from natural-product processing, such as the fermentation of carbohydrates and the reductive cleavage of natural fats and oils, or by chemical synthesis based on the hydrocarbons derived from petroleum or the synthesis gas from coal.

The fermentation of sugars and starches (carbohydrates) to produce alcoholic beverages has been employed at least since history has been recorded. The industrial fermentation process is the biological transformation of a carbohydrate by a highly specialized strain of yeast to produce the desired product, such as ethanol, or 1-butanol and acetone. Fermentation is no longer the major source of 1-butanol, but still accounts for all potable ethanol and a large proportion of the ethanol used industrially worldwide. As sources of hydrocarbons based on petroleum continue to be depleted, fermentation processes based on renewable raw materials are likely to become more important. [P.E.F.]

**Aldehyde** One of a class of organic chemical compounds represented by the general formula RCHO. Formaldehyde, the simplest aldehyde, has the formula HCHO, where R is hydrogen. For all other aldehydes, R is a hydrocarbon radical which may be substituted with other groups such as halogen or hydroxyl (see table). Because of their high chemical reactivity, aldehydes are important intermediates for the manufacture of resins, plasticizers, solvents, dyes, and pharmaceuticals.

At room temperature formaldehyde is a colorless gas. The other low-molecular-weight aldehydes are colorless liquids having characteristic, somewhat acrid odors. The unsaturated aldehydes acrolein and crotonaldehyde are powerful lacrimators. The

**Aldehydes and their formulas**

| Compound | Formula |
|---|---|
| Formaldehyde | $HCHO$ |
| Acetaldehyde | $CH_3CHO$ |
| Acetaldol | $CH_3CHOHCH_2CHO$ |
| Propionaldehyde | $C_2H_5CHO$ |
| *n*-Butyraldhyde | $CH_3(CH_2)_2CHO$ |
| Isobutyraldehyde | $(CH_3)_2CHCHO$ |
| Acrolein | $CH_2{=}CHCHO$ |
| Crotonaldehyde | $CH_3CH{=}CHCHO$ |
| Chloral | $CCl_3CHO$ |
| Chloral hydrate | $CCl_3CH(OH)_2$ |
| Benzaldehyde | $C_6H_5\text{–}C(H){=}O$ |
| Cinnamaldehyde | $C_6H_5\text{–}C(H){=}C(H)\text{–}C(H){=}O$ |

important reactions of aldehydes include oxidation, reduction, aldol condensation, Cannizzaro reaction, and reactions with compounds containing nitrogen.

Because of the importance of aldehydes as chemical intermediates, many industrial and laboratory syntheses have been developed. The more important of these methods include catalytic dehydrogenation of primary alcohols, oxidation of primary alcohols, oxidation of olefins, and hydroformylation of olefins. *See* FORMALDEHYDE. [P.E.F.]

**Alicyclic hydrocarbon** An organic compound that contains one or more closed rings of carbon atoms. The term alicyclic specifically excludes carbocyclic compounds with an array of $\pi$-electrons characteristic of aromatic rings. Compounds with one to five alicyclic rings of great variety and complexity are found in many natural products such as steroids and terpenes. By far the majority of these have six-membered rings. *See* AROMATIC HYDROCARBON; STEROID; TERPENE.

**Structures and nomenclature.** The bonding in cyclic hydrocarbons is much the same as that in open-chain alkanes and alkenes. An important difference, however, is the fact that the atoms in a ring are part of a closed loop. Complete freedom of rotation about a carbon-carbon bond (C—C) is not possible; the ring has faces or sides, like those of a plate.

Simple monocyclic hydrocarbons are usually represented as bond line structures; for example, cyclopropane (**1***a*) is usually represented as structure (**1***b*) and methylcycloheptane as structure (**2**). These hydrocarbons are named by adding the prefix

CH$_3$

CH$_2$

H$_2$C — CH$_2$ ≡

(**1*a***) (**1*b***) (**2**)

cyclo to the stem of the alkane corresponding to the number of atoms in the ring. When two or more substituents are attached to the ring, the relative positions and orientation must be specified: *cis* on the same side and *trans* on the other, as in *cis*-1,2-dimethylcyclopentane (**3**) and *trans*-1,3-dimethylcyclopentane (**4**).

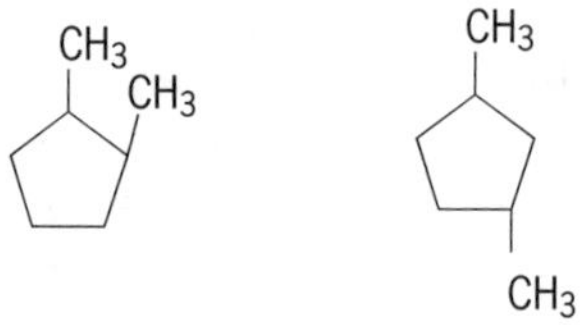

CH$_3$ CH$_3$ CH$_3$

CH$_3$

(**3**) (**4**)

In bicyclic compounds the rings can be joined in three ways: spirocyclic, fused, and bridged, as illustrated in the structures for spiro[4.5]decane (**5**), bicyclo[4.3.0]nonane (**6**), and bicyclo[2.2.1]heptane (**7**). In each case the name indicates the total number

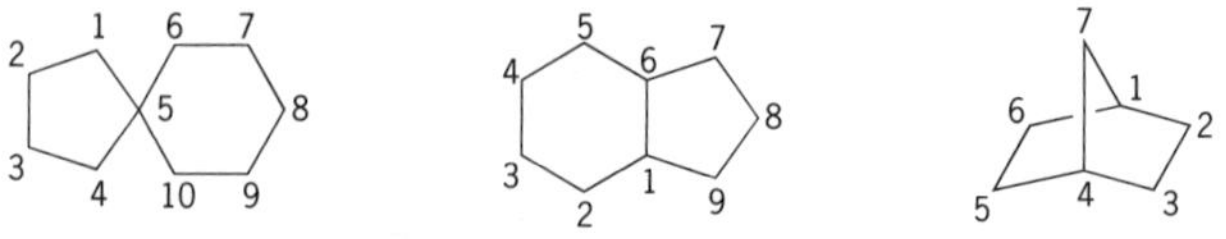

(**5**) (**6**) (**7**)

of carbon atoms, and the number of atoms in each bridge. Atoms are numbered as shown.

Any of these bicyclic systems can be transformed to a tricyclic array by introduction of another bond between nonadjacent carbons, as in structure (**8**), or an additional ring, as in structure (**9**). Cyclic structure gives rise to the possibility of compounds made

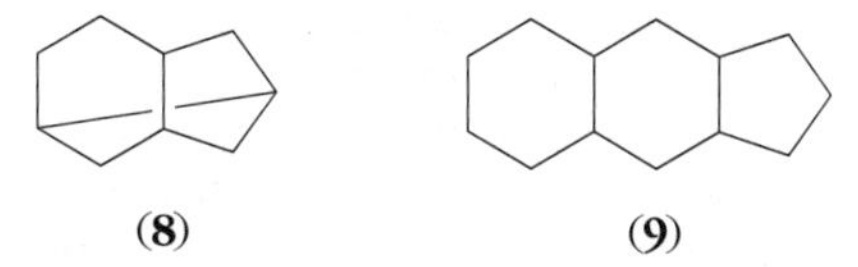

(**8**) (**9**)

up of molecular subunits that are linked mechanically rather than chemically. In rotaxanes (**10**), bulky groups are introduced at the ends of a long chain that is threaded through a large ring ($>C_{30}$). Cyclization of the ends leads to a catenane (**11**). Several examples of compounds with these structures have been prepared.

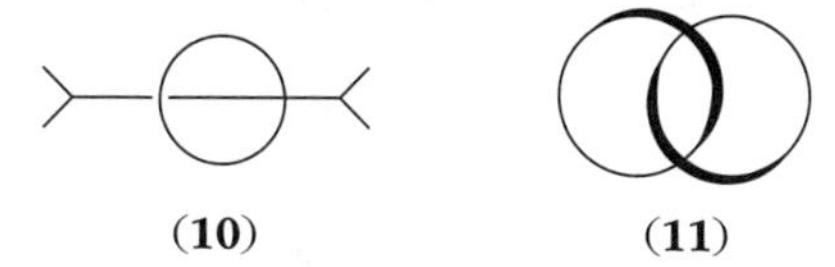

(**10**) (**11**)

The boiling points, melting points, and densities of cycloalkanes are all higher than those of their open-chain counterparts, reflecting the more compact structures and greater association in both liquid and solid. Geometrical constraints in the smaller rings have significant effects on reactions of cycloalkanes and derivatives. Because of ring strain, ring-opening reactions of cyclopropane, such as isomerization to propene, take place under conditions that do not affect alkanes or larger-ring cycloalkanes.

Comparison of reaction rates of compounds with different ring size or cis-trans configuration has provided important insights about reaction mechanism and conformational analysis. *See* ORGANIC REACTION MECHANISM.

**Ring-forming reactions.** A number of useful methods have been devised that lead to alicyclic rings. These reactions are of three types: C-C bond formation between atoms in an open-chain precursor, cycloaddition or cyclooligomerization, and expansion or contraction of a more readily available ring. In cycloaddition, two molecules react with formation of two bonds; in cyclooligomerization, three or more molecules combine to form three or more bonds.

Two alicyclic compounds are manufactured in large volume; both are products of the petroleum industry. Cyclopentadiene (**12**) is formed from various alkylcyclopentanes in naphtha fractions during the refining process. It is a highly reactive diene and spontaneously dimerizes in a 4 + 2 cycloaddition [see reaction below]; it is used as a copolymer in several resins.

(**12**)

Cyclohexane is produced in large quantity by hydrogenation of benzene. The principal use of cyclohexane is conversion by oxidation in air to a mixture of cyclohexanol and the ketone, which is then oxidized further to adipic acid for the manufacture of nylon. *See* ORGANIC SYNTHESIS. [J.A.Mo.]

**Alkali** Any compound having highly basic properties, strong acrid taste, and ability to neutralize acids. Aqueous solutions of alkalies are high in hydroxyl ions, have a pH above 7, and turn litmus paper from red to blue. Caustic alkalies include sodium hydroxide (caustic soda), the sixth-largest-volume chemical produced in the United States, and potassium hydroxide. They are extremely destructive to human tissue; external burns should be washed with large amounts of water. The milder alkalies are the carbonates of the alkali metals; these include the industrially important sodium carbonate (soda ash) and potassium carbonate (potash), as well as the carbonates of lithium, rubidium, and cesium, and the volatile ammonium hydroxide. Sodium bicarbonate is a still milder alkaline material. *See* ACID AND BASE; PH.

About 50% of the caustic soda produced goes into making many chemical products, about 16% into pulp and paper, 6.5% each into aluminum, petroleum, and textiles (including rayon), with smaller percentages into soap and synthetic detergents, and cellophane. For soda ash, about 50% goes to react mainly with sand in making glass, 25% to making miscellaneous chemicals, 6.5% each to alkaline cleaners and pulp and paper, and a few percent to water treatment and other uses. *See* ALKALI METALS; ELECTROCHEMICAL PROCESS; HYDROXIDE; SOAP. [D.F.O.]

**Alkali metals** The elements of group 1 in the periodic table (lithium, sodium, potassium, rubidium, cesium, francium). Of the alkali metals, lithium differs most from the rest of the group, and tends to resemble the alkaline-earth metals (group 2 of the periodic table) in many ways. In this respect lithium behaves as do many other elements that are the first members of groups in the periodic table; these tend to resemble the elements in the group to the right rather than those in the same group. Francium, the heaviest of the alkali-metal elements, has no stable isotopes and exists only in radioactive form.

In general, the alkali metals are soft, low-melting, reactive metals. This reactivity accounts for the fact that they are never found uncombined in nature but are always in chemical combination with other elements. This reactivity also accounts for the fact that they have no utility as structural metals (with the possible exception of lithium in alloys) and that they are used as chemical reactants in industry rather than as metals in the usual sense. The reactivity in the alkali-metal series increases in general with increase in atomic weight from lithium to cesium. *See* CESIUM; ELECTROCHEMICAL SERIES; FRANCIUM; LITHIUM; PERIODIC TABLE; POTASSIUM; RUBIDIUM; SODIUM. [M.Si.]

**Alkaline-earth metals** Usually calcium, strontium, and barium, the heaviest members of group 2 of the periodic table (excepting radium). Other members of the group are beryllium, magnesium, and radium, sometimes included among the alkaline-earth metals. Beryllium resembles aluminum more than any other element, and magnesium behaves more like zinc and cadmium. The gap between beryllium and magnesium and the remainder of the elements of group 2 makes it desirable to discuss these elements separately. Radium is often treated separately because of its radioactivity.

The alkaline earths form a closely related group of highly metallic elements in which there is a regular gradation of properties. The metals, none of which occurs free in nature, are all harder than potassium or sodium, softer than magnesium or beryllium, and about as hard as lead. The metals are somewhat brittle, but are malleable, extrudable, and machinable. They conduct electricity well; the specific conductivity of calcium is 45% of that of silver. The oxidation potentials of the triad are as great as those of the alkali metals.

The elements and their compounds find important industrial uses in low-melting alloys, deoxidizers, and drying agents and as cheap sources of alkalinity. *See* BARIUM; BERYLLIUM; CALCIUM; MAGNESIUM; PERIODIC TABLE; RADIUM; STRONTIUM. [R.F.R.]

**Alkaloid** A cyclic organic compound that contains nitrogen in a negative oxidation state and is of limited distribution among living organisms. Over 10,000 alkaloids of many different structural types are known; and no other class of natural products possesses such an enormous variety of structures. Therefore, alkaloids are difficult to differentiate from other types of organic nitrogen-containing compounds.

Simple low-molecular-weight derivatives of ammonia, as well as polyamines and acyclic amides, are not considered alkaloids because they lack a cyclic structure in some part of the molecule. Amines, amine oxides, amides, and quaternary ammonium salts are included in the alkaloid group because their nitrogen is in a negative oxidation state (the oxidation state designates the positive or negative character of atoms in a molecule). Nitro and nitroso compounds are excluded as alkaloids. The almost-ubiquitous nitrogenous compounds, such as amino acids, amino sugars, peptides, proteins, nucleic acids, nucleotides, prophyrins, and vitamins, are not alkaloids. However, compounds that are exceptions to the classical-type definition (that is, a compound containing nitrogen, usually a cyclic amine, and occurring as a secondary metabolite), such as neutral alkaloids (colchicine, piperine), the $\beta$-phenyl-ethylanines, and the purine bases (caffeine, theophylline, theobromine), are accepted as alkaloids.

Alkaloids often occur as salts of plant acids such as malic, meconic, and quinic acids. Some plant alkaloids are combined with sugars, for example, solanine in potato (*Solanum tuberosum*) and tomatine in tomato (*Lycopersicum esculentum*). Others occur as amides, for example, piperine from black pepper (*Piper nigrum*), or as esters, for example, cocaine from coca leaves (*Erythroxylum coca*). Still other alkaloids occur as quaternary salts or tertiary amine oxides.

While most alkaloids have been isolated from plants, a large number have been isolated from animal sources. They occur in mammals, anurans (frogs, toads), salamanders, arthropods (ants, millipedes, ladybugs, beetles, butterflies), marine organisms, mosses, fungi, and certain bacteria.

Many alkaloids exhibit marked pharmacological activity, and some find important uses in medicine. Atropine, the optically inactive form of hyoscyamine, is used widely in medicine as an antidote to cholinesterase inhibitors such as physostigmine and insecticides of the organophosphate type; it is also used in drying cough secretions. Morphine and codeine are narcotic analgesics, and codeine is also an antitussive agent, less toxic and less habit-forming than morphine. Colchicine, from the corms and seeds of the autumn crocus, is used as a gout suppressant. Caffeine, which occurs in coffee, tea, cocoa, and cola, is a central nervous system stimulant; it is used as a cardiac and respiratory stimulant and as an antidote to barbiturate and morphine poisoning. Emetine, the key alkaloid of ipecac root (*Cephaelis ipecacuanha*), is used in the treatment of amebic dysentery and other protozoal infections. Epinephrine or adrenaline (see structure), produced in most animal species by the adrenal medulla, is used as a

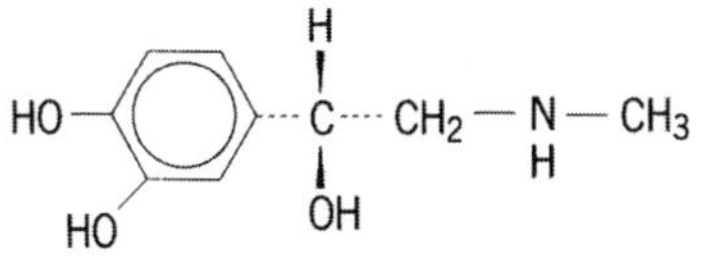

H
HO — C — CH2 — N — CH3
H
HO OH

bronchodilator and cardiac stimulant and to counter allergic reactions, anesthesia, and cardiac arrest. [S.W.Pe.]

**Alkane** A compound with the general formula $C_nH_{2n+2}$. Alkanes are open-chain (aliphatic or noncyclic) hydrocarbons with no multiple bonds or functional groups. They consist of tetrahedral carbon atoms, up to $10^5$ carbons or more in length. The

**Nomenclature and properties of alkanes**

| Name | Structure | Boiling point, °C (°F) | Heat of formation ($\Delta H_f^\circ$), kJ |
|---|---|---|---|
| Methane | $CH_4$ | −162 (−260) | −74.5 |
| Ethane | $CH_3CH_3$ | −89 (−128) | −83.45 |
| Propane | $CH_3CH_2CH_3$ | −42 (−44) | −104.6 |
| Butane | $CH_3(CH_2)_2CH_3$ | −0.5 (33) | −125.7 |
| 2-Methylpropane (isobutane) | $(CH_3)_2CHCH_3$ | −11.7 (10.9) | −134.2 |
| Pentane | $CH_3(CH_2)_3CH_3$ | 36.1 (97.0) | −146.87 |
| 2-Methylbutane (isopentane) | $(CH_3)_2CHCH_2CH_3$ | 29.9 (85.8) | −153.7 |
| 2,2-Dimethyl propane (neopentane) | $(CH_3)_4C$ | 9.4 (49) | −167.9 |
| Hexane | $CH_3(CH_2)_4CH_3$ | 68.7 (156) | −167.0 |

C-C $\sigma$ bonds are formed from $sp^3$ orbitals, and there is free rotation around the bond axis. *See* MOLECULAR ORBITAL THEORY.

Alkanes provide the parent names for all other aliphatic compounds in systematic nomenclature. Alkanes are designated by the ending -ane appended to a stem denoting the chain length. The straight-chain isomer is designated by the prefix C (normal); other isomers are named by specifying the size of the branch and its location (see table). The number of isomers increases enormously in larger molecules; thus there are 75 isomers of $C_{10}H_{22}$ and over 4 billion for $C_{30}H_{62}$.

Alkanes with four or fewer carbons are gases at atmospheric pressure. Higher C alkanes are liquids or, above about 20 carbons, solids known as paraffin wax. Alkanes have densities lower than that of water and have very low water solubility; other properties depend on the degree of branching. The heat of formation ($\Delta H_f^\circ$) is a measure of the energy content of a compound relative to the component elements in standard states. For example, comparison of heat of formation values for three alkane isomers with the molecular formula $C_5H_{12}$ indicates that the relative energy content of the three pentane isomers decreases with increased branching, that is, the branched isomer is thermodynamically most stable. *See* PARAFFIN.

Alkanes are the major components of natural gas and petroleum, which are the only significant sources. Much smaller amounts of alkanes have been produced from coal at various times and locations, either indirectly by the Fischer-Tropsch process or by direct liquefaction. *See* COAL LIQUEFACTION; FISCHER-TROPSCH PROCESS; NATURAL GAS; PETROLEUM.

Individual lower alkanes can be separated from the more volatile distillate fractions of petroleum, but beyond the $C_7$–$C_8$ range the alkanes obtained are mixtures of many isomers. Compounds of a specific structure can be prepared in a laboratory scale by chemical synthesis. Various C-C bond-forming steps such as coupling or condensation are carried out to build up the desired carbon skeleton. The final step is usually removal of a functional group by some type of reduction.

Much of the chemistry of alkanes begins at the petroleum refinery, where several reactions are carried out to adjust the hydrocarbon composition of crude oil to that needed for a constantly changing set of applications. Major reactions are (1) isomerization of straight-chain alkanes to branched compounds; (2) cracking to produce smaller molecules; (3) alkylation, for example, combination of propylene and butane to give 2,3-dimethylpentane; and (4) cyclodehydrogenation (platforming), in which aromatizations occur. An important objective in some of these processes is to increase the yield of highly branched alkanes in the $C_6$–$C_8$ range needed for gasoline. *See* ALKYLATION (PETROLEUM); AROMATIZATION; CRACKING; GASOLINE.

By far the most important end use of alkanes is combustion as fuel to provide heat and electric or motive power. In most cases, complete oxidation is not achieved, and varying amounts of incompletely oxidized fragments, carbon monoxide, and elemental carbon are produced.

Controlled partial oxidation is possible if all the C-H bonds in an alkane are equivalent or if one C-H bond is significantly weaker than all the others. An example of the latter situation is isobutane, which is converted to the hydroperoxide on industrial scale for the manufacture of *t*-butyl alcohol. *See* AUTOXIDATION; COMBUSTION.

Alkanes have been referred to as paraffin hydrocarbons to indicate their low affinity or reactivity. They contain no unshared electron pairs or accessible empty bonding orbitals, and they are unaffected by many reagents that attack $\pi$-bonds or other functional groups. One type of reaction that does occur is substitution by a radical chain process. Examples are chlorination and vapor-phase nitration, involving the odd-electron species $Cl\cdot$ and $NO_2\cdot$, respectively. Neither reaction is selective; when two or more types of C-H bonds are present in the alkane, mixtures of products are usually obtained. Thus propane gives rise to 1- and 2-chloropropanes as well as dichloro compounds. Nitration of propane leads to a mixture of 1- and 2-nitropropane, and also nitromethane and nitroethane by C-C bond cleavage. *See* HALOGENATED HYDROCARBON; HALOGENATION; NITRATION. [J.A.Mo.]

**Alkene** One of the class of acyclic hydrocarbons containing one or more carbon-to-carbon double bonds. Alkenes (also called olefins) and alkynes (also called acetylenes) together constitute the family of organic compounds called unsaturated hydrocarbons, since they contain less than the number of hydrogens found in the corresponding saturated compound, alkane. When the double bond is present in a nonaromatic ring (alicyclic hydrocarbon), the compound is termed a cycloalkene. Hydrocarbons containing more than one double bond are termed dienes, trienes, and so forth, or collectively, polyenes. *See* ALICYCLIC HYDROCARBON; ALKANE; ALKYNE.

In naming alkenes by the system of the International Union of Pure and Applied Chemistry (IUPAC), the longest chain containing the double bond is identified. The presence of the double bond is indicated by changing the "-ane" ending of the alkane having the same number of carbon atoms to "-ene," and the position of the double bond is indicated by a prefixed number. Examples are given in the table, with common or nonsystematic names which are still frequently used given in parentheses.

**Alkenes and dienes (common name given in parentheses)**

| Name | Formula |
|---|---|
| Ethene (ethylene) | $CH_2{=}CH_2$ |
| Propene (propylene) | $CH_2{=}CHCH_3$ |
| 1-Butene | $CH_2{=}CHCH_2CH_3$ |
| 2-Butene | $CH_3CH{=}CHCH_3$ |
| 2-Methylpropene (isobutylene) | $CH_2{=}C(CH_3)CH_3$ (drawn as $CH_2{=}CCH_3$ with $CH_3$ above the C) |
| 1,3-Butadiene | $CH_2{=}CHCH{=}CH_2$ |
| 2-Methyl-l,3-butadiene (isoprene) | $CH_2{=}C(CH_3)CH{=}CH_2$ (drawn as $CH_2{=}CCH{=}CH_2$ with $CH_3$ above the C) |

The lower alkenes and dienes which have up to five carbon atoms are gases at room temperature and pressure. Higher alkenes are colorless liquids or solids. Like other hydrocarbons, alkenes are insoluble in water. Liquid alkenes have specific gravities well below 1.0. Alkenes may undergo polymerization, cyclization, and addition reactions. A major share of structural and elastic polymers are based on homopolymers or copolymers of alkenes and dienes. Alkenes and dienes cyclize readily under various conditions.

Addition reactions of alkenes are among the most important in the entire field of organic chemistry. Industrially, high-octane gasoline is made by the acid-catalyzed alkylation of the three-and four-carbon alkenes. A variety of alkylated aromatics are made by the alkylation of benzene with olefins.

The commercially important alkenes are produced on a large scale in the petroleum industry by thermal or catalytic cracking processes. In the laboratory, most methods for the preparation of alkenes involve some type of elimination reaction, in which atoms or groups on adjacent carbon atoms are removed with concomitant formation of the carbon-carbon double bond. *See* ALKYLATION (PETROLEUM); CRACKING; HALOGENATION; HYDROGENATION. [P.E.F.]

**Alkylation (petroleum)** In the petroleum industry, a chemical process in which an alkene (ethylene, propylene, and so forth) and a hydrocarbon, usually 2-methylpropane, are combined to produce a higher-molecular-weight and higher-carbon-number product. The product has a higher octane rating and is used to improve the quality of gasoline-range fuels. The process was originally developed during World War II to produce high-octane aviation gasoline. Its current main application is in the production of unleaded automotive gasoline. *See* ALKENE; GASOLINE.

The alkylation reaction is initiated by the addition of a proton ($H^+$) to the olefin [reaction (1)]. The protonated olefin (carbonium ion) then reacts with the isobutane by abstraction of a proton from the isobutane to produce the *t*-butyl carbonium ion [reaction (2)]. Reaction of this tertiary carbonium ion with the olefin proceeds by combination of the two species [reaction (3)] to produce a more complex six-carbon carbonium ion which yields a stabilized product by abstraction of a proton from another molecule of isobutane [reaction (4)]. The reaction progresses to more product

$$H_2C{=}CH_2 + H^+ \longrightarrow H_3C{-}CH_2^+ \qquad (1)$$

$$(CH_3)_2CH{-}CH_3 + H_3C{-}\overset{+}{C}H_2 \longrightarrow H_3C{-}\underset{CH_3}{\overset{CH_3}{|}}{C^+} + H_3C{-}CH_3 \qquad (2)$$

$$H_3C{-}\underset{CH_3}{\overset{CH_3}{C^+}} + H_2C{=}CH_2 \longrightarrow H_3C{-}\underset{CH_3}{\overset{CH_3}{C}}{-}\underset{H}{\overset{H}{C}}{-}\overset{+}{C}H_2 \qquad (3)$$

$$H_3C{-}\underset{CH_3}{\overset{CH_3}{C}}{-}\underset{H}{\overset{H}{C}}{-}\overset{+}{C}H_2 + H_3C{-}\underset{CH_3}{\overset{CH_3}{CH}} \longrightarrow \underbrace{H_3C{-}\underset{CH_3}{\overset{CH_3}{C}}{-}\underset{H}{\overset{H}{C}}{-}CH_3}_{\text{Alkylate}} + H_3C{-}\underset{CH_3}{\overset{CH_3}{C^+}} \qquad (4)$$

by reaction of the *t*-butyl and carbonium ion, as already shown in reaction (3). *See* REACTIVE INTERMEDIATES. [J.G.S.]

**Alkyne** One of a group of organic compounds containing a carbon-to-carbon triple-bond linkage (—C≡C—). They are termed acetylenes or alkynes. While exhibiting many of the characteristics of alkenes as regards unsaturation, the acetylenes have many unique properties. Since the bonding in alkyne molecules is linear, R—C≡C—R, cis-trans isomerism is not possible. In the simplest alkyne, acetylene (HC≡CH), or in monosubstituted acetylenes, the hydrogen attached to triply bonded carbon is acidic to such a degree that it is replaceable with metals such as sodium. Structural formulas of several alkynes are as follows:

$$CH{\equiv}CH \qquad CH{\equiv}CH_2{-}CH_2{-}CH_3$$

Ethyne or acetylene — 1-Butyne

$$CH{\equiv}C{-}CH_3 \qquad CH_3{-}C{\equiv}C{-}CH_3$$

Propyne — 2-Butyne

General methods for the preparation of alkynes depend on dehydrohalogenation of $\alpha,\beta$-dihaloparaffins, conversion of aldehydes or ketones to dihaloparaffins with subsequent dehydrohalogenation, and alkylation of metallic acetylides with alkyl halides in liquid ammonia. Grignard reagents of 1-alkynes behave similarly.

The reactions of triply bonded carbon compounds are in general similar to those of compounds containing ethylenic bonds. Addition reactions proceed in two stages to form first a vinyl compound or substituted ethylene, and second the substituted paraffin. In the presence of catalytic quantities of alkoxides, acetylene adds to alcohols and phenols to give vinyl ethers. Reactions of this type, in which addition takes place by replacement of hydrogen with a vinyl group, are termed vinylation. Another form of addition reaction known as ethynylation involves the addition of acetylene to unsaturated compounds. A general reaction of alkynes involves the addition of carbon monoxide and water, or other compounds having an active hydrogen, in the presence of nickel carbonyl.

Polymerization of alkynes may yield acyclic, aromatic, or alicyclic derivatives. In the presence of cuprous chloride, acetylene dimerizes to vinyl acetylene, which adds hydrogen chloride to give chloroprene (2-chloro-l,3-butadiene). Polymerization of chloroprene in the presence of free radical initiators gives the commercially important synthetic rubber, neoprene. Thermal polymerization of acetylene yields benzene as the major product and also a wide variety of polynuclear aromatic compounds. Of great theoretical and practical importance is the polymerization of acetylene to the cyclic tetramer, cyclooctatetraene. *See* ACETYLENE; ALKENE. [C.A.Co./P.E.F.]

**Alum** A colorless to white crystalline substance which occurs naturally as the mineral kalunite and is a constituent of the mineral alunite. Alum is produced as aluminum sulfate by treating bauxite with sulfuric acid to yield alum cake or by treating the bauxite with caustic soda to yield papermaker's alum. Other industrial alums are potash alum, ammonium alum, sodium alum, and chrome alum (potassium chromium sulfate). Major uses of alum are as an astringent, styptic, and emetic. For water purification alum is dissolved. It then crystallizes out into positively charged crystals that attract negatively charged organic impurities to form an aggregate sufficiently heavy to settle out. Alum is also used in sizing paper, dyeing fabrics, and tanning leather. With sodium bicarbonate it is used in baking powder and in some fire extinguishers. *See* ALUMINUM; COLLOID. [F.H.R.]

**Aluminum** A metallic chemical element, symbol Al, atomic number 13, atomic weight 26.98154, in group 13 of the periodic system. Pure aluminum is soft and lacks strength, but it can be alloyed with other elements to increase strength and impart a number of useful properties. Alloys of aluminum are light, strong, and readily formable by many metalworking processes; they can be easily joined, cast, or machined, and accept a wide variety of finishes. Because of its many desirable physical, chemical, and metallurgical properties, aluminum has become the most widely used nonferrous metal. *See* PERIODIC TABLE.

Aluminum is the most abundant metallic element on the Earth and Moon but is never found free in nature. The element is widely distributed in plants, and nearly all rocks, particularly igneous rocks, contain aluminum in the form of aluminum silicate minerals. When these minerals go into solution, depending upon the chemical conditions, aluminum can be precipitated out of the solution as clay minerals or aluminum hydroxides, or both. Under such conditions bauxites are formed. Bauxites serve as principal raw materials for aluminum production.

Aluminum is a silvery metal having a density of 1.56 oz/in.$^3$ at 68°F (2.70 g/cm$^3$ at 20°C). Naturally occurring aluminum consists of a single isotope, $^{27}_{13}Al$. Aluminum crystallizes in the face-centered cubic structure with edge of the unit lattice cube of 4.0495 angstroms (0.40495 nanometer). Aluminum is known for its high electrical and thermal conductivities and its high reflectivity.

The electronic configuration of the element is $1s^2 2s^2 2p^6 3s^2 3p^1$. Aluminum exhibits a valence of +3 in all compounds, with the exception of a few high-temperature monovalent and divalent gaseous species.

Aluminum is stable in air and resistant to corrosion by seawater and many aqueous solutions and other chemical agents. This is due to protection of the metal by a tough, impervious film of oxide. At a purity greater than 99.95%, aluminum resists attack by most acids but dissolves in aqua regia. Its oxide film dissolves in alkaline solutions, and corrosion is rapid.

Aluminum is amphoteric and can react with mineral acids to form soluble salts and to evolve hydrogen.

Molten aluminum can react explosively with water. The molten metal should not be allowed to contact damp tools or containers.

At high temperatures aluminum reduces many compounds containing oxygen, particularly metal oxides. These reactions are used in the manufacture of certain metals and alloys.

Applications in building and construction represent the largest single market of the aluminum industry. Millions of homes use aluminum doors, siding, windows, screening, and down-spouts and gutters. Aluminum is also a major industrial building product. Transportation is the second largest market. Many commercial and military aircraft have become virtually all-aluminum. In automobiles, aluminum is apparent in interior and exterior trim, grilles, wheels, air conditioners, automatic transmissions, and some radiators, engine blocks, and body panels. Aluminum is also found in rapid-transit car bodies, rail cars, forged truck wheels, cargo containers, and in highway signs, divider rails, and lighting standards. In aerospace, aluminum is found in aircraft engines, frames, skins, landing gear, and interiors, often making up 80% of a plane's weight. The food packaging industry is a fast-growing market.

In electrical applications, aluminum wire and cable are major products. Aluminum appears in the home as cooking utensils, cooking foil, hardware, tools, portable appliances, air conditioners, freezers, and refrigerators, and in sporting equipment such as skis, ball bats, and tennis rackets.

There are hundreds of chemical uses of aluminum and aluminum compounds.

Aluminum powder is used in paints, rocket fuels, and explosives, and as a chemical reductant. [A.S.Ru.]

**Americium** A chemical element, symbol Am, atomic number 95. The isotope $^{241}$Am is an alpha emitter with a half-life of 433 years. Other isotopes of americium range in mass from 232 to 247, but only the isotopes of mass 241 and 243 are important. The isotope $^{241}$Am is routinely separated from "old" plutonium and sold for a variety of industrial uses, such as 59-keV gamma sources and as a component in neutron sources. The longer-lived $^{243}$Am (half-life 7400 years) is a precursor in $^{244}$Cm production.

In its most prominent aqueous oxidation state, 3+, americium closely resembles the tripositive rare earths. The formal analogy to the rare earths is also marked in anhydrous compounds of both tripositive and tetrapositive americium. Americium is different in that it is possible to oxidize $Am^{3+}$ to both the 5+ and 6+ states.

Americium metal has a vapor pressure markedly higher than that of its neighboring elements and can be purified by distillation. The metal is nonmagnetic and superconducting at 0.79 K. Under high pressure the metal has been compressed to 80% of its room-temperature volume and displays the $\alpha$-uranium structure. *See* ACTINIDE ELEMENTS; BERKELIUM; CURIUM; PERIODIC TABLE; TRANSURANIUM ELEMENTS. [R.A.Pe.]

**Amide** A derivative of a carboxylic acid with general formula $RCONH_2$, where R is hydrogen or an alkyl or aryl radical. Amides are divided into subclasses, depending on the number of substituents on nitrogen. The simple, or primary, amides are considered to be derivatives formed by replacement of the carboxylic hydroxyl group by the amino group, $NH_2$. They are named by dropping the "-ic acid" or "-oic acid" from the name of the parent carboxylic acid and replacing it with the suffix "amide." In the secondary and tertiary amides, one or both hydrogens are replaced by other groups. The presence of such groups is designated by the prefix capital *N* (for nitrogen).

Except for formamide, all simple amides are relatively low-melting solids, stable, and weakly acidic. They are strongly associated through hydrogen bonding, and hence soluble in hydroxylic solvents, such as water and alcohol. Because of ease of formation and sharp melting points, amides are frequently used for the identification of organic acids and, conversely, for the identification of amines.

Commercial preparation of amides involves thermal dehydration of ammonium salts of carboxylic acids. Thus, slow pyrolysis of ammonium acetate forms water and acetamide. *N, N*-dimethylacetamide may be similarly prepared from dimethylammonium acetate.

Amides are important chemical intermediates since they can be hydrolyzed to acids, dehydrated to nitriles, and degraded to amines containing one less carbon atom by the Hofmann reaction. In pharmacology, acetophenetidin is a popular analgesic. However, the most important commercial application of amides is in the preparation of polyamide resins, also called nylons. *See* ACID ANHYDRIDE; POLYAMIDE RESINS. [P.E.F.]

**Amine** A member of a group of organic compounds which can be considered as derived from ammonia by replacement of one or more hydrogens by organic radicals. Generally amines are bases of widely varying strengths, but a few which are actually acidic are known.

Amines constitute one of the most important classes of organic compounds. The lone pair of electrons on the amine nitrogen enables amines to participate in a large variety of reactions as a base or a nucleophile. Amines play prominent roles in biochemical systems; they are widely distributed in nature in the form of amino acids, alkaloids, and vitamins. Many complex amines have pronounced physiological activity, for example, epinephrine (adrenalin), thiamin or vitamin $B_1$, and Novocaine. The odor of decaying fish is due to simple amines produced by bacterial action. Amines are used to manufacture many medicinal chemicals, such as sulfa drugs and anesthetics. The important synthetic fiber nylon is an amine derivative.

Amines are classified according to the number of hydrogens of ammonia which are replaced by radicals. Replacement of one hydrogen results in a primary amine ($RNH_2$), replacement of two hydrogens results in a secondary amine ($R_2NH$), and replacement of all three hydrogens results in a tertiary amine ($R_3N$). The substituent groups (R) may be alkyl, aryl, or aralkyl. Another group of amines are those in which the nitrogen forms part of a ring (heterocyclic amines). Examples of such compounds are nicotine, which is obtained commercially from tobacco for use as an insecticide, and serotonin, which plays a key role as a chemical mediator in the central nervous system.

Many aromatic and heterocyclic amines are known by trivial names, and derivatives are named as substitution products of the parent amine. Thus, $C_6H_5NH_2$, is aniline and $C_6H_5NHC_2H_5$ is *N*-ethylaniline.

According to the Brönsted-Lowry theory of acids and bases, amines are basic because they accept protons from acids. Stable salts suitable for the identification of amines are in general formed only with strong acids, such as hydrochloric, sulfuric, oxalic, chloroplatinic, or picric.

Commercial preparation of aliphatic amines can be accomplished by direct alkylation of ammonia or by catalytic alkylation of amines with alcohols at elevated temperatures. Reduction of various nitrogen functions carrying the nitrogen in a higher state of oxidation also leads to amines. Such functions are nitro, oximino, nitroso, and cyano. For the preparation of pure primary amines, Gabriel's synthesis and Hofmann's hypohalite reaction are preferred methods. The Bucherer reaction is satisfactory for the preparation of polynuclear primary aromatic amines. *See* AMINO ACIDS. [P.E.F.]

**Amino acids** Organic compounds possessing one or more basic amino groups and one or more acidic carboxyl groups. Of the more than 80 amino acids which have been found in living organisms, about 20 serve as the building blocks for the proteins.

All the amino acids of proteins, and most of the others which occur naturally, are $\alpha$-amino acids, meaning that an amino group ($-NH_2$) and a carboxyl group ($-COOH$) are attached to the same carbon atom. This carbon (the $\alpha$ carbon, being adjacent to the carboxyl group) also carries a hydrogen atom; its fourth valence is satisfied by any of a wide variety of substitutent groups, represented by the letter R in the structural formula below.

```
        R
        |
        CH
       /  \
   H2N    COOH
```

In the simplest amino acid, glycine, R is a hydrogen atom. In all other amino acids, R is an organic radical; for example, in alanine it is a methyl group ($-CH_3$), while in glutamic acid it is an aliphatic chain terminating in a second carboxyl group ($-CH_2-CH-COOH$). Chemically, the amino acids can be considered as falling roughly into nine categories based on the nature of R (see table).

**Occurrence.** Amino acids occur in living tissues principally in the conjugated form. Most conjugated amino acids are peptides, in which the amino group of one amino

**Amino acids of proteins, grouped according to the nature of R**

| Amino acids | R |
|---|---|
| Glycine | Hydrogen |
| Alanine, valine, leucine, isoleucine | Unsubstituted aliphatic chain |
| Serine, threonine | Aliphatic chain bearing a hydroxyl group |
| Aspartic acid, glutamic acid | Aliphatic chain terminating in an acidic carboxyl group |
| Asparagine, glutamine | Aliphatic chain terminating in an amide group |
| Arginine, lysine | Aliphatic chain terminating in a basic amino group |
| Cysteine, cystine, methionine | Sulfur-containing aliphatic chain |
| Phenylalanine, tyrosine | Terminates in an aromatic ring |
| Tryptophan, proline, histidine | Terminates in a heterocyclic ring |

*See articles on the individual amino acids listed in the table.

acid is linked to the carboxyl group of another. Amino acids are capable of linking together to form chains of various lengths, called polypeptides. Proteins are polypeptides ranging in size from about 50 to many thousand amino acid residues. Although most of the conjugated amino acids in nature are proteins, numerous smaller conjugates occur naturally, many with important biological activity. The line between large peptides and small proteins is difficult to draw, with insulin (molecular weight = 7000; 50 amino acids) usually being considered a small protein and adrenocorticotropic hormone (molecular weight = 5000; 39 amino acids) being considered a large peptide.

Free amino acids are found in living cells, as well as the body fluids of higher animals, in amounts which vary according to the tissue and to the amino acid. The amino acids which play key roles in the incorporation and transfer of ammonia, such as glutamic acid, aspartic acid, and their amides, are often present in relatively high amounts, but the concentrations of the other amino acids of proteins are extremely low, ranging from a fraction of a milligram to several milligrams per 100 g wet weight of tissue. The presence of free amino acids in only trace amounts points to the existence of extraordinarily efficient regulation mechanisms. Each amino acid is ordinarily synthesized at precisely the rate needed for protein synthesis.

**General properties.** The amino acids are characterized physically by the following: (1) the $pK_1$, or the dissociation constant of the various titratable groups; (2) the isoelectric point, or pH at which a dipolar ion does not migrate in an electric field; (3) the optical rotation, or the rotation imparted to a beam of plane-polarized light (frequently the D line of the sodium spectrum) passing through 1 decimeter of a solution of 100 grams in 100 milliliters; and (4) solubility. *See* IONIC EQUILIBRIUM; ISOELECTRIC POINT; OPTICAL ACTIVITY.

Since all of the amino acids except glycine possess a center of asymmetry at the $\alpha$ carbon atom, they can exist in either of two optically active, mirror-image forms, or enantiomorphs. All of the common amino acids of proteins appear to have the same configuration about the $\alpha$ carbon; this configuration is symbolized by the prefix L-. The opposite, generally unnatural, form is given the prefix D-. Some amino acids, such as isoleucine, threonine, and hydroxyproline, have a second center of asymmetry and can exist in four stereoisomeric forms. *See* STEREOCHEMISTRY.

At ordinary temperatures, the amino acids are white crystalline solids; when heated to high temperatures, they decompose rather than melt. They are stable in aqueous solution, and with few exceptions can be heated as high as 120°C (248°F) for short periods without decomposition, even in acid or alkaline solution. Thus, the hydrolysis of proteins can be carried out under such conditions with the complete recovery of most of the constituent free amino acids.

**Biosynthesis.** Since amino acids, as precursors of proteins, are essential to all organisms, all cells must be able to synthesize those they cannot obtain from their environment. The selective advantage of being able rapidly to shift from endogenous to exogenous sources of these compounds has led to the evolution of very complex and precise methods of adjusting the rate of synthesis to the available level of the compound. An immediately effective control is that of feedback inhibition. The biosynthesis of amino acids usually requires at least three enzymatic steps. In most cases so far examined, the amino acid end product of the biosynthetic pathway inhibits the first enzyme to catalyze a reaction specific to the biosynthesis of that amino acid. This inhibition is extremely specific; the enzymes involved have special sites for binding the inhibitor. This inhibition functions to shut off the pathway in the presence of transient high levels of the product, thus saving both carbon and energy for other biosynthetic reactions. When the level of the product decreases, the pathway begins to function once more.

The metabolic pathways by which amino acids are synthesized generally are found to be the same in all living cells investigated, whether microbial or animal. Biosynthetic mechanisms thus appear to have developed soon after the origin of life and to have remained unchanged through the divergent evolution of modern organisms.

Biosynthetic pathway diagrams reveal only one quantitatively important reaction by which organic nitrogen enters the amino groups of amino acids: the reductive amination of $\alpha$-ketoglutaric acid to glutamic acid by the enzyme glutamic acid dehydrogenase. All other amino acids are formed either by transamination (transfer of an amino group, ultimately from glutamic acid) or by a modification of an existing amino acid. An example of the former is the formation of valine by transfer of the amino group from glutamic acid to $\alpha$-ketoisovaleric acid; an example of the latter is the reduction and cyclization of glutamic acid to form proline.

**Importance in nutrition.** The nutritional requirement for the amino acids of protein can vary from zero, in the case of an organism which synthesizes them all, to the complete list, in the case of an organism in which all the biosynthetic pathways are blocked. There are 8 or 10 amino acids required by certain mammals; most plants synthesize all of their amino acids, while microorganisms vary from types which synthesize all, to others (such as certain lactic acid bacteria) which require as many as 18 different amino acids. [E.A.Ad.; P.T.M.; R.G.M.]

**Ammine** One of a group of complex compounds formed by the coordination of ammonia molecules with metal ions and, in a few instances, such as calcium, strontium, and barium, with metal atoms. Although ammines are formally analogous to many salt hydrates, the general characteristics of the group of ammines differ considerably from those of the hydrates. For example, hydrated Co(III) salts are strong oxidizing agents whereas Co(II) ammines are strong reducing agents. The ammines of principal interest are those of the transition metals and of the zinc family, but even here there is wide variation in stability or rate of decomposition. Ammines are prepared by treating aqueous solutions of the metal salt with ammonia or, in some instances, by the action of dry gaseous or liquid ammonia on the anhydrous salt. *See* AMMONIA; COORDINATION CHEMISTRY. [H.H.S.]

**Ammonia** The most familiar compound composed of the elements nitrogen and hydrogen, $NH_3$. It is formed as a result of the decomposition of most nitrogenous organic material, and its presence is indicated by its pungent and irritating odor.

Ammonia has a wide range of industrial and agricultural applications. Examples of its use are the production of nitric acid and ammonium salts, particularly the sulfate, nitrate, carbonate, and chloride, and the synthesis of hundreds of organic compounds including many drugs, plastics, and dyes. Its dilute aqueous solution finds use as a household cleansing agent. Anhydrous ammonia and ammonium salts are used as fertilizers, and anhydrous ammonia also serves as a refrigerant, because of its high heat of vaporization and relative ease of liquefaction.

The physical properties of ammonia are analogous to those of water and hydrogen fluoride in that the physical constants are abnormal with respect to those of the binary hydrogen compounds of the other members of the respective periodic families. These abnormalities may be related to the association of molecules through intermolecular hydrogen bonding. Ammonia is highly mobile in the liquid state and has a high thermal coefficient of expansion.

Most of the chemical reactions of ammonia may be classified under three chief groups: (1) addition reactions, commonly called ammonation; (2) substitution reactions, commonly called ammonolysis; and (3) oxidation-reduction reactions.

Ammonation reactions include those in which ammonia molecules add to other molecules or ions. Most familiar of the ammonation reactions is the reaction with water to form ammonium hydroxide. The strong tendency of water and ammonia to combine is evidenced by the very high solubility of ammonia in water. Ammonia reacts readily with strong acids to form ammonium salts. Ammonium salts of weak acids in the solid state dissociate readily into ammonia and the free acid. Ammonation occurs with a variety of molecules capable of acting as electron acceptors (Lewis acids), such as sulfur trioxide, sulfur dioxide, silicon tetrafluoride, and boron trifluoride. Included among ammonation reactions is the formation of complexes (called ammines) with many metal ions, particularly transition metal ions. Ammonolytic reactions include reactions of ammonia in which an amide group ($—NH_2$), an imide group ($=NH$), or a nitride group ($\equiv N$) replaces one or more atoms or groups in the reacting molecule.

Oxidation-reduction reactions may be subdivided into those which involve a change in the oxidation state of the nitrogen atom and those in which elemental hydrogen is liberated. An example of the first group is the catalytic oxidation of ammonia in air to form nitric oxide. In the absence of a catalyst, ammonia burns in oxygen to yield nitrogen. Another example is the reduction with ammonia of hot metal oxides such as cupric oxide.

The physical and chemical properties of liquid ammonia make it appropriate for use as a solvent in certain types of chemical reactions. The solvent properties of liquid ammonia are, in many ways, qualitatively intermediate between those of water and of ethyl alcohol. This is particularly true with respect to dielectric constant; therefore, ammonia is generally superior to ethyl alcohol as a solvent for ionic substances but is inferior to water in this respect. On the other hand, ammonia is generally a better solvent for covalent substances than is water.

The Haber-Bosch synthesis is the major source of industrial ammonia. In a typical process, water gas ($CO$, $H_2$, $CO_2$) mixed with nitrogen is passed through a scrubber cooler to remove dust and undecomposed material. The $CO_2$ and $CO$ are removed by a $CO_2$ purifier and ammoniacal cuprous solution, respectively. The remaining $H_2$ and $N_2$ gases are passed over a catalyst at high pressures (up to 1000 atm or 100 megapascals) and high temperatures (approx. 1300°F or 700°C). Other industrial sources of ammonia include its formation as a by-product of the destructive distillation of coal, and its synthesis through the cyanamide process. In the laboratory, ammonia is usually formed by its displacement from ammonium salts (either dry or in solution) by strong bases. Another source is the hydrolysis of metal nitrides. *See* AMIDE; NITROGEN. [H.H.S.]

**Ammonium salt** A product of a reaction between ammonia, $NH_3$, and various acids. The general reaction for formation is $NH_3 + HX \rightarrow NH_2X$. Examples of ammonium salts are ammonium chloride, $NH_4Cl$, ammonium nitrate, $NH_4NO_3$, ammonium sulfate, $(NH_4)_2SO_4$, and ammonium carbonate, $(NH_4)_2CO_3$. These compounds are addition products of ammonia and the acid. For this reason, their formulas are sometimes written as $[H(NH_3)]X$.

All ammonium salts decompose into ammonia and the acid when heated. Their stability, however, varies according to the nature of the acid. Salts of weak acids decompose at lower temperatures than do salts of strong acids.

Ammonium chloride is made by absorbing ammonia in hydrochloric acid. This salt, sometimes called sal ammoniac, is used in galvanizing iron, in textile dyeing, and in manufacturing dry cell batteries.

Ammonium nitrate is prepared from ammonia and nitric acid. It is used as a source of nitrous oxide, $N_2O$, or laughing gas, and in the manufacture of explosives. A mixture of ammonium nitrate and trinitrotoluene is known as amatol.

Ammonium sulfate, obtained from ammonia and sulfuric acid, is prepared commercially by passing ammonia and carbon dioxide, $CO_2$, into a suspension of finely ground calcium sulfate, $CaSO_4$. Large quantities are also produced as a byproduct of coke ovens and coal-gas works. The chief use of ammonium sulfate is as a fertilizer.

Ammonium carbonate may be prepared by bringing ammonia and carbon dioxide together in aqueous solution. It is also obtained by heating a mixture of ammonium sulfate and a fine suspension of calcium carbonate. *See* AMMONIA; HYDROLYSIS. [F.J.J.]

**Analytical chemistry** The science of chemical characterization and measurement. Qualitative analysis is concerned with the description of chemical composition in terms of elements, compounds, or structural units, whereas quantitative analysis is concerned with the measurement of amount.

Analytical chemistry, once limited to the determination of chemical composition in terms of the relative amounts of elements or compounds in a sample, has been expanded to involve the spatial distribution of elements or compounds in a sample, the distinction between different crystalline forms of a given element or compound, the distinction between different chemical forms (such as the oxidation state of an element), the distinction between a component on the surface or in the interior of a particle, and the detection of single atoms on a surface. To permit these more detailed questions to be answered, as well as to improve the speed, accuracy, sensitivity, and selectivity of traditional analysis, a large variety of physical measurements are used. These methods are based on spectroscopic, electrochemical, chromatographic, chemical, and nuclear principles.

Modern analysis has also placed significant demands on sampling techniques. It has become necessary, for example, to handle very small liquid samples [in the nanoliter ($10^{-9}$ liter) range or less] as part of the analysis of complex mixtures such as biological fluids and to simultaneously determine many different components. The sample may be a solid that must be converted through vaporization into a form suitable for analysis.

Spectroscopy includes the measurement of emission, absorption, reflection, and scattering phenomena resulting from interaction of a sample with gamma rays and x-rays at the high-energy end of the spectrum and with the less energetic ultraviolet, visible, infrared, and microwave radiation. *See* SPECTROSCOPY.

Lower-energy forms of excitation such as ultraviolet, visible, or infrared radiation are used in molecular spectroscopy. Ultraviolet radiation and visible radiation, which are reflective of the electronic structure of molecules, are used extensively for quantitative analysis. The radiation absorbed by the sample is measured. It is also possible to measure the radiation emitted (fluorescence). The absorption of infrared radiation is controlled by the properties of bonds between atoms, and it is accordingly most widely used for structure identification and determination. It is not widely used for quantitative analysis except for gases such as carbon monoxide (CO) and hydrocarbons. X-rays are used through emission of characteristic radiation, absorption, or diffraction. In the last case, characteristic diffraction patterns reveal information about specific structural entities, such as a particular crystalline form. Extended x-ray absorption fine structure (EXAFS) is based on the use of x-rays from a synchrotron source to reveal structural details such as interatomic distances. *See* EMISSION SPECTROCHEMICAL ANALYSIS; EXTENDED X-RAY ABSORPTION FINE STRUCTURE (EXAFS); X-RAY FLUORESCENCE ANALYSIS.

Though not strictly a spectroscopic technique, mass spectrometry is an important and increasingly applied method of analysis, especially for organic and biological samples. Among the applications are the analysis of more than 70 elements (spark-source mass spectrometry), surface analysis (secondary ion mass spectrometry and ion-probe mass spectrometry), and the determination of the structure of organic molecules and of proteins and peptides (high-resolution mass spectrometry). *See* MASS SPECTROMETRY; SECONDARY ION MASS SPECTROMETRY (SIMS).

Nuclear magnetic resonance measures the magnetic environment around individual atoms and provides one of the most important means for deducing the structure of a molecule. Atoms possessing nuclear spin are probed by monitoring the interaction between their nuclear spin and an applied external magnetic field. For large molecules these interactions are complex, and a variety of nuclear excitation techniques have been developed that permit establishment of the connectivity between the various atoms in a molecule. Since the technique is nondestructive, it can be used to monitor living systems. *See* NUCLEAR MAGNETIC RESONANCE (NMR).

Several forms of spectroscopy are especially useful for surface analysis. The scanning electron microscope (SEM) involves a finely collimated electron beam that sweeps across the surface to produce an image. At the same time the surface atoms are excited to emit characteristic x-rays, thus making it possible to obtain an image of the surface along with its spatially resolved elemental composition. The resolution of this technique (electron microprobe) is in the micrometer ($10^{-4}$ cm) range. Images with a resolution of angstroms ($10^{-8}$ cm) have been obtained by using the techniques of atomic force microscopy (AFM) and scanning tunneling microscopy (STM), which correspond to the dimensions of individual atoms. A significant advantage of the latter two techniques is that a high vacuum is not required, so samples can be analyzed at atmospheric pressure. *See* ELECTRON-PROBE MICROANALYSIS; ELECTRON SPECTROSCOPY.

Potentiometry is the most widely applied electrochemical technique, since it includes a variety of ion-selective electrodes, the most important of which is the glass electrode used to measure pH. Other important ion-selective electrodes measure ions of sodium, potassium, calcium, sulfide, chloride, and fluoride. When the electrodes are used in conjunction with gas-permeable membranes, gases such as ammonia, carbon dioxide, and hydrogen sulfide can be measured. *See* ELECTROCHEMISTRY; ION-SELECTIVE MEMBRANES AND ELECTRODES; PH; POLAROGRAPHIC ANALYSIS. [H.A.L.; G.S.W.]

Separation techniques include the various forms of chromatography and electrophoresis. They are based on the separation of a mixture of species in a sample due to differential migration. Two forces act in opposition: a stationary phase acts to retard a migrating species, while the mobile phase tends to promote migration. The mobile phase may be liquid (liquid chromatography) or gaseous (gas chromatography), while the stationary phase may be a solid or a solid covered with a thin film of liquid. The stationary phase is typically packed in a column through which the mobile phase is pumped. High-performance liquid chromatography (HPLC) has become especially important for the separation of complex mixtures of nonvolatile materials. Separations may often be accomplished in a matter of several minutes. The stationary phase can preferentially interact with the migrating species according to charge, size, hydrophobicity, or in some cases because of the special affinity which a species has for the stationary phase (affinity chromatography). The stationary phase can also be a thin layer of solid support deposited on a plate (thin-layer chromatography). *See* CHROMATOGRAPHY; GAS CHROMATOGRAPHY; LIQUID CHROMATOGRAPHY.

Alternatively, the driving force for separation will be the migration of charged species in an electric field (electrophoresis). The stationary phase may be a gel on a plate or in a tube, or a solution maintained in a capillary through which the analytes move. The important techniques in this area are capillary electrophoresis, isotachophoresis, and isoelectric focusing. *See* ELECTROPHORESIS.

Thermal methods are based on the heating of a sample over a range of temperatures. This approach may result in absorption of heat by the sample or in evolution of heat due to physical or chemical changes. Thermogravimetry involves the measurement of mass; differential thermal analysis involves a detection of chemical or physical processes through a measurement of the difference in temperature between a sample and a stable reference material; differential thermal calorimetry evaluates the heat evolved in such processes. A variety of calorimetric techniques are used to measure the extent of reactions that are otherwise difficult to evaluate. *See* CALORIMETRY. [G.S.W.]

**Antimony** A chemical element, symbol Sb, atomic number 51. Antimony is not a naturally abundant element; it is occasionally found native, often in isomorphous mixture with arsenic, as allemonite. The symbol Sb is derived from the Latin name stibium. *See* PERIODIC TABLE.

The element is dimorphic, existing as a yellow, metastable form composed of $Sb_4$ molecules, as in antimony vapor and the structural unit in yellow antimony; and a gray, metallic form, which crystallizes with a layered rhombohedral structure. Antimony differs from normal metals in having a lower electrical conductivity as a solid than as a liquid (as does its congener, bismuth). Metallic antimony is quite brittle, bluish-white with a typical metallic luster, but a flaky appearance. Although stable in air at normal temperatures, it burns brilliantly when heated, with the formation of a white smoke of $Sb_2O_3$. Vaporization of the metal gives molecules of $Sb_4O_6$, which break down to $Sb_2O_3$ above the transition temperature.

Antimony occurs in nature mainly as $Sb_2S_3$ (stibnite, antimonite); $Sb_2O_3$ (valentinite) occurs as a decomposition product of stibnite. Antimony is commonly found in ores of copper, silver, and lead. The metal antimonides NiSb (breithaupite), NiSbS (ullmannite), and $Ag_2Sb$ (dicrasite) also are found naturally; there are numerous thioantimonates such as $Ag_3SbS_3$ (pyrargyrite).

Antimony is produced either by roasting the sulfide with iron, or by roasting the sulfide and reducing the sublimate of $Sb_4O_6$ thus produced with carbon; high-purity antimony is produced by electrolytic refining.

Commercial-grade antimony is used in many alloys (1–20%), especially lead alloys, which are much harder and mechanically stronger than pure lead; batteries, cable sheathing, antifriction bearings, and type metal consume almost half of all the antimony produced. The valuable property of Sn-Sb-Pb alloys, that they expand on cooling from the melt, thus enabling the production of sharp castings, makes them especially useful as type metal. [J.L.T.W.]

**Antioxidant** A substance that, when present at a lower concentration than that of the oxidizable substrate, significantly inhibits or delays oxidative processes, while being itself oxidized. In primary antioxidants, such as polyphenols, this antioxidative activity is implemented by the donation of an electron or hydrogen atom to a radical derivative, and in secondary antioxidants by the removal of an oxidative catalyst and the consequent prevention of the initiation of oxidation.

Antioxidants have diverse applications. They are used to prevent degradation in polymers, weakening in rubber and plastics, autoxidation and gum formation in gasoline, and discoloration of synthetic and natural pigments. They are used in foods, beverages, and cosmetic products to inhibit deterioration and spoilage. Interest is increasing in the application of antioxidants to medicine relating to human diseases attributed to oxidative stress.

The autoxidation process is shown in reactions (1), (2), and (3). Lipids, mainly those containing unsaturated fatty acids, such as linoleic acid [RH in reaction (1)], can

$$\mathrm{RH} + \text{initiator (L)} \rightarrow \mathrm{R\cdot} + \mathrm{LH} \tag{1}$$

$$\mathrm{R\cdot} + \mathrm{O_2} \rightarrow \mathrm{ROO\cdot} \tag{2}$$

$$\mathrm{ROO\cdot} + \mathrm{RH} \rightarrow \mathrm{ROOH} + \mathrm{R\cdot} \tag{3}$$

undergo autoxidation via a free-radical chain reaction, which is unlikely to take place with atmospheric oxygen (ground state) alone. A catalyst (L) is required, such as light, heat, heavy-metal ions (copper or iron), or specific enzymes present in the biological system [reaction (1)]. The catalyst allows a lipid radical to be formed (alkyl radical R·) on a carbon atom next to the double bond of the unsaturated fatty acid. This radical is very unstable and reacts with oxygen [reaction (2)] to form a peroxyl radical (ROO·), which in turn can react with an additional lipid molecule to form a hydroperoxide [ROOH in reaction (3)] plus a new alkyl radical, and hence to start a chain reaction. Reactions (2) and (3), the propagation steps, continue unless a decay reaction takes place (a termination step), which involves the combination of two radicals to form stable products. *See* AUTOXIDATION; CATALYSIS; CHAIN REACTION (CHEMISTRY).

When lipid autoxidation occurs in food, it can cause deterioration, rancidity, bad odor, spoilage, reduction in nutritional value, and possibly the formation of toxic byproducts. Oxidation stress in a lipid membrane in a biological system can alter its structure, affect its fluidity, and change its function, causing disease.

An antioxidant can eliminate potential initiators of oxidation and thus prevent reaction (1). It can also stop the process by donating an electron and reducing one of the radicals in reaction (2) or (3), thus halting the propagation steps. A primary antioxidant can be effective if it is able to donate an electron (or hydrogen atom) rapidly to a lipid radical and itself become more stable then the original radical. The ease of electron donation depends on the molecular structure of the antioxidant, which dictates the stability of the new radical. Many naturally occurring polyphenols, such as flavonoids, anthocyanins, and saponins, which can be found in wine, fruit, grain, vegetables, and almost all herbs and spices, are effective antioxidants that operate by this mechanism.

A secondary antioxidant can prevent reaction (1) from taking place by absorbing ultraviolet light, scavenging oxygen, chelating transition metals, or inhibiting enzymes involved in the formation of reactive oxygen species, for example, NADPH oxidase and xanthine oxidase (reducing molecular oxygen to superoxide and hydrogen peroxide), dopamine-$\beta$-hydroxylase, and lipoxygenases. The common principle of action in the above examples is the removal of the component acting as the catalyst that initiates and stimulates the free-radical chain reaction. *See* ENZYME.

Among antioxidants, the synthetic compounds butylated hydroxyanisole (BHA), propyl gallate, ethoxyquin, and diphenylamine are commonly used as food additives. Quercetin belongs to a large natural group of antioxidants, the flavonoid family, with more than 6000 known members, many acting through both mechanisms described above. Ascorbic acid is an important water-soluble plasma antioxidant; it and the tocopherols, the main lipid soluble antioxidants, represent the antioxidants in biological systems. $\beta$-Carotene belongs to the carotenoid family, which includes lycopene, the red pigment in tomatoes; the family is known to be very effective in reacting with singlet oxygen ($^1O_2$), a highly energetic species of molecular oxygen. *See* ASCORBIC ACID; CAROTENOID. [J.Va.; L.P.]

**Aqua regia** A mixture of one part by volume of concentrated nitric acid and three parts of concentrated hydrochloric acid. Aqua regia was so named by the alchemists because of its ability to dissolve platinum and gold. Either acid alone will not dissolve these noble metals. [E.E.W.]

**Argon** A chemical element, Ar, atomic number 18, and atomic weight 39.948. Argon is the third member of group 18 in the periodic table. The gaseous elements in this group are called the noble, inert, or rare gases, although argon is not actually rare. The Earth's atmosphere is the only natural argon source; however, traces of this gas are found in minerals and meteorites. Argon constitutes 0.934% by volume of the Earth's atmosphere. Of this argon, 99.6% is the argon-40 isotope; the remainder is argon-36 and argon-38. There is good evidence that all the argon-40 in the air was produced by the radioactive decay of the radioisotope potassium-40. *See* INERT GASES; PERIODIC TABLE.

Argon is colorless, odorless, and tasteless. The element is a gas under ordinary conditions, but it can be liquefied and solidified readily. Some salient properties of the gas are listed in the table. Argon does not form any chemical compounds in the ordinary sense of the word, although it does form some weakly bonded clathrate compounds with water, hydroquinone, and phenol. There is one atom in each molecule of gaseous argon.

The oldest large-scale use for argon is in filling electric light bulbs. Welding and cutting metal consumes the largest amount of argon. Metallurgical processing constitutes the most rapidly growing application. Argon and argon-krypton mixtures are used, along with a little mercury vapor, to fill fluorescent lamps. Argon mixed with a little neon is used to fill luminous electric-discharge tubes employed in advertising signs (similar to

**Properties of argon**

| Property | Value |
|---|---|
| Atomic number | 18 |
| Atomic weight (atmospheric argon) | 39.948 |
| Melting point (triple point), °C | −189.4 |
| Boiling point at 1 atm pressure, °C | −185.9 |
| Gas density at 0°C and 1 atm (101.325 kPa) pressure, g/liter | 1.7840 |
| Liquid density at normal boiling point, g/ml | 1.3998 |
| Solubility in water at 20°C, ml argon (STP) per 1000 g water at 1 atm (101.325 kPa) partial pressure of argon | 33.6 |

neon signs) when a blue or green color is desired instead of the red color of neon. Argon is also used in gas-filled thyratrons, Geiger-Müller radiation counters, ionization chambers which measure cosmic radiation, and electron tubes of various kinds. Argon atmospheres are used in dry boxes during manipulation of very reactive chemicals in the laboratory and in sealed-package shipments of such materials.

Most argon is produced in air-separation plants. Air is liquefied and subjected to fractional distillation. Because the boiling point of argon is between that of nitrogen and oxygen, an argon-rich mixture can be taken from a tray near the center of the upper distillation column. The argon-rich mixture is further distilled and then warmed and catalytically burned with hydrogen to remove oxygen. A final distillation removes hydrogen and nitrogen, yielding a very high-purity argon containing only a few parts per million of impurities. [A.W.F.]

**Aromatic hydrocarbon** A hydrocarbon with a chemistry similar to that of benzene. Aromatic hydrocarbons are either benzenoid or nonbenzenoid. Benzenoid aromatic hydrocarbons contain one or more benzene rings and are by far the more common and the more important commercially. Nonbenzenoid aromatic hydrocarbons have carbon rings that are either smaller or larger than the six-membered benzene ring. Their importance arises mainly from a theoretical interest in understanding those structural features that impart the property of aromaticity.

Benzenoid aromatic hydrocarbons are also called arenes. Benzene itself is the prototypical arene. The properties associated with aromaticity have little to do with aroma, although the aromatic hydrocarbons were first studied in connection with naturally occurring fragrances. Instead, these compounds possess special stability; take part in certain types of reactions; and exhibit persistence of the structural integrity of aromatic rings during chemical reactions, while groups attached to those rings are chemically altered or manipulated.

**Benzene.** With molecular formula (**1**), benzene is highly unsaturated; it has three double bonds, alternating with single bonds. The double bonds in the benzene structure

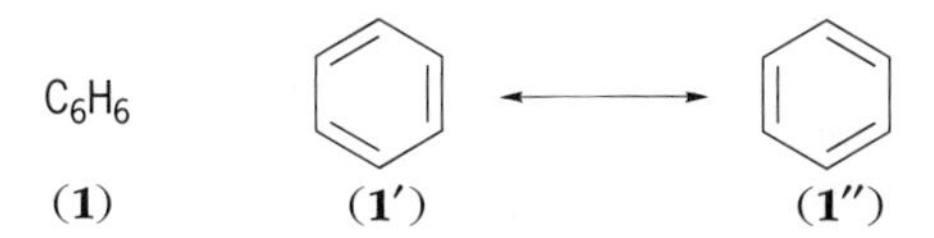


can be arranged in two ways, (**1′**) and (**1″**). Benzene is a resonance hybrid of these two structures, called Kekulé structures; the double-headed arrow is used to signify that the benzene structure is neither (**1′**) nor (**1″**), but a single structure that is a hybrid of the two. That is, the bonds between adjacent carbon atoms are neither double nor single, but of some intermediate or hybrid type.

Each carbon atom in benzene is connected to three atoms, two adjacent carbon atoms and a hydrogen atom. These three bonds lie in a single plane and use three of the carbon's four valence electrons. The fourth valence electron of each carbon is located in a *p* orbital, extending perpendicularly above and below the plane of the other three bonds. These electrons, one from each carbon atom and called $\pi$ electrons, form three molecular orbitals located above and below, but parallel to, the plane of the ring.

The symbol of a hexagon with an inscribed circle (**2**) is often used to express the

(**2**)

delocalized nature of the $\pi$ electrons in benzene and other arenes. There is physical evidence that the $\pi$ electrons circulate around the ring carbons, as implied by this formula. For example, in the nuclear magnetic resonance (NMR) spectra of arenes, the chemical shifts of arene hydrogen atoms (protons) are characteristically at lower magnetic fields than those of protons attached to carbon-carbon double bonds. This difference is due to an induced magnetic field caused by circulation of the $\pi$ electrons in the molecular orbitals above and below the arene ring plane. Indeed, this chemical shift difference, due to a diamagnetic ring current, is sometimes used as evidence for aromaticity in nonbenzenoid aromatic hydrocarbons. *See* DELOCALIZATION; MOLECULAR ORBITAL THEORY.

**Other arenes.** Besides benzene itself, several alkylbenzenes are commercially important and produced on a large scale—millions of pounds annually. Production is commonly by the cyclodehydrogenation of alkanes at high temperatures over metallic catalysts such as platinum.

Benzene, toluene, and the xylenes are added to unleaded gasoline to raise the octane number. These arenes are also essential to the petrochemical industry. Products derived from them include polyesters, polyurethanes, polystyrene, and synthetic rubber; alkylbenzenesulfonate detergents; phenol and acetone; pharmaceuticals, flavors, and perfumes; plasticizers; and many others. *See* PETROCHEMICAL.

Arenes with fused rings are also known as polynuclear aromatic hydrocarbons. Rings are said to be fused when they share two carbon atoms. The simplest example is naphthalene (**3**), a colorless crystalline compound found in coal tar, best known as a moth repellent.

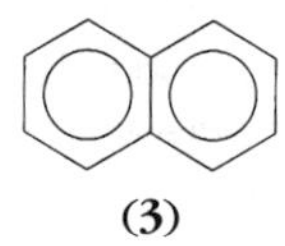

(3)

Additional arene rings can be fused. For example, anthracene, tetracene, and pentacene are linearly fused, while phenanthrene, triphenylene, and pyrene are angularly fused (see illustration). In general, angular fusion results in more stable systems than linear fusion. Phenanthrene, for example, is about 6 kcal/mol more stable than its linear isomer anthracene. Stability falls off sharply in the linearly fused series, and compounds with more than seven such rings are unknown.

**Hückel rule.** From molecular orbital theory, E. Hückel derived the rule that planar, cyclic conjugated (alternate single and double bonds) systems with $4n + 2\pi$ electrons

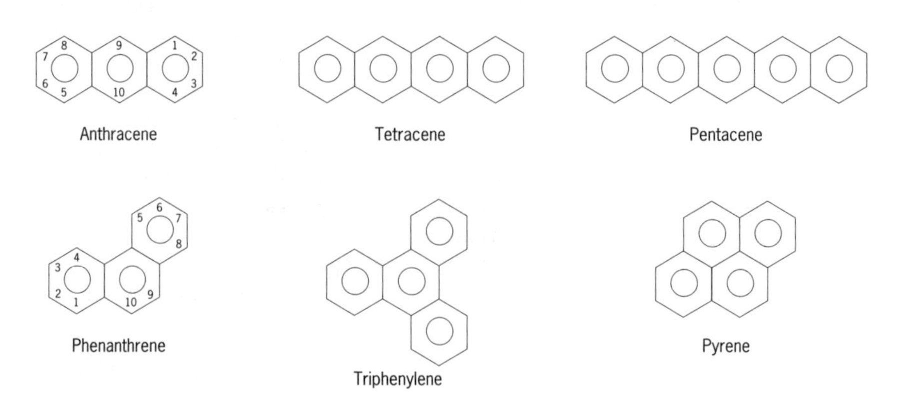


**Structures of some fused-ring (polynuclear) aromatic hydrocarbons.**

($n$ is an integer, 0, 1, 2, . . . ) will be aromatic and have substantial resonance energy, whereas those with $4n$ such electrons will not; indeed, it was later shown that $4n$ systems are often destabilized, hence antiaromatic. Benzene is a $4n + 2$ system ($n = 1$) and aromatic. As striking confirmation of these ideas, pentalene (**4**), a planar analog of cyclooctatetraene, is exceptionally reactive, unstable, and antiaromatic (a $4n$ system, $n = 2$), whereas the purple hydrocarbon azulene (**5**; a $4n = 2$ system, $n = 2$, and

(**4**) (**5**)

an isomer of naphthalene) is stable and undergoes substitution reactions analogous to those of benzenoid arenes. [H.Ha.]

**Aromatization** The conversion of any nonaromatic hydrocarbon structures, especially those found in petroleum, to aromatic hydrocarbons. There are numerous routes and means to accomplish this transformation, the simplest and most important of which are direct dehydrogenation of naphthenes to aromatics, reaction (1); dehydroisomerization of naphthenes to aromatics, reaction (2); dehydrocyclization of aliphatics to aromatics, reaction (3); and high-temperature condensation of hydrocarbons to aromatics, reaction (4).

$$\text{Cyclohexane} \longrightarrow \text{Benzene} + 3H_2 \text{ (Hydrogen)} \quad (1)$$

$$\text{Methylcyclopentane} \longrightarrow \text{Benzene} + 3H_2 \text{ (Hydrogen)} \quad (2)$$

$$H_3C-CH_2-CH_2-CH_2-CH_2-CH_2-CH_3 \text{ (} n\text{-Heptane)} \longrightarrow \text{Toluene} + 4H_2 \text{ (Hydrogen)} \quad (3)$$

$$3C_3H_8 \text{ (Propane)} \longrightarrow \text{Benzene} + 3CH_4 \text{ (Methane)} + 3H_2 \text{ (Hydrogen)} \quad (4)$$

Reforming of naphthas with catalysts comprising small amounts of platinum on an acidified alumina support accomplishes reactions (1), (2), and (3) readily and simultaneously. It is a major process for benzene, toluene, and other aromatics from petroleum sources.

Reaction (4) illustrates one type of reaction that may occur in the high-temperature (600–800°C or 1100–1500°F) thermal cracking of petroleum fractions. *See* PETROLEUM PROCESSING AND REFINING. [B.S.G.; M.Sou.]

**Arsenic** A chemical element, symbol As, atomic number 33. Arsenic is found widely distributed in nature (approximately $5 \times 10^{-4}$ of the Earth's crust). It is one of the 22 known elements composed of only one stable nuclide, ${}^{75}_{33}As$; the atomic weight is 74.92158. There are 17 other radioactive arsenic nuclides known.

There are three polymorphic modifications of arsenic. The yellow cubic $\alpha$-form is made by condensing the vapor at very low temperatures. The black $\beta$-polymorph is isostructural with black phosphorus. Both these modifications revert to the stable $\gamma$-form, gray or metallic, rhombohedral arsenic, on heating or exposure to light. The metallic form is a moderately good thermal and electric conductor and is brittle, easily fractured, and of low ductility.

Arsenic is found native as the mineral scherbenkobalt, but generally occurs among surface rocks combined with sulfur or metals such as Mn, Fe, Co, Ni, Ag, or Sn. The principal arsenic mineral is FeAsS (arsenopyrite, mispickel); other metal arsenide ores are $FeAs_2$ (löllingite), NiAs (nicolite), CoAsS (cobalt glance), NiAsS (gersdorffite), and $CoAs_2$ (smaltite). Naturally occurring arsenates and thioarsenates are common, and most sulfide ores contain arsenic. $As_4S_4$ (realgar) and $As_4S_6$ (orpiment) are the most important sulfur-containing minerals. The oxide, arsenolite, $As_4O_6$ is found as the product of the weathering of other arsenical minerals, and is also recovered from flue dusts collected during the extraction of Ni, Cu, and Sn from their ores; it also results when the arsenides of Fe, Co, or Ni are roasted in air or oxygen. The element may be obtained by roasting FeAsS or $FeAs_2$ in the absence of air or by reduction of $As_4O_6$ carbon, when $As_4$ may be sublimed away.

Elemental arsenic has few uses. It is one of the few minerals available in 99.9999+% purity, which is largely used in the laser material GaAs and as a doping agent in the manufacture of various solid-state devices. Arsenic oxide is used in glass manufacture. The arsenic sulfides are used as pigments and in pyrotechnics. Dihydrogen arsenate is used in medicine, as are several other arsenic compounds. Most of the medicinal uses of arsenic compounds depend on their toxic nature. *See* ANTIMONY; PERIODIC TABLE; PHOSPHORUS. [J.L.T.W.]

**Art conservation chemistry** The application of chemistry to the technical examination, authentication, and preservation of cultural property. Chemists working in museums engage in a broad range of investigations, most frequently studying the chemical composition and structure of artifacts, their corrosion products, and the materials used in their repair, restoration, and conservation. The effects of the museum environment, including air pollutants, fluctuations in temperature and relative humidity, biological activity, and ultraviolet and visible illumination, represent a second major area of research. A third area of interest is the evaluation of the effectiveness, safety, and long-term stability of materials and techniques for the conservation of works of art. Though analytical techniques appear to dominate, many other areas of chemistry, biology, physics, and engineering, including polymer chemistry, kinetic studies, imaging methodologies, biodegradation studies, dating methods, computer modeling, metallography, and corrosion engineering, play active roles in conservation science.

Methods of examination may be divided into two classes: those that provide an image of the entire object (holistic examination) or a section of it; and those that provide an analysis at a point on the object, with or without sampling. Nondestructive methods, not requiring sampling, are always preferable. However, modern methods of analysis can be employed on such minute samples that they are in effect nondestructive. In some cases, samples must be taken for methods that are in principle nondestructive because the object is too large to fit into a sample chamber. The ability to analyze minute samples introduces the serious concern that the sample may not be representative of the composition of the artifact but may be an inclusion or contaminant introduced by the experimentalist. With specimens from painted surfaces, great care must be taken to identify areas of restoration.

A further concern arises from the differing depths from which signals originate. On a metal surface, ion scattering spectrometry (ISS) would see the initial fraction-of-a-nanometer, predominantly adsorbed species and contaminants. Secondary ion mass spectrometry (SIMS) would begin to penetrate the oxidized area; Auger electron spectrometry (AES) would examine the bulk of the oxidized layer; and x-ray-induced photoelectron spectrometry (XPS) would give data on the bulk sample some 10 nm below the specimen surface. *See* ACTIVATION ANALYSIS; ANALYTICAL CHEMISTRY; AUGER EFFECT; SECONDARY ION MASS SPECTROMETRY (SIMS).

The most commonly employed holistic method is x-ray radiography, where variations in the density and average atomic number of the sample attenuate an x-ray beam, leaving a negative image on film. Other methods, such as ultraviolet and infrared reflectance and fluorescence, are used to show areas of compositional difference indicating restoration or variation in the pigments used by the artist. *See* LUMINESCENCE ANALYSIS.

In the examination of paintings, small samples are taken under the binocular microscope, embedded in transparent resin, and polished to produce a cross section for microscopic examination. This permits a study of the artist's painting technique and shows how several layers may have been built up to achieve a desired effect. Conservation studies of the composition and technique embrace the entire spectrum of modern chemical analysis. *See* CHEMICAL MICROSCOPY.

The separation of the fake from the authentic is a small but often spectacular aspect of the technical examination of artifacts. In some cases, direct age determination (dendrochronology for panel paintings, fission track dating for uranium glass, radiocarbon dating for organic materials, thermoluminescence dating for ceramics) is possible.

More commonly, the issue of authenticity turns upon anachronisms in composition or technique when the artifact in question is compared to accepted artifacts of the period. Thus, the greater part of the work in the conservation laboratory concerns the building of databases of analyses of composition, trace-element distributions, and studies of technique.

Many artifacts are sensitive to destructive agents in the museum atmosphere. Rapid changes in relative humidity will cause dimensional changes in wood furniture, polychrome sculpture, and panel paintings, leading to cracking and splitting of the wood with loss of painted surface decoration. High relative humidity can lead to mold growth and foxing on books and prints, while low relative humidity will cause photographic prints and films to become brittle.

Oxidation of iron objects, tarnishing of silver plate, and the development of corrosion products on lead artifacts by the action of formic and acetic acids emitted by wooden display cases have regularly been observed in museums.

The common air pollutants sulfur dioxide ($SO_2$) and ozone ($O_3$) have been monitored at elevated levels in museums, libraries, and archives. These pollutants cause the degradation of leather, spotting of photographic prints, and fading of dyes and pigments. Chemical methods of analysis are used to identify degradation products and to study the kinetics of degradation mechanisms. Specialists in air-pollution monitoring use analytical instrumentation to measure ambient pollution levels in museums. [N.S.B.]

**Ascorbic acid** A white, crystalline compound, also known as vitamin C. It is highly soluble in water, which is a stronger reducing agent than the hexose sugars, which it resembles chemically. Vitamin C deficiency in humans has been known for centuries as scurvy. The compound has the structural formula shown below.

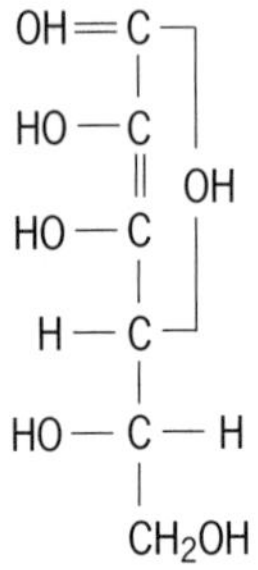

```
OH═C ─┐
    │  │
HO ─C  │
    ║  OH
HO ─C  │
    │  │
 H ─C ─┘
    │
HO ─C ─ H
    │
   CH2OH
```

The stability of ascorbic acid decreases with increases in temperature and pH. This destruction by oxidation is a serious problem in that a considerable quantity of the vitamin C content of foods is lost during processing, storage, and preparation.

While vitamin C is widespread in plant materials, it is found sparingly in animal tissues. Of all the animals studied, only a few, including humans, require a dietary source of vitamin C. The other species are capable of synthesizing the vitamin in such tissues as liver and kidneys. Some drugs, particularly the terpene-like cyclic ketones, stimulate the production of ascorbic acid by rat tissues.

Vitamin C–deficient animals suffer from defects in their mesenchymal tissues. Their ability to manufacture collagen, dentine, and osteoid, the intercellular cement substances, is impaired. This may be related to a role of ascorbic acid in the formation of hydroxy-proline, an amino acid found in structural proteins, particularly collagen. People with scurvy lose weight and are easily fatigued. Their bones are fragile, and their joints sore and swollen. Their gums are swollen and bloody, and in advanced stages their teeth fall out. They also develop internal and subcutaneous hemorrhages.

There is evidence that vitamin C may play roles in stress reactions, in infectious disease, or in wound healing. Therefore, many nutritionists believe that the human intake of ascorbic acid should be many times more than that intake level which produces deficiency symptoms. The recommended dietary allowances of the Food and Nutrition Board of the National Research Council are 30 mg per day for 1- to 3-month infants, 80 mg per day for growing boys and girls, and 100 mg per day for pregnant and lactating women. These values represent an intake which tends to maintain tissue and plasma concentrations in a range similar to that of other well-nourished species of animals. [S.N.G.; W.A.Li.]

**Aspartame** A white, crystalline compound, 1-aspartyl-1-phenylalanine methyl ester (APM), with formula (**1**). It is slightly soluble in water. Its sweetening properties were discovered accidentally in 1965 when the compound, a dipeptide, was

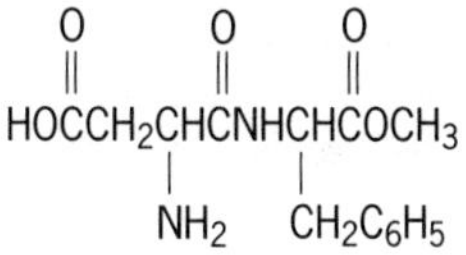

$$HOC(=O)CH_2CH(NH_2)C(=O)NHCH(CH_2C_6H_5)C(=O)OCH_3$$

**(1)**

produced as an intermediate in the synthesis of the C-terminal tetrapeptide of gastrin. Aspartame is the L,L-diastereoisomer; the three other possible diastereoisomers are not sweet. The taste of aspartame would not have been predictable based on its component amino acids, aspartic acid and phenylalanine.

The sweetness of aspartame relative to sucrose is a function of the latter's concentration, and is also dependent upon the presence of other flavors and materials. In a number of applications, such as chewing gum and various fruit-flavored products, aspartame favorably extends and enhances the flavor perception, and it shows synergy with other sweeteners. The sweetness perception may also last longer with aspartame than with sucrose or other sweeteners. *See* SUCROSE.

Aspartame is metabolized to its component amino acids, which are further metabolized by the usual metabolic pathways. Under certain conditions of heat and pH in aqueous solution, aspartame is transformed into its diketopiperazine derivative, 3,6-dioxo-5-benzyl-2-piperazineacetic acid (**2**), which is tasteless.

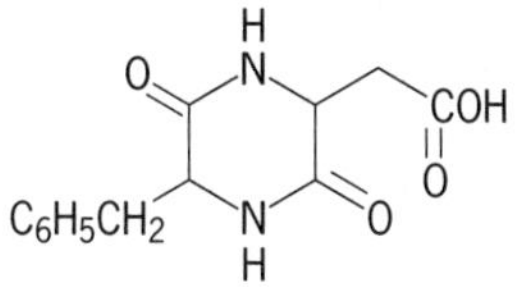

**(2)**

This property limits the use of aspartame when it is exposed to high temperatures, such as in baking. The stability of aspartame in aqueous solution is pH-dependent; it is most stable at a pH of approximately 4. The rate of conversion (its half-life is 262 days at 77°F or 25°C) is sufficiently slow under the conditions of normal use that aspartame has found an increasing number of applications in various food products, and is particularly successful in soft drinks. The safety of aspartame has been established by studies in animals and human beings. Aspartame has been approved in many countries for uses in both dry and wet applications. [D.L.A.]

**Aspirin** The acetyl ester of salicylic acid, also known as 2-(acetyloxy)-benzoic acid and acetylsalicylic acid (see structure below). Aspirin is prepared by the acetylation of salicylic acid with acetic anhydride.

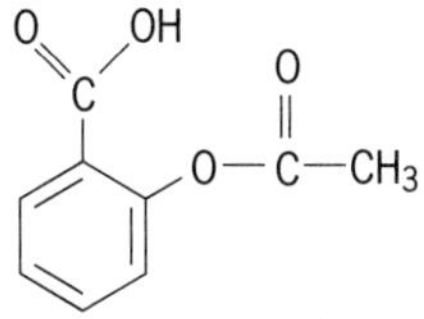

Aspirin is effective as an analgesic, antipyretic, and anti-inflammatory drug. It prevents the aggregation of platelets, and there is some evidence that it can prevent stroke. Aspirin, if tolerated, is the preferred drug for the treatment of rheumatoid arthritis, and it has been used in the treatment of osteoarthritis. Aspirin lowers fever, probably by acting on the hypothalamus. Salicylates inhibit aldose reductase in the lens; it has been suggested that they might retard the development of cataracts. Aspirin might encourage the development of Reye's syndrome, an acute encephalopathy which occurs in children who recover from viral disease, but this cause-and-effect relationship remains to be confirmed. *See* ANALGESIC.

Intolerance to aspirin is not uncommon. It tends to develop in middle age and involve the skin or the respiratory tract, or both. Death rarely ensues because people rapidly become aware of their intolerance. [M.S.]

**Astatine** A chemical element, At, atomic number 85. Astatine is the heaviest of the halogen groups, filling the place immediately below iodine in group 17 of the periodic table. Astatine is a highly unstable element existing only in short-lived radioactive forms. About 25 isotopes have been prepared by nuclear reactions of artificial transmutation. The longest-lived of these is $^{210}$At, which decays with a half-life of only 8.3 h. It is unlikely that a stable or long-lived form will be found in nature or prepared artificially. The most important isotope, used for tracer studies, is $^{211}$At. Astatine exists in nature in uranium minerals, but only in the form of trace amounts of shortlived isotopes, continuously replenished by the slow decay of uranium, The total amount of astatine in the Earth's crust is less than 1 oz (28 g).

In aqueous solution, astatine resembles iodine except for differences attributable to the fact that astatine solutions are of necessity extremely dilute. Like the halogen iodine, when astatine exists as a free element in solution, it is extracted by benzene. The element in solution is reduced by agents such as sulfur dioxide and is oxidized by bromine. It is more electropositive than the other halogens. It has oxidation states with coprecipitation characteristics similar to those of the iodide ion, free iodine, and the iodate ion. Powerful oxidizing agents produce an astatate ion, but not a perastatate ion. The free state is most readily obtained and is characterized by high volatility and high extractability into organic solvents. *See* HALOGEN ELEMENTS; PERIODIC TABLE. [E.K.H.]

**Asymmetric synthesis** A reaction or series of reactions leading to predominant or exclusive formation of a single enantiomer, that is, a stereoisomer that is not superimposable on its mirror image. Among the organic compounds that are usually the target of asymmetric synthesis, the most common structural element that makes one exist as an enantiomer is a carbon atom with a single bond to four different atoms or groups (a stereogenic center), as the two enantiomers of 3-methylhexane, (R)-3-methylhexane (**1**) and (S)-3-methylhexane (**2**).

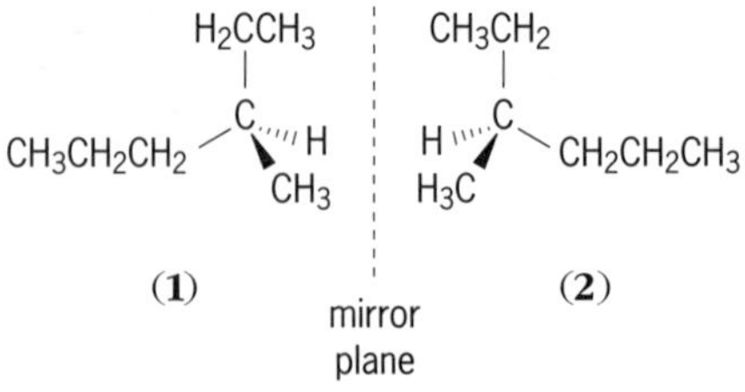

$H_2CCH_3$ $CH_3CH_2$
$CH_3CH_2CH_2$ C H H C $CH_2CH_2CH_3$
$CH_3$ $H_3C$
(1) (2)
mirror plane

Enantiomers are said to be chiral. They are asymmetric; symmetric molecules, possessing a plane or point of symmetry, are superimposable on their mirror images. Not all molecules exist as enantiomers. *See* PROCHIRALITY.

Other structural elements can give rise to asymmetry, for example substituted allene functional groups, as in (S)-2,3-pentadiene (**3**) and (R)-2,3-pentadiene (**4**), and binaphthyl systems, as in (R)-1,1′-binaphthyl (**5**) and (S)-1,1′-binaphthyl (**6**), which

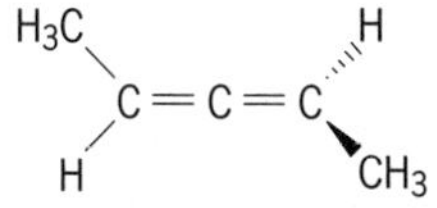

(**3**) 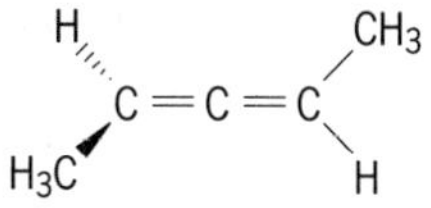 (**4**)

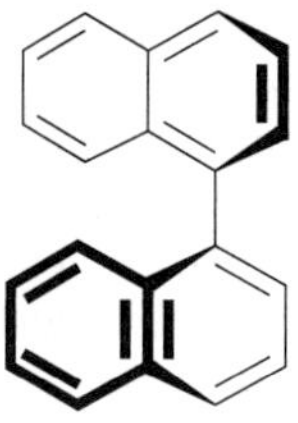

(**5**) 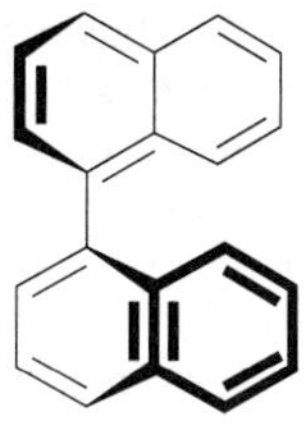 (**6**)

are said to have axial chirality. Rules exist for unambiguously designating the three-dimensional orientations (configurations) of the atoms attached to such structural elements; these designations are the R versus S terms. *See* STEREOCHEMISTRY.

Asymmetry in molecules is very important to the biological activity of the molecule. Because almost all of the molecules in an organism, such as occur in cell membranes, enzymes, receptors, and nucleic acids (which mediate all life processes) are asymmetric, they interact differently with different enantiomers. For example, the S enantiomer of asparagine (**7**; a common amino acid) has a bitter flavor, while the R enantiomer (**8**)

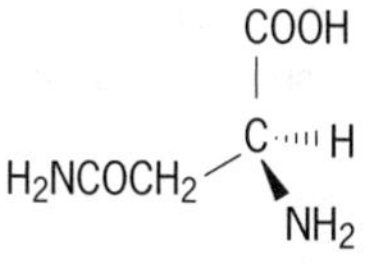

(**7**) 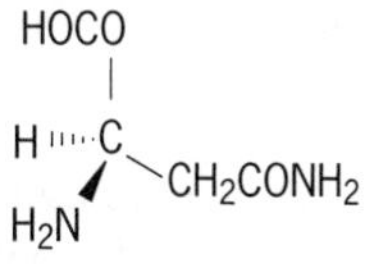 (**8**)

has a sweet flavor, due to the fact that each of the two enantiomers binds differently to chemoreceptors in the tongue. Thalidomide, a drug once prescribed to counteract pregnancy-related morning sickness, is an effective sedative as the R enantiomer (**9**), but the S enantiomer (**10**) is a potent teratogen (it causes birth defects). In general,

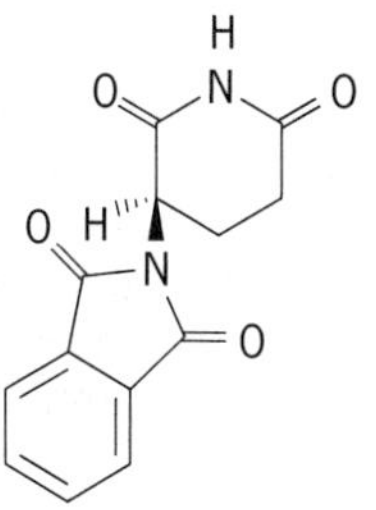

(**9**) 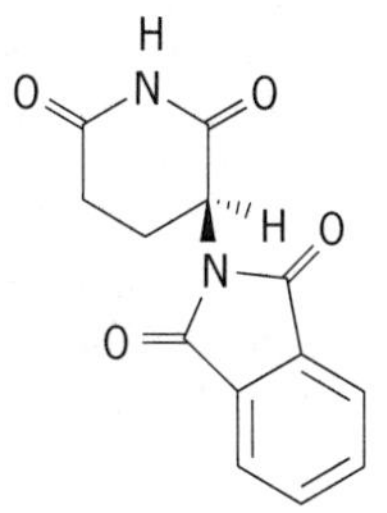 (**10**)

only one enantiomer of a drug, agrochemical (herbicide, pesticide), flavoring agent, or other molecule (when asymmetric) has the desired biological effect, while the other enantiomer has very different effects or, at least, places a metabolic burden on the body.

For this reason, asymmetric synthesis to produce only one enantiomer of a molecule for such uses is extremely important. *See* AMINO ACIDS; CHEMORECEPTION; PROTEIN.

The two enantiomers of an asymmetric molecule have identical physical properties, except that they rotate plane-polarized light in opposite directions. The ability to rotate plane-polarized light (referred to as optical activity) is a property that only asymmetric molecules possess: one pure enantiomer will rotate the plane of polarization in one direction [clockwise, thus behaving as a *d* (dextrorotatory) or + enantiomer], and the opposite enantiomer will rotate the plane of polarization the same number of degrees but in the opposite direction [counterclockwise, thus an *l* (levorotatory) or − enantiomer]. A 50:50 mixture of two enantiomers of a molecule is called a racemic mixture (designated *dl* or ±); it will not rotate the plane of plane-polarized light. By knowing the specific rotation of a pure enantiomer, it is possible to calculate the relative amounts of each enantiomer (the so-called optical purity) in an unequal mixture. *See* OPTICAL ACTIVITY; RACEMIZATION.

Another distinguishing property of two enantiomers is that each will react with a single enantiomer of another chiral molecule at a different rate. This process is related to the existence of diastereomers, which are stereoisomers that are not enantiomers. Diastereomers can occur in many forms. One common manifestation is the case where a molecule possesses two (or more) stereogenic carbon centers. Diastereomers, unlike enantiomers, possess different physical and chemical properties; they have different free energies while enantiomers are identical in energy. Therefore, if a reaction is designed so that it passes through two possible pathways, each involving transition states which are diastereomeric, to produce two possible stereoisomers of the product, then the pathway which involves the lower-energy transition state will proceed faster; thus one stereoisomer of the product will predominate in the product mixture. The greater the difference between the energies of the transition states, the greater the predominance of one product stereoisomer. This is the basis of asymmetric synthesis. *See* FREE ENERGY.

A common strategy to achieve asymmetric synthesis is to place a chiral center in proximity to the location where the new stereogenic center is to be introduced. When the reaction proceeds, the configuration of the new stereogenic centers being formed are influenced by the chirality of the chiral reactant; the chiral reactant "induces" chirality at the newly formed stereogenic centers.

In some cases, a chiral solvent or a chiral catalyst is used to induce chirality. In all cases, the existing chiral entity in the reaction (reactant or solvent or catalyst) is involved in the transition state, resulting in diastereomeric transition states of which the lower-energy one is favored.

Another strategy for synthesizing predominantly one enantiomer of a product is to react a racemic mixture of a starting material with a chiral reagent or catalyst that reacts faster with one of the enantiomers of the starting material than the other so that one enantiomer is consumed and the other is not. Such processes are known as kinetic resolutions. A kinetic resolution strategy for asymmetric synthesis is not as desirable as an asymmetric reaction strategy, because half of the starting material is left behind as the unwanted stereoisomer. *See* ENZYME; MOLECULAR ISOMERISM. [R.D.Wa.]

**Atom** A constituent of matter consisting of $z$ negatively charged electrons bound predominantly by the Coulomb force to a tiny, positively charged nucleus consisting of $Z$ protons and $(A - Z)$ neutrons. $Z$ is the atomic number, and $A$ is the mass or nucleon number. The atomic mass unit is $u = 1.6605397 \times 10^{-24}$ g. Electrically neutral atoms ($z = Z$) with the range $Z = 1$ (hydrogen) to $Z = 92$ (uranium) make up the periodic table of the elements naturally occurring on Earth. Isotopes of a given element have different values of $A$ but nearly identical chemical properties, which are fixed by the value of $Z$. Certain isotopes are not stable; they decay by various processes called radioactivity. Atoms with $Z$ greater than 92 are all radioactive but may be synthesized, either naturally in stellar explosions or in the laboratory using accelerator techniques. *See* ATOMIC MASS UNIT; ATOMIC NUMBER; ELECTRON; MASS NUMBER; RADIOACTIVITY; TRANSURANIUM ELEMENTS.

Atoms with $Z - z$ ranging from 1 to $Z - 1$ are called positive ions. Those having $z - Z = 1$ are called negative ions; none has been found with $z - Z$ greater than 1. *See* ION. [P.M.K.]

**Atom cluster** Clusters are aggregates of atoms (or molecules) containing between three and a few thousand atoms that have properties intermediate between those of the isolated monomer (atom or molecule) and the bulk or solid-state material. The study of such species has been an increasingly active research field since about 1980. This activity is due to the fundamental interest in studying a completely new area that can bridge the gap between atomic and solid-state physics and also shows many analogies to nuclear physics. However, the research is also done for its potential technological interest in areas such as catalysis, photography, and epitaxy. A characteristic of clusters which is responsible for many of their interesting properties is the large number of atoms at the surface compared to those in the cluster interior. For many kinds of atomic clusters, all atoms are at the surface for sizes of up to 12 atoms. As the clusters grow further in size, the relative number of atoms at the surface scales as approximately $4N^{-1/3}$, where $N$ is the total number of atoms. Even in a cluster as big as $10^5$ atoms, almost 10% of the atoms are at the surface. Clusters can be placed in the following categories:

1. Microclusters have from 3 to 10–13 atoms. Concepts and methods of molecular physics are applicable.

2. Small clusters have from 10–13 to about 100 atoms. Many different geometrical isomers exist for a given cluster size with almost the same energies. Molecular concepts lose their applicability.

3. Large clusters have from 100 to 1000 atoms. A gradual transition is observed to the properties of the solid state.

4. Small particles or nanocrystals have at least 1000 atoms. These bodies display some of the properties of the solid state.

The most favored geometry for rare-gas (neon, argon, and krypton) clusters of up to a few thousand atoms is icosahedral. However, the preferred cluster geometry depends critically on the bonding between the monomers in the clusters. For example, ionic clusters such as those of sodium chloride [$(NaCl)_N$] very rapidly assume the cubic form of the bulk crystal lattice, and for metallic clusters it is the electronic structure rather than the geometric structure which is most important. *See* CHEMICAL BONDING.

There are two main types of sources for producing free cluster beams. In a gas-aggregation source, the atoms or molecules are vaporized into a cold, flowing rare-gas atmosphere. In a jet-expansion source, a gas is expanded under high pressure through a small hole into a vacuum.

In most situations, the valence electrons of the atoms making up the clusters can be regarded as being delocalized, that is, not attached to any particular atom but with a certain probability of being found anywhere within the cluster. The simplest and most widely used model to describe the delocalized electrons in metallic clusters is that of a free-electron gas, known as the jellium model. The positive charge is regarded as being smeared out over the entire volume of the cluster, while the valence electrons are free to move within this homogeneously distributed, positively charged background. [E.Ca.]

**Atomic mass** The mass of an atom or molecule on a scale where the mass of a carbon-12 ($^{12}C$) atom is exactly 12.0. The mass of any atom is approximately equal to the total number of its protons and neutrons multiplied by the atomic mass unit, $u = 1.6605397 \times 10^{-24}$ gram. (Electrons are much lighter, about 0.0005486 u.) No atom differs from this simple formula by more than 1%, and stable atoms heavier than helium all lie within 0.3%. *See* ATOMIC MASS UNIT.

This simplicity of nature led to the confirmation of the atomic hypothesis—the idea that all matter is composed of atoms, which are identical and chemically indivisible for each chemical element. In 1802, G. E. Fischer noticed that the weights of acids needed to neutralize various bases could be described systematically by assigning relative weights to each of the acids and bases. A few years later, John Dalton proposed an atomic theory in which elements were made up of atoms that combine in simple ways to form molecules.

In reality, nature is more complicated, and the great regularity of atomic masses more revealing. Two fundamental ideas about atomic structure come out of this regularity: that the atomic nucleus is composed of charged protons and uncharged neutrons, and that these particles have approximately equal mass. The number of protons in an atom is called its atomic number, and equals the number of electrons in the neutral atom. The electrons, in turn, determine the chemical properties of the atom. Adding a neutron or two does not change the chemistry (or the name) of an atom, but does give it an atomic mass which is 1 u larger for each added neutron. Such atoms are called isotopes of the element, and their existence was first revealed by careful study of radioactive elements. Most naturally occurring elements are mixtures of isotopes, although a single isotope frequently predominates. Since the proportion of the various isotopes is usually about the same everywhere on Earth, an average atomic mass of an element can be defined, and is called the atomic weight. Atomic weights are routinely used in chemistry in order to determine how much of one chemical will react with a given weight of another. *See* ATOMIC STRUCTURE AND SPECTRA; RELATIVE ATOMIC MASS.

In contrast to atomic weights, which can be defined only approximately, atomic masses are exact constants of nature. All atoms of a given isotope are truly identical; they cannot be distinguished by any method. This is known to be true because the quantum mechanics treats identical objects in special ways, and makes predictions that depend on this assumption. One such prediction, the exclusion principle, is the reason that the chemical behavior of atoms with different numbers of electrons is so different.

[F.L.P.; D.E.P.]

**Atomic mass unit** An arbitrarily defined unit in terms of which the masses of individual atoms are expressed. One atomic mass unit is defined as exactly $^1/_{12}$ of the mass of an atom of the nuclide $^{12}C$, the predominant isotope of carbon. The unit, also known as the dalton, is often abbreviated amu, and is designated by the symbol u. The relative atomic mass of a chemical element is the average mass of its atoms expressed in atomic mass units. *See* RELATIVE ATOMIC MASS. [J.F.We.]

**Atomic nucleus** The central region of an atom. Atoms are composed of negatively charged electrons, positively charged protons, and electrically neutral neutrons. The protons and neutrons (collectively known as nucleons) are located in a small central region known as the nucleus. The electrons move in orbits which are large in comparison with the dimensions of the nucleus itself. Protons and neutrons possess approximately equal masses, each roughly 1840 times that of an electron. The number of nucleons in a nucleus is given by the mass number *A* and the number of protons by the atomic number $Z$. Nuclear radii $r$ are given approximately by $r = 1.2 \times 10^{-15}\ \mathrm{m}\ A^{1/3}$. [H.E.D.]

**Atomic number** The number of elementary positive charges (protons) contained within the nucleus of an atom. It is denoted by the letter $Z$. Correspondingly, it is also the number of planetary electrons in the neutral atom.

The concept of atomic number emerged from the work of G. Moseley, done in 1913–1914. He measured the wavelengths of the most energetic rays ($K$ and $L$ lines) produced by using the elements calcium to zinc as targets in an x-ray tube. The square root of the frequency, $\nu$, of these x-rays increased by a constant amount in passing from one target to the next. These data, when extended, gave a linear plot of atomic number versus $\nu$ for all elements studied, using 13 as the atomic number for aluminum and 79 for that of gold. *See* X-RAY SPECTROMETRY.

Moseley's atomic numbers were quickly recognized as providing an accurate sequence of the elements, which the chemical atomic weights had sometimes failed to do. Additionally, the atomic number sequence indicated the positions of elements that had not yet been discovered.

The atomic number not only identifies the chemical properties of an element but facilitates the description of other aspects of atoms and nuclei. Thus, atoms with the same atomic number are isotopes and belong to the same element, while nuclear reactions may alter the atomic number.

When specifically written, the atomic number is placed as a subscript preceding the symbol of the element, while the mass number ($A$) precedes as a superscript, for example, $^{27}_{13}\mathrm{Al}$, $^{238}_{92}\mathrm{U}$. *See* ATOMIC STRUCTURE AND SPECTRA; ELEMENT (CHEMISTRY); MASS NUMBER. [H.E.D.]

**Atomic spectrometry** A branch of chemical analysis that seeks to determine the composition of a sample in terms of which chemical elements are present and their quantities or concentrations. Unlike other methods of elemental analysis, however, the sample is decomposed into its constituent atoms which are then probed spectroscopically.

In routine atomic spectrometry, a device called the atom source or atom cell is responsible for producing atoms from the sample; there are many different kinds of atom sources. After atomization of the sample, any of several techniques can determine which atoms are present and in what amounts, but the most common are atomic absorption, atomic emission, atomic fluorescence (the least used of these four alternatives), and mass spectrometry.

Most atomic spectrometric measurements (all those just mentioned except mass spectrometry) exploit the narrow-line spectra characteristic of gas-phase atoms. Because the atom source yields atomic species in the vapor phase, chemical bonds are disrupted, so valence electronic transitions are unperturbed by bonding effects. As a result, transitions among atomic energy levels yield narrow spectral lines, with spectral bandwidths commonly in the 1–5-picometer wavelength range. Moreover, because each atom possesses its unique set of energy levels, these narrow-band transitions can be measured individually, with little mutual interference. Thus, sodium, potassium, and scandium can all be monitored simultaneously and with minimal spectral influence on each other. This lack of spectral overlap remains one of the most attractive features of atomic spectrometry. *See* ATOMIC STRUCTURE AND SPECTRA.

In atomic absorption spectrometry, light from a primary source is directed through the atom cell, where a fraction of the light is absorbed by atoms from the sample.

The amount of radiation that remains can then be monitored on the far side of the cell. The concentration of atoms in the path of the light beam can be determined by Beer's law, which can be expressed as the equation below, where $P_0$ is the light inten-

$$\log\frac{P_0}{P} = kC$$

sity incident on the atom cell, $P$ is the amount of light which remains unabsorbed, $C$ is the concentration of atoms in the cell, and $k$ is the calibration constant, which is determined by means of standard samples having known concentrations.

The two most common kinds of atom cells employed in atomic absorption spectrometry are chemical flames and electrical furnaces. Chemical flames are usually simple to use, but furnaces offer higher sensitivity.

The most common primary light source employed in atomic absorption spectrometry is the hollow-cathode lamp. Conveniently, the hollow-cathode lamp emits an extremely narrow line spectrum of one, two, or three elements of interest. As a result, the atomic absorption spectrometry measurement is automatically tuned to the particular spectral lines of interest.

In atomic emission spectrometry, atomic species are measured by their emission spectra. For such spectra to be produced, the atoms must first be excited by thermal or nonthermal means. Therefore, the atom sources employed in atomic emission spectrometry are hotter or more energetic than those commonly used in atomic absorption spectrometry. Although several such sources are in common use, the dominant one is the inductively coupled plasma. From the simplest standpoint, the inductively coupled plasma is a flowing stream of hot, partially ionized (positively charged) argon. Power is coupled into the plasma by means of an induction coil.

There are two common modes for observing emission spectra from an inductively coupled plasma. The less expensive and more flexible approach employs a so-called slew-scan spectrometer, which accesses spectral lines in rapid sequence, so that a number of chemical elements can be measured rapidly, one after the other. Moreover, because each viewed elemental spectral line can be scanned completely, it is possible to subtract spectral emission background independently for each element. The alternative approach is to view all spectral lines simultaneously, either with a number of individual photo-detectors keyed to particular spectral lines or with a truly multichannel electronic detector driven by a computer. This approach enables samples to be analyzed more rapidly and permits transient atom signals (as from a furnace-based atomizer) to be recorded. *See* EMISSION SPECTROCHEMICAL ANALYSIS.

Elemental mass spectrometry has been practiced for many years in the form of spark-source mass spectrometry and, more recently, glow-discharge-lamp mass spectrometry. However, a hybrid technique that combines the inductively coupled plasma with a mass spectrometer has assumed a prominent place.

At the high temperatures present in an inductively coupled plasma, many atomic species occur in an ionic form. These ions can be readily extracted into a mass spectrometer.

The advantages of the combination of inductively coupled plasma and mass spectrometry are substantial. The system is capable of some of the best detection limits in atomic spectrometry, typically $10^{-3}$ to $10^{-2}$ ng/ml for most elements. Also, virtually all elements in the periodic table can be determined during a single scan. The method is also capable of providing isotopic information, unavailable by any other atomic spectrometric method for such a broad range of elements. *See* MASS SPECTROMETRY.

[G.M.H.]

**Atomic structure and spectra** The idea that matter is subdivided into discrete and further indivisible building blocks called atoms dates back to the Greek philosopher Democritus, whose teachings of the 5th century B.C. are commonly accepted as the earliest authenticated ones concerning what has come to be called atomism by students of Greek philosophy. The weaving of the philosophical thread of atomism into the analytical fabric of physics began in the late 18th and the 19th centuries. Robert Boyle is generally credited with introducing the concept of chemical elements, the irreducible units of which are now recognized as individual atoms of a given element. In the early 19th century John Dalton developed his atomic theory, which postulated that matter consists of indivisible atoms as the irreducible units of Boyle's elements, that each atom of a given element has identical attributes, that differences among elements are due to fundamental differences among their constituent atoms, that chemical reactions proceed by simple rearrangement of indestructible atoms, and that chemical compounds consist of molecules which are reasonably stable aggregates of such indestructible atoms. *See* CHEMISTRY.

The work of J. J. Thomson in 1897 clearly demonstrated that atoms are electromagnetically constituted and that from them can be extracted fundamental material units bearing electric charge that are now called electrons. The electrons of an atom account for a negligible fraction of its mass. By virtue of overall electrical neutrality of every atom, the mass must therefore reside in a compensating, positively charged atomic component of equal charge magnitude but vastly greater mass. *See* ELECTRON.

Thomson's work was followed by the demonstration by Ernest Rutherford in 1911 that nearly all the mass and all of the positive electric charge of an atom are concentrated in a small nuclear core approximately 10,000 times smaller in extent than an atomic diameter. Niels Bohr in 1913 and others carried out some remarkably successful attempts to build solar system models of atoms containing planetary pointlike electrons orbiting around a positive core through mutual electrical attraction (though only certain "quantized" orbits were "permitted"). These models were ultimately superseded by nonparticulate-matter wave quantum theories of both electrons and atomic nuclei.

The modern picture of condensed matter (such as solid crystals) consists of an aggregate of atoms or molecules which respond to each other's proximity through attractive electrical interactions at separation distances of the order of 1 atomic diameter (approximately $10^{-10}$ m) and repulsive electrical interactions at much smaller distances. These interactions are mediated by the electrons, which are in some sense shared and exchanged by all atoms of a particular sample, and serve as a kind of interatomic glue which binds the mutually repulsive, heavy, positively charged atomic cores together.

The hydrogen atom is the simplest atom, and its spectrum (or pattern of light frequencies emitted) is also the simplest. The regularity of its spectrum had defied explanation until Bohr solved it with three postulates, these representing a model which is useful, but quite insufficient, for understanding the atom.

Postulate 1: The force that holds the electron to the nucleus is the Coulomb force between electrically charged bodies.

Postulate 2: Only certain stable, nonradiating orbits for the electron's motion are possible, those for which the angular momentum is an integral multiple of $h/2\pi$ (Bohr's quantum condition on the orbital angular momentum). Each stable orbit represents a discrete energy state.

Postulate 3: Emission or absorption of light occurs when the electron makes a transition from one stable orbit to another, and the frequency $\upsilon$ of the light is such that the difference in the orbital energies equals $h\upsilon$ (A. Einstein's frequency condition for the photon, the quantum of light).

Here the concept of angular momentum, a continuous measure of rotational motion in classical physics, has been asserted to have a discrete quantum behavior, so that its quantized size is related to Planck's constant $h$, a universal constant of nature.

Modern quantum mechanics has provided justification for Bohr's quantum condition on the orbital angular momentum. It has also shown that the concept of definite orbits cannot be retained except in the limiting case of very large orbits. In this limit, the frequency, intensity, and polarization can be accurately calculated by applying the classical laws of electrodynamics to the radiation from the orbiting electron. This fact illustrates Bohr's correspondence principle, according to which the quantum results must agree with the classical ones for large dimensions. The deviation from classical theory that occurs when the orbits are smaller than the limiting case is such that one may no longer picture an accurately defined orbit. Bohr's other hypotheses are still valid.

According to Bohr's theory, the energies of the hydrogen atom are quantized (that is, can take on only certain discrete values). These energies can be calculated from the electron orbits permitted by the quantized orbital angular momentum. The orbit may be circular or elliptical, so only the circular orbit is considered here for simplicity. Let the electron, of mass $m$ and electric charge $-e$, describe a circular orbit of radius $r$ around a nucleus of charge $+e$ and of infinite mass. With the electron velocity $v$, the angular momentum is $mvr$, and the second postulate becomes Eq. (1). The integer $n$ is called

$$mvr = n(h/2\pi) \quad (n = 1, 2, 3, \ldots) \tag{1}$$

the principal quantum number. The possible energies of the nonradiating states of the atom are given by Eq. (2). Here $\varepsilon_0$ is the permittivity of free space, a constant included

$$E = -\frac{me^4}{8\epsilon_0{}^2h^2} \cdot \frac{1}{n^2} \tag{2}$$

to give the correct units according to the mks statement of Coulomb's law.

The same equation for the hydrogen atom's energy levels, except for some small but significant corrections, is obtained from the solution of the Schrödinger equation for the hydrogen atom.

The frequencies of electromagnetic radiation or light emitted or absorbed in transitions are given by Eq. (3) where $E'$ and $E''$ are the energies of the initial and final states

$$v = E' - E''/h \tag{3}$$

of the atom. Spectroscopists usually express their measurements in wavelength $\lambda$ or in wave number $\sigma$ in order to obtain numbers of a convenient size. The wave number of a transition is shown in Eq. (4). If $T = -E/hc$, then Eq. (5) results. Here $T$ is called the spectral term.

$$\sigma = \frac{v}{c} = \frac{E'}{hc} - \frac{E'}{hc} \tag{4}$$

$$\sigma = T'' - T' \tag{5}$$

The allowed terms for hydrogen, from Eq. (2), are given by Eq. (6). The quantity

$$T = \frac{me^4}{8\epsilon_0{}^2ch^3} \cdot \frac{1}{n^2} = \frac{R}{n^2} \tag{6}$$

$R$ is the important Rydberg constant. Its value, which has been accurately measured by laser spectroscopy, is related to the values of other well-known atomic constants, as shown in Eq. (6).

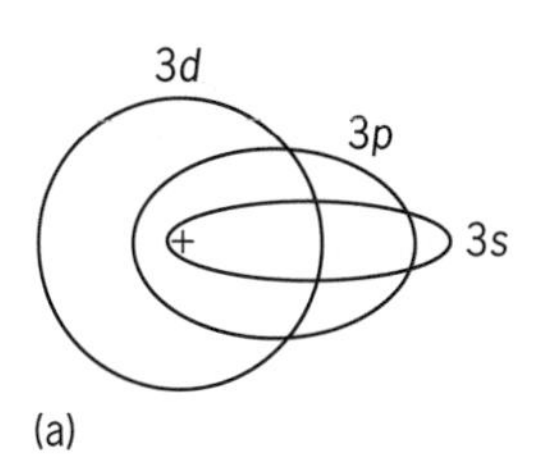


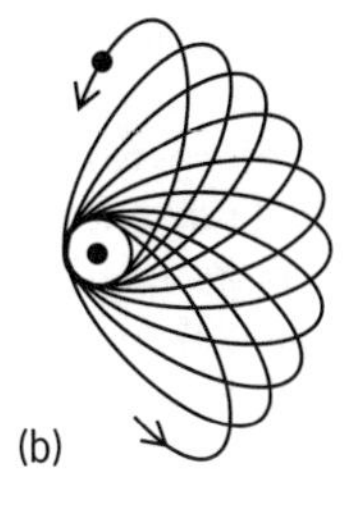


**Possible elliptical orbits, according to the Bohr-Sommerfeld theory. (*a*) The three permitted orbits for $n = 3$. (*b*) Precession of the 3*s* orbit caused by the relativistic variation of mass. (*After A. P. Arya, Fundamentals of Atomic Physics, Allyn and Bacon, pp. 281 and 286, 1971*).**

The effect of finite nuclear mass must be considered, since the nucleus does not actually remain at rest at the center of the atom. Instead, the electron and nucleus revolve about their common center of mass. This effect can be accurately accounted for and requires a small change in the value of the effective mass $m$ in Eq. (6).

In addition to the circular orbits already described, elliptical ones are also consistent with the requirement that the angular momentum be quantized. A. Sommerfeld showed that for each value of $n$ there is a family of $n$ permitted elliptical orbits, all having the same major axis but with different eccentricities. Illustration *a* shows, for example, the Bohr-Sommerfeld orbits for $n = 3$. The orbits are labeled $s$, $p$, and $d$, indicating values of the azimuthal quantum number $l = 0$, 1, and 2. This number determines the shape of the orbit, since the ratio of the major to the minor axis is found to be $n/(l + 1)$. To a first approximation, the energies of all orbits of the same $n$ are equal. In the case of the highly eccentric orbits, however, there is a slight lowering of the energy due to precession of the orbit (illus. *b*). According to Einstein's theory of relativity, the mass increases somewhat in the inner part of the orbit, because of greater velocity. The velocity increase is greater as the eccentricity is greater, so the orbits of higher eccentricity have their energies lowered more. The quantity $l$ is called the orbital angular momentum quantum number or the azimuthal quantum number.

In attempting to extend Bohr's model to atoms with more than one electron, it is logical to compare the experimentally observed terms of the alkali atoms, which contain only a single electron outside closed shells, with those of hydrogen. A definite similarity is found but with the striking difference that all terms with $l > 0$ are double. This fact was interpreted by S. A. Goudsmit and G. E. Uhlenbeck as due to the presence of an additional angular momentum of $^1/_2(h/2\pi)$ attributed to the electron spinning about its axis. The spin quantum number of the electron is $s = {}^1/_2$.

The relativistic quantum mechanics developed by P. A. M. Dirac provided the theoretical basis for this experimental observation.

Implicit in much of the following discussion is W. Pauli's exclusion principle, first enunciated in 1925, which when applied to atoms may be stated as follows: no more than one electron in a multielectron atom can possess precisely the same quantum numbers. In an independent, hydrogenic electron approximation to multielectron atoms, there are $2n^2$ possible independent choices of the principal ($n$), orbital ($l$), and magnetic ($m_l$, $m_s$) quantum numbers available for electrons belonging to a given $n$, and no more. Here $m_l$ and $m_s$ refer to the quantized projections of $l$ and $s$ along some chosen direction. The organization of atomic electrons into shells of increasing radius (the Bohr radius scales as $n^2$) follows from this principle.

The energy of interaction of the electron's spin with its orbital angular momentum is known as spin-orbit coupling. A charge in motion through either "pure" electric or "pure" magnetic fields, that is, through fields perceived as "pure" in a static laboratory, actually experiences a combination of electric and magnetic fields, if viewed in the frame of reference of a moving observer with respect to whom the charge is momentarily at rest. For example, moving charges are well known to be deflected by magnetic fields. But in the rest frame of such a charge, there is no motion, and any acceleration of a charge must be due to the presence of a pure electric field from the point of view of an observer analyzing the motion in that reference frame.

A spinning electron can crudely be pictured as a spinning ball of charge, imitating a circulating electric current. This circulating current gives rise to a magnetic field distribution very similar to that of a small bar magnet, with north and south magnetic poles symmetrically distributed along the spin axis above and below the spin equator. This representative bar magnet can interact with external magnetic fields, one source of which is the magnetic field experienced by an electron in its rest frame, owing to its orbital motion through the electric field established by the central nucleus of an atom. In multielectron atoms, there can be additional, though generally weaker, interactions arising from the magnetic interactions of each electron with its neighbors, as all are moving with respect to each other, and all have spin. The strength of the bar magnet equivalent to each electron spin, and its direction in space are characterized by a quantity called the magnetic moment, which also is quantized essentially because the spin itself is quantized. Studies of the effect of an external magnetic field on the stages of atoms show that the magnetic moment associated with the electron spin is equal in magnitude to a unit called the Bohr magneton.

The energy of the interaction between the electron's magnetic moment and the magnetic field generated by its orbital motion is usually a small correction to the spectral term, and depends on the angle between the magnetic moment and the magnetic field, or equivalently, between the spin angular momentum vector and the orbital angular momentum vector (a vector perpendicular to the orbital plane whose magnitude is the size of the orbital angular momentum). Since quantum theory requires that the quantum number $j$ of the electron's total angular momentum shall take values differing by integers, while $l$ is always an integer, there are only two possible orientations for $s$ relative to $l$: $s$ must be either parallel or antiparallel to $l$.

For the case of a single electron outside the nucleus, the Dirac theory gives Eq. (7)

$$\Delta T = \frac{R\alpha^2 Z^4}{n^3} \cdot \frac{j(j+1) - l(l+1) - s(s+1)}{l(2l+1)(l+1)} \tag{7}$$

for the spin-orbit correction to the spectral terms. Here $\alpha = e^2/2\varepsilon_0 hc \cong 1/137$ is called the fine structure constant.

In atoms having more than one electron, this fine structure becomes what is called the multiplet structure. The doublets in the alkali spectra, for example, are due to spin-orbit coupling; Eq. (7), with suitable modifications, can be applied.

When more than one electron is present in the atom, there are various ways in which the spins and orbital angular momenta can interact. Each spin may couple to its own orbit, as in the one-electron case; other possibilities are orbit–other orbit, spin-spin, and so on. The most common interaction in the light atoms, called $LS$ coupling or Russell-Saunders coupling, is described schematically in Eq. (8). This notation indicates that

$$\{(l_1, l_2, l_3, \ldots)(s_1, s_2, s_3, \ldots)\} = \{L, S\} = J \tag{8}$$

the $l_i$ are coupled strongly together to form a resultant $L$, representing the total orbital angular momentum. The $s_i$, are coupled strongly together to form a resultant $S$, the total spin angular momentum. The weakest coupling is that between $L$ and $S$ to form $J$, the total angular momentum of the electron system of the atom in this state.

Coupling of the *LS* type is generally applicable to the low-energy states of the lighter atoms. The next commonest type is called *jj* coupling, represented in Eq. (9). Each

$$\{(l_1, s_1)(l_2, s_2)(l_3, s_3)\ldots\} = \{j_1, j_2, J_3, \ldots\} = J \tag{9}$$

electron has its spin coupled to its own orbital angular momentum to form a *ji* for that electron. The various *ji* are then more weakly coupled together to give *J*. This type of coupling is seldom strictly observed. In the heavier atoms it is common to find a condition intermediate between *LS* and *jj* coupling; then either the *LS* or *jj* notation may be used to describe the levels, because the number of levels for a given electron configuration is independent of the coupling scheme.

Most atomic nuclei also possess spin, but rotate about 2000 times slower than electrons because their mass is on the order of 2000 or more times greater than that of electrons. Because of this, very weak nuclear magnetic fields, analogous to the electronic ones that produce fine structure in spectral lines, further split atomic energy levels. Consequently, spectral lines arising from them are split according to the relative orientations, and hence energies of interaction, of the nuclear magnetic moments with the electronic ones. The resulting pattern of energy levels and corresponding spectral-line components is referred to as hyperfine structure.

The enormous capabilities of tunable lasers have allowed observations which were impossible previously. For example, high-resolution saturation spectroscopy, utilizing a saturating beam and a probe beam from the same laser, has been used to measure the hyperfine structure of the sodium resonance lines (called the $D_1$ and $D_2$ lines). The smallest separation resolved was less than 0.001 $\text{cm}^{-1}$ which was far less than the Doppler width of the lines. *See* LASER SPECTROSCOPY.

Nuclear properties also affect atomic spectra through the isotope shift. This is the result of the difference in nuclear masses of two isotopes, which results in a slight change in the Rydberg constant. There is also sometimes a distortion of the nucleus.

It would be misleading to think that the most probable fate of excited atomic electrons consists of transitions to lower orbits, accompanied by photon emission. In fact, for at least the first third of the periodic table, the preferred decay mode of most excited atomic systems in most states of excitation and ionization is the electron emission process first observed by P. Auger in 1925 and named after him. For example, a singly charged neon ion lacking a 1*s* electron is more than 50 times as likely to decay by electron emission as by photon emission. In the process, an outer atomic electron descends to fill an inner vacancy, while another is ejected from the atom to conserve both total energy and momentum in the atom. The ejection usually arises because of the interelectron Coulomb repulsion. *See* AUGER EFFECT. [I.A.S.]

**Atropine** An alkaloid, $C_{17}H_{23}NO_3$, with the chemical structure below. The systematic chemical name is endo-(±)-$\alpha$-(hydroxymethyl)phenylacetic acid 8-methyl-

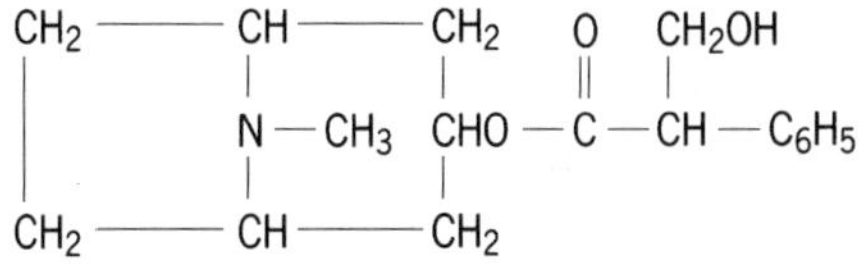

```
CH2 ——— CH ——— CH2   O   CH2OH
 |       |       |    ||    |
 |      N—CH3   CHO — C — CH — C6H5
 |       |       |
CH2 ——— CH ——— CH2
```

8-azabicyclo[3.2.1]oct-3-yl ester, and in phamacy it is sometimes known as *dl*-hyoscyamine. It occurs in minute amounts in the leaves of *Atropa belladonna, A. betica, Datura stramonium, D. innoxia,* and *D. sanguinea*, as well as many related plants. It is chiefly manufactured by racemization of *l*-hyoscyamine, which is isolated from the leaves and stems of the henbane, *Hyoscyamus niger*. It melts at 114–116°C

(237–241°F) and is poorly soluble in water. The nitrate and sulfate are used in medicine instead of the free base.

Atropine is used clinically as a mydriatic (pupil dilator). Dilation is produced by paralyzing the iris and ciliary muscles. Atropine is also administered in small doses before general anesthesia to lessen oral and air-passage secretions. Its ability to reduce these secretions is also utilized in several preparations commonly used for symptomatic relief of colds. *See* ALKALOID. [F.W.]

**Auger effect** One of the two principal processes for the relaxation of an inner-shell electron vacancy in an excited or ionized atom. The Auger effect is a two-electron process in which an electron makes a discrete transition from a less bound shell to the vacant, but more tightly bound, electron shell. The energy gained in this process is transferred, via the electrostatic interaction, to another bound electron which then escapes from the atom. This outgoing electron is referred to as an Auger electron and is labeled by letters corresponding to the atomic shells involved in the process. For example, a $KL_IL_{III}$ Auger electron corresponds to a process in which an $L_I$ electron makes a transition to the $K$ shell and the energy is transferred to an $L_{III}$ electron (illus. *a*). By the conservation of energy, the Auger electron kinetic energy $E$ is given by $E = E(K) - E(L_I) - E(L_{III})$ where $E(K,L)$ is the binding energy of the various electron shells. Since the energy levels of atoms are discrete and well understood, the Auger energy is a signature of the emitting atom. *See* ELECTRON CONFIGURATION.

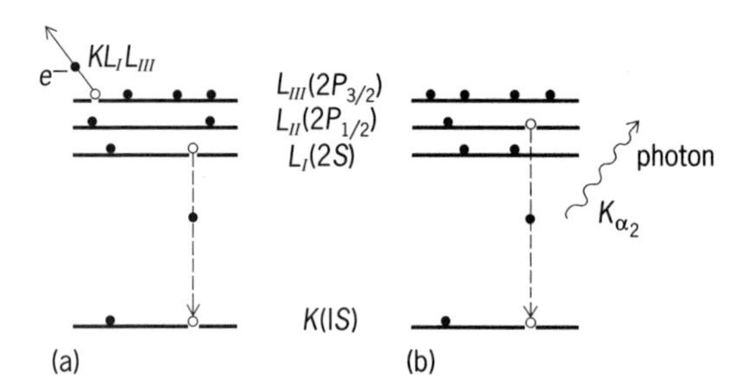


**Two principal processes for the filling of an inner-shell electron vacancy. (*a*) Auger emission; a $KL_IL_{III}$ Auger process in which an $L_I$ electron fills the *K*-shell vacancy with the emission of a $KL_IL_{III}$ Auger electron from the $L_{III}$ shell. (*b*) Photon emission; a radiative process in which an $L_{II}$ electron fills the *K*-shell vacancy with the emission of a $K_{\alpha_2}$ photon.**

The other principal process for the filling of an inner-shell hole is a radiative one in which the transition energy is carried off by a photon (illus. *b*). Inner-shell vacancies in elements with large atomic number correspond to large transition energies and usually decay by such radiative processes; vacancies in elements with low atomic number or outer-shell vacancies with low transition energies decay primarily by Auger processes. *See* ATOMIC STRUCTURE AND SPECTRA. [L.C.F.]

**Auger electron spectroscopy** Auger electron spectroscopy (AES) is a widely used technique that detects the elements in the first atomic layers of a solid surface. Although many elements can be detected, hydrogen usually cannot be observed. Excellent spatial resolution can be achieved. Auger electron spectroscopy is important in many areas of science and technology, such as catalysis, electronics, lubrication, and new materials, and also understanding chemical bonding in the surface region. Auger spectra can be observed with gas-phase species.

**Basic principles.** In Auger electron spectroscopy an electron beam, usually 2–20 kV in energy, impinges on a surface, and a core-level electron in an atom is ejected. An electron from a higher level falls into the vacant core level. Two deexcitation processes are possible: An x-ray can be emitted, or a third electron is ejected from the atom (see illustration). This electron is an Auger electron; the effect is named after its discoverer, Pierre Auger. Auger electrons for surface analysis can be created from other sources, such as, x-rays, ion beams, and positrons. $K$ capture (radioactive decay) is another source of Auger electrons. The Auger yield (the ratio of Auger electrons to the number of core holes created) depends upon the element, the initial core level, and the excitation conditions. *See* AUGER EFFECT.

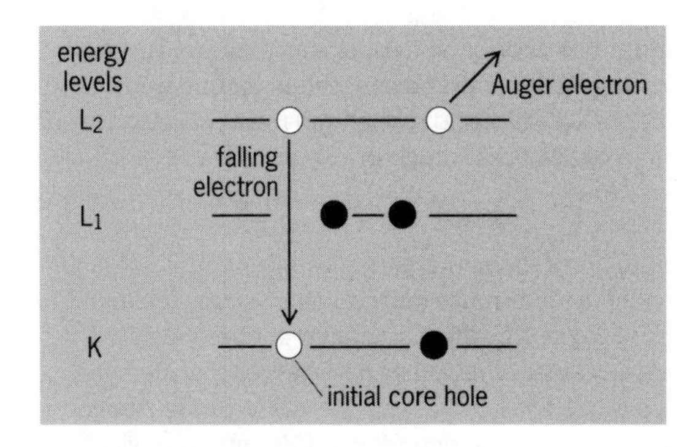


**Schematic diagram of an Auger electron being ejected from a carbon atom for a $KL_2L_2$ transition. Round forms are electrons.**

The surface sensitivity of Auger electron spectroscopy is based upon the fact that electrons with kinetic energy in the range of 30–2500 eV have an inelastic mean free path (IMFP) of about 0.5–3 nanometers for most materials. The inelastic mean free path is defined as the average distance that an electron will travel before it undergoes an inelastic collision with another electron or nucleus of the matrix constituent atoms. Very few Auger electrons created at depths greater than two to three times the IMFP will leave the bulk with their initial kinetic energy. Those electrons formed at greater depths undergo collisions with atoms and electrons in the bulk and lose part or all of the initial creation energy. In this energy range, all elements with an atomic number of 3 or more have detectable Auger electrons. Most elements can be detected to 1 atomic percent in the analysis volume, and relative atomic amounts usually can be determined.

X-ray notation is used to signify Auger transitions. For example, if a $K$ ($1s$) electron in silicon is ejected and an $L_3$ ($2p_{3/2}$) electron falls into the $K$-level hole and another $L_3$ is ejected as the Auger electron, the transition is $KL_3L_3$. If a valence electron is involved, it often is denoted with a $V$. [N.H.T.]

**Autoxidation** The slow, flameless combustion of materials by reaction with oxygen; it is sometimes spelled autooxidation. Autoxidation is important because it is a useful reaction for converting compounds to oxygenated derivatives, and also because it occurs in situations where it is not desired (as in the destructive cracking of the rubber in automobile tires). *See* COMBUSTION; OXIDATION PROCESS.

Although virtually all types of organic materials can undergo air oxidation, certain types are particularly prone to autoxidation, including unsaturated compounds that have allylic hydrogens or benzylic hydrogens; these materials are converted to hydroperoxides by autoxidation.

Autoxidation is a free-radical chain process. Such reactions can be divided into three stages: initation, propagation, and termination. In the initiation process, some event causes free radicals to be formed. For example, free radicals can be produced purposefully by the decomposition of a free-radical initiator, such as benzoyl peroxide.

In some cases, initiation occurs by a process that is not well understood but is thought to be the spontaneous reaction of oxygen with a material with a readily abstractable hydrogen. Destructive autoxidation processes also are initiated by pollutants such as those in smog.

Once free radicals are formed, they react in a chain to convert the material to a hydroperoxide. The chain is ended by termination reactions in which free radicals collide and combine their odd electrons to form a new bond. *See* CHAIN REACTION (CHEMISTRY); FREE RADICAL; ORGANIC REACTION MECHANISM.

Autoxidation is a process of enormous economic impact, since all foods, plastics, gasolines, oils, rubber, and other materials that must be exposed to air undergo continuous destructive reactions of this type. All plastics and rubber and most processed foods contain antioxidants to protect them against the attack of oxygen. *See* ANTIOXIDANT; PLASTICS PROCESSING; RUBBER. [Wi.A.P.]

**Avogadro number** The number of elementary entities in one mole of a substance. A mole is defined as an amount of a substance that contains as many elementary entities as there are atoms in exactly 12 g of $^{12}C$; the elementary entities must be specified and may be atoms, molecules, ions, electrons, other particles, or specified groups of such particles. Experiments give $6.0221367 \times 10^{23}$ as the value of the Avogadro number. In most calculations the coefficient is rounded off to 6.02. Thus, a mole of $^{12}C$ atoms has $6.02 \times 10^{23}$ carbon atoms, a mole of water molecules contains $6.02 \times 10^{23}$ $H_2O$ molecules, a mole of electrons contains $6.02 \times 10^{23}$ electrons, and so forth. *See* MOLE (CHEMISTRY).

The atomic weight (relative atomic mass) of $^{12}C$ is exactly 12, by definition. Consider 12 g of $^{12}C$ (which is one mole and contains the Avogadro number of atoms) compared with 4 g of He, whose atomic weight is 4. The 12 g to 4 g ratio of the masses of the two samples is the same as the 12 to 4 ratio of the masses of the atoms of $^{12}C$ and He. Therefore the two samples must contain the same number of atoms, and 4 g of He contains the Avogadro number of atoms. The same argument holds for any element. Thus, for an element with atomic weight $x$, a sample with mass $x$ grams contains the Avogadro number of atoms. Similarly, for a substance with molecular weight $y$, a sample whose mass is $y$ grams must contain the Avogadro number of molecules. For example, 18 g of water contains $6.02 \times 10^{23}$ $H_2O$ molecules. *See* RELATIVE ATOMIC MASS.

The Avogadro number is a dimensionless number. The Avogadro constant is defined as the Avogadro number divided by the unit "mole." The Avogadro constant is usually symbolized by $N_A$, $N_0$, or $L$. Since $N_A$ gives the number of molecules per mole, $N_A = N/n$, where $N$ is the number of molecules present in $n$ moles of a substance.

The Avogadro number relates the mass of a mole of a substance to the mass of a single molecule. For example, for $H_2O$ (whose molecular weight is 18) the mass of one mole is 18 g and the mass of one molecule is $(18\text{ g})/(6.02 \times 10^{23}) \approx 3 \times 10^{-23}$ g. The mass $m$ of one molecule of a substance with molar mass $M$ is $m = M/N_A$.

The Avogadro constant $N_A$ is related to other fundamental physical constants. The Faraday constant $F$ is the absolute value of the charge on one mole of electrons. Therefore $F = N_A e$, where $e$ is the absolute value of the charge on one electron. Also, $R = N_A k$, where $R$ is the gas constant and $k$ is the Boltzmann constant. *See* GAS CONSTANT.

Widespread use of the mole concept began only around 1900. The nineteenth-century concept most closely related to the Avogadro number is the number of molecules per unit volume in a gas at 0°C and 1 atm. [The ideal-gas law $PV = nRT = (N/N_A)RT$ gives $N/V = N_A P/RT$, so $N/V$, the number of gas molecules per unit volume, is proportional to the Avogadro constant $N_A$ at fixed pressure $P$ and temperature $T$.] Avogadro hypothesized in 1811 that at a fixed temperature and pressure the number of molecules per unit volume is the same for different gases, but he had no way of estimating this number. [I.N.L.]

**Avogadro's law** The principle that equal volumes of all gases and vapors, under the same conditions of temperature and pressure, contain identical number of molecules; also known as Avogadro's hypothesis. From Avogadro's law the converse follows that equal numbers of molecules of any gases under identical conditions occupy equal volumes. Therefore, under identical physical conditions the gram-molecular weights of all gases occupy equal volumes. *See* GAS. [T.C.W.]

**Azeotropic distillation** Any of several processes by which liquid mixtures containing azeotropes may be separated into their pure components with the aid of an additional substance (called the entrainer, the solvent, or the mass separating agent) to facilitate the distillation. Distillation is a separation technique that exploits the fact that when a liquid is partially vaporized the compositions of the two phases are different. By separating the phases, and repeating the procedure, it is often possible to separate the original mixture completely. However, many mixtures exhibit special states, known as azeotropes, at which the composition, temperature, and pressure of the liquid phase become equal to those of the vapor phase. Thus, further separation by conventional distillation is no longer possible. By adding a carefully selected entrainer to the mixture, it is often possible to "break" the azeotrope and thereby achieve the desired separation. *See* AZEOTROPIC MIXTURE; DISTILLATION.

Entrainers fall into at least four distinct categories that may be identified by the way in which they make the separation possible. These categories are: (1) liquid entrainers that do not induce liquid-phase separation, used in homogeneous azeotropic distillations, of which classical extractive distillation is a special case; (2) liquid entrainers that do induce a liquid-phase separation, used in heterogeneous azeotropic distillations; (3) entrainers that react with one of the components; and (4) entrainers that dissociate ionically, that is, salts. *See* SALT-EFFECT DISTILLATION.

Within each of these categories, not all entrainers will make the separation possible, that is, not all entrainers will break the azeotrope. In order to determine whether a given entrainer is feasible, a schematic representation known as a residue curve map for a mixture undergoing simple distillation is created. The path of liquid compositions starting from some initial point is the residue curve. The collection of all such curves for a given mixture is known as a residue curve map (see illustration). These maps contain exactly the same information as the corresponding phase diagram for the mixture, but they represent it in such a way that it is more useful for understanding and designing distillation systems.

Mixtures that do not contain azeotropes have residue curve maps that all look the same. The presence of even one binary azeotrope destroys the structure. If the mixture contains a single minimum-boiling binary azeotrope, three residue curve maps are possible, depending on whether the azeotrope is between the lowest- and highest-boiling components, between the intermediate- and highest-boiling components, or between the intermediate- and lowest-boiling components.

Nonazeotropic mixtures may be separated into their pure components by using a sequence of distillation columns because there are no distillation boundaries to get in the way. The situation is quite different when azeotropes are present, as can be seen from the illustration. It is possible to separate mixtures that have residue curve maps similar to those shown in illus. *a* and *c* by straightforward sequences of distillation

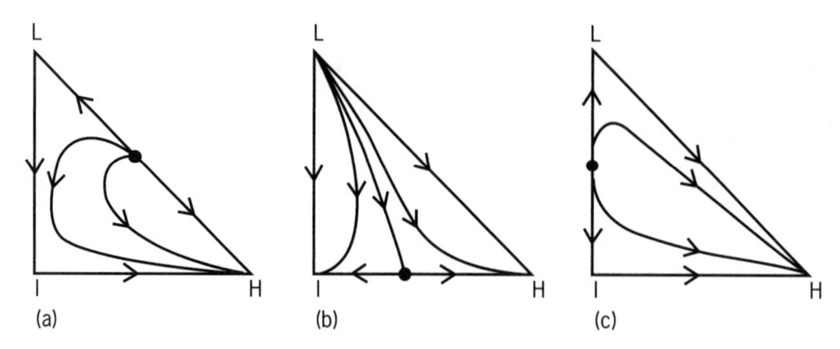


**Schematic representation of the residue curve maps for ternary mixtures with one minimum-boiling binary azeotrope. (*a*) Azeotrope between the lowest- (L) and highest-boiling (H) pure components. (*b*) Azeotrope between the intermediate- (I) and highest-boiling components. (*c*) Azeotrope between the intermediate- and lowest-boiling components.**

columns. This is because these maps do not have any distillation boundaries. These, and other feasible separations for more complex mixtures, are referred to collectively as homogeneous azeotropic distillations. Without exploiting some other effect (such as changing the pressure from column to column), it is impossible to separate mixtures that have residue curve maps like illus. *b*.

A large number of mixtures have residue curve maps similar to illus. *c*, and therefore the corresponding distillation is given the special name extractive distillation.

Heterogeneous entrainers cause liquid-liquid phase separations to occur in such a way that the composition of each phase lies on either side of a distillation boundary. In this way, the entrainer allows the separation to "jump" over a boundary that would otherwise be impassable. [M.F.D.]

**Azeotropic mixture** A solution of two or more liquids, the composition of which does not change upon distillation. The composition of the liquid phase at the boiling point is identical to that of the vapor in equilibrium with it, and such mixtures or azeotropes form constant-boiling solutions. The exact composition of the azeotrope changes if the boiling point is altered by a change in the external pressure. A solution of two components which form an azeotrope may be separated by distillation into one pure component and the azeotrope, but not into two pure components. *See* DISTILLATION; SOLUTION. [F.J.J.]

**Azide** A compound containing the group $—N_3$, which can be represented as a resonance hybrid of two structures, as shown in expression (1). Sodium azide, from

$$-\overset{\ominus}{N}-\overset{\oplus}{N}\equiv N \longleftrightarrow -N=\overset{\oplus}{N}=\overset{\ominus}{N} \qquad (1)$$

which most other azides are prepared, is manufactured by passing nitrous oxide over heated sodium amide, reaction (2). It is a water-soluble, stable compound. Heavy-

$$2NaNH_2 + N_2O \rightarrow NaN_3 + NaOH + NH_3 \qquad (2)$$

metal azides are highly explosive and very shock sensitive; lead azide, $Pb(N_3)_2$, is used as a detonator to set off explosives. Sodium azide, in combination with an oxidizing agent, may be used as a gas generator in motor-vehicle passive-restraint systems.

A variety of organic azides are known. The most important of these are aryl azides ($ArN_3$), azidoformates ($ROCON_3$), and sulfonyl azides ($RSO_2N_3$). These lose $N_2$ when heated or exposed to ultraviolet light to generate species known as nitrenes, which are so reactive that they will react with almost any organic compound to form amine derivatives, as shown in reaction (3). Aryl azides are widely used to probe the active

$$RN_3 \xrightarrow[-N_2]{\text{heat or light}} R\ddot{\underset{..}{N}} \xrightarrow{R'CH_3} RNHCH_2R' \quad (3)$$

sites of biological targets by photoaffinity labeling. Difunctional aryl azides are used commercially to prepare photoresists, the nitrenes reacting with a polymer containing double bonds to insolubilize the polymer in the light-struck areas. The insoluble polymer protects the underlying metal from being attacked by an etching solution. Difunctional azido-formates and sulfonyl azides can be used to cross-link polymers, to prepare polymeric foams, and to adhere tire cord to rubber in the manufacture of automobile tires. Compounds containing a sulfonyl azide group and a hydrolyzable silane group, such as $—Si(OCH_3)_3$, in the same molecule are used to bond silaceous fillers (glass fibers, silica, mica, and so forth) to almost any organic polymer. Such coupled systems have much superior properties to simple mechanical mixtures.

Simple acyl azides (R = alkyl or aryl) undergo a Curtius rearrangement on heating to form an isocyanate, reaction (4).

$$RCON_3 \xrightarrow{\text{heat}} RN{=}C{=}O + N_2 \quad (4)$$

*See* NITROGEN; POLYMERIZATION; RESONANCE (MOLECULAR STRUCTURE); RUBBER. [D.S.Br.]

# B

**Balance** An instrument used for the precise measurement of small weights or masses in amounts ranging from micrograms up to a few kilograms.

Balances are differentiated according to design, weighing principle, and metrological criteria (see table). For a given weighing task, a balance is selected primarily for its maximum weighing load (Max) and for the finest graduation or division (*d*) of its weight-reading device (scale dial, digital display, readout).

**Classification of balances**

| Type | Division (*d*) | Typical capacity (Max) |
|---|---|---|
| Ultramicroanalytical | 0.1 μg | 3 g |
| Microanalytical | 1 μg | 3 g |
| Semimicroanalytical | 0.01 mg | 30 g |
| Macroanalytical | 0.1 mg | 160 g |
| Precision | ≥1 mg | 160 g–60 kg |

Balances can be roughly differentiated from scales by their resolution or number of scale divisions, $n = \text{Max}/d$. Balances typically have a resolution of more than 10,000 divisions, and scales for the most part have less.

A traditional mechanical balance consists of a symmetric lever called a balance beam, two pans suspended from its ends, and a pivotal axis (fulcrum) at its center (see illustration). The object to be weighed is placed on one pan, whereupon the balance is brought into equilibrium by placing the required amount of weights on the opposite pan. Thus the weight of an object is defined as the amount represented by the calibrated standard masses that will exactly counterbalance the object on a classic equal-arm balance. Although this is not self-evident with modern balances and scales, the measurement of weight continues to be based on this original understanding.

The substitution principle represented the conclusive step in the evolution of the mechanical balance. Substitution balances have only one hanger assembly, incorporating both the load pan and a built-in set of weights on a holding rack. The hanger assembly is balanced by a counterpoise which is rigidly connected to the other side of the beam. The weight of an object is determined by lifting weights off the holding rack until the balance returns to an equilibrium position within its angular, differential weighing range. Small increments of weight in between the discrete dial weight steps are read from the projected screen image of a graduated optical reticle which is rigidly connected to the balance beam.

The evolution of electronic (more accurately, electromechanical) balances started in the late 1960s and has extended over several generations of electronic technology.

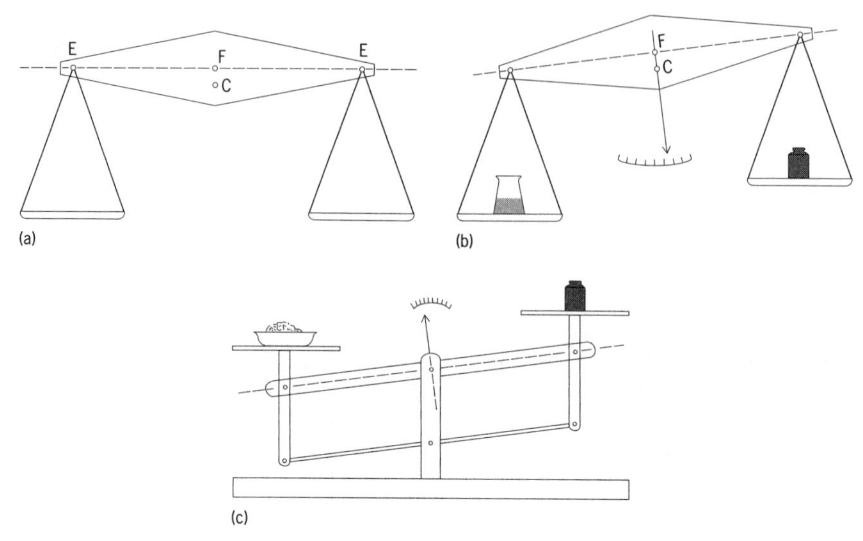


**Mechanical balance design. (*a*) Critical design aspects for an equal-arm balance. (*b*) Weighing small weight differentials with an equal-arm balance. (*c*) Top-loading equal-arm balance. F, fulcrum; E, end pivot; C, center of gravity.**

Among a number of technical possibilities, one operating principle, electromagnetic force compensation, emerged early as the standard in high-precision weighing. First described by K. Ångström in 1895, the principle of electromagnetic force compensation became feasible for technical application as a result of the advancements in solid-state electronic components.

In every electromechanical weighing system, there are three basic functions: (1) The load-transfer mechanism, composed of the weighing platform or pan, levers, and guides, receives the weighing load on the pan as a randomly distributed pressure force and translates it into a measurable single force. (2) The electromechanical force transducer, often called load cell, converts the mechanical input force into an electrical output, for example, voltage, current, or frequency. (3) The electronic signal-processing part of the balance receives the output signal, converts it to numbers, performs computation, and displays the final weight data on the readout.

Besides improved accuracy, reliability, and speed of operation, the main benefits from this technology are human-engineered design for optimized interaction between operator and instrument, and numerous operating conveniences such as push-button zero setting, automatic calibration, built-in computing capabilities for frequently used work procedures, and data output to printers and computers. [W.E.Kup.]

**Barium** A chemical element, Ba, with atomic number 56 and atomic weight of 137.34. Barium is eighteenth in abundance in the Earth's crust, where it is found to the extent of 0.04%, making it intermediate in amount between calcium and strontium, the other alkaline-earth metals. Barium compounds are obtained from the mining and conversion of two barium minerals. Barite, barium sulfate, is the principal ore and contains 65.79% barium oxide. Witherite, sometimes called heavy spar, is barium carbonate and is 72% barium oxide. *See* PERIODIC TABLE.

The metal was first isolated by Sir Humphry Davy in 1808 by electrolysis. Industrially, only small amounts are prepared by aluminum reduction of barium oxide in large retorts. These are used in barium-nickel alloys for spark-plug wire (the barium increases the emissivity of the alloy) and in frary metal, which is an alloy of lead, barium, and calcium used in place of babbitt metal because it can be cast.

The metal reacts with water more readily than do strontium and calcium, but less readily than sodium; it oxidizes quickly in air to form a surface film that inhibits further reaction, but in moist air it may inflame. The metal is sufficiently active chemically to react with most nonmetals. Freshly cut pieces have a lustrous gray-white appearance, and the metal is both ductile and malleable. The physical properties of the elementary form are given in the table.

For the manufacture of barium compounds, soft (easily crushable) barite is preferred, but crystalline varieties may be used. Crude barite is crushed and then mixed with pulverized coal. The mixture is roasted in a rotary reduction furnace, and the barium sulfate is thus reduced to barium sulfide or black ash. Black ash is roughly 70% barium sulfide and is treated with hot water to make a solution used as the starting material for the manufacture of many compounds.

**Properties of barium**

| Property | Value |
|---|---|
| Atomic number | 56 |
| Atomic weight | 137.34 |
| Isotopes (stable) | 130, 132, 134, 135, 136, 137, 138 |
| Atomic volume | 36.2 $cm^3$/g-atom |
| Crystal structure | Face-centered cubic |
| Electron configuration | 2 8 18 18 8 2 |
| Valence | 2+ |
| Ionic radius (A) | 1.35 |
| Boiling point, °C | 1140(?) |
| Melting point, °C | 850(?) |
| Density | 3.75 $g/cm^3$ at 20°C |
| Latent heat of vaporization at boiling point, kj/g-atom | 374 |

Lithopone, a white powder consisting of 20% barium sulfate, 30% zinc sulfide, and less than 3% zinc oxide, is widely used as a pigment in white paints. Blanc fixe is used in the manufacture of brilliant coloring compounds. It is the best grade of barium sulfate for paint pigments. Because of the large absorption of x-rays by barium, the sulfate is used to coat the alimentary tract for x-ray photographs in order to increase the contrast. Barium carbonate is useful in the ceramic industry to prevent efflorescence on claywares. It is used also as a pottery glaze, in optical glass, and in rat poisons. Barium chloride is used in purifying salt brines, in chlorine and sodium hydroxide manufacture, as a flux for magnesium alloys, as a water softener in boiler compounds, and in medicinal preparations. Barium nitrate, or the so-called baryta saltpeter, finds use in pyrotechnics and signal flares (to produce a green color), and to a small extent in medicinal preparations. Barium oxide, known as baryta or calcined baryta, finds use both as an industrial drying agent and in the case-hardening of steels. Barium peroxide is sometimes used as a bleaching agent. Barium chromate, lemon chrome or chrome yellow, is used in yellow pigments and safety matches. Barium chlorate finds use in the manufacture of pyrotechnics. Barium acetate and cyanide are used industrially as a chemical reagent and in metallurgy, respectively. [R.F.R.]

**Base (chemistry)** In the Brønsted- Lowry classification, any chemical species, ionic or molecular, capable of accepting or receiving a proton (hydrogen ion) from another substance. The other substance acts as an acid in giving up the proton. A substance may act as a base, then, only in the presence of an acid. The greater the tendency to accept a proton, the stronger the base. The hydroxyl ion acts as a strong base. Substances that ionize in aqueous solutions to produce the hydroxyl ion (OH), such as potassium hydroxide (KOH) and barium hydroxide [$Ba(OH)_2$], are also conventionally called bases.

Anions of weak acids such as acetic and formic, act as bases in reacting with solvent water to form the molecular acid and hydroxyl ion, for example, the acetate ion ($CH_3COO^-$). Ammonia ($NH_3$) and amines react similarly in aqueous solutions. In these examples, the acetate ion and acetic acid ($CH_3COOH$) and $NH_3$ and the ammonium ion are conjugate base-acid pairs. The basicity constant, $K_b$, is the equilibrium constant for the proton transfer reaction, and it is a quantitative measure of base strength.

The Lewis classification involves the concept of a base as a substance that donates an electron pair to an acid aceptor. In the gas phase, $NH_3$ acts as a base contributing an electron pair to the formation of a convalent bond with the boron trifluoride ($BF_3$) molecule. *See* ACID AND BASE. [F.J.J.]

**Benzene** A colorless, liquid, inflammable, aromatic hydrocarbon of chemical formula $C_6H_6$ which boils at 80.1°C (176.2°F) and freezes at 5.4–5.5°C (41.7–41.9°F). In the older American and British technical literature benzene is designated by the German name benzol. In current usage the term benzol is commonly reserved for the less pure grades of benzene.

Benzene is used as a solvent and particularly in Europe as a constituent of motor fuel. In the United States the largest uses of benzene are for the manufacture of styrene and phenol. Other important outlets are in the production of dodecylbenzene, aniline, maleic anhydride, chlorinated benzenes (used in making DDT and as moth flakes), and benzene hexachloride, an insecticide.

The six carbon atoms of benzene, each with a hydrogen atom attached, are arranged symmetrically in a plane, forming a regular hexagon. The hexagon symbol, commonly used to represent the structural formula for benzene, implies the presence of a carbon atom at each of the six angles and, unless substituents are attached, a hydrogen at each carbon atom. Whereas the three double bonds usually included in the formula are convenient in accounting for the addition reactions of benzene, present evidence is that all the carbon-to-carbon bonds are identical.

Nearly all commercial benzene is a product of petroleum technology. The gasoline fractions obtained by reforming or steam cracking of feedstocks from petroleum contain benzene and toluene which can be separated economically. Benzene may also be produced by the dealkylation of toluene.

Benzene is a toxic substance, and prolonged exposure to concentrations in excess of 35–100 parts per million in air may lead to symptoms ranging from nausea and excess fatigue to anemia and leukopenia. *See* AROMATIC HYDROCARBON. [C.K.B.]

**Benzoic acid** An organic acid, also known as benzene carboxylic acid, with the formula below. Melting point is 250.2°F (121.2°C), and the acid sublimes at 212°F

$$C_6H_5-\overset{\displaystyle O}{\overset{\|}{C}}-OH$$

(100°C). Benzoic acid is only slightly soluble in water but is soluble in most organic solvents, and reacts with bases to form the corresponding benzoate salts. Benzoic acid was first obtained by sublimation from gum benzoin. It occurs both free and combined in nature, being found in many berries (cranberries, prunes, cloves) and as the end product of phenylalanine metabolism.

Benzoic acid is prepared in the laboratory by the Grignard reaction, hydrolysis of benzonitrile ($C_6H_5CN$), or prolonged oxidation of alkyl benzenes with potassium permanganate regardless of the length of the alkyl group. Commercially it was previously prepared by the chlorination of toluene ($C_6H_5CH_3$) with the subsequent hydrolysis of the benzotrichloride ($C_6H_5CCl_3$), and by the monodecarboxylation of phthalic anhydride (from naphthalene). Modern preparation is by the catalytic oxidation of toluene at elevated temperatures with a cobalt catalyst, and purification by sublimation. *See* GRIGNARD REACTION.

Benzoic acid undergoes the normal reactions of the aromatic ring (nitration, sulfonation, halogenation, alkylation). Groups are inserted in the meta position due to the directive influence of the carboxyl group. Substitution occurs less readily than with ortho- or para-directing groups due to the deactivating effect of the meta-directing group. Ortho or para derivatives can be obtained with some starting materials other than the acid. Benzoic acid also undergoes the usual reactions of the carboxyl group, forming acyl halides, anhydrides, amides, esters, and salts. *See* HALOGENATION; NITRATION; SUBSTITUTION REACTION; SULFONATION AND SULFATION.

Sodium benzoate is the only salt of importance. It is water-soluble, has antipyretic and antiseptic properties, is useful as a corrosion inhibitor with sodium nitrite if used for iron, and is also used to modify alkyd resins by increasing hardness, adhesion, and gloss.

Esters of benzoic acid are also found in nature. They are almost universally fragrant. Methyl benzoate is the fragrant principle in tuberose. Some esters of benzoic acid are used in the perfume industry, for example, benzyl ester as a fixative. The butyl ester is used as a dye carrier because of its desirable biodegradable properties; and glycol esters are used as plasticizers. *See* ESTER.

Uses for both benzoic acid and its derivatives include the pharmaceuticals and synthetic polymers. Benzoic acid is used in preservatives and many cosmetics. The derivatives are used in the dyeing industry, with some applications in the cosmetic industry. Pure benzoic acid is a standard for bomb calorimetry because of its ease of purification by sublimation. *See* CALORIMETRY; CARBOXYLIC ACID. [E.H.H.]

**Benzoyl peroxide** A chemical compound, sometimes called dibenzoyl peroxide, with the formula below. It is a colorless, crystalline solid, melting point 106–108°C

$$C_6H_5\overset{\overset{\displaystyle O}{\|}}{C}-O-O-\overset{\overset{\displaystyle O}{\|}}{C}-C_6H_5$$

(223–226°F), that is virtually insoluble in water but very soluble in most organic solvents. Benzoyl peroxide is an initiator, a type of material that decomposes at a controlled rate at moderate temperatures to give free radicals. *See* FREE RADICAL.

Benzoyl peroxide has a half-life (10 h at 73°C or 163°F in benzene) for decomposition that is very convenient for many laboratory and commercial processes. This, plus its relative stability among peroxides, makes it one of the most frequently used initiators. Its primary use is as an initiator of vinyl polymerization. It also is the preferred bleaching agent for flour; is used to bleach many commercial waxes and oils; and is the active ingredient in commercial acne preparations. *See* POLYMERIZATION.

Benzoyl peroxide itself is neither a carcinogen nor a mutagen. However, when coapplied with carcinogens to mouse skin it is a potent promoter of tumor development. All peroxides should be treated as potentially explosive, but benzoyl peroxide is one of the least prone to detonate. [Wi.A.P.]

**Berkelium** Element number 97, symbol Bk, the eighth member of the actinide series of elements. In this series the 5*f* electron shell is being filled, just as the 4*f* shell is being filled in the lanthanide (rare-earth) elements. These two series of elements are very similar in their chemical properties, and berkelium, aside from small differences in ionic radius, is especially similar to its homolog terbium. *See* PERIODIC TABLE; RARE-EARTH ELEMENTS; TERBIUM.

Berkelium does not occur in the Earth's crust because it has no stable isotopes. It must be prepared by means of nuclear reactions using more abundant target elements. These reactions usually involve bombardments with charged particles, irradiations with neutrons from high-flux reactors, or production in a thermonuclear device.

Berkelium metal is chemically reactive, exists in two crystal modifications, and melts at 986°C (1806°F). Berkelium was discovered in 1949 by S. G. Thompson, A. Ghiorso, and G. T. Seaborg at the University of California in Berkeley and was named in honor of that city. Nine isotopes of berkelium are known, ranging in mass from 243 to 251 and in half-life from 1 hour to 1380 years. The most easily produced isotope is $^{249}Bk$, which undergoes beta decay with a half-life of 314 days and is therefore a valuable source for the preparation of the isotope $^{249}Cf$. The berkelium isotope with the longest half-life is $^{247}Bk$ (1380 years), but it is difficult to produce in sufficient amounts to be applied to berkelium chemistry studies. *See* ACTINIDE ELELMENTS; TRANSURANIUM ELEMENTS. [G.T.S.]

**Beryllium** A chemical element, Be, atomic number 4, with an atomic weight of 9.0122. Beryllium, a rare metal, is one of the lightest structural metals, having a density about one-third that of aluminum. Some of the important physical and chemical properties of beryllium are given in the table. Beryllium has a number of unusual and even unique properties. *See* PERIODIC TABLE.

The largest volume uses of beryllium metal are in the manufacture of beryllium-copper alloys and in the development of beryllium-containing moderator and reflector materials for nuclear reactors. Addition of 2% beryllium to copper forms a nonmagnetic alloy which is six times stronger than copper. These beryllium-copper alloys find numerous applications in industry as nonsparking tools, as critical moving parts in aircraft engines, and in the key components of precision instruments, mechanical computers, electrical relays, and camera shutters. Beryllium-copper hammers, wrenches, and other tools are employed in petroleum refineries and other plants in which a spark from steel against steel might lead to an explosion or fire. *See* ALKALINE-EARTH METALS.

Beryllium has found many special uses in nuclear energy because it is one of the most efficient materials for slowing down the speed of neutrons and acting as a neutron reflector. Consequently, much beryllium is used in the construction of nuclear reactors as a moderator and as a support or alloy with the fuel elements.

The following list shows some of the principal compounds of beryllium. Many of the compounds listed are useful as intermediates in the processes for the preparation of ceramics, beryllium oxide, and beryllium metal. Other compounds are useful in analysis and organic synthesis.

| | |
|---|---|
| Acetylacetonate | Fluoride, $BeF_2$ |
| Ammonium beryllium fluoride | Hydroxide |
| Aurintricarboxylate | Nitrate, $Be(NO_3)_2 \cdot 4H_2O$ |
| Basic acetate, $BeO \cdot Be_3(CH_3COO)_6$ | Nitride, $Be_3N_2$ |
| Basic beryllium carbonate | Oxide, BeO |
| Beryllate, $BeO_2^{2-}$ | Perchlorate, $Be(ClO_4)_2 \cdot 4H_2O$ |
| Beryllium ammonium phosphate | Plutonium-beryllium, $PuBe_{13}$ |
| Bromide, $BeB_2$ | Salicylate |
| Carbide, $Be_2C$ | Silicates(emerald) |
| Chloride, $BeCl_2$ | Sulfate, $BeSO_4 \cdot 4H_2O$ |
| Dimethyl, $Be(CH_3)_2$ | Uranium-beryllium, $UBe_{13}$ |

Beryllium is surprisingly rare for a light element, constituting about 0.005% of the Earth's crust. It is about thirty-second in order of abundance, occurring in concentrations approximating those of cesium, scandium, and arsenic. Actually, the

**Physical and chemical properties of beryllium**

| | |
|---|---|
| *Atomic and mass properties* | |
| Mass number of stable isotopes | 9 |
| Atomic number | 4 |
| Outer electronic configuration | $1s^2 2s^2$ |
| Atomic weight | 9.0122 |
| Atomic diameter | 0.221 nm |
| Atomic volume | 4.96 $cm^3$/mole |
| Crystal structure | Hexagonal close-packed |
| Lattice parameters | $a = 0.2285$ nm<br>$c = 0.3583$ nm |
| Axial ratio | $c/a = 1.568$ |
| Field of cation (charge/radius$^2$) | 17 |
| Density*, 25°, x-ray (theoretical) | 1.8477±0.0007 g/$cm^3$ |
| Density, 1000°, x-ray | 1.756 g/$cm^3$ |
| Radius of atom ($Be^0$) | 0.111 nm |
| Radius of ion, $Be^{2+}$ | 0.034 nm |
| Ionization energy ($Be^0 \rightarrow Be^{2+}$) | 27.4 eV |
| *Thermal properties* | |
| Melting point | 1285°C (2345°F) |
| Boiling point† | 2970°C (5378°F) |
| Vapor pressure ($T = K°$) | $\log P$ (atm) $= 6.186 + 1.454\ 10^{-4}T - (16{,}700/T)$ |
| Heat of fusion | 250–275 cal/g |
| Heat of vaporization | 53,490 cal/mole |
| Specific heat (20–100°) | 0.43–0.52 cal/(g)(°C) |
| Thermal conductivity (20°) | 0.355 cal/($cm^2$)(cm)(s)(°C) (42% of copper) |
| Heat of oxidation | 140.15 cal |
| *Electrical properties* | |
| Electrical conductivity | 40–44% of copper |
| Electrical resistivity | 4 microhms/cm (0°C)<br>6 microhms/cm (100°C) |
| Electrolytic solution potential, Be/Be† | $E^0 = -1.69$ volts |
| Electrochemical equivalent | 0.04674 mg/coulomb |

*Measured values vary from 1.79 to 1.86, depending on purity and method of fabrication.

†Obtained by extrapolation of vapor pressure data, not considered very reliable.

abundances of beryllium and its neighbors, lithium and boron, are about $10^{-5}$ times those of the next heavier elements, carbon, nitrogen, and oxygen. At least 50 different beryllium-bearing minerals are known, but in only about 30 is beryllium a regular constituent. The only beryllium-bearing mineral of industrial importance is beryl. Bertrandite substitutes for beryl as a domestic source in the United States. [J.Sch.]

**Biochemical engineering** The application of engineering principles to conceive, design, develop, operate, or use processes and products based on biological and biochemical phenomena. Biochemical engineering, a subset of chemical engineering, impacts a broad range of industries, including health care, agriculture, food, enzymes, chemicals, waste treatment, and energy. Historically, biochemical engineering has been distinguished from biomedical engineering by its emphasis on biochemistry and microbiology and by the lack of a health care focus. However, now there is increasing participation of biochemical engineers in the direct development of health care products. Biochemical engineering has been central to the development of the biotechnology industry, especially with the need to generate prospective products (often using genetically engineered microorganisms) on scales sufficient for testing, regulatory evaluation, and subsequent sale.

In the discipline's initial stages, biochemical engineers were chiefly concerned with optimizing the growth of microorganisms under aerobic conditions at scales of up to thousands of liters. While the scope of the discipline has expanded, this focus remains. Often the aim is the development of an economical process to maximize biomass production (and hence a particular chemical, biochemical, or protein), taking into consideration raw-material and other operating costs. The elemental constituents of biomass (carbon, nitrogen, oxygen, hydrogen, and to a lesser extent phosphorus, sulfur, mineral salts, and trace amounts of certain metals) are added to the biological reactor (often called a fermentor) and consumed by the bacteria as they reproduce and carry out metabolic processes. Sufficient amounts of oxygen (usually supplied as sterile air) are added to the fermentor in such a way as to promote its availability to the growing culture. *See* BIOMASS; CHEMICAL REACTOR; TRANSPORT PROCESSES.

In some situations, microorganisms may be cultivated whose activity is adversely affected by the presence of dissolved oxygen. Anaerobic cultures are typical of fermentations in which organic acids and solvents are produced; these systems are usually characterized by slower growth rates and lower biomass yields. The largest application of anaerobic microorganisms is in waste treatment, where anaerobic digesters containing mixed communities of anaerobic microorganisms are used to reduce the quantity of solids in industrial and municipal wastes.

While the operation and optimization of large-scale, aerobic cultures of microorganisms is still of major importance in biochemical engineering, the capability of cultivating a wide range of cell types has become important also. Biochemical engineers are often involved in the culture of plant cells, insect cells, and mammalian cells, as well as the genetically engineered versions of these cell types. Metabolic engineering uses the tools of molecular genetics, often coupled with quantitative models of metabolic pathways and bioreactor operation, to optimize cellular function for the production of specific metabolites and proteins. Enzyme engineering focuses on the identification, design, and use of biocatalysts for the production of useful chemicals and biochemicals. Tissue engineering involves material, biochemical, and medical aspects related to the transplant of living cells to treat diseases. Biochemical engineers are also actively involved in many aspects of bioremediation, immunotechnology, vaccine development, and the use of cells and enzymes capable of functioning in extreme environments. [R.M.Ke.]

**Biochemistry** The study of the substances and chemical processes which occur in living organisms. It includes the identification and quantitative determination of the substances, studies of their structure, determining how they are synthesized and degraded in organisms, and elucidating their role in the operation of the organism.

Substances studied in biochemistry include carbohydrates (including simple sugars and large polysaccharides), proteins (such as enzymes), ribonucleic acid (RNA) and deoxyribonucleic acid (DNA), lipids, minerals, vitamins, and hormones. *See* CARBOHYDRATE; ENZYME; PROTEIN.

**Metabolism and energy production.** Many of the chemical steps involved in the biological breakdown of sugars, lipids (fats), and amino acids are known. It is well established that living organisms capture the energy liberated from these reactions by forming a high-energy compound, adenosine triphosphate (ATP). In the absence of oxygen, some organisms and tissues derive ATP from an incomplete breakdown of glucose, degrading the sugar to an alcohol or an acid in the process. In the presence of oxygen, many organisms degrade glucose and other foodstuff to carbon dioxide and water, producing ATP in a process known as oxidative phosphorylation.

**Structure and function studies.** The relationship of the structure of enzymes to their catalytic activity is becoming increasingly clear. It is now possible to visualize atoms and groups of atoms in some enzymes by x-ray crystallography. Some enzyme-catalyzed processes can now be described in terms of the spatial arrangement of the groups on the enzyme surface and how these groups influence the reacting molecules to promote the reaction. It is also possible to explain how the catalytic activity of an enzyme may be increased or decreased by changes in the shape of the enzyme molecule. An important advance has been the development of an automated procedure for joining amino acids together into a predetermined sequence. This technology will permit the synthesis of slightly altered enzymes and will improve the understanding of the relationship between the structure and the function of enzymes. In addition, this procedure permits the synthesis of medically important polypeptides (short chains of amino acids) such as some hormones and antibiotics.

**Molecular genetics.** A subject of intensive investigation has been the explanation of genetics in molecular terms. It is now well established that genetic information is encoded in the sequence of nucleotides of DNA and that, with the exception of some viruses which utilize RNA, DNA is the ultimate repository of genetic information. The sequence of amino acids in a protein is programmed in DNA; this information is first transferred by copying the nucleotide sequence of DNA into that of messenger RNA, from which this sequence is translated into the specific sequence of amino acids of the protein.

The biochemical basis for a number of genetically inherited diseases, in which the cause has been traced to the production of a defective protein, has been determined. Sickle cell anemia is a striking example; it is well established that the change of a single amino acid in hemoglobin has resulted in a serious abnormality in the properties of the hemoglobin molecule.

**Regulation.** Increased understanding of the chemical events in biological processes has permitted the investigation of the regulation of these proceses. An important concept is the chemical feedback circuit: the product of a series of reactions can itself influence the rates of the reactions. For example, the reactions which lead to the production of ATP proceed vigorously when the supply of ATP within the cell is low, but they slow down markedly when ATP is plentiful. These observations can be explained, in part, by the fact that ATP molecules bind to some of the enzymes involved, changing the surface features of the enzymes sufficiently to decrease their effectiveness as catalysts. It is also possible to regulate these reactions by changing the amounts of the enzymes; the amount of an enzyme can be controlled by modulating the synthesis of its specific messenger RNA or by modulating the translation of the information of the RNA molecule into the enzyme molecule. Another level of regulation involves the interaction of cells and tissues in multicellular organisms. For instance, endocrine glands can sense certain tissue activities and appropriately secrete hormones which control these activities. The chemical events and substances involved in cellular and tissue "communication" have become subjects of much investigation.

**Photosynthesis and nitrogen fixation.** Two subjects of substantial interest are the processes of photosynthesis and nitrogen fixation. In photosynthesis, the chemical reactions whereby the gas carbon dioxide is converted into carbohydrate are understood, but the reactions whereby light energy is trapped and converted into the chemical energy necessary for the synthesis of carbohydrate are unclear. The process of nitrogen fixation involves the conversion of nitrogen gas into a chemical form which can be utilized for the synthesis of numerous biologically important substances; the chemical events of this process are not fully understood. [A.S.L.H.]

**Bioinorganic chemistry** The field at the interface between biochemistry and inorganic chemistry; also known as inorganic biochemistry or metallobiochemistry. This field involves the application of the principles of inorganic chemistry to problems of biology and biochemistry. Because most biological components are organic, that is, they involve the chemistry of carbon compounds, the combination of the prefix bio- and inorganic may appear contradictory. However, organisms require a number of other elements to carry out their basic functions. Many of these elements are present as metal ions that are involved in crucial biological processes such as respiration, metabolism, cell division, muscle contraction, nerve impulse transmission, and gene regulation. The characterization of the interactions between such metal centers and biological components is the heart of bioinorganic chemistry. *See* BIOCHEMISTRY; INORGANIC CHEMISTRY.

Metal ions influence biological phenomena by interacting with organic functional groups on biomolecules, forming metal complexes. From this perspective, much of bioinorganic chemistry may be considered as coordination chemistry applied to biological questions. In general, bioinorganic chemists tackle such problems by first focusing on the elucidation of the structure of the metal complex of interest and then correlating structure with function. The attainment of solutions usually requires a combination of physical, chemical, and biological approaches. Biochemistry and molecular biology are often used to provide sufficient amounts of the system for investigation. Physical approaches such as crystallography and spectroscopy are useful in defining structural properties of the metal site. Synthetic methods can be used for the design and assembly of structural, spectroscopic, and functional models of the metal site. All these approaches then converge to elucidate how such a site functions. *See* COORDINATION CHEMISTRY; CRYSTALLOGRAPHY; SPECTROSCOPY.

**Low-molecular-weight compounds.** A number of coordination compounds found in organisms have relatively low molecular weights. Ionophores, molecules that are able to carry ions across lipid barriers, are polydentate ligands designed to bind alkali and alkaline-earth metal ions; they span membranes and serve to transport such ions across these biological barriers. Molecular receptors known as siderophores are also polydentate ligands; they have a very high affinity for iron. *See* IONOPHORE.

Other low-molecular-weight compounds are metal-containing cofactors that interact with macromolecules to promote important biological processes. Perhaps the most widely studied of the metal ligands found in biochemistry are the porphyrins; iron protoporphyrin IX (see illustration) is an example of the all-important complex

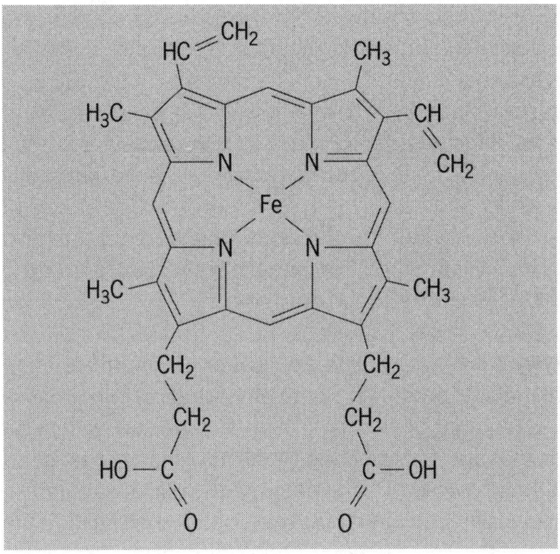

**Iron complex of protoporphyrin IX, or heme.**

in biology known as heme. Chlorophyll and vitamin $B_{12}$ are chemically related to the porphyrins. Magnesium is the central metal ion in chlorophyll, which is the green pigment in plants used to convert light energy into chemical energy. Cobalt is the central metal ion in vitamin $B_{12}$; it is converted into coenzyme $B_{12}$ in cells, where it participates in a variety of enzymatic reactions. *See* PORPHYRIN.

**Metalloproteins and metalloenzymes.** These are metal complexes of proteins. In many cases, the metal ion is coordinated directly to functional groups on amino acid residues. In some cases, the protein contains a bound metallo-cofactor such as heme. In metalloproteins with more than one metal-binding site, the metal ions may be found in clusters. Examples include ferredoxins, which contain iron-sulfur clusters ($Fe_2S_2$ or $Fe_4S_4$), and nitrogenase, which contains both $Fe_4S_4$ units and a novel $MoFe_7S_8$ cluster. *See* PROTEIN.

Some metalloproteins are designed for the storage and transport of the metal ions themselves—for example, ferritin and transferrin for iron and metallothionein for zinc. Others, such as the yeast protein Atx1, act as metallochaperones that aid in the insertion of the appropriate metal ion into a metalloenzyme. Still others function as transport agents. Cytochromes and ferredoxins facilitate the transfer of electrons in various metabolic processes.

Many metalloproteins catalyze important cellular reactions and are thus more specifically called metalloenzymes. For example, cytochrome oxidase is the respiratory enzyme in mitochondria responsible for disposing of the electrons generated by mammalian metabolism; it does so by reducing $O_2$ to water with the help of both heme and copper centers. In contrast, the conversion of water to $O_2$ is carried out in the photosynthetic apparatus by manganese centers. Other metalloenzymes are involved in the transformation of organic molecules in cells. For example, tyrosine hydroxylase (an iron enzyme) and dopamine $\beta$-hydroxylase (a copper enzyme) carry out oxidation reactions important for the biosynthesis of neurotransmitters. Alternatively, the metal center can serve as a Lewis acidic site to activate substrates for nucleophilic displacement reactions (that is, hydrolysis).

**Metals in medicine.** Metal complexes have also been found to be useful as therapeutic or diagnostic agents. Prominent among metal-based drugs is cisplatin, which is particularly effective in the treatment of testicular and ovarian cancers. Gold, gallium, and bismuth compounds are used for the treatment of rheumatoid arthritis, hypercalcemia, and peptic ulcers, respectively.

In clinical diagnosis, metal complexes can be used as imaging agents. The convenient half-life and radioemission properties of technetium-99 make its complexes very useful for a number of applications; by varying the ligands bound to the metal ion, diagnostic agents have been developed for imaging the heart, brain, and kidneys. Complexes of paramagnetic metal ions such as gadolinium(III), iron(III), and manganese(II) are also used as contrast agents to enhance images obtained from magnetic resonance imaging (MRI). *See* COORDINATION COMPLEXES; ORGANOMETALLIC COMPOUND. [L.Q.]

**Biomedical chemical engineering** The application of chemical engineering principles to the solution of medical problems due to physiological impairment. A knowledge of organic chemistry is required of all chemical engineers, and many also study biochemistry and molecular biology. This training at the molecular level gives chemical engineers a unique advantage over other engineering disciplines in communication with life scientists and clinicians in medicine. Practical applications include the development of tissue culture systems, the construction of three-dimensional scaffolds of biodegradable polymers for cell growth in the laboratory, and the design of artificial organs. *See* BIOCHEMISTRY.

Cell transplantation is explored as a means of restoring tissue function. With this approach, individual cells are harvested from a healthy section of donor tissue, isolated, expanded in culture, and implanted at the desired site of the functioning tissue. Isolated cells cannot form new tissues on their own and require specific environments that often include the presence of supporting material to act as a template for growth. Three-dimensional scaffolds can be used to mimic their natural counterparts, the extracellular matrices of the body. These scaffolds serve as both a physical support and an adhesive substrate for isolated parenchymal cells during cell culture and subsequent implantation. The scaffold must be made of biocompatible materials. As the transplanted cell population grows and the cells function normally, they will begin to secrete their own extracellular matrix support. The need for an artificial support will gradually diminish; and thus if the implant is biodegradable, it will be eliminated as its function is replaced. The development of processing methods to fabricate reproducibly three-dimensional scaffolds of biodegradable polymers that will provide temporary scaffolding to transplanted cells will be instrumental in engineering tissues.

Chemical engineers have made significant contributions to the design and optimization of many commonly used devices for both short-term and long-term organ replacement. Examples include the artificial kidney for hemodialysis and the heart-lung machine employed in open heart surgery. The artificial kidney removes waste metabolites (such as urea and creatinine) from blood across a polymeric membrane that separates the flowing blood from the dialysis fluid. The mass transport properties and biocompatibility of these membranes are crucial to the functioning of hemodialysis equipment. The heart-lung machine replaces both the pumping function of the heart and the gas exchange function of the lung in one fairly complex device. While often life saving, both types of artificial organs only partially replace real organ function. Long-term use often leads to problems with control of blood coagulation mechanisms to avoid both excessive clotting initiated by blood contact with artificial surfaces and excessive bleeding due to platelet consumption or overuse of anticoagulants. *See* DIALYSIS; MEMBRANE SEPARATIONS.

Other chemical engineering applications include methodology for development of artificial bloods, utilizing fluorocarbon emulsions or encapsulated or polymerized hemoglobin, and controlled delivery devices for release of drugs or of specific molecules (such as insulin) missing in the body because of disease or genetic alteration. *See* POLYMER. [L.V.M.; A.G.M.]

**Bioorganic chemistry** The science that describes the structure, interactions, and reactions of organic compounds of biological significance at the molecular level. It represents the meeting of biochemistry, as attempts are made to describe the structure and physiology of organisms on an ever smaller scale, with organic chemistry, as attempts are made to synthesize and understand the behavior of molecules of ever-increasing size and complexity. Areas of research include enzymatic catalysis, the structure and folding of proteins, the structure and function of biological membranes, the chemistry of poly(ribonucleic acids) and poly(deoxyribonucleic acids), biosynthetic pathways, immunology, and mechanisms of drug action.

Being at the interface of two disciplines, bioorganic chemistry utilizes experimental techniques and theoretical concepts drawn from both. Important experimental techniques include organic synthesis, kinetics, structure-activity relationships, the use of model systems, methods of protein purification and manipulation, genetic mutation, cloning and overexpression (engineered enhancement of gene transcription), and the elicitation of monoclonal antibodies. Theoretical concepts important to bioorganic chemistry include thermodynamics, transition-state theory, acid-base theory, concepts of hydrophobicity and hydrophilicity, theories of stereocontrol, and theories of adaptation of organisms to selective pressures.

Historically, a major focus of bioorganic research has been the study of catalysis by enzymes. Enzymes have a dramatic ability to increase the rates at which reactions occur. One of the ways that enzymes increase the rates of bimolecular reactions is to overcome the entropic barrier associated with bringing two particles together to form one. *See* ENZYME.

Enzymes also catalyze reactions by facilitating proton transfers. Many of the reactions catalyzed by enzymes, such as the formation and hydrolysis of esters and amides, require the deprotonation of a nucleophile (base catalysis) or the protonation of an electrophile (acid catalysis).

Enzymes may act as preorganized solvation shells for transition states. If the active site of an enzyme has just the right size, shape, and arrangement of functional groups to bind a transition state, it will automatically bind the reactants less well. Selective binding of the transition state lowers the energy of the transition state relative to that of the reactants. Because the energy barrier between reactants and transition state is reduced, the reaction proceeds more rapidly.

The selectivity of enzymes makes them useful as catalysts for organic synthesis. Surprisingly, enzymes are able to catalyze not only the reactions that they mediate in living systems but also similar, selective transformations of unnatural substrates. Because of their selectivity, several enzyme-catalyzed reactions may run simultaneously in the same vessel. Thus, a reactant can undergo several reactions in series without the need for isolation of intermediates. Enzymes can be combined so as to reconstitute within a reaction vessel naturally occurring metabolic pathways or to create new, artificial metabolic pathways. Sequences of up to 12 serial reactions have been executed successfully in a single reaction vessel. *See* CATALYSIS.

The biological activity of a protein, whether binding, catalytic, or structural, depends on its full three-dimensional or conformational structure. The linear sequence of amino acids that make up a protein constitutes its primary structure. Local regions of highly organized conformation ($\alpha$-helices, $\beta$-pleats, $\beta$-turns, and so on) are called secondary structure. Further folding of the protein causes regions of secondary structure to associate or come into correct alignment. This action establishes the tertiary structure of the native (active) protein. A goal of bioorganic chemistry is to achieve an understanding of the process of protein folding. *See* AMINO ACIDS; BIOCHEMISTRY; ORGANIC CHEMISTRY; PROTEIN. [H.K.C.]

**Biopolymer** A macromolecule derived from natural sources; also known as biological polymer. Some biopolymers are used as structural materials, food sources, or catalysts. Others have evolved as entities for information storage and transfer. Examples of biopolymers include polypeptides; polysaccharides; polypeptide/polysaccharide hybrids; polynucleotides, which are polymers derived from ribonucleic acid (RNA) and deoxyribonucleic acid (DNA); polyhydroxybutyrates, a class of polyesters produced by certain bacteria; and *cis*-1,4-polyisoprene, the major component of rubber tree latex. *See* POLYMER.

Amino acids are the monomers from which polypeptides are derived. Polypeptides alone, as well as multipolypeptide complexes or complexes with other molecules, are known as proteins, and each has a specific biological function. There are numerous examples of biopolymers having a polysaccharide and polypeptide in the same molecule, usually with a polysaccharide as a side chain in a polypeptide, or vice versa. *See* PEPTIDE; POLYSACCHARIDE; PROTEIN. [G.E.W.]

**Biosensor** An integrated device consisting of a biological recognition element and a transducer capable of detecting the biological reaction and converting it into a signal which can be processed. Ideally, the sensor should be self-contained, so that it is not necessary to add reagents to the sample matrix to obtain the desired response. There are a number of analytes (the target substances to be detected) which are measured in biological media: pH, partial pressure of carbon dioxide ($pCO_2$), partial pressure of oxygen ($pO_2$), and the ionic concentrations of sodium, potassium, calcium, and chloride. However, these sensors do not use biological recognition elements, and are considered chemical sensors. Normally, the biological recognition element is a protein or protein complex which is able to recognize a particular analyte in the presence of many other components in a complex biological matrix. This definition has since been expanded to include oligonucleotides. The recognition process involves a chemical or biological reaction, and the transducer must be capable of detecting not only the reaction but also its extent. An ideal sensor should yield a selective, rapid, and reliable response to the analyte, and the signal generated by the sensor should be proportional to the analyte concentration.

Biosensors are typically classified by the type of recognition element or transduction element employed. A sensor might be described as a catalytic biosensor if its recognition element comprised an enzyme or series of enzymes, a living tissue slice (vegetal or animal), or whole cells derived from microorganisms such as bacteria, fungi, or yeast. The sensor might be described as a bioaffinity sensor if the basis of its operation were a biospecific complex formation. Accordingly, the reaction of an antibody with an antigen or hapten, or the reaction of an agonist or antagonist with a receptor, could be employed. In the former case, the sensor might be called an immunosensor.

Since enzyme-based sensors measure the rate of the enzyme-catalyzed reaction as the basis for their response, any physical measurement which yields a quantity related to this rate can be used for detection. The enzyme may be immobilized on the end of an optical fiber, and the spectroscopic properties (absorbance, fluorescence, chemiluminescence) related to the disappearance of the reactants or appearance of products of the reaction can be measured. Since biochemical reactions can be either endothermic (absorbing heat) or exothermic (giving off heat), the rate of the reaction can be measured by microcalorimetry. Miniaturized thermistor-based calorimeters, called enzyme thermistors, have been developed and widely applied, especially for bioprocess monitoring.

As in the case of the catalytic biosensors, many physical techniques can be used to detect affinity binding: microcalorimetry (thermometric enzyme-linked immunosorbent assay, or TELISA), fluorescence energy transfer, fluorescence polarization, or bioluminescence.

The quality of the results obtained from sensors based on biological recognition elements depends most heavily on their ability to react rapidly, selectively, and with high affinity. Antibodies and receptors frequently react with such high affinity that the analyte does not easily become unbound. To reuse the sensor requires a time-consuming regeneration step. Nonetheless, if this step can be automated, semicontinuous monitoring may be possible. [G.S.W.]

**Bismuth** The metallic element, Bi, of atomic number 83 and atomic weight 208.980 belonging in the periodic table to group 15. Bismuth is the most metallic element in this group in both physical and chemical properties. The only stable isotope

**Physical and mechanical properties of bismuth**

| Property | Value | Temperature |
|---|---|---|
| Melting point, °C | 271.4 | |
| Boiling point, °C | 1559 | |
| Heat of fusion, kcal/mole | 2.60 | |
| Heat of vaporization, kcal/mole | 36.2 | |
| Vapor pressure, mm Hg | 1 | 917°C |
| | 10 | 1067°C |
| | 100 | 1257°C |
| Density, g/cm$^3$ | 9.80 | 20 (solid) |
| | 10.03 | 300 (liquid) |
| | 9.91 | 400 (liquid) |
| | 9.66 | 600 (liquid) |
| Mean specific heat, cal/g | 0.0294 | 0–270°C |
| | 0.0373 | 300–1000°C |
| Coefficient of linear expansion | $13.45 \times 10^{-6}$/°C | |
| Thermal conductivity, cal/(s)(cm$^2$)(°C) | 0.018 | 100 (solid) |
| | 0.041 | 300 (liquid) |
| | 0.037 | 400 (liquid) |
| Electrical resistivity, $\mu$ohm-cm | 106.5 | 0 (solid) |
| | 160.2 | 100 (solid) |
| | 267.0 | 269 (solid) |
| | 128.9 | 300 (liquid) |
| | 134.2 | 400 (liquid) |
| | 145.3 | 600 (liquid) |
| Surface tension, dynes/cm | 376 | 300°C |
| | 370 | 400°C |
| | 363 | 500°C |
| Viscosity, centipoise | 1.662 | 300°C |
| | 1.280 | 450°C |
| | 0.996 | 600°C |
| Magnetic susceptibility, cgs units | $-1.35 \times 10^{-6}$ | |
| Crystallography | Rhombohedral, $a_0 = 0.47457$ nm | |
| Thermal-neutron absorption cross section, barns | $0.032 \pm 0.003$ | |
| Modulus of elasticity, lb/cm$^2$ | $4.6 \times 10^6$ | |
| Shear modulus, lb/cm$^2$ | $1.8 \times 10^6$ | |
| Poisson's ratio | 0.33 | |
| Hardness, Brinell | 4–8 | |

is that of mass 209. It is estimated that the Earth's crust contains about 0.00002% bismuth. It occurs in nature as the free metal and in ores. The principal ore deposits are in South America. However, the primary source of bismuth in the United States is as a by-product in refining of copper and lead ores. *See* PERIODIC TABLE.

The main use of bismuth is in the manufacture of low-melting alloys which are used in fusible elements in automatic sprinklers, special solders, safety plugs in compressed gas cylinders, and automatic shutoffs for gas and electric water-heating systems. Some bismuth alloys, which expand on freezing, are used in castings and in type metal. Another important use of bismuth is in the manufacture of pharmaceutical compounds.

Bismuth is a gray-white, lustrous, hard, brittle, coarsely crystalline metal. It is one of the few metals which expand on solidification. The thermal conductivity of bismuth is lower than that of any metal, with the exception of mercury. The table cites the chief physical and mechanical properties of bismuth. Bismuth is inert in dry air at room temperature, although it oxidizes slightly in moist air. It rapidly forms an oxide film at temperatures above its melting point, and it burns at red heat, forming the yellow oxide, $Bi_2O_3$. The metal combines directly with halogens and with sulfur, selenium, and tellurium; however, it does not combine directly with nitrogen or phosphorus. Bismuth is not attacked at ordinary temperatures by air-free water, but it is slowly oxidized at red heat by water vapor.

Almost all compounds of bismuth contain trivalent bismuth. However, bismuth can occasionally be pentavalent or monovalent. Sodium bismuthate and bismuth pentafluoride are perhaps the most important compounds of Bi(V). The former is a powerful oxidizing agent, and the latter a useful fluorinating agent for organic compounds. [S.J.Y.]

**Bohrium** A chemical element, symbol Bh, atomic number 107. Bohrium was synthesized and identified in 1981 by using the Universal Linear Accelerator (UNILAC) of the Gesellschaft für Schwerionenforschung (GSI) at Darmstadt, West Germany, by a team led by P. Armbruster and G. Müzenberg. The reaction used to produce the element was proposed and applied in 1976 by Y. T. Oganessian and colleagues at Dubna Laboratories in Russia. A $^{209}$Bi target was bombarded by a beam of $^{54}$Cr projectiles.

The best technique to identify a new isotope is its genetic correlation to known isotopes through a radioactive decay chain. These decay chains are generally interrupted by spontaneous fission. In order to apply decay chain analysis, those isotopes that are most stable against spontaneous fission should be produced, that is, isotopes with odd numbers of protons and neutrons. Not only does the fission barrier govern the spontaneous fission of a species produced, but also, in the deexcitation of the virgin nucleus, fission competing with neutron emission determines the final production probability. To keep the fission losses small, a nucleus should be produced with the minimum excitation energy possible. In this regard, reactions using relatively symmetric collision partners and strongly bound closed-shell nuclei, such as $^{209}$Bi and $^{208}$Pb as targets and $^{48}$Ca and $^{50}$Ti as projectiles, are advantageous.

Six decay chains were found in the Darmstadt experiment. All the decays can be attributed to $^{262}$Bh, an odd nucleus produced in a one-neutron reaction. The isotope $^{262}$Bh undergoes alpha-particle decay (10.38 MeV) with a half-life of about 5 ms.

Experiments at Dubna, performed in 1983 using the 157-in. (400-cm) cyclotron, established the production of $^{262}$Bh in the reaction $^{209}$Bi $^{54}$Cr. *See* PERIODIC TABLE; TRANSURANIUM ELEMENTS. [P.Ar.]

**Boiling point** The boiling point of a liquid is the temperature at which the liquid and vapor phases are in equilibrium with each other at a specified pressure. Therefore, the boiling point is the temperature at which the vapor pressure of the liquid is equal to the applied pressure on the liquid. The boiling point at a pressure of 1 atmosphere is called the normal boiling point.

For a pure substance at a particular pressure $P$, the stable phase is the vapor phase at temperatures immediately above the boiling point and is the liquid phase at temperatures immediately below the boiling point. The liquid-vapor equilibrium line on the phase diagram of a pure substance gives the boiling point as a function of pressure. Alternatively, this line gives the vapor pressure of the liquid as a function of temperature. The vapor pressure of water is 1 atm (101.325 kilopascals) at 100°C (212°F), the normal boiling point of water. The vapor pressure of water is 3.2 kPa (0.031 atm) at 25°C (77°F), so the boiling point of water at 3.2 kPa is 25°C. The liquid-vapor equilibrium line on the phase diagram of a pure substance begins at the triple point (where solid, liquid, and vapor coexist in equilibrium) and ends at the critical point, where the densities of the liquid and vapor phases have become equal. For pressures below the triple-point pressure or above the critical-point pressure, the boiling point is meaningless. Carbon dioxide has a triple-point pressure of 5.11 atm (518 kPa), so carbon dioxide has no normal boiling point. *See* TRIPLE POINT; VAPOR PRESSURE.

The normal boiling point is high for liquids with strong intermolecular attractions and low for liquids with weak intermolecular attractions. Helium has the lowest normal boiling point, 4.2 K (−268.9°C). Some other normal boiling points are 111.1 K (−162°C) for $CH_4$, 450°C (842°F) for $n$-$C_{30}H_{62}$, 1465°C (2669°F) for NaCl, and 5555°C (10031°F) for tungsten.

The rate of change of the boiling-point absolute temperature $T_b$ of a pure substance with pressure is given by the equation below. $\Delta H_{\mathrm{vap,m}}$ is the molar enthalpy (heat) of

$$\frac{dT_b}{dP} = \frac{T_b \Delta V_{\mathrm{vap,m}}}{\Delta H_{\mathrm{vap,m}}}$$

vaporization, and $\Delta V_{\mathrm{vap,m}}$ is the molar volume change on vaporization.

The quantity $\Delta H_{\mathrm{vap,m}}/T_b$ is $\Delta S_{\mathrm{vap,m}}$, the molar entropy of vaporization. The molar entropy of vaporization at the normal boiling point (nbp) is given approximately by Trouton's rule: $\Delta S_{\mathrm{vap,m,nbp}} \approx 87$ J/mol K (21 cal/mol K). Trouton's rule fails for highly polar liquids (especially hydrogen-bonded liquids). It also fails for liquids boiling at very low or very high temperatures, because the molar volume of the vapor changes with temperature and the entropy of a gas depends on its volume.

When a pure liquid is boiled at fixed pressure, the temperature remains constant until all the liquid has vaporized. When a solution is boiled at fixed pressure, the composition of the vapor usually differs from that of the liquid, and the change in liquid composition during boiling changes the boiling point. Thus the boiling process occurs over a range of temperatures for a solution. An exception is an azeotrope, which is a solution that boils entirely at a constant temperature because the vapor in equilibrium with the solution has the same composition as the solution. In fractional distillation, the variation of boiling point with composition is used to separate liquid mixtures into their components. *See* AZEOTROPIC MIXTURE; DISTILLATION; PHASE EQUILIBRIUM. [I.N.L.]

**Bond angle and distance** The angle between two bonds sharing a common atom is known as the bond angle. The distance between the nuclei of bonded atoms is known as bond distance. The geometry of a molecule can be characterized by bond angles and distances. The angle between two bonds sharing a common atom through a third bond is known as the torsional or dihedral angle (see illustration).

Certain pairs of atoms in a molecule are held together at distances of about 0.075–0.3 nanometer, with energies of about 150–1000 kilojoules/mol, because of a balance between electrostatic attraction and repulsion among the electrons and nuclei, subject

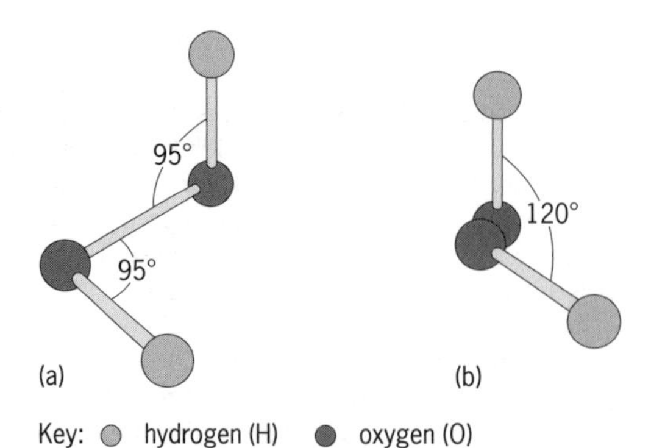


**Bond angle versus torsional angle. (*a*) Two O—O—H bond angles in hydrogen peroxide ($H_2O_2$). (*b*) The same molecule when viewed along the O—O bond, revealing the torsional angle, that is, the angle between the two O—H bonds.**

to the electron distributions allowed by quantum mechanics. Such interactions are called chemical bonds. *See* CHEMICAL BONDING.

Bond angles and distances are important because they determine the shape and size of a molecule and therefore affect many physical properties of a substance. The shape of a molecule influences its charge distribution, often determining whether a molecule is polar or nonpolar; polar molecules exhibit stronger intermolecular attraction and thus have higher boiling points. Molecular shapes also influence the packing density of molecules in solids and liquids. *See* MOLECULAR STRUCTURE AND SPECTRA; POLAR MOLECULE.

Bond angles and distances depend on the identity of the atoms bonded together and the type of chemical bonding involved. For example, carbon-hydrogen bonds are remarkably similar in length in a wide range of compounds. Bonds vary in length depending on the bond multiplicity: they are shorter for a triple bond than for a single bond. Single bonds vary in length slightly, depending on whether they are adjacent to multiple bonds. Bond angles vary considerably from molecule to molecule; the variation in bond angles and distances depends primarily on the electronic structures of the molecules.

The bond lengths and angles of an electronically excited molecule usually differ from the ground-state values. For example, carbon dioxide is linear in its ground state but is bent in its lowest excited state; that is, the O—C—O angle is $180^\circ$ in the ground state but is $122 \pm 2^\circ$ in the first excited state. The C═O bond distance is 0.116 nm in the ground state but is 0.125 nm in the first excited state. [B.A.Ga.]

**Borane** One of a class of binary compounds of boron and hydrogen, often referred to as boron hydrides. The term borane is sometimes used to denote substances which may be considered to be derivatives of the boron-hydrogen compounds, such as boron trichloride ($BCl_3$), and diiododecaborane ($B_{10}H_{12}I_2$).

The simplest borane is diborane ($B_2H_6$); other boranes of increasingly higher molecular weight are known, one of the least volatile of which is an apparently polymeric solid of composition $(BH)_x$. Certain boranes, such as $BH_3$, and $B_3H_7$, are not known as such, but can be prepared in the form of adducts with electron-donor molecules.

The most spectacular projected large-scale use of the boranes and their derivatives is in the field of high-energy fuels for jet planes and rockets. The thermal decomposition of diborane (6) [$B_2H_6$] has been used to produce coatings of pure elementary boron for neutron-detecting devices and for applications requiring hard, corrosion-resistant surfaces. Boranes can also be used as vulcanizing agents for natural and synthetic rubbers, and are especially effective in the preparation of silicone rubbers.

The molecular structures possessed by the boranes are exhibited by no other class of substances. Because of the lack of sufficient electrons for the formation of the requisite number of covalent bonds, normal covalently bonded structures of the hydrocarbon type are not possible. The boranes are sometimes referred to as electron-deficient substances. In no case are the simple chain and ring configurations of carbon chemistry encountered in the more complex boranes. Instead, the boron atoms are situated at the corners of polyhedrons. An example of such a structure is that of pentaborane (9) [$B_5H_9$], shown below. Boron nomenclature uses a prefix to designate the number

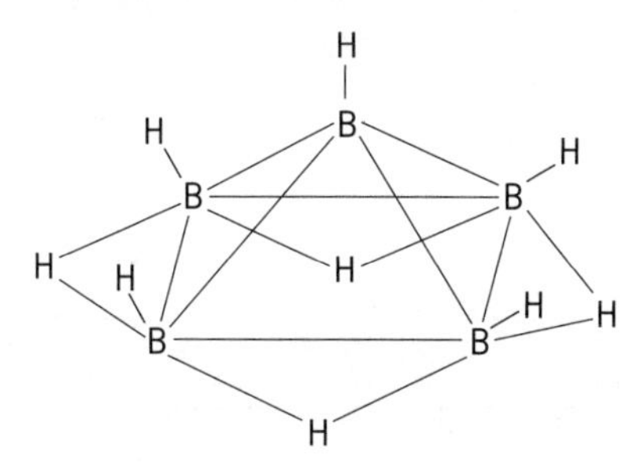


of boron atoms in the molecule and a numeral suffix in parentheses to indicate the number of hydrogen atoms.

As a class, the boranes are quite reactive substances and are generally decomposed, at times explosively, on contact with air. Their reactivities with air and water decrease with increasing molecular weight. Because boranes react readily with air, laboratory investigations are almost invariably carried out in all-glass vacuum apparatus or in inert-atmosphere dry boxes. With the possible exception of decaborane (14) [$B_{10}H_{14}$], the known boranes are not indefinitely stable at room temperature. They decompose more or less rapidly to yield elementary hydrogen and boranes richer in boron.

The known derivatives of the boranes (other than $BH_3$) are relatively few in number. Several halo, alkyl, and amino boranes have been reported but, in general, these have not been extensively characterized. *See* BORON; CARBORANE; METAL HYDRIDES. [T.W.]

**Boron** A chemical element, B, atomic number 5, atomic weight 10.811, in group III of the periodic table. It has three valence electrons and is nonmetallic in behavior. It is classified as a metalloid and is the only nonmetallic element which has fewer than four electrons in its outer shell. The free element is prepared in crystalline or amorphous form. The crystalline form is an extremely hard, brittle solid. It is of jet-black to silvery-gray color with a metallic luster. One form of crystalline boron is bright red. The amorphous form is less dense than the crystalline and is a dark-brown to black powder. In the naturally occurring compounds, boron exists as a mixture of two stable isotopes with atomic weights of 10 and 11. *See* PERIODIC TABLE.

Many properties of boron have not been sufficiently established experimentally as a result of the questionable purity of some sources of boron, as well as of the variations in the methods and temperatures of preparation. A summary of the physical properties is shown in the table.

Boron and boron compounds have numerous uses in many fields, although elemental boron is employed chiefly in the metal industry. Its extreme reactivity at high temperatures, particularly with oxygen and nitrogen, makes it a suitable metallurgical degasifying agent. It is used to refine the grain of aluminum castings and to facilitate the heat treatment of malleable iron. Boron considerably increases the high-temperature

**Physical properties of boron**

| Property | Temp., °C | Value |
|---|---|---|
| Density | | |
| Crystalline | 25–27 | 2.31 g/cm$^3$ |
| Amorphous | 25–27 | 2.3 g/cm$^3$ |
| Mohs hardness | | |
| Crystalline | | 9.3 |
| Melting point | | 2100°C |
| Boiling point | | 2500°C |
| Resistivity | 25 | $1.7 \times 10^{-6}$ ohm-cm |
| Coefficient of thermal expansion | 20–750 | $8.3 \times 10^{-6}$ cm/°C |
| Heat of combustion | 25 | $302.0 \pm 3.4$ kcal/mole |
| Entropy | | |
| Crystalline | 25 | 1.403 cal/(mole)(deg) |
| Amorphous | 25 | 1.564 cal/(mole)(deg) |
| Heat capacity | | |
| Gas | 25 | 4.97 cal/(mole)(deg) |
| Crystalline | 25 | 2.65 cal/(mole)(deg) |
| Amorphous | 25 | 2.86 cal/(mole)(deg) |

strength characteristics of alloy steels. Elemental boron is used in the atomic reactor and in high-temperature technologies. The physical properties that make boron attractive as a construction material in missile and rocket technology are its low density, extreme hardness, high melting point, and remarkable tensile strength in filament form. When boron fibers are used in an epoxy (or other plastic) carrier material or matrix, the resulting composite is stronger and stiffer than steel and 25% lighter than aluminum. Refined borax, $Na_2B_4O_7 \cdot 10H_2O$, is an important ingredient of a variety of detergents, soaps, water-softening compounds, laundry starches, adhesives, toilet preparations, cosmetics, talcum powder, and glazed paper. It is also used in fireproofing, disinfecting of fruit and lumber, weed control, and insecticides, as well as in the manufacture of leather, paper, and plastics.

Boron makes up 0.001% of the Earth's crust. It is never found in the uncombined or elementary state in nature. Besides being present to the extent of a few parts per million in sea water, it occurs as a trace element in most soils and is an essential constituent of several rock-forming silicate minerals, such as tourmaline and datolite. The presence of boron in extremely small amounts seems to be necessary in nearly all forms of plant life, but in larger concentrations, it becomes quite toxic to vegetation. Only in a very limited number of localities are high concentrations of boron or large deposits of boron minerals to be found in nature; the more important of these seem to be primarily of volcanic origin. [F.H.M./V.V.L.]

**Boyle's law** A law of gases which states that at constant temperature the volume of a gas varies inversely with its pressure. This law, formulated by Robert Boyle (1627–1691), can also be stated thus: The product of the volume of a gas times the pressure exerted on it is a constant at a fixed temperature. The relation is approximately true for most gases, but is not followed at high pressure. The phenomenon was discovered independently by Edme Mariotte about 1650 and is known in Europe as Mariotte's law. *See* GAS. [F.H.R.]

**Branched polymer** A polymer chain having branch points that connect three or more chain segments. Examples of branched polymers include long chains having occasional and usually short branches comprising the same repeat units as the

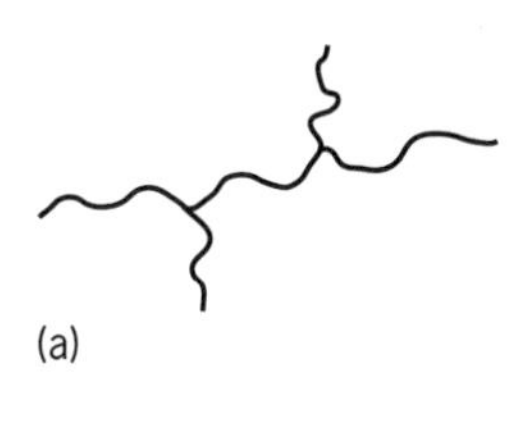

(a)

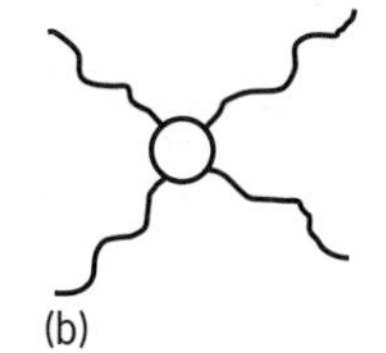

(b)

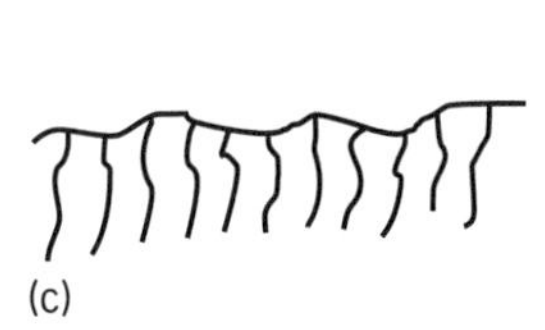

(c)

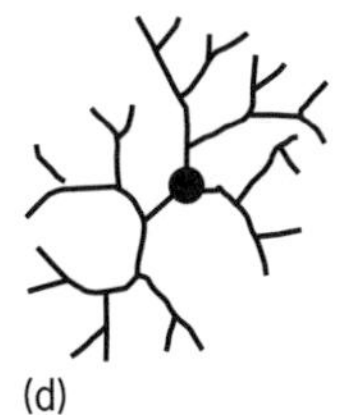

(d)

**Examples of branched polymers. (a) Branched polymer (if arms are of composition similar to backbone) or graft polymer (if compositions are different). (b) Star polymer. (c) Comb polymer. (d) Dendritic polymer.**

main chain (nominally termed a branched polymer); long chains having occasional branches comprising repeat units different from those of the main chain (termed graft copolymers); main chains having one long branch per repeat unit (referred to as comb polymers); and small core molecules with branches radiating from the core (star polymers). Starburst or dendritic polymers are a special class of star polymer in which the branches are multifunctional, leading to further branching with polymer growth. Star, comb, and starburst polymers (see illustration), especially the last, represent interesting molecular structures that may lead to unusual supramolecular structures (for example, micelles and liposomes) that mimic the functions of complex biomolecules. *See* BIOPOLYMER; POLYMER; POLYMERIZATION. [G.E.W.]

**Bromine** A chemical element, Br, atomic number 35, atomic weight 79.909, which normally exists as $Br_2$, a dark-red, low-boiling but high-density liquid of intensely irritating odor. This is the only nonmetallic element that is liquid at normal temperature and pressure. Bromine is very reactive chemically; one of the halogen group of elements, it has properties intermediate between those of chlorine and iodine. *See* HALOGEN ELEMENTS; PERIODIC TABLE.

The most stable valence states of bromine in its salts are $-1$ and $+5$, although $+1$, $+3$, and $+7$ are known. Within wide limits of temperature and pressure, molecules of the liquid and vapor are diatomic, $Br_2$, with a formula weight of 159.818. There are two stable isotopes ($^{79}Br$ and $^{81}Br$) that occur naturally in nearly equal proportion, so that the atomic weight is 79.909. A number of radioisotopes are also known.

The solubility of bromine in water at 20°C (68°F) is 3.38 g/100 g (3.38 oz/100 oz) solution, but its solubility is increased tremendously in the presence of its salts and in hydrobromic acid. The ability of this inorganic element to dissolve in organic solvents is of considerable importance in its reactions. The table summarizes the physical properties of bromine.

Although it is estimated that from $10^{15}$ to $10^{16}$ tons of bromine are contained in the Earth's crust, the element is widely distributed and found only in low concentrations in the form of its salts. The bulk of the recoverable bromine, however, is found in the hydrosphere. Sea water contains an average of 65 parts per million (ppm) of bromine. The other major sources of bromine in the United States are underground brines and salt lakes, with commercial production in Michigan, Arkansas, and California.

**Physical properties of bromine**

| Property | Value |
|---|---|
| Flash point | None |
| Fire point | None |
| Freezing point, °C | −7.27 |
| Density, 20°C | 3.1226 |
| Pounds per gallon, 25°C | 25.8 |
| Boiling point, 760 mm Hg, °C | 58.8 |
| Refractive index, 20°C | 1.6083 |
| Latent heat of fusion, cal/g | 15.8 |
| Latent heat of vaporization, cal/g, bp | 44.9 |
| Vapor density, g/liter, standard conditions (0°C, 1 atm) | 7.139 |
| Viscosity, centistokes, 20°C | 0.314 |
| Surface tension, dynes/cm, 20°C | 49.5 |
| 30°C | 47.3 |
| 40°C | 45.2 |
| Dielectric constant, $10^5$ freq, 25°C | 3.33 |
| Compressibility, vapors, 25°C | 0.998 |

Thermodynamic data, cal/(mole K)

| | *T*,K | *Entropy* | *Heat capacity* |
|---|---|---|---|
| Solid | 265.9 | 24.786 | 14.732 |
| Liquid | 265.9 | 34.290 | 18.579 |

While many inorganic bromides have found industrial use, the organic bromides have even wider application. Because of the ease of reaction of bromine with organic compounds and the ease of its subsequent removal or replacement, organic bromides have been much studied and used as chemical intermediates. In addition, any of the bromine reactions are so clean-cut that they can be used for the study of reaction mechanisms without complication of side reactions. The ability of bromine to add into unusual places on organic molecules has added to its value as a research tool.

Bromine and its compounds have found acceptance as disinfection and sanitizing agents in swimming pools and potable water. Certain bromine-containing compounds are safer to use than the analogous chlorine compounds due to certain persistent residuals found in the chlorine-containing materials. Other bromine chemicals are used as a working fluid in gages, as hydraulic fluids, as chemical intermediates in the manufacture of organic dyes, in storage batteries, and in explosion-suppressant and fire-extinguishing systems. Bromine compounds, because of their density, also find use in the gradation of coal and other minerals where separations are effected by density gradients. The versatility of bromine compounds is illustrated by the commercial use of over 100 compounds that contain bromine.

Bromine is almost instantaneously injurious to the skin, and it is difficult to remove quickly enough to prevent a painful burn that heals slowly. Bromine vapor is extremely toxic, but its odor gives good warning; it is difficult to remain in an area of sufficient concentration to be permanently damaging. Bromine can be handled safely, but the recommendations of the manufacturers should be respected. [R.C.S.]

**Buffers (chemistry)** A solution selected or prepared to minimize changes in hydrogen ion concentration which would otherwise tend to occur as a result of a chemical reaction. In general, chemical buffers are systems which, once constituted, tend to resist further change due to external influences. Thus it is possible, for example, to make buffers resistant to changes in temperature, pressure, volume, redox potential, or acidity. The commonest buffer in chemical solution systems is the acid-base buffer.

Chemical reactions known or suspected to be dependent on the acidity of the solution, as well as on other variables, are frequently studied by measurements in comixture with an appropriate buffer. For example, it may be desirable to investigate how the rate of a chemical reaction depends upon the hydrogen ion activity (pH). This is accomplished by measurements in several buffer systems, each of which provides a nearly constant, different pH. Alternatively, it may be desirable to measure the effects of other variables on a pH-sensitive system, by stabilizing the pH at a convenient value with a particular buffer. *See* PH.

Buffer action depends upon the fact that, if two or more reactions coexist in a solution, then the chemical potential of any species is common to all reactions in which it takes part, and may be defined by specification of the chemical potentials of all other species in any one of the reactions. To be effective, a buffer must be able to respond to an increase as well as a decrease of the species to be buffered. In order to do so, it is necessary that the proton transfer step of the buffer be reversible with respect to the species involved, in the reaction to be buffered. In aqueous solution the proton transfer between most acids, their conjugate bases, and water, is so rapid and reversible that the dominant direct source of protons for a chemical reaction is $H_3O^+$, the hydronium ion.

Buffers are particularly effective in water, because of the unusual properties of water as a solvent. Its high dielectric constant tends to promote the existence of formally charged ions (ionization). Because it has both an acidic (H) and a basic (O) group, it may form bonds with ionic species leading to an organized sheath of solvent surrounding an ion (solvation). Water also tends to self-ionize to form its own conjugate acid-base system. *See* ACID AND BASE; ACID-BASE INDICATOR; IONIC EQUILIBRIUM; SOLVATION. [A.M.H.]

# C

**C—H activation** The cleavage of carbon-hydrogen bonds in organic compounds, leading to subsequent functionalization to introduce useful chemical groups. For example, petroleum and natural gas, important energy resources for the modern world, are both alkanes, $C_nH_{2n+2}$, with many C—H bonds. One problem is to convert these alkanes to useful alcohols, $C_nH_{2n+1}OH$. This is particularly difficult because of the lack of reactivity of alkanes as indicated by their older name, paraffins (from Latin, meaning low affinity). In C—H activation, metal catalysts are often used to mediate the hydrocarbon conversion reactions. *See* ALCOHOL; ALKANE; CHEMICAL BONDING; ORGANIC CHEMISTRY.

A variety of $ML_n$ fragments, consisting of a metal M and its associated ligands $L_n$, are capable of reaction with an alkane to give C—H bond breaking as shown in the first step of reaction (1). In principle, it should be possible to follow this with a second

$$CH_4 + ML_n \xrightarrow[\text{activation step}]{} \begin{matrix} CH_3 \\ H \end{matrix} \!\!>\! ML_n \xrightarrow[\text{functionalization step}]{\frac{1}{2}O_2} CH_3OH + ML_n \qquad (1)$$

step to give the functionalized alkane, but this has proved difficult in practice. R. G. Bergman showed as early as 1982 that $(C_5Me_5)Ir(PMe_3)$, formed photochemically, is suitable for the first reaction, for example. *See* COORDINATION CHEMISTRY; LIGAND.

Other cases are known where the first step of reaction (1) is followed by a functionalization reaction. H. Felkin, R. H. Crabtree, and A. S. Goldman showed reactions of this kind where the intermediate alkyl hydride decomposes to give alkene and free $H_2$ or, in the presence of a second alkene as sacrificial oxidant, alkene and hydrogenated sacrificial oxidant [reaction (2)].

-H₂ → (alkene)
CO → (aldehyde, C(=O)H)
$(RO)_2B—B(OR)_2$ → (alkyl)$B(OR)_2$ (2)

Y. Saito and M. Tanaka showed that the intermediate alkyl hydride can be trapped by carbon monoxide (CO) to give aldehyde (RCHO) as final product, and J. F. Hartwig trapped the alkyl with diborane derivatives to give alkyl boronic esters [reaction (2)]. In all of these cases, C—H activation preferentially occurs at the least hindered C—H bond, leading to products that are quite different from those formed in radical and acid pathways where the most substituted and most hindered C—H bonds are most reactive. In each case, appropriate transition-metal compounds are present and catalyze the reactions. These reactions do not yet form the basis of any practical process, but produce terminally functionalized products (the most desirable type) rather than the mixtures commonly found in other reactions. [R.H.Cr.]

**Cadmium** A relatively rare chemical element, symbol Cd, atomic number 48, closely related to zinc, with which it is usually associated in nature. It is a silvery-white ductile metal with a faint bluish tinge. It is softer and more malleable than zinc, but slightly harder than tin. It has an atomic weight of 112.40 and a specific gravity of 8.65 at 20°C (68°F). Its melting point of 321°C (610°F) and boiling point of 765°C (1410°F) are lower than those of zinc. There are eight naturally occurring stable isotopes, and eleven artificial unstable radio isotopes have been reported. Cadmium is the middle member of group 12 (zinc, cadmium, and mercury) in the periodic table, and its chemical properties generally are intermediate between zinc and mercury. The cadmium ion is displaced by zinc metal in acidic sulfate solutions. Cadmium is bivalent in all its stable compounds, and its ion is colorless. *See* PERIODIC TABLE; TIN; ZINC.

Cadmium does not occur uncombined in nature, and the one true cadmium mineral, greenockite (cadmium sulfide), is not a commercial source of the metal. Almost all of the cadmium produced is obtained as a by-product of the smelting and refining of zinc ores, which usually contain 0.2–0.4% cadmium. The United States, Canada, Mexico, Australia, Belgium-Luxembourg, and the Republic of Korea are principal sources, although not all are producers.

At one time an important commercial use of cadmium was as an electrodeposited coating on iron and steel for corrosion protection. Nickel-cadmium batteries are the second-largest application, with pigment and chemical uses third. Sizable amounts are used in low-melting-point alloys, similar to Wood's metal, and in automatic fire sprinklers, and relatively smaller uses are in brazing alloys, solders, and bearings. Cadmium compounds are used as stabilizers in plastics and the production of cadmium phosphors. Because of its great neutron-absorbing capacity, especially the isotope 113, cadmium is used in control rods and shielding for nuclear reactors. [W.H.]

**Caffeine** An alkaloid, formerly synthesized by methylation of theobromine isolated from cacao, but now recovered from the solvents used in the manufacture of decaffeinated coffee. Chemically, caffeine is 1,3,7-trimethylxanthine, and has the formula below. It is widely used in medicine as a stimulant for the central nervous system

and as a diuretic. It occurs naturally in tea, coffee, and yerba maté, and small amounts are found in cola nuts and cacao. Caffeine crystallizes into long, white needlelike crystals that slowly lose their water of hydration to give a white solid that melts at 235–237.2°C (455–459.0°F). It sublimes without decomposition at lower temperatures. Caffeine has an intensely bitter taste, though it is neutral to litmus. *See* ALKALOID. [F.W.]

**Cage hydrocarbon** A compound that is composed of only carbon and hydrogen atoms and contains three or more rings arranged topologically so as to enclose a volume of space. In general, the "hole" within a cage hydrocarbon is too small to accommodate even a proton. The carbon frameworks of many cage hydrocarbons are quite rigid. Consequently, the geometric relationships between substituents on the cage are well defined. This quality makes these compounds exceptionally valuable for testing concepts concerning bonding, reactivity, structure-activity relationships, and structure-property relationships. *See* CHEMICAL BONDING.

The carbocyclic analogs of the platonic solids that are tenable are tetrahedrane (structure **1**, where X = –H), cubane (**2**), and dodecahedrane. *See* ALICYCLIC HYDROCARBON.

An unsubstituted prismane has the general formula of $(CH)_n$, and the carbon atoms are located at the corners of a regular prism. Prismane (**3**), cubane (**2**), pentaprismane, and hexaprismane are the simplest members of this family of cage hydrocarbons.

The monomer of the diamond carbon skeleton is adamantane (**4**, where X = –H).

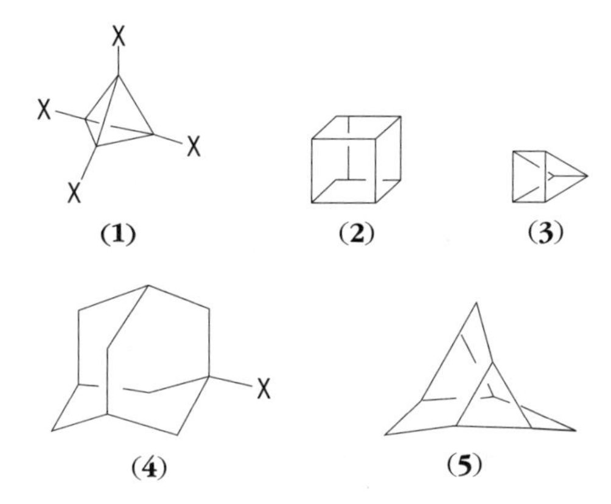


Amantadine (**4**, where X = $-NH_2$) was developed commercially as the first orally active antiviral drug for the prevention of respiratory illness due to influenza A2-Asian viruses.

The other simple diamondoid hydrocarbons are diamantane and triamantane.

Organic chemists have prepared a wide variety of cage hydrocarbons that do not occur in nature. Among these compounds are triasterane (**5**), iceane or wurtzitane, and pagodane. [R.K.Mu.]

**Calcium** A chemical element, Ca, of atomic number 20, fifth among elements and third among metals in abundance in the Earth's crust. Calcium compounds make up 3.64% of the Earth's crust. The physical properties of calcium metal are given in the table. The metal is trimorphous and is harder than sodium, but softer than aluminum. Like beryllium and aluminum, but unlike the alkali metals, it will not cause burns on the skin. It is less reactive chemically than the alkali metals and the other alkaline-earth metals. *See* PERIODIC TABLE.

Occurrence of calcium is very widespread; it is found in every major land area of the world. This element is essential to plant and animal life, and is present in bones, teeth, eggshell, coral, and many soils. Calcium chloride is present in sea water to the extent of 0.15%.

Calcium metal is prepared industrially by the electrolysis of molten calcium chloride. Calcium chloride is obtained either by treatment of a carbonate ore with hydrochloric acid or as a waste product from the Solvay carbonate process. The pure metal may be machined in a lathe, threaded, sawed, extruded, drawn into wire, pressed, and hammered into plates.

**Properties of calcium metal**

| Property | Value |
|---|---|
| Atomic number | 20 |
| Atomic weight | 40.08 |
| Isotopes (stable) | 40, 42, 43, 44, 46, 48 |
| Atomic volume, $cm^3$/g-atom | 25.9 |
| Crystal form | Face-centered cubic |
| Valence | 2+ |
| Ionic radius, nm | 0.099 |
| Electron configuration | 2882 |
| Boiling point, °C | 1487(?) |
| Melting point, °C | 810(?) |
| Density, g/$cm^3$ at 20°C | 1.55 |
| Latent heat of vaporization at boiling point, kilojoules/g-atom | 399 |

In air, calcium forms a thin film of oxide and nitride, which protects it from further attack. At elevated temperatures, it burns in air to form largely the nitride. The commercially produced metal reacts easily with water and acids, yielding hydrogen that contains noticeable amounts of ammonia and hydrocarbons as impurities.

The metal is employed as an alloying agent for aluminum-bearing metal, as an aid in removing bismuth from lead, and as a controller for graphitic carbon in cast iron. It is also used as a deoxidizer in the manufacture of many steels, as a reducing agent in preparation of such metals as chromium, thorium, zirconium, and uranium, and as a separating material for gaseous mixtures of nitrogen and argon.

Calcium oxide, $CaO$, is made by the thermal decomposition of carbonate minerals in tall kilns using a continuous-feed process. The oxide is used in high-intensity arc lights (limelights) because of its unusual spectral features and as an industrial dehydrating agent. The metallurgical industry makes wide use of the oxide during the reduction of ferrous alloys.

Calcium hydroxide, $Ca(OH)_2$, is used in many applications where hydroxide ion is needed. During the slaking process for producing calcium hydroxide, the volume of the slaked lime [$Ca(OH)_2$] produced expands to twice that of quicklime ($CaO$), and because of this, it can be used for the splitting of rock or wood. Slaked lime is an excellent absorbent for carbon dioxide to produce the very insoluble carbonate.

Calcium silicide, $CaSi$, an electric-furnace product made from lime, silica, and a carbonaceous reducing agent, is useful as a steel deoxidizer. Calcium carbide, $CaC_2$, is produced by heating a mixture of lime and carbon to 5432°F (3000°C) in an electric furnace. The compound is an acetylide which yields acetylene upon hydrolysis. Acetylene is the starting material for a great number of chemicals important in the organic chemicals industry.

Pure calcium carbonate exists in two crystalline forms: calcite, the hexagonal form, which possesses the property of birefringence, and aragonite, the rhombohedral form. Naturally occurring carbonates are the most abundant of the calcium minerals. Iceland spar and calcite are essentially pure carbonate forms, whereas marble is a somewhat impure and much more compact variety which, because it may be given a high polish, is much in demand as a construction stone. Although calcium carbonate is quite insoluble in water, it has considerable solubility in water containing dissolved carbon dioxide, because in these solutions it dissolves to form the bicarbonate. This fact accounts for cave formation in which limestone deposits have been leached away by the acidic ground waters.

The halides of calcium include the phosphorescent fluoride, which is the most widely distributed calcium compound and which has important applications in spectroscopy. Calcium chloride has in the anhydrous form important deliquescent properties which make it useful as an industrial drying agent and as a dust quieter on roads. Calcium chloride hypochlorite (bleaching powder) is produced industrially by passing chlorine into slaked lime, and has been used as a bleaching agent and a water purifier. *See* CHLORINE.

Calcium sulfate dihydrate is the mineral gypsum. It constitutes the major portion of portland cement, and has been used to help reduce soil alkalinity. A hemihydrate of calcium sulfate, produced by heating gypsum at elevated temperatures, is sold under the commercial name plaster of paris.

Calcium is an invariable constituent of all plants because it is essential for their growth. It is contained both as a structural constituent and as a physiological ion. Calcium is found in all animals in the soft tissues, in tissue fluid, and in the skeletal structures. The bones of vertebrates contain calcium as calcium fluoride, as calcium carbonate, and as calcium phosphate. [R.F.R.]

**Californium** A chemical element, Cf, atomic number 98, the ninth member of the actinide series of elements. Its discovery and production have been based upon artificial nuclear transmutation of radioactive isotopes of lighter elements. All isotopes of californium are radioactive, with half-lives ranging from a minute to about 1000 years. Because of its nuclear instability, californium does not exist in the Earth's crust. *See* ACTINIDE ELEMENTS; BERKELIUM; PERIODIC TABLE.

The chemical properties are similar to those observed for other 3+ actinide elements: a water-soluble nitrate, sulfate, chloride, and perchlorate. Californium is precipitated as the fluoride, oxalate, or hydroxide. Ion-exchange chromatography can be used for the isolation and identification of californium in the presence of other actinide elements. Californium metal is quite volatile and can be distilled at temperatures of the order of 1100–1200°C (2010–2190°F). It is chemically reactive and appears to exist in three different crystalline modifications between room temperature and its melting point, 900°C (1600°F).

The most easily produced isotope for many purposes is $^{252}Cf$, which is obtained in gram quantities in nuclear reactors and has a half-life of 2.6 years. It decays partially by spontaneous fission, and has been very useful for the study of fission. It has also had an important influence on the development of counters and electronic systems with applications not only in nuclear physics but in medical research as well. *See* TRANSURANIUM ELEMENTS. [G.T.S.]

**Calomel** Mercury(I) chloride, $Hg_2Cl_2$, a covalent compound which is insoluble in water. The substance sublimes when heated. The formula weight is 472.086 and the specific gravity is 7.16 at 20°C (68°F). The material is a white, impalpable powder consisting of fine tetragonal crystals.

Calomel is used in preparing insecticides and medicines. It is well known in the laboratory as the constituent of the calomel reference electrodes which are commonly used in conjunction with a glass electrode to measure pH. *See* MERCURY. [E.E.W.]

**Calorimetry** The measurement of the quantity of heat energy involved in processes such as chemical reactions, changes of state, and mixing of substances, or in the determination of heat capacities of substances. The unit of energy in the International

System of Units is the joule. Another unit still being used is the calorie, defined as 4.184 joules.

A calorimeter is an apparatus for measuring the quantity of heat energy released or absorbed during a process. Since there are many processes that can be studied over a wide range of temperature and pressure, a large variety of calorimeters have been developed.

Nonisothermal calorimeters measure the temperature change that occurs during the process. An aneroid-type nonisothermal calorimeter is normally constructed of a material having a high thermal conductivity, such as copper, so that there is rapid temperature equilibration. It is isolated from its surroundings by a high vacuum to reduce heat leaks. This type of calorimeter can be used for determining the heat capacity of materials when measurements involve low temperatures. Aneroid-type nonisothermal calorimeters have also been developed for measuring the energy of combustion for small samples of rare materials.

With most nonisothermal calorimeters, it is necessary to relate the temperature rise to the quantity of energy released in the process. This is done by determining the calorimeter constant, which is the amount of energy required to increase the temperature of the calorimeter itself by 1°. This value can be determined by electrical calibration or by measurement on a well-defined test system. For example, in bomb calorimetry the calorimeter constant is often determined from the temperature rise which occurs when a known mass of a very pure standard sample of benzoic acid is burned.

Isothermal calorimeters make measurements at constant temperature. The simplest example is a calorimeter containing an outer annular space filled with a liquid in equilibrium with a crystalline solid at its melting point, arranged so that any volume change will displace mercury along a capillary tube. The Bunsen ice calorimeter operates at 0°C (32°F) with a mixture of ice and water. Changes as a result of the process being studied cause the ice to melt or the water to freeze, and the consequent volume change is determined by measurement of the movement of the mercury meniscus in the capillary tube. While these calorimeters can yield accurate results, they are limited to operation at the equilibrium temperature of the two-phase system. Other types of isothermal calorimeters use the addition of electrical energy to achieve exact balance of the heat absorption that occurs during an endothermic process.

All calorimeters consist of the calorimeter proper and a jacket or a bath, which is used to control the temperature of the calorimeter and the rate of heat leak to the environment. For temperatures not too far removed from room temperature, the jacket or bath contains liquid at a controlled temperature. For measurements at extreme temperatures, the jacket usually consists of a metal block containing a heater to control the temperature. With nonisothermal calorimeters, where the jacket is kept at a constant temperature, there will be some heat leak to the jacket when the temperature of the calorimeter changes. It is necessary to correct the temperature change observed to the value it would have been if there were no leak. This is achieved by measuring the temperature of the calorimeter for a time period both before and after the process and applying Newton's law of cooling. This correction can be avoided by using the technique of adiabatic calorimetry, where the temperature of the jacket is kept equal to the temperature of the calorimeter as a change occurs. This technique requires more elaborate temperature control, and its primary use is for accurate heat capacity measurements at low temperatures.

In calorimetric experiments it is necessary to measure temperature differences accurately; in some cases the temperature itself must be accurately known. Modern calorimeters use resistance thermometers to measure both temperatures and temperature differences, while thermocouples or thermistors are used to measure smaller temperature differences.

Heat capacities of materials and heats of combustion are processes that are routinely measured with calorimeters. Calorimeters are also used to measure the heat involved in phase changes, for example, the change from a liquid to a solid (fusion) or from a liquid to a gas (vaporization). Calorimetry has also been applied to the measurement of heats of hydrogenation of unsaturated organic compounds, the heat of dissolution of a solid in a liquid, or the heat change on mixing two liquids. [K.M.M.]

**Camphor** A bicyclic, saturated terpene ketone. It exists in the optically active dextro and levo forms, and as the racemic mixture of the two forms. All of these melt within a degree of 178°C (352°F). The principal form is *dextro*-camphor, which occurs in the wood and leaves of the camphor tree (*Cinnamomum camphora*). Camphor is also synthesized commercially on a large scale from pinene which yields mainly the racemic variety. The structural formula of the molecule is shown below.

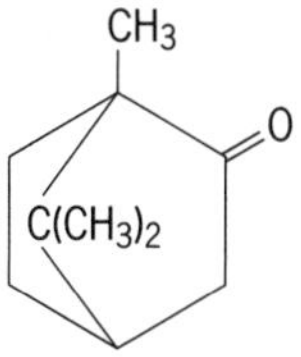

$CH_3$

O

$C(CH_3)_2$

Camphor has a characteristic odor; it crystallizes in thin plates and sublimes readily at ordinary temperatures.

Camphor has use in liniments and as a mild rubefacient, analgesic, and antipruritic. It has a local action on the gastrointestinal tract, producing a feeling of warmth and comfort in the stomach. It is also used in photographic film and as a plasticizer in the manufacture of plastics. *See* KETONE; PINE TERPENE; TERPENE. [E.L.S.]

**Carbohydrate** A term applied to a group of substances which include the sugars, starches, and cellulose, along with many other related substances. This group of compounds plays a vitally important part in the lives of plants and animals, both as structural elements and in the maintenance of functional activity. Plants are unique in that they alone in nature have the power to synthesize carbohydrates from carbon dioxide and water in the presence of the green plant chlorophyll through the energy derived from sunlight, by the process of photosynthesis. This process is responsible not only for the existence of plants but for the maintenance of animal life as well, since animals obtain their entire food supply directly or indirectly from the carbohydrates of plants.

The term carbohydrate originated in the belief that naturally occurring compounds of this class, for example, D-glucose ($C_6H_{12}O_6$), sucrose ($C_{12}H_{22}O_{11}$), and cellulose $(C_6H_{10}O_5)_n$, could be represented formally as hydrates of carbon, that is, $C_x(H_2O)_y$. Later it became evident that this definition for carbohydrates was not a satisfactory one. New substances were discovered whose properties clearly indicated that they had the characteristics of sugars and belonged in the carbohydrate class, but which nevertheless showed a deviation from the required hydrogen-to-oxygen ratio. Examples of these are the important deoxy sugars, D-deoxyribose, L-fucose, and L-rhamnose, the uronic acids, and such compounds as ascorbic acid (vitamin C). The retention of the term carbohydrate is therefore a matter of convenience rather than of exact definition. A carbohydrate is usually defined as either a polyhydroxy aldehyde (aldose) or ketone (ketose), or as a substance which yields one of these compounds on hydrolysis. However, included within this class of compounds are substances also containing nitrogen and sulfur.

The properties of many carbohydrates differ enormously from one substance to another. The sugars, such as D-glucose or sucrose, are easily soluble, sweet-tasting, and crystalline; the starches are colloidal and paste-forming; and cellulose is completely insoluble. Yet chemical analysis shows that they have a common basis; the starches and cellulose may be degraded by different methods to the same crystalline sugar, D-glucose.

The carbohydrates usually are classified into three main groups according to complexity: monosaccharides, oligosaccharides, and polysaccharides. Monosaccharides are simple sugars that consist of a single carbohydrate unit which cannot be hydrolyzed into simpler substances. These are characterized, according to their length of carbon chain, as trioses ($C_3H_6O_3$), tetroses ($C_4H_8O_4$), pentoses ($C_5H_{10}O_5$), hexoses ($C_6H_{12}O_6$), heptoses ($C_7H_{14}O_7$), and so on. Oligosaccharides are compound sugars that are condensation products of two to five molecules of simple sugars and are subclassified into disaccharides, trisaccharides, tetrasaccharides, and pentasaccharides, according to the number of monosaccharide molecules yielded upon hydrolysis. Polysaccharides comprise a heterogeneous group of compounds which represent large aggregates of monosaccharide units, joined through glycosidic bonds. They are tasteless, nonreducing, amorphous substances that yield a large and indefinite number of monosaccharide units on hydrolysis. Their molecular weight is usually very high, and many of them, like starch or glycogen, have molecular weights of several million. They form colloidal solutions, but some polysaccharides, of which cellulose is an example, are completely insoluble in water. On account of their heterogeneity they are difficult to classify. *See* POLYSACCHARIDE.

The sugars are also classified into two general groups, the reducing and nonreducing. The reducing sugars are distinguished by the fact that because of their free, or potentially free, aldehyde or ketone groups they possess the property of readily reducing alkaline solutions of many metallic salts, such as those of copper, silver, bismuth, mercury, and iron. The most widely used reagent for this purpose is Fehling's solution. The reducing sugars constitute by far the larger group. The monosaccharides and many of their derivatives reduce Fehling's solution. Most of the disaccharides, including maltose, lactose, and the rarer sugars cellobiose, gentiobiose, melibiose, and turanose, are also reducing sugars. The best-known nonreducing sugar is the disaccharide sucrose. Among other nonreducing sugars are the disaccharide trehalose, the trisaccharides raffinose and melezitose, the tetrasaccharide stachyose, and the pentasaccharide verbascose.

The sugars consist of chains of carbon atoms which are united to one another at a tetrahedral angle of $109°28'$. A carbon atom to which are attached four different groups is called asymmetric. A sugar, or any other compound containing one or more asymmetric carbon atoms possesses optical activity; that is, it rotates the plane of polarized light to the right or left. *See* OPTICAL ACTIVITY. [W.Z.H.]

**Carbon** A chemical element, C, with an atomic number of 6 and an atomic weight of 12.01115. Carbon is unique in chemistry because it forms a vast number of compounds, larger than the sum total of all other elements combined. By far the largest group of these compounds are those composed of carbon and hydrogen. It has been estimated that there are at least 1,000,000 known organic compounds, and this number is increasing rapidly each year. Although the classification is not rigorous, carbon forms another series of compounds, classified as inorganic, comprising a much smaller number than the organic compounds. *See* ORGANIC CHEMISTRY; PERIODIC TABLE.

Elemental carbon exists in two well-defined crystalline allotropic forms, diamond and graphite. Other forms, which are poorly developed in crystallinity, are charcoal, coke, and carbon black. Chemically pure carbon is prepared by the thermal decomposition of sugar (sucrose) in the absence of air. The physical and chemical properties of carbon are dependent on the crystal structure of the element. The density varies from 2.25 g/cm$^3$ (1.30 oz/in.$^3$) for graphite to 3.51 g/cm$^3$ (2.03 oz/in.$^3$) for diamond. For graphite, the melting point is 3500°C (6332°F) and the extrapolated boiling point is 4830°C (8726°F). Elemental carbon is a fairly inert substance. It is insoluble in water, dilute acids and bases, and organic solvents. At elevated temperatures, it combines with oxygen to form carbon monoxide or carbon dioxide. With hot oxidizing agents, such as nitric acid and potassium nitrate, mellitic acid, $C_6(CO_2H)_6$, is obtained. Of the halogens, only fluorine reacts with elemental carbon. A number of metals combine with the element at elevated temperatures to form carbides.

Carbon forms three gaseous compounds with oxygen: carbon monoxide, $CO$; carbon dioxide, $CO_2$; and carbon suboxide, $C_3O_2$. The first two oxides are the more important from an industrial standpoint. Carbon forms compounds with the halogens which have the general formula $CX_4$, where X is fluorine, chlorine, bromine, or iodine. At room temperature, carbon tetrafluoride is a gas, carbon tetrachloride is a liquid, and the other two compounds are solids. Mixed carbon tetrahalides are also known. Perhaps the most important of them is dichlorodifluoromethane, $CCl_2F_2$, commonly called Freon. *See* CARBON DIOXIDE; HALOGENATED HYDROCARBON.

Carbon and its compounds are found widely distributed in nature. It is estimated that carbon makes up 0.032% of the Earth's crust. Free carbon is found in large deposits as coal, an amorphous form of the element which contains additional complex carbon-hydrogen-nitrogen compounds. Pure crystalline carbon is found as graphite and as diamonds.

Extensive amounts of carbon are found in the form of its compounds. In the atmosphere, carbon is present in amounts of up to 0.03% by volume as carbon dioxide. Various minerals such as limestone, dolomite, marble, and chalk all contain carbon in the form of carbonate. All plant and animal life is composed of complex organic compounds containing carbon combined with hydrogen, oxygen, nitrogen, and other elements. The remains of past plant and animal life are found as deposits of petroleum, asphalt, and bitumen. Deposits of natural gas contain compounds that are composed of carbon and hydrogen.

The free element has many uses, ranging from ornamental applications of the diamond in jewelry to the black-colored pigment of carbon black in automobile tires and printing inks. Another form of carbon, graphite, is used for high-temperature crucibles, arc-light and dry-cell electrodes, lead pencils, and as a lubricant. Charcoal, an amorphous form of carbon, is used as an absorbent for gases and as a decolorizing agent. *See* CHARCOAL; GRAPHITE.

The compounds of carbon find many uses. Carbon dioxide is used for the carbonation of beverages, for fire extinguishers, and in the solid state as a refrigerant. Carbon monoxide finds use as a reducing agent for many metallurgical processes. Carbon tetrachloride and carbon disulfide are important solvents for industrial uses. Freon is used in refrigeration devices. Calcium carbide is used to prepare acetylene, which is used for the welding and cutting of metals as well as for the preparation of other organic compounds. Other metal carbides find important uses as refractories and metal cutters.

[E.E.W.]

**Carbon dioxide** A colorless, odorless, tasteless gas, formula $CO_2$, about 1.5 times as heavy as air. Under normal conditions, it is stable, inert, and nontoxic. The decay (slow oxidation) of all organic materials produces $CO_2$. Fresh air contains approximately 0.033% $CO_2$ by volume. In the respiratory action (breathing) of all animals and humans, $CO_2$ is exhaled.

Carbon dioxide gas may be liquefied or solidified. Solid $CO_2$ is known as dry ice. Carbon dioxide is obtained commercially from four sources: gas wells, fermentation, combustion of carbonaceous fuels, and as a by-product of chemical processing. Applications include use as a refrigerant, in either solid or liquid form, inerting medium, chemical reactant, neutralizing agent for alkalies, and pressurizing agent.

Most $CO_2$ is obtained as a by-product from steam-hydrocarbon reformers used in the production of ammonia, gasoline, and other chemicals; other sources include fermentation, deep gas wells, and direct production from carbonaceous fuels. Whatever the source, the crude $CO_2$ (containing at least 90% $CO_2$) is compressed in either two or three stages, cooled, purified, condensed to the liquid phase, and placed in insulated storage vessels. Carbon dioxide is distributed in three ways; in high-pressure uninsulated steel cylinders; as a low-pressure liquid in insulated truck trailers or rail tank cars; and as dry ice in insulated boxes, trucks, or boxcars. [J.S.L.]

**Carbonyl** A functional group found in organic compounds in which a carbon atom is doubly bonded to an oxygen atom:

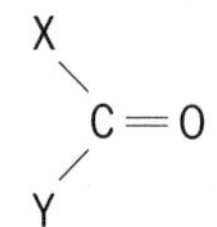


Depending upon the nature of the other groups attached to carbon, the most common compounds containing the carbonyl group are aldehydes (X and Y = H; X = H, Y = alkyl or aryl), ketones (X and Y = alkyl or aryl), carboxylic acids (X = OH, Y = H, alkyl, or aryl), esters (X = O-alkyl or aryl; Y = H, alkyl, or aryl), and amides (X = N—H, N-alkyl, or N-aryl; Y = H, alkyl, or aryl). Other compounds that contain the carbonyl group are acid halides, acid anhydrides, lactones, and lactams. *See* ACID ANHYDRIDE; ACID HALIDE; ALDEHYDE; AMIDE; ESTER; KETONE.

All the compounds containing this functional group are referred to in a general way as carbonyl compounds. It is important, however, to distinguish these compounds from a large group formed from metals and carbon monoxide, which are known as metal carbonyls. In these latter compounds, there is only one group attached to the carbon in addition to the oxygen, and the carbon atom is viewed as triply bonded to the oxygen. [J.P.Fr.]

**Carborane** A cluster compound containing both carbon (C) and boron (B) atoms as well as hydrogen (H) atoms external to the framework of the cluster. A cluster compound is one with insufficient electrons to allow for classical two-center two-electron bonds between all adjacent atoms. Sometimes the term carborane is used as a synonym for *closo*-1,2-$C_2B_{10}H_{12}$, commonly referred to as *ortho*-carborane. Carboranes are of interest because of their nonclassical bonding, their relatively high thermal stability, and their ability, when containing the $^{10}B$ isotope, to capture neutrons efficiently. *See* BORANE.

The structures of carboranes are based upon a series of three-dimensional, cage-like geometric shapes possessing triangulated faces; such shapes are termed delta polyhedra. The structure for any given carborane may be predicted by determining the framework electrons, by determining the number of electrons involved in bonding the boron and carbon atoms of the cluster framework together, and by using Wade's rule. Wade's rule states that a cluster containing $n$ framework electrons will be derived from a delta polyhedron containing $(n - 2)/2$ vertices, the parent cluster. Once this parent cluster has been determined, the geometry of the cluster framework may be predicted by clipping off vertices from the parent cluster until a polyhedron whose number of vertices is equivalent to the sum of boron and carbon atoms in the cluster framework is obtained.

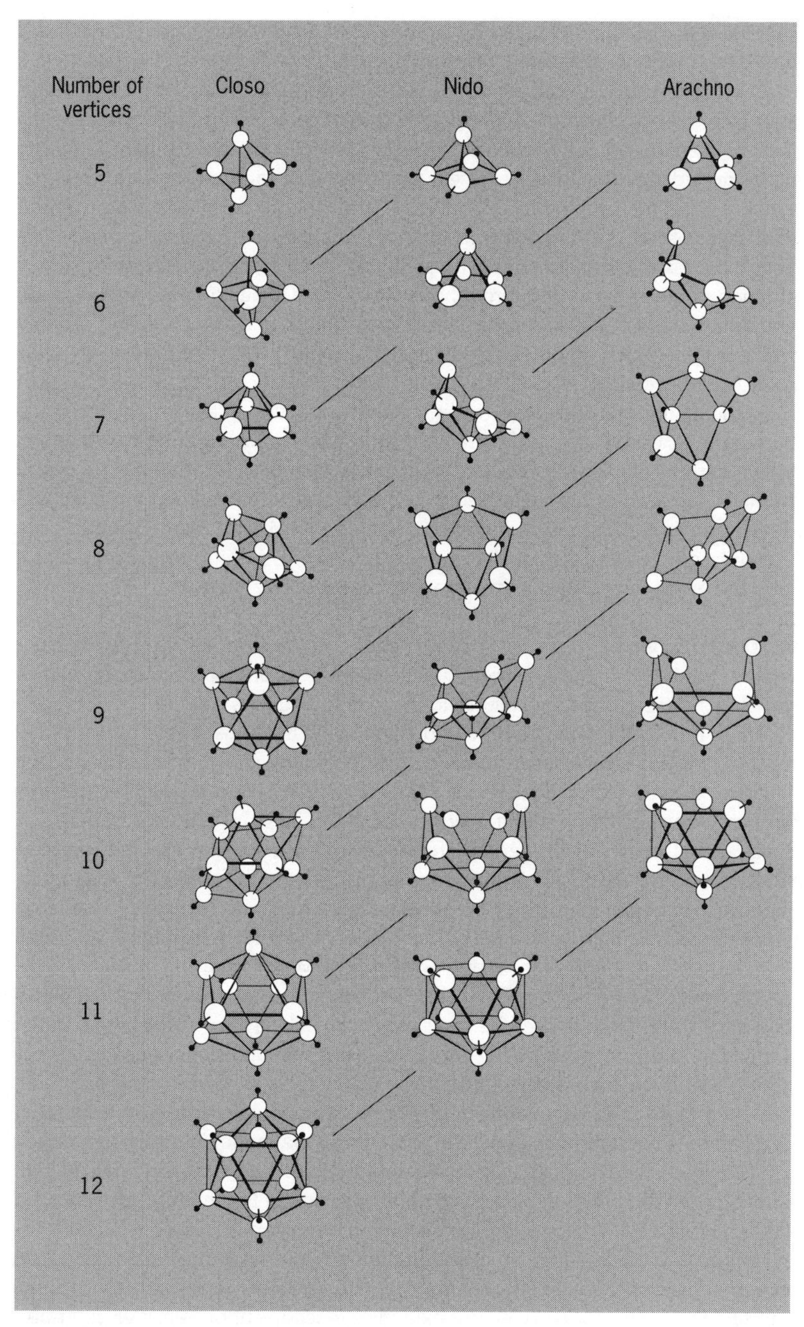


**Parent clusters from which carborane structures are determined. The diagonal lines define series of related closo, nido, and arachno structures.**

Carboranes are placed, according to their structure, into several classifications. The most common classifications are closo (closed), nido (nestlike), and arachno (cobweb) (see illustration). If a carborane's framework structure is that of a closed delta polyhedron, the carborane is said to be a *closo*-carborane. If a carborane's framework structure is that of a closed delta polyhedron minus one or two vertices, the carborane is said to be a *nido*- or *arachno*-carborane, respectively.

The bonding within a carborane can be thought of in terms of localized atomic orbitals forming both classical two-center two-electron bonds and nonclassical three-center two-electron bonds. Each vertex boron and carbon atom can be thought of as being $sp^3$ hybridized, with three of these hybrid orbitals of each vertex atom employed in framework bonding. Employing this simple approach, the bonding within carboranes can be represented by employing resonance structures. For example, *nido*-$C_2B_4H_8$ exhibits three resonance structures. The existence of resonance structures implies a delocalization of electron density throughout the cluster framework. Indeed, carboranes exhibit a high degree of electron density delocalization, and their bonding can be described more accurately by employing molecular orbital theory. *See* CHEMICAL BONDING; DELOCALIZATION; MOLECULAR ORBITAL THEORY; RESONANCE (MOLECULAR STRUCTURE).

The typical synthesis of a carborane involves the reaction of a boron hydride cluster, containing only boron and hydrogen, with an alkyne. The resulting carborane contains two carbon atoms in its skeletal structure, a dicarbon carborane. The dicarbon carboranes, because of their relative ease of preparation, have been the most widely studied group of carboranes. In particular, *closo*-1,2-$C_2B_{10}H_{12}$, the most readily available carborane, has been extensively studied. Other common groups of carboranes include the monocarbon and tetracarbon carboranes. *See* ALKYNE. [T.D.G.]

**Carboxylic acid** One of a large family of organic substances widely distributed in nature, and characterized by the presence of one or more carboxyl groups (—COOH). These groups typically yield protons in aqueous solution. In the type formula, $R(CXY)_nCOOH$, symbols R, X, and Y can be hydrogen, saturated or unsaturated groups, carboxyl, alicyclic, or aromatic groups, halogens, or other substituents, and *n* may vary from zero (formic acid, HCOOH) to more than 100, provided that the normal carbon covalence of four is maintained.

Physical and chemical properties of carboxylic acids are represented, grossly, by the resultant of the various chemical groupings present in the molecule. A short-chain aliphatic acid, wherein the carboxyl is dominant, is a pungent, corrosive, water-soluble liquid of abnormally high boiling point (because of molecular association), with specific gravity close to 1 (higher for formic and acetic acids). With increasing molecular weight, the hydrocarbon grouping overbalances the carboxyl; sharpness of odor diminishes, boiling and melting points rise, the specific gravity falls toward that of the parent hydrocarbon, and the water solubility decreases. Thus the typical high-molecular-weight saturated acid is a bland, waxlike solid.

Acids are used in large quantities in the production of esters, acid halides, acid amides, and acid anhydrides. They find wide use in the manufacture of soaps and detergents, in thickening lubricating greases (stearate soaps), in modifying rigidity in plastics, in compounding buffing bricks and abrasives, and in the manufacture of crayons, dictaphone cylinders, and phonograph records. The solvent action of acids finds use in manufacture of carbon paper, inks, and in the compounding of synthetic and natural rubber. Because of the stability of saturated fatty acids toward oxidation, these are often used as solvents for carrying out oxidation reactions upon sensitive compounds. [E.B.R.]

**Catalysis** The phenomenon in which a relatively small amount of foreign material, called a catalyst, augments the rate of a chemical reaction without itself being consumed. A catalyst is material, and not light or heat. It increases a reaction rate. *See* ANTIOXIDANT; INHIBITOR (CHEMISTRY).

If the reaction A + B → D occurs very slowly but is catalyzed by some catalyst (Cat), the addition of Cat must open new channels for the reaction. In a very simple case shown below, the two propagation processes, which are fast compared to the

$$\left.\begin{array}{l} A + Cat \longrightarrow ACat \\ ACat + B \longrightarrow D + Cat \end{array}\right\} \text{Chain propagation}$$

$$A + B \longrightarrow D \qquad \text{Overall reaction}$$

uncatalyzed reactions, A + B → D, provide the new channel for the reaction. The catalyst reacts in the first step, but is regenerated in the second step to commence a new cycle. A catalytic reaction is thus a kind of chain reaction.

If a reaction is in chemical equilibrium under some fixed conditions, the addition of a catalyst cannot change the position of equilibrium without violating the second law of thermodynamics. Therefore, if a catalyst augments the rate of A + B → D, it must also augment the reverse rate, D → A + B. *See* CHEMICAL THERMODYNAMICS.

Catalysis is conventionally divided into three categories: homogeneous, heterogeneous, and enzyme. In homogeneous catalysis, reactants, products, and catalyst are all present molecularly in one phase, usually liquid. Homogeneous catalysis is important in the petrochemical and chemical industries. In heterogeneous catalysis, the catalyst is in a separate phase; usually the reactants and products are in gaseous or liquid phases and the catalyst is a solid. Heterogeneous catalysis plays a dominant role in chemical processes in the petroleum, petrochemical, and chemical industries. Transformations of matter in living organisms occur by an elaborate sequence of reactions, most of which are catalyzed by biocatalysts called enzymes. Enzyme catalysis plays a key role in all metabolic processes and in some industries, such as the fermentation industry. The mechanisms of these categories involve chemical interaction between the catalyst and one or more reactants. In phase-transfer catalysis the interaction is physical. Electrocatalysis and photocatalysis are more specialized forms of catalysis. *See* ENZYME; HETEROGENEOUS CATALYSIS; HOMOGENEOUS CATALYSIS; PHASE-TRANSFER CATALYSIS.

In most cases of catalysis, a given set of reactants could react in two or more ways. The degree to which just one of the possible reactions is favored over the other is called selectivity. The fraction of reactants that react by a specific path is called the selectivity by that path, and it will vary from catalyst to catalyst. Selectivity is a key property of a catalyst in any practical application of the catalyst.

The number of sets of molecules that react consequent to the presence of a catalytic site (heterogeneous or enzyme catalysis) or molecule of catalyst (homogeneous catalysis), the turnover number, is substantially greater than unity and may be very large. The turnover frequency is the number of sets of molecules that react per site or catalyst molecule per second. It must be greater than one or the reaction is stoichiometric, not catalytic.

In practical applications, catalyst life, that is, the time or number of turnovers before the reaction rate becomes uselessly low, is important. It will be reduced by the presence of molecules that adsorb at (react with) and block the active site (poisons). *See* ADSORPTION. [R.L.Bu.; G.L.H.]

**Catenanes** Compounds that are made of interlocking macrocycles. The interlocking rings are said to be mechanically rather than chemically bound. Catenanes are named according to the number of interlocking rings. The simplest catenane, containing two interlocking rings, is called [2]-catenane. The most common catenanes consist of a linear arrangement of interlocking rings, so [2]-catenane is the first member of this series. Next is [3]-catenane, and so on. Catenanes involving necklace arrangements of

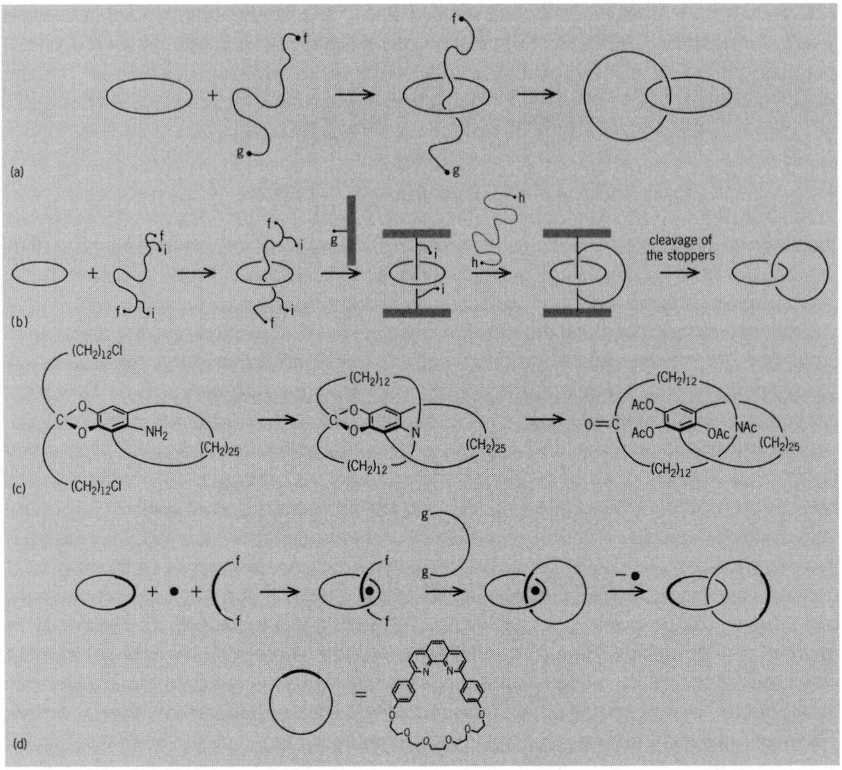

Synthetic routes to catenanes. (*a*) Statistical; f and g are complementary functions that react to close the ring. (*b*) Statistical, using a rotaxane intermediate; f and g are complementary functions that react to anchor the stoppers; i and h are complementary functions that react to form the second, interlocked ring. (*c*) Directed. (*d*) Transition-metal-templated; the thick ring portions represent the coordination sites, and the black disk is the metal; f and g are complementary functions that react to close the second, interlocked ring.

beadlike rings have also been synthesized. [7]-Catenane combines linear and necklace topologies; it is the highest-order molecular catenane isolated so far.

Molecular catenanes are compounds of synthetic origin. However, natural deoxyribonucleic acid (DNA) macromolecules were shown to assemble, in certain conditions, into catenated structures, which were studied by electron microscopy.

The rings of the catenanes may be purely organic macrocycles or metallomacrocycles, that is, macrocycles including transition-metal ions in their bond sequences. Actually, the chemical nature of the ring is dictated by the method of synthesis. Basically, three methods have been developed for the synthesis of catenanes (see illustration): statistical, directed, and template syntheses.

Statistical synthesis relies on the probability that a macrocycle can be threaded onto a molecular string to afford an intermediate that will undergo the cyclization react (illus. *a*). Convincing approaches to the statistical method used the trick of stabilizing the threaded complex by stoppering the extremities of the string with bulky groups: a so-called rotaxane species is obtained (illus. *b*). Subsequently, conventional macrocycle synthesis is used to prepare the catenane, which is obtained after removal of the stoppers. This method produced the first hydrocarbon catenane, made of interlocked $(CH_2)_{28}$ and $(CH_2)_{46}$ macrocycles.

Directed synthesis uses a catechol-based acetal incorporated in a macrocycle (illus. *c*). Two pendent arms, as precursors to the second macrocycle, are anchored perpendicularly to the plane of the first macrocycle. Their functionalized extremities are compelled to react with a complementary function localized inside the first macrocycle, so that the construction of the second macrocycle is directed to take place inside the first one. The last key step is the cleavage of the bonds linking the two macrocyclic sequences of atoms. This multistep synthetic method was used to prepare [2]- and [3]-catenanes.

Template synthesis methods are highly directed and economical in terms of numbers of steps. In these methods, metal cations or molecules gather and preorganize reactive molecular fragments in a spatially controlled manner, so that the desired structure will be obtained preferentially over many others. In one method, the transition-metal-templated synthesis of catenanes, the metal [generally Cu(I)] gathers a macrocycle incorporating a chelating subunit and a linear fragment made up of the complementary chelate such that both components are more or less at right angles to each other (illus. *d*). Cyclization of the linear fragment affords the metallocatenane, or catenate. Removal of the metal template by competitive complexation provides the catenane as a free ligand, or catenand. In the one-step alternative strategy, two open-chain chelates are assembled orthogonally at a metal center. A double cyclization reaction provides the catenate in one step.

Applications of catenanes are a research field in its infancy, and therefore they are more or less speculative. Promising approaches include incorporation of catenane structures into polymeric species to endow the polymers with peculiar rheological properties, because of the mechanical linking; and use of catenanes as elements of molecular machines—for example, the mechanical link of catenanes could be used for making a primitive rotary motor at the molecular level. *See* SUPRAMOLECULAR CHEMISTRY.

[J.C.Ch.; C.O.D.B.; J.P.Sa.]

**Cellophane** A clear, flexible film made from cellulose. It first appeared commercially in the United States in 1924, and it revolutionized the packaging industry, which had been using opaque waxed paper or glassine as wrapping materials. Cellophane was also the first transparent mending tape. By 1960, petrochemical-based polymers (polyolefins) such as polyethylene had surpassed cellophane for use as a packaging film. Nevertheless, cellophane is still often used for packaging because it is stiffer and more easily imprinted than are polyolefin films.

Cellophane is manufactured in a process that is very similar to that for rayon. Special wood pulp, known as dissolving pulp, which is white like cotton and contains 92–98% cellulose, is treated with strong alkali in a process known as mercerization. The mercerized pulp is aged for several days.

The aged, shredded pulp is then treated with carbon disulfide, which reacts with the cellulose and dissolves it to form a viscous, orange solution of cellulose xanthate known as viscose. Rayon fibers are formed by forcing the viscose through a small hole into an acid bath that regenerates the original cellulose while carbon disulfide is given off. To make cellophane, the viscose passes through a long slot into a bath of ammonium sulfate which causes it to coagulate. The coagulated viscose is then put into an acidic bath that returns the cellulose to its original, insoluble form. The cellophane is now clear.

The cellophane is then treated in a glycerol bath and dried. The glycerol acts like a plasticizer, making the dry cellophane less brittle. The cellophane may be coated with nitrocellulose or wax to make it impermeable to water vapor; it is coated with polyethylene or other materials to make it heat sealable for automated wrapping machines. Cellophane is typically 0.03 mm (0.001 in.) thick, is available in widths to 132 cm (52 in.), and can be made to be heat sealable from 82 to 177°C (180 to 350°F). *See* POLYMER. [C.J.Bi.]

**Centrifugation** A mechanical method of separating immiscible liquids or solids from liquids by the application of centrifugal force. This force can be very great, and separations which proceed slowly by gravity can be speeded up enormously in centrifugal equipment.

Centrifugal force is generated inside stationary equipment by introducing a high-velocity fluid stream tangentially into a cylindrical-conical chamber, forming a vortex of considerable intensity. Cyclone separators based on this principle remove liquid drops or solid particles from gases, down to 1 or 2 $\mu$m in diameter. Smaller units, called liquid cyclones, separate solid particles from liquids. The high velocity required at the inlet of a liquid cyclone is obtained with standard pumps. Much higher centrifugal forces than in stationary equipment are generated in rotating equipment (mechanically driven bowls or baskets, usually of metal, turning inside a stationary casing). Rotating a cylinder at high speed induces a considerable tensile stress in the cylinder wall. This limits the centrifugal force which can be generated in a unit of a given size and material of construction. Very high forces, therefore, can be developed only in very small centrifuges.

There are two major types of centrifuges: sedimenters and filters. A sedimenting centrifuge contains a solid-wall cylinder or cone rotating about a horizontal or vertical axis. An annular layer of liquid, of fixed thickness, is held against the wall by centrifugal force; because this force is so large compared with that of gravity, the liquid surface is essentially parallel with the axis of rotation regardless of the orientation of the unit. Heavy phases "sink" outwardly from the center, and less dense phases "rise" inwardly. Heavy solid particles collect on the wall and must be periodically or continuously removed.

A filtering centrifuge operates on the same principle as the spinner in a household washing machine. The basket wall is perforated and lined with a filter medium such as a cloth or a fine screen; liquid passes through the wall, impelled by centrifugal force, leaving behind a cake of solids on the filter medium. The filtration rate increases with the centrifugal force and with the permeability of the solid cake. Some compressible solids do not filter well in a centrifuge because the particles deform under centrifugal force and the permeability of the cake is greatly reduced. The amount of liquid adhering to the solids after they have been spun also depends on the centrifugal force applied; in general, it is substantially less than in the cake from other types of filtration devices. *See* MECHANICAL SEPARATION TECHNIQUES. [J.C.Sm.]

**Cerium** A chemical element, Ce, atomic number 58, atomic weight 140.12. It is the most abundant metallic element of the rare-earth group in the periodic table. The naturally occurring element is made up of the isotopes $^{136}$Ce, $^{138}$Ce, $^{140}$Ce, and $^{142}$Ce. A radioactive $\alpha$-emitter, $^{142}$Ce has a half-life of $5 \times 10^{15}$ years. Cerium occurs mixed with other rare earths in many minerals, particularly monazite and blastnasite, and is found among the products of the fission of uranium, thorium, and plutonium.

Although the common valence of cerium is 3, it also forms a series of quadrivalent compounds and is the only rare earth which occurs as a quadrivalent ion in aqueous solution. Although it can be separated from the other rare earths in high purity by ion-exchange methods, it is usually separated chemically by taking advantage of its quadrivalent state. *See* PERIODIC TABLE; RARE-EARTH ELEMENTS. [F.H.Sp.]

**Cesium** A chemical element, Cs, with an atomic number of 55 and an atomic weight of 132.905, the heaviest of the alkali metals in group 1 of the periodic table (except for francium, the radioactive member of the alkali metal family). Cesium is a soft, light, very low-melting metal. It is the most reactive of the alkali metals and indeed is the most electropositive and the most reactive of all the elements. *See* PERIODIC TABLE.

Cesium reacts vigorously with oxygen to form a mixture of oxides. In moist air, the heat of oxidation may be sufficient to melt and ignite the metal. Cesium does not appear to react with nitrogen to form a nitride, but does react with hydrogen at high temperatures to form a fairly stable hydride. Cesium reacts violently with water and even with ice at temperatures as low as −116°C (−177°F). Cesium reacts with the halogens, ammonia, and carbon monoxide. In general, cesium undergoes some of the same type of reactions with organic compounds as do the other alkali metals, but it is much more reactive. *See* SODIUM.

The physical properties of cesium metal are summarized in the table.

**Physical properties of cesium metal**

| Property | Temp., °C | Value |
|---|---|---|
| Density | 20 | 1.9 g/cm$^3$ |
| Melting point | 28.5 | |
| Boiling point | 705 | |
| Heat of fusion | 28.5 | 3.8 cal/g |
| Heat of vaporization | 705 | 146 cal/g |
| Viscosity | 100 | 4.75 millipoises |
| Vapor pressure | 278 | 1 mm |
| | 635 | 400 mm |
| Thermal conductivity | 28.5 | 0.044 cal/(s)(cm$^2$)(°C) |
| Heat capacity | 28.5 | 0.06 cal/(g)(°C) |
| Electrical resistivity | 30 | 36.6 microhm-cm |

Cesium is not very abundant in the Earth's crust, there being only 7 parts per million (ppm) present. Like lithium and rubidium, cesium is found as a constituent of complex minerals and not in relatively pure halide form as are sodium and potassium. Indeed, lithium, rubidium, and cesium frequently occur together in lepidolite ores.

Cesium metal is used in photoelectric cells, spectrographic instruments, scintillation counters, radio tubes, military infrared signaling lamps, and various optical and detecting devices. Cesium compounds are used in glass and ceramic production, as absorbents in carbon dioxide purification plants, as components of getters in radio tubes, and in microchemistry. Cesium salts have been used medicinally as antishock agents after administration of arsenic drugs. The isotope cesium-137 is supplanting cobalt-60 in the treatment of cancer. *See* ALKALI METALS. [M.Si.]

**Chain reaction (chemistry)** A chemical reaction in which many molecules undergo chemical reaction after one molecule becomes activated. In ordinary chemical reactions, every molecule that reacts must first become activated by collision with other rapidly moving molecules. The number of these violent collisions per second is so small that the reaction is slow. After a chain reaction is started, it is not necessary to wait for more collisions with activated molecules to accelerate the reaction because the reaction now proceeds spontaneously.

A typical chain reaction is the photochemical reaction between hydrogen and chlorine as described by the following reactions.

$$Cl_2 + \text{light} \rightarrow Cl + Cl$$
$$Cl + H_2 \rightarrow HCl + H$$
$$H + Cl_2 \rightarrow HCl + Cl$$
$$Cl + H_2 \rightarrow HCl + H$$

The light absorbed by a chlorine molecule dissociates the molecule into chlorine atoms; these in turn react rapidly with hydrogen molecules to give hydrogen chloride and hydrogen atoms. The hydrogen atoms react with chlorine molecules to give hydrogen chloride and chlorine atoms. The chlorine atoms react further with hydrogen and continue the chain until some other reaction uses up the free atoms of chlorine or hydrogen. The chain-stopping reaction may be the reaction between two chlorine atoms to give chlorine molecules, or between two hydrogen atoms to give hydrogen molecules. Again the atoms may collide with the walls of the containing vessel, or they may react with some impurity which is present in the vessel only as a trace.

Certain oxidations in the gas phase are known to be chain reactions. The carbon knock which occurs at times in internal combustion engines is caused by a too-rapid combustion rate caused by chain reactions. This chain reaction is reduced by adding tetraethyllead, which acts as an inhibitor.

The polymerization of styrene to give polystyrene and the polymerization of other organic materials to give industrial plastics involve chain reactions. The spoilage of foods, the precipitation of insoluble gums in gasoline, and the deterioration of certain plastics in sunlight involve chain reactions, which can be minimized with inhibitors. *See* CHEMICAL DYNAMICS; PHOTOCHEMISTRY. [F.D.]

**Charcoal** A porous solid product containing 85–98% carbon produced by heating carbonaceous materials such as cellulose, wood, peat, and coals of bituminous or lower rank at 930–1100°F (500–600°C) in the absence of air.

Chars or charcoals from cellulose or wood are soft and friable. They are used chiefly for decolorizing solutions of sugar and other foodstuffs and for removing objectionable tastes and odors from water. Chars from nutshells and coal are dense, hard carbons. They are used in gas masks and in chemical manufacturing for many mixture separations. Another use is for the tertiary treatment of waste water. Residual organic matter is adsorbed effectively to improve the water quality. *See* ADSORPTION; CARBON. [J.H.Fi.]

**Charles' law** A thermodynamic law, also known as Gay-Lussac's law, which states that at constant pressure the volume of a fixed mass or quantity of gas varies directly with the absolute temperature. Conversely, at constant volume the gas pressure varies directly with the absolute temperature. *See* GAS. [F.H.R.]

**Chelation** A chemical reaction or process involving chelate ring formation and characterized by multiple coordinate bonding between two or more of the electron-pair-donor groups of a multidentate ligand and an electron-pair-acceptor metal ion. The multidentate ligand is usually called a chelating agent, and the product is known as a metal chelate compound or metal chelate complex. Metal chelate chemistry is a subdivision of coordination chemistry and is characterized by the special properties resulting from the utilization of ligands possessing bridged donor groups, two or more of which coordinate simultaneously to a metal ion. *See* COORDINATION CHEMISTRY.

Many of the functional groups of both synthetic and naturally occurring organic compounds can form coordinate bonds to metal ions, producing metal-organic complexes or chelates, many of which are biologically active. Thus chelate compounds are frequently found in an interdisciplinary field of science called bioinorganic chemistry. The biological significance of chelates is demonstrated by the large number of biologically important compounds that are either metal chelates or chelating agents. Included in this group are the alpha amino acids, peptides, proteins, enzymes, porphyrins (such as hemoglobin), corrins (such as vitamin $B_{12}$), catechols, hydroxypolycarboxylic acids (such as citric acid), ascorbic acid (vitamin C), polyphosphates, nucleosides and other genetic compounds, pyridoxal phosphate (vitamin $B_6$), and sugars. The ubiquitous green plant pigment, chlorophyll, is a magnesium chelate of a tetradentate ligand formed from a modified porphin compound, and similarly the oxygen transport heme of red blood cells contains an Fe(II) chelate. *See* BIOINORGANIC CHEMISTRY; ORGANOMETALLIC COMPOUND.

The ability of chelating agents to reduce the chemical activity of metal ions has found extensive application in many areas of science and industry. Ethylenediaminetetraacetic acid (EDTA), a hexadentate chelating agent, has been employed commercially for water softening, boiler scale removal, industrial cleaning, soil metal micronutrient transport, and food preservation. Nitrilotriacetic acid (NTA) is a tetradentate chelating agent which, because of lower cost, has taken over some of the commercial applications of EDTA. Many commercially important dyes and pigments, such as copper phthalocyanines, are chelate compounds. Humic and fulvic acids are plant degradation products in lake and sea-water sediments that have been suggested as important chelating agents which regulate metalion balance in natural waters. By virtue of its abundance, low toxicity, low cost, and good chelating tendencies for metal ions that produce water hardness, the tripolyphosphate ion (as its sodium salt) is used in large quantities as a builder in synthetic detergents. Both synthetic ion exchangers and the mineral zeolites are chelating ion-exchange resins which are used in analytical and water-softening applications. As final examples, less conventional chelating agents are the multidentate, cyclic ligands, termed collectively crown ethers, which are particularly suited for the complexation of the alkali and alkaline-earth metals. *See* ETHYLENEDIAMINETETRAACETIC ACID.
[A.E.M.; R.J.Mo.]

**Chemical bonding** The force that holds atoms together in molecules and solids. Chemical bonds are very strong. To break one bond in each molecule in a mole of material typically requires an energy of many tens of kilocalories.

It is convenient to classify chemical bonding into several types, although all real cases are mixtures of these idealized cases. The theory of the various bond types has been well developed and tested by theoretical chemists. *See* COMPUTATIONAL CHEMISTRY; MOLECULAR ORBITAL THEORY; QUANTUM CHEMISTRY.

The simplest chemical bonds to describe are those resulting from direct coulombic attractions between ions of opposite charge, as in most crystalline salts. These are termed ionic bonds. *See* STRUCTURAL CHEMISTRY.

Other chemical bonds include a wide variety of types, ranging from the very weak van der Waals attractions, which bind neon atoms together in solid neon, to metallic bonds or metallike bonds, in which very many electrons are spread over a lattice of positively charged atom cores and give rise to a stable configuration for those cores. *See* INTERMOLECULAR FORCES.

The covalent bond, in which two electrons bind two atoms together, as in

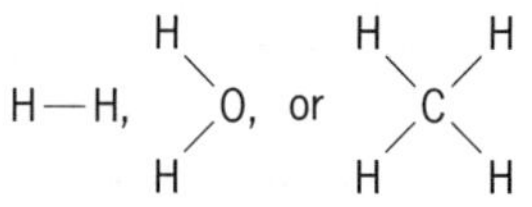

```
                H            H     H
                 \            \   /
       H—H,       O,   or       C
                 /            /   \
                H            H     H
```

is the most characteristic link in chemistry. The theory that accounts for it is a cornerstone of chemical science. The physical and chemical properties of any molecule are direct consequences of its particular detailed electronic structure. Yet the theory of any one covalent chemical bond, for example, the H—H bond in the hydrogen molecule, has much in common with the theory of any other covalent bond, for example, the O—H bond in the water molecule. The current theory of covalent bonds both treats their qualitative features and quantitatively accounts for the molecular properties which are a consequence of those features. The theory is a branch of quantum theory. *See* QUANTUM CHEMISTRY.

The problem of the proper description of chemical bonds in molecules that are more complicated than $H_2$ has many inherent difficulties. The qualitative theory of chemical bonding in complex molecules preserves the use of many chemical concepts that predate quantum chemistry itself; among these are electrostatic and steric factors, tautomerism, and electronegativity. The quantitative theory is highly computational in nature and involves extensive use of computers and supercomputers.

The number of covalent bonds which an atom can form is called the covalence and is determined by the detailed electron configuration of the atom. An extremely important case is that of carbon. In most of its compounds, carbon forms four bonds. When these connect it to four other atoms, the directions of the bonds to these other atoms normally make angles of about 109° to one another, unless the attached atoms are crowded or constrained by other bonds. That is, covalent bonds have preferred directions. However, in accord with the idea that carbon forms four bonds, it is necessary to introduce the notion of double and triple bonds. Thus in the structural formula of ethylene, $C_2H_4$ (**1**), all lines denote covalent bonds, the double line connecting the

```
H         H
 \       /
  C  =  C
 /       \
H         H          H—C≡C—H
    (1)                (2)
```

carbon atoms being a double bond. Such double bonds are distinctly shorter, almost twice as stiff, and require considerably more energy to break completely than do single bonds. However, they do not require twice as much energy to break as a single bond. Similarly, acetylene (**2**) is written with a triple bond, which is still shorter than a double bond. A carbon-carbon single bond has a length close to $1.54 \times 10^{-8}$ cm, whereas the triple bond is about $1.21 \times 10^{-8}$ cm long. *See* BOND ANGLE AND DISTANCE; VALENCE.

Many substances have some bonds which are covalent and others which are ionic. Thus in crystalline ammonium chloride, $NH_4Cl$, the hydrogens are bound to nitrogen by electron pairs, but the $NH_4$ group is a positive ion and the chlorine is a negative ion.

Both electrons of a covalent bond may come from one of the atoms. Such a bond is called a coordinate or dative covalent bond or semipolar double bond, and is one example of the combination of ionic and covalent bonding.

The hydrogen bond is a special bond in which a hydrogen atom links a pair of other atoms. The linked atoms are normally oxygen, fluorine, chlorine, or nitrogen. These four elements are all quite electronegative, a fact which favors a partially ionic interpretation of this kind of bonding. *See* ELECTRONEGATIVITY; HYDROGEN BOND. [R.G.P.]

**Chemical conversion** A chemical manufacturing process in which chemical transformation takes place, that is, the product differs chemically from the starting materials. Most chemical manufacturing processes consist of a sequence of steps, each of which involves making some sort of change in either chemical makeup, concentration, phase state, energy level, or a combination of these, in the materials passing through the particular step. If the changes are of a strictly physical nature (for example, mixing, distillation, drying, filtration, adsorption, condensation), the step is referred to as a unit operation. If the changes are of a chemical nature, where conversion from one chemical species to another takes place (for example, combustion, polymerization, chlorination, fermentation, reduction, hydrolysis), the step is called a unit process. Some steps involve both, for example, gas absorption with an accompanying chemical reaction in the liquid phase. The term chemical conversion is used not only in describing overall processes involving chemical transformation, but in certain contexts as a synonym for the term unit process. The chemical process industry as a whole has tended to favor the former usage, while the petroleum industry has favored the latter. *See* CHEMICAL PROCESS INDUSTRY; UNIT PROCESSES.

Another usage of the term chemical conversion is to define the percentage of reactants converted to products inside a chemical reactor or unit process. This quantitative usage is expressed as percent conversion per pass, in the case of reactors where unconverted reactants are recovered from the product stream and recycled to the reactor inlet. *See* CHEMICAL ENGINEERING. [W.F.F.]

**Chemical dynamics** That branch of physical chemistry which seeks to explain time-dependent phenomena, such as energy transfer and chemical reaction, in terms of the detailed motion of the nuclei and electrons which constitute the system.

In principle, it is possible to prepare two reagents in specific quantum states and to determine the quantum-state distribution of the products. In practice, this is very difficult, and experiments have mostly been limited to preparing one reagent or to determining some aspect of the product distribution. This approach yields data concerning the detailed aspects of the dynamics.

**Energy distribution.** An important question regarding the dynamics of chemical reactions has to do with the product energy distribution in exothermic reactions. For example, because the hydrogen fluoride (HF) molecule is more strongly bound than the $H_2$ molecule, reaction (1) releases a considerable amount of energy (more than

$$F + H_2 \longrightarrow HF + H \qquad (1)$$

30 kcal/mol or 126 kJ/mol). The two possible paths for this energy release to follow are into translations, that is, with HF and H speeding away from each other, or into vibrational motion of HF.

In this case it is vibration, and this has rather dramatic consequences; the reaction creates a population inversion among the vibrational energy levels of HF—that is, the higher vibrational levels have more population than the lower levels—and the emission of infrared light from these excited vibrational levels can be made to form a chemical laser. A number of other reactions also give a population inversion among the vibrational energy levels, and can thus be used to make lasers.

**Most effective energy.** The rates of most chemical reactions are increased if they are given more energy. In macroscopic kinetics this corresponds to increasing the temperature, and most reactions are faster at higher temperatures. It seems reasonable, though, that some types of energy will be more effective in accelerating the reaction than others. For example, in reaction (2), where potassium (K) reacts with hydrogen

$$K + HCl \longrightarrow KCl + H \qquad (2)$$

chloride (HCl) to form potassium chloride (KCl), studies have shown that if HCl is vibrationally excited (by using a laser), this reaction is found to proceed approximately 100 times faster, while the same amount of energy in translational kinetic energy has a smaller effect. Here, therefore, vibrational energy is much more effective than translational energy in accelerating the reaction.

For reaction (1), however, translational energy is more effective than vibrational energy in accelerating the reaction. The general rule of thumb is that vibrational energy is more effective for endothermic reactions (those for which the new molecule is less stable than the original molecule), while translational energy is most effective for exothermic reactions.

**Lasers.** Lasers are also important for probing the dynamics of chemical reactions. Because they are light sources with a very narrow wavelength, they are able to excite molecules to specific quantum states (and also to detect what states molecules are in), an example of which is reaction (2). For polyatomic molecules—that is, those with more than two atoms—there is the even more interesting question of how the rate of reaction depends on which vibration is excited.

For example, when the molecule allyl isocyanide, $CH_2{=}CH{-}CH_2{-}NC$, is given sufficient vibrational energy, the isocyanide part (—NC) will rearrange to the cyanide (—CN) configuration. A laser can be used to excite a C-H bond vibrationally. An interesting question is whether the rate of the rearrangement process depends on which C-H bond is excited. Only with a laser is it possible to excite different C-H bonds and begin to answer such questions. This question of mode-specific chemistry—whether excitation of specific modes of a molecule causes specific chemistry to result—has been a subject of great interest. (For the example above, the reaction is fastest if the C-H bond closest to the NC group is excited.) Mode-specific chemistry would allow much greater control over the course of chemical reactions, and it would be possible to accelerate the rate of some reactions (or reactions at one part of a molecule) and not others.

**Theoretical methods.** The goal of chemical dynamics is to understand kinetic phenomena from the basic laws of molecular mechanics, and it is thus a field which sees close interplay between experimental and theoretical research. Many different theoretical models and methods have been useful in understanding and analyzing the phenomena described above. Probably the single most useful approach has been the calculation of classical trajectories. Assuming that the potential energy function or a reasonable approximation is known for the three atoms in reaction (1), for example, it is possible by use of electronic computers to calculate the classical motion of the three atoms. It is thus an easy matter to give the initial molecule more or less vibrational

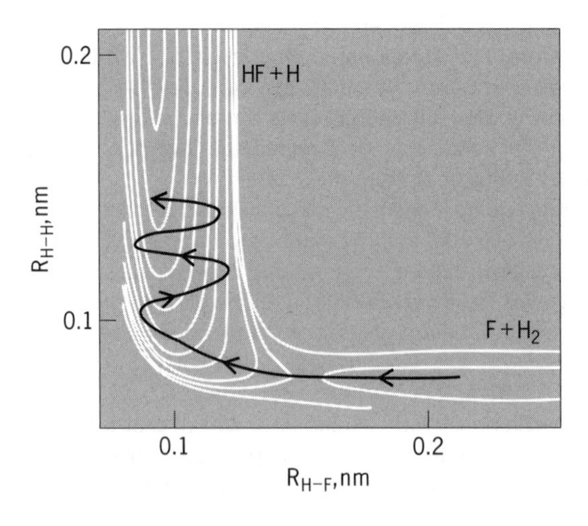


**Contour plot of the potential energy surface for the reaction $F + H_2 = HF + H$, with a typical reactive trajectory indicated.**

or translational energy, and then compute the probability of reaction. Similarly, the final molecule and atom can be studied to see where the energy appears, that is, as translation or as vibration.

It is thus a relatively straightforward matter theoretically to answer the questions and to see whether or not mode-specific excitation leads to significantly different chemistry than simply increasing the temperature under bulk conditions.

The most crucial step in carrying out these calculations is obtaining the potential energy surface—that is, the potential energy as a function of the positions of the atoms—for the system. The illustration shows a plot of the contours of the potential energy surface for reaction (1). Even without carrying out classical trajectory calculations, it is possible to deduce some of the dynamical features of this reaction; for example, the motion of the system first surmounts a small potential barrier, and then it slides down a steep hill, turning the corner at the bottom of the hill. It is evident that such motion will cause much of the energy released in going down the hill to appear in vibrational motion of HF.

This and other theoretical methods have interacted strongly with experimental research in helping to understand the dynamics of chemical reactions. *See* CHEMICAL KINETICS; INORGANIC PHOTOCHEMISTRY. [P.R.B.; W.H.M.]

**Chemical energy** A useful but obsolescent term for the energy available from elements and compounds when they react, as in a combustion reaction. In precise terminology, there is no such thing as chemical energy, since all energy is stored in matter as either kinetic energy or potential energy. *See* COMBUSTION; ENERGY.

When a chemical reaction takes place, the atoms of the reactants change their bonding pattern and become products. The breaking of bonds in the reactants requires energy, and the formation of bonds in the products releases energy. The net change in energy is commonly referred to as chemical energy. To be more precise, when a reaction takes place, there is an overall change in the enthalpy $H$ of the system as bonds are broken and new bonds are formed. This change in enthalpy is denoted $\Delta H$. Under standard conditions [a pressure of 1 bar (100 kilopascals) and all substances pure], the change is noted $\Delta H^\circ$ and called the standard enthalpy of reaction. Provided the pressure is constant, the standard enthalpy can be identified with the energy released as heat (when $\Delta H^\circ < 0$) or gained as heat ($\Delta H^\circ > 0$) when the reaction takes place. Reactions for which $\Delta H^\circ < 0$ are classified as exothermic; those for which $\Delta H^\circ > 0$ are classified as endothermic. All combustions are exothermic, the released heat being used either to provide warmth or to raise the temperature of a working fluid in an engine of some kind. There are very few common endothermic reactions; one example is the dissolution of ammonium nitrate in water (a process utilized in medical cold packs). *See* CHEMICAL EQUILIBRIUM; CHEMICAL THERMODYNAMICS; ENTHALPY; THERMOCHEMISTRY.

The "chemical energy" available from a typical fuel (that is, the enthalpy change accompanying the combustion of the fuel, when carbon-hydrogen bonds are replaced by stronger carbon-oxygen and hydrogen-oxygen bonds) is commonly reported as either the specific enthalpy or the enthalpy density. The specific enthalpy is the standard enthalpy of combustion divided by the mass of the reactant. The enthalpy density is the standard enthalpy of combustion divided by the volume of the reactant. The former is of primary concern when mass is an important consideration, as in raising a rocket into orbit. The latter is of primary concern when storage space is a limitation. The specific enthalpy of hydrogen gas is relatively high (142 kilojoules/g), but its enthalpy density is low (13 kJ/L). The values for octane, a compound representative of gasoline, are¥break 48 kJ/g and 38 MJ/L, respectively (note the change in units). The high enthalpy density of octane means that a gasoline tank need not be large to store a lot of "chemical energy." [P.W.A.]

**Chemical engineering** The application of engineering principles to conceive, design, develop, operate, or use processes and products based on chemical and physical phenomena. The chemical engineer is considered an engineering generalist because of a unique ability (among engineers) to understand and exploit chemical change. Drawing on the principles of mathematics, physics, and chemistry and familiar with all forms of matter and energy and their manipulation, the chemical engineer is well suited for working in a wide range of technologies.

Although chemical engineering was conceived primarily in England, it underwent its main development in America, propelled at first by the petroleum and heavy-chemical industries, and later by the petrochemical industry with its production of plastics, synthetic rubber, and synthetic fibers from petroleum and natural-gas starting materials. In the early twentieth century, chemical engineering developed the physical separations such as distillation, absorption, and extraction, in which the principles of mass transfer, fluid dynamics, and heat transfer were combined in equipment design. The chemical and physical aspects of chemical engineering are known as unit processes and unit operations, respectively.

Chemical engineering now is applied in biotechnology, energy, environmental, food processing, microelectronics, and pharmaceutical industries, to name a few. In such industries, chemical engineers work in production, research, design, process and product development, marketing, data processing, sales, and, almost invariably, throughout top management. *See* BIOCHEMICAL ENGINEERING; BIOMEDICAL CHEMICAL ENGINEERING; CHEMICAL CONVERSION; CHEMICAL PROCESS INDUSTRY; ELECTROCHEMICAL PROCESS; UNIT OPERATIONS; UNIT PROCESSES. [W.F.F.]

**Chemical equilibrium** In a dynamic or kinetic sense, chemical equilibrium is a condition in which a chemical reaction is occurring at equal rates in its forward and reverse directions, so that the concentrations of the reacting substances do not change with time. In a thermodynamic sense, it is the condition in which there is no tendency for the composition of the system to change; no change can occur in the system without the expenditure of some form of work upon it. From the viewpoint of statistical mechanics, the equilibrium state places the system in a condition of maximum freedom (or minimum restraint) compatible with the energy, volume, and composition of the system. The statistical approach has been merged with thermodynamics into a field called statistical thermodynamics; this merger has been of immense value for its intellectual stimulus, as well as for its practical contributions to the study of equilibria. *See* CHEMICAL THERMODYNAMICS.

Of the three viewpoints, the thermodynamic approach is by far the most powerful and fruitful in treating the quantitative relationships between the position of equilibrium and the factors which govern it. Since thermodynamics is concerned with relationships among observable properties, such as temperature, pressure, concentration, heat, and work, the relationships possess general validity, independent of theories of molecular behavior.

**Chemical potential.** Thermodynamics attributes to each chemical substance a property called the chemical potential, which may be thought of as the tendency of the substance to enter into chemical (or physical) change. Although the chemical potential of a substance cannot be directly measured (except on a relative basis), differences in chemical potential are measurable. (The units are those of energy per mole.)

The importance of the chemical potential lies in its relation to the affinity or driving force of a chemical reaction. Consider general reaction (1). Let $\mu_A$ be the chemical po-

$$aA + bB \rightleftharpoons gG + hH \tag{1}$$

tential per mole of substance A, $\mu_B$ be the chemical potential per mole of B, and so on. Then, according to one of the fundamental principles of thermodynamics (the second law), the reaction will be spontaneous when the total chemical potential of the reactants is greater than that of the products. Thus, for spontaneous change (naturally occurring processes) notation (2) applies When equilibrium is reached, the total chemical potentials of products and reactants become equal; thus Eq. (3) holds at equilibrium. The

$$[g\mu_G + h\mu_H] - [a\mu_A + b\mu_B] < 0 \tag{2}$$

$$[g\mu_G + h\mu_H] - [a\mu_A + b\mu_B] = 0 \tag{3}$$

difference in chemical potentials in Eqs. (2) and (3) is called the driving force or affinity of the process or reaction; naturally, it is zero when the chemical system is in chemical equilibrium.

For reactions at constant temperature and pressure (the usual restraints in a chemical laboratory), the difference in chemical potentials becomes equal to the free energy change $\Delta G$ for the process in Eq. (4). The decrease in free energy represents the

$$\Delta G = [g\mu_G + h\mu_H] - [a\mu_A + b\mu_B] \tag{4}$$

maximum net work obtainable from the process. When no more work is obtainable, the system is at equilibrium. Conversely, if the value of $\Delta G$ for a process is positive, some useful work will have to be expended upon the process, or reaction, in order to make it proceed; the process cannot proceed naturally or spontaneously. (The term spontaneously as used here implies only that a process can occur. It does not imply that the reaction will be rapid or instantaneous. Thus, the reaction between hydrogen and oxygen is a spontaneous process in the sense of the term as used here, even though a mixture of hydrogen and oxygen can remain unchanged for years unless ignited or exposed to a catalyst.)

Since by definition a catalyst remains unchanged chemically through a reaction, its chemical potential does not appear in Eqs. (2), (3), and (4). A catalyst, therefore, can contribute nothing to the driving force of a reaction, nor can it, in consequence, alter the position of the chemical equilibrium in a system. *See* CATALYSIS.

In addition to furnishing a criterion for the equilibrium state of a chemical system, the thermodynamic method goes much further. In many cases, it yields a relation between the change in chemical potentials (or change in free energy) and the equilibrium concentrations of the substances involved in the reaction. To do this, the chemical potential must be expressed as a function of concentration (and other properties of the substance). *See* CONCENTRATION SCALES.

**Activity and standard states.** It is often convenient to utilize the product $fx$, called the activity of the substance and defined by $a = fx$. The activity may be looked upon as an effective concentration of the substance, measured in the same units as the concentration $x$ with which it is associated. The standard state of the substance is then defined as the state of unit activity (where $a = 1$) and is characterized by the standard chemical potential $\mu^\circ$. Clearly, the terms $\mu^\circ$, $f$, and $x$ are not independent; the choice of the activity scale serves to fix the standard state. For example, for an aqueous solution of hydrochloric acid, the standard state for the solute (HCl) would be an (hypothetical) ideal 1 molar (or molal) solution, and for the solvent ($H_2O$) the standard state would be pure water (mole fraction = 1). The reference state would be an infinitely dilute solution; here the activity coefficients would be unity for both solute and solvent. For the vapor of HCl above the solution, the standard state would be the ideal gaseous state at 1 atm (101.325 kilopascals) partial pressure; the reference state would be a state of zero pressure. (For gases, the term fugacity is used instead of activity.)

It should be noted that the reference state is a limiting state which in many cases can be reached only through an extrapolation from observed behavior. *See* FUGACITY.

**Equilibrium constant.** If general reaction (1) occurs at constant temperature $T$ and pressure $P$ when all of the substances involved are in their standard states of unit activity, Eq. (4) would become Eq. (5). The quantity $\Delta G^\circ$ is known as the standard

$$\Delta G^\circ = [g\mu_G^\circ + h\mu_H^\circ] - [a\mu_A^\circ + b\mu_B^\circ] \tag{5}$$

free energy change for the reaction at that temperature and pressure for the chosen standard states. (Standard state properties are commonly designated by a superscript, $\Delta G^\circ$, $\mu^\circ$.) Since each of the standard chemical potentials ($\mu^\circ$) is a unique property determined by the temperature, pressure, standard state, and chemical identity of the substance concerned, the standard free energy change $\Delta G^\circ$ is a constant (parameter) characteristic of the particular reaction for the chosen temperature, pressure, and standard states.

When the system has come to chemical equilibrium at constant temperature and pressure, $\Delta G = 0$, and $\Delta G^\circ$ is given by Eq. (6), where the value of $K^\circ$ is shown as Eq. (7), and the activities are the equilibrium values. The ratio of the activities at equi-

$$\Delta G^\circ = -RT \ln K^\circ \tag{6}$$

$$K^\circ = \left[\frac{a_G{}^g a_H{}^h}{a_A{}^a a_B{}^b}\right] \tag{7}$$

librium, $K^\circ$, is called the equilibrium constant or, more precisely, the thermodynamic equilibrium constant. (The terms $K^\circ$ and $Q^\circ$ are written with superscripts to emphasize that they represent ratios of activities.) The equilibrium constant is a characteristic property of the reaction system, since it is determined uniquely in terms of the standard free energy change. The term $-\Delta G^\circ$ represents the maximum net work which the reaction could make available when carried out at constant temperature and pressure with the substances in their standard states.

The kinetic concept of chemical equilibrium introduced by C. M. Guldberg and P. Waage (1864) led to the formulation of the equilibrium constant in terms of concentrations. Although the concept is correct in terms of the dynamic picture of opposing reactions occurring at equal speeds, it has not been successful in coping with the problems of activity coefficients. Conversely, the thermodynamic approach yields no relationship between the driving force of the reaction and the rate of approach to equilibrium. *See* CHEMICAL DYNAMICS.

The influence of temperature upon the chemical potentials, and hence upon the equilibrium constant, is given by the Gibbs-Helmholtz equation, Eq. (8). The derivative

$$\left[\frac{d \ln K^\circ}{dT}\right]_P = \frac{\Delta H^\circ}{RT^2} \tag{8}$$

on the left represents the slope of the curve obtained when values of $\ln K^\circ$ for a reaction, obtained at different temperatures but always at the same pressure $P$, are plotted against temperature. The standard heat of reaction $\Delta H^\circ$ for the temperature $T$ at which the slope is measured is the heat effect which could also be observed by carrying out the reaction involving the standard states in a calorimeter at the corresponding temperature and pressure. *See* HEAT CAPACITY; THERMOCHEMISTRY.

**Homogeneous equilibria.** These involve single-phase systems: gaseous, liquid, and solid solutions. In most cases, solid solutions are so far from ideal that equilibrium constants cannot be evaluated, and such systems are treated in terms of the phase rule. A typical gas-phase equilibrium is the ammonia synthesis shown in reaction (9). A typical liquid-phase equilibrium is the dissociation of acetic acid in water, reaction (10).

$$N_2 + 3H_2 \rightleftharpoons 2NH_3 \tag{9}$$

$$HC_2H_3O_2 + H_2O \rightleftharpoons H_2O^+ + C_2H_3O_2^- \tag{10}$$

The solvent appears to be inert, since its chemical potential remains practically unchanged over the useful concentration range. As a result of this apparent inertness of the solvent, it is not possible to determine the extent of hydration of any dissolved species from equilibrium studies. Thus, whether the actual ion is $H^+$, $H_3O^+$, or $H_9O_4{}^+$, it is the total stoichiometric concentration that is measured and used. *See* IONIC EQUILIBRIUM.

**Heterogeneous equilibria.** These are usually studied at constant pressure, since at least one of the phases will be a solid or liquid. The imposed pressure may be that of an equilibrium gaseous phase, or it may be an externally controlled pressure.

In describing such systems, the nature of each phase must be specified. In the following example, the terms s, *l*, and *g* identify solid, liquid, and gaseous phases, respectively. For solutions or mixtures, the composition is needed, in addition to the temperature and pressure, to complete the specification of the system. If not obvious, the identity of the solvent must be given.

In the equilibrium shown as reaction (11), the relationship of Eq. (12) holds. Here

$$H_2O(l) \rightleftharpoons H_2O(g) \tag{11}$$

$$\Delta G^\circ = \mu_g^\circ - \mu_l^\circ = -RT \ln \frac{p}{N} \tag{12}$$

$K^\circ = p/N$, the ratio of the vapor pressure $p$ to the liquid mole fraction $N$. For pure water, the equilibrium constant is simply the standard vapor pressure $p^\circ$, and the Clausius-Clapeyron equation is just a special case of the Gibbs-Helmholtz equation, Eq. (8). Now when a small amount of solute is added, decreasing the mole fraction of solvent, the vapor pressure $p$ must be lowered to maintain equilibrium (Raoult's law). The effect of the total applied pressure $P$ upon the vapor pressure $p$ of the liquid is given by the Gibbs-Poynting equation, Eq. (13). Here $V_l$ and $V_g$ are the molar volumes of liquid

$$\left[\frac{dp}{dP}\right]_T = \frac{V_l}{V_g} \tag{13}$$

and vapor. The vapor pressure will increase as external pressure is applied (activity increases with pressure). If the external pressure is applied to a solution by a semipermeable membrane, an applied pressure can be found which will restore the vapor pressure (or activity) of the solvent to its standard state value. *See* OSMOSIS.

Other heterogeneous equilibria are solubility equilibria, reactions involving two immiscible phases, and reactions involving condensed and immiscible phases. *See* EXTRACTION; SOLUBILITY PRODUCT CONSTANT. [C.E.V.]

**Chemical fuel** The principal fuels used in internal combustion engines (automobiles, diesel, and turbojet) and in the furnaces of stationary power plants are organic fossil fuels. These fuels, and others derived from them by various refining and separation processes, are found in the earth in the solid (coal), liquid (petroleum), and gas (natural gas) phases.

Special fuels to improve the performance of combustion engines are obtained by synthetic chemical procedures. These special fuels serve to increase the specific impulse of the engine or to increase the heat of combustion available to the engine per unit mass or per unit volume of the fuel. A special fuel which possesses a very high heat of combustion per unit mass is liquid hydrogen. It has been used along with liquid oxygen in rocket engines. Because of its low liquid density, liquid hydrogen is not too useful in systems requiring high heats of combustion per unit volume of fuel ("volume-limited" systems).

A special fuel which produces high flame temperatures of the order of 5000°F (2800°C) is gaseous cyanogen. This is used with gaseous oxygen as the oxidizer. The liquid fuel hydrazine, and other hydrazine-based fuels, with the liquid oxidizer nitrogen tetroxide are used in many space-oriented rocket engines. The boron hydrides, such as diborane and pentaborane, are high-energy fuels which are used in advanced rocket engines.

For air-breathing propulsion engines (turbojets and ramjets), hydrocarbon fuels are most often used. For some applications, metal alkyl fuels which are pyrophoric (that is, ignite spontaneously in the presence of air), and even liquid hydrogen, are being used.

Fuels which liberate heat in the absence of an oxidizer while decomposing either spontaneously or because of the presence of a catalyst are called monopropellants and have been used in rocket engines. Examples of these monopropellants are hydrogen peroxide and nitro-methane.

Liquid fuels and oxidizers are used in most large-thrust rocket engines. When thrust is not a consideration, solid-propellant fuels and oxidizers are frequently employed because of the lack of moving parts such as valves and pumps, and the consequent simplicity of this type of rocket engine. Solid fuels fall into two broad classes, double-base and composites. Double-base fuels are compounded of nitroglycerin (glycerol trinitrate) and nitrocellulose, with no separate oxidizer required. The double-base propellant is generally formed in a mold into the desired shape (called a grain) required for the rocket case. Composite propellants are made of a fuel and an oxidizer. The latter could be an inorganic perchlorate or a nitrate. Fuels for composite propellants are generally the asphalt-oil-type, thermosetting plastics or several types of synthetic rubber and gumlike substances. Metal particles such as boron, aluminum, and beryllium have been added to solid propellants to increase their heats of combustion and to eliminate certain types of combustion instability. *See* HYDROGEN PEROXIDE. [W.Ch.]

**Chemical kinetics** A branch of physical chemistry that seeks to measure the rates of chemical reactions, describe them in terms of elementary steps, and understand them in terms of the fundamental interactions between molecules.

**Reaction kinetics.** Although the ultimate state of a chemical system is specified by thermodynamics, the time required to reach that equilibrium state is highly dependent upon the reaction. For example, diamonds are thermodynamically unstable with respect to graphite, but the rate of transformation of diamonds to graphite is negligible. As a consequence, determining the rate of chemical reactions has proved to be important for practical reasons. Rate studies have also yielded fundamental information about the details of the nuclear rearrangements which constitute the chemical reaction.

Traditional chemical kinetic investigations of the reaction between species X and Y to form Z and W, reaction (1), sought a rate of the form given in Eq. (2), where $d[\mathrm{Z}]/dt$

$$\mathrm{X} + \mathrm{Y} \rightarrow \mathrm{Z} + \mathrm{W} \tag{1}$$

$$d[\mathrm{Z}]/dt = kf([X], [Y], [Z], [W]) \tag{2}$$

is the rate of appearance of product Z, $f$ is some function of concentrations of X, Y, Z, and W which are themselves functions of time, and $k$ is the rate constant. Chemical reactions are incredibly diverse, and often the function $f$ is quite complicated, even for seemingly simple reactions such as that in which hydrogen and bromine combine directly to form hydrogen bromide (HBr). This is an example of a complex reaction which proceeds through a sequence of simpler reactions, called elementary reactions. For reaction (3*d*), the sequence of elementary reactions is a chain mechanism known to involve a series of steps, reactions (3*a*)–(3*c*). This sequence of elementary reactions

$$Br_2 \rightarrow 2Br \tag{3a}$$

$$Br + H_2 \rightarrow HBr + H \tag{3b}$$

$$H + Br_2 \rightarrow HBr + Br \tag{3c}$$

$$H_2 + Br_2 \rightarrow 2HBr \tag{3d}$$

was formerly known as the reaction mechanism, but in the chemical dynamical sense the word mechanism is reserved to mean the detailed motion of the nuclei during a collision.

An elementary reaction is considered to occur exactly as written. Reaction (3*b*) is assumed to occur when a bromine atom hits a hydrogen molecule. The products of the collision are a hydrogen bromide molecule and a hydrogen atom. On the other hand, the overall reaction is a sequence of these elementary steps and on a molecular basis does not occur as reaction (3*d*) is written. With few exceptions, the rate law for an elementary reaction A + B → C + D is given by $d[C]/dt = k[A][B]$. The order (sum of the exponents of the concentrations) is two, which is expected if the reaction is bimolecular (requires only species A to collide with species B). The rate constant $k$ for such a reaction depends very strongly on temperature, and is usually expressed as $k = Z_{AB}\rho \exp(-E_a/RT)$. $Z_{AB}$ is the frequency of collision between A and B calculated from molecular diameters and temperature; $\rho$ is an empirically determined steric factor which arises because only collisions with the proper orientation of reagents will be effective; and $E_a$, the experimentally determined activation energy, apparently reflects the need to overcome repulsive forces before the reagents can get close enough to react.

In some instances, especially for decompositions, AB → A + B, the elementary reaction step is first-order, Eq. (4), which means that the reaction is unimolecular. The

$$d[A]/dt = d[B]/dt = k[AB] \tag{4}$$

species AB does not spontaneously dissociate; it must first be given some critical amount of energy, usually through collisions, to form an excited species AB*. It is the species AB* which decomposes unimolecularly. [P.R.B.]

**Relaxation methods.** Considerable use has been made of perturbation techniques to measure rates and determine mechanisms of rapid chemical reactions. These methods provide measurements of chemical reaction rates by displacing equilibria. In situations where the reaction of interest occurs in a system at equilibrium, perturbation techniques called relaxation methods have been found most effective for determining reaction rate constants.

A chemical system at equilibrium is one in which the rate of a forward reaction is exactly balanced by the rate of the corresponding back reaction. Examples are chemical reactions occurring in liquid solutions, such as the familiar equilibrium in pure water, shown in reaction (5). The molar equilibrium constant at 25°C (77°F) is given by Eq. (6), where bracketed quantities indicate molar concentrations. It arises naturally from the equality of forward and backward reaction rates, Eq. (7). Here $k_f$ and $k_b$

are the respective rate constants that depend on temperature but not concentrations. Furthermore, the combination of Eqs. (6) and (7) gives rise to Eq. (8). Thus a reasonable

$$H_2O \underset{k_b}{\rightleftharpoons} H^+(aq) + OH^-(aq) \tag{5}$$

$$K_{eq} = \frac{[H^+][OH^-]}{[H_2O]} = \frac{10^{-14}}{55.5} = 1.8 \times 10^{-16} \tag{6}$$

$$k_f[H_2O] = k_b[H^+][OH^-] \tag{7}$$

$$K_{eq} = k_f / k_b = 1.8 \times 10^{-16} \tag{8}$$

question might be what the numerical values of $k_f$ in units of $s^{-1}$ and $k_h$ in units of $dm^3\ mol^{-1}\ s^{-1}$ must be to satisfy Eqs. (6) through (8) in water at room temperature. Stated another way, when a liter of 1 *M* hydrochloric acid is poured into a liter of 1 *M* sodium hydroxide (with considerable hazardous sputtering), how rapidly do the hydronium ions, $H^+(aq)$, react with hydroxide ions, $OH^-(aq)$, to produce a warm 0.5 *M* aqueous solution of sodium chloride? In the early 1950s it was asserted that such a reaction is instantaneous. Turbulent mixing techniques were (and still are) insufficiently fast (mixing time of the order of 1 ms) for this particular reaction to occur outside the mixing chamber. The relaxation techniques were conceived by M. Eigen, who accepted the implied challenge of measuring the rates of seemingly immeasurably fast reactions. *See* ULTRAFAST MOLECULAR PROCESSES.

The essence of any of the relaxation methods is the perturbation of a chemical equilibrium (by a small change in temperature, pressure, electric-field intensity, or solvent composition) in so sudden a fashion that the chemical system, in seeking to reachieve equilibrium, is forced by the comparative slowness of the chemical reactions to lag behind the perturbation. [E.M.Ey.]

**Gas-phase reactions.** The rates of thermal gas-phase chemical reactions are important in understanding processes such as combustion and atmospheric chemistry.

Chemical conversion of one stable, gas-phase molecule into another is an apparently simple process; yet it is highly unlikely to occur in just a single step, but as a web of sequential and parallel reactions involving many species. The oxidation of methane ($CH_4$) to carbon dioxide ($CO_2$) and water provides an excellent example. It occurs in combustion (for example, in burning natural gas, which is mostly methane) as well as in the atmosphere. In both cases, the net process may be written down as single reaction (9).

$$CH_4 + 2O_2 \rightarrow 2H_2O + CO_2 \tag{9}$$

The reaction does not, however, result from collision of two oxygen ($O_2$) molecules with one methane molecule. Rather, it involves many separate steps.

All elementary reactions fundamentally require a collision between two molecules. Even in the case of a unimolecular reaction, in which a single molecule breaks apart or isomerizes to another form, the energy required for the process comes from collision with other molecules. The species involved in many gas-phase elementary reactions are free radicals, molecules that have one or more unpaired electrons. Such species tend to be highly reactive, and they are responsible for carrying out most gas-phase chemistry.

A reaction rate is the rate at which the concentration of one of the reactants or products changes with time. The objective of a kinetics experiment is not to measure the reaction rate itself but to measure the rate coefficient, an intrinsic property of the reaction that relates the reactant concentrations to their time rates of change. For example, the mathematical expression for the rate of a biomolecular reaction, $A + B \rightarrow$ products, is differential equation (10). The square brackets denote the concentrations of A and B,

$$-d[A]/dt = -d[B]/dt = k[A][B] \tag{10}$$

and $k$ is the rate coefficient described above. The dependence of the rate expression on reactant concentrations is determined experimentally, and it also arises from a fundamental tenet of chemistry known as the law of mass action. Once the rate constant is known, the rate of a reaction can be computed for any given set of concentrations.

Rate constants usually change with temperature because of the change in the mean energy of colliding molecules. The temperature dependence often follows an Arrhenius expression, $k = A \exp(-E_A/RT)$, where $A$ is a preexponential factor that is related to the gas-phase collision rate, $R$ is the universal gas constant, and $T$ is the absolute temperature (in kelvins). The key quantity is the activation energy, $E_A$, the amount of energy required to induce a reaction. Pressure dependences are usually important only for association reactions, $A + B \rightarrow AB$, since collision of A with B will form an energized complex, $AB^*$, that will simply redissociate unless a subsequent collision carries away enough energy to stabilize the AB product. The probability of a stabilizing collision increases with the collision frequency and thus the total pressure.

In the simple case of a unimolecular reaction, $A \rightarrow P$, the rate expression is Eq. (11).

$$-d[A]/dt = d[P]/dt = k[A] \tag{11}$$

Equation (11) is first-order since the rate is proportional to the reactant concentration to the first power, and it leads to an expression for the change in the concentration of A or P with time (called an integrated rate expression), as in Eqs. (12). The subscripts

$$[A]_t = [A]_0 \exp(-kt) \tag{12a}$$

$$[P]_t = [A]_0\{1 - \exp(-kt)\} \tag{12b}$$

denote the concentrations at zero time (initial concentration) and an arbitrary time $t$. The concentration of A decreases (because it is reacting away) as a function of time, while the concentration of P increases with time such that the sum of the concentrations of A and P is always constant and equal to the initial concentration of A. Because first-order reactions are mathematically simple, kineticists try to reduce all studied reactions (if at all possible) to this form. A second-order reaction, for example $A + B \rightarrow$ products, has the rate expression given in Eq. (10). To reduce the second-order expression to the first-order expression, Eq. (11), one chooses one of the concentrations to be in large excess, for example $[B] \gg [A]$. The concentration of B is then approximately constant during the course of the reaction, and it may be combined with the rate constant to give an expression identical to Eq. (11) that depends on the concentration of A alone.

[A.R.Ra.; S.S.B.]

**Chemical microscopy** A scientific discipline in which microscopes are used to solve chemical problems. The unique ability to form a visual image of a specimen, to select a small volume of the specimen, and to perform a chemical or structural analysis on the material in the selected volume makes chemical microscopy indispensable to modern chemical analysis. *See* ANALYTICAL CHEMISTRY.

Microscopes can be combined with most analytical instruments. For example, a light microscope can be combined with a spectroscope, making it possible to determine the molecular composition of microscopic objects or structures. Similarly, an x-ray spectrometer can be combined with an electron microscope to determine the elemental composition of small objects. *See* SPECTROSCOPY; X-RAY SPECTROMETRY.

Phase analyses can also be made microscopically. The boundaries of amorphous phases can usually be distinguished in the microscope, and an elemental or physical analysis can be used to identify the phase. An example of a physical analysis is the measurement of refractive index. Crystalline phases are even more amenable to microscopical analysis. For example, a polarizing microscope can be used to measure the optical properties of a crystalline phase and thus identify it. Or a transmission electron microscope can be used to select a tiny area of a crystalline phase and identify the crystal structure by means of electron diffraction.

The minimum volume that can be analyzed varies widely with the instrument used. Light microscopes can be used to identify particles as small as 1 micrometer in diameter and weighing about 1 picogram. A field ion microscope has been combined with a mass spectrometer and used to identify single atoms extracted from the surface of a specimen. *See* MASS SPECTROMETRY.

After a portion of a specimen has been selected microscopically, it can be analyzed in many ways. An experienced microscopist may learn to recognize various structures by studying known materials, using published atlases, or an atlas that the individual microscopist has constructed. If the object or structure cannot be recognized, many means of analysis are available. For example, a polarizing microscope may be employed to identify the object by using optical crystallographic methods. Other light microscopes useful for chemical analysis include phase-contrast and interference-contrast microscopes, microspectrophotometers, the confocal scanning laser microscope, and the laser Raman microscope. Physicochemical methods may be used to measure melting points, or mixed-melt phenomena and dispersion staining may also be used. *See* LASER SPECTROSCOPY.

Microscopes using other types of image-forming beams serve for chemical analysis. Scanning or transmission electron microscopes are powerful tools for chemical microscopy. Scanning electron microscopes are often fitted with x-ray spectrometers which are capable of both qualitative and quantitative analysis for most of the elements. Other electron microscopes capable of chemical analysis are the Auger electron microscope, field electron microscope, scanning tunneling microscope, and cathodoluminescence microscope. Microscopes which use ion beams, neutron beams, and x-ray beams also have analytical capabilities. *See* AUGER EFFECT; AUGER ELECTRON SPECTROSCOPY.

[G.C.]

**Chemical process industry** An industry, abbreviated CPI, in which the raw materials undergo chemical conversion during their processing into finished products, as well as (or instead of) the physical conversions common to industry in general. In the chemical process industry the products differ chemically from the raw materials as a result of undergoing one or more chemical reactions during the manufacturing process. The chemical process industries broadly include the traditional chemical industries, both organic and inorganic; the petroleum industry; the petrochemical industry, which produces the majority of plastics, synthetic fibers, and synthetic rubber from petroleum and natural-gas raw materials; and a series of allied industries in which chemical processing plays a substantial part. While the chemical process industries are primarily the realm of the chemical engineer and the chemist, they also involve a wide range of other scientific, engineering, and economic specialists.

For a discussion of the more prominent chemical process industries, *see* ADHESIVE; BIOMEDICAL CHEMICAL ENGINEERING; BIOCHEMICAL ENGINEERING; COAL CHEMICALS; COAL GASIFICATION; COAL LIQUEFACTION; ELECTROCHEMICAL PROCESS; FAT AND OIL; GRAPHITE;

HYDROCRACKING; NUCLEAR CHEMICAL ENGINEERING; PETROCHEMICAL; PETROLEUM PROCESSING AND REFINING; PLASTICS PROCESSING; POLYMER; RUBBER. [W.F.F.]

**Chemical reactor** A vessel in which chemical reactions take place. A combination of vessels is known as a chemical reactor network. Chemical reactors have diverse sizes, shapes, and modes and conditions of operation based on the nature of the reaction system and its behavior as a function of temperature, pressure, catalyst properties, and other factors.

Laboratory chemical reactors are used to obtain reaction characteristics. Therefore, the shape and mode of operation of a reactor on this scale differ markedly from that of the large-scale industrial reactor, which is designed for efficient production rather than for gathering information. Laboratory reactors are best designed to achieve well-defined conditions of concentrations and temperature so that a reaction model can be developed which will prove useful in the design of a large-scale reactor model.

Chemical reactions may occur in the presence of a single phase (liquid or gas), in which case they are called homogeneous, or they may occur in the presence of more than one phase and are referred to as heterogeneous. In addition, chemical reactions may be catalyzed. Examples of homogeneous reactions are gaseous fuel combustion (gas phase) and acid-base neutralization (liquid phase). Examples of heterogeneous systems are carbon dioxide absorption into alkali (gas-liquid); coal combustion and automobile exhaust purification (gas-solid); water softening (liquid-solid); coal liquefaction and oil hydrogenation (gas-liquid-solid); and cake reduction of iron ore (solid-solid).

Chemical reactors may be operated in batch, semibatch, or continuous modes. When a reactor is operated in a batch mode, the reactants are charged, and the vessel is closed and brought to the desired temperature and pressure. These conditions are maintained for the time needed to achieve the desired conversion and selectivity, that is, the required quantity and quality of product. At the end of the reaction cycle, the entire mass is discharged and another cycle is begun. Batch operation is labor-intensive and therefore is commonly used only in industries involved in limited production of fine chemicals, such as pharmaceuticals. In a semibatch reactor operation, one or more reactants are in the batch mode, while the coreactant is fed and withdrawn continuously. In a chemical reactor designed for continuous operation, there is continuous addition to, and withdrawal of reactants and products from, the reactor system.

There are a number of different types of reactors designed for gas-solid heterogeneous reactions. These include fixed beds, tubular catalytic wall reactors, and fluid beds. Many different types of gas-liquid-solid reactors have been developed for specific reaction conditions. The three-phase trickle-bed reactor employs a fixed bed of solid catalyst over which a liquid phase trickles downward in the presence of a cocurrent gas phase. An alternative is the slurry reactor, a vessel within which coreactant gas is dispersed into a liquid phase bearing suspended catalyst or coreactant solid particles. At high ratios of reactor to diameter, the gas-liquid-solid reactor is often termed an ebulating-bed (high solids concentration) or bubble column reactor (low solids concentration). Gas-liquid reactors assume a form virtually identical to the absorbers utilized in physical absorption processes. Solid-solid reactions are often conducted in rotary kilns which provide the necessary intimacy of contact between the solid coreactants. *See* GAS ABSORPTION OPERATIONS. [J.J.Ca.]

**Chemical separation techniques** Methods used in chemistry to purify substances or to isolate them from other substances, for either preparative or analytical purposes. In industrial applications the ultimate goal is the isolation of a product of given purity, whereas in analysis the primary goal is the determination of the amount or concentration of that substance in a sample. There are three factors of importance to be considered in all separations: (1) the completeness of recovery of the substance being isolated, (2) the extent of separation from associated substances, and (3) the efficiency of the separation.

There are many types of separations based on a variety of properties of materials. Among the most commonly used properties are those involving solubility, volatility, adsorption, and electrical and magnetic effects, although others have been used to advantage. The most efficient separation will obviously be obtained under conditions for which the differences in properties between two substances undergoing separation are at a maximum.

The common aspect of all separation methods is the need for two phases. The desired substance will partition or distribute between the two phases in a definite manner, and the separation is completed by physically separating the two phases. The ratio of the concentrations of a substance in the two phases is called its partition or distribution coefficient. If two substances have very similar distribution coefficients, many successive steps may be required for a separation. The resulting process is called a fractionation.

Based on the nature of the second phase, the more commonly used methods of separation are classified as follows:

1. Methods involving a solid second phase include precipitation, electrodeposition, chromatography (adsorption), ion exchange, and crystallization.

2. The outstanding method involving a liquid second phase is solvent extraction, in which the original solution is placed in contact with another liquid phase immiscible with the first.

3. Methods involving a gaseous second phase include gas evolution, distillation, sublimation, and gas chromatography. Mixtures of volatile substances can often be separated by fractional distillation. *See* EXTRACTION. [G.H.Mo.]

**Chemical symbols and formulas** A system of symbols and notation for the chemical elements and the combinations of these elements which form numerous chemical compounds. This system consists of letters, numerals, and marks that are designed to denote the chemical element, formula, or structure of the molecule or compound. These symbols give a concise and instantly recognizable description of the element or compound. In many cases, through the efforts of international conferences, the symbols are recognized throughout the scientific world, and they greatly simplify the universal language of chemistry.

**Elements.** At the present time, 109 chemical elements have been given symbols, usually derived from the name of the element. Examples of names and symbols are chlorine, Cl; fluorine, F; beryllium, Be; aluminum, Al; oxygen, O; and carbon, C. However, symbols for some elements are derived from Latin or other names for the element. Examples are Au, gold (from *aurum*); Fe, iron (from *ferrum*); Pb, lead (from *plumbum*); Na, sodium (from *natrium*); and K, potassium (from *kalium*). The symbols consist of one or, more commonly, two letters. The first letter is a capital, followed by a lowercase letter.

**Inorganic molecules and compounds.** Simple diatomic molecules of a single element are designated by the element symbol with a subscript 2, indicating that the molecule contains two atoms. Thus the hydrogen molecule is $H_2$; the nitrogen molecule, $N_2$; and the oxygen molecule, $O_2$. Polyatomic molecules of a single element are designated by the element symbol with a subscript corresponding to the number of atoms in the molecule. Examples are the phosphorus molecule, $P_4$; the sulfur molecule, $S_8$; and the arsenic molecule, $As_4$.

Diatomic covalent molecules containing unlike elements are given a similar designation. The formula for hydrogen chloride is HCl; for iodine monochloride, ICl; and for hydrogen iodide, HI. The more electropositive element is always designated first in the formula.

For polyatomic covalent molecules containing unlike elements, subscripts designate the number of atoms of each element that are present in the molecule. Examples are arsine, $AsH_3$; ammonia, $NH_3$; and water, $H_2O$. Again, the more electropositive element is placed first in the formula.

Ionic inorganic compounds are designated by a similar notation. The positive ion is given first in the formula, followed by the negative ion; subscripts denote the number of ions of each element present in the compound. The formulas for several common compounds are sodium chloride, NaCl; ammonium nitrate, $NH_4NO_3$; and aluminum sulfate, $Al_2(SO_4)_3$.

More complex inorganic compounds are designated in a similar manner. The positive ion is given first, but may contain attached or coordinated groups, and this is followed by the negative ion. Examples are hexammine-cobalt(III) chloride, $[Co(NH_3)_6]Cl_3$; and potassium trioxalatoferrate(III), $K_3[Fe(C_2O4)_3]$. Hydrates of inorganic compounds, such as copper(II) sulfate pentahydrate, are designated by the formula of the compound followed by the formula for water, the number of water molecules being designated by a prefix. Thus the symbol for the last compound is $CuSO_4 \cdot 5H_2O$.

**Organic compounds.** Because there are many more organic than inorganic compounds, the designation or notation for the organic group becomes complex. Many different types of organic compounds are known; in the case of hydrocarbons, there are aromatic and aliphatic, saturated and unsaturated, cyclic and polycyclic, and so on. The system of notation must distinguish between the various hydrocarbons themselves as well as setting this group of compounds apart from others such as alcohols, ethers, amines, esters, and phenols. *See* CHEMISTRY; COORDINATION COMPLEXES; INORGANIC CHEMISTRY; ORGANIC CHEMISTRY. [W.W.We.]

**Chemical thermodynamics** The application of thermodynamic principles to systems involving physical and chemical transformations in order to (1) develop quantitative relationships among the identifiable forms of energy and their conjugate variables, (2) establish the criteria for spontaneous change, for equilibrium, and for thermodynamic stability, and (3) provide the macroscopic base for the statistical-mechanical bridge to atomic and molecular properties. The thermodynamic principles applied are the conservation of energy as embodied in the first law of thermodynamics, the principle of internal entropy production as embodied in the second law of thermodynamics, and the principle of absolute entropy and its statistical thermodynamic formulation as embodied in the third law of thermodynamics.

The basic goal of thermodynamics is to provide a description of a system of interest in order to investigate the nature and extent of changes in the state of that system as it undergoes spontaneous change toward equilibrium and interacts with its surroundings. This goal implicitly carries with it the concept that there are measurable properties of the system which can be used to adequately describe the state of the system and that the system is enclosed by a boundary or wall which separates the system and its surroundings. Properties that define the state of the system can be classified as extensive and intensive properties. Extensive properties are dependent upon the mass of the system, whereas intensive properties are not. Typical extensive properties are the energy, volume, and numbers of moles of each component in the system, while typical intensive properties are temperature, pressure, density, and the mole fractions or concentrations of the components.

The concept of a boundary enclosing the system and separating it from the surroundings requires specification of the nature of the boundary and of any constraints the boundary places upon the interaction of the system and its surroundings. Boundaries that restrain a system to a particular value of an extensive property are said to be restrictive with respect to that property. A boundary which restrains the system to a given volume is a fixed wall. A boundary which is restrictive to one component of a system but not to the other components is a semipermeable wall or membrane. A system whose boundaries are restrictive to energy and to mass or moles of components is said to be an isolated system. A system whose boundaries are restrictive only to mass or moles of components is a closed system, whereas an open system has nonrestrictive walls and hence can exchange energy, volume, and mass with its surroundings. Boundaries can be restrictive with respect to specific forms of energy, and two important types are those restrictive to thermal energy but not work (adiabatic walls) and those restrictive to work but not thermal energy (diathermal walls).

Changes in the state of the system can result from processes taking place within the system and from processes involving exchange of mass or energy with the surroundings. After a process is carried out, if it is possible to restore both the system and the surroundings to their original states, the process is said to be reversible; otherwise the process is irreversible. All naturally occurring spontaneous processes are irreversible. The first law defines the internal energy as a state function or property of the state of a system, and restricts the system and its surroundings to those processes which conserve energy. The second law establishes which of the permissible processes can occur spontaneously.

According to the first law of thermodynamics, the total energy $E$ of a system is the sum of its kinetic energy $T$, its potential energy $V$, and its internal energy $U$, Eq. (1).

$$E = T + V + U \tag{1}$$

If a system has constant mass and its center of mass is moving with uniform velocity in a uniform potential, then changes in the total energy of the system $\delta E$ are equal to changes in its internal energy $\delta U$. Chemical thermodynamics concentrates on the internal energy of the system, but kinetic and potential energy changes of the system as a whole can be important for chemical systems. The principle of conservation of energy requires that the change in the internal energy of a system be the result of energy transfer between the system and its surroundings. The internal energy $U$ is a function of the set of extensive variables associated with the various forms of internal energy. Each form of internal energy is manifest by the product of an extensive variable and its conjugate intensive variable.

Thermal energy exchange or heat (that form of energy transferred as a result of temperature differences between a system and its surroundings) plays a central role in thermodynamics, and is singled out from the other forms of energy or work. This is expressed by Eq. (2), where $\delta q$ is the differential thermal energy (heat) absorbed by the

$$dU = \delta q + \delta w \tag{2}$$

system from the surroundings and $\delta \omega$ is the differential work performed on the system by the surroundings. It is convenient to write Eqs. (3), where $T$ is temperature, $S$ is the

$$\delta q = TdS - \delta a \tag{3a}$$

$$dU = TdS + \delta w - \delta a \tag{3b}$$

entropy, and $(-\delta a)$ is a sum of the nonthermal differential work terms. The term $\delta a$ can be either zero or nonzero. If it is zero, the heat absorbed by the system is equal to $TdS$. In an adiabatic process $\delta q$ is zero and $TdS - \delta a$, and hence if $\delta a$ is nonzero, it must correspond to an internally generated thermal energy. This is frequently referred to as the uncompensated heat of a process, since it does not result from the transfer of heat from the surroundings.

The heat capacity of a system is of particular importance in such thermochemical calculations. The heat capacity is the amount of thermal energy that can be absorbed by a system for a unit rise in temperature. This is defined by Eq. (4), where $C_{\text{process}}$ is

$$\delta q = C_{\text{process}} dT \tag{4}$$

the heat capacity of a system for a given type of process. *See* HEAT CAPACITY.

There are many possible and essentially equivalent statements of the second law of thermodynamics. It will suffice to state the empirical result that in all spontaneous processes the uncompensated heat $\delta a$ in Eqs. (3) is always positive. Equation (3*a*) can be rewritten as Eq. (5), where the term $\delta q/T$ is the contribution to the entropy due to

$$dS = \delta q/T + \delta a/T \tag{5}$$

heat exchange with the surrounding ($d_eS$), while $\delta a/T$ is the contribution to the entropy produced as a result of the interconversion of work terms ($d_iS$). The second law can then be summarized as Eqs. (6), where $d_iS$ greater than zero applies to irreversible process. When $d_iS = 0$, that is, for a reversible process, Eq. (7) holds. This is the

$$dS = d_eS + d_iS \tag{6a}$$

$$d_iS \geq 0 \tag{6b}$$

$$dS = \delta q_{\text{rev}}/T \tag{7}$$

basic equation for establishing the thermodynamic temperature scale based upon the theoretical limits of reversible cycles. The requirement that $d_iS > 0$ for spontaneous processes provides the criteria for examining the specific conditions for spontaneous paths, and the criteria for establishing the equilibrium state of a system.

Many chemical systems can be considered closed systems in which a single parameter $\xi$ can be defined as a measure of the extent of the reaction or the degree of achievement of a process. If the reaction proceeds or the process advances spontaneously, entropy must be produced according to the second law and $\delta a$ must be positive. In terms of the advancement parameter $\xi$, this uncompensated heat $\delta a$ can be given by Eq. (8), where

$$\delta a = \underline{A}\, d\xi = T d_iS \tag{8}$$

$\underline{A}$ is the affinity of the process or reaction. The affinity is related to internal entropy production by Eq. (9). The condition that the entropy production is zero represents

$$\underline{A} = T d_iS/d\xi \geq 0 \tag{9}$$

equilibrium, and hence $\underline{A} = 0$ is an equivalent condition for equilibrium in a closed system. For spontaneous processes, since the signs of $\underline{A}$ and $d\xi$ must be the same, for positive $\underline{A}$ the process must advance or go in a forward direction in the usual sense of chemical reactions or physical processes, while for negative $\underline{A}$ the process must proceed in the reverse direction.

The affinity of a chemical reaction establishes the spontaneous direction of the reaction, and consequently methods for determining the affinity are important in thermochemical studies. The affinity is simply related to the stoichiometric coefficients of the reaction and the chemical potentials of the reactants and products in the reaction.

Classical equilibrium thermodynamics is primarily concerned with calculations for reversible processes, and deals with irreversibility in terms of inequalities. In the case of irreversible processes in systems slightly removed from equilibrium, the rate of internal entropy production $d_iS/dt$ is related to the fluxes $J_i$ associated with thermal, concentration, or other differences in intensive parameters or potentials $X_i$. This entropy production is then given by Eq. (10). The fluxes include heat conduction, diffusion, electric conduction, and other direct effects.

$$d_iS/dt = \sum_i J_i X_i \geq 0 \tag{10}$$

In addition, a flux of one type may be coupled to a potential difference of another type. For example, a thermal gradient can result in a mass flux (thermal diffusion), or a concentration gradient in any energy flux. Thermal conductivity, thermo-osmosis, and thermoelectric effects are all coupled effects.

Far removed from equilibrium, thermodynamics must be formulated somewhat differently and more cautiously. The interplay of thermodynamic stability and kinetics can give rise to macroscopic structures with both temporal and spatial coherence called dissipative structures. Much theoretical effort is being directed to these studies because of their apparent relevance to biological structures, but it is still too early to assess how far-reaching these theories will be in the future. *See* THERMODYNAMIC PRINCIPLES; THERMODYNAMIC PROCESSES. [R.A.Pi.]

**Chemiluminescence** The type of luminescence wherein a chemical reaction supplies the energy responsible for the emission of light (ultraviolet, visible, or infrared) in excess of that of a blackbody (thermal radiation) at the same temperature and within the same spectral range. Below 900°F (500°C), the emission of any light during a chemical reaction is a chemiluminescence. The blue inner cone of a bunsen burner or the Coleman gas lamp are examples.

Many chemical reactions generate energy. Usually this exothermicity appears as heat, that is, translational, rotational, and vibrational energy of the product molecules; whereas, for a visible chemiluminescence to occur, one of the reaction products must be generated in an excited electronic state (designated by an asterisk) from which it can undergo deactivation by emission of a photon. Hence a chemiluminescent reaction, as shown in reactions (1) and (2), can be regarded as the reverse of a photochemical reaction.

$$\mathrm{A + B \rightarrow C^* + D} \tag{1}$$

$$\mathrm{C^* \rightarrow C} + h\nu \tag{2}$$

The energy of the light quantum $h\upsilon$ (where $h$ is Planck's constant, and $\upsilon$ is the light frequency) depends on the separation between the ground and the first excited electronic state of C; and the spectrum of the chemiluminescence usually matches the fluorescence spectrum of the emitter. Occasionally, the reaction involves an additional step, the transfer of electronic energy from C* to another molecule, not necessarily otherwise involved in the reaction. Sometimes no discrete excited state can be specified, in which case the chemiluminescence spectrum is a structureless continuum associated with the formation of a molecule, as in the so-called air afterglow: $NO + O \rightarrow NO_2 + h\upsilon$ (green light). *See* PHOTOCHEMISTRY.

Only very exothermic, or "exergonic," chemical processes can be expected to be chemiluminescent. Partly for this reason, most familiar examples of chemiluminescence involve oxygen and oxidation processes; the most efficient examples of these are the enzyme-mediated bioluminescences. [T.Wi.]

Electrogenerated chemiluminescence, also known as electrochemiluminescence, is a luminescent chemical reaction in which the reactants are formed electrochemically.

Electrochemical reactions are electron-transfer reactions occurring in an electrochemical cell. In such a reaction, light emission may occur as with chemiluminescence; however, the excitation is from the application of a voltage to an electrode. In chemiluminescence, the luminophore is excited to a higher energetic state by means of a chemical reaction initiated by mixing of the reagents. In electrogenerated chemiluminescence, the emitting luminophore is excited to a higher energy state by reactions of species that are generated at an electrode surface by the passage of current through the working electrode. Upon decay to the electronic ground state, light emitted by the luminophore (fluorescent or phosphorescent) can be detected. The luminophore is typically a polycyclic hydrocarbon, an aromatic heterocycle, or certain transition-metal chelates. *See* ELECTRON-TRANSFER REACTION.

Chief among the developments since the early research and discovery of electrogenerated chemiluminescence was the construction of instrumentation for detection of electrogenerated chemiluminescence. These instruments made it possible for the methodology to be used by practitioners other than electrochemists.

Measurement of the light intensity of electrogenerated chemiluminescence is very sensitive and is proportional to the luminophore concentration. Trace amounts of luminophore as low as $10^{-13}$ mol/liter can be detected, making electrogenerated chemiluminescence very useful in analytical and diagnostic applications.

A commercial application of the phenomenon forms the basis of a highly sensitive technique for detection of biological analytes such as deoxyribonucleic acid (DNA), ribonucleic acid (RNA), proteins, antibodies, haptens, and therapeutic drugs in the clinical laboratory. The technique combines a binding assay method and a system for detecting electrogenerated chemiluminescence. [J.K.Le.; L.Na.; H.Ya.]

**Chemistry** The science that embraces the properties, composition, and structure of matter, the changes in structure and composition that matter undergoes, and the accompanying energy changes. It is important to distinguish chemical change, implicit in this definition, and changes in physical form. An example of the latter is the conversion of liquid water to solid or gas by cooling or heating; the water substance is unchanged. In chemical change, such as the rusting of iron, the metal is consumed as it reacts with air in the presence of water to form the new substance, iron oxide.

Modern chemistry grew out of the alchemy of the Middle Ages, and the attempts to transmute base metals into gold. Seminal observations were made in the early eighteenth century on the changes in volume of air during combustion in a closed vessel, and the French chemist Antoine Lavoisier in the 1770s interpreted these phenomena in essentially modern terms.

**Atoms and elements.** Underlying all of chemistry is the concept of elementary units of matter which cannot be subdivided. This idea was adumbrated in classical Greek writings, and was clearly expressed by the Englishman John Dalton in 1803, who called these units atoms. Different kinds of atoms were recognized, each corresponding to one of the chemical elements such as oxygen, sulfur, tin, iron, and a few other metals. By the midnineteenth century, about 80 elements had been characterized, and these were organized on the basis of regularities in behavior and properties, into a periodic table. *See* ELEMENT (CHEMISTRY); PERIODIC TABLE.

In the early twentieth century, observations of radiation from various sources and its impact on solid targets led to the recognition of three fundamental particles that are common to all elements; the electron, with negative charge; the proton, with positive charge; and the neutron, with zero charge. An atom consists of a nucleus containing protons and neutrons, and a diffuse cloud of electrons, equal in number to the number of protons and arranged in orbitals of progressively higher energy levels as the distance from the nucleus increases. The atomic number of an element ($Z$) is defined as the number of protons in the nucleus; this is the sequence of ordering in the periodic table. The mass number corresponds to the total number of protons and neutrons. *See* ATOMIC NUMBER; ELECTRON; NEUTRON.

**Isotopes.** Most elements exist as isotopes, which have differing numbers of neutrons. All isotopes of an element exhibit the same chemical behavior, although isotopes can be separated on the basis of differences in atomic mass. The known elements total 116; of these, 88 have been detected in one or more isotopic forms in the Earth's crust. The other elements, including all but one of those with atomic number above 92, are synthetic isotopes produced in nuclear reactions that take place in nuclear piles or particle accelerators. Most of the isotopes of these heavier elements and also some lighter ones are radioactive; that is, the nuclei are unstable and decay, resulting in the emission of radiation.

**Molecules and chemical reactions.** Molecules are combinations of two or more atoms, bonded together in definite proportions and specific geometric arrangements. These entities are chemical compounds; a molecule is the smallest unit. The bonding of atoms in compounds involves the distribution of electrons, and is the central concern of chemistry.

Compounds result from chemical reactions of atoms or molecules. The process involves formation and breaking of bonds, and may be either exothermic, in which the net bond charges lead to a more stable (lower-energy) system and heat is evolved, or endothermic, in which energy must be added to overcome a net loss of bonding energy.

A simple case is the reaction of hydrogen and oxygen to give water, which can be expressed as reaction (1). The equation is balanced; no atoms are gained or lost in a

$$2H_2 + O_2 \rightarrow 2H_2O \quad \Delta H = 572 \text{ kilojoules} \qquad (1)$$

chemical reaction. The symbols represent the nature of the initial and final materials and also the relative amounts. Thus $H_2O$ represents a molecule of water or a mole, which is the quantity in grams (or other mass units) equivalent to the molecular weight. The symbol $\Delta H$ indicates the energy (enthalpy) change for the process. The reaction of hydrogen and oxygen is highly exothermic, and the sign of the energy charge is therefore negative since the system has lost heat to the surroundings. *See* ENTHALPY; MOLE (CHEMISTRY); STOICHIOMETRY.

**Bonds.** Bonds can be broadly classified as ionic or covalent. An ion is an atom or molecule which has an electric charge. Ionic compounds can be illustrated by salts such as sodium chloride, NaCl, in which a positive sodium ion, $Na^+$, and negative chloride ion, $Cl^-$, are associated by electrostatic attractions in regular locations of a crystal lattice. In solution the ions are solvated by water molecules and can conduct an electric current.

In covalent molecules, bonds are formed by the presence of pairs of electrons in overlapping orbitals between two atoms. Thus when two hydrogen atoms (H·) come within bonding distance, a molecule of hydrogen is formed in an exothermic reaction, by formation of a covalent bond. In this case the heat of reaction represents the energy of the H—H bond [reaction (2)].

$$2H\cdot \rightarrow H\text{—}H \quad \Delta H = -435 \text{ kilojoules} \qquad (2)$$

*See* CHEMICAL BONDING.

**Chemical compounds.** A compound is specified by the elements it contains, the number of atoms of each element, the bonding arrangement, and the characteristic properties. The number of unique compounds that have been isolated from natural sources or prepared by synthesis is enormous; as of 2000, over 15 million substances were registered in the file maintained by Chemical Abstracts (American Chemical Society). Most of these are organic compounds, containing from a few to many hundred carbon atoms. The element carbon, unlike any others, can form long chains of covalently bonded atoms. Moreover, there can be many compounds, called isomers, with the same atomic composition. Thus a molecular formula such as $C_8H_{16}O$ can represent many thousand different compounds. *See* CARBON; MOLECULAR ISOMERISM.

**Branches of chemistry.** Traditionally, five main subdivisions are designated for the activities, professional organizations, and literature of chemistry and chemists.

Analytical chemistry deals with the determination of the composition of matter and the amount of each component in mixtures of any kind. Analytical measurements are an integral and indispensable part of all chemical endeavor. Originally, analytical chemistry involved detection, separation, and weighing of the substances present in a mixture. Determination of the atomic ratio and thence the molecular formula of a compound is a prerequisite for any other investigation; the development of balances and techniques for doing this on milligram quantities of material had an enormous impact on organic chemistry. Advances have involved increasingly sophisticated instrumentation; mass spectrometers are a notable example. Other important methods include high-resolution chromatography and various applications of electrochemistry. A constant goal in analytical chemistry is the development of methods and instruments of greater sensitivity. It is now possible to detect trace compounds such as environmental pollutants at the picogram level. *See* ANALYTICAL CHEMISTRY.

Biochemistry is the study of living systems from a chemical viewpoint; thus it is concerned with the compounds and reactions that occur in plant and animal cells. Most of the substances in living tissues, including carbohydrates, lipids, proteins, nucleic acids, and hormones, are well-defined organic substances. However, the metabolic and regulatory processes of these compounds and their biological function are the special province of biochemistry. One of the major areas is the characterization of enzymes and their cofactors, and the mechanism of enzyme catalysis. Other topics of interest include the transport of ions and molecules across cell membranes, and the target sites of neurotransmitters and other regulatory molecules. Biochemical methods and thinking have contributed extensively to the fields of endocrinology, genetics, immunology, and virology. *See* BIOCHEMISTRY.

Inorganic chemistry is concerned with any material in which metals and metalloid elements are of primary interest. Inorganic chemistry is therefore concerned with the structure, synthesis, and bonding of a very diverse range of compounds. One of the early interests was the composition of minerals and the discovery of new elements; from this has grown the specialized area of geochemistry. Early synthetic work emphasized compounds of the main group elements, and particularly in the twentieth century, complex compounds of the transition metals. These studies have led to soluble transition-metal catalysts, and a greatly increased understanding of catalytic processes and the pivotal role of metal atoms in major biochemical processes, such as oxygen transport in blood, photosynthesis, and biological nitrogen fixation. Other contributions of inorganic chemistry are seen in advanced ceramics, high-performance composite materials, and the growing number of high-temperature superconductors. *See* CATALYSIS; INORGANIC CHEMISTRY.

Organic chemistry is centered on compounds of carbon. Originally these were the compounds isolated from plant and animal sources, but the term was early broadened to include all compounds in which a linear or cyclic carbon chain is the main feature. Two of the major thrusts have been the elucidation of new structures and their preparation by synthesis; another long-standing interest has been study of the reaction mechanisms and rearrangements of organic compounds. Structure work on naturally occurring compounds progressed over a 150-year period from simple straight-chain compounds with 2–10 carbon atoms, hydrogen, and 1 or 2 oxygen atoms to antibiotics and toxins with many rings and as many as 100 carbon atoms. In modern work, nuclear magnetic resonance spectroscopy and x-ray diffraction have become indispensable tools. Paralleling structural studies has been the synthesis of increasingly complex target molecules. Synthetic work is directed also to the preparation of large numbers of compounds for screening as potential drugs and agricultural chemicals. Plastics, synthetic fibers, and other high polymers are other products of organic chemistry. *See* NUCLEAR MAGNETIC RESONANCE (NMR); ORGANIC CHEMISTRY; X-RAY DIFFRACTION.

Physical chemistry deals with the interpretation of chemical phenomena and the underlying physical processes. One of the classical topics of physical chemistry involves the thermodynamic and kinetic principles that govern chemical reactions. Another is a description of the physical states of matter in molecular terms. Experimentation and theoretical analysis have been directed to the understanding of equilibria, solution behavior, electrolysis, and surface phenomena. One of the major contributions has been quantum chemistry, and the applications and insights that it has provided. The methods and instruments of physical chemistry, including such hardware as spectrometers and magnetic resonance and diffraction instruments, are an integral part of every other area. *See* CHEMICAL THERMODYNAMICS; PHYSICAL CHEMISTRY.

Each broad area of chemistry embraces many specialized topics. There are also a number of hybrid areas, such as bioorganic and bioinorganic chemistry, analytical biochemistry, and physical organic chemistry. Each of these areas has borrowed extensively from and contributed to every other one. *See* BIOINORGANIC CHEMISTRY.

[J.A.Mo.]

**Chemometrics** A chemical discipline that uses mathematical and statistical methods to design or select optimal measurement procedures and experiments and to provide maximum chemical information by analyzing chemical data. Chemometrics is actually a collection of procedures, mathematics, and statistics that can help chemists perform well-designed experiments and proceed rapidly from data, to information, to knowledge of chemical systems and processes.

Medicinal chemists use chemometrics to relate measured or calculated properties of candidate drug molecules to their biological function; this subdiscipline is known as quantitative structure activity relations (QSAR). Environmental chemists use chemometrics to find pollution sources or understand the effect of point pollution sources on regional or global ecosystems by analyzing masses of environmental data. Forensic chemists analyze chemical measurements made on evidence (for example, gasoline in an arson case) or contraband to determine its source. Experimental physical chemists use chemometrics to unravel and identify physical or chemical states from spectral data acquired during the course of an experiment. *See* FORENSIC CHEMISTRY; PHYSICAL CHEMISTRY.

In analytical chemistry, chemometrics has seen rapid growth and widespread application, primarily due to the computerization of analytical instrumentation. Automation provides an opportunity to acquire enormous amounts of data on chemical systems. Virtually every branch of analytical chemistry has been impacted significantly by chemometrics; commercial software implementing chemometrics methods has become commonplace in analytical instruments. *See* ANALYTICAL CHEMISTRY.

Whether the analyst is concerned with a single sample or, as in process analytical chemistry, an entire chemical process (for example, the human body, a manufacturing process, or an ecosystem), chemometrics can assist in the experimental design, instrument response, optimization, standardization, and calibration as well as in the various steps involved in going from measurements (data), to chemical information, to knowledge of the chemical system under study. [B.Ko.]

**Chlorine** A chemical element, Cl, atomic number 17 and atomic weight 35.453. Chlorine exists as a greenish-yellow gas at ordinary temperatures and pressures. It is second in reactivity only to fluorine among the halogen elements, and hence is never found free in nature, except at the elevated temperatures of volcanic gases. It is estimated that 0.045% of the Earth's crust is chlorine. It combines with metals, nonmetals, and organic materials to form hundreds of chlorine compounds, the most important of which are discussed here. *See* PERIODIC TABLE.

Chlorine and its common acid derivative, hydrochloric (or muriatic) acid, were probably noted by experimental investigators as early as the thirteenth century. C. W. Scheele identified chlorine as "dephlogisticated muriatic acid" in 1774, and H. Davy proved that a new element had been found in 1810. Extensive production started 100 years later. During the twentieth century, the amount of chlorine used has been considered a measure of industrial growth.

**Physical properties.** The atomic weight of naturally occurring chlorine is 35.453 (based on carbon at 12). It is formed of stable isotopes of mass 35 and 37; radioactive isotopes have been made artificially. The diatomic gas has a molecular weight of 70.906. The boiling point of liquid chlorine (golden-yellow in color) is −33.97°C (−29.15°F) at 760 mm Hg ($10^2$ kilopascals) and the melting point of solid chlorine (tetragonal crystals) is −100.98°C (−149.76°F). The critical temperature is 144°C (292°F); the critical pressure is 78.7 atm; the critical volume is 1.745 ml/g; and density at the critical point is 0.573 g/ml. Thermodynamic properties include heat of sublimation at $7370 \pm 10$ cal/mole at 0 K, heat of evaporation at 4882 cal/mole at −33.97°C, heat of fusion at 1531 cal/mole, vapor heat capacity at a constant pressure of 1 atm of 8.32 cal/(mole °C) at 0°C (32°F) and 8.46 cal/(mole °C) at 100°C (212°F). Chlorine forms solid hydrates, $Cl_2 \cdot 6H_2O$ (pale-green crystals) and $Cl_2 \cdot 8H_2O$. It hydrolyzes in water as shown in reaction (1).

$$Cl_2 + H_2O \rightarrow HClO + HCl \quad (1)$$

**Chemical properties.** Chlorine is one of four closely related chemical elements which have been called the halogen elements. Fluorine is more active chemically, and bromine and iodine are less active. Chlorine replaces iodine and bromine from their salts. It enters into substitution and addition reactions with both organic and inorganic materials. Dry chlorine is somewhat inert, but moist chlorine unites directly with most of the elements. *See* HALOGEN ELEMENTS.

**Compounds.** Sodium chloride, NaCl, is used directly as mined (rock salt), or as found on the surface, or as brine. It may also be dissolved, purified, and reprecipitated for use in foods or when chemical purity is required. Its main uses are in the production of soda ash and chlorine products. Farm use, refrigeration, dust and ice control, water treatment, food processing, and food preservation are other uses. Calcium chloride, $CaCl_2$, is usually obtained from brines or as a by-product of chemical processing. Its main uses are in road treatment, coal treatment, concrete conditioning, and refrigeration.

Wet chlorine reacts with metals to form chlorides, most of which are soluble in water. It also reacts with sulfur and phosphorus and with other halogens as in reactions (2).

$$
\begin{aligned}
H_2 + Cl_2 &\rightarrow 2HCl \\
2Fe + 3Cl_2 &\rightarrow 2FeCl_3 \\
2S + Cl_2 &\rightarrow S_2Cl_2 \\
S + Cl_2 &\rightarrow SCl_2 \\
S + 2Cl_2 &\rightarrow SCl_4 \\
P_4 + 6Cl_2 &\rightarrow 4PCl_3 \\
P_4 + 10Cl_2 &\rightarrow 4PCl_5 \\
Br_2 + Cl_2 &\rightarrow 2BrCl \\
2F_2 + Cl_2 &\rightarrow ClF + ClF_3
\end{aligned}
\tag{2}
$$

The oxides of chlorine, dichlorine monoxide, $Cl_2O$, chlorine monoxide, ClO, chlorine dioxide, $ClO_2$, chlorine hexoxide, $Cl_2O_6$, and chlorine heptoxide, $Cl_2O_7$, are all made indirectly. $Cl_2O$ is commonly called chlorine monoxide also. Chlorine dioxide, a green gas, has become increasingly important in commercial bleaching of cellulose, water treatment, and waste treatment.

Hydrogen chloride, HCl, is a colorless, pungent, poisonous gas which liquefies at 82 atm at 51°C (124°F). It boils at −85°C (−121°F) at 1 atm ($10^2$ kPa). Its major production is as the by-product of many organic chlorinations. It can be made by direct reaction of chlorine and hydrogen in an open combustion chamber submerged in cooled, aqueous hydrochloric acid solution. It is used as a strong acid and as a reducing agent.

Aluminum chloride, $AlCl_3$, is an anhydrous, white, deliquescent, hexagonal crystalline substance. Either scrap aluminum or the oxide (bauxite) may be chlorinated. Aluminum chloride is a catalyst for production of cumene, styrene, and isomerized butane. Of the aluminum chloride uses in anhydrous form, ethylbenzene production uses 25%, dyes 30%, detergents 15%, ethyl chloride 10%, drugs 8%, and miscellaneous production 12%. Hydrated and liquid forms are also available, 50% of which are used in drug and cosmetic production.

Ferric chloride, $FeCl_3$, is a solid composed of dark, hexagonal crystals. Much chlorine from chemical processes is converted to ferric chloride, which is then used for the manufacture of salts, pigments, pharmaceuticals, and dyes and for photoengraving, preparation of catalysts, and waste and sewage treatment.

**Natural occurrence.** Because many inorganic chlorides are quite soluble in water, they are leached out of land areas by rain and ground water to accumulate in the sea or in lakes that have no outlets. Seawater contains 18.97 g of chloride ion per kilogram (3% sodium chloride). Solar evaporation produces large deposits of salts in landlocked areas. Similar evaporation in the past is responsible for vast underground deposits of rock salt and brines in Michigan, central New York, the Gulf Coast of Texas, Stassfurt in Germany, and elsewhere. These deposits are mainly of sodium chloride, the supply of which is unlimited for practical purposes. Other rocks and minerals in the Earth's surface average slightly over 0.03% chloride.

**Manufacture.** The first electrolytic process was patented in 1851 by Charles Watt in Great Britain. In 1868 Henry Deacon produced chlorine from hydrochloric acid and oxygen at 400°C (750°F) with copper chloride absorbed in pumice stone as a catalyst. The electrolytic cells now used may be classified for the most part as diaphragm and mercury types. Both make caustic (NaOH or KOH), chlorine, and hydrogen. The economics of the chlor-alkali industry mainly involve the balanced marketing or internal use of caustic and chlorine in the same proportions as obtained from the electrolytic cell process.

**Uses.** Chlorine is an excellent oxidizing agent. Historically, the use of chlorine as a bleaching agent in the paper, pulp, and textile industries and as a germicide for drinking water preparation, swimming pool purification, and hospital sanitation has made community living possible. Chlorine is used to produce bromine from bromides found in brines and seawater. The automotive age increased the production of bromine tremendously for the manufacture of ethylene dibromide for use in gasoline. Compounds of chlorine are used as bleaching agents, oxidizing agents, solvents, and intermediates in the manufacture of other substances. [J.Do.; F.W.Ko.; R.W.B.]

**Cholesterol** A cyclic hydrocarbon alcohol commonly classified as a lipid because it is insoluble in water but soluble in a number of organic solvents. It is the major sterol in all vertebrate cells and the most common sterol of eukaryotes. In vertebrates, the highest concentration of cholesterol is in the myelin sheath that surrounds nerves and in the plasma membrane that surrounds all cells.

Cholesterol can exist either in the free (unesterified) form (see structure below) or in

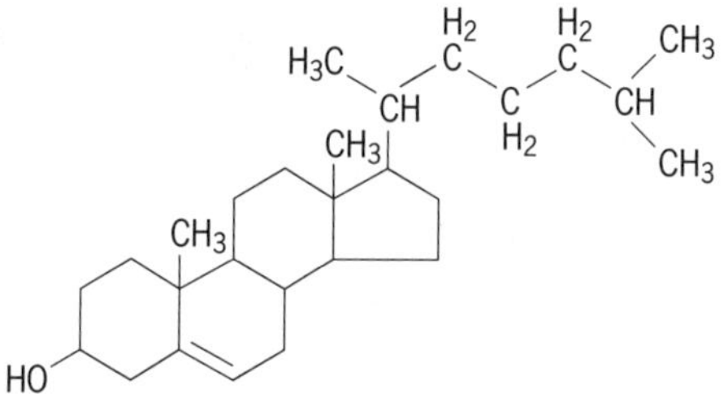

the esterified form, in which a fatty acid is bound to the hydroxyl group of cholesterol by an ester bond. The free form is found in membranes. Cholesteryl esters are normally found in lipid droplets either within the cells of steroidogenic tissues, where it can be converted to free cholesterol and then to steroid hormones, or in the middle of spherical lipid-protein complexes, called lipoproteins, that are found in blood.

Cholesterol, together with phospholipids and proteins, is important in the maintenance of normal cellular membrane fluidity. At physiological temperatures, the cholesterol molecule interacts with the fatty acids of the membrane phospholipids and causes increased packing of the lipid molecules and hence a reduction of membrane fluidity. Thus, all vertebrate cells require cholesterol in their membranes in order for the cell to function normally. Cholesterol is also important as a precursor for a number of other essential compounds, including steroid hormones, bile acids, and vitamin D. *See* STEROID.

Cellular cholesterol is obtained both from the diet, following its absorption in the intestine, and from synthesis within all cells of the body. Foods that are particularly high in cholesterol include eggs, red meat, and organs such as liver and brain. About 40–50% of the dietary cholesterol is absorbed from the intestine per day. In contrast, plant sterols are very poorly absorbed. Cholesterol synthesis occurs in all vertebrate cells but is highest in the liver, intestine, and skin, and in the brain at the time of myelination.

Cholesterol and cholesteryl esters are essentially insoluble in water. In order to transport these compounds around the body in the blood, the liver and intestine produce various lipid-protein complexes, called lipoproteins, which serve to solubilize them. Lipoproteins are large, complex mixtures of cholesterol, cholesteryl esters, phospholipids, triglycerides (fats), and various proteins. The major lipoproteins include chylomicrons, very low density lipoprotein (VLDL), low-density lipoprotein, and high-density lipoprotein (HDL).

Total plasma cholesterol levels of less than 200 mg/100 ml are considered desirable. Values of 200–239 or greater than 239 mg per 100 ml are considered, respectively, borderline high or high risk values, indicating the potential for a heart attack. High levels of low-density lipoprotein in the plasma are associated with increased risk of atherosclerosis, ("hardening of the arteries"), which involves deposition of cholesterol and other lipids in the artery wall. Diets low in cholesterol and saturated fats often result in a reduction in total plasma and LDL cholesterol levels. Such changes in blood cholesterol levels are thought to be beneficial and to reduce the incidence of heart attacks. [P.A.E.]

**Chromatography** A physical separation method in which the components of a mixture are separated by differences in their distribution between two phases, one of which is stationary (stationary phase) while the other (mobile phase) moves through it in a definite direction. The substances must interact with the stationary phase to be retained and separated by it.

Retention results from a combination of reversible physical interactions that can be characterized as adsorption at a surface, absorption in an immobilized solvent layer, and electrostatic interactions between ions. When the stationary phase is a porous medium, accessibility to its regions may be restricted and a separation can result from size differences between the sample components. More than one interaction may contribute simultaneously to a separation mechanism. The general requirements are that all interactions must be reversible, and that the two phases can be separated (two immiscible liquids, a gas and a solid, and so forth) in such a way that a distribution of sample components between phases and mass transport by one phase can be established. *See* ABSORPTION; ADSORPTION.

Reversibility of the interactions can be achieved by purely physical means, such as by a change in temperature or by competition; the latter condition is achieved by introducing substances into the mobile phase that have suitable properties to ensure reversibility for the interactions responsible for retention of the sample components. Since this competition with the sample components is itself selective, it provides a general approach to adjusting the outcome of a chromatographic experiment to obtain a desired separation. It is an absolute requirement that a difference in the distribution constants for the sample components in the chromatographic system exist for a separation to be possible.

**Methods.** A distinction between the principal chromatographic methods can be made in terms of the properties of the mobile phase and configuration of the stationary phase. In gas chromatography the mobile phase is an inert gas, in supercritical fluid chromatography the mobile phase is a fluid (dense gas above its critical pressure and temperature), and in liquid chromatography the mobile phase is a liquid of low viscosity. The stationary phase can be a porous, granular powder with a narrow particle-size distribution packed into a tube (called a column) as a dense homogeneous bed. This configuration is referred to as a packed column and is nearly always used in liquid chromatography and is commonly used in supercritical fluid and gas chromatography. Alternatively, the stationary phase can be distributed as a thin film or layer on the wall of an open tube of capillary dimensions, leaving an open space through the center of the column. This configuration is referred to as an open tubular column (or incorrectly as a capillary column); and it is commonly used in gas chromatography, frequently used in supercritical fluid chromatography, but rarely used in liquid chromatography.

Thin-layer chromatography is a form of liquid chromatography in which the stationary phase is spread as a thin layer over the surface of a glass or plastic supporting structure. The stationary phase must be immobilized on the support by using a binder to impart the desired mechanical strength and stability to the layer. The samples are applied to the layer as spots or bands near the bottom edge of the plate. The separation is achieved by contacting the bottom edge of the plate below the line of samples with the mobile phase, which proceeds to ascend the layer by capillary action. This process is called development and is performed in a chamber, with the lower edge of the layer in contact with the mobile phase and the remaining portion of the layer in contact with solvent vapors from the mobile phase. The chamber may be a simple device such as a covered jar or beaker or a more elaborate device providing control of the mobile-phase velocity and elimination or control of the vapor phase. Thin-layer chromatography is the most popular form of planar chromatography having virtually replaced paper chromatography in laboratory practice. *See* GAS CHROMATOGRAPHY; GEL PERMEATION CHROMATOGRAPHY; LIQUID CHROMATOGRAPHY; SUPERCRITICAL-FLUID CHROMATOGRAPHY.

**Uses.** Chromatographic methods provide a means of analyzing samples (to determine component identity and relative amount), of isolating significant quantities of purified material for further experimentation or commerce, and for determining fundamental physical properties of either the samples or the mobile or stationary phases (for example, diffusion coefficients, solubilities, or thermodynamic properties). There are virtually no boundaries to the sample types that can be separated. Examples include organic and inorganic compounds in the form of fixed gases, ions, polymers, as well as other species. Applications are found in all areas of technological development, making chromatography one of the most widely used laboratory procedures in chemistry. Depending on intent, chromatography can be applied to trace quantities at the limit of detector response (for example, $10^{-15}$ g) or to kilogram amounts in preparative separations.

**Instrumentation.** Modern chromatographic methods are instrumental techniques in which the optimal conditions for the separation are set and varied by electromechanical devices external to the column or layer. Separations are largely automated, with important features of the instrumentation being control of the flow and composition of the mobile phase, introduction of the sample onto the stationary phase, and on-line detection of the separated components. In column chromatography the sample components are detected in the presence of the mobile phase after they have exited the stationary phase. In thin-layer chromatography the sample components are detected in the presence of the stationary phase, resulting in different detection strategies.

Instrument requirements differ by the needs of the method employed. Gas chromatography, for example, employs a mobile phase of constant composition at a few atmospheres of column inlet pressure and variation in the temperature of the column to effect a separation. Liquid chromatography uses a pump to select or vary the composition of the mobile phase with a high column inlet pressure (typically a few hundred atmospheres) and a constant temperature for the separation. These differences in optimized separation conditions result in different equipment configurations for each chromatographic method.

**Interpretation.** The results of a chromatographic experiment are summarized in a chromatogram (see illustration), a two-dimensional record of the detector response to the sample components ($y$ axis) plotted against the residence time of the components in column chromatography or migration distance in planar chromatography ($x$ axis). Individual compounds or mixtures of unseparated compounds appear as peaks in the chromatogram. These peaks are ideally symmetrical and occur at positions in the chromatogram that are characteristic of their identity, with a distribution around the

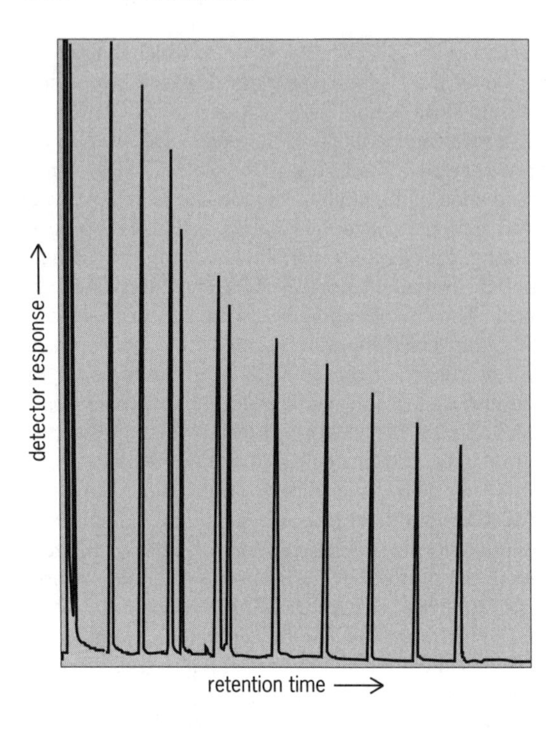


**Typical chromatogram obtained by gas chromatography.**

mean position (apex of the peak) that is characteristic of the kinetic properties of the chromatographic system. The area inscribed by the peak is proportional to the amount of substance separated in the chromatographic system.

Information readily extracted from the chromatogram includes an indication of sample complexity (the number of observed peaks), qualitative substance identification (determined by peak position), relative composition of the sample (peak dimensions; area or height), and a summary of the kinetic characteristics of the chromatographic system (peak shapes). [C.F.P.]

**Chromium** A chemical element, Cr, atomic number 24, and atomic weight 51.996, which is the weighted average for several isotopes weighing 50 (4.31%), 52 (83.76%), 53 (9.55%), and 54 (2.38%). The orbital arrangement of the electrons is $1s^2$, $2s^2$, $2p^6$, $3s^2$, $3p^6$, $3d^5$, $4s^1$. The stability of the half-filled *d* shell doubtless accounts for this rather unusual arrangement. In the crust of the Earth, chromium is the twenty-first element in abundance, which ranks it along with vanadium, zinc, nickel, and copper. Traces of chromium are present in the human body; in fact, it is essential to life. *See* PERIODIC TABLE.

The element was discovered in 1797 and isolated the following year by the French chemist L. N. Vauquelin. It was named chromium because of the many colors of its compounds. It occurs in nature largely as the mineral chromite ($FeO \cdot Cr_2O_3$), which is a spinel, but the ore is usually contaminated with $Al^{3+}$, $Fe^{3+}$, $Mn^{2+}$, and $Mg^{2+}$. Smaller quantities are found as the yellow mineral crocoite ($PbCrO_4$).

As a transition metal, chromium exists in all oxidation states from 2− to 6+. The chemistry of its aqueous solutions, at least in the 3+ (chromic) state, is complicated by the fact that the compounds exist in many isomeric forms, which have quite different chemical properties.

Pure chromium metal has a bluish-white color, reflects light well, and takes a high polish. When pure, it is ductile, but even small amounts of impurities render it brittle. The metal melts at about 1900°C (3452°F) and boils at 2642°C (4788°F). Chromium shows a wide range of oxidation states; the compounds in which the metal is in a low oxidation state are powerful reducing agents, whereas those in which it shows a high oxidation state are strong oxidizing agents.

The bright color and resistance to corrosion make chromium highly desirable for plating plumbing fixtures, automobile radiators and bumpers, and other decorative pieces. Unfortunately, chrome plating is difficult and expensive. It must be done by electrolytic reduction of dichromate in sulfuric acid solution. This requires the addition of six electrons per chromium ion. This reduction does not take place in one step, but through a series of steps, most of which are not clearly understood. The current efficiency is low (maybe 12%), and the chromium plate contains microscopic cracks and other flaws, and so it does not adequately protect the metal under it from corrosion. It is customary, therefore, to first plate the object with copper, then with nickel, and finally, with chromium.

In alloys with iron, nickel, and other metals, chromium has many desirable properties. Chrome steel is hard and strong and resists corrosion to a marked degree. Stainless steel contains roughly 18% chromium and 8% nickel. Some chrome steels can be hardened by heat treatment and find use in cutlery; still others are used in jet engines. Nichrome and chromel consist largely of nickel and chromium; they have low electrical conductivity and resist corrosion, even at red heat, so they are used for heating coils in space heaters, toasters, and similar devices. Other important alloys are Hastelloy C (Cr, Mo, W, Fe, Ni), used in chemical equipment which is in contact with HCl, oxidizing acids, and hypochlorite. Stellite [Co, Cr, Ni, C, W (or Mo)], noted for its hardness and abrasion resistance at high temperatures, is used for lathes and engine valves, and Inconel (Cr, Fe, Ni) is used in heat treating and in corrosion-resistant equipment in the chemical industry.

Several chromium compounds are used as paint pigments—chrome oxide green ($Cr_2O_3$), chrome yellow ($PbCrO_4$), chrome orange ($PbCrO_4 \cdot PbO$), molybdate orange (a solution of $PbSO_4$, $PbCrO_4$, and $PbMoO_4$), chrome green (a mixture of $PbCrO_4$ and Prussian blue), and zinc yellow (potassium zinc chromate). Several of these, particularly zinc yellow, are used to inhibit corrosion. The gems ruby, emerald, and alexandrite owe their colors to traces of chromium compounds. *See* PAINT.

Dichromates are widely used as oxidizing agents, as rust inhibitors on steel, and as wood preservatives. In the last application, they kill fungi, termites, and boring insects. The wood can still be painted and glued, and retains its strength. Other chromium compounds find use as catalysts, as drilling muds, and in photochemical reactions. The last are important in the printing industry. A metal plate is coated with a colloidal material (for example, glue, shellac, or casein) containing a dichromate. On exposure to strong light under a negative image, the dichromate is reduced to $Cr^{3+}$, which reacts with the colloid, hardening it and making it resistant to removal by washing. The unexposed material is washed off, and the metal plate is etched with acid to give a printing plate.

Chromium is essential to life. A deficiency (in rats and monkeys) has been shown to impair glucose tolerance, decrease glycogen reserve, and inhibit the utilization of amino acids. It has also been found that inclusion of chromium in the diet of humans sometimes, but not always, improves glucose tolerance. Certain chromium(III) compounds enhance the action of insulin.

On the other hand, chromates and dichromates are severe irritants to the skin and mucous membranes, so workers who handle large amounts of these materials must be protected against dusts and mists. Continued breathing of the dusts finally leads to ulceration and perforation of the nasal septum. Contact of cuts or abrasions with chromate may lead to serious ulceration. Even on normal skin, dermatitis frequently results. Cases of lung cancer have been observed in plants where chromates are manufactured.

[J.C.Ba.]

**Citric acid** A hydroxytricarboxylic acid, general formula $C_6H_8O_7$, with the structure shown below. It is available primarily as anhydrous material but also as the

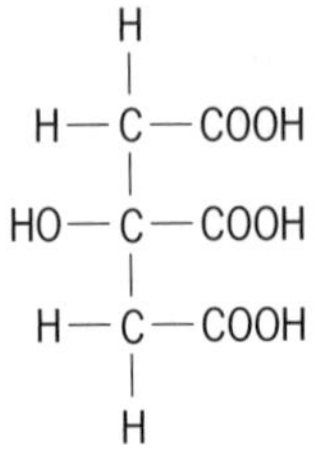

```
       H
       |
   H — C — COOH
       |
  HO — C — COOH
       |
   H — C — COOH
       |
       H
```

monohydrate. The major commercial salts are sodium and potassium, with calcium, diammonium, and ferric ammonium (complex) also available. *See* ACID AND BASE.

Citric acid is a relatively strong organic acid, and is very soluble in water. Citric acid and its salts are widely used because they are nontoxic, safe to handle, and easily biodegraded.

Citric acid occurs in relatively large quantities in citrus fruits. It also occurs in other fruits, in vegetables, and in animal tissues and fluids either as the free acid or as citrate ion. It is an integral part of the Krebs (citric acid) cycle involving the metabolic conversion of carbohydrates, fats, and proteins in most living organisms.

Today, essentially all of the commercial citric acid is produced by fermentation. Processes employed are surface or submerged fermentation by mold (*Aspergillus niger*) and submerged fermentation by yeast (*Candida guilliermondii, C. lipolytica*), using a variety of substrates including sucrose, molasses, corn syrup, enzyme-treated starch, and normal paraffins. Citric acid is recovered from the fermentation broth by solvent extraction or more commonly by precipitation as calcium citrate, followed by treatment with sulfuric acid to convert the calcium citrate to calcium sulfate and citric acid. The calcium sulfate is removed by filtration, and the citric acid solution is further purified. Crystallization of citric acid from a hot aqueous solution (above the transition temperature of 36.6°C or 97.9°F) yields anhydrous citric acid; crystallization from a cold solution yields the monohydrate. Although total chemical syntheses for citric acid have been published, they have never achieved commercial success.

Citric acid is widely used in the food and pharmaceutical industries. In foods it is used primarily to produce a tart taste and to complement fruit flavors in carbonated beverages, beverage powders, fruit-flavored drinks, jams and jellies, candy, sherbets, water ices, and wine. It is also used to reduce pH in certain canned foods to make heat treatment more effective, and in conjunction with antioxidants to chelate trace metals and retard enzymatic activity.

In pharmaceuticals, citric acid provides the acid source in effervescent tablets in addition to being used to adjust pH, impart a tart taste, and chelate trace metals. It is also used as a blood anticoagulant. *See* PHARMACEUTICAL CHEMISTRY.

Citric acid, because of its low toxicity, relative noncorrosiveness, and biodegradability, is also being used for applications normally reserved for the strong mineral acids. These include preoperational and operational cleaning of iron and copper oxides from boilers, nuclear reactors, and heat exchangers; passivation of stainless steel tanks and equipment; and etching of concrete floors prior to coating. It is also used as a dispersant to retard settling of titanium dioxide slurries and as a sequestering and pH control agent in the textile industry. [F.Sa.]

**Clarification** The removal of small amounts of fine, particulate solids from liquids. The purpose is almost invariably to improve the quality of the liquid, and the removed solids often are discarded. The particles removed by a clarifier may be as large as 100 micrometers or as small as 2 micrometers. Clarification is used in the manufacture of pharmaceuticals, beverages, and fiber and film polymers; in the reconditioning of electroplating solutions; in the recovery of dry-cleaning solvent; and for the purification of drinking water and waste water. The filters in the feed line and lubricating oil system of an internal combustion engine are clarifiers.

The methods of clarification include gravity sedimentation, centrifugal sedimentation, filtration, and magnetic separation. Clarification differs from other applications of these mechanical separation techniques by the low solid content of the suspension to be clarified (usually less than 0.2%) and the substantial completion of the particle removal. *See* FILTRATION; MECHANICAL SEPARATION TECHNIQUES; SEDIMENTATION (INDUSTRY). [S.A.M.]

**Clathrate compounds** Well-defined addition compounds formed by inclusion of molecules in cavities existing in crystal lattices or present in large molecules. The constituents are bound in definite ratios, but these are not necessarily integral. The components are not held together by primary valence forces, but instead are the consequence of a tight fit which prevents the smaller partner, the guest, from escaping from the cavity of the host. Consequently, the geometry of the molecules is the decisive factor.

Inclusion compounds can be subdivided into (1) lattice inclusion compounds (inclusion within a lattice which, as such, is built up from smaller single molecules); (2) molecular inclusion compounds (inclusion into larger ring molecules with holes); and (3) inclusion compounds of macromolecules. The best-known lattice inclusion compounds are the urea and thiourea channel inclusion compounds, which are formed by mixing hydrocarbons, carboxylic acids, or long-chain fatty alcohols with solutions of urea. Other representatives of lattice inclusion compounds are the choleic acids, which are adducts of deoxycholic acid with fatty acids, and other lipoic substances. Some aromatic compounds form an open crystal lattice which can accommodate smaller gas and solvent molecules (clathrates in the stricter sense of the word). The gas hydrates are inclusion compounds of gases in a somewhat expanded ice lattice. The gas or solvent molecules are inserted into definite places within the ice lattice and are surrounded by water molecules on all sides.

Crown ether compounds are cyclic or polycyclic polyether compounds capable of including another atom in the center of the ring. In this way, sodium or potassium compounds can be solubilized in organic solvents. Similarly, a series of ionophore antibiotics can complex inorganic cations.

Some clay minerals are made up of distinct silicate layers. Between these layers some free space may exist in the shape of channels. Smaller hydrocarbon molecules can be accommodated reversibly within these channels. This phenomenon is used in some technical separation processes for separating hydrocarbons (molecular sieves). Furthermore, ion-exchange processes used for water deionization are based on similar minerals. *See* MOLECULAR SIEVE.

Enzymes are believed to accommodate their substrates in active sites, pockets, or clefts prior to the chemical reaction which then changes the chemical structures of the substrates. These binding processes are identical to those of low-molecular-weight inclusion compounds. [F.Cr.; W.S.; D.G.]

**Coal chemicals** For about 100 years, chemicals obtained as by-products in the primary processing of coal to metallurgical coke have been the main source of aromatic compounds used as intermediates in the synthesis of dyes, drugs, antiseptics, and solvents. Although some aromatic hydrocarbons, such as toluene and xylene, are now obtained largely from petroleum refineries, the main source of others, such as benzene, naphthalene, anthracene, and phenanthrene, is still the by-product coke oven. Heterocyclic nitrogen compounds, such as pyridines and quinolines, are also obtained largely from coal tar. Although much phenol is produced by hydrolysis of monochlorobenzene and by decomposition of cumene hydroperoxide, much of the phenol, cresols, and xylenols are still obtained from coal tar.

**Coal tar chemicals**

| Compound | Fraction of whole tar, % | Use |
|---|---|---|
| Naphthalene | 10.9 | Phthalic acid |
| Monomethylnaphthalenes | 2.5 | |
| Acenaphthenes | 1.4 | Dye intermediates |
| Fluorene | 1.6 | Organic syntheses |
| Phenanthrene | 4.0 | Dyes, explosives |
| Anthracene | 1.0 | Dye intermediates |
| Carbazole (and other similar compounds) | 2.3 | Dye intermediates |
| Phenol | 0.7 | Plastics |
| Cresols and xylenols | 1.5 | Antiseptics, organic syntheses |
| Pyridine, picolines, lutidines, quinolines, acridine, and other tar bases | 2.3 | Drugs, dyes, antioxidants |

Coke oven by-products are gas, light oil, and tar. Coke oven gas is a mixture of methane, carbon monoxide, hydrogen, small amounts of higher hydrocarbons, ammonia, and hydrogen sulfide. Most of the coke oven gas is used as fuel. Although several hundred chemical compounds have been isolated from coal tar, a relatively small number are present in appreciable amounts. These may be grouped as in the table. All the compounds in the table except the monomethylnaphthalenes are of some commercial importance.

The direct utilization of coal as a source of bulk organic chemicals has been the objective of much research and development. Oxidation of aqueous alkaline slurries of coal with oxygen under pressure yields a mixture of aromatic carboxylic acids. Because of the presence of nitrogen compounds and hydroxy acids, this mixture is difficult to refine. Hydrogenation of coal at elevated temperatures and pressures yields much larger amounts of tar acids and aromatic hydrocarbons of commercial importance than are obtained by carbonization. However, this operation is more costly than other sources of these chemicals. *See* COKE; DESTRUCTIVE DISTILLATION; PYROLYSIS. [H.W.W.]

**Coal gasification** The conversion of coal or coal char to gaseous products by reaction with steam, oxygen, air, hydrogen, carbon dioxide, or a mixture of these.

Products consist of carbon monoxide, carbon dioxide, hydrogen, methane, and some other gases in proportions dependent upon the specific reactants and conditions (temperatures and pressures) employed within the reactors, and the treatment steps which the gases undergo subsequent to leaving the gasifier. Similar chemistry can also be applied to the gasification of coke derived from petroleum and other sources. The reaction of coal or coal char with air or oxygen to produce heat and carbon dioxide could be called gasification, but it is more properly classified as combustion. The principal purposes of such conversion are the production of synthetic natural gas as a substitute gaseous fuel and synthesis gases for production of chemicals and plastics. *See* COMBUSTION.

In all cases of commercial interest, gasification with steam, which is endothermic, is an important chemical reaction. The necessary heat input is typically supplied to the gasifier by combusting a portion of the coal with oxygen added along with the steam. From the industrial viewpoint, the final product is either chemical synthesis gas (CSG), medium-Btu gas (MBG), or a substitute natural gas (SNG).

Each of the gas types has potential industrial applications. In the chemical industry, synthesis gas from coal is a potential alternative source of hydrogen and carbon monoxide. This mixture is obtained primarily from the steam reforming of natural gas, natural gas liquids, or other petroleum liquids. Fuel users in the industrial sector have studied the feasibility of using medium-Btu gas instead of natural gas or oil for fuel applications. Finally, the natural gas industry is interested in substitute natural gas, which can be distributed in existing pipeline networks.

There has also been some interest by the electric power industry in gasifying coal by using air to provide the necessary heat input. This could produce low-Btu gas (because of the nitrogen present), which can be burned in a combined-cycle power generation system.

In nearly all of the processes, the general process is the same. Coal is prepared by crushing and drying, pretreated if necessary to prevent caking, and then gasified with a mixture of air or oxygen and steam. The resulting gas is cooled and cleaned of char fines, hydrogen sulfide, and carbon dioxide before entering optional processing steps to adjust its composition for the intended end use. [W.R.Ep.]

**Coal liquefaction** The conversion of most types of coal (with the exception of anthracite) primarily to petroleumlike hydrocarbon liquids which can be substituted for the standard liquid or solid fuels used to meet transportation, residential, commercial, and industrial fuel requirements. Coal liquids contain less sulfur, nitrogen, and ash, and are easier to transport and use than the parent (solid) coal. These liquids are suitable refinery feedstocks for the manufacture of gasoline, heating oil, diesel fuel, jet fuel, turbine fuel, fuel oil, and petrochemicals.

Liquefying coal involves increasing the ratio of hydrogen to carbon atoms (H:C) considerably—from about 0.8 to 1.5–2.0. This can be done in two ways: (1) indirectly, by first gasifying the coal to produce a synthesis gas (carbon monoxide and hydrogen) and then reconstructing liquid molecules by Fischer-Tropsch or methanol synthesis reactions; or (2) directly, by chemically adding hydrogen to the coal matrix under conditions of high pressure and temperature. In either case (with the exception of methanol synthesis), a wide range of products is obtained, from light hydrocarbon gases to heavy liquids. Even waxes, which are solid at room temperature, may be produced, depending on the specific conditions employed. [W.R.Ep.]

**Cobalt** A lustrous, silvery-blue metallic chemical element, Co, with an atomic number of 27 and an atomic weight of 58.93. Metallic cobalt was isolated in 1735 by the

Swedish scientist G. Brandt, who called the impure metal cobalt rex, after the ore from which it was extracted. The metal was shown to be a previously unknown element by T. O. Bergman in 1780. *See* PERIODIC TABLE.

Cobalt is a transition element in the same group as rhodium and iridium. In the periodic table it occupies a position between iron and nickel in the third period. Cobalt resembles iron and nickel in both its free and combined states, possessing similar tensile strength, machinability, thermal properties, and electrochemical behavior. Constituting 0.0029% of the Earth's crust, cobalt is widely distributed in nature, occurring in meteorites, stars, lunar rocks, seawater, fresh water, soils, plants, and animals. *See* PERIODIC TABLE; TRANSITION ELEMENTS.

Cobalt and its alloys resist wear and corrosion even at high temperatures. The most important commercial uses are in making alloys for heavy-wear, high-temperature, and magnetic applications. Small amounts of the element are required by plants and animals. The artificially produced radioactive isotope of cobalt, $^{60}Co$, has many medical and industrial applications.

Cobalt, with a melting point of 1495°C (2723°F) and a boiling point of 3100°C (5612°F), has a density (20°C; 68°F) of 8.90 $g{\cdot}cm^{-3}$, an electrical resistivity (20°C) of 6.24 microhm·cm, and a hardness (diamond pyramid, Vickers; 20°C) of 225. It is harder than iron and, although brittle, it can be machined. The latent heat of fusion is 259.4 joules/g, and the latent heat of vaporization is 6276 J/g; the specific heat (15–100°C; 59–212°F) is 0.442 J/g · °C. Cobalt is ferromagnetic, with the very high Curie temperature of 1121°C (2050°F). The electronic configuration is $1s^2 2s^2 2p^6 3s^2 3p^6 3d^7 4s^2$. At normal temperatures the stable crystal form of cobalt is hexagonal close-packed, but above 417°C (783°F) face-centered cubic is the stable structure. Although the finely divided metal is pyrophoric in air, cobalt is relatively unreactive and stable to oxygen in the air, unless heated. It is attacked by sulfuric, hydrochloric, and nitric acids, and more slowly by hydrofluoric and phosphoric acids, ammonium hydroxide, and sodium hydroxide. Cobalt reacts when heated with the halogens and other nonmetals such as boron, carbon, phosphorus, arsenic, antimony, and sulfur. Dinitrogen, superoxo, peroxo, and mixed hydride complexes also exist. In its compounds, cobalt exhibits all the oxidation states from −I to IV, the most common being II and III. The highest oxidation state is found in cesium hexafluorocobaltate(IV), $Cs_2CoF_6$, and a few other compounds.

There are over 200 ores known to contain cobalt; traces of the metal are found in many ores of iron, nickel, copper, silver, manganese, and zinc. However, the commercially important cobalt minerals are the arsenides, oxides, and sulfides. Zaire is the chief producer, followed by Zambia. Russia, Canada, Cuba, Australia, and New Caledonia produce most of the rest. Zaire and Zambia together account for just over 50% of the world's cobalt reserves. Nickel-containing laterites (hydrated iron oxides) found in the soils of the Celebes, Cuba, New Caledonia, and many other tropical areas are being developed as sources of cobalt. The manganese nodules found on the ocean floor are another large potential reserve of cobalt. They are estimated to contain at least 400 times as much cobalt as land-based deposits.

Since cobalt production is usually subsidiary to that of copper, nickel, or lead, extraction procedures vary according to which of these metals is associated with the cobalt. In general, the ore is roasted to remove stony gangue material as a slag, leaving a speiss of mixed metal and oxides, which is then reduced electrolytically, reduced thermally with aluminum, or leached with sulfuric acid to dissolve iron, cobalt, and nickel, leaving metallic copper behind. Lime is used to precipitate iron, and sodium hypochlorite is used to precipitate cobalt as the hydroxide. The cobalt hydroxide can be heated to give the oxide, which in turn is reduced to the metal by heating with charcoal.

Cobalt ores have long been used to produce a blue color in pottery, glass, enamels, and glazes. Cobalt is contained in Egyptian pottery dated as early as 2600 B.C. and in the blue and white porcelain ware of the Ming Dynasty in China (1368–1644).

An important modern industrial use involves the addition of small quantities of cobalt oxide during manufacture of ceramic materials to achieve a white color. The cobalt oxide counteracts yellow tints resulting from iron impurities. Cobalt oxide is also used in enamel coatings on steel to improve the adherence of the enamel to the metal. Cobalt arsenates, phosphates, and aluminates are used in artists' pigments, and various cobalt compounds are used in inks for full-color jet printing and in reactive dyes for cotton. Cobalt blue (Thenard's blue), one of the most durable of all blue pigments, is essentially cobalt aluminate. Cobalt linoleates, naphthenates, oleates, and ethylhexoates are used to speed up the drying of paints, lacquers, varnishes, and inks by promoting oxidation. In all, about a third of the world's cobalt production is used to make chemicals for the ceramic and paint industries. *See* DYE.

Cobalt catalysts are used throughout the chemical industry for various processes. These include hydrogenations and dehydrogenations, halogenations, aminations, polymerizations (for example, butadiene), oxidation of xylenes to toluic acid, production of hydrogen sulfide and carbon disulfide, carbonylation of methanol to acetic acid, olefin synthesis, denitrogenation and desulfurization of coal tars, reductions with borohydrides, and nitrile syntheses, and such important reactions as the Fisher-Tropsch method for synthesizing liquid fuels and the hydroformylation process. Cobalt catalysts have also been used in the oxidation of poisonous hydrogen cyanide in gas masks and in the oxidation of carbon monoxide in automobile exhausts.

Although cobalt was not used in its metallic state until the twentieth century, the principal use of cobalt is as a metal in the production of alloys, chiefly high-temperature and magnetic types. Superalloys needed to stand high stress at high temperatures, as in jet engines and gas turbines, typically contain 20–65% cobalt along with nickel, chromium, molybdenum, tungsten, and other elements.

In parts of the world where soil and plants are deficient in cobalt, trace amounts of cobalt salts [for example, the chloride and nitrate of Co(II)] are added to livestock feeds and fertilizers to prevent serious wasting diseases of cattle and sheep, such as pining, a debilitating disease especially common in sheep. Symptoms of cobalt deprivation in animals include retarded growth, anemia, loss of appetite, and decreased lactation.

The principal biological role of cobalt involves corrin compounds (porphyrin-like macrocycles). The active forms contain an alkyl group (5′-deoxyadenosine or methyl) attached to the cobalt as well as four nitrogens from the corrin and a nitrogen from a heterocycle, usually 5,6-dimethylbenzimidazole. These active forms act in concert with enzymes to catalyze essential reactions in humans. However, the corrin compounds are not synthesized in the body; they must be ingested in very small quantities. Vitamin $B_{12}$, with cyanide in place of the alkyl, prevents pernicious anemia but is itself inactive. The body metabolizes the vitamin into the active forms. Although the cobalt in corrins is usually Co(III), both Co(II) and Co(I) are involved in enzymic processes. Roughly one-third of all enzymes are metalloenzymes. Cobalt(II) substitutes for zinc in many of these to yield active forms. Such substitution of zinc may account, in part, for the toxicity of cobalt. *See* ENZYME. [L.Ma.; P.A.Mar.]

**Coenzyme** An organic cofactor or prosthetic group (nonprotein portion of the enzyme) whose presence is required for the activity of many enzymes. The prosthetic groups attached to the protein of the enzyme (the apoenzyme) may be regarded as dissociable portions of conjugated proteins. Neither the apoenzyme nor the coenzyme moieties can function singly. In general, the coenzymes function as acceptors of electrons or functional groupings, such as the carboxyl groups in $\alpha$-keto acids, which are removed from the substrate. *See* PROTEIN.

Well-known coenzymes include the pyridine nucleotides, nicotinamide adenine dinucleotide (NAD) and nicotinamide adenine dinucleotide phosphate (NADP); thiamine pyrophosphate (TPP); flavin mononucleotide (FMN) and flavinadenine dinucleotide (FAD); iron protoporphyrin (hemin); uridine diphosphate (UDP) and UDP-glucose; and adenosine triphosphate (ATP), adenosine diphosphate (ADP), and adenosine monophosphate (AMP). Coenzyme A (CoA), a coenzyme in certain condensing enzymes, acts in acetyl or other acyl group transfer and in fatty acid synthesis and oxidation. Folic acid coenzymes are involved in the metabolism of one carbon unit. Biotin is the coenzyme in a number of carboxylation reactions, where it functions as the actual carrier of carbon dioxide. *See* ADENOSINE TRIPHOSPHATE (ATP); ENZYME; NICOTINAMIDE ADENINE DINUCLEOTIDE PHOSPHATE (NADP). [M.B.McC.]

**Coke** A coherent, cellular, carbonaceous residue remaining from the dry (destructive) distillation of a coking coal. It contains carbon as its principal constituent, together with mineral matter and residual volatile matter. The residue obtained from the carbonization of a noncoking coal, such as subbituminous coal, lignite, or anthracite, is normally called a char. Coke is produced chiefly in chemical-recovery coke ovens, but a small amount is also produced in beehive or other types of nonrecovery ovens.

Coke is used predominantly as a fuel reductant in the blast furnace, in which it also serves to support the burden. As the fuel, it supplies the heat as well as the gases required for the reduction of the iron ore. It also finds use in other reduction processes, the foundry cupola, and house heating.

Coke is formed when coal is heated in the absence of air. During the heating in the range of 660–930°F (350–500°C), the coal softens and then fuses into a solid mass. The degree of softening attained during heating determines to a large extent the character of the coke produced. In order to produce coke having desired properties, two or more coals are blended before charging into the coke oven. In addition to the types of coals blended, the carbonizing conditions in the coke oven influence the characteristics of the coke produced. Oven temperature is the most important of these and has a significant effect on the size and the strength of the coke. In general, for a given coal, the size and shatter strength of the coke increase with decrease in carbonization temperature.

The important properties of coke that are of concern in metallurgical operations are its chemical composition, such as moisture, volatile-matter, ash, and sulfur contents, and its physical character, such as size, strength, and density. The moisture and the volatile-matter contents are a function of manner of oven operation and quenching, whereas ash and sulfur contents depend upon the composition of the coal charged. *See* CHARCOAL; DESTRUCTIVE DISTILLATION. [M.P.]

**Coking (petroleum)** In the petroleum industry, a process for converting nondistillable fractions (residua) of crude oil to lower-boiling-point products and coke. Coking is often used in preference to catalytic cracking because of the presence of metals and nitrogen components that poison catalysts. *See* CRACKING.

The liquid products from the coker, after cleanup via commercially available hydrodesulfurization technology, can provide large quantities of low-sulfur liquid fuels (less than 0.2 wt% sulfur). Another major application for the processes is upgrading heavy low-value crude oils into lighter products.

Petroleum coke is used principally as a fuel or, after calcining, for carbon electrodes. The feedstock from which the coke is produced controls the coke properties, especially in terms of sulfur, nitrogen, and metals content. A concentration effect tends to deposit the majority of the sulfur, nitrogen, and metals in the coke. Cokes exceeding around 2.5% sulfur content and 200 parts per million vanadium are mainly used, environmental regulations permitting, for fuel or fuel additives. The properties of coke for nonfuel use include low sulfur, metals, and ash as well as a definable physical structure. [J.G.S.]

**Colloid** A state of matter characterized by large specific surface areas, that is, large surfaces per unit volume or unit mass. The term colloid refers to any matter, regardless of chemical composition, structure (crystalline or amorphous), geometric form, or degree of condensation (solid, liquid, or gas), as long as at least one of the dimensions is less than approximately 1 micrometer but larger than about 1 nanometer. Thus, it is possible to distinguish films (for example, oil slick), fibers (spider web), or colloidal particles (fog) if one, two, or three dimensions, respectively, are within the submicrometer range.

A colloid consists of dispersed matter in a given medium. In the case of finely subdivided particles, classification of a number of systems is possible, as given in the table. In addition to the colloids listed in the table, there are systems that do not fit into any of the listed categories. Among these are gels, which consist of a network-type internal structure loaded with larger or smaller amounts of fluid. Some gels may have the consistency of a solid, while others are truly elastic bodies that can reversibly deform. Another colloid system that may occur is termed coacervate, and is identified as a liquid phase separated on coagulation of hydrophilic colloids, such as proteins. *See* GEL.

**Types of colloid dispersions**

| Medium | Dispersed matter | Technical name | Examples |
|---|---|---|---|
| Gas | Liquid | Aerosol | Fog, sprays |
| | Solid | Aerosol | Smoke, atmospheric or interstellar dust |
| Liquid | Gas | Foam | Head on beer, lather |
| | Liquid | Emulsion | Milk, cosmetic lotions |
| | Solid | Sol | Paints, muddy water |
| Solid | Gas | Solid foam | Foam rubber |
| | Liquid | Solid emulsion | Opal |
| | Solid | Solid sol | Steel |

It is customary to distinguish between hydrophobic and hydrophilic colloids. The former are assumed to be solvent-repellent, while the latter are solvent-attractant (dispersed matter is said to be solvated). In reality there are various degrees of hydrophilicity for which the degree of solvation cannot be determined quantitatively.

Certain properties of matter are greatly enhanced in the colloidal state due to the large specific surface area. Thus, finely dispersed particles are excellent adsorbents; that is, they can bind various molecules or ions on their surfaces. This property may be used for removal of toxic gases from the atmosphere (in gas masks), for elimination of soluble contaminants in purification of water, or decolorization of sugar, to give just a few examples. *See* ADSORPTION.

Colloids are too small to be seen by the naked eye or in optical microscopes. However, they can be observed and photographed in transmission or scanning electron microscopes. Owing to their small size, they cannot be separated from the medium (liquid or gas) by simple filtration or normal centrifugation. Special membranes with exceedingly small pores, known as ultrafilters, can be used for collection of such finely dispersed particles. The ultracentrifuge, which spins at very high velocities, can also be employed to promote colloid settling. *See* ULTRACENTRIFUGE; ULTRAFILTRATION.

Colloids show characteristic optical properties. They strongly scatter light, causing turbidity such as in fog, milk, or muddy water. Scattering of light (recognized by the Tyndall beam) can be used for the observation of tiny particles in the ultramicroscope. Colloidal state of silica is also responsible for iridescence, the beautiful effect observed with opals.

Since the characteristic dimensions of colloids fall between those of simple ions or molecules and those of coarse systems, there are in principle two sets of techniques available for their preparation: dispersion and condensation. In dispersion methods the starting materials consist of coarse units which are broken down into finely dispersed particles, drawn into fibers, or flattened into films. For example, colloid mills grind solids to colloid sizes, nebulizers can produce finely dispersed droplets from bulk liquids, and blenders are used to prepare emulsions from two immiscible liquids (such as oil and water). *See* EMULSION.

In condensation methods, ions or molecules are aggregated to give colloidal particles, fibers, or films. Thus, insoluble monolayer films can be developed by spreading onto the surface of water a long-chain fatty acid (for example, stearic acid) from a solution in an organic liquid (such as benzene or ethyl ether). Colloidal aggregates of detergents (micelles) form by dissolving the surface-active material in an aqueous solution in amounts that exceed the critical micelle concentration. *See* MICELLE; MONOMOLECULAR FILM.

The most common procedure to prepare sols is by homogeneous precipitation of electrolytes. Thus, if aqueous silver nitrate and potassium bromide solutions are mixed in proper concentrations, colloidal dispersions of silver bromide will form, which may remain stable for a long time. Major efforts have focused on preparation of monodispersed sols, which consist of colloidal particles that are uniform in size, shape, and composition. *See* PRECIPITATION (CHEMISTRY). [E.Ma.]

**Colloidal crystals** Periodic arrays of suspended colloidal particles. Common colloidal suspensions (colloids) such as milk, blood, or latex are polydisperse; that is, the suspended particles have a distribution of sizes and shapes. However, suspensions of particles of identical size, shape, and interaction, the so-called monodisperse colloids, do occur. In such suspensions, a new phenomenon that is not found in polydisperse systems, colloidal crystallization, appears: under appropriate conditions, the particles can spontaneously arrange themselves into spatially periodic structures. This ordering is analogous to that of identical atoms or molecules into periodic arrays to form atomic or molecular crystals. However, colloidal crystals are distinguished from molecular crystals, such as those formed by very large protein molecules, in that the individual particles do not have precisely identical internal atomic or molecular arrangements. On the other hand, they are distinguished from periodic stackings of macroscopic objects like cannonballs in that the periodic ordering is spontaneously adopted by the system through the thermal agitation (brownian motion) of the particles. These conditions limit the sizes of particles which can form colloidal crystals to the range from about 0.01 to about 5 micrometers.

The most spectacular evidence for colloidal crystallization is the existence of naturally occurring opals. The ideal opal structure is a periodic close-packed three-dimensional array of silica microspheres with hydrated silica filling the spaces not occupied by particles. Opals are the fossilized remains of an earlier colloidal crystal suspension. Another important class of naturally occurring colloidal crystals are found in concentrated suspensions of nearly spherical virus particles, such as *Tipula* iridescent virus and tomato bushy stunt virus. Colloidal crystals can also be made from the synthetic monodisperse colloids, suspensions of plastic (organic polymer) microspheres. Such suspensions have become important systems for the study of colloidal crystals, by virtue of the controllability of the particle size and interaction. *See* COLLOID. [N.A.C.]

**Colorimetry** Any technique by which an unknown color is evaluated in terms of known colors. Colorimetry may be visual, photoelectric, or indirect by means of spectrophotometry. These techniques are widely used in scientific studies involving the appearance of objects and lights, but are of greatest importance in the color specification of the raw materials and finished products of industry.

In visual colorimetry, the unknown color is presented beside a comparison field into which may be introduced any one of a range of known colors from which the operator chooses the one matching the unknown. To be generally applicable, the comparison field must not only cover a sufficient color range but must also be continuously adjustable in color.

In indirect colorimetry, the light leaving the unknown specimen is split into its component spectral parts by means of a prism or diffraction grating, and the amount of each component part is separately measured by a photometer. The quantity evaluated is spectral radiance of a light source, spectral transmittance of a filter (glass, plastic, gelatin, or liquid), or spectral reflectance of an opaque body.

In photoelectric colorimetry, the light leaving the specimen is measured separately by three photocells. The spectral sensitivity of these photocells is adjusted, usually by color filters, to conform as closely as possible to the three color-mixture functions for the average normal human eye (CIE standard observer). The responses of the photocells give directly the amounts of red, green, and blue primaries required to produce the color of the unknown specimen for the kind of vision represented by the three photocells.

If two objects have the same color because the light leaving one of them toward the eye is spectrally identical to that leaving the other, any type of colorimetry serves reliably to establish the fact of color match. If, however, the two lights are spectrally dissimilar, they may still color-match for any one observer; such pairs of lights are called metamers. Normal color vision differs sufficiently from person to person so that a metameric color match for one observer may be seriously mismatched for another. On this account, the question of color match of spectrally dissimilar lights can be reliably settled only by the indirect method which uses spectrophotometry combined with a precisely defined standard observer. [D.B.J./J.L.L.]

**Combinatorial chemistry** A method in which very large numbers of chemical entities are synthesized by condensing a small number of reagents together in all combinations defined by a small set of reactions. The main objective of combinatorial chemistry is synthesis of arrays of chemical or biological compounds called libraries. These libraries are screened to identify useful components, such as drug candidates. Synthesis and screening are often treated as separate tasks because they require different conditions, instrumentation, and scientific expertise. Synthesis involves the development of new chemical reactions to produce the compounds, while screening aims to identify the biological effect of these compounds, such as strong binding to proteins and other biomolecular targets.

Combinatorial chemistry is sometimes referred to as matrix chemistry. If a chemical synthesis consists of three steps, each employing one class of reagent to accomplish the conversion, then employing one type of each reagent class will yield $1 \times 1 \times 1 = 1$ product as the result of $1 + 1 + 1 = 3$ total reactions. Combining 10 types of each reagent class will yield $10 \times 10 \times 10 = 1000$ products as the result of as few as

10 + 10 + 10 = 30 total reactions; 100 types of each reagent will yield 1,000,000 products as the result of as few as 300 total reactions. While the concept is simple, considerable strategy is required to identify 1,000,000 products worth making and to carry out their synthesis in a manner that minimizes labor and maximizes the value of the resulting organized collection, called a chemical library.

The earliest work was motivated by a desire to discover novel ligands (that is, compounds that associate without the formation of covalent bonds) for biological macromolecules, such as proteins. Such ligands can be useful tools in understanding the structure and function of proteins; and if the ligand meets certain physiochemical constraints, it may be useful as a drug. For this reason, pharmaceutical applications provided early and strong motivation for the development of combinatorial chemistry. *See* LIGAND. [A.W.Cz.]

In combinatorial chemistry, attention has been focused on the problem of how to identify the set of molecules that possess a desired combination of properties. In a drug-discovery effort, the library members that strongly bind to a particular biological receptor are of interest. In a search for new materials that behave as superconductors at relatively high temperatures, the special combination of elements yielding the best electrical properties is a goal. In each case, the library might consist of up to a million members, while the subset of target molecules might consist of several thousand contenders or just a single highly selective binder. This subset could then be studied in more detail by conventional means.

Several emerging strategies promise to address this problem. In the first case, a library is constructed in a spatial array such that the chemical composition of each location in the array is noted during the construction. The binding molecules, usually labeled with a fluorescent tag, are exposed to the entire assay. The locations that light up can then be immediately identified from their spatial location. This approach is being actively developed for libraries of proteins and nucleotides. A problem is that the chemistry required to attach various molecules to the solid surface, usually silicon, is quite tricky and difficult to generalize. The assaying strategy is intertwined with the available procedures for synthesizing the libraries themselves.

A conceptually straightforward approach is to first synthesize the library by using polystyrene beads as the solid support. The product molecules are then stripped from the support and pooled together into a master solution. This complex mixture consisting of a potentially large selection of ligand molecules could then be exposed to an excess of a target receptor. The next step is to devise a method for identifying the ligand-receptor pairs that point to molecularly specific binding. One approach is to examine a part of the mixture en masse by using affinity capillary electrophoresis. With this technique, the migration times of the ligand-receptor pair are significantly longer than the unreactive ligands, and can be interrogated by electrospray mass spectrometry.

The mass spectrometric method often provides a direct structural identification of the ligand, either by determination of its molecular weight or by collision-induced dissociation experiments. In the latter case, the molecular ion is selected by a primary mass spectrometer and is driven into a region of high-pressure inert gas for fragmentation. The fragment ions are then used to reconstruct the original molecular structure. This direct approach to screening and assaying has the advantage that the screening is carried out in solution rather than on a solid support, and it avoids steric problems associated with resin-bound molecules. At present the approach seems limited to libraries of about 1000 compounds because of interference from unbound ligands, and limited by sensitivity issues. New strategies using mass spectrometry may eliminate this limit.

A different tack involves assaying the polystyrene beads one by one after the resin-bound molecules are exposed to a receptor. With this approach, active beads may be identified by color or by fluorescence associated with the receptor, and are subsequently indexed in standard 96-well titer plates. Identification is then possible by using a variety of spectroscopic techniques; at present, the most popular methods are electrospray mass spectrometry and matrix-assisted laser desorption ionization mass spectrometry. *See* MASS SPECTROMETRY. [N.W.]

Dynamic combinatorial chemistry integrates library synthesis and screening in one process, potentially accelerating the discovery of useful compounds. In the dynamic approach the libraries are not created as arrays of individual compounds, but are generated as mixtures of components, similar to natural pools of antibodies. One important requirement is that the mixture components exist in dynamic equilibrium with each other. According to basic laws of thermodynamics (Le Chatelier's principle), if one of the components ($A_i$) is removed from the equilibrated mixture, the system will respond by producing more of the removed component to maintain the equilibrium balance in the mixture. *See* CHEMICAL EQUILIBRIUM.

The dynamic mixture, as any other combinatorial library, is so designed that some of the components have potentially high affinity to a biomolecular target. These high-affinity (effective) components can form strong complexes with the target. If the target is added to the equilibrated mixture, when the effective components form complexes with the target they are removed from the equilibrium. This forces the system to make more of these components at the expense of other ones that bind to the target with less strength. As a result of such an equilibrium shift, the combinatorial library reorganizes to increase the amount of strong binders and decrease the amount of the weaker ones. This reorganization leads to enrichment of the library with the effective components and simplifies their identification. [A.V.E.]

**Combinatorial synthesis** A method for preparing a large number of chemical compounds, commonly known as a combinatorial library, which are then screened to identify compounds having a desired function, such as a particular biological or catalytic activity. Combinatorial synthesis is an aspect of combinatorial chemistry, which allows for the simultaneous generation and rapid testing for a desired property of large numbers of chemically related compounds. One could regard combinatorial chemistry as the scientist's attempt to mimic the natural principles of random mutation and selection of the fittest. Combinatorial chemistry has already become an invaluable tool in the areas of molecular recognition, materials science, drug discovery and optimization, and catalyst development. *See* COMBINATORIAL CHEMISTRY; ORGANIC SYNTHESIS.

Combinatorial synthesis was developed to prepare libraries of organic compounds containing from a few dozen to several million members simultaneously, in contrast to traditional organic synthesis where one target compound is prepared at a time in one reaction (see illustration). Thus, a target compound AB, for example, would be prepared

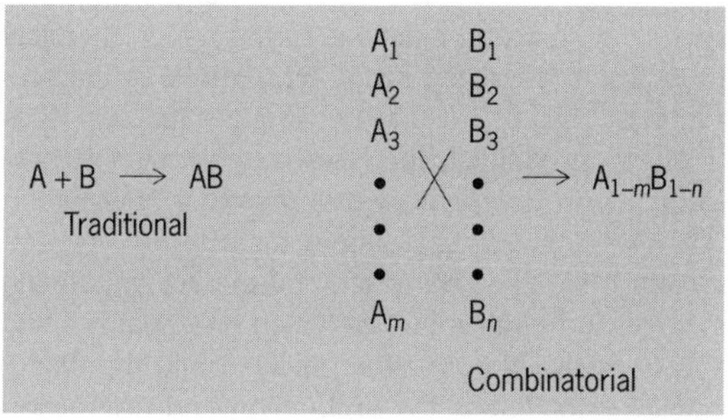

$A + B \longrightarrow AB$ Traditional

$A_1, A_2, A_3, \ldots, A_m \times B_1, B_2, B_3, \ldots, B_n \longrightarrow A_{1-m}B_{1-n}$ Combinatorial

**Traditional synthesis versus combinatorial synthesis.**

by coupling of the substrates A and B in a traditional (orthodox) synthesis and would be isolated after reaction processing (workup) and purification (such as crystallization, chromatography, or distillation). Combinatorial synthesis offers the potential to prepare every combination of substrates, type $A_{1-m}$ and type $B_{1-n}$, providing a set of compounds, $A_{(1-m)}B_{(1-n)}$. [Y.U.]

**Combining volumes, law of** The principle that when gases take part in chemical reactions the volumes of the reacting gases and those of the products, if gaseous, are in the ratio of small whole numbers, provided that all measurements are made at the same temperature and pressure. The law is illustrated by the following reactions:

1. One volume of chlorine and one volume of hydrogen combine to give two volumes of hydrogen chloride.
2. Two volumes of hydrogen and one volume of oxygen combine to give two volumes of steam.
3. One volume of ammonia and one volume of hydrogen chloride combine to give solid ammonium chloride.
4. One volume of oxygen when heated with solid carbon gives one volume of carbon dioxide.

The law of combining volumes is similar to the other gas laws in that it is strictly true only for an ideal gas, though most gases obey it closely at room temperatures and atmospheric pressure. Under high pressures used in many large-scale industrial operations, such as the manufacture of ammonia from hydrogen and nitrogen, the law ceases to be even approximately true. *See* GAS. [T.C.W.]

**Combustion** The burning of any substance, in gaseous, liquid, or solid form. In its broad definition, combustion includes fast exothermic chemical reactions, generally in the gas phase but not excluding the reaction of solid carbon with a gaseous oxidant. Flames represent combustion reactions that can propagate through space at subsonic velocity and are accompanied by the emission of light. The flame is the result of complex interactions of chemical and physical processes whose quantitative description must draw on a wide range of disciplines, such as chemistry, thermodynamics, fluid dynamics, and molecular physics. In the course of the chemical reaction, energy is released in the form of heat, and atoms and free radicals, all highly reactive intermediates of the combustion reactions, are generated. *See* FLAME; FREE RADICAL; REACTIVE INTERMEDIATES.

The physical processes involved in combustion are primarily transport processes: transport of mass and energy and, in systems with flow of the reactants, transport of momentum. The reactants in the chemical reaction are normally a fuel and an oxidant. In practical combustion systems the chemical reactions of the major chemical species, carbon and hydrogen in the fuel and oxygen in the air, are fast at the prevailing high temperatures (greater than 1200 K or 1700°F) because the reaction rates increase exponentially with temperature. In contrast, the rates of the transport processes exhibit much smaller dependence on temperature are, therefore, lower than those of the chemical reactions. Thus in most practical flames the rate of evolution of the main combustion products, carbon dioxide and water, and the accompanying heat release depends on the rates at which the reactants are mixed and heat is being transferred from the flame to the fresh fuel-oxidant mixture injected into the flame. However, this generalization cannot be extended to the production and destruction of minor species in the flame, including those of trace concentrations of air pollutants such as nitrogen oxides, polycyclic aromatic hydrocarbons, soot, carbon monoxide, and submicrometer-size inorganic particulate matter. *See* TRANSPORT PROCESSES.

Combustion applications are wide ranging with respect to the fields in which they are used and to their thermal input, extending from a few watts for a candle to hundreds of megawatts for a utility boiler. Combustion is the major mode of fuel utilization in domestic and industrial heating, in production of steam for industrial processes and for electric power generation, in waste incineration, and in propulsion in internal combustion engines, gas turbines, or rocket engines. [J.M.Be.]

**Computational chemistry** A branch of theoretical chemistry that uses a digital computer to model systems of chemical interest. In this discipline, the computer itself is the primary instrument of research. The use of computers for analysis of experimental data, and for the storage and display of results obtained with other tools, is distinct from computational chemistry. The latter permits calculation of quantities which can be measured experimentally, such as molecular geometries of ground and excited states, heats of formation, and ionization potentials. Alternatively, quantities not readily accessible by existing experimental techniques, such as geometries of transition states and detailed structure of liquids, may be evaluated.

Because of the increasing power and availability of computers, and the simultaneous development of well-tested and reliable theoretical methods,the use of computational chemistry as an adjunct to experimental research has increased rapidly. Calculations ranging from a few seconds to many hours of computer time can serve as a guide to exclude less favorable reactions or unstable products, or to select several more fruitful procedures from the many possible ones. In addition, modeling of chemical systems with a computer enables the researcher to examine them on a scale of space or time as yet unmeasurable by experimental techniques. This can give insight into a chemical system beyond that provided by experiment. Examples are examination of the dynamics of a chemical reaction or of detailed changes in conformation of a polymer in solution. Examination of the molecular orbitals occupied by the electrons of the molecule can provide insight into chemical bonding and the electronic interactions which determine specific geometric configurations. Thus computational chemistry can yield information which may not be experimentally available. *See* CHEMICAL DYNAMICS.

Computational chemistry may be the application of existing theory and numerical methods to new molecules, or it may be the development of new computational methods. The latter may include incorporating more physics into the mathematical model in order to provide a better theoretical description of the system being studied, for example, inclusion of interactions between individual electrons in molecular orbital calculations. These more complete studies, for "large" molecules, usually need to be performed on a supercomputer, or on a mid-size computer with an array processor. Another approach is to devise simpler methods which will approximate the accuracy of more complex calculations. These include semiempirical methods in which values of hard-to-calculate terms are derived from experiment. Such an approach allows fruitful work with a mid-size or even a desk-top computer, avoiding the need for expensive computer resources. [Z.R.W.]

**Concentration scales** Concentration is a very important property of mixtures, because it defines the quantitative relation of the components. In solutions the concentration is expressed as the mass, volume, or number of moles of solute present in proportion to the amount of solvent or of total solution.

The simplest scale to measure is percentage; hence it is often used for medicinal or household solutions. Weight percent is the number of parts of weight of solute per hundred parts of solution (total). For example, a 10% saline solution contains 10 g of salt in 90 g of water, that is, 100 g total weight. Gaseous mixtures, being difficult to weigh, are often expressed as volume percent. Thus, air is said to contain 78% nitrogen by volume. Solutions of liquids in liquids (say, alcohol in water) may also be expressed in volume percent.

To the chemist, the number of moles of solute is of more significance than the number of grams. The molarity (abbreviated *M*) is the number of moles of solute per liter of total solution. Thus, 12 *M* HCl means that the solution contains 12 formula weights ($12 \times 36.5$), or 438 g, of HCl/liter. (As used here, the mole is an amount of substance whose weight in grams is numerically the same as the molecular weight.)

Molality (abbreviated *m*) relates the number of moles of solute to the weight of solvent rather than to the volume of solution. This scale indicates the number of moles of solute/1000 g of solvent. Thus 34.2 g of sucrose ($C_{12}H_{22}O_{11}$, mol wt 342), if dissolved in 200 g of water, has the concentration of 0.5 mole of sucrose/1000 g of water, and hence is 0.5 *m*.

When it is important to know the reactive capacities of reagents, as in volumetric analysis, the normality scale is used. Normality (abbreviated *N*) is found by multiplying molarity by the number of active units in the formula.

In recent years many chemists have sought to avoid the molarity scale lest it imply that electrolyte solutes exist as molecules rather than ions. The formality scale (abbreviated *F*) represents formula weights per liter. Its values are identical with the molarities of un-ionized solutes.

Many properties of solutions (for example, vapor pressure of one component) are dependent on the ratio of the number of moles of solute to the number of moles of solvent, rather than on the ratios of respective volumes or masses. The mole fraction (abbreviated $N_A$ or $X_A$ for component *A*) is the ratio of the number of moles of solute to the total number of moles of all components. Thus for 16 g of methanol (0.5 mole) dissolved in 18 g of water (1 mole), the mole fraction of methanol is 0.5/1.5, or 1/3; the mole percent is 33.3. For gases the mole percent is identical with the volume percent. *See* GRAM-MOLECULAR WEIGHT; SOLUTION; TITRATION. [A.L.H.]

**Conformational analysis** The study of the energies and structures of conformations of organic molecules and their chemical and physical properties. Organic molecules are not static entities. The constituent atoms vibrate and groups rotate about the bond axes. *See* STEREOCHEMISTRY.

**Linear structures.** Rotation about the C-C bond in ethane, for example, results in an infinite number of slightly different structures called conformations; one of these is indicated below. In conformation (**1**) of ethane, the pairs of C-H bonds on the

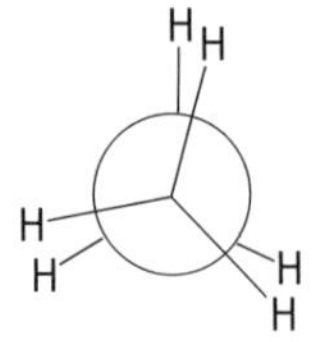

H H

H

H H

H

(1)

two carbon atoms reside in a plane, and are termed eclipsed. (The circle represents the front-most carbon atom, and the long bonds those from that carbon atom to the hydrogen atoms. The bonds from the circle represent those bonds from the rearward carbon to its hydrogen atoms.) Conformation (**1**) is called the eclipsed conformation. In conformation (**2**), the C-H bonds on one carbon reside between the C-H bonds on the other carbon. Conformation (**2**) is called the staggered conformation. An infinite number of conformations between (**1**) and (**2**) are possible; however, conformations (**1**) and (**2**) are of greatest interest because they are the maximum and minimum energy structures. In conformation (**1**), the electrons in the C-H bonds and the nuclear charges of the hydrogen atoms repel each other, resulting in a higher energy state. In conformation (**2**), the electrons and hydrogen atoms are at their greatest possible separation and the repulsion is at a minimum. The conversion of (**2**) to (**3**) by

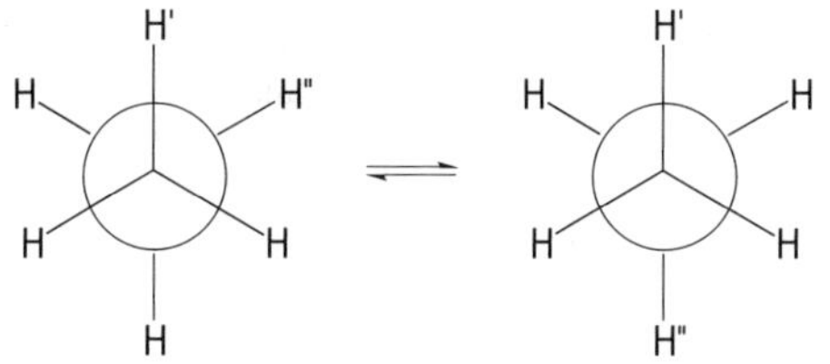

(**2**) (**3**)

rotation of one of the methyl groups by 120° requires the input of energy, approximately 3.0 kcal (13 kilojoules) per mole, in order to pass through the eclipsed conformation. This amount is small compared to the thermal energy at room temperature, and the rotation about the C-C bond in ethane occurs at about $10^9$–$10^{10}$ times per second.

Butane provides a more complicated case. There are two different eclipsed (**4** and **6**) and two staggered (**5** and **7**) conformations which are designated by the names below

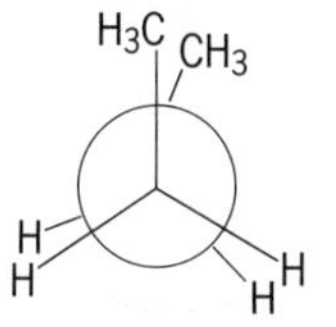

(**4**) synperiplanar

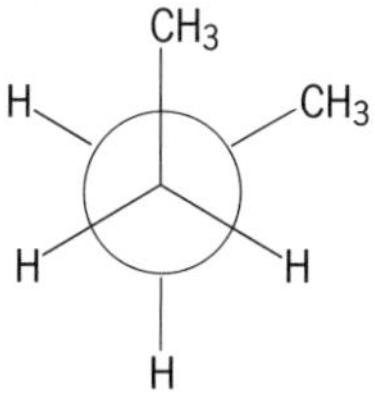

(**5**) synclinal

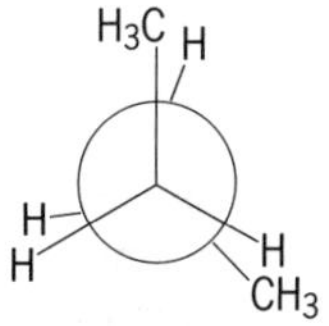

(**6**) anticlinal

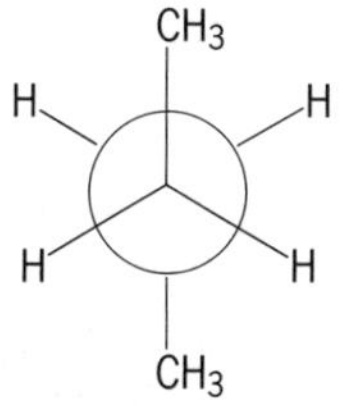

(**7**) antiperiplanar

the structures. As the methyl group is larger than a hydrogen atom, (**4**) is higher in energy than (**6**), and (**5**) is higher in energy than (**7**). As the molecular weight of an alkane increases, the number of molecular conformations (combinations of all possible individual bond conformations) increases dramatically, although the antiperiplanar conformations are always favored.

**Cyclic structures.** These also exist in various conformations. Cyclopropane, a planar structure, can exist in only one conformation. In cyclobutane, slight twisting about the C-C bonds can occur which relieves some of the C-H eclipsing strain energy. Cyclobutane exists in rapidly interconverting so-called butterfly conformations (**8**).

(8)

Cyclopentane exists in two types of nonplanar conformations: the envelope conformation (**9**) and the half-chair (or twist) conformation (**10**). The flap atom of (**9**), the

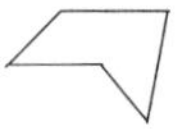

(9)  (10)

atom out of the plane of the other four, can migrate around the ring, as can also the twist in (**10**). These motions, called pseudorotation, require very little energy, and cyclopentane presents a very complex conformational system.

Cyclohexane exists predominantly in the chair conformation (**11**), in which all C-C bond conformations are of the staggered type. Interconversion between chair conformations occurs rapidly at room temperature, passing through the high-energy, boat conformation (**12**). In the chair conformation, there are two distinctly different types of

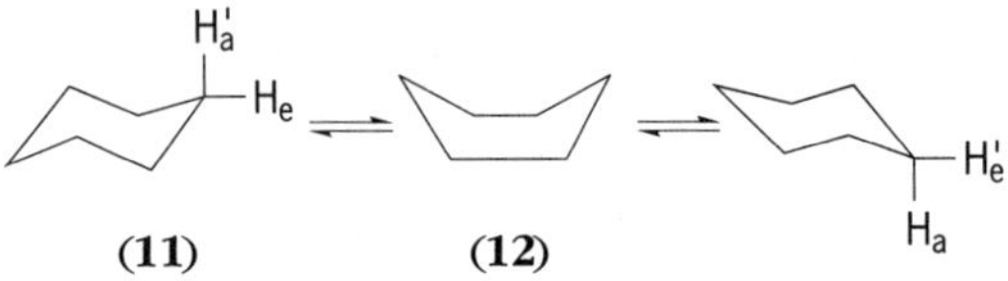

$H_a'$ $H_e$ $H_e'$ $H_a$

(11) (12)

hydrogen atoms; one set oriented perpendicular to the general plane of the ring, called axial (a) hydrogens, and one set oriented parallel to the plane of the ring, called equatorial (e) hydrogens. These hydrogens interchange orientation on chair interconversion. Axial and equatorial hydrogens possess different chemical and physical properties, although at room temperature the interconversion occurs very rapidly ($\sim 10^5$ per second), and it is in general not possible to detect the differences. It is easy to design molecules in which the interconversion is not possible and the differences in properties become readily apparent. [D.J.Pa.]

**Conjugation and hyperconjugation** A higher-order bonding interaction between electron orbitals on three or more contiguous atoms in a molecule, which leads to characteristic changes in physical properties and chemical reactivity. One participant in this interaction can be the electron pair in the $\pi$-orbital of a multiple (that is, double or triple) bond between two atoms, or a single electron or electron pair or electron vacancy on a single atom. The second component will be the pair of $\pi$-electrons in an adjacent multiple bond in the case of conjugation, and in the case of hyperconjugation it will be the pair of electrons in an adjacent polarized $\sigma$-bond (that is, a bond where the electrons are held closer to one atom than the other due to electronegativity differences between the two atoms). *See* CHEMICAL BONDING.

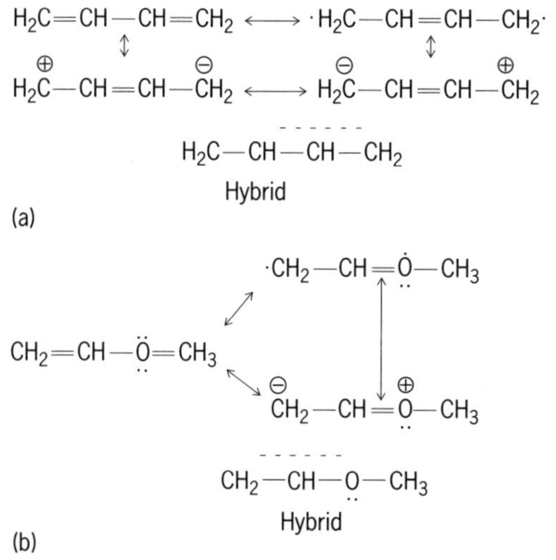

$$H_2C{=}CH{-}CH{=}CH_2 \longleftrightarrow \cdot H_2C{-}CH{=}CH{-}CH_2\cdot$$

$$\overset{\oplus}{H_2C}{-}CH{=}CH{-}\overset{\ominus}{CH_2} \longleftrightarrow \overset{\ominus}{H_2C}{-}CH{=}CH{-}\overset{\oplus}{CH_2}$$

$$H_2C{-}\overline{CH{-}CH}{-}CH_2$$

Hybrid

(a)

$$\cdot CH_2{-}CH{=}\dot{\ddot{O}}{-}CH_3$$

$$CH_2{=}CH{-}\ddot{\underset{..}{O}}{=}CH_3$$

$$\overset{\ominus}{CH_2}{-}CH{=}\overset{\oplus}{\underset{..}{O}}{-}CH_3$$

$$\overline{CH_2{-}CH{-}O}{-}CH_3$$

Hybrid

(b)

**Fig. 1. Conjugated molecules. Broken overbars indicate the effects of conjugation. (*a*) 1,3-Butadiene. (*b*) Methyl vinyl ether.**

The conjugated orbitals reside on atoms that are separated by a single bond in the classical valence-bond molecular model, and the conjugation effect is at a maximum when the axes of the component orbitals are aligned in a parallel fashion because this allows maximum orbital overlap. Conjugation thus has a stereoelectronic requirement, or a restriction on how the participating orbitals must be oriented with respect to each other. Two simple examples are shown in Fig. 1; in 1,3-butadiene ($H_2C{=}H\underline{C{-}C}H{=}CH_2$) conjugation occurs between the *p* orbitals ($\pi$-bonds) of the two double bonds, and in methyl vinyl ether ($H_2C{=}H\underline{C{-}O}{-}CH_3$) the nonbonding $sp^3$ orbital on oxygen is conjugated with the *p* orbitals of the double bond. This interaction is manifest in an effective bond order between single and double for the underlined single bond.

Hyperconjugation, the conjugation of polarized $\sigma$-bonds with adjacent $\pi$-orbitals, was introduced in the late 1930s by R. S. Mulliken. This rationale was used to explain successfully a wide variety of chemical phenomena; however, the confusing adaptation of the valence-bond model necessary to depict it and its inappropriate extension to some phenomena led to difficulties. The advent of the molecular orbital treatment has eliminated many of these difficulties.

Early on, hyperconjugation was used to explain the stabilization by alkyl groups of carbocations, or positively charged trivalent carbon. Figure 2 shows how the orbitals of the $\sigma$-bonds of a methyl group ($H_3C$) exert a hyperconjugative stabilizing effect upon a neighboring *p* orbital on a methylene group ($CH_2$). The bonding molecular orbitals of the methyl group utilize aspects of its carbon *p* orbitals; in both of the conformations or rotations the methyl group has such an orbital which overlaps in phase with the neighboring *p* orbital on $CH_2$. As hydrogen is an electropositive atom, the methyl group acts as an electron donor to stabilize an empty *p* orbital and to destabilize a filled *p* orbital on the right-hand carbon. This is in accord with empirical observations. If $H^1$ is replaced with some other atom X, the hyperconjugative effect of X will be at a maximum in conformation A, where there is orbital density on X, and it will be at a minimum in conformation B, where X is in the nodal or null plane. If X is a more electropositive atom such as silicon, the C-X bond is a donor bond and will stabilize an adjacent empty *p* orbital, such as in a cation. If X is a more electronegative element such as fluorine, the C-X bond is an acceptor bond and will stabilize an adjacent filled *p*

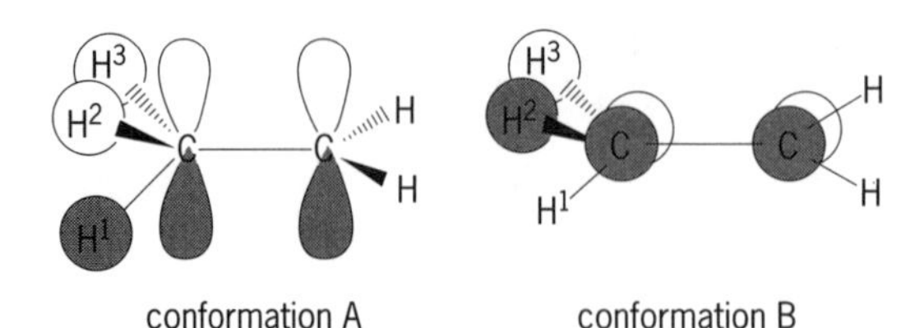


**Fig. 2. Diagram showing hyperconjugative effect of orbitals of the $\sigma$-bonds of a methyl group ($H_3C$) on a neighboring *p* orbital of a methylene group ($CH_2$). Algebraic phrase is indicated by the presence or absence of tint.**

orbital such as in an anion; this result accords with experiment as well. The donor and acceptor effects are also important where the single *p* orbital is replaced by a multiple bond. *See* MOLECULAR ORBITAL THEORY; REACTIVE INTERMEDIATES.

A number of scientific phenomena depend on the properties of conjugated systems; these include vision (the highly tuned photoreceptors are triggered by molecules with extended conjugation), electrical conduction (organic semiconductors such as polyacetylenes are extended conjugated systems), color (most dyes are conjugated molecules designed to absorb particular wavelengths of light), and medicine (a number of antibiotics and cancer chemotherapy agents contain conjugated systems which trap enzyme sulfhydryl groups by conjugate addition). *See* VALENCE. [M.F.S.]

**Coordination chemistry** A field which, in its broadest usage, is acid-base chemistry as defined by G. N. Lewis. However, the term coordination chemistry is generally used to describe the chemistry of metals and metal ions in their interactions with other molecules or ions. For example, reactions (1)–(3) show acid-base-type reactions;

$$Mg^{2} + 6H_2O \rightarrow Mg(H_2O)_6{}^{2+} \quad (1)$$

$$Ni + 4CO \rightarrow Ni(CO)_4 \quad (2)$$

$$Fe^{2+} + 6CN^- \rightarrow Fe(CN)_6{}^{4-} \quad (3)$$

the products formed are coordination ions or compounds, and this area of chemistry is known as coordination chemistry.

Thus, it follows that coordination compounds are compounds that contain a central atom or ion and a group of ions or molecules surrounding it. Such a compound tends to retain its identity, even in solution, although partial dissociation may occur. The charge on the coordinated species may be positive, zero, or negative, depending on the charges carried by the central atom and the coordinated groups. These groups are called ligands, and the total number of attachments to the central atom is called the coordination number. Other names commonly used for these compounds include complex compounds, complex ions, Werner complexes, coordinated complexes, chelate compounds, or simply complexes. *See* ACID AND BASE; CHELATION.

Experimental observations as early as the middle of the 18th century reported the isolation of coordination compounds, but valence theory could not adequately account for such materials. The correct interpretation of these compounds was given by Alfred Werner in 1893. He introduced the concept of residual or secondary valence, and suggested that elements have this type of valence in addition to their normal or primary valence. Thus, platinum(IV) has a normal valence of 4 but a secondary valence or coordination number of 6. This then led to the formulaton of $PtCl_4 \cdot 6NH_3$ as $[Pt(NH_3)_6]^{4+}$, $4Cl^-$ and of $PtCl_4 \cdot 5NH_3$ as $[Pt(NH_3)_5Cl]^{3+}$, $3Cl^-$. The compound with five ammonias has only three ionic chlorides,the fourth is inside the coordination sphere, and therefore is not readily precipitated upon the addition of silver ion. Although the exact nature of the coordinate bond between metal and ligand remains the subject of considerable discussion, it is agreed that the formulations of Werner are essentially correct.

Three theories have been used to explain the nature of the coordinate bond. These are the valence bond theory, the electrostatic theory, including crystal field corrections, and the molecular orbital theory. Currently, the theory used almost exclusively is the molecular orbital theory. The valence bond theory for metal complexes considers that the pair of electrons on the ligand enter the hybridized atomic orbitals of the metal and that the bond is either essentially covalent or essentially ionic. Several of the properties of these substances can be explained on the basis of this theory. The electrostatic theory, plus the crystal field theory for the transition metals, assumes that the metal-ligand bond is caused by electrostatic interactions between point charges and dipoles and that there is no sharing of electrons. In addition to explaining the structure and magnetic properties, the crystal field theory affords an adequate interpretation of the visible spectra of metal complexes. The molecular orbital theory assumes that the electrons move in molecular orbitals which extend over all the nuclei of the metalligand system. In this manner, it serves to make use of both the valence bond theory and crystal field theory. The molecular orbital theory is therefore the best approximation to the nature of the coordinate bond because it is sufficiently flexible to permit both covalent and ionic bonding as well as the splitting of *d* orbitals into various energy levels. *See* MOLECULAR ORBITAL THEORY.

The stability of metal complexes depends both on the metal ion and the ligand. In general the stability of metal complexes increases if the central ion increases in charge, decreases in size, and increases in electron affinity. Several characteristics of the ligand are known to influence the stability of complexes: (1) basicity of the ligand, (2) the number of metal-chelate rings per ligand, (3) the size of the chelate ring, (4) steric effects, (5) resonance effects, and (6) the ligand atom. Since coordination compounds are formed as a result of acid-base reactions where the metal ion is the acid and the ligand is the base, it follows that generally the more basic ligand will tend to form the more stable complex. The size of the chelate ring is likewise an important factor. For saturated ligands such as ethylenediamine, five-membered rings are the most stable for chelates containing one or more double bonds.

Steric factors often have a very large effect on the stability of metal complexes. This is most frequently observed with ligands having a large group attached to the ligand atom or near it. Thus complexes, of the type shown in the illustration, with alkyl groups R in the position designated are much less stable than the parent complex where R = H. This results from the steric strain introduced by the size of the alkyl group on or adjacent to the ligand atom. In contrast to this, alkyl substitution at any other position results in the formation of more stable complexes because the ligand becomes more basic, and the bulky group is now removed from a position near the coordination site.

Finally, the ligand atom itself plays a significant role in controlling the stability of metal complexes. For most of the metal ions, the smallest ligand atom with the largest electron density will form the most stable complex.

Often the most stable complex is also the least reactive or most inert. Several factors, such as the electronic configuration of the central metal ion, its coordination number,

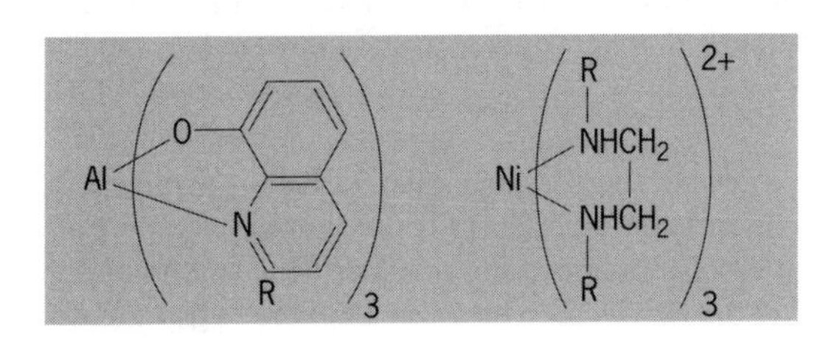


**Structural formulas of metal complexes which are affected by steric factors.**

and the extent of chelation, all have a marked effect on the rate of reaction of a given compound. *See* CHEMICAL BONDING; MAGNETOCHEMISTRY; SOLID-STATE CHEMISTRY; STEREOCHEMISTRY. [F.Ba.]

**Coordination complexes** A group of chemical compounds in which a part of the molecular bonding is of the coordinate covalent type. For a discussion of the nature of the coordinate bond, and the stability and reactivity of complex compounds *see* CHELATION; COORDINATION CHEMISTRY.

Coordination complexes contain a central atom or ion and a group of ions or molecules surrounding it. Many simple hydrates, such as $MgCl_2 \cdot 6H_2O$, are best formulated as $[Mg(H_2O)_6]Cl_2$ because it is known that the 6 molecules of water surround the central magnesium ion. Therefore, $[Mg(H_2O)_6]^{2+}$ is a complex ion, and $[Mg(H_2O)_6]Cl_2$ is a complex compound. The charge on this complex ion is +2, because this is the charge on the magnesium ion and the coordinated water molecules are neutral. However, if the coordinated groups are charged, then the charge on the complex is represented by the sum of the charge on the metal and that of the coordinated ions. See, for example, the progression of charges on the platinum(IV) complexes listed below.

| | |
|---|---|
| $[Pt(NH_3)_6]Cl_4$ | Hexaammineplatinum(IV) chloride |
| $[Pt(NH_3)_5Cl]Cl_3$ | Chloropentaammineplatinum(IV) chloride |
| $[Pt(NH_3)_4Cl_2]Cl_2$ | Dichlorotetraammineplatinum(IV) chloride |
| $[Pt(NH_3)_3Cl_3]Cl$ | Trichlorotriammineplatinum(IV) chloride |
| $[Pt(NH_3)_2Cl_4]$ | Tetrachlorodiammineplatinum(IV) |
| $K[Pt(NH_3)Cl_5)]$ | Potassium pentachloroammineplatinate(IV) |
| $K_2[PtCl_6]$ | Potassium hexachloroplatinate(IV) |

Thus the charge is +4 for $[Pt(NH_3)_6]^{4+}$ because Pt is +4 and $NH_3$ is neutral. But the charge is −2 for $[PtCl_6]^{2-}$ because of the 6 for $Cl^-$, that is, $+4 - 6 = -2$.

Metal complexes exhibit various types of isomerism. In many ways, inorganic stereochemistry is similar to that observed with organic compounds. Geometrical isomers are common among the inert complexes of coordination numbers 4 and 6. *See* STEREOCHEMISTRY.

The synthesis of metal complexes containing only one kind of ligand generally involves just the reaction of the metal salt in aqueous solution with an excess of the ligand reagent, as shown below.

$$[Ni(H_2O)_6](NO_3)_2 + 6NH_3 \rightarrow [Ni(NH_3)_6](NO_3)_2 + 6H_2O$$

The desired complex salt can then be isolated by removal of water until it crystallizes, or by addition of a water-miscible organic solvent to cause it to separate. For the inert complexes (those slow to react), prolonged treatment at more drastic conditions is often necessary. The preparation of geometrical isomers is much more difficult, and in most cases, the approach used is rather empirical. Generally, reactions yield a mixture of cis-trans products, and these are separated on the basis of their differences in solubility. *See* AMMINE; HYDRATE. [F.Ba.]

**Copolymer** A macromolecule in which two or more different species of monomer are incorporated into a polymer chain. The properties of copolymer depend on both the nature of the monomers and their distribution in the polymer. Thus, monomers A and B can polymerize randomly to form ABBAABA; they can alternate to give ABAB;

they can form blocks AAABBB; or one monomer can be grafted onto a polymer of the other:

AAAA
B
B

Copolymers can be prepared by all the known methods of polymerization: addition polymerization of vinyl monomers (by free-radical, anionic, cationic, or coordination catalysis), ring-opening polymerization, or condensation polymerization. *See* POLYMERIZATION.

In the polymerization of two monomers, $M_1$ and $M_2$, the monomers can add to a growing chain ending in either monomer, designated as $M_1^*$ and $M_2^*$. To a first approximation, only the terminal group of the growing chain is important, so two reactivity ratios can be defined: $r_1$ is the relative reactivity of chain end $M_1^*$ to monomers $M_1$ and $M_2$, and $r_2$ is the relative reactivity of chain end $M_2^*$ to monomers $M_2$ and $M_1$. The reactivity ratios can be determined experimentally by analyzing polymer compositions at different monomer feeds. *See* ACRYLONITRILE; POLYACRYLONITRILE RESINS; POLYMER.

[D.S.Br.]

**Copper** A chemical element, Cu, atomic number 29, atomic weight 63.546. Copper, a nonferrous metal, is the twentieth most abundant element present in the Earth's crust, at an average level of 68 parts per million (0.22 lb/ton or 0.11 kg/metric ton). Copper metal and copper alloys have considerable technological importance due to their combined electrical, mechanical, and physical properties. The discoveries that mixed-valence Cu(II)/Cu(III) oxides exhibit superconductivity (zero electrical resistance) at temperatures as high as 125 K (−234°F; liquid nitrogen, a cheap coolant, boils at 90 K or −297°F) have generated intense international competition to understand these new materials and to develop technological applications. Although some pure copper metal is present in nature, commercial copper is obtained by reduction of the copper compounds in ores followed by electrolytic refining. The rich chemistry of copper is restricted mostly to the valence states Cu(I) and Cu(II); compounds containing Cu(0), Cu(III), and Cu(IV) are uncommon. Soluble copper salts are potent bacteriocides and algicides at low levels and toxic to humans in large doses. Yet copper is an essential trace element that is present in various metalloproteins required for the survival of plants and animals.

Copper is located in the periodic table between nickel and zinc in the first row of transition elements and in the same subgroup as the other so-called coinage metals, silver and gold. The electronic configuration of elemental copper is $[1s^2 2s^2 2p^6 3s^2]\ 3d^{10} 4s^1$ or $[\text{argon}]3d^{10}4s^1$. At first glance, the sole 4*s* electron might suggest chemical similarity to potassium, which has the $[\text{argon}]4s^1$ configuration. However, metallic copper, in sharp contrast to metallic potassium, is relatively unreactive. The higher nuclear charge of copper relative to that of potassium is not fully shielded by the 10 additional *d* electrons, with the result that the copper 4*s* electron has a higher ionization potential than that of potassium (745.5 versus 418.9 kilojoules/mole, respectively). Moreover, the second and third ionization potentials of copper (1958.1 and 3554 J/mole, respectively) are considerably lower than those of potassium, and account for the higher valence-state accessibility associated with transition-metal chemistry as opposed to alkali-metal chemistry. *See* ELECTRON CONFIGURATION; PERIODIC TABLE; TRANSITION ELEMENTS; VALENCE.

Copper is a comparatively heavy metal. The density of the pure solid is 8.96 g/cm$^3$ (5.18 oz/in.$^3$) at 20°C (68°F). The density of commercial copper varies with method of manufacture, averaging 8.90–8.92 g/cm$^3$ (5.14–5.16 oz/in.$^3$) in cast refinery shapes, 8.93 g/cm$^3$ (5.16 oz/in.$^3$) for annealed tough-pitch copper, and 8.94 g/cm$^3$ (517 oz/in.$^3$) for oxygen-free copper. The density of liquid copper is 8.22 g/cm$^3$ (4.75 oz/in.$^3$) near the freezing point.

The melting point of copper is 1083.0 ∓ 0.1°C (1981.4 ∓ 0.2°F). Its normal boiling point is 2595°C (4703°F).

The coefficient of linear expansion of copper is $1.65 \times 10^{-5}$/°C at 20°C.

The specific heat of the solid is 0.092 cal/g at 20°C (68°F). The specific heat of liquid copper is 0.112 cal/ g, and of copper in the vapor state about 0.08 cal/g.

The electrical resistivity of copper in the usual volumetric unit, that of a cube measuring 1 cm in each direction, is $1.6730 \times 10^{-6}$ ohm · cm at 20°C (68°F). Only silver has a greater volumetric conductivity than copper. On a relative basis in which silver is rated 100, copper is 94, aluminum 57, and iron 16.

The mass resistivity of pure copper for a length of 1 m weighing 1 g at 20°C (68°F) is 0.14983 ohm. The conductivity of copper on the mass basis is surpassed by several light metals, notably aluminum. The relative values are 100 for aluminum, 50 for copper, and 44 for silver.

By far the largest use of copper is in the electrical industry, and therefore high electrical conductivity is its most important single property, although for industrial use this property must be accompanied by suitable characteristics in other respects. *See* CONDUCTOR (ELECTRICITY).

Copper-containing proteins provide diverse biochemical functions, including copper uptake and transport (ceruloplasmin), copper storage (metallothionen), protective roles (superoxide dismutase), catalysis of substrate oxygenation (dopamine $\beta$-monooxygenase), biosynthesis of connective tissue (lysyl oxidase), terminal oxidases for oxygen metabolism (cytochrome *c* oxidase), oxygen transport (hemocyanin), and electron transfer in photosynthetic pathways (plastocyanin). *See* ENZYME. [H.Sc.]

**Cotton effect** The characteristic wavelength dependence of the optical rotatory dispersion curve or the circular dichroism curve or both in the vicinity of an absorption band.

When an initially plane-polarized light wave traverses an optically active medium, two principal effects are manifested: a change from planar to elliptic polarization, and a rotation of the major axis of the ellipse through an angle relative to the initial direction of polarization. Both effects are wavelength dependent. The first effect is known as circular dichroism, and a plot of its wavelength (or frequency) dependence is referred to as a circular dichroism (CD) curve. The second effect is called optical rotation and, when plotted as a function of wavelength, is known as an optical rotatory dispersion (ORD) curve. In the vicinity of absorption bands, both curves take on characteristic shapes, and this behavior is known as the Cotton effect, which may be either positive or negative (see illustration). There is a Cotton effect associated with each absorption

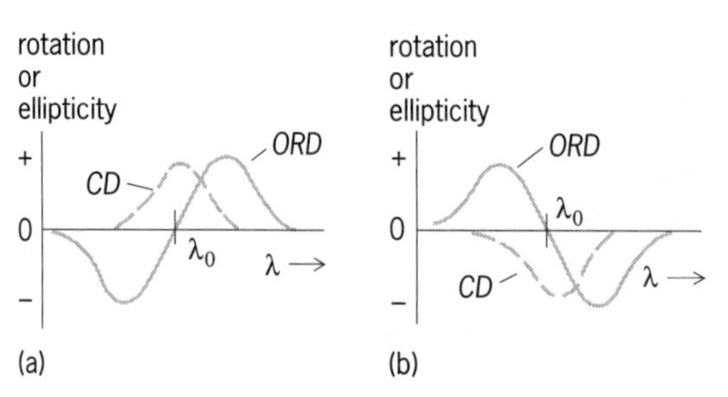


**Behavior of the ORD and CD curves in the vicinity of an absorption band at wavelength $\lambda_0$ (idealized). (*a*) Positive Cotton effect. (*b*) Negative Cotton effect.**

process, and hence a partial CD curve or partial ORD curve is associated with each particular absorption band or process. *See* OPTICAL ROTATORY DISPERSION.

The rotational strengths actually observed vary over quite a few orders of magnitude; this variation in magnitude is useful in stereochemical interpretation of molecular structure. [A.Mo.]

**Coulometer** Electrolysis cell in which a product is obtained with 100% efficiency as a result of an electrochemical reaction. The quantity of electricity, that is, the number of coulombs of electricity ($Q$), can be determined very accurately by weighing the product that is deposited on an electrode in the course of the electrochemical reaction. The relationship between the weight of the product formed in the coulometer and the quantity of electricity used is given by Faraday's laws of electrolysis. When a constant current of $i$ amperes flows through the electrolyte in the coulometer for $t$ seconds, the number of coulombs passed is given by Eq. (1).

$$Q = it \tag{1}$$

If the current varies in the course of the electrolysis, the simple current-time product in Eq. (1) is replaced by the current-time integral, Eq. (2).

$$Q = \int_0^t i \, dt \tag{2}$$

When $Q$ coulombs of electricity are passed through the electrolyte, the weight in grams of the material that is deposited on the electrode ($w$) is given by Eq. (3), where $n$

$$w = \frac{QM}{Fn} \tag{3}$$

is the number of electrons transferred per mole of material deposited, $M$ is its molecular weight, and $F$ is the Faraday constant, 96,487 $\pm$ 1.6 coulombs.

Equation (3) is fundamental in coulometry and is a mathematical statement of Faraday's laws. This equation is used for the accurate determination of $Q$, the current-time integral, by weighing or measuring a product that is formed at an electrode by an electrochemical reaction that occurs with 100% current efficiency. The electrolysis cell that is used for this purpose is a coulometer.

Only a few electrode reactions proceed with the 100% current efficiency that is required for the use of Eq. (3). The deposition of silver or copper (in a silver or copper coulometer), the evolution of oxygen and hydrogen (in a gas coulometer), and the oxidation of iodide to iodine (in an iodine coulometer) are examples of electrode reactions that have been successfully employed. One coulomb of electricity will deposit 1.1180 mg of silver at the cathode in a silver coulometer or liberate 1.315 mg of iodine at the anode in an iodine coulometer. Although these classical chemical coulometers are capable of measuring the quantity of electricity with high precision and accuracy, their use is time-consuming and inconvenient; and they have been largely replaced by operational amplifier integrator circuits or digital circuits that display in a direct readout the number of coulombs passed during electrolysis. *See* ELECTROCHEMICAL EQUIVALENT; ELECTROLYSIS. [Q.F.]

**Countercurrent transfer operations** Industrial processes in chemical engineering or laboratory operations in which heat or mass or both are transferred from one fluid to another, with the fluids moving continuously in very nearly steady state or constant manner and in opposite directions through the unit. Other geometrical arrangements for transfer operations are the parallel or concurrent flow, where the two fluids enter at the same end of the apparatus and flow in the same direction to the other end, and the cross-flow apparatus, where the two fluids flow at right angles to each other through the apparatus.

In heat transfer there can be almost complete transfer in countercurrent operation. The limit is reached when the temperature of the colder fluid becomes equal to that of the hotter fluid at some point in the apparatus. At this condition the heat transfer is zero between the two fluids. Most heat transfer equipment has a solid wall between the hot fluid and the cold fluid, so the fluids do not mix. Heat is transferred from the hot fluid through the wall into the cold fluid. Another type of equipment does use direct contact between the two fluids—for example, the cooling towers used to remove heat from a circulating water stream.

Mass transfer involves the changing compositions of mixtures, and is done usually by physical means. A material is transferred within a single phase from a region of high concentration to one of lower concentration by processes of molecular diffusion and eddy diffusion. In typical mass transfer processes, at least two phases are in direct contact in some state of dispersion, and mass (of one or more substances) is transferred from one phase across the interface into the second phase. Mass transfer takes place between two immiscible phases until equilibrium between the two phases is attained. In mass transfer there is seldom an equality of concentration in the two equilibrium phases. This means that a component may be transferred from a phase at low concentration (but at a concentration higher than that at equilibrium) to a second phase of greater concentration. The approach to equilibrium is controlled by diffusion transport across phase boundaries.

Although the two phases may be in concurrent flow or cross-flow, usual arrangements have the phases moving in Countercurrent directions. The more dense phase enters near the top of a vertical cylinder and moves downward under the influence of gravity. The less dense phase enters near the bottom of the cylinder and moves upward under the influence of a small pressure gradient. *See* ADSORPTION; CHEMICAL SEPARATION TECHNIQUES ; DISTILLATION; ELECTROPHORESIS. [F.J.L.]

**Cracking** A process used in the petroleum industry to reduce the molecular weight of hydrocarbons by breaking molecular bonds. Cracking is carried out by thermal, catalytic, or hydrocracking methods. Increasing demand for gasoline and other middle distillates relative to demand for heavier fractions makes cracking processes important in balancing the supply of petroleum products.

Thermal cracking depends on a free-radical mechanism to cause scission of hydrocarbon carbon-carbon bonds and a reduction in molecular size, with the formation of olefins, paraffins, and some aromatics. Side reactions such as radical saturation and polymerization are controlled by regulating reaction conditions. In catalytic cracking, carbonium ions are formed on a catalyst surface, where bond scissions, isomerizations, hydrogen exchange, and so on, yield lower olefins, isoparaffins, isoolefins, and aromatics. Hydrocracking is based on catalytic formation of hydrogen radicals to break carbon-carbon bonds and saturate olefinic bonds. It converts intermediate- and high-boiling distillates to middle distillates high in paraffins and low in cyclics and olefins. *See* HYDROCRACKING. [E.C.L.]

**Critical phenomena** The unusual physical properties displayed by substances near their critical points. The study of critical phenomena of different substances is directed toward a common theory.

Ideally, if a certain amount of water ($H_2O$) is sealed inside a transparent cell and heated to a high temperature $T$, for instance, $T > 647$ K (374°C or 705°F), the enclosed water exists as a transparent homogeneous substance. When the cell is allowed to cool

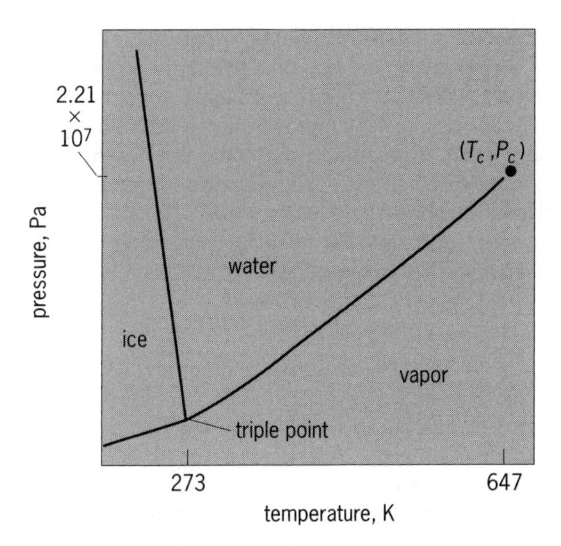


**Phase diagram of water ($H_2O$) on pressure-temperature (*P-T*) plane.**

down gradually and reaches a particular temperature, namely the boiling point, the enclosed water will go through a phase transition and separate into liquid and vapor phases. The liquid phase, being more dense, will settle into the bottom half of the cell. This sequence of events takes place for water at most moderate densities. However, if the enclosed water is at a density close to 322.2 kg · m$^{-3}$, rather extraordinary phenomena will be observed. As the cell is cooled toward 647 K (374°C or 705°F), the originally transparent water will become increasingly turbid and milky, indicating that visible light is being strongly scattered. Upon slight additional cooling, the turbidity disappears and two clear phases, water and vapor, are found. This phenomenon is called the critical opalescence, and the water sample is said to have gone through the critical phase transition. The density, temperature, and pressure at which this transition happens determine the critical point and are called respectively the critical density $\rho_c$, the critical temperature $T_c$, and the critical pressure $P_c$. For water $\rho_c = 322.2$ kg · m$^{-3}$, $T_c = 647$ K (374°C or 705°F), and $P_c = 2.21 \times 10^7$ pascals. *See* OPALESCENCE.

Different fluids, as expected, have different critical points. Although the critical point is the end point of the vapor pressure curve on the pressure-temperature (*P-T*) plane (see illustration), the critical phase transition is qualitatively different from that of the ordinary boiling phenomenon that happens along the vapor pressure curve. In addition to the critical opalescence, there are other highly unusual phenomena that are manifested near the critical point; for example, both the isothermal compressibility and heat capacity diverge to infinity as the fluid approaches $T_c$. *See* THERMODYNAMIC PROCESSES.

Many other systems, for example, ferromagnetic materials such as iron and nickel, also have critical points. The ferromagnetic critical point is also known as the Curie point. As in the case of fluids, a number of unusual phenomena take place near the critical point of ferromagnets, including singular heat capacity and divergent magnetic susceptibility. The study of critical phenomena is directed toward describing the various anomalous and interesting types of behavior near the critical points of these diverse and different systems with a single common theory. *See* CURIE TEMPERATURE; FERROMAGNETISM. [M.H.W.C.]

**Crystallization** The formation of a solid from a solution, melt, vapor, or a different solid phase. Crystallization from solution is an important industrial operation because of the large number of materials marketed as crystalline particles. Fractional crystallization is one of the most widely used methods of separating and purifying chemicals. In fractional crystallization it is desired to separate several solutes present in the same solution. This is generally done by picking crystallization temperatures and solvents such that only one solute is supersaturated and crystallizes out. By changing conditions, other solutes may be crystallized subsequently. Repeated crystallizations are necessary to achieve desired purities when many inclusions are present or when the solid solubility of other solutes is significant. For solubility and other relationships between solid and liquid phases *see* NUCLEATION; PHASE EQUILIBRIUM; SOLUTION. [W.R.W.]

**Curium** A chemical element, Cm, in the actinide series, with an atomic number of 96. Curium does not exist in the terrestrial environment, but may be produced artificially. The chemical properties of curium are so similar to those of the typical rare earths that, if it were not for its radioactivity, it might easily be mistaken for one of these elements. The known isotopes of curium include mass numbers 238–250. The isotope $^{244}Cm$ is of particular interest because of its potential use as a compact, thermoelectric power source, through conversion to electrical power of the heat generated by nuclear decay. *See* PERIODIC TABLE.

Metallic curium may be produced by the reduction of curium trifluoride with barium vapor. The metal has a silvery luster, tarnishes in air, and has a specific gravity of 13.5. The melting point has been determined as 2444 ± 72°F (1340 ± 40°C). The metal dissolves readily in common mineral acids with the formation of the tripositive ion.

A number of solid compounds of curium have been prepared and their structures determined by x-ray diffraction. These include $CmF_4$, $CmF_3$, $CmCl_3$, $CmBr_3$, $CmI_3$, $Cm_2O_3$, and $CmO_2$. Isostructural analogs of the compounds of curium are observed in the lanthanide series of elements. *See* ACTINIDE ELEMENTS; TRANSURANIUM ELEMENTS. [G.T.S.]

**Cyanamide** A term used to refer to the free acid hydrogen cyanamide, $H_2NCN$, or commonly to the calcium salt of this acid, calcium cyanamide, $CaCN_2$. Calcium cyanamide is manufactured by the cyanamide process, in which nitrogen gas is passed through finely divided calcium carbide at a temperature of 1832°F (1000°C). Most calcium cyanamide is used in agriculture as a fertilizer; some is used as a weed killer, as a pesticide, and as a cotton defoliant.

Hydrogen cyanamide is prepared from calcium cyanamide by treating the salt with acid. It is used in the manufacture of dicyandiamide and thiourea. [E.E.W.]

**Cyanate** A compound containing the —OCN group, typically a salt or ester of cyanic acid (HOCN). The cyanate ion is ambidentate, that is, it has two reactive sites, because it can bind through the oxygen (O) or the nitrogen (N). Cyanate is commonly N-bonded with most nonmetallic elements, presumably because of the small charge density on the oxygen. Cyanic acid has the structure H—O=C≡N, but it may exist in an isomeric form known as isocyanic acid, H—N=C=O. The cyanates are isomeric with fulminates, where the carbon and the nitrogen are transposed (—ONC).

Cyanic acid is a volatile liquid, and it polymerizes upon standing to form cyamelide acid and cyanuric acid. In water cyanic acid undergoes hydrolysis to ammonia and carbon dioxide. However, dilute solutions in ice water may be kept for several hours. Both cyanic acid and isocyanic acid are prepared from cyanuric acid. The linear —OCN ion may be prepared by mild oxidation of aqueous cyanide ion ($CN^-$) using lead oxide ($PbO$) and potassium cyanide ($KCN$).

The primary use for cyanates is in the synthesis of a number of organic compounds such as unsubstituted carbamates (R—NH=C=OOR′). These compounds are used as derivatives of alcohols. Other important reactions include addition to amines to give substituted ureas. *See* CARBON; CYANIDE; NITROGEN; THIOCYANATE. [T.J.Me.]

**Cyanide** A compound containing the —CN group, for example, potassium cyanide, KCN; calcium cyanide, $Ca(CN)_2$; and hydrocyanic (or prussic) acid, HCN. Chemically, the simple inorganic cyanides resemble chlorides in many ways. Organic compounds containing this group are called nitriles, for example, acrylonitrile, $CH_2CHCN$. *See* ACRYLONITRILE.

HCN is a weak acid. In the pure state, it is a highly volatile liquid, boiling at 26°C (78.8°F). HCN and the cyanides are highly toxic to animals and humans.

The cyanide ion forms a variety of coordination complexes with transition-metal ions, a property responsible for several of the commercial uses of cyanides. The cyanide process T is the most widely used method for extracting gold and silver from the ores. In silver-plating, a smooth adherent deposit is obtained on a metal cathode when electrolysis is carried out in the presence of an excess of cyanide ion.

$Ca(CN)_2$ is extensively used in pest control and as a fumigant in the storage of grain. In finely divided form, it reacts slowly with the moisture in the air to liberate HCN.

In case hardening of metals, an iron or steel article is immersed in a bath of molten sodium or potassium cyanide containing sodium chloride or carbonate. The cyanide decomposes at the surface, forming a deposit of carbon which combines with and penetrates the metal. *See* COORDINATION CHEMISTRY. [F.J.J.]

**Cyanocarbon** A derivative of hydrocarbon in which all of the hydrogen atoms are replaced by the —C≡N group. However, the term cyanocarbon has been applied to compounds which do not strictly follow the above definition: tetracyanoethylene oxide, tetracyanothiophene, tetracyanofuran, tetracynopyrrole, tetracyanobenzoquinone, tetracyanoquinodimethane, tetrayanodithiin, pentacyanopyridine, diazomalononitrile, and diazotetracyanocyclopentadiene.

Tetracyanoethylene (with the formula below), the simplest olefinic cyanocarbon, is

```
H≡C         C≡N
   \       /
    C  =  C
   /       \
H≡C         C≡C
```

a colorless, thermally stable solid, having a melting range of 198–200°C (388–392°F), and readily forming stable complexes with most aromatic systems. Cyanocarbon acids are among the strongest protonic organic acids known and are usually isolated only as the cyanocarbon anion salt. [D.W.W.]

**Cyanogen** A colorless, highly toxic gas having the molecular formula $C_2N_2$. Structurally, cyanogen is written N≡C—C≡N. Cyanogen belongs to a class of compounds known as pseudohalogens, because of the similarity of their chemical behavior to that of the halogens. Liquid cyanogen boils at −21.17°C (−6.11°F) and freezes at −27.9°C (−18.2°F) at 1 atm (101.325 kilopascals).

Cyanogen reacts with hydrogen at elevated temperatures in a manner analogous to the halogens, forming hydrogen cyanide, HCN. With hydrogen sulfide, $H_2S$, cyanogen forms thiocyanoformamide or dithiooxamide. Cyanogen burns in oxygen, producing one of the hottest flames known from a chemical reaction. It is considered to be a promising component of high-energy fuels. *See* CYANIDE. [F.J.J.]

**Cyclic nucleotides** Derivatives of nucleic acids that control the activity of several proteins within cells to regulate and coordinate metabolism. They are members of a group of molecules known as intracellular second messengers; their levels are regulated by hormones and neurotransmitters, which are the extracellular first messengers in a regulatory pathway. Cyclic nucleotides are found naturally in all living cells.

Two major forms of cyclic nucleotides are characterized: 3′,5′-cyclic adenosine monophosphate (cyclic AMP or cAMP) and 3′,5′-cyclic guanosine monophosphate (cyclic GMP or cGMP). Like all nucleotides, cAMP and cGMP contain three functional groups: a nitrogenous aromatic base (adenine or guanine), a sugar (ribose), and a phosphate. Cyclic nucleotides differ from other nucleotides in that the phosphate group is linked to two different hydroxyl (3′ and 5′) groups of the ribose sugar and hence forms a cyclic ring. This cyclic conformation allows cAMP and cGMP to bind to proteins to which other nucleotides cannot.

An increase in cAMP or cGMP triggered by hormones and neurotransmitters can have many different effects on any individual cell. The type of effect is dependent to some extent on the cellular proteins to which the cyclic nucleotides may bind. Three types of effector proteins are able to bind cyclic nucleotides: protein kinases, ion channels, and cyclic nucleotide phosphodiesterases.

Protein kinases are enzymes which are able to transfer a phosphate group to (phosphorylate) individual amino acids of other proteins. This action often changes the function of the phosphorylated protein. Ion channels are proteins found in the outer plasma membrane of some cells; binding of cyclic nucleotides to them can alter the flow of sodium ions across the cell membranes. Cyclic nucleotide phosphodiesterases are enzymes responsible for the degradation of cyclic nucleotides.

In bacteria, cAMP can bind to a fourth type of protein, which can also bind to deoxyribonucleic acid (DNA). This catabolite gene activator protein (CAP) binds to specific bacterial DNA sequences, stimulating the rate at which DNA is copied into ribonucleic acid (RNA) and increasing the amount of key metabolic enzymes in the bacteria.

In humans, cyclic nucleotides acting as second messengers play a key role in many vital processes and some diseases. For example, in the brain, cAMP and possibly cGMP are critical in the formation of both long-term and short-term memory. In the liver, cAMP coordinates the function of many metabolic enzymes to control the level of glucose and other nutrients in the bloodstream. *See* ADENOSINE TRIPHOSPHATE (ATP); ENZYME; NUCLEIC ACID; NUCLEOPROTEIN; NUCLEOTIDE; PROTEIN. [M.D.U.]

**Cyclophane** A molecule composed of two building blocks: an aromatic ring (most frequently a benzene ring) and an aliphatic unit forming a bridge between two (or more) positions of the aromatic ring. Those cyclophanes that contain heteroatoms in the aromatic part are generally called phanes, as "cyclo" in cyclophanes more strictly means that only benzene rings are present in addition to aliphatic bridges. Phane molecules containing hetero atoms in the aromatic ring (for example, nitrogen or sulfur) are called heterophanes, while in heteraphanes they are part of the aliphatic bridge.

Cyclophanes usually are not at all planar molecules; they exhibit an interesting stereochemistry (arrangement of atoms in three-dimensional space); molecular parts are placed and sometimes fixed in unusual orientations toward each other, and often the molecules are not rigid but (conformationally) flexible. Ring strain and, as a consequence, deformation of the benzene rings out of planarity are often encountered. Electronic interactions between aromatic rings fixed face to face can take place. In addition, influence on substitution reactions in the aromatic rings results; that is, substituents in one ring induce transannular electronic effects in the other ring, often leading to unexpected products. *See* CHEMICAL BONDING; CONFORMATIONAL ANALYSIS; STEREOCHEMISTRY.

Cyclophanes have become important in host-guest chemistry and supramolecular chemistry as they constitute host compounds for guest particles because of their cavities. Recognition processes at the molecular level are understood to mean the ability to design host molecules to encompass or attach selectively to smaller, sterically complementary guests in solution—by analogy to biological receptors and enzymes. For this reason, water-soluble cyclophanes were synthesized; they have the ability to include nonpolar guest molecules in aqueous solution. *See* CAGE HYDROCARBON; MOLECULAR RECOGNITION; POLAR MOLECULE.

Cyclophanes also exist in nature as alkaloids, cytotoxic agents, antibiotics, and many cyclopeptides. With the possibility of locating groups precisely in space, cyclophane chemistry has provided building units for molecular niches, nests, hollow cavities, multifloor structures, helices, macropolycycles, macro-hollow tubes, novel ligand systems, and so on. Cyclophane chemistry has become a major component of supramolecular chemistry, molecular recognition, models for intercalation, building blocks for organic catalysts, receptor models, and variation of crown ethers and cryptands. *See* MACROCYCLIC COMPOUND; SUPRAMOLECULAR CHEMISTRY. [F.Vo.; M.Boh.]

**Cytochrome** Any of a group of proteins that carry as prosthetic groups various iron porphyrins called hemes. Hemes also constitute prosthetic groups for other proteins, but the function of prosthetic groups in the cytochromes is largely restricted to oxidation to the ferric heme, with the iron in the $3^+$ valence state, and reduction to ferrous heme with a $2^+$ iron. Thus, by alternate oxidation and reduction the cytochromes can transfer electrons to and from each other and other substances, and can operate in the oxidation of substrates. The energy released in their oxidation reactions is conserved by using it to drive the formation of the energy-rich compound adenosine triphosphate (ATP) from adenosine diphosphate (ADP) and inorganic phosphate. This process of coupling the oxidation of substrates to phosphorylation of ADP is called oxidative phosphorylation. In cells of eukaryotic organisms, the cytochromes have rather uniform properties; they are part of the respiratory chain and are located in the mitochondria. In contrast, prokaryotes exhibit much more varied cytochromes. Cytochromes are found even in metabolic pathways that employ oxidants other than oxygen. *See* ADENOSINE TRIPHOSPHATE (ATP); PROTEIN.

**Respiratory chain.** There are four cytochromes in the respiratory chain of eukaryotes, termed respectively $aa_3$, $b$, $c$, and $c_1$. Cytochrome $aa_3$, also called cytochrome oxidase, functions by oxidizing reduced cytochrome $c$ (ferrocytochrome $c$) to the ferric form. It then transfers the reducing equivalents acquired in this reaction to molecular oxygen, reducing it to water. The cytochrome oxidase reaction is probably the most important reaction in biology since it drives the entire respiratory chain and takes up over 95% of the oxygen employed by organisms, thus providing nearly all of the energy needed for living processes.

The energy released during oxidation is utilized to actively pump protons ($H^+$) from the matrix of the mitochondrion through the inner membrane into the intermembrane space. This creates a proton gradient across the membrane, with the matrix space having a lower proton concentration and the outside having a higher proton concentration. This chemical and potential gradient can be released by allowing protons to flow down the gradient and back into the mitochondrial matrix, thereby driving the formation of ATP. A pair of electrons flowing down the respiratory chain yields three molecules of ATP, a remarkable feat of energy conservation. This is called the chemiosmotic mechanism of oxidative phosphorylation, which is generally considered a true picture of respiratory chain function.

**Cytochrome oxidase.** The cytochrome oxidase of eukaryotes is a very complex protein assembly containing from 8 to 13 polypeptide subunits, two hemes, *a* and $a_3$, and two atoms of copper. The two hemes are chemically identical but are placed in different protein environments, so that heme *a* can accept an electron from cytochrome *c* and heme $a_3$ can react with oxygen. When cytochrome oxidase has accepted four electrons, one from each of four molecules of reduced cytochrome *c*, both its hemes and both its copper atoms are in reduced form, and it can transfer the electrons in a series of reactions to a molecule of oxygen to yield two molecules of water.

Cytochrome oxidase straddles the inner membrane of mitochondria, part of it on the matrix side, part within the membrane, and part on the outer surface or cytochrome *c* side of the inner membrane.

**Cytochrome c.** Cytochrome *c* is the only protein member of the respiratory chain that is freely mobile in the mitochondrial intermembrane space. It is a small protein consisting of a single polypeptide chain of 104 to 112 amino acid residues, wrapped around a single heme prosthetic group. The cytochromes *c* of eukaryotes are all positively charged proteins, with strong dipoles, while the systems from which cytochrome *c* accepts electrons, cytochrome reductase, and to which cytochrome *c* delivers electrons, cytochrome oxidase, are negatively charged. There is good evidence that this electrostatic arrangement correctly orients cytochrome *c* as it approaches the reductase or the oxidase, so that electron transfer can take place very efficiently, even though the surface area at which the reaction occurs is less than 1% of the total surface of the protein.

The amino acid sequences of the cytochromes *c* of eukaryotes have been determined for well over 100 different species, from yeast to humans, and have provided some very interesting correlations between protein structure and the evolutionary relatedness of different taxonomic groups. The extensive degree of similarity over the entire range of extant organisms has been taken as evidence that this is an ancient structure, developed long before the divergence of plants and animals, which in the course of its evolutionary descent has been adapted to serve a variety of electron transfer functions in different organisms.

**Cytochrome reductase.** Like cytochrome oxidase, the cytochrome reductase complex is an integral membrane protein system. There are numerous subunits, consisting of two molecules of cytochrome *b*, one molecule of a nonheme iron protein, and one molecule of cytochrome $c_1$. As in the case of the oxidase, the two cytochrome *b* hemes are chemically identical, but are present in somewhat different protein environments. The reductase complex is reduced by reaction with the reduced form of the fat-soluble coenzyme Q, dissolved within the inner mitochondrial membrane, which is itself reduced by the succinate dehydrogenase, the NADH dehydrogenase, and other systems. *See* COENZYME.

**Other cytochromes.** In addition to the mitochondrial respiratory chain cytochromes, animals have a heme protein, termed cytochrome P450, located in the liver and adrenal gland cortex. In the liver it is part of a mono-oxygenase system that can utilize oxygen and the reduced coenzyme NADPH, to hydroxylate a large variety of foreign substances and drugs and thus detoxify them; in the adrenal it functions in the hydroxylation of steroid precursors in the normal biosynthesis of adrenocortical hormones.

Two varieties of cytochrome *b*, termed $b_{563}$ and $b_{559}$, and one of cytochrome *c*, $c_{552}$, are involved in the photosynthetic systems of plants. Other plant cytochromes occur in specialized tissues and certain species. [E.Ma.]

# D

**Dalton's law** The total pressure of a mixture of gases is the sum of the partial pressures of each gas in the mixture. The law was established by John Dalton (1766–1844). In his original formulation, the partial pressure of a gas is the pressure of the gas if it alone occupied the container at the same temperature. Dalton's law may be expressed as $P = P_A + P_B + \cdots$, where $P_J$ is the partial pressure of the gas $J$, and $P$ is the total pressure of the mixture; this formulation is strictly valid only for mixtures of ideal gases. For real gases, the total pressure is not the sum of the partial pressures (except in the limit of zero pressure) because of interactions between the molecules.

In modern physical chemistry the partial pressure is defined as $P_J = x_J P$, where $x_J$ is the mole fraction of the gas $J$, the ratio of its amount in moles to the total number of moles of gas molecules present in the mixture. With this definition, the total pressure of a mixture of any kind of gases is the sum of their partial pressures. However, only for an ideal gas is the partial pressure (as defined here) the pressure that the gas would exert if it alone occupied the container. *See* AVOGADRO NUMBER; GAS; KINETIC THEORY OF MATTER; THERMODYNAMIC PRINCIPLES. [P.W.A.]

**Darmstadtium** The eighteenth of the synthetic transuranium elements. Element 110 should be a heavy homolog of the elements platinum, palladium, and nickel (group 10). It is the eighth element in the 6*d* shell. The various isotopes of element 110 have been reliably documented by the observation of at least two atoms with half-lives in the range of milliseconds, a time too short for the application of chemical methods, verifying its position in the periodic system. *See* HALF-LIFE; NICKEL; PALLADIUM; PERIODIC TABLE; PLATINUM; RADIOISOTOPE.

Searches for this element began in 1985. With the finding of the isotope $^{269}110$ (the isotope of element 110 with mass number 269) in late 1994 at Gesellschaft für Schwerionenforschung (GSI), Darmstadt, Germany, its discovery became conclusive.

After an extensive search for the optimum bombarding energy on November 9, 1994, a decay chain was observed which proved the existence of the isotope $^{269}110$. The isotopes were produced in a fusion reaction of a nickel-62 projectile with a lead-208 target nucleus. The fused system, with an excitation energy of 13 MeV, cooled down by emitting one neutron and forming $^{269}110$, which by sequential alpha decays transformed to hassium-265, seaborgium-261, rutherfordium-257, and nobelium-253. All of these daughter isotopes were already known, and four decay chains observed in the following 12 days corroborated without any doubt the discovery of the element. The illustration shows the first decay chain observed, which ended in $^{257}Rf$. The isotope $^{269}110$ has a half-life of 0.2 ms, and is

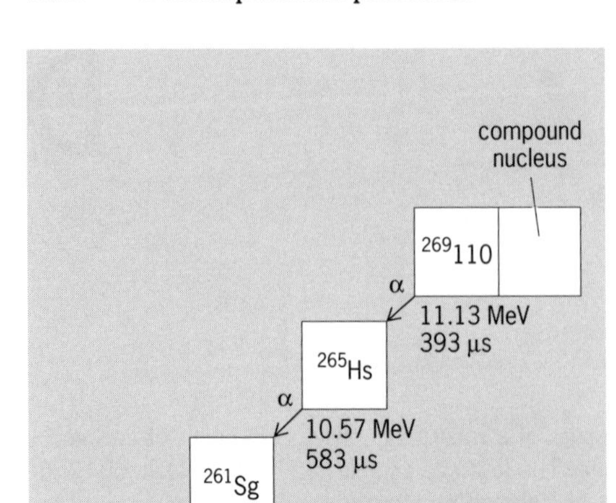


**Sequence of decay chains that document the discovery of element 110. Numbers below boxes are alpha energies and correlation times. Element 110 is produced in the reaction $^{62}Ni + {}^{208}Pb \rightarrow {}^{269}110 + 1n$.**

produced with a cross section of about $3 \times 10^{-36}$ cm$^2$. *See* NEUTRON; RUTHERFORDIUM. [P.Ar.]

**Decomposition potential** The electrode potential at which the electrolysis current begins to increase appreciably. Decomposition potentials are used as an approximate characteristic of industrial electrode processes. *See* ELECTROCHEMICAL PROCESS; ELECTROLYSIS.

Decomposition potentials are obtained by extrapolation of current-potential curves. Extrapolation is not precise because there is a progressive increase of current as the electrode potential is varied. The decomposition potential, for a given element, depends on the range of currents being considered. The cell voltage at which electrolysis becomes appreciable is approximately equal to the algebraic sum of the decomposition potentials of the reactions at the two electrodes and the ohmic drop, or voltage drop, in the electrolytic cell. The ohmic drop term is quite negligible for electrolytes with high conductance. For a discussion of current-potential curves *see* OVERVOLTAGE. [P.De.]

**Definite composition, law of** The law that a given chemical compound always contains the same elements in the same fixed proportions by weight. Thus, whatever its source, silver chloride always contains 100 g (3.53 oz) of silver to every 32.85 g (1.159 oz) of chlorine. If a compound is formed by the union of $m$ atoms of one element, each weighing $a$, with $n$ atoms of another element, each weighing $b$, the composition by weight of one molecule of the compound is in the ratio $ma{:}nb$. This must be the composition of any mass of the compound, provided that all atoms of the same kind have the same weight. It is now known that this is not usually the case but that the atoms of an element may consist of a number of isotopes, having different masses. However, as long as any sample of the element always contains the same relative proportions of the isotopes, the law still holds. Much more widespread and serious departures from the law of definite composition occur in a large variety of solid compounds (the nonstoichiometric compounds). *See* NONSTOICHIOMETRIC COMPOUNDS; STOICHIOMETRY. [T.C.W.]

**Dehydrogenation** A reaction in which hydrogen is detached from a molecule. The reaction is strongly endothermic, and therefore heat must be supplied to maintain the reaction temperature. When the detached hydrogen is immediately oxidized, two benefits accrue: (1) the conversion of reactants to products is increased because the equilibrium concentration is shifted toward the products (law of mass action); and (2) the added exothermic oxidation reaction supplies the needed heat of reaction. This process is called oxidative dehydrogenation. On the other hand, excess hydrogen is sometimes added to a dehydrogenation reaction in order to diminish the complete breakup of the molecule into many fragments.

The primary types of dehydrogenation reactions are vapor-phase conversion of primary alcohols to aldehydes, vapor-phase conversion of secondary alcohols to ketones, dehydrogenation of a side chain, and catalytic reforming of naphthas and naphthenes in the presence of a platinum catalyst. All four of these types of dehydrogenation reactions are of major industrial importance. They account for the production of billions of pounds of organic compounds that enter into the manufacture of lubricants, explosives, plastics, plasticizers, and elastomers. *See* HYDROGENATION; OXIDATION PROCESS. [J.W.Fu.]

**Deliquescence** The absorption of atmospheric water vapor by a crystalline solid until the crystal eventually dissolves into a saturated solution. This behavior is well known for certain salts such as hydrated calcium chloride, $CaCl_2 \cdot 6H_2O$, and zinc chloride, $ZnCl_2$, but it is a property of all soluble salts in air of sufficiently high humidity.

Thermodynamically, the condition for deliquescence is that the partial pressure of the water vapor in the air exceeds the vapor pressure (aqueous tension) of the water in the saturated solution of the salt. The speed at which the process takes place depends upon the rate of diffusion of water vapor into the crystal lattice, crystal size, and other factors. The process will stop when the water vapor in the atmosphere is depleted to the point at which its partial pressure equals that of the saturated solution.

Crystalline solids also may absorb water by increasing their water of hydration if the dissociation pressure of the hydrated species to be formed is less than the partial pressure of the water vapor. It is this process, not deliquescence, which is the opposite of efflorescence.

Deliquescent substances can be used to remove water vapor from air, although they have no special advantage over substances which merely add water of hydration and remain crystalline. *See* DESICCANT; EFFLORESCENCE; VAPOR PRESSURE. [R.L.S.]

**Delocalization** A phenomenon in which the most loosely held bonding electrons of some molecules serve to bind not two but several atoms. This contrasts with localization, a characteristic of ordinary single bonds, as in the normal paraffins, for example, methane ($CH_4$); in ethane, ($C_2H_6$); or in the molecules water ($H_2O$) and ammonia ($NH_3$). *See* CHEMICAL BONDING.

The prototype completely delocalized system is the ideal crystalline metal, in which the valence electrons are spaced uniformly over a periodic lattice of positive-ion cores. This confers a special stability to the metal, and it also accounts for its high electrical conductivity and other metallic properties.

The aromatic and conjugated molecules of organic chemistry contain delocalized electronic systems. The benzene molecule ($C_6H_6$) is considered the archetypical aromatic molecule. Benzene possesses an underlying single-bonded planar framework $(C_6H_6)^{6+}$ plus six additional electrons. Available for these electrons are six 2*p* orbitals, one on every carbon atom, each perpendicular to the molecular plane. The six electrons, called pi electrons, are ascribable to the six carbon atoms in such a way that neither structure (**1**) nor structure (**2**) is an accurate representation, but instead a structure

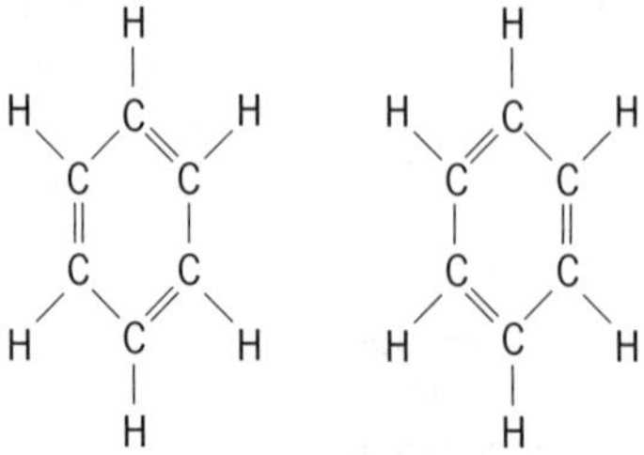

(1) (2)

that can be described in the valence bond language as a resonance hybrid of the two, schematically written as (**3**), where the circle stands for the delocalized electrons. All

(3)

molecules containing such rings possess an extra stability associated with the delocalization phenomenon. As a consequence, they also have a low propensity for chemical reactivity. *See* MOLECULAR ORBITAL THEORY; RESONANCE (MOLECULAR STRUCTURE). [R.G.P.]

**Dendritic macromolecule** A large molecule having a well-defined three-dimensional structure. Dendritic macromolecules play a crucial role in the chemistry of living systems. In contrast to the high level of structural precision that characterizes many biologically active macromolecules, the sizes and shapes of macromolecules made by polymer chemists are usually far less controlled. Most synthetic polymers are best described as statistical mixtures. However, chemists have sought to develop ways to prepare large molecules with more control over their architecture. If properly designed, such molecules might be capable of performing chemical or physical functions reminiscent of the macromolecules found in living systems.

Dendritic macromolecules are characterized by a highly branched molecular connectivity, whereby each repeat unit forms a branch juncture. The monomers used to prepare dendrimers possess three or more functional groups, and are of the type $AB_2$, $AB_3$, and so forth, where A and B represent a functionality (a site of chemical activity) that can combine to form a new covalent bond. Monomer chemistry is thus similar to that used to make condensation polymers except that the functionality is higher (there are more sites of chemical activity). *See* POLYMER.

The architectures of dendritic macromolecules are dramatically different from those of conventional macromolecules. Variations in molecular size, shape, and flexibility are all possible, depending on the monomer's chemical structure and geometry, the branch-point multiplicity, and the number of repetitive cycles carried out. The architecture most typical of dendritic macromolecules is a globular shape where segment density increases in going from the core to the periphery. The unusual structure of dendritic macromolecules, yet to be fully investigated, results in rheological and solubility characteristics that are dramatically different from linear macromolecules. [J.S.Mo.]

**Density** The mass per unit volume of a material. The term is applicable to mixtures and pure substances and to matter in the solid, liquid, gaseous, or plasma state. Density of all matter depends on temperature; the density of a mixture may depend on its composition, and the density of a gas on its pressure. Common units of density are grams per cubic centimeter, and slugs or pounds per cubic foot. The specific gravity of a material is defined as the ratio of its density to the density of some standard material, such as water at a specified temperature, for example, 60°F (15.6°C), or, for gases the basis may be air at standard temperature and pressure. Another related concept is weight density, which is defined as the weight of a unit volume of the material. *See* DENSITY MEASUREMENT; WEIGHT. [L.N.]

**Density measurement** Determination of the mass per unit volume of a substance. The term density is equally applicable to solids (including powders), liquids, and gases. Usually values of density are given in terms of grams per cubic centimeter or pounds per cubic foot. The density of all substances depends on temperature; in the case of gases, on temperature and pressure. The temperature used as a base for determining or reporting values of density is not the same for all substances. For solids 32°F (0°C) is the preferred temperature, although some tables give values at average room temperature because the effect of temperature on density is relatively small for most solids. The effect of temperature on density is more pronounced in liquids, so that the temperature must always be stated along with the density value. For many liquids the reference temperature is 60°F (15.6°C). For gases 32°F and a pressure of 29.921 in. of mercury (or 0°C and 760 mm of mercury or 101.325 kilopascals) are used for most scientific work and for tables of gas data. For fuel gases 60°F (15.6°C) and a pressure of 14.73 lb/in.$^2$ absolute (29.99 in. mercury or 101.56 kPa) are the values used in the United States.

In the case of a solid, if the sample is of regular shape, such as a cube or a cylinder, its volume may be determined by linear measurement. The mass of the sample is determined by weighing it on a suitable scale or balance; then this weight divided by the volume gives the density. Ordinarily the weighing is done in air, and the density value is the density in air, or apparent density. By adjusting for the buoyant effect of the air upon the weight of the sample, the real density is obtained.

A second procedure, applicable to irregular as well as regular shaped samples, is to weigh the sample in air and then to suspend it in a liquid of known density. The volume of the sample is equal to its loss of weight in the liquid divided by the density of the liquid. This is the method of hydrostatic weighing.

One method to determine the density of a gas is to completely evacuate a light but strong vessel of suitable size, the interior volume of which is known. The evacuated vessel is weighed, filled with a sample of the gas, and then weighed again. Of course the pressure and temperature of this sample of gas must be obtained.

A densitometer or gravitometer may be used to indicate and record the density of a flowing stream of a liquid or a gas. In a densitometer, a spinning propeller produces a pressure difference between inlet and outlet chambers of the device, which is proportional to the density of the gas flowing through it. [H.S.B.]

Two precision methods, the oscillator or vibrator method and the magnetic method, have emerged which allow more rapid and accurate determinations on liquid systems.

In the oscillator method, the density of a sample is related to the change in resonance frequency of a laterally vibrating tube. This frequency is inversely proportional to the square root of the mass of the tube and its contents. By calibrating the tube with media of known density at a given temperature, the density of unknown solutions may be determined if the volumes are strictly indentical. It is now established, that the accuracy of this method decreases as the viscosity of the medium increases. Hence, accurate viscosity measurements must accompany the density calibrations for a given instrument.

The instruments used in the magnetic method are called magnetic densimeters. This densimeter is a device whereby a tiny ferromagnetic cylinder, encased in a glass or plastic jacket, is held at a precise height within a medium by virtue of a solenoid controlled by a servo system in circuit with a height sensor. The jacket and ferromagnetic material constitute a buoy or float. The solenoid induces a magnetic moment $M$ at the buoy which is proportional to the electric current $I$ to the solenoid. The total magnetic force on the buoy is the product of this moment and the field gradient, $dH/dz$, where $H$ is the magnetic intensity and $z$ is the vertical distance from the center of the solenoid. The field gradient varies with $z$ and is also proportional to the current. Thus the total magnetic force at a particular distance $z$ in the solution which compensates for the difference in the opposing forces of gravity (downward) and the buoyancy (upward) exerted by the medium, through Archimedes' principle, is $M\,(dH/dz)$. The magnetic force, under proper conditions, is directly proportional to the square of the current, which can be measured very accurately. Thus $M(dH/dz) = kI^2$, where $k$ is a constant. If the buoyant force on the buoy is sufficient to make it float on the liquids of interest, the force generated by the solenoid must be downward to add to the force of gravity. The equation relating these forces is shown below, where, $V_B$ is the volume of the buoy, $g$ is

$$M\left(\frac{dH}{dz}\right) = gV_B\rho - gV_B\rho_B = gV_B(\rho - \rho_B)$$

the acceleration of gravity, and $\rho$ and $\rho_B$ are the densities of the solution and the buoy, respectively. By means of a precision resistor and an accurate differential voltmeter, the measurements consist simply of reading or recording the voltage, which is a parabolic function of the density. [D.W.Ku.]

**Desiccant** A substance (adsorbant) used to withdraw moisture from other materials. Although the removal of large quantities of water is done by evaporation, aided by moving air currents and by elevated temperature, the last traces of moisture are often held very tightly and do not evaporate readily. Furthermore, evaporation ceases when the moisture content of the material is reduced to that of the drying-air current. For final drying, a desiccant is used. It may react with water chemically or retain water through capillarity of adsorption. The drying agent is placed directly into the gas or liquid to be dried; solid materials are placed in a desiccator, a closed vessel in which moisture diffuses to the desiccant through the dry desiccator atmosphere. A desiccant loses potency as it takes on water; often it can be renewed by heating. Desiccants which form hydrates can be selected to maintain certain levels of low humidity in a closed vessel. *See* ADSORPTION; DELIQUESCENCE.

Among the more important types of solid desiccants are silica gel, activated alumina, anhydrous calcium sulfate, magnesium perchlorate, oxides (of barium and calcium), and activated carbon. [A.L.H.]

**Desorption** A process in which atomic and molecular species residing on the surface of a solid leave the surface and enter the surrounding gas or vacuum. In stimulated desorption studies, species residing on a surface are made to desorb by incident electrons or photons. Measurements of these species provide insight into the ways that radiation affects matter, and are useful analytical probes of surface physics and chemistry. In thermal desorption studies, adsorbed surface species are caused to desorb as the sample is heated under controlled conditions. These measurements can provide information on surface-bond energies, the species present on the surface and their coverage, the order of the desorption process, and the number of bonding states or sites.

**Stimulated desorption.** Stimulated desorption from surfaces is initiated by electronic excitation of the surface bond by incident electrons or photons. The classical model of desorption is an adaptation of the theory of gas-phase dissociation, in which desorption results from excitation from a bonding state to an antibonding state.

Another model which is more applicable to the phenomenon of ion desorption was first observed in studies of the desorption of positively ionized oxygen ($O^+$) from the surface of titanium(IV) oxide ($TiO_2$). Here it is found that $O^+$ is desorbed not by valence level excitation, but by ionization of the titanium and oxygen core levels. These levels, of course, have little to do with bonding. Furthermore, the fact that the oxygen is desorbed as an $O^+$ ion (whereas it is nominally at $O^{2-}$ on the surface) implies a large (three-electron) charge-transfer preceding desorption. This mechanism for desorption can also be effective for covalently bonded surface species.

Stimulated desorption studies are finding wide use. First, they can show the ways in which radiation affects the structure of solids. This will have important applications in the areas of radiation-induced damage and chemistry. Second, as an analytical tool, they offer a unique new way to study the physics and chemistry of atoms on surfaces which, when combined with the many other surface techniques based largely on electron spectroscopy, can provide new insight. Finally, models of the surface bond are put to a much sterner test in attempting to explain desorption phenomena.

An additional important discovery is that ion angular distributions from stimulated desorption are not isotropic, but show that ions are emitted in relatively narrow cones which project along the nominal ground-state bond directions. Thus this technique provides a direct display of the surface-bonding geometry.

**Thermal desorption.** Thermal desorption mass spectroscopy is possibly the oldest technique for the study of adsorbates on surfaces. Three primary forms of the thermal desorption experiment involve measurement of (1) the rate of desorption from a surface during controlled heating (temperature-programmed thermal desorption), (2) the rate of desorption at constant temperature (isothermal desorption), and (3) surface lifetimes and diffusion under exposure to a pulsed beam of adsorbates (molecular-beam experiments). Of the three, temperature-programmed thermal desorption is by far the most widely applied. The most straightforward information provided is the nature of the desorbed species from mass analysis, and a determination of the absolute coverage by the adsorbate, which is very difficult to obtain with other techniques. The technique can also provide important kinetic parameters of the desorption process.

While the thermal desorption techniques are among the simplest of surface probes, they remain indispensable because of their directness and the variety of information they convey. Thus while surface science moves to detailed methods involving extremely sophisticated apparatus, the simple thermal desorption methods remain an important part of the overall picture. [M.L.K.]

**Destructive distillation** The primary chemical processing of materials such as wood, coal, oil shale, and some residual oils from refining of petroleum. It consists in heating material in an inert atmosphere at a temperature high enough for chemical decomposition. The principal products are (1) gases containing carbon monoxide, hydrogen, hydrogen sulfide, and ammonia, (2) oils, and (3) water solutions of organic acids, alcohols, and ammonium salts.

Crude shale oil may be obtained by destructive distillation of carboniferous shales. It may be subjected to a destructive, or coking, distillation to reduce its viscosity and increase its hydrogen content. Residual oils from petroleum refinery operations are subjected to coking distillation to reduce the carbon content. The coke is used for the manufacture of electrode carbon. The main product of the destructive distillation of wood is 40–45% charcoal used in metallurgical processes in which the low content of ash, sulfur, and phosphorus is important. *See* CHARCOAL; COAL CHEMICALS; COKE; COKING (PETROLEUM); PYROLYSIS; WOOD CHEMICALS. [H.H.St./H.W.W.]

**Deuterium** The isotope of the element hydrogen with atomic weight 2.0144 and symbols $^2H$ or D. The terrestrial natural abundance of deuterium is 1 part in 6700 parts of ordinary hydrogen (protium). Small variations in natural sources are found as a result of fractionation by geological processes.

Deuterium is a gas ($D_2$) at room temperature. It is prepared from heavy water, $D_2O$, either by electrolysis or by reaction of $D_2O$ with metals such as zinc, iron, calcium, and uranium. It is also prepared directly by the fractional distillation of liquid hydrogen.

Deuterium is used mainly in the form of heavy water. In the uncombined state it finds uses as a research tool. Liquid deuterium is used in bubble chambers to study the reactions of elementary particles with the deuterium nucleus, the deuteron. Deuterons are frequently accelerated in cyclotrons to study their reactions with other nuclei and also to produce radioactive nuclides. Deuterium gas is used in the direct synthesis of organic compounds for tracer studies. *See* HEAVY WATER; HYDROGEN; TRITIUM. [J.Big.]

**Dewaxing of petroleum** The process of separating hydrocarbons which solidify readily (waxes) from petroleum fractions. Removal of wax is usually necessary to produce lubricating oil which will remain fluid down to the lowest temperature of use. The wax removed may be purified further to produce commercial paraffin or microcrystalline waxes.

Most commercial dewaxing processes utilize solvent dilution, chilling to crystallize the wax, and filtration. Wax crystals are formed by chilling through the walls of scraped surface chillers, and wax is separated from the resultant wax-oil-solvent slurry by using fully enclosed rotary vacuum filters. In a process modification, most of the chilling is accomplished by multistage injection of very cold solvent into the waxy oil with vigorous agitation, resulting in more uniform and compact wax crystals which filter faster.

Complex dewaxing requires no refrigeration, but depends upon the formation of a solid urea-*n*-paraffin complex which is separated by filtration and then decomposed. The catalytic dewaxing process is based on selective hydrocracking of the normal paraffins; it uses a molecular sieve-based catalyst in which the active hydrocracking sites are accessible only to the paraffin molecules. *See* WAX, PETROLEUM. [S.F.P.]

**Dextran** A polyglucose biopolymer characterized by preponderance of $\alpha$-1,6 linkage, and generally produced by enzymes from certain strains of *Leuconostoc* or *Streptococcus*. While formerly its principal utility was as blood plasma substitute, dextran is also employed as packing material in column chromatography and as pharmaceutical agents; its average molecular weight determines usage to a great extent. Dextran's chemical and physical properties depend upon the strain of microorganism employed and the environmental conditions imposed upon the bacterium during growth, or the reaction conditions where an enzymatic method of dextran production is employed. *Leuconostoc* and *Streptococcus* species primarily convert sucrose to dextran and fructose. *Acetobacter* species convert dextrin to dextran. *See* POLYMERIZATION; POLYSACCHARIDE. [H.M.Ts.]

**Dialysis** A process of selective diffusion through a membrane by dissolved solutes in liquid solution. As dialysis is usually carried out, the membrane permits the diffusion of low-molecular-weight solutes (crystalloids) but prevents the passage of colloidal and high-molecular-weight solutes (macro-molecules). Membranes suitable for this purpose include vegetable parchment, animal parchment, goldbeater's skin (peritoneal membranes of cattle), fish bladders, dialyzing cellophane (Visking sausage casing), and collodion (nitrocellulose deposited from alcohol-ether solution).

The solution is contained within such a membrane. The low-molecular-weight solutes are removed by placing pure solvent outside the membrane. This solvent is changed periodically or continuously until the concentration of diffusible solutes in the solution is reduced to near zero. The technique is used extensively in separating and purifying macro-molecules of biological origin. [Q.V.W.]

**Diazotization** The process by which an aromatic primary amine is converted to a diazonium compound. The preparation and reactions of diazonium salts were discovered in 1858 and were the basis of the synthetic dye industry and the development of other industrial chemistry in Europe. In diazotization, sodium nitrite is added to a solution of the amine in aqueous acid solution at 0–5°C (32–41°F). Reaction of the amine with nitrous acid gives a nitrosamine. Tautomerization and loss of water lead to the diazonium ion, which is stabilized by delocalization of the positive charge at the ortho and para carbon atoms of the ring, as in the reaction below.

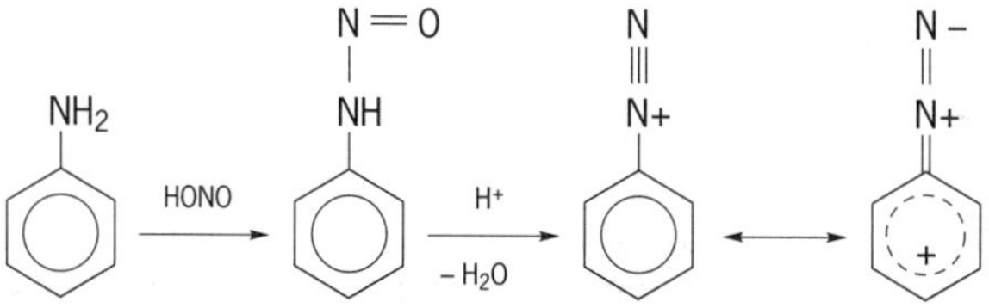

Nitrosamine Diazonium ion

*See* AMINE; AROMATIC HYDROCARBON; DELOCALIZATION; TAUTOMERISM.

The overall reaction is simple and very general. Substituents of all types—alkyl, halogen, nitro, hydroxyl, sulfonic acid—can be present at any position. Heterocyclic amines such as aminothiazole or aminopyridines can also be diazotized. Aromatic diamines are converted to *bis*-diazonium compounds. Diazonium salts are generally used and handled in aqueous solution; they are explosive if isolated and dried.

The great importance of diazonium compounds in dye technology lies in the coupling reactions that occur with an activated aromatic ring, such as that in phenols or aromatic amines. Coupling, or electrophilic substitution by $ArN_2^+$, gives compounds with an arylazo group at the position para or ortho to —OH or $—NH_2$. Reaction with amines occurs in weak acid solution. With phenols the phenoxide ion is the reactive species, and slightly basic solution is used. *See* CHEMICAL EQUILIBRIUM.

The azo dyes obtained in these coupling reactions are one of the important types of synthetic dyes. The color of the dye can be varied widely by choice of diazonium and coupling components. The coupling reaction lends itself to an important method of applying the dye to fabrics. In this process the coupling reagent, such as a naphthol-sulfonic acid, is absorbed onto the fiber, and the coupling reaction is then carried out directly on the fiber by passing the fabric through a bath of the diazonium solution. *See* DYE. [J.A.Mo.]

**Diels-Alder reaction** The 1,4-addition of an alkene (the dienophile) to a conjugated diene. The reaction, also known as the diene synthesis, is one of the most valuable and versatile methods for the preparation of compounds containing a six-membered ring, and proceeds most rapidly when the dienophile is substituted by electron-attracting groups. An example is

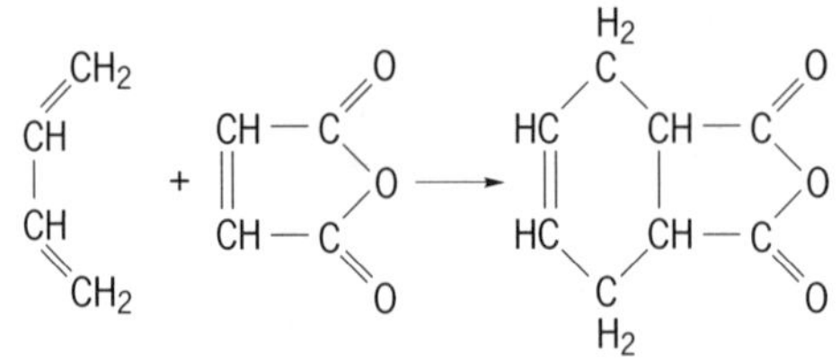

Diene(1,3-butadiene) + Dienophile (maleic anhydride) → Adduct

The Diels-Alder reaction does not require a catalyst, nor is the reaction retarded by the presence of oxidation inhibitors with which dienes are commonly treated to prevent formation of peroxides.

Industrially the diene synthesis is used in the production of the insecticides aldrin and dieldrin. The adduct of butadiene and maleic anhydride is used in the synthesis of the important fungicide captan. [P.E.F.]

**Diffusion** The transport of matter from one point to another by random molecular motions. It occurs in gases, liquids, and solids.

Diffusion plays a key role in processes as diverse as permeation through membranes, evaporation of liquids, dyeing textile fibers, drying timber, doping silicon wafers to make semiconductors, and transporting of thermal neutrons in nuclear power reactors. Rates of important chemical reactions are limited by how fast diffusion can bring reactants together or deliver them to reaction sites on enzymes or catalysts. The forces between molecules and molecular sizes and shapes can be studied by making diffusion measurements. *See* EVAPORATION.

Molecules in fluids (gases and liquids) are constantly moving. Even in still air, for example, nitrogen and oxygen molecules ricochet off each other at bullet speeds. Molecular diffusion is easily demonstrated by pouring a layer of water over a layer of ink in a narrow glass tube. The boundary between the ink and water is sharp at first, but it slowly blurs as the ink diffuses upward into the clear water. Eventually, the ink spreads evenly along the tube without any help from stirring.

**Gases.** A number of techniques are used to measure diffusion in gases. In a two-bulb experiment, two vessels of gas are connected by a narrow tube through which diffusion occurs. Diffusion is followed by measuring the subsequent changes in the composition of gas in each vessel. Excellent results are also obtained by placing a lighter gas mixture on top of a denser gas mixture in a vertical tube and then measuring the composition along the tube after a timed interval.

Rates of diffusion in gases increase with the temperature ($T$) approximately as $T^{3/2}$ and are inversely proportional to the pressure. The interdiffusion coefficients of gas mixtures are almost independent of the composition.

Kinetic theory shows that the self-diffusion coefficient of a pure gas is inversely proportional to both the square root of the molecular weight and the square of the molecular diameter. Interdiffusion coefficients for pairs of gases can be estimated by taking averages of the molecular weights and collision diameters. Kinetic-theory predictions are accurate to about 5% at pressures up to 10 atm (1 megapascal). Theories which take into account the forces between molecules are more accurate, especially for dense gases.

**Liquids.** The most accurate diffusion measurements on liquids are made by layering a solution over a denser solution and then using optical methods to follow the changes in refractive index along the column of solution. Excellent results are also obtained with cells in which diffusion occurs between two solution compartments through a porous diaphragm. Many other reliable experimental techniques have been devised.

Room-temperature liquids usually have diffusion coefficients in the range 0.5–5 $\times 10^{-5}$ cm$^2$ s$^{-1}$. Diffusion in liquids, unlike diffusion in gases, is sensitive to changes in composition but relatively insensitive to changes in pressure. Diffusion of high-viscosity, syrupy liquids and macromolecules is slower. The diffusion coefficient of aqueous serum albumin, a protein of molecular weight 60,000 atomic mass units, is only 0.06 $\times 10^{-5}$ cm$^2$ s$^{-1}$ at 25°C (77°F).

When solute molecules diffuse through a solution, solvent molecules must be pushed out of the way. For this reason, liquid-phase interdiffusion coefficients are inversely proportional to both the viscosity of the solvent and the effective radius of the solute molecules. Accurate theories of diffusion in liquids are still under development.

[D.G.L.]

**Solids.** Diffusion in solids is an important topic of physical metallurgy and materials science since diffusion processes are ubiquitous in solid matter at elevated temperatures. They play a key role in the kinetics of many microstructural changes that occur during the processing of metals, alloys, ceramics, semiconductors, glasses, and polymers. Typical examples of such changes include nucleation of new phases, diffusive phase transformations, precipitation and dissolution of a second phase, recrystallization, high-temperature creep, and thermal oxidation. Direct technological applications concern diffusion doping during the fabrication of microelectronic devices, solid electrolytes for battery and fuel cells, surface hardening of steels through carburization or nitridation, diffusion bonding, and sintering. *See* FUEL CELL; PHASE TRANSITIONS.

The atomic mechanisms of diffusion are closely connected with defects in solids. Point defects such as vacancies and interstitials are the simplest defects and often mediate diffusion in an otherwise perfect crystal. Dislocations, grain boundaries, phase boundaries, and free surfaces are other types of defects in a crystalline solid. They can act as diffusion short circuits because the mobility of atoms along such defects is usually much higher than in the lattice.

[H.Me.]

**Dimethyl sulfoxide** A versatile solvent (formula $C_2H_6OS$) abbreviated DMSO, used industrially and in chemical laboratories as a medium for carrying out chemical reactions. Its uses have been extended to that of a chemical reagent where DMSO itself is involved in a chemical change. It is the simplest member of a class of organic compounds which are typified by the polar sulfur-oxygen bond represented in the resonance hybrid shown below. The molecule is pyramidal in shape with the oxygen and the carbons at the corners.

$$\overset{O}{\overset{\|}{CH_3SCH_3}} \longleftrightarrow \overset{\bar{O}}{\overset{|}{CH_3\underset{+}{S}CH_3}}$$

DMSO is a colorless, odorless (when pure), and very hygroscopic stable liquid (bp 189°C, mp 19.5°C). It is manufactured commercially by reacting the black liquor from digestion in the kraft pulp process, with molten sulfur to form dimethyl sulfide which is then oxidized with nitrogen tetroxide. This highly polar aprotic solvent is water- and alcohol-miscible and will dissolve most polar organic compounds and many inorganic salts.

As an aprotic solvent, DMSO strongly solvates cations, leaving a highly reactive anion. Thus in DMSO, basicity and nucleophilicity is enhanced, and it is a superior solvent for many elimination, nucleophilic substitution, and solvolysis reactions in which nucleophile and base strength are important. Nucleophilic substitution reactions in which halogens or sulfonate esters are displaced by anions such as cyanide, alkoxide, thiocyanate, azide, and others are accelerated 1000 to 10,000 times in DMSO over the reaction in aqueous alcohol. Dimethyl sulfoxide is also superior to protic solvents as a media for elimination reactions. Its high dielectric constant makes DMSO a useful solvent for dissolving resins, polymers, and carbohydrates. It is employed commercially as a spinning solvent in the manufacture of synthetic fibers.

The apparent low toxicity and high skin permeability have led to extensive studies in numerous biological systems, including humans. Inorganic salts or small-molecular-weight organic compounds dissolved in DMSO can be transported across skin membrane, indicating a potential hazard in commercial use. [W.W.E./F.W.Sw.]

**Dipole-dipole interaction** The interaction of two atoms, molecules, or nuclei by means of their electric or magnetic dipole moments. This is the first term of the multipole-multipole series of invariants. More precisely, the interaction occurs when one dipole is placed in the field of another dipole. The interaction energy depends on the strength and relative orientation of the two dipoles, as well as on the distance between the centers and the orientation of the radius vector connecting the centers with respect to the dipole vectors. The electric dipole-dipole interaction and magnetic dipole-dipole interaction must be distinguished.

The center of the negative charge distribution of a molecule may fail to coincide with its center of gravity, thus creating a dipole moment. An example is the water molecule. If such molecules are close together, there will be a (electric) dipole-dipole interaction between them. Atoms do not have permanent dipole moments, but a dipole moment may be induced by the presence of another atom nearby; this is called the induced dipole-dipole interaction. Induced dipole-dipole forces between atoms and molecules are known by many different names: van der Waals forces, London forces, or dispersion forces. These induced dipole-dipole forces are responsible for cohesion and surface tension in liquids. They also act between unlike molecules, resulting in the adsorption of atoms on macroscopic objects. *See* ADSORPTION; INTERMOLECULAR FORCES; VAN DER WAALS EQUATION.

The magnetic dipole-dipole interaction is found both on a macroscopic and on a microscopic scale. Two compass needles within reasonable proximity of each other illustrate clearly the influence of the dipole-dipole interaction. In quantum mechanics, the magnetic moment is partially due to a current arising from the motion of the electrons in their orbits, and partially due to the intrinsic moment of the spin. The same interaction exists between nuclear spins. Magnetic dipole-dipole forces are particularly important in low-temperature solid-state physics, the interaction between the spins of the ions in paramagnetic salts being a crucial element in the use of such salts as thermometers and as cooling substances. *See* ELECTRON. [P.H.E.M.]

**Distillation** A method for separating homogeneous mixtures based upon equilibration of liquid and vapor phases. Substances that differ in volatility appear in different proportions in vapor and liquid phases at equilibrium with one another. Thus, vaporizing part of a volatile liquid produces vapor and liquid products that differ in composition. This outcome constitutes a separation among the components in the original liquid. Through appropriate configurations of repeated vapor-liquid contactings, the degree of separation among components differing in volatility can be increased many fold. *See* PHASE EQUILIBRIUM.

Distillation is by far the most common method of separation in the petroleum, natural gas, and petrochemical industries. Its many applications in other industries include air fractionation, solvent recovery and recycling, separation of light isotopes such as hydrogen and deuterium, and production of alcoholic beverages, flavors, fatty acids, and food oils.

**Simple distillations.** The two most elementary forms of distillation are a continuous equilibrium distillation and a simple batch distillation.

In a continuous equilibrium distillation, a continuously flowing liquid feed is heated or reduced in pressure (flashed) so as to cause partial vaporization. The vapor and liquid disengage while flowing through an open drum, and the products emerge as vapor and liquid streams. The vapor product can be condensed to form a liquid distillate. It is also possible to use a vapor feed, subjected to cooling and thereby partial condensation, again followed by disengagement of the resultant vapor and liquid in an open drum.

In a simple batch distillation, an entire batch of liquid is initially charged to a vessel and is then heated, typically by condensation of steam inside a metal coil within the vessel. Vapor is thereby continuously generated, and may be condensed to form a liquid distillate, which is collected. In the batch distillation, increments of vapor are formed in equilibrium with all liquid compositions ranging from the original to the final, whereas the continuous equilibrium distillation gives vapor in equilibrium with only the final liquid composition. Since the distillate consists primarily of the more volatile components and the feed liquid contains more of these substances than does the final liquid, the simple batch distillation gives a more enriched distillate than does the continuous equilibrium distillation.

**Fractional distillation.** Unless the vapor pressures of the species being separated are very dissimilar, a simple distillation does not produce highly purified products. Product purities can be increased by repeated partial vaporizations and condensations. The liquid from an initial continuous equilibrium distillation can be partially vaporized by additional heating. The remaining liquid can again be heated and partially vaporized, forming another liquid and so forth. Each liquid is progressively enriched in the less volatile substances. Similarly, successive partial condensations of the vapor fraction from the initial continuous equilibrium distillation produce vapor products successively enriched in the more volatile components. *See* VAPOR PRESSURE.

The process involved in successive partial vaporizations and condensations would lead to only very small amounts of the most enriched products, along with numerous streams having intermediate compositions. A logical step is to recycle each of these intermediate streams to the prior vessel in the sequence of contactors. Recycling of the intermediate vapors and liquids has another highly beneficial effect in that it negates the need for intermediate heaters and coolers. The resultant process is known as continuous fractional distillation. It is usually carried out in a distillation column, which is a simpler, more compact form of equipment than the cascade of vessels used in the process of recycling intermediate vapors and liquids. *See* DISTILLATION COLUMN.

**Vapor-liquid equilibria.** The separation accomplished in a distillation relates to the difference in composition of vapor and liquid phases at equilibrium. These relationships are the subject matter of phase-equilibrium thermodynamics. *See* CHEMICAL THERMODYNAMICS; GAS; LIQUID.

The dependence of liquid-phase activity coefficients upon liquid composition is complex; consequently vapor-liquid equilibrium relationships can have quite different

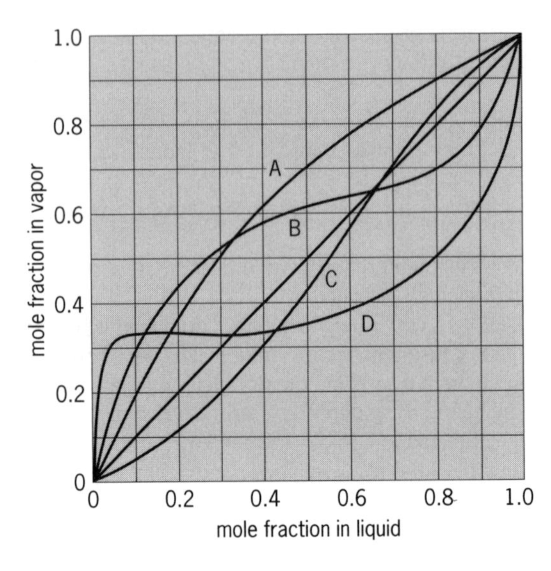


**Types of binary vapor-liquid equilibrium relationship; equilibrium vapor mole fraction of the more volatile (first named) component as a function of the liquid mole fraction of the more volatile component. The pressure is constant at 1 atm (101 kilopascals) for all systems. Curve A, benzene-toluene; B, acetone-carbon disulfide; C, acetone-chloroform; D, isobutyl alcohol-water.**

characteristics. For example, the two-component benzene-toluene system (curve A in the illustration) is a relatively ideal system, for which the activity coefficients are close to 1.0 over the liquid composition range. The mole fraction of benzene in the vapor is therefore always greater than that in the liquid, and the relative volatility has a nearly constant value of 2.4. Such a system can be separated readily into products of high purity by continuous fractional distillation.

In the acetone-carbon disulfide system (curve B), there are positive deviations from ideality, meaning that the activity coefficients are greater than 1.0 and increase for either component as that component becomes more dilute in the mixture. The result is that the relative volatility is greatest at low concentrations of the more volatile component (acetone) and decreases at higher concentrations. Because of hydrogen bonding between the two components, the acetone-chloroform system (curve C) exhibits negative deviations from ideality, meaning that the activity coefficients are less than 1.0. The isobutanol-water system (curve D) exhibits sufficiently extreme positive deviations from ideality so that two immiscible liquid phases form over much of the composition range. Continuous fractional distillation cannot produce two products of high purity for such a system.

**Distillation pressure.** The pressure under which a distillation is performed is a matter of choice, although operation at pressures more removed from atmospheric becomes more costly. Altering the pressure of a distillation can serve to alter the vapor-liquid equilibrium relationship, a feature that can be used to advantage.

Because of the relationship between vapor pressure and temperature, the temperatures within a distillation column are lower for lower pressures. Vacuum distillation is an effective means of maintaining lower temperatures for separations involving heat-sensitive materials.

Steam distillation is an alternative to vacuum distillation for separations of organic substances. In this process, steam is fed directly to the bottom of a column and passes upward, composing a substantial fraction of the vapor phase. The combined partial pressures of the organic substances being distilled are thereby lessened, giving the lower temperatures characteristic of a vacuum distillation. [C.J.Ke.]

**Distillation column** An apparatus used widely for countercurrent contacting of vapor and liquid to effect separations by distillation or absorption. In general, the apparatus consists of a cylindrical vessel with internals designed to obtain multiple contacting of ascending vapor and descending liquid, together with means for introducing or generating liquid at the top and vapor at the bottom.

In a column that can be applied to distillation (see illustration), a vapor condenser is used to produce liquid (reflux) which is returned to the top, and a liquid heater (reboiler) is used to generate vapor for introduction at the bottom. In a simple absorber, the absorption oil is the top liquid and the feed gas is the bottom vapor. In all cases, changes in composition produce heat effects and volume changes, so that there is a temperature gradient and a variation in vapor, and liquid flows from top to bottom of the column. These changes affect the internal flow rates from point to point throughout the column and must be considered in its design.

Distillation columns used in industrial plants range in diameter from a few inches to 40 ft (12 m) and in height from 10 to 200 ft (3 to 60 m). They operate at pressures as low as a few millimeters of mercury and as high as 3000 lb/in.$^2$ (2 megapascals) at temperatures from −300 to 700°F (−180 to 370°C). They are made of steel and other metals, of ceramics and glass, and even of such materials as bonded carbon and plastics.

A variety of internal devices have been used to obtain more efficient contacting of vapor and liquid. The most widely used devices are the bubble-cap plate, the perforated or sieve plate, and the packed column.

The bubble-cap plate is a horizontal deck with a large number of chimneys over which circular or rectangular caps are mounted to channel and distribute the vapor

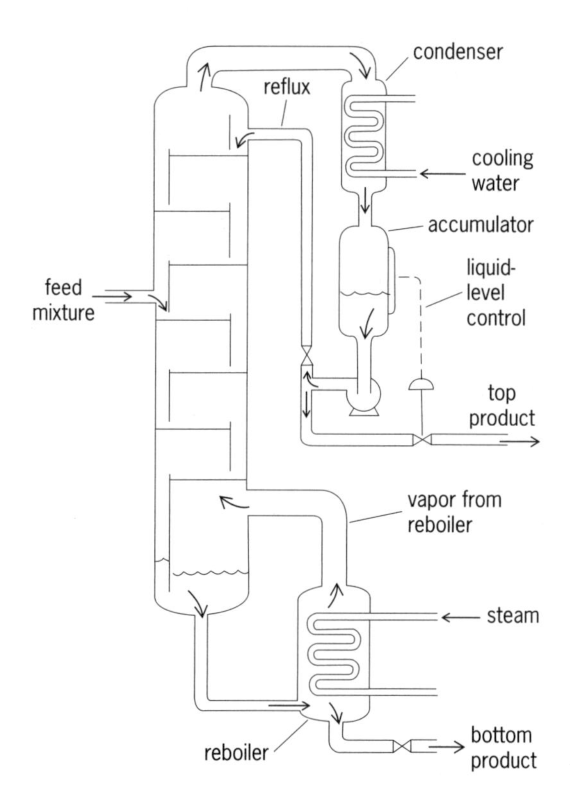


**Elements of a distillation column.**

through the liquid. Liquid flows by gravity downward from plate to plate through separate passages known as downcomers.

The perforated or sieve plate is a horizontal deck with a multiplicity of round holes or rectangular slots for distribution of vapor through the liquid. The sieve plate can be designed with downcomers similar to those used for bubble-cap trays, or it can be made without downcomers so that both liquid and vapor flow through the perforations in the deck.

The packed column is a bed or succession of beds made up of small solid shapes over which liquid and vapor flow in tortuous countercurrent paths. Expanded metal or woven mats are also used as packing. The packed column is used without downcomers, but in larger sizes usually has horizontal redistribution decks to collect and redistribute the liquid over the bed at successive intervals of height. The packed column is widely used in laboratories. It is often used in small industrial plants, especially where corrosion is severe and ceramic or glass materials must be chosen. *See* DISTILLATION; GAS ABSORPTION OPERATIONS. [M.Sou.]

**Donnan equilibrium** The distribution of diffusible ions on either side of a semipermeable membrane in the presence of macro-ions that are too large to pass through the membrane. The Donnan equilibrium thus is the result of (1) external constraints (boundary conditions) that enforce an unequal distribution of mobile ions and (2) a corresponding electrical potential on the membrane as a balance. This equilibrium is named for F. G. Donnan, the first to describe it and formulate the theory.

Typically, within a solution system that consists of two communicating compartments (for example, an inner and outer compartment), the polyions of a macromolecular polyelectrolyte (a colloidal electrolyte) are confined to one compartment (for example, the inner one) by the semipermeable membrane, which allows the exchange of solvent and low-molecular-weight electrolytes by diffusion between the compartments. Since electrical neutrality must obtain in either compartment, and since the macro-ions are confined to one side, the concentration of the diffusible ions in the two compartments cannot be the same. At thermodynamic equilibrium—assuming for simplicity an electrolyte of which both ions are small, carry a charge of 1, and diffuse readily throughout the system in which they are dissolved—the chemical potential (activities) for the outer ionic pair must equal that of the diffusible ion pair on the inside. This condition is fulfilled when the products of the activities are equal, as shown in Eq. (1), where $m_o$

$$m_o^+ m_o^- \gamma_o^{\pm} = m_i^+ m_i^- \gamma_i^{\pm} \qquad (1)$$

and $m_i$ are the molalities and $\gamma_o$ and $\gamma_i$ are the activity coefficients within the outer and inner compartments, respectively. In dilute solution, the molalities are approximately equal to the molarities, and the activity coefficients are practically equal to 1. The equilibrium of the solvent (whether the nondiffusible species is ionized or not) follows upon flux of the solvent into the (closed) cell that holds the macromolecular solute, until osmotic pressure equalizes the activities of the solvent in both cells. *See* ACTIVITY (THERMODYNAMICS).

The inequality of the individual ionic concentrations on either side of the membrane means that, at equilibrium, the membrane must carry a charge whose free energy is equal in magnitude and opposite in sign to the free energy that results from the unequal individual activities of the diffusible ions. This is shown in Eq. (2) for a negatively

$$E = \frac{RT}{F} \ln \frac{[\mathrm{Na}_i^+]}{[\mathrm{Na}_o^-]} = \frac{RT}{F} \ln \frac{[\mathrm{Cl}_o^-]}{[\mathrm{Cl}_i^-]} \qquad (2)$$

charged nondiffusible polyelectrolyte, where $E$ is the potential difference. In other words, the presence of the nondiffusing species on one side of the membrane and the resulting differences in salt concentration enforce a polarization of the membrane (which is independent of the nature of the diffusible ions).

Equation (1) permits the determination of the concentration (activity) of the ions of the same charge as the macro-ions within the inside cell, which is equal to the activity of the ions that have diffused in from the outer cell. It can also be seen that the osmotic pressure of a closed inner cell must be larger than the osmotic pressure that would be caused by the macro-ions alone, by an amount that corresponds to the number of ions that compensate the charges on the macro-ions, plus that of the ions that diffused in; that is, the osmotic pressure depends on the average molecular weight of the nonsolvent components of the inner cell. Therefore, if the salt concentration of the outer cell, and along with it the concentration in the inner cell, is raised, the contribution to the osmotic pressure by the macro-ions becomes relatively smaller and eventually becomes constant, amounting to the osmotic pressure of the macro-ions in their uncharged state. At the same time, the concentrations of the diffusible ions inside and outside approach one another. The same situation obtains in the absence of a membrane, when microscopic gel particles possess covalently bound ionized groups in their interior. Suspended microscopic gel particles, or dense polyelectrolyte coils, may thus carry a charge due to the effect of the Donnan equilibrium. *See* OSMOSIS.

The Donnan equilibrium is a frequent contributor to membrane potentials, but differential adsorption or, as in living systems, differences in the rates of passive or active ion transport through biological membranes are usually the main sources of membrane potentials of living cells. *See* COLLOID; ION-SELECTIVE MEMBRANES AND ELECTRODES. [F.R.E.]

**Dry ice** A solid form of carbon dioxide, $CO_2$, which finds its largest application as a cooling agent in the transportation of perishables. It is nontoxic and noncorrosive and sublimes directly from a solid to a gas, leaving no residue. At atmospheric pressure it sublimes at −109.6°F (−78.7°C). Slabs of dry ice can easily be cut and used in shipping containers for frozen foods, in refrigerated trucks, and as a supplemental cooling agent in refrigerator cars. *See* CARBON DIOXIDE. [C.F.K.]

**Drying** An operation in which a liquid, usually water, is removed from a wet solid in equipment termed dryers. The use of heat to remove liquids distinguishes drying from mechanical dewatering methods such as centrifugation, decantation or sedimentation, and filtration, in which no change in phase from liquid to vapor is experienced. Drying is preferred to the term dehydration, which usually implies removal of water accompanied by a chemical change. Drying is a widespread operation in the chemical process industries. It is used for chemicals of all types, pharmaceuticals, biological materials, foods, detergents, wood, minerals, and industrial wastes. Drying processes may evaporate liquids at rates varying from only a few ounces per hour to 10 tons per hour in a single dryer. Drying temperatures may be as high as 1400°F (760°C), or as low as −40°F (−40°C) in freeze drying. Dryers range in size from small cabinets to spray dryers with steel towers 100 ft (30 m) high and 30 ft (9 m) in diameter. The materials dried may be in the form of thin solutions, suspensions, slurries, pastes, granular materials, bulk objects, fibers, or sheets. Drying may be accomplished by convective heat transfer, by conduction from heated surfaces, by radiation, and by dielectric heating. In general, the removal of moisture from liquids (that is, the drying of liquids) and the drying of gases are classified as distillation processes and adsorption processes, respectively, and they are performed in special equipment usually termed distillation columns (for liquids) and adsorbers (for gases and liquids). Gases also may be dried by compression. *See* ADSORPTION; DISTILLATION.

**Drying of solids.** In the drying of solids, the desirable end product is in solid form. Thus, even though the solid is initially in solution, the problem of producing this solid in dry form is classed under this heading. Final moisture contents of dry solids are usually less than 10%, and in many instances, less than 1%.

The mechanism of the drying of solids is reasonably simple in concept. When drying is done with heated gases, in the most general case, a wet solid begins to dry as though the water were present alone without any solid, and hence evaporation proceeds as it would from a so-called free water surface, that is, as water standing in an open pan. The period or stage of drying during this initial phase, therefore, is commonly referred to as the constant-rate period because evaporation occurs at a constant rate and is independent of the solid present. The presence of any dissolved salts will cause the evaporation rate to be less than that of pure water. Nevertheless, this lower rate can still be constant during the first stages of drying.

A fundamental theory of drying depends on a knowledge of the forces governing the flow of liquids inside solids. Attempts have been made to develop a general theory of drying on the basis that liquids move inside solids by a diffusional process. However, this is not true in all cases. In fact, only in a limited number of types of solids does true diffusion of liquids occur. In most cases, the internal flow mechanism results from a combination of forces which may include capillarity, internal pressure gradients caused by shrinkage, a vapor-liquid flow sequence caused by temperature gradients, diffusion, and osmosis. Because of the complexities of the internal flow mechanism, it has not been possible to evolve a generalized theory of drying applicable to all materials. Only in the drying of certain bulk objects such as wood, ceramics, and soap has a significant understanding of the internal mechanism been gained which permits control of product quality.

Most investigations of drying have been made from the so-called external viewpoint, wherein the effects of the external drying medium such as air velocity, humidity, temperature, and wet material shape and subdivision are studied with respect to their influence on the drying rate. The results of such investigations are usually presented as drying rate curves, and the natures of these curves are used to interpret the drying mechanism.

When materials are dried in contact with hot surfaces, termed indirect drying, the air humidity and air velocity may no longer be significant factors controlling the rate. The "goodness" of the contact between the wet material and the heated surfaces, plus the surface temperature, will be controlling. This may involve agitation of the wet material in some cases.

Drying equipment for solids may be conveniently grouped into three classes on the basis of the method of transferring heat for evaporation. The first class is termed direct dryers; the second class, indirect dryers; and the third class, radiant heat dryers. Batch dryers are restricted to low capacities and long drying times. Most industrial drying operations are performed in continuous dryers. The large numbers of different types of dryers reflect the efforts to handle the larger numbers of wet materials in ways which result in the most efficient contacting with the drying medium. Thus, filter cakes, pastes, and similar materials, when preformed in small pieces, can be dried many times faster in continuous through-circulation dryers than in batch tray dryers. Similarly, materials which are sprayed to form small drops, as in spray drying, dry much faster than in through-circulation drying.

**Drying of gases.** The removal of 95–100% of the water vapor in air or other gases is frequently necessary. Gases having a dew point of −40°F (−40°C) are considered commercially dry. The more important reasons for the removal of water vapor from air are (1) comfort, as in air conditioning; (2) control of the humidity of manufacturing atmospheres; (3) protection of electrical equipment against corrosion, short circuits, and electrostatic discharges; (4) requirement of dry air for use in chemical processes where moisture present in air adversely affects the economy of the process; (5) prevention of water adsorption in pneumatic conveying; and (6) as a prerequisite to liquefaction.

Gases may be dried by the following processes: (1) absorption by use of spray chambers with such organic liquids as glycerin, or aqueous solutions of salts such as lithium chloride, and by use of packed columns with countercurrent flow of sulfuric acid, phosphoric acid, or organic liquids; (2) adsorption by use of solid adsorbents such as activated alumina, silica gel, or molecular sieves; (3) compression to a partial pressure of water vapor greater than the saturation pressure to effect condensation of liquid water; (4) cooling below dew point of the gas with surface condensers or coldwater sprays; and (5) compression and cooling, in which liquid desiccants are used in continuous processes in spray chambers and packed towers—solid desiccants are generally used in an intermittent operation that requires periodic interruption for regeneration of the spent desiccant.

Desiccants are classified as solid adsorbents, which remove water vapor by the phenomena of surface adsorption and capillary condensation (silica gel and activated alumina); solid absorbents, which remove water vapor by chemical reaction (fused anhydrous calcium sulfate, lime, and magnesium perchlorate); deliquescent absorbents, which remove water vapor by chemical reaction and dissolution (calcium chloride and potassium hydroxide); or liquid absorbents, which remove water vapor by absorption (sulfuric acid, lithium chloride solutions, and ethylene glycol).

The mechanical methods of drying gases, compression and cooling and refrigeration, are used in large-scale operations, and generally are more expensive methods than those using desiccants. Such mechanical methods are used when compression or cooling of the gas is required.

Liquid desiccants (concentrated acids and organic liquids) are generally liquid at all stages of a drying process. Soluble desiccants (calcium chloride and sodium hydroxide) include those solids which are deliquescent in the presence of high concentrations of water vapor.

Deliquescent salts and hydrates are generally used as concentrated solutions because of the practical difficulties in handling, replacing, and regenerating the wet corrosive solids. The degree of drying possible with solutions is much less than with corresponding solids; but, where only moderately low humidities are required and large volumes of air are dried, solutions are satisfactory. *See* DESICCANT; EVAPORATION; FILTRATION; UNIT OPERATIONS; VAPOR PRESSURE. [W.R.M.]

**Dubnium** A chemical element, symbol Db, atomic number 105. It was synthesized and identified unambiguously in March 1970 at the heavy-ion linear accelerator (HILAC) at the Lawrence Radiation Laboratory, Berkeley, University of California. The discovery team consisted of A. Ghiorso and colleagues. *See* PERIODIC TABLE.

The dubnium isotope, with a half-life of 1.6 s, decayed by emitting alpha particles with energies of 9.06 (55%), 9.10 (25%), and 9.14 (20%) MeV. It was shown to be of mass 260 by identifying lawrencium-256 as its daughter by two different methods.

Previous work on dubnium was reported in 1968 by G. N. Flerov and colleagues at Dubna Laboratories in Russia. They claimed to have discovered two isotopes of dubnium produced by the bombardment of $^{243}$Am by $^{22}$Ne ions. However, the Lawrence Radiation Laboratory work did not confirm these findings (due to energy and decay differences). *See* NUCLEAR CHEMISTRY. [A.Gh.]

**Dye** A colored substance, also called a dyestuff, which imparts more or less permanent color to other materials.

Customarily, colored water-insoluble substances are called pigment. Dyes are generally water-soluble, although some are soluble only during application, after which they become insoluble.

The mechanism by which soluble colored substances enter the internal structure of fibers and there become fixed has been variously explained in terms of the physical and chemical concepts of the times when the explanations were given. It is said to be an adsorption phenomenon, a salt formation, a quasi-chemical union caused by hydrogen bonding, or an ether linkage, and in some cases it is considered to be a true solution effect. The end result, however, is that the dye has imparted a color (not necessarily that of the solid dye itself) to the fiber which is more or less resistant to washing or removal by similar mechanical operations. The dye is said to be fixed on and to have affinity for the material it has colored. The material is designated as the substrate. If the color is quite resistant to washing and light, it is called a fast color; if the color is easily removed or fades quickly, it is a fugitive dye.

Dyestuffs may be classified in various ways: according to color (blue, red, and so on); origin (natural—from vegetable and animal matter—or synthetic); chemical structure (the most precise); kinds of material to which they are applied (cloth, paper, leather, plastic, food, and biological specimens); and method of application (used most frequently by the practical dyer). [W.Mi.]

Dyes that form a covalent bond with the substrate during the dyeing process are known as reactive dyes. They offer improved washfastness, improved brightness, and improved rubbing fastness and shade range. Reactive dyes are one of the more important single types of cellulosic dye. To a lesser degree, they are used for polyamide fibers. The simplicity of continuous application and the brightness of reactive dyes make them useful in printing as well as dyeing. [D.R.B.; D.H.A.]

**Dysprosium** A metallic rare-earth element, Dy, atomic number 66 and atomic weight 162.50. The naturally occurring element is composed of seven stable isotopes. Dysprosium forms a white oxide, $Dy_2O_3$, which dissolves in acid to give a yellowish-green solution. *See* PERIODIC TABLE; RARE-EARTH ELEMENTS.

The metal is attacked readily by air at high tempratures, but at room temperatures, in massive blocks, it is fairly satble in the armosphere and remains shiny for long periods of time. Dysprosium is partamagnetic, but as the temperature is lowered, it becomes antiferromagnetic at the Néel point (178 K or $-139^{\circ}$F) and ferromagnetic at the Curie point (85 K or $-306.4^{\circ}$F). At very low temperatures, the metal shows strong anisotropic magentic properties. [F.H.Sp.]

# E

**Efflorescence** The spontaneous loss of water (as vapor) from hydrated crystalline solids. The thermodynamic requirement for efflorescence is that the partial pressure of water vapor at the surface of the solid (its dissociation pressure) exceed the partial pressure of water vapor in the air. A typical efflorescent substance is Glauber's salt, $Na_2SO_4 \cdot 10H_2O$. The spontaneous loss of water normally requires that the crystal structure be rearranged, and consequently, efflorescent salts usually go to microcrystalline powders when they lose their water of hydration. *See* PHASE EQUILIBRIUM; VAPOR PRESSURE. [R.L.S.]

**Einsteinium** A chemical element. Es, atomic number 99, a member of the actinide series in the periodic table. It is not found in nature but is produced by artificial nuclear transmutation of lighter elements. All isotopes of einsteinium are radioactive, decaying with half-lives ranging from a few seconds to about 1 year. *See* ACTINIDE ELEMENTS; PERIODIC TABLE.

Einsteinium is the heaviest actinide element to be isolated in weighable form. The metal is chemically reactive, is quite volatile, and melts at 860°C (1580°F); one crystal structure is known. *See* TRANSURANIUM ELEMENTS. [S.G.T.; G.T.S.]

**Electrochemical equivalent** The mass of a substance, according to Faraday's law, produced or consumed by electrolysis with 100% current efficiency during the flow of a quantity of electricity equal to 1 faraday or 96,487 coulombs (1 coulomb corresponds to a current of 1 ampere during 1 second). Electrochemical equivalents are essential in the calculation of the current efficiency of an electrode process.

The electrochemical equivalent of a substance is equal to the gram-atomic or gram-molecular mass of this substance divided by the number of electrons involved in the electrode reaction. For example, the electrochemical equivalent of zinc, for which two electrons are required in order to deposit one atom, is Zn/2 or 65.37/2 g. Thus, the faraday is equal to the product of the charge of the electron times the number of electrons (the Avogadro number) required to react with 1 atom- or molecule-equivalent of substance. *See* COULOMETER. [P.De.]

**Electrochemical process** The principles of electrochemistry may be adapted for use in the preparation of commercially important quantities of certain substances, both inorganic and organic in nature. *See* ELECTROCHEMISTRY.

**Inorganic processes.** Inorganic chemical processes can be classified as electrolytic, electrothermic, and miscellaneous processes including electric discharge through gases and separation by electrical means. In electrolytic processes, chemical and electrical energy are interchanged. Current passed through an electrolytic cell causes chemical reactions at the electrodes. Voltaic cells convert chemicals into electricity. Electrothermic processes use electricity to attain the necessary temperature for reaction. *See* ELECTROCHEMISTRY; ELECTROLYSIS; ELECTROLYTIC CONDUCTANCE; ELECTROMOTIVE FORCE (CELLS).

*Electrolysis in aqueous solutions.* The electrolysis of water to form hydrogen and oxygen, according to the reaction $2H_2O \rightarrow 2H_2 + O_2$, may be considered as the simplest process for aqueous electrolytes. It does not compete with hydrogen from propane or from natural gas and with oxygen from liquid air, except in small installations. While simplicity, high hydrogen purity requirement, and lower capital cost (in small plants) have justified electrolytic plants, severely rising energy costs have limited such applications. Heavy water, or deuterium oxide, used in moderating nuclear reactors is also a by-product of the electrolysis of water. *See* DEUTERIUM; HEAVY WATER; HYDROGEN; OXYGEN.

*Metallurgical applications.* Protective or decorative coatings on a base metal such as steel are obtained by electroplating. Plating may also be used to replace worn metal or to provide a wear-resistant surface. Electrogalvanizing is preferred over hot dipping for applying zinc to steel. Tin plate for containers is electrolytic.

Electroforming is a method of forming or reproducing articles by electrodeposition. In contrast to electroplating, the product is removed from the base surface or mold. Electrodeposition of metal powders is used to produce particles in the 1- to 1000-micrometer range for use in powder metallurgy and metallic pigments. Electrolytic polishing of metals is accomplished by making the article anodic in an electrolyte of mixed acids. Electrolytic machining of metals is accomplished by making the metal part anodic in a suitable electrolyte. Electrorefining is a process for purifying metals and recovering their impurities, which at times are more valuable than the original metal. Electrowinning, sometimes termed aqueous electrometallurgy, involves processing of metallic ores by leaching solutions to obtain metal-containing electrolytes which can be processed with insoluble anodes and metal cathodes.

*Alkali-chlorine processes.* Electrolysis of alkali halides is the basis of the alkali-chlorine and chlorate industries. Chlorine, $Cl_2$, and caustic soda, NaOH (or caustic potash, KOH), are made by electrolysis of brine, a solution of sodium chloride, NaCl, in water. Hydrochloric acid electrolysis is of interest for recovery of chlorine from HCl resulting as a by-product from organic chlorinations. *See* CHLORINE.

*Oxidations and reductions.* These reactions occur in all cells, but in a narrower sense oxidation reactions are those in which oxygen or chlorine at the anode oxidizes some material to form a new compound; reduction reactions are those in which hydrogen, liberated at the cathode, reduces a material to a new product. There are no commercial applications of inorganic electrochemical reductions by this narrow definition.

*Ion-permeable membrane cells.* These utilize diaphragms made of ion-exchange resins. Cation-permeable membranes permit cations to pass through but not anions, whereas the reverse holds for anion-permeable membranes. Purification of sea water is the most important application. Salt has been recovered from sea water which has been concentrated in this way. *See* ION-SELECTIVE MEMBRANES AND ELECTRODES.

*Fused-salt electrolysis.* Aluminum, barium, beryllium, cerium and misch metal, fluorine, lithium, magnesium, sodium, molybdenum, thorium, titanium, uranium, and zirconium are obtained by electrolysis of fused salts, because water interferes with the desired reaction. Raw materials must all be purified before addition to fused-salt cells, because purification of the electrolyte is not economical as in aqueous electrolytes. Metallizing is a process of depositing a metal as an alloy on a substrate from a fused complex metal salt.

*Electrothermics.* The manufacture of many products requires temperatures higher than can be obtained by combustion methods. Electric heat can usually be developed at, or close to, the point where it is required, so that it is relatively quick. It permits easy control of the atmosphere for oxidizing, reducing, or neutral conditions.

Products of the electric furnace include iron and steel; ferroalloys; nonferrous metals and alloys; the exotic metals titanium, zirconium, hafnium, thorium, and uranium; and nonmetallic products such as calcium carbide, calcium cyanamide, sodium cyanide, silicon carbide, boron carbide, and graphite.

Zone refining of metals for the electronics industry, such as silicon for diodes and transistors, is accomplished by induction melting of the metal in a narrow zone and slow movement of the molten zone in the metal ingot from one end to the other in an evacuated or inert gas–filled enclosure. Impurities move toward the end of the ingot. The operation is repeated until the desired purity is obtained.

*Electrodialysis.* This is the separation of low-molecular-weight electrolytes from aqueous solutions by migration of the electrolyte through semipermeable membranes in an electric field. It is used on an industrial scale for deashing starch hydrolyzates and whey, and in many municipalities for producing potable water from saline water. Its uses also include the concentration of liquid foods such as dairy products and citrus juices, the recovery of sulfite pulp waste and pickling acid, and the isolation of proteins. *See* COLLOID; DIALYSIS.

*Electrophoretic deposition.* This is the deposition of a non-conductive material in a finely divided state from a suspension in an inert medium. Electrophoresis is the migration of colloidal particles, which acquire positive or negative charges in an electric field. The process is useful in electropainting; for instance, electropainting of automobile bodies and other objects has now been adopted on a large scale. Rubber latex is an example of a negatively charged colloid which can be plated on an anode. Electronic components can be coated with inorganic salts, oxides, and ceramics suspended in organic media. *See* ELECTROPHORESIS.

*Electroendosmosis.* This is the movement of a liquid with respect to an immobilized colloid in an electric field. The process is used in the dehydration of peat, dye pastes, and clay. It is also used commercially for dewatering soils in mining, road building construction, and other civil engineering works.

*Electrostatic technique.* The deposition of charged particles from suspension in gases has many useful applications. The Cottrell electrostatic precipitator removes dusts and mists from gases. *See* ELECTROSTATIC PRECIPITATOR.

Spray painting with a high voltage between the spray gun and the work is particularly effective in providing an even coating with an economical use of paint on irregular and open surfaces, such as a screen.

In xerography a sheet of plain paper is electrically sensitized in those areas corresponding to an original so that colored resin particles carrying an opposite charge are attracted and retained only on the sensitized areas, thus producing a visible image corresponding to the original.

Abrasive paper and cloth are coated with an adhesive and abrasive powders attracted to the base material in an electrostatic field. Pile fabrics can be produced in a similar manner, with the short fibers oriented by the electric field.

**Organic processes.** Organic electrochemistry was once regarded as a tantalizing area with many important laboratory achievements but few successes in commercial practice. This situation is changing, however, in that electroorganic processes are likely to prove commercially advantageous if they can fulfill either of two conditions: (1) performance under conditions of voltage corresponding thermodynamically to the conversion of an organic group to a reduced or oxidized group, with the cell products relatively easy to isolate and purify; (2) performance of a highly selective, specific technique to make an addition at a double bond, or to split a particular bond (for example, between carbon atoms 17 and 18 of a complex molecule having 25 carbon atoms).

Selectivity and specificity are highly important in electroorganic processes for the manufacture of complicated molecules of vitamins and hormones—as well as for the medicinal products whose action on pathogenic organisms is a function of their spatial arrangement, steric forms, and resonance.

The electrolytic approach can also be competitive for some low-cost, tonnage products. Here continuous processing is important, and only a single phase should be present, that is, a solution rather than an emulsion, dispersion, or mechanical mixture. Only for fairly valuable products is it practical to find a conducting solvent and then to engineer around it.

The electrolytic oxidation and reduction of organic compounds differ from the corresponding and more familiar inorganic reactions only in that organic reactions tend to be more complex and have low yields. The electrochemical principles are precisely those of inorganic reactions, while the procedures for handling the chemicals are precisely those of organic chemistry. [C.L.M.]

**Electrochemical series** A series in which the metals are listed in the order of their chemical reactivity, the most active at the top and the less reactive or more "noble" metals at the bottom. In a broader sense such an activity series need not be limited to the metals but may be carried on through the electronegative (nonmetallic) elements as well. *See* the table for a list of common elements.

The electrochemical series as it applies to metals was first established by laboratory experiments in which the purpose was to determine which metals would displace others from solutions of their salts. By exhaustive experiments it becomes possible to draw up a complete list in the order of chemical activity, in which the metals at the top of the list are those which are found to give up their electrons most readily (that is, are the most electropositive elements). Such a list is shown in the table, where lithium exhibits the most reactivity as a metal.

**Electrochemical series of the elements***

| | | | | | |
|---|---|---|---|---|---|
| Lithium | Li | Aluminum | Al | Molybdenum | Mo |
| Potassium | K | Titanium | Ti | Tin | Sn |
| Rubidium | Rb | Zirconium | Zr | Lead | Pb |
| Cesium | Cs | Manganese | Mn | Germanium | Ge |
| Radium | Ra | Vanadium | V | Tungsten | W |
| Barium | Ba | Niobium | Nb | **Hydrogen** | **H** |
| Strontium | Sr | Boron | B | | |
| Calcium | Ca | Silicon | Si | Copper | Cu |
| Sodium | Na | Tantalum | Ta | Mercury | Hg |
| Lanthanum | La | Zinc | Zn | Silver | Ag |
| Cerium | Ce | Chromium | Cr | Gold | Au |
| Magnesium | Mg | Gallium | Ga | Rhodium | Rh |
| Scandium | Sc | Iron | Fe | Platinum | Pt |
| Plutonium | Pu | Cadmium | Cd | Palladium | Pd |
| Thorium | Th | Indium | In | Bromine | Br |
| Beryllium | Be | Thallium | Tl | Chlorine | Cl |
| Uranium | U | Cobalt | Co | Oxygen | O |
| Hafnium | Hf | Nickel | Ni | Fluorine | F |

*According to standard oxidation potentials $E^{\circ}$ at 25°C (77°F).

To obtain an accurate and reproducible activity series, it is best to use the electrode potential, or oxidation-reduction potential, which is defined as the voltage developed by a sample of pure metal immersed in a solution of one of its salts (at unit activity and at 25°C or 77°F) versus a hydrogen electrode immersed in hydrochloric or sulfuric acid of equivalent concentration. *See* ELECTROCHEMISTRY; ELECTRODE POTENTIAL; OXIDATION-REDUCTION. [E.G.Ro.]

**Electrochemical techniques** Experimental methods developed to study the physical and chemical phenomena associated with electron transfer at the interface of an electrode and solution. The objective is to obtain either analytical or fundamental information regarding electroactive species in solution. Fundamental electrode characteristics may be investigated also.

The physical and chemical phenomena important in electrode processes generally occur very close to the electrode surface (usually within a few micrometers). Mass transfer of species involved in an electrode process to and from the bulk of solution is one important aspect. Inclusion of a large excess of inert electrolyte in most electrochemical systems eliminates electrical migration as an important means of mass transfer for electroactive species, and only convection and diffusion are considered.

Important chemical aspects of electrode processes include the oxidation or reduction occurring as a result of electron transfer, and coupled chemical reactions. Coupled reactions are initiated by production or depletion of the primary products or reactants at the electrode surface. Identification of the nature and mechanism of such coupled reactions is of particular importance in studies of electrode reactions of organic compounds, where multireaction cascades are often found to be initiated by electron transfer to or from an electrode.

The primary experimental variables involved in electrochemical techniques are the potential $E$, the current $I$, and the time $t$. Either the potential or current at the working electrode is controlled and the other observed as a function of time. The many ways in which either may be controlled give rise to a wide variety of controlled-potential or controlled-current techniques. In all such techniques it is necessary to specify whether mass transport of the electroactive species to the electrode is by convection or diffusion, since mathematical treatments of these two processes are quite different. Mass transport by convection is more efficient than diffusion by several orders of magnitude but is much more difficult to model mathematically.

The general scheme for electron transfer at an electrode in solution is shown in reaction (1), where $O$ and $R$ are the oxidized and reduced forms of the electroactive

$$O + ne^- \underset{k_b}{\overset{k_f}{\rightleftharpoons}} R \tag{1}$$

species and $n$ is the number of electrons transferred. When $k_f$ and $k_b$, the rate constants for the forward and back reaction, respectively, are very fast and $O$ and $R$ are not involved in preceding or following chemical reactions, the system is called reversible and the Nernst equation (2) holds. In this relation $E$ is the electrode potential, $E^{\circ\prime}$ is the

$$E = E^{\circ\prime} + \frac{0.059}{n} \log \frac{C_O}{C_R} \tag{2}$$

formal standard potential for the redox couple, and $C_O$ and $C_R$ are concentrations at the electrode surface. In the following discussions, only reversible reduction processes will be considered, although oxidations are equally applicable. *See* ELECTROCHEMICAL PROCESS; ELECTRODE; ELECTRODE POTENTIAL; ELECTROLYSIS.

**Controlled potential.** A variety of methods of this type have been developed, depending on whether the electrode potential is held constant or varied during the experiment and on whether mass transport is by convection (stirring) or diffusion.

In constant potential with convection, known as controlled potential electrolysis, the electrode potential is held constant as the solution is stirred or the electrode is rotated at a constant rate. The current is controlled by the concentration of electroactive substance and the stirring rate. Complete conversion of the electroactive substance takes place at a first-order rate whose constant is determined by cell geometry and stirring efficiency. Under ideal circumstances, electrolysis is complete in a manner of minutes.

In constant potential with diffusion (chronoamperometry), when a reducing potential is imposed instantaneously on a stationary working electrode in quiescent solution, current will rise sharply and then decay as the electroactive species in the electrode vicinity is depleted by electrolysis. The magnitude of the current is proportional to the bulk concentration of electroactive species, and if the imposed potential is sufficiently negative of $E^{\circ\prime}$, the Nernst equation demands complete conversion to the reduced form. Under these conditions the current for this so-called potential-step experiment is diffusion-controlled and decays with $1/t^{1/2}$.

An important variant on this experiment is double-potential-step chronoamperometry, in which a second potential step is applied. During the initial step, electrolysis occurs, depleting the oxidized form $O$, but producing the reduced form $R$ in the immediate vicinity of the electrode. If the potential is instantaneously switched back to the initial value after a time $\tau$, species $R$ will be reoxidized. A wide dynamic range of rate constants can be measured by varying $\tau$.

**Variable potential.** Several procedures have been developed in this area. In linear sweep voltammetry (LSV), the potential of a stationary electrode in quiescent solution is varied by applying a linear voltage ramp to the electrode. The resulting signal is recorded as a plot of current versus potential.

The most widely used electrochemical technique is cyclic voltammetry, which bears the same relation to linear sweep voltammetry as double-potential-step chronoamperometry does to the single-step experiment. That is, at the end of the first linear ramp, the potential is swept linearly back to the initial potential. Typical cyclic voltammograms are shown in illustration for the case of reduction of $O$ to $R$ and subsequent decomposition

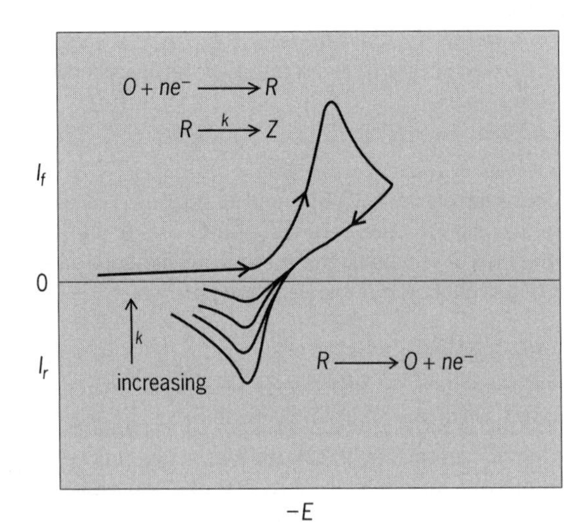


**Current-voltage curves in cyclic voltammetry.**

of $R$ to $Z$. The figure shows a series of cyclic voltammograms corresponding to different values of the rate constant $k$. As in the double-step experiment, the appearance of the forward sweep is unaffected by the coupled chemical process. However, as $k$ increases, the height of the peak on the reverse sweep due to reoxidation of $R$ decreases. Sweep rates may be varied over a range of roughly $10^{-2}$ to $10^{2}$ V/s in conventional cyclic voltammetry. With proper experimental design this may be extended considerably, as high as $10^{5}$ V/s or greater by using microelectrodes (electrodes with diameters in the micrometer range).

**Controlled current.** Controlled current techniques, although easier to implement than those in which the potential is controlled, are not as selective and therefore have fallen out of favor in recent years. In chronopotentiometry, a constant current is imposed at the working electrode, and its potential is monitored with time. Coulostatic analysis involves application of a very large, short pulse of current to the electrode, after which the cell circuit is opened. The current pulse charges up the electrode-solution interface to a new potential, and the electrode discharges, and returns to its original potential, by reducing the electroactive species in solution. [A.J.Fr.]

**Electrochemistry** The science dealing with the chemical changes accompanying the passage of an electric current, or the reverse process in which a chemical reaction is used as the source of energy to produce an electric current, as in a battery. Ionic conduction in electrolytes (liquid solutions, molten salts, and certain ionically conductive solids) is a phase of electrochemistry. Conduction in metals, semiconductors, and gases is generally considered a portion of physics. Other aspects of electrochemistry are described below. *See* ELECTROLYTIC CONDUCTANCE.

**Galvanic cells.** These are better known as electric batteries. Many chemical reactions can be arranged to produce electrical energy by physically separating the reaction into two half-reactions, one supplying electrons to an electrode forming the negative terminal of the cell, and the other removing the electrons from the positive terminal. *See* FUEL CELL.

**Electrodeposition.** The most important type of chemical reaction brought about by the passage of electric current is the deposition of a metal at a cathode from a solution of its ions. Electroforming is a variety of electrodeposition in which an article to be reproduced is rendered conductive by spraying a thin metallic coating, then electroplated with a metallic deposit that is stripped from its substrate and filled with backing to reproduce the original article. Electrowinning is used for the commercial production of active metals, such as aluminum, magnesium, and sodium, from molten salts and others, such as copper, manganese, and antimony, from aqueous solution. Electrorefining is commonly used to purify metals such as silver, lead, and copper. The impure metal is used as the anode, and purified metal is deposited at the cathode.

**Electrolytic processes.** Many electrode reactions other than metal deposition are of commercial or scientific use. Electrolysis of brine to yield chlorine at the anode, hydrogen at the cathode, and sodium hydroxide in the electrolyte is an important industrial process. Many organic compounds can be prepared electrolytically. *See* ELECTROLYSIS.

**Electroanalytical chemistry.** Many electrochemical measurements are useful for analytical purposes. Electrodes that are commonly used for analytical purposes through measurement of their potentials include the glass electrode for pH measurements, and ion-selective electrodes for certain ions, such as sodium or potassium ion (special glass compositions), calcium ion (liquid membrane), and fluoride ion (doped lanthanum fluoride single crystals). Polarography involves the use of a dropping mercury electrode as one electrode of an electrolytic cell. Qualitative analysis is carried out by measurement of characteristic potentials (half-wave potentials) for electrode processes, and quantitative analysis by measurement of diffusion-controlled currents. Coulometry involves the application of Faraday's law for analytical purposes.

Other analytical applications of electrochemistry include chronopotentiometry (measurement of potential-time transients under constant current conditions), and linear sweep and cyclic voltammetry (measurement of currents with linear voltage scan). Several titration methods involve electrochemical measurements, for example, conductometric, potentiometric, and amperometric titrations. *See* POLAROGRAPHIC ANALYSIS; TITRATION.

**Miscellaneous phenomena.** Electrochemical transport of ions through synthetic or natural membranes is important for processes, such as desalination of water and electrodialysis. In biological systems, the transmittal of nerve impulses and the generation of electrical signals, such as brain waves, are basically of electrochemical origin. A set of related phenomena can be grouped together under electrokinetic behavior, including the motion of colloidal particles in an electric field (electrophoresis), the motion of the liquid phase relative to a stationary solid under the influence of a potential gradient (electroosmosis), and the inverse generation of a potential gradient caused by a flowing liquid (streaming potential). Alternating-current phenomena, such as dielectric behavior, double-layer charging, and faradaic rectification, may also be included in a general definition of electrochemistry. Corrosion and passivation of metals are electrochemical in nature. [H.A.L.]

**Organic electrochemistry.** Organic electrochemistry involves the study of the chemical reactions that take place when an electric current is passed through a solution containing one or more organic compounds. It is a highly interdisciplinary science. Understanding and fully developing a given organic electrochemical reaction may involve techniques of synthesis, purification, and identification of organic compounds, as well as theory and practice of a variety of sophisticated electroanalytical techniques, surface science, cell design, electronics, engineering scaleup, and materials modification. Organic electrochemistry has been of increasing interest for industrial applications in recent years because the costs of electrochemistry have been rising more slowly than the costs of conventional chemical reagents, and because electrochemical procedures can be environmentally less intrusive than other chemical processes. A number of so-called fine chemicals are made electrochemically on a scale ranging from several kilograms to several tons per day. A few chemicals are made on much larger scale; the best known is adiponitrile [$NC(CH_2)_4CN$], a key intermediate in the synthesis of nylon. *See* ORGANIC CONDUCTOR.

Cathodic organic reactions fall into several categories: cleavage of single bonds, reduction of functional groups, and reduction of large conjugated systems such as activated alkenes and aromatic compounds. Anodic reactions are equally diverse. The oldest and best-known anodic reaction is the Kolbe reaction, in which electrochemical oxidation of carboxylate ions yields dimeric products with evolution of carbon dioxide. A wide range of functional groups can be accommodated in this versatile reaction.

The powerful yet precisely controllable oxidizing and reducing conditions that can be achieved electrochemically are useful for generating novel organic intermediates, including carbocations, carbanions, radicals, radical ions, carbenes, nitrenes, and arynes. These reactive intermediates are frequently trapped by other organic substances added to the medium to extend the synthetic utility of an electrode reaction.

Voltammetric methods are commonly used to provide detailed mechanistic information. In these, the electrode potential is varied in a controlled fashion in the vicinity of the redox potential of the substrate, and the current response is measured as a function of experimental variables such as scan rate, concentration, and added reagents. Other methods include identification of the products from a preparative electrolysis (generally carried out at controlled potential in mechanistic experiments); coulometry, or measurement of the actual amount of current passed in the electrolysis; comparison of experimental responses with computer simulations of the theoretical behavior for various mechanisms; studies of changes in the structure of the electrode surface during electrolysis; and spectroscopic identification of intermediates. Microelectrodes, which have micrometer diameters, extend the speeds with which electrochemical measurements can be made, and thus permit measurement of even very fast rates of chemical reactions associated with electron transfers at electrodes. *See* ORGANIC REACTION MECHANISM; REACTIVE INTERMEDIATES. [A.J.Fr.]

**Electrode** An electrical conductor through which an electric current enters or leaves a conducting medium, whether it be an electrolytic solution, solid, molten mass, gas, or vacuum. For electrolytic solutions, many solids, and molten masses, an electrode is an electric conductor at the surface of which a change occurs from conduction by electrons to conduction by ions. For gases and vacuum, the electrodes merely serve to conduct electricity to and from the medium. *See* ELECTRODE POTENTIAL; ELECTROLYSIS; ELECTROMOTIVE FORCE (CELLS). [W.J.H.]

**Electrode potential** The equilibrium potential difference between two conducting phases in contact, most often an electronic conductor such as a metal or semiconductor on the one hand, and an ionic conductor such as an electrolyte solution (a solution containing ions) on the other. Electrode potentials are not experimentally accessible, but the differences in potential between two electronic conductors making contact with the same ionic conductor (that is, the difference between two electrode potentials) can be measured. A useful scale of electrode potentials can therefore be obtained when a particular electrode potential is set equal to zero by definition. There are several conventions, based on different definitions of the zero point on the scale of electrode potential, but all tables use the so-called standard hydrogen convention. *See* ELECTRODE; REFERENCE ELECTRODE.

The interfacial potential difference is usually the consequence of the transfer of some charge carriers from one conducting phase to the other. For example, when a piece of silver, which contains silver ions and free, so-called conduction electrons, is in contact with an aqueous solution of silver nitrate, the only species common to the two phases are the silver ions. Their concentration (volume density) is constant in the metal but variable (from zero to the solubility limit of the silver salt used) in the solution. When more silver ions transfer from the solution to the metal than in the opposite direction, an excess of negatively charged nitrate ions remains in the solution, which therefore acquires a negative charge. However, the metal gains more silver ions than it loses, and therefore acquires a positive charge. Such a charge separation leads to a potential difference across the boundary between the two phases. The continued buildup of such charges makes the potential of the metal more and more positive with respect to that of the solution. This effect in turn leads to electrostatic repulsion of the silver ions in the solution phase immediately adjacent to the metal; these are the very metal ions that are candidates for transfer across the boundary. Consequently, the electrostatic repulsion decreases the tendency of silver ions to move from the solution to the metal, and eventually the process reaches equilibrium, at which point the tendency of ions to transfer is precisely counterbalanced by the repulsion of the candidate ions by the existing potential difference. At that potential, there is no further net transfer of charges between the contacting phases, although individual charges can still exchange across the phase boundary, a process which gives rise to the exchange current.

For a metal in contact with its metal ions of valence $z$, the potential difference $E$ can be expressed in terms of a standard potential $E^\circ$ (describing the affinity of the metal for its ions) and the concentration $c$ of these ions in solution through the Nernst equation (1), where $R$ is the gas constant, $T$ is the absolute temperature, and $F$ is the Faraday.

$$E = E^\circ + (RT/zF)\ln c \qquad (1)$$

When the metal ions in solution form a sparingly soluble salt, the solubility equilibrium can be used to convert a metal electrode responding to its own metal cations (positive ions) into an electrode responding to the concentration of anions (negative ions). Typical examples are the silver/silver chloride electrode, based on the low solubility of silver chloride (AgCl), and the calomel electrode, based on the poor solubility of calomel ($Hg_2Cl_2$).

An equilibrium potential difference between a metal and an electrolyte solution can also be established when the latter contains a redox couple, that is, a pair of chemical components that can be converted into each other by the addition or withdrawal of electrons, by reduction or oxidation respectively. In that case the metal often merely acts as the supplier or acceptor of electrons. When metal electrons are donated to the solution, the oxidized form of the redox couple is reduced; when the metal withdraws electrons from the redox couple, its reduced component is oxidized. Again, the buildup of a charge separation generates a potential difference, which counteracts the electrochemical charge transfer and eventually brings the process to equilibrium, a state in which the rate of oxidation is exactly equal to the rate of reduction. The dependence of the equilibrium electrode potential on the concentrations of $c_{ox}$ and $c_{red}$ of the oxidized and reduced forms respectively is described by a Nernst equation of the form (2),

$$E = E^\circ + (RT/nF)\ln(c_{ox}/c_{red}) \qquad (2)$$

where $n$ denotes the number of electrons ($e^-$) transferred between the oxidized species (Ox) and the reduced species (Red) in the reactions $Ox + ne^- \rightleftarrows Red$. A typical example is the reduction of hydrogen ions $H^+$ to dissolved hydrogen molecules $H_2$, and vice versa, in which case the reactions are $2H^+ + 2e^- \rightleftarrows H_2$, and for which the Nernst equation is of the form (3).

$$E = E^\circ + (RT/2F)\ln(c_{H^+}^2/c_{H_2}) \qquad (3)$$

Redox potentials involving a gas are often established slowly, if at all. For determinations of such a redox potential, platinum is often used as the metal, because it is chemically and electrochemically stable. *See* OXIDATION-REDUCTION.

Electrode potentials can also be established at double phase boundaries, such as that between two aqueous solutions separated by a glass membrane. This glass electrode is commonly used for measurements of the pH, a measure of the acidity or basicity of solutions. The mechanism by which the glass electrode operates involves ion exchange of hydrogen ions at the two glass-solution interfaces. *See* ION EXCHANGE.

In all the above examples, the two contacting phases can have only one type of charge carrier in common. Usually, no equilibrium potential difference is established when more than one type of charge carrier can cross the interface, but often (depending on the nature of the metal and of the chemical components of the solution, and sometimes also depending on the geometry of the contact region) an apparently stable potential can still be obtained, which corresponds to zero net charge transfer. This can be a so-called mixed potential, important in metal corrosion, or a junction potential, which figures in most measurements of electrochemical potentials and usually limits the accuracy and precision of such measurements, including that of the pH. *See* ELECTROCHEMICAL SERIES.

In determining electrode potentials, there are several complications. In the first place, it follows from thermodynamics that the Nernst equation should be written in terms of activities rather than concentrations. The difference between these two parameters is often small but seldom completely negligible. *See* ACTIVITY (THERMODYNAMICS).

Second, measurements of potential differences always involve the potential difference between two metals rather than that between a metal and a solution. Therefore, electrode potentials as defined above cannot be measured either. Because these measurements involve a potential difference, there has been considerable confusion about the definition of that difference; that is, whether it is defined as the potential of the metal minus that of the solution, or the other way around. This is a matter of a sign convention. The problem is usually framed in terms of oxidation potentials versus reduction potentials.

There are four main applications for measurements of electrode potentials: (1) in the establishment of the oxidative and reductive power of redox systems, the so-called electromotive series; (2) as concentration probes, such as in pH measurements; (3) as sources of chemical equilibrium data; and (4) as the primary (or independent) variable in studies of electrode reactions. [R.DeD.]

**Electrokinetic phenomena** Phenomena associated with the movement of charged particles through a continuous medium or with the movement of a continuous medium over a charged surface. The four principal electrokinetic phenomena are electrophoresis, electroosmosis, streaming potential, and sedimentation potential, or Dorn effect. These phenomena are related to one another through the zeta potential $\zeta$ of the electrical double layer which exists in the neighborhood of the charged surface. *See* ELECTROPHORESIS; STREAMING POTENTIAL. [Q.V.W.]

**Electrolysis** A means of producing chemical changes through reactions at electrodes in contact with an electrolyte by the passage of an electric current. Electrolysis cells, also known as electrochemical cells, generally consist of two electrodes connected to an external source of electricity (a power supply or battery) and immersed in a liquid that can conduct electricity through the movement of ions. Reactions occur at both electrode-solution interfaces because of the flow of electrons. Reduction reactions, where substances add electrons, occur at the electrode called the cathode; oxidation reactions, where species lose electrons, occur at the other electrode, the anode. In the cell shown in the illustration, water is reduced at the cathode to produce hydrogen gas and hydroxide ion; chloride ion is oxidized at the anode to generate chlorine gas. Electrodes are typically constructed of metals (such as platinum or steel) or carbon. Electrolytes usually consist of salts dissolved in either water or a nonaqueous solvent, or they are molten salts. *See* ELECTROCHEMISTRY; ELECTRODE; ELECTROLYTE; OXIDATION-REDUCTION.

There are many industrial applications for the production of important inorganic chemicals. Chlorine and alkali are produced by the large-scale electrolysis of brine (the chloralkali process) in cells carrying out the same reactions as those shown in the illustration. Other chemicals produced include hydrogen and oxygen (via water electrolysis), chlorates, peroxysulfate, and permanganate.

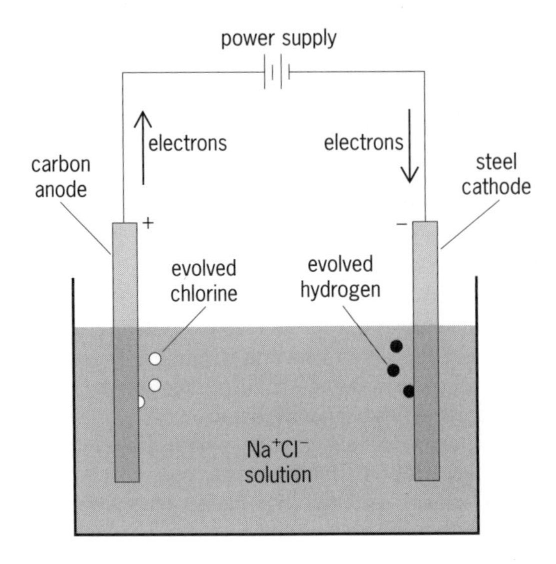


**Schematic diagram of an electrolysis cell in which the electrolyte is a solution of sodium chloride.**

The major electrolytic processes involving organic compounds are the hydrodimerization of acrylonitrile to produce adiponitrile and the production of tetraethyllead. Many other organic compounds have been studied on the laboratory scale.

Electroplating involves the electrochemical deposition of a thin layer of metal on a conductive substrate, for example, to produce a more attractive or corrosion-resistant surface. Chromium, nickel, tin, copper, zinc, cadmium, lead, silver, gold, and platinum are the most frequently electroplated metals. Metal surfaces can also be electrolytically oxidized (anodized) to form protective oxide layers. This surface-finishing technique is most widely used for aluminum but is also used for titanium, copper, and steel.

Metals can be purified by electrorefining. Here, the impure metal is used as the anode, which dissolves during the electrolysis. The metal is plated, in purer form, on the cathode. Copper, nickel, cobalt, lead, and tin are all purified by this technique.

Electroanalysis involves the use of electrolytic processes to identify and quantitate a species. Coulometric methods are based on measuring the quantity of electricity used for a desired process. Voltammetric methods allow characterization of species through an analysis of the effect of potential and electrolysis conditions on the observed currents. [A.J.Ba.]

**Electrolyte** A material that conducts an electric current when it is fused or dissolved in a solvent, usually water. Electrolytes are composed of positively charged species, called cations, and negatively charged species, called anions. For example, sodium chloride (NaCl) is an electrolyte composed of sodium cations ($Na^+$) and chlorine anions ($Cl^-$). The ratio of cations to anions is always such that the substance is electrically neutral. If two wires connected to a light bulb and to a power source are placed in a beaker of water, the light bulb will not glow. If an electrolyte, such as sodium chloride, is dissolved in the water, the light bulb will glow because the solution can now conduct electricity. The amount of electric current that can be carried by an electrolyte solution is proportional to the number of ions dissolved. Thus, the bulb will glow more brightly if the amount of sodium chloride in the solution is increased. *See* Ion.

Any substance that produces ions when dissolved is an electrolyte. These substances include ionic materials composed of simple monatomic ions, such as sodium chloride, or substances composed of polyatomic ions, such as ammonium nitrate ($NH_4NO_3$). When these substances are dissolved, hydrated ions are generated, as in reactions (1) and (2).

$$NaCl \rightarrow Na^+ + Cl^- \quad (1)$$

$$NH_4NO_3 \rightarrow NH_4{}^+ + NO_3{}^- \quad (2)$$

A special type of electrolyte is an acid, in which the cation is $H^+$. When acids are dissolved in water, $H^+$ ions are produced along with an anion, which can be either a monatomic ion, as in hydrochloric acid (HCl), or a polyatomic ion, as in nitric acid ($HNO_3$) or acetic acid ($HC_2H_3O_2$), as in reactions (3)–(5).

$$HCl \rightarrow H^+ + Cl^- \quad (3)$$

$$HNO_3 \rightarrow H^+ + NO_3{}^- \quad (4)$$

$$HC_2H_3O_2 \rightarrow H^+ + C_2H_3O_2{}^- \quad (5)$$

Electrolytes such as sodium hydroxide (NaOH) that yield the hydroxyl ($OH^-$) anion when dissolved in water are called bases [reaction (6)]. Some molecules that do not

$$NaOH \rightarrow Na^+ + OH^- \quad (6)$$

contain ions, such as ammonia ($NH_3$), generate $OH^-$ ions when dissolved in water as in reaction (7); these bases are also electrolytes. Polar covalent molecules, such as

$$NH_3 + H_2O \rightarrow NH_4{}^+ + OH^- \quad (7)$$

ethanol, dissolve in water but do not generate any ions and are called nonelectrolytes.

Many electrolytic substances completely dissociate into ions when they are dissolved in water. In these cases, the reactions shown above would proceed completely to the right, leaving only dissolved ions and no associated, electrically neutral molecules. For example, when sodium chloride is dissolved in water, all of the dissolved material is present as $Na^+$ and $Cl^-$ ions, with no dissolved NaCl molecules. Substances that completely dissociate are called strong electrolytes, because every molecule dissolved generates ions that contribute to the electrical conductivity. Ionic substances, such as NaCl and $NH_4NO_3$, are strong electrolytes. Acids and bases that completely dissociate are called strong acids and strong bases, and these substances are also strong electrolytes. Hydrochloric acid (HCl) and $HNO_3$ are strong acids, so every molecule of HCl or $HNO_3$ that dissolves generates a $H^+$ cation and a $Cl^-$ or $NO_3{}^-$ anion, respectively.

Some substances dissolve in water but do not dissociate completely. For example, when acetic acid is dissolved in water, some molecules dissociate to form $H^+$ and $C_2H_3O_2{}^-$ ions, while others remain associated as $C_2H_3O_2H$ units, which contain polar O-H bonds and are therefore readily soluble in water. Acids such as acetic acid that do not dissociate completely are called weak acids. Similarly, the reaction of ammonia with water in reaction (7) proceeds to only a small extent. Thus, some molecules of ammonia react with water to generate $OH^-$ and $NH_4{}^+$ ions, but many simply remain as $NH_3$. Since each dissolved molecule of ammonia does not generate an $OH^-$ anion, ammonia is said to be a weak base. *See* IONIC EQUILIBRIUM.

In biological systems, electrolytes play important roles in regulating kidney function and the retention of water. Electrolytes are also vital for providing the electric current needed for nerve impulses in neurons. [H.H.T]

**Electrolytic conductance** The transport of electric charges, under electric potential differences, by particles of atomic or larger size. This phenomenon is distinguished from metallic conductance, which is due to the movement of electrons. The charged particles that carry the electricity are called ions.

Positively charged ions are termed cations; the sodium ion, $Na^+$, is an example. The negatively charged chloride ion, $Cl^-$, is typical of anions. The negative charges are identical with those of electrons or integral multiples thereof. The unit positive charges have the same magnitude as those of electrons but are of opposite sign. Colloidal particles, which may have relatively large weights, may be ions, and may carry many positive or negative charges. Electrolytic conductors may be solids, liquids, or gases. Semiconductors have properties that are intermediate between the metallic and electrolytic types. *See* ELECTROCHEMISTRY; ELECTROMOTIVE FORCE (CELLS).

Conductances are usually reported as specific conductances $\kappa$, which are the reciprocals of the resistances of cubes of the materials, 1 cm (0.39 in.) in each dimension, placed between electrodes 1 cm square, on opposite sides. These units are sometimes called mhos, that is, ohm spelled backward. Conductances of solutions are usually measured by a method in which a Wheatstone bridge is employed. [H.A.L.]

**Electromotive force (cells)** The voltage or electric potential difference across the terminals of a cell when no current is drawn from it. The electromotive force (emf) is the sum of the electric potential differences produced by a separation of charges (electrons or ions) that can occur at each phase boundary (or interface) in the cell. The magnitude of each potential difference depends on the chemical nature of the two contacting phases. Thus, at the interface between two different metals, some electrons will have moved from the metal with a higher free energy of electrons to the metal with a lower free energy of electrons. The resultant charge separation will produce a potential difference, just as charge separation produces a voltage across a capacitor; at equilibrium this exactly opposes further electron flow. Similarly, potential differences can be produced when electrons partition across a metal|solution interface or metal|solid interface, and when ions partition across a solution|membrane|solution interface. *See* CHEMICAL THERMODYNAMICS.

**Origin.** How a cell emf is composed of the sum of interfacial potential differences is shown by the Daniell cell (see illustration), where aq soln denotes an aqueous solution, a solid line indicates a phase boundary, and the broken line a porous barrier permeable to all the ions in the adjacent solutions. The copper connector to the zinc electrode is denoted Cu′. The barrier prevents physical mixing of the zinc sulfate ($ZnSO_4$) and copper sulfate ($CuSO_4$) solutions.

The cell emf is the open-circuit (that is, zero current) potential difference measured between the two Cu leads (any potential-measuring device must ultimately measure the potential difference between two chemically identical phases, in this example the Cu and Cu′ phases).

It is convenient to describe any electrochemical cell in terms of half cells. A half cell consists of an oxidant (Ox) and reductant (Red) such that $\mathrm{Ox} + ne^- \rightleftharpoons \mathrm{Red}$; species Ox and Red are commonly referred to as a redox couple. The Daniell cell, for example, comprises the two half cells $Cu^{2+} + 2e^- \rightleftharpoons Cu$ (redox couple is $Cu^{2+} \mid Cu$) and $Zn^{2+} + 2e^- \rightleftharpoons Zn$ (redox couple is $Zn^{2+} \mid Zn$) with the half-cell potentials $E_{Cu^{2+}|Cu}$ and $E_{Zn^{2+}|Zn}$

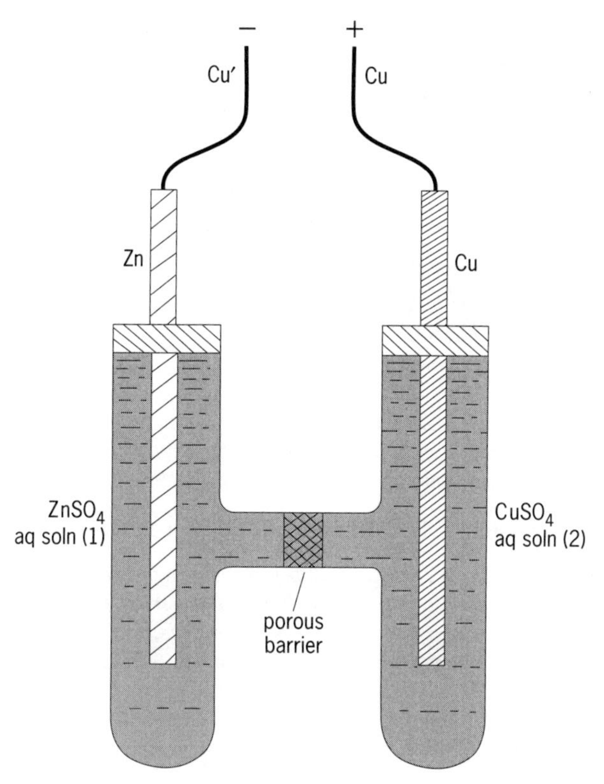


**Cross-section diagram of Daniell cell. (*After W. J. Moore, Physical Chemistry, 5th ed., Longman, 1974*)**

given by Eqs. (1) and (2), where $F$ is the Faraday constant (96485.3 coulombs for

$$E_{Cu^{2+}|Cu} = E^{\circ}_{Cu^{2+}|Cu} + \frac{RT}{2F} \ln a_{Cu^{2+}} \quad (1)$$

$$E_{Zn^{2+}|Zn} = E^{\circ}_{Zn^{2+}|Zn} + \frac{RT}{2F} \ln a_{Zn^{2+}} \quad (2)$$

the Avogadro number of electrons), $R$ is the gas constant (8.3144 joules/mole/degree Kelvin), $T$ is the temperature in Kelvin, and $a$ is the activity of the species shown. Such expressions are useful only if the standard electrode potential $E^{\circ}$ value for each half cell is known. Values of $E^{\circ}$ can be assigned to any given half cell by arbitrarily specifying that $E^{\circ}_{H^+}$ $H_2$ [the $E^{\circ}$ for the standard hydrogen electrode (SHE) half cell, $H^+$(aq, $a = 1) + e^- \rightleftharpoons {}^1\!/_2H_2$(g, 1 atm)] is zero. The temperature dependence $dE^{\circ}_{H^+|H^2}/dT$ is also specified as zero. $E^{\circ}$ values versus the SHE for selected half cells are given in the table. In principle, any $E^{\circ}$ (and its temperature dependence) can be measured directly versus the SHE, or another half cell whose electrode potential and temperature dependence have been determined. Such an electrode is termed a reference electrode. The reference electrode is a half cell designed so that its potential is stable, and reproducible, and it neither contaminates nor is contaminated by the medium in which it is immersed. Two convenient reference electrodes commonly used in aqueous systems are the saturated calomel and the silver|silver chloride electrodes (see table). *See* ELECTROCHEMISTRY; ELECTRODE POTENTIAL; OXIDATION-REDUCTION; REFERENCE ELECTRODE.

**Selected standard electrode potentials at 25°C**

| Electrode reaction | $E^\circ$/V |
|---|---|
| $Li^+(aq) + e^- \rightleftharpoons Li(s)$ | −3.045 |
| $Zn^{2+}(aq) + 2e^- \rightleftharpoons Zn(s)$ | −0.763 |
| $2H^+(aq) + 2e^- \rightleftharpoons H_2(g)$ | 0 |
| $Hg_2Cl_2(s) + 2e^- \rightleftharpoons 2Hg(s) + 2Cl^-$ (sat aq KCl) | 0.246 |
| $AgCl(s) + e^- \rightleftharpoons Ag(s) + Cl^-(aq)$ | 0.222 |
| $Cu^{2+}(aq) + 2e^- \rightleftharpoons Cu(s)$ | 0.337 |
| $Fe(CN)_6^{3-}(aq) + e^- \rightleftharpoons Fe(CN)_6^{4-}(aq)$ | 0.69 |
| $O_2(g) + 4H^+(aq) + 4e^- \rightleftharpoons 2H_2O$ | 1.223 |
| $F_2(g) + 2H^+ + 2e^- \rightleftharpoons 2HF(aq)$ | 3.06 |

The emf produced by a single cell or a set of cells in series (a battery) is used as a dc power source for a wide array of applications, ranging from powering wristwatches to emergency power supplies. The emf of the saturated Weston cell (scheme 3;

$$\mathrm{Hg(liquid)} \mid \mathrm{Hg_2SO_4(solid)} \mid \mathrm{CdSO_4\ (sat\ aq)} \mid \mathrm{Cd(Hg)} \qquad (3)$$

sat aq = saturated aqueous solution) still serves as a high-level voltage reference for the National Institute of Standards and Technology. However, Josephson arrays are now considered the most precise voltage references, and Zener diodes are used to produce reference voltages for many laboratory and field voltage measurement devices.

There are also a few photoelectrochemical cells with possible application in solar energy conversion, such as the Grätzel cell (scheme 4; nonaq = nonaqueous solution),

$$\mathrm{TiO_2(solid)} \mid \mathrm{adsorbed\ Ru^{2+}\ dye} \mid \mathrm{I_2,\ I\ (nonaq)} \mid \mathrm{C} \qquad (4)$$

where the cell emf is produced by absorption of visible light.

The emf of a cell can also be used as an indicator of chemical composition. Devices that depend on the measurement of an open circuit-cell potential work well when the device is sensitive to a single analyte, but they are notoriously sensitive to interferences. An example is the so-called alkaline error that occurs when glass electrodes are used to measure very high pH values. For complex systems containing several species that can undergo electrochemical reaction, individual $E^\circ$ values can be determined using electroanalytical techniques that involve controlling the potential applied to a cell with measurement of the resultant current. Techniques such as polarography and cyclic voltammetry, for example, involve changing the potential of an indicator electrode and observing a wave or peak in the current at the redox potential of the species in solution; the height of the wave or peak indicates the concentration of that species. *See* ELECTROCHEMICAL TECHNIQUES; FUEL CELL; POLAROGRAPHIC ANALYSIS. [M.D.Ar.; S.W.F.]

**Electron** An elementary particle which is the negatively charged constituent of ordinary matter. The electron is the lightest known particle which possesses an electric charge. Its rest mass is $m_e \cong 9.1 \times 10^{-28}$ g, about $^1/_{1836}$ of the mass of the proton or neutron, which are, respectively, the positively charged and neutral constituents of ordinary matter. Discovered in 1895 by J. J. Thomson in the form of cathode rays, the electron was the first elementary particle to be identified.

The charge of the electron is $-e \cong -4.8 \times 10^{-10}$ esu $= -1.6 \times 10^{-19}$ coulomb. The sign of the electron's charge is negative by convention, and that of the equally charged proton is positive. This is a somewhat unfortunate convention, because the flow of electrons in a conductor is thus opposite to the conventional direction of the current.

Electrons are emitted in radioactivity (as beta rays) and in many other decay processes; for instance, the ultimate decay products of all mesons are electrons, neutrinos, and photons, the meson's charge being carried away by the electrons. The electron itself is completely stable. Electrons contribute the bulk to ordinary matter; the volume of an atom is nearly all occupied by the cloud of electrons surrounding the nucleus, which occupies only about $10^{-13}$ of the atom's volume. The chemical properties of ordinary matter are determined by the electron cloud.

The electron obeys the Fermi-Dirac statistics, and for this reason is often called a fermion. One of the primary attributes of matter, impenetrability, results from the fact that the electron, being a fermion, obeys the Pauli exclusion principle; the world would be completely different if the lightest charged particle were a boson, that is, a particle that obeys Bose-Einstein statistics. [C.J.G.]

**Magnetic moment.** The electron has magnetic properties by virtue of (1) its orbital motion about the nucleus of its parent atom and (2) its rotation about its own axis. The magnetic properties are best described through the magnetic dipole moment associated with 1 and 2. The classical analog of the orbital magnetic dipole moment is the dipole moment of a small current-carrying circuit. The electron spin magnetic dipole moment may be thought of as arising from the circulation of charge, that is, a current, about the electron axis; but a classical analog to this moment has much less meaning than that to the orbital magnetic dipole moment. The magnetic moments of the electrons in the atoms that make up a solid give rise to the bulk magnetism of the solid.

**Spin.** That property of an electron which gives rise to its angular momentum about an axis within the electron. Spin is one of the permanent and basic properties of the electron. Both the spin and the associated magnetic dipole moment of the electron were postulated by G. E. Uhlenbeck and S. Goudsmit in 1925 as necessary to allow the interpretation of many observed effects, among them the so-called anomalous Zeeman effect, the existence of doublets (pairs of closely spaced lines) in the spectra of the alkali atoms, and certain features of x-ray spectra.

The spin quantum number is $s$, where $s$ is always $^1/_2$. This means that the component of spin angular momentum along a preferred direction, such as the direction of a magnetic field, is $\pm\ ^1/_2\ \hbar$, where $\hbar$ is Planck's constant $h$ divided by $2\pi$. The spin angular momentum of the electron is not to be confused with the orbital angular momentum of the electron associated with its motion about the nucleus. In the latter case the maximum component of angular momentum along a preferred direction is $l\hbar$, where $l$ is the angular momentum quantum number and may be any positive integer or zero.

The electron has a magnetic dipole moment by virtue of its spin. The approximate value of the dipole moment is the Bohr magneton $\mu_0$ which is equal to $eh/4\pi mc = 9.27 \times 10^{-21}$ erg/oersted, where $e$ is the electron charge measured in electrostatic units, $m$ is the mass of the electron, and $c$ is the velocity of light. (In SI units, $\mu_0 = 9.27 \times 10^{-24}$ joule/tesla.) The orbital motion of the electron also gives rise to a magnetic dipole moment $\mu_l$, that is equal to $\mu_0$ when $l = 1$. [Ar.R.]

**Electron affinity** The amount of energy released when an electron at rest is captured by a species M, producing the negative ion $M^-$. The electron affinity of a species M can also be thought of as the ionization potential of the negative ion $M^-$. Stated in terms of a chemical equation, the electron affinity of a species M is equal to the exothermocity of the reaction $e + \mathrm{M} \rightarrow \mathrm{M}^-$, where the negative ion $\mathrm{M}^-$ is left in its lowest electronic, vibrational, and rotational state. *See* IONIZATION POTENTIAL.

If the electron affinity of M is negative, the $\mathrm{M}^-$ ion is unstable with respect to decomposition into $\mathrm{M} + e$. Most atoms have positive electron affinities, even though there is no net Coulomb attraction between the electron and the atom until the electron is close enough to be "a part of the atom." The simple rules of chemical valency provide a qualitative guide to the magnitude of electron affinities. Thus the noble gases, which have a filled outer electronic shell and are chemically inert, are not capable of binding an additional electron to form a negative ion. The largest electron affinities are possessed by the halogens, atoms which require only one additional electron to fill the valence shell. *See* VALENCE.

The major exception to this concept is that multiply charged negative ions—for example, $O^{2-}$, one of many multiply charged negative ions which are stable in solution—are not stable in the gas phase. The ability to place more than one additional electron in the valence shell of a neutral atom or molecule appears to come from the medium: the solvent shell surrounding the ion in liquid solutions and the amorphous or crystalline region surrounding the ion in solids. [W.C.L.]

**Electron configuration** The orbital arrangement of an atom's electrons. Negatively charged electrons are attracted to a positively charged nucleus to form an atom or ion. Although such bound electrons exhibit a high degree of quantum-mechanical wavelike behavior, there still remain particle aspects to their motion. Bound electrons occupy orbitals that are somewhat concentrated in spatial shells lying at different distances from the nucleus. As the set of electron energies allowed by quantum mechanics is discrete, so is the set of mean shell radii. Both these quantized physical quantities are primarily specified by integral values of the principal, or total, quantum number $n$. The full electron configuration of an atom is correlated with a set of values for all the quantum numbers of each and every electron. In addition to $n$, another important quantum number is $l$, an integer representing the orbital angular momentum of an electron in units of $h/2\pi$, where $h$ is Planck's constant. The values 1, 2, 3, 4, 5, 6, 7 for $n$ and 0, 1, 2, 3 for $l$ together suffice to describe the electron configurations of all known normal atoms and ions, that is, those that have their lowest possible values of total electronic energy. The first seven shells are also given the letter designations $K$, $L$, $M$, $N$, $O$, $P$, and $Q$ respectively. Electrons with $l$ equal to 0, 1, 2, and 3 are designated $s$, $p$, $d$, and $f$, respectively.

In any configuration the number of equivalent electrons (same $n$ and $l$) is indicated by an integral exponent (not a quantum number) attached to the letters $s$, $p$, $d$, and $f$. According to the Pauli exclusion principle, the maximum is $s^2$, $p^6$, $d^{10}$, and $f^{14}$.

An electron configuration is categorized as having even or odd parity, according to whether the sum of $p$ and $f$ electrons is even or odd. Strong spectral lines result only from transitions between configurations of unlike parity.

Insofar as they are known from spectroscopic investigations, the electron configurations characteristic of the normal or ground states of the first 103 chemical elements are shown in the following table:

**Distribution of electrons in the atoms**

| Element and atomic number | | K | L | | M | | | N | | | | O | | | | Ground term | Ionization potential, eV |
|---|---|---|---|---|---|---|---|---|---|---|---|---|---|---|---|---|---|
| | | 1,0 1s | 2,0 2s | 2,1 2p | 3,0 3s | 3,1 3p | 3,2 3d | 4,0 4s | 4,1 4p | 4,2 4d | 4,3 4f | 5,0 5s | 5,1 5p | 5,2 5d | 5,3 5f | | |
| H | 1 | 1 | — | — | — | — | — | — | — | — | — | — | — | — | — | $^2S_{1/2}$ | 13.5981 |
| He | 2 | 2 | — | — | — | — | — | — | — | — | — | — | — | — | — | $^1S_0$ | 24.5868 |
| Li | 3 | 2 | 1 | — | — | — | — | — | — | — | — | — | — | — | — | $^2S_{1/2}$ | 5.3916 |
| Be | 4 | 2 | 2 | — | — | — | — | — | — | — | — | — | — | — | — | $^1S_0$ | 9.322 |
| B | 5 | 2 | 2 | 1 | — | — | — | — | — | — | — | — | — | — | — | $^2P^\circ_{1/2}$ | 8.298 |
| C | 6 | 2 | 2 | 2 | — | — | — | — | — | — | — | — | — | — | — | $^3P_0$ | 11.260 |
| N | 7 | 2 | 2 | 3 | — | — | — | — | — | — | — | — | — | — | — | $^4S^\circ_{3/2}$ | 14.534 |
| O | 8 | 2 | 2 | 4 | — | — | — | — | — | — | — | — | — | — | — | $^3P_2$ | 13.618 |
| F | 9 | 2 | 2 | 5 | — | — | — | — | — | — | — | — | — | — | — | $^2P^\circ_{3/2}$ | 17.422 |
| Ne | 10 | 2 | 2 | 6 | — | — | — | — | — | — | — | — | — | — | — | $^1S_0$ | 21.564 |
| Na | 11 | Neon configuration | | | 1 | — | — | — | — | — | — | — | — | — | — | $^2S_{1/2}$ | 5.139 |
| Mg | 12 | | | | 2 | — | — | — | — | — | — | — | — | — | — | $^1S_0$ | 7.646 |
| Al | 13 | | | | 2 | 1 | — | — | — | — | — | — | — | — | — | $^2P^\circ_{1/2}$ | 5.986 |
| Si | 14 | | | | 2 | 2 | — | — | — | — | — | — | — | — | — | $^3P_0$ | 8.151 |
| P | 15 | | | | 2 | 3 | — | — | — | — | — | — | — | — | — | $^4S^\circ_{3/2}$ | 10.486 |
| S | 16 | | | | 2 | 4 | — | — | — | — | — | — | — | — | — | $^3P_2$ | 10.360 |
| Cl | 17 | | | | 2 | 5 | — | — | — | — | — | — | — | — | — | $^2P^\circ_{3/2}$ | 12.967 |
| Ar | 18 | | | | 2 | 6 | — | — | — | — | — | — | — | — | — | $^1S_0$ | 15.759 |
| K | 19 | Argon configuration | | | | | — | 1 | — | — | — | — | — | — | — | $^2S_{1/2}$ | 4.341 |
| Ca | 20 | | | | | | — | 2 | — | — | — | — | — | — | — | $^1S_0$ | 6.113 |
| Sc | 21 | | | | | | 1 | 2 | — | — | — | — | — | — | — | $^2D_{3/2}$ | 6.54 |
| Ti | 22 | | | | | | 2 | 2 | — | — | — | — | — | — | — | $^3F_2$ | 6.82 |
| V | 23 | | | | | | 3 | 2 | — | — | — | — | — | — | — | $^4F_{3/2}$ | 6.74 |
| Cr | 24 | | | | | | 5 | 1 | — | — | — | — | — | — | — | $^7S_3$ | 6.765 |
| Mn | 25 | | | | | | 5 | 2 | — | — | — | — | — | — | — | $^6S_{5/2}$ | 7.432 |
| Fe | 26 | | | | | | 6 | 2 | — | — | — | — | — | — | — | $^5D_4$ | 7.870 |
| Co | 27 | | | | | | 7 | 2 | — | — | — | — | — | — | — | $^4F_{9/2}$ | 7.86 |
| Ni | 28 | | | | | | 8 | 2 | — | — | — | — | — | — | — | $^3F_4$ | 7.635 |
| Cu | 29 | | | | | | 10 | 1 | — | — | — | — | — | — | — | $^2S_{1/2}$ | 7.726 |
| Zn | 30 | | | | | | 10 | 2 | — | — | — | — | — | — | — | $^1S_0$ | 9.394 |
| Ga | 31 | | | | | | 10 | 2 | 1 | — | — | — | — | — | — | $^2P^\circ_{1/2}$ | 5.999 |
| Ge | 32 | | | | | | 10 | 2 | 2 | — | — | — | — | — | — | $^3P_0$ | 7.899 |
| As | 33 | | | | | | 10 | 2 | 3 | — | — | — | — | — | — | $^4S^\circ_{3/2}$ | 9.81 |
| Se | 34 | | | | | | 10 | 2 | 4 | — | — | — | — | — | — | $^3P_2$ | 9.752 |
| Br | 35 | | | | | | 10 | 2 | 5 | — | — | — | — | — | — | $^2P^\circ_{3/2}$ | 11.814 |
| Kr | 36 | | | | | | 10 | 2 | 6 | — | — | — | — | — | — | $^1S_0$ | 13.999 |
| Rb | 37 | Krypton configuration | | | | | | | | — | — | 1 | — | — | — | $^2S_{1/2}$ | 4.177 |
| Sr | 38 | | | | | | | | | — | — | 2 | — | — | — | $^1S_0$ | 5.693 |
| Y | 39 | | | | | | | | | 1 | — | 2 | — | — | — | $^2D_{3/2}$ | 6.38 |
| Zr | 40 | | | | | | | | | 2 | — | 2 | — | — | — | $^3F_2$ | 6.84 |
| Nb | 41 | | | | | | | | | 4 | — | 1 | — | — | — | $^6D_{1/2}$ | 6.88 |
| Mo | 42 | | | | | | | | | 5 | — | 1 | — | — | — | $^7S_3$ | 7.10 |
| Tc | 43 | | | | | | | | | 5 | — | 2 | — | — | — | $^6S_{5/2}$ | 7.28 |
| Ru | 44 | | | | | | | | | 7 | — | 1 | — | — | — | $^5F_5$ | 7.366 |
| Rh | 45 | | | | | | | | | 8 | — | 1 | — | — | — | $^4F_{9/2}$ | 7.46 |
| Pd | 46 | | | | | | | | | 10 | — | — | — | — | — | $^1S_0$ | 8.33 |

(*continued on next page*)

**Distribution of electrons in the atoms (cont.)**

| Element and atomic number | | Configuration of inner shells | N | O | | | | P | | | Q | Ground term | Ionization potential, eV |
|---|---|---|---|---|---|---|---|---|---|---|---|---|---|
| | | | 4,3 4*f* | 5,0 5*s* | 5,1 5*p* | 5,2 5*d* | 5,3 5*f* | 6,0 6*s* | 6,1 6*p* | 6,2 6*d* | 7,0 7*s* | | |
| Ag | 47 | | — | 1 | — | — | — | — | — | — | — | $^2S_{1/2}$ | 7.576 |
| Cd | 48 | | — | 2 | — | — | — | — | — | — | — | $^1S_0$ | 8.993 |
| In | 49 | | — | 2 | 1 | — | — | — | — | — | — | $^2P^\circ_{1/2}$ | 5.786 |
| Sn | 50 | Palladium | — | 2 | 2 | — | — | — | — | — | — | $^3P_0$ | 7.344 |
| Sb | 51 | configuration | — | 2 | 3 | — | — | — | — | — | — | $^4S^\circ_{3/2}$ | 8.641 |
| Te | 52 | | — | 2 | 4 | — | — | — | — | — | — | $^3P_2$ | 9.01 |
| I | 53 | | — | 2 | 5 | — | — | — | — | — | — | $^2P^\circ_{3/2}$ | 10.457 |
| Xe | 54 | | — | 2 | 6 | — | — | — | — | — | — | $^1S_0$ | 12.130 |
| Cs | 55 | | — | | | — | — | 1 | — | — | — | $^2S_{1/2}$ | 3.894 |
| Ba | 56 | | — | | | — | — | 2 | — | — | — | $^1S_0$ | 5.211 |
| La | 57 | | — | | | 1 | — | 2 | — | — | — | $^2D_{3/2}$ | 5.5770 |
| Ce | 58 | | 1 | | | 1 | — | 2 | — | — | — | $^1G^\circ_4$ | 5.466 |
| Pr | 59 | | 3 | | | — | — | 2 | — | — | — | $^4I^\circ_{9/2}$ | 5.422 |
| Nd | 60 | | 4 | | | — | — | 2 | — | — | — | $^5I_4$ | 5.489 |
| Pm | 61 | | 5 | | | — | — | 2 | — | — | — | $^6H^\circ_{5/2}$ | 5.554 |
| Sm | 62 | The shells 1*s* to 4*d* contain 46 electrons | 6 | The shells 5*s* to 5*p* contain 8 electrons | | — | — | 2 | — | — | — | $^7F_0$ | 5.631 |
| Eu | 63 | | 7 | | | — | — | 2 | — | — | — | $^8S^\circ_{7/2}$ | 5.666 |
| Gd | 64 | | 7 | | | 1 | — | 2 | — | — | — | $^9D^\circ_2$ | 6.141 |
| Tb | 65 | | 9 | | | — | — | 2 | — | — | — | $^6H^\circ_{15/2}$ | 5.852 |
| Dy | 66 | | 10 | | | — | — | 2 | — | — | — | $^5I_8$ | 5.927 |
| Ho | 67 | | 11 | | | — | — | 2 | — | — | — | $^4I^\circ_{15/2}$ | 6.018 |
| Er | 68 | | 12 | | | — | — | 2 | — | — | — | $^3H_6$ | 6.101 |
| Tm | 69 | | 13 | | | — | — | 2 | — | — | — | $^2F^\circ_{7/2}$ | 6.184 |
| Yb | 70 | | 14 | | | — | — | 2 | — | — | — | $^1S_0$ | 6.254 |
| Lu | 71 | | 14 | | | 1 | — | 2 | — | — | — | $^2D_{3/2}$ | 5.426 |
| Hf | 72 | | | | | 2 | — | 2 | — | — | — | $^3F_2$ | 6.865 |
| Ta | 73 | | | | | 3 | — | 2 | — | — | — | $^4F_{3/2}$ | 7.88 |
| W | 74 | | | | | 4 | — | 2 | — | — | — | $^5D_0$ | 7.98 |
| Re | 75 | The shells 1*s* to 5*p* contain 68 electrons | | | | 5 | — | 2 | — | — | — | $^6S_{5/2}$ | 7.87 |
| Os | 76 | | | | | 6 | — | 2 | — | — | — | $^5D_4$ | 8.5 |
| Ir | 77 | | | | | 7 | — | 2 | — | — | — | $^4F_{9/2}$ | 9.1 |
| Pt | 78 | | | | | 9 | — | 1 | — | — | — | $^3D_3$ | 9.0 |
| Au | 79 | | | | | | — | 1 | — | — | — | $^2S_{1/2}$ | 9.22 |
| Hg | 80 | | | | | | — | 2 | — | — | — | $^1S_0$ | 10.43 |
| Tl | 81 | | | | | | — | 2 | 1 | — | — | $^2P^\circ_{1/2}$ | 6.108 |
| Pb | 82 | | | | | | — | 2 | 2 | — | — | $^3P_0$ | 7.417 |
| Bi | 83 | | | | | | — | 2 | 3 | — | — | $^4S^\circ_{3/2}$ | 7.289 |
| Po | 84 | | | | | | — | 2 | 4 | — | — | $^3P_2$ | 8.43 |
| At | 85 | | | | | | — | 2 | 5 | — | — | $^2P^\circ_{3/2}$ | |
| Rn | 86 | | | | | | — | 2 | 6 | — | — | $^1S_0$ | 10.749 |
| Fr | 87 | | | | | | — | 2 | 6 | — | 1 | $^2S_{1/2}$ | |
| Ra | 88 | | | | | | — | 2 | 6 | — | 2 | $^1S_0$ | 5.278 |
| Ac | 89 | | | | | | — | 2 | 6 | 1 | 2 | $^2D_{3/2}$ | 5.17 |
| Th | 90 | The shells 1*s* to 5*d* contain 78 electrons | | | | | — | 2 | 6 | 2 | 2 | $^3F_2$ | 6.08 |
| Pa | 91 | | | | | | 2 | 2 | 6 | 1 | 2 | $^4K_{11/2}$ | 5.89 |
| U | 92 | | | | | | 3 | 2 | 6 | 1 | 2 | $^5L_6$ | 6.05 |
| Np | 93 | | | | | | 4 | 2 | 6 | 1 | 2 | $^6L_{11/2}$ | 6.19 |
| Pu | 94 | | | | | | 6 | 2 | 6 | — | 2 | $^7F_0$ | 6.06 |
| Am | 95 | | | | | | 7 | 2 | 6 | — | 2 | $^8S^\circ_{7/2}$ | 5.993 |
| Cm | 96 | | | | | | 7 | 2 | 6 | 1 | 2 | $^9D^\circ_2$ | 6.02 |
| Bk | 97 | | | | | | 9 | 2 | 6 | 0 | 2 | $^6He^\circ_{5/2}$ | 6.23 |
| Cf | 98 | | | | | | 10 | 2 | 6 | 0 | 2 | $^5I_8$ | 6.30 |
| Es | 99 | | | | | | 11 | 2 | 6 | 0 | 2 | $^4I^\circ_{15/2}$ | 6.42 |
| Fm | 100 | | | | | | 12 | 2 | 6 | 0 | 2 | $^3H_6$ | 6.50 |
| Md | 101 | | | | | | 13 | 2 | 6 | 0 | 2 | $^2F^\circ_{7/2}$ | 6.58 |
| No | 102 | | | | | | 14 | 2 | 6 | 0 | 2 | $^1S_0$ | 6.65 |
| Lw | 103 | | | | | | (14) | 2 | 6 | (1) | (2) | | |

In the next-to-last column of the table, the spectral term of the energy level with lowest total electronic energy is shown. The main part of the term symbol is a capital letter, $S$, $P$, $D$, $F$, and so on, that represents the total electronic orbital angular momentum. Attached to this is a superior prefix, 1, 2, 3, 4, and so on, that indicates the multiplicity, and an anterior suffix, 0, $^1/_2$, 1, $^3/_2$, 2, $^5/_2$, and so on, that shows the total angular momentum, or $J$ value, of the atom in the given state. A sign $^\circ$ above the $J$ value signifies that the spectral term and electron configuration have odd parity.

The last column of the table presents the first ionization potential of the atom, the energy required to remove from an atom its least firmly bound electron and transform a neutral atom into a singly charged ion. *See* ATOMIC STRUCTURE AND SPECTRA; IONIZATION POTENTIAL. [J.E.B.]

**Electron paramagnetic resonance (EPR) spectroscopy** The study of the resonant response to microwave- or radio-frequency radiation of paramagnetic materials placed in a magnetic field. It is sometimes referred to as electron spin resonance (ESR). Paramagnetic substances normally have an odd number of electrons or unpaired electrons, but sometimes electron paramagnetic resonance (EPR) is observed for ions or biradicals with an even number of electrons. EPR spectra are normally presented as plots of the first derivative of the energy absorbed from an oscillating magnetic field at a fixed microwave frequency versus the magnetic field strength. The dispersion may also be detected.

To overcome the intrinsic low sensitivity of the magnetic dipole transitions responsible for EPR, samples are placed in resonant cavities. Routine experiments are carried out in the steady state at a fixed microwave frequency of approximately 9 gigahertz by slowly sweeping the magnetic field through resonance. Free electrons resonate in a magnetic field of 3250 gauss (325 millitesla) at the microwave frequency of 9.1081 GHz, whereas organic free radicals resonate at slightly different magnetic fields characteristic of each particular molecule. *See* ELECTRON SPIN.

The observation of EPR spectra depends on spin-lattice relaxation, which is the exchange of magnetic energy with the thermal motion of the crystal or molecule. For transition-metal ions and rare-earth ions, experiments often require operation at or near liquid helium temperature (4 K; −269°C; −452°F). Organic free radicals can usually be studied successfully at room temperature.

**Applications.** EPR spectroscopy is used to determine the electronic structure of free radicals as well as transition-metal and rare-earth ions in a variety of substances, to study interactions between molecules, and to measure nuclear spins and magnetic moments. It is applied in the fields of physics, chemistry, biology, archeology, geology, and mineralogy. It is also used in the investigation of radiation-damaged materials and in radiation dosimetry.

The basic physics of transition-metal ions and rare-earth ions present in low concentrations in diamagnetic host crystals has provided a theoretical basis for how electronic structure is modified by the surrounding atoms. Particular applications include probing phase transitions in solids and studies of pairs and triads of magnetically interacting ions.

Applications of EPR in chemistry include characterization of free radicals, studies of organic reactions, and investigations of the electronic properties of paramagnetic inorganic molecules. Information obtained is used in the investigation of molecular structure. EPR is used widely in biology in the study of metal proteins, for nitroxide spin labeling, and in the investigation of radicals produced during reaction processes in proteins and other biomacromolecules. EPR has proved to be an important technique for interdisciplinary investigations of photosynthetic systems. By means of EPR, more than 20 proteins that function in the mitochondrial respiratory chains of mammals have been identified, and details regarding their electron transfer processes have been elucidated.

**Solids.** EPR spectra from single crystals clearly provide the greatest amount of information. These include crystals containing small concentrations of paramagnetic ions substituting for the regular ions in the crystal or, for organic molecules, small fractions of free radicals produced by ionizing radiation. Spectra which may contain up to several hundred lines are often highly anisotropic; that is, they change with the orientation of the magnetic field direction in the crystal. Transition-metal-ion and rare-earth-ion EPR spectra in crystals are generally much more anisotropic than free radicals due to the intrinsic anisotropy of the electron magnetic moments, and of other effects that are important when there is more than one unpaired electron.

The occurrence of many lines is due to interactions of the orbital motions of electrons with the electric potential of the local surrounding atoms, and to hyperfine interactions between the paramagnetic electrons and nuclear magnetic moments of the paramagnetic ion and surrounding atoms. In the case of free radicals, symmetric or nearly symmetric characteristic hyperfine patterns are observed. From knowledge of hyperfine interactions with nuclei whose spins and magnetic moments are known, the electron distribution throughout a molecule may be determined. Since hyperfine interactions vary as the reciprocal of the cube of the distance between the center of the free radical and the nucleus, structural information may be obtained in addition to electron densities.

**Liquids and motional averaging.** Spectra in solution due to free radicals are often quite simple as a result of motional averaging, and this clearly gives less information than would be obtained from a single-crystal investigation. Linewidths are very narrow (approximately 0.1 gauss or smaller). By varying the temperature above and below room temperature, EPR spectra range from the frozen solution at low temperatures, with a powderlike spectrum, to rapid motional averaging at room temperature where anisotropies are averaged out. The intermediate region can provide information about slow molecular motions, which is especially important for nitroxide spin labels selectively attached to different parts of macromolecules such as the components of natural and synthetic phospholipid membranes, liquid crystals, and proteins. Such measurements have revealed important structural and functional information. [J.R.P.]

**Electron-probe microanalysis** A method used for determining the elemental composition of materials, based on the x-rays emitted by different elements when bombarded with high-energy electrons. It is a micro method that can detect x-ray photons emitted by the atoms within a small volume excited by an electron beam focused to 10 nanometers diameter or less. In biology electron-probe microanalysis can be used to determine the composition of cell organelles without isolating them and therefore altering the distribution of diffusible elements.

High-energy electrons can ionize atoms, ejecting an inner-shell electron. To fill the resultant vacancy, an outer-shell electron falls into the ionized shell; the atom remains in the higher-energy excited state. The emission of a characteristic x-ray by the ionized atom is one of the mechanisms for releasing its excess energy. Another mechanism is the emission of Auger electrons; the probability of ejecting an x-ray instead of an Auger electron is the fluorescence yield. In the case of electron-probe microanalysis the source of the exciting electrons is the electron gun, which, in modern electron microscopes equipped with field-emission guns, can produce a focused beam narrower than 1 nm. The x-rays emitted as the result of atomic ionization are called characteristic x-rays, because their energy is characteristic of the core shell of the ionized element and of the shell from which the electron relaxed into the vacancy. *See* AUGER EFFECT. [A.P.S.]

**Electron spectroscopy** A form of spectroscopy which deals with the emission and recording of the electrons which constitute matter—solids, liquids, or gases. The usual form of spectroscopy concerns the emission or absorption of photons (x-rays, ultraviolet rays, visible or microwave wavelengths, and so on). Electron spectra can be excited by x-rays, which is the basis for electron spectroscopy for chemical analysis (ESCA), or by ultraviolet photons, or by ions (electrons). By means of ESCA, complete sets of photoelectron lines can be excited from the internal (core) levels as well as from the external (valence) region. Also, complete sequences of the Auger electron lines are automatically obtained in this mode.

The electron lines in an ESCA spectrum are extremely sharp and well suited for precision measurements. With a high-resolving ESCA spectrometer which has a magnetic or electrostatic focusing dispersive system, the electron lines have widths which are set by the limit caused by the uncertainty principle (the "inherent" widths of atomic levels). With a suitable choice of radiation, electron spectroscopy reproduces directly the electronic level structure from the innermost shells (core electrons) to the atomic surface (valence or conduction band). Furthermore, all elements from hydrogen to the heaviest ones can be studied even if the element occurs together with several other elements and even if the element represents only a small part of the chemical compound.

When applied to solid materials, ESCA is a typical surface spectroscopy with applications to problems such as chemical surface reactions, for example, corrosion or heterogeneous catalysis. ESCA also reproduces bulk matter properties such as valence electron band structures. Electron spectroscopy can supply a detailed knowledge of the valence orbital structure for all molecules which can be brought into gaseous form with pressures of $10^{-5}$ torr ($10^{-3}$ pascal) or more. Under certain conditions, liquids and solutions of various compositions can be studied by ESCA techniques.

A unique feature of ESCA is that, if the exact position of the electron lines characteristic of the various elements in the molecule is measured, the area of inspection can be moved from one atomic species to another in the molecular structure. If the structure of the molecule is known, the charge distribution can be estimated in a simple way by using, for example, the electronegativity concept and assuming certain resonance structures. More sophisticated quantum-chemical treatments can also be applied. Conversely, if, by means of ESCA, the approximate charge distribution is known, conclusions concerning the structure of the molecule can be drawn. *See* ATOMIC STRUCTURE AND SPECTRA; ELECTRON CONFIGURATION; MOLECULAR ORBITAL THEORY; SPECTROSCOPY. [K.S.]

**Electron spin** That property of an electron which gives rise to its angular momentum about an axis within the electron. Spin is one of the permanent and basic properties of the electron. Both the spin and the associated magnetic dipole moment of the electron were postulated by G. E. Uhlenbeck and S. Goudsmit in 1925 as necessary to allow the interpretation of many observed effects, among them the so-called anomalous Zeeman effect, the existence of doublets (pairs of closely spaced lines) in the spectra of the alkali atoms, and certain features of x-ray spectra.

The spin quantum number is $s$, which is always $^1/_2$. This means that the component of spin angular momentum along a preferred direction, such as the direction of a magnetic field, is $\pm{}^1/_2\hbar$ where $\hbar = h/2\pi$ and $h$ is Planck's constant. The spin angular momentum of the electron is not to be confused with the orbital angular momentum of the electron associated with its motion about the nucleus. In the latter case the maximum component of angular momentum along a preferred direction is $l\hbar$, where $l$ is the angular momentum quantum number and may be any positive integer or zero.

**Electron magnetic moment.** The electron has a magnetic dipole moment by virtue of its spin. The approximate value of the dipole moment is the Bohr magneton $\mu_0$ which is equal, in SI units, to $eh/4\pi m = 9.27 \times 10^{-24}$ joule/tesla, where $e$ is the electron charge measured in coulombs, and $m$ is the mass of the electron. The orbital motion of the electron also gives rise to a magnetic dipole moment $\mu_l$ that is equal to $\mu_0$ when $l = 1$.

The orbital magnetic moment of an electron can readily be deduced with the use of the classical statements of electromagnetic theory in quantum-mechanical theory; the simple classical analog of a current flowing in a loop of wire describes the magnetic effects of an electron moving in an orbit. The spin of an electron and the magnetic properties associated with it are, however, not possible to understand from a classical point of view.

In the Landé $g$ factor, $g$ is defined as the negative ratio of the magnetic moment, in units of $\mu_0$, to the angular momentum, in units of $\hbar$. For the orbital motion of an electron, $g_l = 1$. For the spin of the electron the appropriate $g$ value is $g_s \simeq 2$; that is, unit spin angular momentum produces twice the magnetic moment that unit orbital angular momentum produces. The total electronic magnetic moment of an atom depends on the state of coupling between the orbital and spin angular momenta of the electron.

**Atomic beam measurements.** With the development of spectroscopy by the atomic beam method, a new order of precision in the measurement of the frequencies of spectral lines became possible. By using the atomic-beam techniques, it became possible to measure $g_s/g_l$ directly, with the result $g_s/g_l = 2\ (1.001168 \pm 0.000005)$. The magnetic moment of the electron therefore is not $\mu_0$ but $1.001168\mu_0$, or equivalently the $g$ factor of the electron departs from 2 by the so-called $g$ factor anomaly defined as $a = (g_2 - 2)/2$ so that $\mu = (1 + a)_0$. Thus the first molecular beam work gave $a = 0.001168$.

**Calculation of g-factor anomaly.** It is not possible to give a qualitative description of the effects which give rise to the $g$-factor anomaly of the electron. The detailed theoretical calculation of the quantity is in the domain of quantum electrodynamics, and involves the interaction of the zero-point oscillation of the electromagnetic field with the electron. Comparison of theoretical determination of $a$ with its experimental measurement constitutes the most accurate and direct existing test of the theory of quantum electrodynamics. *See* ATOMIC STRUCTURE AND SPECTRA. [A.Ri.; T.Ki.]

**Electron-transfer reaction** A reaction in which one electron is transferred from one molecule or ion to another molecule or ion. Electron-transfer reactions are ubiquitous in nature. Some are deceptively simple [for example, reaction (1), where the

$$^{*}Fe(H_2O)_6{}^{3+} + Fe(H_2O)_6{}^{2+} \rightarrow {}^{*}Fe(H_2O)_6{}^{2+} + Fe(H_2O)_6{}^{3+} \quad (1)$$

asterisk is used to identify a specific isotope]; others look very complicated (for example, the long-range electron transfers found in biology). The widespread occurrence of electron-transfer reactions has stimulated much theoretical and experimental work.

The simplest reactions in solution chemistry are electron self-exchange reactions (2),

$$^{*}A_{ox} + A_{red} \rightarrow {}^{*}A_{red} + A_{ox} \quad (2)$$

in which the reactants and products are the same (the asterisk is used to identify a specific isotope). The only way to determine chemically that a reaction has taken place is to introduce an isotopic label. There is no change in the free energy ($\Delta G^\circ = 0$) for this type of reaction.

Much more common are cross reactions (3), where $A_{ox}$ is the oxidized reactant, $B_{red}$

$$A_{ox} + B_{red} \rightarrow A_{red} + B_{ox} \qquad (3)$$

is the reduced reactant, $A_{red}$ is the reduced product, and $B_{ox}$ is the oxidized product. For these reactions, $\Delta G^\circ \neq 0$.

Both types of electron-transfer reactions (self-exchange and cross reactions) can be classified broadly as inner sphere or outer sphere. In an inner-sphere reaction, a ligand is shared between the oxidant and reductant in the transition state. An outer-sphere reaction, on the other hand, is one in which the inner coordination shells of both the oxidant and reductant remain intact in the transition state. There is no bond breaking or bond making, and no shared ligands between redox centers. Long-range electron transfers in biology are all of the outer-sphere type. *See* OXIDATION-REDUCTION.

**Electron-transfer theory.** The simplest electron transfer occurs in an outer-sphere reaction. The changes in oxidation states of the donor and acceptor centers result in a change in their equilibrium nuclear configurations. This process involves geometrical changes, the magnitudes of which vary from system to system. In addition, changes in the interactions of the donor and acceptor with the surrounding solvent molecules will occur. The Franck-Condon principle governs the coupling of the electron transfer to these changes in nuclear geometry: during an electronic transition, the electronic motion is so rapid that the nuclei (including metal ligands and solvent molecules) do not have time to move. Hence, electron transfer occurs at a fixed nuclear configuration. In a self-exchange reaction, the energies of the donor and acceptor orbitals (hence, the bond lengths and bond angles of the donor and acceptor) must be the same before efficient electron transfer can take place.

**Long-range electron transfer.** The rate of long-range electron transfer between an electron donor ($B_{red}$) and an electron acceptor ($A_{ox}$) depends on both the electronic coupling between $A_{ox}$ and $B_{red}$ (which is a function of the intersite $A_{ox}//B_{red}$ distance $d$, the nature of the intervening medium, and the relative $A_{ox}//B_{red}$ orientation, where // represents the protein medium that separates the donor and the acceptor) and an activation energy term. A standard theoretical rate equation (4) expresses $k_{et}$ in terms of

$$k_{et} = \nu[\exp(-\beta d)]\{\exp[-(\lambda + \Delta G^\circ)^2/4\lambda RT]\} \qquad (4)$$

these factors; here, $\nu$ is a frequency factor, $\beta$ is a medium- and orientation-dependent quantity, $d$ is the intersite distance, $\lambda$ is the reorganization energy, $\Delta G^\circ$ is the reaction free energy of the electron-transfer process, $R$ is the universal gas constant, and $T$ is the absolute temperature of the system. Experiments in several laboratories have been designed to estimate the values of $\lambda$ and $\beta$ in modified metalloproteins, rigid organic molecules, and protein–protein complexes. [B.E.Bo.; W.R.El.; H.B.Gr.; T.J.Me.]

**Electronegativity** According to L. Pauling, "the power of an atom in a molecule to attract electrons to itself." Quantitative definitions and scales of electronegativity have been based not on electron distribution itself but on properties which were assumed to reflect electronegativity.

The electronegativity of an element depends upon its valence state and thus is not an invariant atomic property. As an example, the electron-withdrawing ability of an $sp^n$ hybrid orbital centered on carbon and directed toward hydrogen increases as the percentage of $s$ character in the orbital increases in the series ethane < ethylene < acetylene. Thus, according to this concept of orbital electronegativity, each element exhibits a range of electronegativity values.

The original scale, proposed by Pauling in 1932, is based upon the difference between the energy of the A-B bond in the compound $AB_n$ and the mean of the energies

**Average electronegativities from thermochemical data**

| Element | Value | Element | Value |
|---|---|---|---|
| H | 2.20 | Al | 1.61 |
| Li | 0.98 | Ga | 1.81 |
| Na | 0.93 | In | 1.78 |
| K | 0.82 | Tl | 2.04 |
| Rb | 0.82 | C | 2.55 |
| Cs | 0.79 | Si | 1.90 |
| Be | 1.57 | Ge | 2.01 |
| Mg | 1.31 | Sn | 1.96 |
| Ca | 1.00 | Pb | 2.33 |
| Sr | 0.95 | N | 3.04 |
| Ba | 0.89 | P | 2.19 |
| Sc | 1.36 | As | 2.18 |
| Ti | 1.54 | Sb | 2.05 |
| V | 1.63 | Bi | 2.02 |
| Cr | 1.66 | O | 3.44 |
| Mn | 1.55 | S | 2.58 |
| Fe | 1.83 | Se | 2.55 |
| Co | 1.88 | F | 3.98 |
| Ni | 1.91 | Cl | 3.16 |
| Cu | 1.90 | Br | 2.96 |
| Zn | 1.65 | I | 2.66 |
| B | 2.04 | | |

of the homopolar bonds A-A and B-B (see table). R. S. Mulliken proposed that the electronegativity of an element is given by the average of the valence-state ionization potential and electron affinity. The Mulliken approach is consistent with Pauling's original definition and gives orbital electronegativities, not invariant atomic electronegativities. Electronegativity was defined by A. L. Allred and E. G. Rochow as the force of attraction between a nucleus and an electron from a bonded atom. A quantum-defect electronegativity scale has been developed from potentials based on atomic spectral data, and a nonempirical scale has been calculated by an ab initio method using floating gaussian orbitals.

Other methods for calculating electronegativities utilize such observables as bond-stretching force constants, electrostatic potentials, spectra, and covalent radii. The measurement of electronegativities involves observations of properties dependent upon electron distribution. Close agreement of electronegativity values obtained from measurements of several diverse properties lends confidence and utility to the concept. [A.L.A.]

**Electrophilic and nucleophilic reagents** Electrophilic reagents are chemical species which, in the course of chemical reactions, acquire electrons, or a share in electrons, from other molecules or ions. Although this definition embraces all oxidizing agents and all Lewis acids, electrophilic reagents are ordinarily thought of as cationic species, such as $H^+$, $NO_2^+$, $Br^+$, or $SO_3$ (or carriers of these species such as HCl, $CH_3COONO_2$, or $Br_2$), which can form stable covalent bonds with carbon atoms. Electrophilic reagents frequently are positively charged ions (cations). *See* ACID AND BASE.

Nucleophilic reagents are the opposite of electrophilic reagents. Nucleophilic reagents give up electrons, or a share in electrons, to other molecules or ions in the course of chemical reactions. Nucleophilic reagents frequently are negatively charged ions (anions). Typical nucleophilic reagents are hydroxide ion ($OH^-$), halide ions ($F^-$, $Cl^-$,

$Br^-$, and $I^-$), cyanide ion ($CN^-$), ammonia ($NH_3$), amines, alkoxide ions (such as $CH_3O^-$), and mercaptide ions (such as $C_6H_5S^-$). *See* SUBSTITUTION REACTION. [J.F.B.]

**Electrophoresis** The migration of electrically charged particles in solution or suspension in the presence of an applied electric field. Each particle moves toward the electrode of opposite electrical polarity. For a given set of solution conditions, the velocity with which a particle moves divided by the magnitude of the electric field is a characteristic number called the electrophoretic mobility. The electrophoretic mobility is directly proportional to the magnitude of the charge on the particle, and is inversely proportional to the size of the particle. An electrophoresis experiment may be either analytical, in which case the objective is to measure the magnitude of the electrophoretic mobility, or preparative, in which case the objective is to separate various species which differ in their electrophoretic mobilities under the experimental solution conditions.

**Gel techniques.** Electrophoresis was first employed as an experimental technique by Arne Tiselius in 1937. The apparatus of Tiselius detected electrophoretic motion by the moving-boundary method, in which a boundary is created between the solution of particles to be examined and a sample of pure solvent. As the particles migrate in an electric field, the boundary between solution and solvent can be observed to move, and if there are a number of species in the solution with different electrophoretic mobilities, a series of boundaries of various shapes and magnitudes can be detected. The moving-boundary method was used for three decades to separate complex mixtures of charged macromolecules in solution and to study the physical characteristics of solutions of proteins and other macromolecules of biological and industrial importance.

The resolving power of electrophoresis was greatly improved by the introduction of the use of gel supporting media. The gel matrix prevents thermal convection caused by the heat which results from the passage of electric current through the sample. The absence of convection reduces greatly the mixing of the various parts of the sample, and therefore allows for more stable separation. The dimensions of the cross-links of the gel may also provide a molecular sieving effect, which increases the resolving power of the electrophoretic separation of molecules of different size. In addition, the gel media may support a gradient of a separate reagent, which assists in the separation of macromolecules. Gradients of pH and of reagents of various types may be combined in two-dimensional arrays for even greater resolving power. A very successful derivative of the gel technique is the determination of the molecular weights of protein molecules by electrophoresis of the molecules in a gel medium which contains substantial amounts of detergent. The detergent denatures the protein molecules, changing them from globular, compact structures to long, flexible polymers which are coated with detergent molecules. These polymers move in the electric field through the gel medium with a velocity which is determined by the length of the polymer, and therefore by the molecular weight of the protein unit. This method is the most common technique for the determination of molecular weights of proteins in biochemical studies. *See* GEL; PH; PROTEIN.

**Isoelectric focusing.** An important variation of the electrophoresis technique is isoelectric focusing. In this technique the medium supports a pH gradient which includes the isoelectric pH of the species being studied. Many charged macromolecules have both positive and negative charges on their surfaces, and the electrophoretic mobility is related to the net excess of charge of one type or the other. As the pH becomes more acidic, the number of positive charges increases, and as the pH becomes more basic, the number of negative charges increases. For each molecule of this type, there is one pH at which the net charge on the surface is zero, so that the molecule does not move when an electric field is applied and thus has an electrophoretic mobility of zero. This pH is called the isoelectric pH. If the molecule is introduced into a pH gradient which includes its isoelectric pH, it will migrate to the position of the isoelectric pH and then become stationary. In this way, all molecules of a given isoelectric pH will migrate to the same region—hence the term isoelectric focusing. The method of isoelectric focusing is particularly good for the analysis of microheterogeneity of protein species and other species which may differ slightly in their chemical content. *See* ISOELECTRIC POINT.

**Laser applications.** Application of the optical laser to electrophoretic detection resulted in the development of a technique which can be used for analytical electrophoresis experiments on particles of all sizes. The basic principle is that the highly monochromatic (single-frequency) laser light impinges upon the particles and is scattered from the particles in all directions. When observing the laser light which has been scattered from a moving particle, one can detect that there is a slight shift in the frequency of the light as a result of the motion of the particle. The application of the laser Doppler principle to electrophoresis experiments, often called electrophoretic light scattering (ELS), is an important method for the rapid determination of electrophoretic velocities. Electrophoretic light scattering has been used for the study of many types of living cells, cell organelles, viruses, proteins, nucleic acids, and synthetic polymers. *See* ELECTROLYTIC CONDUCTANCE . [B.R.W.]

**Capillary electrophoresis.** Electrophoresis can be performed in a capillary format. A typical system consists of two reservoirs and a capillary filled with a buffer solution. A high voltage is applied across the capillary by using a high-voltage power supply. The very small diameter capillaries (typically 5–100 micrometers) employed in this technique allow for efficient heat dissipation. Therefore, much higher voltages can be employed than those used in slab gel electrophoresis, leading to faster, more efficient separations. Compounds are separated on the basis of their net electrophoretic mobilities.

Most often, the detector is placed on line and analytes are detected as they flow past the detector. Spectroscopic detection (ultraviolet and laser-based fluorescence) is usually performed in this manner by using the capillary itself as the optical cell. Alternatively, detectors can be placed off line (after the column). In this case, the detector is isolated from the applied electric field through the use of a grounding joint. Electrochemical detection and mass spectroscopic detection are generally accomplished in this manner, since the electric field can interfere with the performance of these detectors.

Capillary zone electrophoresis is the simplest and most widely used form of capillary electrophoresis. The capillary is filled with a homogeneous buffer, and compounds are separated on the basis of their relative charge and size. Most often, fused silica capillaries are employed. In this case, an electrical double layer is produced at the capillary surface due to the attraction of positively charged cations in the buffer to the ionized silanol groups on the capillary wall. In the presence of an electric field, the cations in the diffuse portion of this double layer move toward the cathode and drag the solvent with them, producing an electroosmotic flow. The resulting flow profile is flat rather than the parabolic shape characteristic of liquid chromatography. This flat flow profile causes analytes to migrate in very narrow bands and leads to highly efficient separations. The electroosmotic flow is also pH dependent, and it is highest at alkaline pH values.

In most cases, the electroosmotic flow is the strongest driving force in the separation, and all analytes, regardless of charge, migrate toward the cathode. Therefore, it is possible to separate and detect positive, negative, and neutral molecules in the same electrophoretic run, if the detector is placed at the cathodic end. Negatively charged compounds are attracted to the anode but are swept up by the electroosmotic flow and

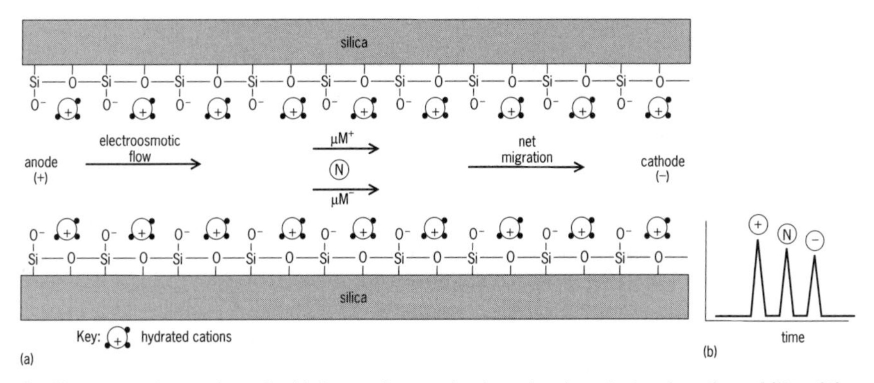


**Capillary zone electrophoresis. (*a*) Separation mechanism showing electrophoretic mobility of the positive ion ($\mu M^+$) and negative ion ($\mu M^-$); N is a neutral molecule. (*b*) Migration order of the ions.**

elute last. Neutral molecules, which are not separated from each other in capillary zone electrophoresis, elute as a single band with the same velocity as the electroosmotic flow. Positive compounds have positive electrophoretic mobilities in the same direction as the electroosmotic flow, and they elute first (see illustration). Capillary zone electrophoresis is generally employed for the separation of small molecules, including amino acids, peptides, and small ions, and for the separation of drugs, their metabolites, and degradation products.

All capillary electrophoresis methods have the advantage of the ability to analyze small volumes (typical injection volumes are 1–50 nanoliters). This makes it possible to analyze very small samples or to use the same sample for several different analyses. One unique application of this technique is the determination of amino acids and neutrotransmitters in single cells. [S.Lu.]

**Alternating-field electrophoresis.** Alternating-field agarose gel electrophoresis is a technique for separating very large molecules of deoxyribonucleic acid (DNA); fragments of DNA ranging in size from 30 to 10,000 kilobasepairs (kb) can be resolved. For the molecular biologist, this is a considerable advance over conventional agarose gel electrophoresis, which is limited to the resolution of less than 50 kb.

Conventional gel electrophoresis employs a single pair of electrodes to generate an electric field that is constant in both time and direction and that is uniform across the gel. DNA molecules are negatively charged with a uniform charge-to-mass ratio and thus migrate steadily toward the positive electrode. Although DNA is a linear molecule, in solution it tends to collapse into a random coil configuration. Agarose is a porous material that acts like a sieve, retarding the movement of the DNA; the larger the molecule, the more the retardation, and thus the molecules separate on the basis of size. However, above approximately 50 kb, the dimensions of the random coil are larger than the pore size of the agarose. The DNA can no longer be sieved through the gel, and resolution is lost.

Contrary to conventional electrophoresis, alternating-field electrophoresis does not use a constant electric field but one which regularly alternates in direction. With each change in field direction, the DNA molecules attempt to reorient themselves. When this happens, an end or a small loop of the molecule, which has dimensions much smaller than those of the random coil of the entire molecule, may find itself positioned by a pore in the agarose. The electric field can then pull the DNA by the end through the hole. When the field regularly alternates from one direction to another, the DNA regularly reorients and is pulled through an adjacent hole.

Not all molecules in the system will make equal progress under these conditions, because not all molecules can reorient themselves with equal speed. The larger the molecule, the more time it requires in a given field strength (determined by the applied voltage) to change directions and the less time it has left to move. [K.Ga.]

**Electrostatic precipitator** A device used to remove liquid droplets or solid particles from a gas in which they are suspended. The process depends on two steps. In the first step the suspension passes through an electric discharge (corona discharge) area where ionization of the gas occurs. The ions produced collide with the suspended particles and confer on them an electric charge. The charged particles drift toward an electrode of opposite sign and are deposited on the electrode where their electric charge is neutralized. The phenomenon would be more correctly designated as electrodeposition from the gas phase.

The use of electrostatic precipitators has become common in numerous industrial applications. Among the advantages of the electrostatic precipitator are its ability to handle large volumes of gas, at elevated temperatures if necessary, with a reasonably small pressure drop, and the removal of particles in the micrometer range. Some of the usual applications are: (1) removal of dirt from flue gases in steam plants; (2) cleaning of air to remove fungi and bacteria in establishments producing antibiotics and other drugs, and in operating rooms; (3) cleaning of air in ventilation and air conditioning systems; (4) removal of oil mists in machine shops and acid mists in chemical process plants; (5) cleaning of blast furnace gases; (6) recovery of valuable materials such as oxides of copper, lead, and tin; and (7) separation of rutile from zirconium sand. [G.S.M.; W.O.M.]

**Element (chemistry)** An element is a substance made up of atoms with the same atomic number. Some common elements are oxygen, hydrogen, iron, copper, gold, silver, nitrogen, chlorine, and uranium. Approximately 75% of the elements are metals and the others are nonmetals. Most of the elements are solids at room temperature, two of them (mercury and bromine) are liquids, and the rest are gases.

A few of the elements are found in nature in the free (uncombined) state. Some of these are oxygen, nitrogen, the noble gases (helium, neon, argon, krypton, xenon, and radon), sulfur, copper, silver, and gold. Most of the elements in nature are combined with other elements in the form of compounds. The most abundant element on the Earth is oxygen; the next most abundant is silicon. The most abundant element in the universe is hydrogen and the next most abundant is helium.

The elements are classified in families or groups in the periodic table. Elements are also frequently classified as metals and nonmetals. A metallic element is one whose atoms form positive ions in solution, and a nonmetallic element is one whose atoms form negative ions in solution. *See* PERIODIC TABLE.

Atoms of a given element have the same atomic number, but may not all have the same atomic weight. Atoms with identical atomic numbers but different atomic weights are called isotopes. Oxygen, for example, is made up of atoms whose atomic weights are 16, 17, and 18. Hydrogen is made up of isotopes 1, 2, and 3; the isotopes of masses 2 and 3 are called deuterium and tritium, respectively. Carbon is made up of isotopes 11, 12, 13, and 14. Carbon-14 is radioactive and is used as a tracer in many chemical experiments.

All the elements have isotopes, although in certain cases only synthetic isotopes are known. Thus, fluorine exists in nature as $^{19}F$, but the artificial radioactive isotope $^{18}F$ can be prepared. Many of the isotopes of the different elements are unstable, or radioactive, and hence they disintegrate to form stable atoms either of that element or of some other element. *See* ATOMIC MASS.

The origin of the chemical elements is believed to be the result of the synthesis by fusion processes at very high temperatures (in the order of 100,000,000°C or 180,000,000°F and higher) of the simple nuclear particles (protons and neutrons) first to heavier atomic nuclei such as those of helium and then on to the heavier and more complex nuclei of the light elements (lithium, boron, and so on). The helium atoms bombard the atoms of the light elements and produce neutrons. The neutrons are captured by the nuclei of elements and produce heavier elements.

A number of elements that are found in only very slight traces or not at all in nature, such as technetium, promethium, astatine, francium, and all the elements with atomic numbers above 92, have been synthesized by a variety of nuclear reactions that involve transmuting atoms of one element into atoms of another by bombarding that element with neutrons or fast-moving particles (protons, deuterons, and alpha particles) which will change the atomic number to that of the new element. *See* ATOMIC STRUCTURE AND SPECTRA; TRANSURANIUM ELEMENTS. [A.B.G.]

**Element 111** Element 111 was discovered in late 1994. It should be a homolog of the elements gold, silver, and copper. It is expected to be the ninth element in the 6*d* shell, but the half-life of 1.5 ms of the only isotope known today is too short to allow chemical studies. *See* COPPER; GOLD; HALF-LIFE; PERIODIC TABLE; SILVER.

The element was discovered on December 17, 1994, at GSI (Gesellschaft für Schwerionenforschung), Darmstadt, Germany, by detection of the isotope $^{272}111$ (the isotope of element 111 with mass number 272), which was produced by fusion of a nickel-64 projectile and a bismuth-209 target nucleus after the fused system was cooled by emission of one neutron. Sequential alpha decays to meitnerium-268, bohrium-264, dubnium-260, and lawrencium-256 allowed identification from the known decay properties of $^{260}$Db and $^{256}$Lr. In the decay chain (see illustration), the first three

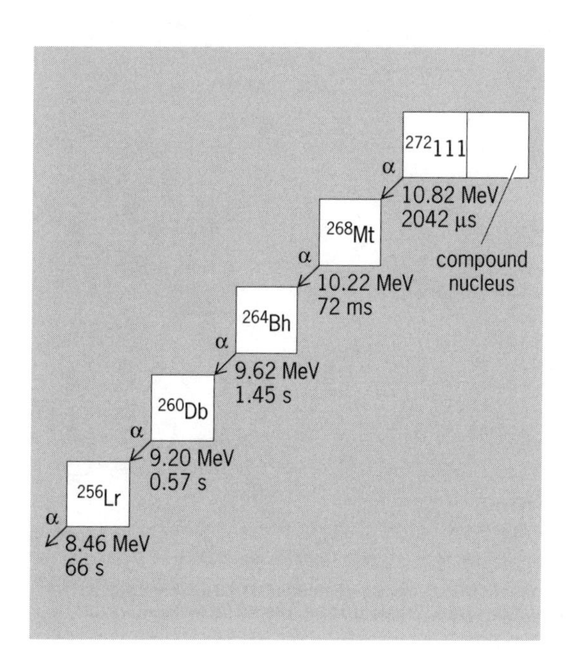


**Sequence of decay chains that document the discovery of element 111. Numbers below boxes are alpha energies and correlation times. Element 111 is produced in the reaction $^{64}\text{Ni} + {}^{209}\text{Bi} \rightarrow {}^{272}111 + 1n$.**

members are new isotopes. The isotope $^{272}111$ has a half-life of 1.5 ms, and is produced with a cross section of $3.5 \times 10^{-36}$ cm$^2$. Altogether, three chains were observed. The new isotopes meitnerium-268 and bohrium-264 are the heaviest isotopes of these elements currently known. Their half-lives of 70 ms and 0.4 s, respectively, are longer than those of the previously known isotopes of these elements. *See* NEUTRON.

[P.Ar.]

**Element 112** Element 112 was discovered in 1996. It should be a heavy homolog of the elements mercury, cadmium, and zinc. It is expected to be the last element in the 6*d* shell. No chemistry is possible in the near future as all cross sections are in the range of $10^{-36}$ cm$^2$ and the half-lives are short. *See* CADMIUM; HALF-LIFE; MERCURY; PERIODIC TABLE; ZINC.

Element 112 was discovered on February 9, 1996, at GSI (Gesellschaft für Schwerionenforschung), Darmstadt, Germany, by detection of the isotope $^{277}112$, which was produced by fusion of a zinc-70 projectile and a lead-208 target nucleus following the cooling down of the fused system by emission of a single neutron. Sequential alpha decays to $^{273}110$, hassium-269, seaborgium-265, rutherfordium-261, and nobelium-257 allowed unambiguous identification by using the known decay properties of the last three members of the chain. In the decay chain (see illustration), the first three members are new isotopes. Isotope $^{277}112$ has a half-life of 0.24 ms, and it is produced

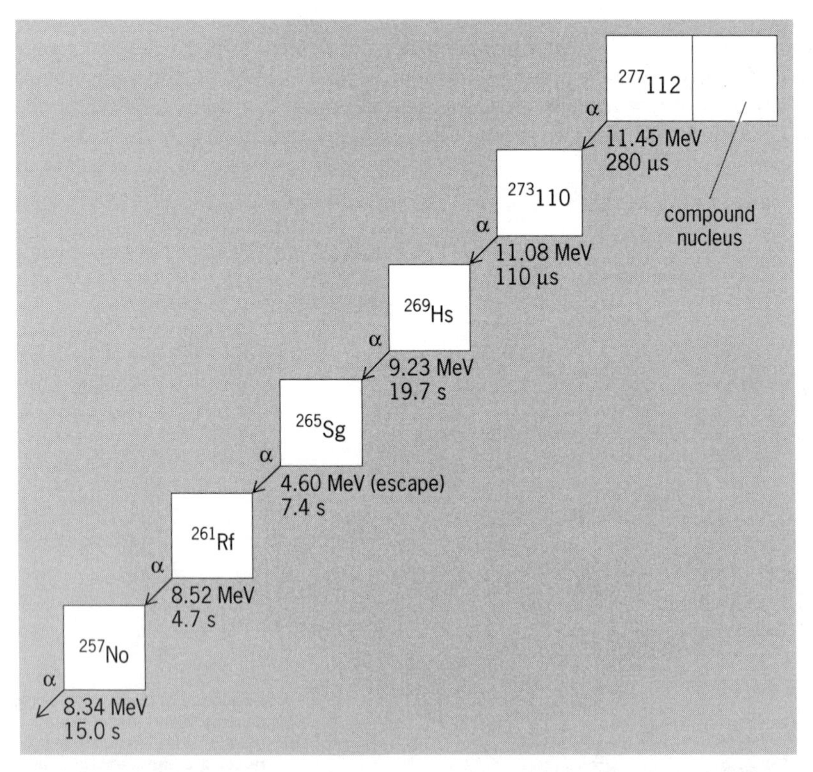


**Sequence of decay chains that document the discovery of element 112. Numbers below boxes are alpha energies and correlation times. Element 112 is produced in the reaction $^{70}Zn + {}^{208}Pb \rightarrow {}^{277}112 + 1n$.**

with a cross section of $1.0 \times 10^{-36}$ $cm^2$ (the smallest observed in the production of heavy elements). The new isotopes of darmstadtium and hassium are of special interest. Their half-lives and alpha energies are very different, as is characteristic of a closed-shell crossing. At the neutron number $N = 162$, a closed shell was theoretically predicted, and this closed shell is verified in the decay chain observed. Isotope $^{269}Hs$ has a half-life of 9 s, which is long enough to allow studies on the chemistry of this element. The crossing of the neutron shell at $N = 162$ is an important achievement in the field of research on superheavy elements. The stabilization of superheavy elements is based on high fission barriers, which are due to corrections in the binding energies found near closed shells. The shell at $N = 162$ is the first such shell predicted, and is now verified. *See* DARMSTADTIUM; ELEMENT 111; HASSIUM; LEAD; NEUTRON; NOBELIUM; RUTHERFORDIUM; SEABORGIUM. [P.Ar.]

**Emission spectrochemical analysis** A technique conducted by monitoring and measuring the spectrum of light caused to be emitted by the material to be analyzed. In general, there are many ways in which to conduct an emission spectrometric measurement; the differences among approaches result mainly from the choice of location within the electromagnetic spectrum at which to observe emitted radiation. However, emission spectrochemical analysis traditionally refers to those analytical determinations based on radiation in the visible through vacuum ultraviolet region of the electromagnetic spectrum (wavelengths about 800 to 100 nanometers). The technique is used principally to detect (qualitative analysis) and determine (quantitative analysis) metals and some nonmetals. Under optimum conditions, as little as $10^{-10}$ gram of an element per gram of sample can be determined.

The steps in emission spectrochemical analysis are: vaporization and atomization of sample; excitation of atomic vapor; resolution of emitted radiation; and observation and measurement of resolved radiation. *See* SPECTROSCOPY. [A.T.Z.]

**Emulsion** A dispersion of one liquid in a second immiscible liquid. Since the majority of emulsions contain water as one of the phases, it is customary to classify emulsions into two types: the oil-in-water (O/W) type consisting of droplets of oil dispersed in water, and the water-in-oil (W/O) type in which the phases are reversed. The continuous liquid is referred to as the dispersion medium, and the liquid which is in the form of droplets is called the disperse phase.

A stable emulsion consisting of two pure liquids cannot be prepared; to achieve stability, a third component, an emulsifying agent, must be present. Generally, the introduction of an emulsifying agent will lower the interfacial tension of the two phases. A large number of emulsifying agents are known; they can be classified broadly into several groups. The largest group is that of the soaps, detergents, and other compounds whose basic structure is a paraffin chain terminating in a polar group. Some solid powders can act as emulsifiers by being wetted more by one phase than by the other. Whichever phase shows the greater wetting power will become the dispersion medium. Many naturally occurring emulsions, such as milk or rubber latex, are stabilized by proteins. Egg yolk proteins stabilize mayonnaise and salad dressing. Certain hydrophilic colloids such as gum arabic or gelatin also stabilize water-in-oil emulsions by a similar mode of action.

Emulsions may be prepared readily by shaking together the two liquids or by adding one phase drop by drop to the other phase with some form of agitation, such as irradiation by ultrasonic waves of high intensity. In industry, emulsification is accomplished by means of emulsifying machines.

The breaking of emulsions is necessary in many industrial operations, for example, in the separation of water-in-oil emulsions in the petroleum industry and in product recovery from emulsions produced by the steam distillation of organic liquids. Emulsions may be broken by (1) addition of multivalent ions of charge opposite to the emulsion droplet, (2) chemical action (addition of acids to emulsions stabilized by soaps), (3) freezing, (4) heating, (5) aging, (6) centrifuging, (7) application of high-potential alternating electric fields, and (8) treatment with ultrasonic waves of low intensity. *See* COLLOID; SOAP. [G.S.M.; W.O.M.]

**Enthalpy** For any system, that is, the volume of substance under discussion, enthalpy is the sum of the internal energy of the system plus the system's volume multiplied by the pressure exerted by the system on its surroundings. The sum is given the symbol $H$ primarily as a matter of convenience because this sum appears repeatedly in thermodynamic discussion. Previously, enthalpy was referred to as total heat or heat content, but these terms are misleading and should be avoided. Enthalpy is, from the viewpoint of mathematics, a point function, as contrasted with heat and work, which are path functions. Point functions depend only on the initial and final states of the system undergoing a change; they are independent of the paths or character of the change. *See* ENTROPY; THERMODYNAMIC PRINCIPLES; THERMODYNAMIC PROCESSES. [H.C.W.; W.A.S.]

**Entropy** A function first introduced in classical thermodynamics to provide a quantitative basis for the common observation that naturally occurring processes have a particular direction. Subsequently, in statistical thermodynamics, entropy was shown to be a measure of the number of microstates a system could assume. Finally, in communication theory, entropy is a measure of information. Each of these aspects will be considered in turn. Before the entropy function is introduced, it is necessary to discuss reversible processes.

**Reversible processes.** Any system under constant external conditions is observed to change in such a way as to approach a particularly simple final state called an equilibrium state. For example, two bodies initially at different temperatures are connected by a metal wire. Heat flows from the hot to the cold body until the temperatures of both bodies are the same. It is common experience that the reverse processes never occur if the systems are left to themselves; that is, heat is never observed to flow from the cold to the hot body. Max Planck classified all elementary processes into three categories: natural, unnatural, and reversible. Natural processes do occur, and proceed in a direction toward equilibrium. Unnatural processes move away from equilibrium and never occur. A reversible process is an idealized natural process that passes through a continuous sequence of equilibrium states.

**Entropy function.** The state function entropy $S$ puts the foregoing discussion on a quantitative basis. Entropy is related to $q$, the heat flowing into the system from its surroundings, and to $T$, the absolute temperature of the system. The important properties for this discussion are:

1. $dS > q/T$ for a natural change.
   $dS = q/T$ for a reversible change.
2. The entropy of the system $S$ is made up of the sum of all the parts of the system so that $S = S_1 + S_2 + S_3 \cdots$.

*See* THERMODYNAMIC PRINCIPLES.

**Nonconservation.** In his study of the first law of thermodynamics, J. P. Joule caused work to be expended by rubbing metal blocks together in a large mass of water. By this and similar experiments, he established numerical relationships between heat and work. When the experiment was completed, the apparatus remained unchanged except for a slight increase in the water temperature. Work ($W$) had been converted into heat ($Q$) with 100% efficiency. Provided the process was carried out slowly, the temperature difference between the blocks and the water would be small, and heat transfer could be considered a reversible process. The entropy increase of the water at its temperature $T$ is $\Delta S = Q/T = W/T$. Since everything but the water is unchanged, this equation also represents the total entropy increase. The entropy has been created from the work input, and this process could be continued indefinitely, creating more and more entropy. Unlike energy, entropy is not conserved. *See* THERMODYNAMIC PROCESSES.

**Degradation of energy.** Energy is never destroyed. But in the Joule friction experiment and in heat transfer between bodies, as in any natural process, something is lost. In the Joule experiment, the energy expended in work now resides in the water bath. But if this energy is reused, less useful work is obtained than was originally put in. The original energy input has been degraded to a less useful form. The energy transferred from a high-temperature body to a lower-temperature body is also in a less useful form. If another system is used to restore this degraded energy to its original form, it is found that the restoring system has degraded the energy even more than the original system had. Thus, every process occurring in the world results in an overall increase in entropy and a corresponding degradation in energy. [W.F.J.]

**Measure of information.** The probability characteristic of entropy leads to its use in communication theory as a measure of information. The absence of information about a situation is equivalent to an uncertainty associated with the nature of the situation. This uncertainty is the entropy of the information about the particular situation. [F.H.R.]

**Enzyme** A catalytic protein produced by living cells. The chemical reactions involved in the digestion of foods, the biosynthesis of macromolecules, the controlled release and utilization of chemical energy, and other processes characteristic of life are all catalyzed by enzymes. In the absence of enzymes, these reactions would not take place at a significant rate. Several hundred different reactions can proceed simultaneously within a living cell, and the cell contains a comparable number of individual enzymes, each of which controls the rate of one or more of these reactions. The potentiality of a cell for growing, dividing, and performing specialized functions, such as contraction or transmission of nerve impulses, is determined by the complement of enzymes it possesses. Some representative enzymes, their sources, and reaction specificities are shown in the table.

**Characteristics.** Enzymes can be isolated and are active outside the living cell. They are such efficient catalysts that they accelerate chemical reactions measurably, even at concentrations so low that they cannot be detected by most chemical tests for protein. Like other chemical reactions, enzyme-catalyzed reactions proceed only when accompanied by a decrease in free energy; at equilibrium the concentrations of reactants and products are the same in the presence of an enzyme as in its absence. An enzyme can catalyze an indefinite amount of chemical change without itself being diminished or altered by the reaction. However, because most isolated enzymes are relatively unstable, they often gradually lose activity under the conditions employed for their study.

**Chemical nature.** All enzymes are proteins. Their molecular weights range from about 10,000 to more than 1,000,000. Like other proteins, enzymes consist of chains of amino acids linked together by peptide bonds. An enzyme molecule may contain

**Some representative enzymes, their sources, and reaction specificities**

| Enzyme | Some sources | Reaction catalyzed |
|---|---|---|
| Pepsin | Gastric juice | Hydrolysis of proteins to peptides and amino acids |
| Urease | Jackbean, bacteria | Hydrolysis of urea to ammonia and carbon dioxide |
| Amylase | Saliva, pancreatic juice | Hydrolysis of starch to maltose |
| Phosphorylase | Muscle, liver, plants | Reversible phosphorolysis of starch or glycogen to glucose-1-phosphate |
| Transaminases | Many animal and plant tissues | Transfer of an amino group from an amino acid to a keto acid |
| Phosphohexose isomerase | Muscle, yeast | Interconversion of glucose-6-phosphate and fructose-6-phosphate |
| Pyruvic carboxylase | Yeast, bacteria, plants | Decarboxylation of pyruvate to acetaldehyde and carbon dioxide |
| Catalase | Erythrocytes, liver | Decomposition of hydrogen peroxide to oxygen and water |
| Alcohol dehydrogenase | Liver | Oxidation of ethanol to acetaldehyde |
| Xanthine oxidase | Milk, liver | Oxidation of xanthine and hypoxanthine to uric acid |

one or more of these polypeptide chains. The sequence of amino acids within the polypeptide chains is characteristic for each enzyme and is believed to determine the unique three-dimensional conformation in which the chains are folded. This conformation, which is necessary for the activity of the enzyme, is stabilized by interactions of amino acids in different parts of the peptide chains with each other and with the surrounding medium. These interactions are relatively weak and may be disrupted readily by high temperatures, acid or alkaline conditions, or changes in the polarity of the medium. Such changes lead to an unfolding of the peptide chains (denaturation) and a concomitant loss of enzymatic activity, solubility, and other properties characteristic of the native enzyme. Enzyme denaturation is sometimes reversible. *See* AMINO ACIDS; PROTEIN.

Many enzymes contain an additional, nonprotein component, termed a coenzyme or prosthetic group. This may be an organic molecule, often a vitamin derivative, or a metal ion. The coenzyme, in most instances, participates directly in the catalytic reaction. For example, it may serve as an intermediate carrier of a group being transferred from one substrate to another. Some enzymes have coenzymes that are tightly bound to the protein and difficult to remove, while others have coenzymes that dissociate readily. When the protein moiety (the apoenzyme) and the coenzyme are separated from each other, neither possesses the catalytic properties of the original conjugated protein (the holoenzyme). By simply mixing the apoenzyme and the coenzyme together, the fully active holoenzyme can often be reconstituted. The same coenzyme may be associated with many enzymes which catalyze different reactions. It is thus primarily the nature of the apoenzyme rather than that of the coenzyme which determines the specificity of the reaction. *See* COENZYME.

The complete amino acid sequence of several enzymes has been determined by chemical methods. By x-ray crystallographic methods even the exact three-dimensional molecular structure of a few enzymes has been deduced. *See* X-RAY CRYSTALLOGRAPHY.

**Classification and nomenclature.** Enzymes are usually classified and named according to the reaction they catalyze. The principal classes are as follows.

Oxidoreductases catalyze reactions involving electron transfer, and play an important role in cellular respiration and energy production. Some of them participate in the process of oxidative phosphorylation, whereby the energy released by the oxidation of carbohydrates and fats is utilized for the synthesis of adenosine triphosphate (ATP) and thus made directly available for energy-requiring reactions.

Transferases catalyze the transfer of a particular chemical group from one substance to another. Thus, transaminases transfer amino groups, transmethylases transfer methyl groups, and so on. An important subclass of this group are the kinases, which catalyze the phosphorylation of their substrates by transferring a phosphate group, usually from ATP, thereby activating an otherwise metabolically inert compound for further transformations.

Hydrolases catalyze the hydrolysis of proteins (proteinases and peptidases), nucleic acids (nucleases), starch (amylases), fats (lipases), phosphate esters (phosphatases), and other substances. Many hydrolases are secreted by the stomach, pancreas, and intestine and are responsible for the digestion of foods. Others participate in more specialized cellular functions. For example, cholinesterase, which catalyzes the hydrolysis of acetylcholine, plays an important role in the transmission of nervous impulses. *See* ACETYLCHOLINE.

Lyases catalyze the nonhydrolytic cleavage of their substrate with the formation of a double bond. Examples are decarboxylases, which remove carboxyl groups as carbon dioxide, and dehydrases, which remove a molecule of water. The reverse reactions are catalyzed by the same enzymes.

Isomerases catalyze the interconversion of isomeric compounds.

Ligases, or synthetases, catalyze endergonic syntheses coupled with the exergonic hydrolysis of ATP. They allow the chemical energy stored in ATP to be utilized for driving reactions uphill.

**Specificity.** The majority of enzymes catalyze only one type of reaction and act on only one compound or on a group of closely related compounds. There must exist between an enzyme and its substrate a close fit, or complementarity. In many cases, a small structural change, even in a part of the molecule remote from that altered by the enzymatic reaction, abolishes the ability of a compound to serve as a substrate. An example of an enzyme highly specific for a single substrate is urease, which catalyzes the hydrolysis of urea to carbon dioxide and ammonia. On the other hand, some enzymes exhibit a less restricted specificity and act on a number of different compounds that possess a particular chemical group. This is termed group specificity.

A remarkable property of many enzymes is their high degree of stereospecificity, that is, their ability to discriminate between asymmetric molecules of the right-handed and left-handed configurations. An example of a stereospecific enzyme is L-amino acid oxidase. This enzyme catalyzes the oxidation of a variety of amino acids of the type R—CH($NH_2$)COOH. The rate of oxidation varies greatly, depending on the nature of the R group, but only amino acids of the L configuration react. *See* STEREOCHEMISTRY.

[D.W.]

**Epoxide** A member of a class of three-membered ring cyclic ethers that are also known as oxiranes or alkylene oxides. The basic structure of an epoxide is analogous to that of the first member, ethylene oxide, shown below.

$$H_2C \overset{O}{\text{——}} CH_2$$

*See* ETHYLENE OXIDE.

Epoxides are made primarily by select oxidation reactions of alkenes; however, another classical preparation results from the ring closure of halohydrins by way of intramolecular nucleophilic displacement; that is, the reaction occurs within the same molecule, the alkoxide ion displacing the halogen to form a ring. The interest in epoxides results from their ease in preparation, and their usefulness as a reactive functional group that can give a variety of products after treatment with either electrophilic or nucleophilic reagents, or on occasion after treatment with oxidizing (for example, periodic acid) and reducing (for example, titanocene dichloride) agents. The ease of opening of the strained three-membered ring epoxides with attack of reagents in a stereospecific manner gives one or two stereochemical products (when applicable), usually in good yield. Epoxides are also used to prepare monomers, prepolymers, polymers, and copolymers. *See* ALKENE; ELECTROPHILIC AND NUCLEOPHILIC REAGENTS; HETEROCYCLIC COMPOUNDS.

Poly(ethylene oxide), $(CH_2CH_2O)_n$, was one of the first polymers to be prepared in the laboratory. The use of numerous epoxides (as monomers, or for the synthesis of other epoxide monomers or prepolymers) in polymer synthesis has been quite extensive, and these polyethers have been prepared by both cationic and anionic polymerization techniques. The molecular weights of resulting polyethers generally range from 500 to 10,000, usually because of interfering chain-transfer reactions. An especially interesting synthesis is the opening of propylene oxide to give optically active poly(propylene oxides). Epoxides are used in the preparation of copolymers. *See* COPOLYMER; FURAN; POLYETHER RESINS; POLYMERIZATION.

Epoxides are promoters; that is, they are easy to polymerize, and their polymerization causes more difficult cyclic ethers such as tetrahydrofuran to polymerize. Epichlorohydrin has also been used for the preparation of prepolymers by condensations with the sodium salt of bisphenols. The epichlorohydrin/bisphenol A and other epoxide prepolymers have been used for the preparation of epoxy resins (thermosets), which are crossed-linked (3D) polymers, linear polymers that are connected by cross-linked molecular units. They are insoluble, infusible, and intractable, and sometimes are called epoxy resins. For example, a low-molecular-weight epoxy prepolymer can be treated and cross-linked with a polyamine or smaller molecules such as diethylenetriamine, or crossed-linked by the addition of carboxylic acid anhydrides, such as maleic anhydride. Other polymers can also be incorporated or can participate in the cross-linking process, and include phenolics, ureas, and melamines. The products resulting from these prepolymer techniques are used for surface coating materials, molding, pipes, laminating, repair of damaged automobile bodies, manufacture of articles reinforced with glass fibers, durable and tough epoxy resin adhesives (glues), and many other applications. [C.F.B.]

**Equivalent weight** The number of parts by weight of an element or compound which will combine with or replace, directly or indirectly, 1.008 parts by weight of hydrogen, 8.00 parts of oxygen, or the equivalent weight of any other element or compound. For all elements, the atomic weight is equal to the equivalent weight times a small whole number, called the valence of the element. An element can have more than one valence and therefore more than one equivalent weight. *See* VALENCE.

The concept of equivalent weight, together with that of gram-equivalent weight, tends to have been abandoned, and relations are expressed in terms of balanced stoichiometric chemical equations and relative numbers of moles reacting. *See* ELECTROCHEMICAL EQUIVALENT; MOLE (CHEMISTRY); STOICHIOMETRY. [T.C.W.]

**Erbium** A chemical element, Er, atomic number 68, atomic weight 167.26, belonging to the rare-earth group. The naturally occurring element is made up of the six stable isotopes. The rose-pink oxide, $Er_2O_3$, dissolves in mineral acids to give rose-colored solutions. The salts are paramagnetic and the ions are trivalent. At low temperatures the metal is antiferromagnetic and at still lower temperatures becomes strongly ferromagnetic. *See* PERIODIC TABLE; RARE-EARTH ELEMENTS. [F.H.Sp.]

**Ester** The product of a condensation reaction (esterification) in which a molecule of an acid unites with a molecule of alcohol with elimination of a molecule of water as shown in the following reaction.

$$\underset{\text{Acid}}{R\overset{O}{\overset{\|}{C}}OH} + \underset{\text{Alcohol}}{HOR'} \longrightarrow \underset{\text{Ester}}{R\overset{O}{\overset{\|}{C}}OR'} + H_2O$$

At one time it was thought that esterification was analogous to neutralization, and esters are still named as though they are "alkyl salts" of carboxylic acids.

Esters are generally insoluble in water and have boiling points slightly higher than hydrocarbons of similar molecular weight.

Ethyl and butyl acetates are volatile industrial solvents, used particularly in the formulation of lacquers. Higher-boiling esters such as butyl phthalate are used as softening agents (plasticizers) in the compounding of plastics. The natural waxes of biological origin are largely simple esters. For example, a principal component of beeswax is myricyl palmitate. *See* SOLVENT.

Esters of cellulose (cellulose triacetate) are used in photographic film, as a textile fiber (acetate rayon), and several have become important as thermoplastic materials. Cellulose nitrate, called celluloid pyroxylin, forms celluloid, dynamite cotton, and gun cotton. Cordite and ballistite are made from gun cotton. Dimethyl and diethyl sulfates (esters of sulfuric acid) are excellent agents for alkylating organic molecules that contain labile hydrogen atoms, for example, starch and cellulose.

Esters of unsaturated acids, for example, acrylic or methacrylic acid, are reactive and polymerize rapidly, yielding resins; thus, methyl methacrylate yields a polymethyl methacrylate resin (Lucite). Analogously, esters of unsaturated alcohols are reactive and readily react with themselves; thus, vinyl acetate polymerizes to polyvinyl acetate.

Many low-molecular-weight esters have characteristic, fruit-like odors: banana (isoamyl acetate), rum (isobutyl propionate), and pineapple (butyl butyrate). These esters are used to some extent in compounding synthetic flavors and perfumes. *See* ALCOHOL; CARBOXYLIC ACID; FAT AND OIL; POLYESTER RESINS. [P.E.F.]

**Ether** One of a class of organic compounds characterized by the structural feature of an oxygen atom linking two hydrocarbon groups, R—O—R′. Ethers are used widely as solvents, both in chemical manufacture and in the research laboratory.

The hydrocarbon radicals R and R′ may be identical (simple ether) or different (mixed ether). They may be aromatic or aliphatic, and the names of the ethers correspond to the hydrocarbon groups present. Thus, $CH_3—O—CH_3$ is methyl ether, rarely dimethyl ether, and $C_6H_5—O—CH_3$ is phenyl methyl ether.

Simple ethers may be considered to be the anhydrides of alcohols and are manufactured from alcohols by catalytic dehydration, as in reaction (1), or from olefins by controlled catalytic hydration, as in reaction (2).

$$2ROH \rightarrow ROR + H_2O \qquad (1)$$

$$2CH_3CH{=}CH_2 + HOH \rightarrow (CH_3)_2CH{-}O{-}CH(CH_3)_2 \qquad (2)$$

Ethers are less soluble in water than are the corresponding alcohols, but are miscible with most organic solvents. Low-molecular-weight ethers have a lower boiling point than the corresponding alcohols, but for those ethers containing radicals larger than butyl, the reverse is true. Inertness at moderate temperatures, an outstanding chemical characteristic of the saturated alkyl ethers, leads to their wide use as reaction media.

The best known of the ethers is ethyl ether, sometimes called diethyl ether or simply ether, $CH_3CH_2OCH_2CH_3$. It is used in industry as a solvent and in medicine as an anesthetic. When ethyl ether is used as a solvent, its high volatility can cause loss. However this volatility is advantageous in that the ether can be readily removed from the concentrated or crystallized product. The toxicity to humans is low, and recovery from overexposure is rapid and complete. It readily forms explosive mixtures with air, and on standing in containers which have been opened, it forms dangerous peroxides.

Several cyclic ethers (epoxides) are of special importance and interest. The simplest of these is ethylene oxide or oxirane (**1**), made industrially by the oxidation of ethylene with air over a silver catalyst. Dioxane or 1,4-dioxane (**2**) is prepared by the catalytic dimerization of ethylene oxide. It is used extensively as an industrial solvent. Furan (**3**), made by the decarbonylation of furfural, is the most important ether obtained from an agricultural source. Most of it is hydrogenated to form the useful solvent tetrahydrofuran (**4**). *See* ETHYLENE; FURAN; OXIDE.

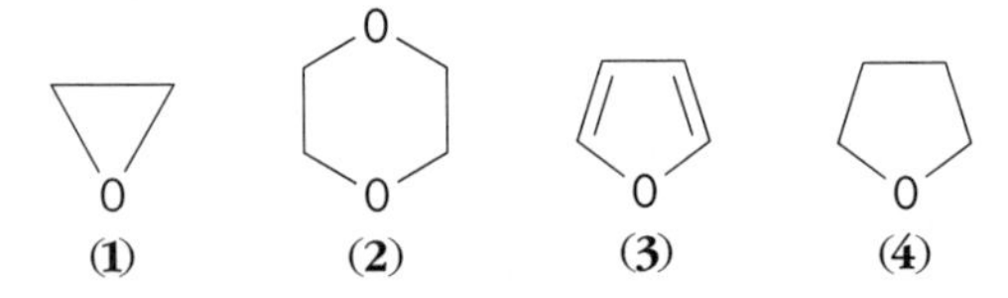


Certain large-ring polyethers, the crown ethers, are able to increase the solubility of alkali metal salts in nonpolar organic solvents. Specific metal complexes are formed by these crown ethers with alkali metal cations, the specificity for a given cation depending upon the hole in the middle of the crown ether structure. *See* HETEROCYCLIC COMPOUNDS.

[P.E.F.]

**Ethyl alcohol** Probably the best known of the alcohols, ethyl alcohol, formula $CH_3CH_2OH$, is also called alcohol, ethanol, grain alcohol, industrial alcohol, fermentation alcohol, cologne spirits, ethyl hydroxide, and methylcarbinol. Pure ethyl alcohol is a colorless, limpid, volatile liquid which is flammable and toxic and has a pungent taste. It boils at 78.4°C (173°F) and melts at −112.3°C (−170.1°F), has a specific gravity of 0.7851 at 20°C (68°F), and is soluble in water and most organic liquids. It is one of the most important industrial organic chemicals. Ethyl alcohol is produced by chemical synthesis and by fermentation or biosynthetic processes. *See* ORGANIC SYNTHESIS.

**Uses.** Ethyl alcohol is used as a solvent, extractant, antifreeze, and intermediate in the synthesis of innumerable organic chemicals. It is also an essential ingredient of alcoholic beverages.

Various grades of ethyl alcohol are produced, depending on their intended use. U.S. Pharmaceutical (USP XV) grade is the water azeotrope of ethyl alcohol and is 95% ethyl alcohol by volume. National Formulary (NFX) grade is 99+% ethyl alcohol by weight; it is also called absolute, or anhydrous, alcohol. This grade is generally prepared by azeotropic dehydration with benzene and therefore usually contains about 0.5% benzene. Denatured alcohol contains a small amount of a malodorous or obnoxious material to prevent the use of this grade of ethyl alcohol for beverage purposes. *See* AZEOTROPIC MIXTURE.

The major use of ethyl alcohol is as a starting material for various organic syntheses. Bimolecular dehydration of ethyl alcohol gives diethyl ether, which is employed as a solvent, extractant, and anesthetic. Dehydrogenation of ethyl alcohol yields acetaldehyde, which is the precursor of a vast number of organic chemicals, such as acetic acid, acetic anhydride, chloral, butanol, crotonaldehyde, and ethylhexanol. Reaction with carboxylic acids or anhydrides yields esters which are useful in many applications. The hydroxyl group of ethyl alcohol may be replaced by halogen to give the ethyl halides. Treatment with sulfuric acid gives ethyl hydrogen sulfate and diethyl sulfate, a useful ethylating agent. Reaction of ethyl alcohol with aldehydes gives the respective diethyl acetals, and reaction with acetylene produces the acetals, as well as ethyl vinyl ether. Treatment of ethyl alcohol with ammonia produces acetonitrile, which may be reduced to ethylamine. These and other ethyl alcohol-derived chemicals are used in dyes, drugs, synthetic rubber, solvents, extractants, detergents, plasticizers, lubricants, surface coatings, adhesives, moldings, cosmetics, explosives, pesticides, and synthetic fiber resins. [H.J.P.; E.M.M.]

**Ethylene** A colorless gas, formula $CH_2{=}CH_2$, with a boiling point of −103.8°C (−155°F) and a melting point of −169.4°C (−273°F). Ethylene is the most important synthetic organic chemical in terms of volume, sales value, and number of derivatives. About half of the ethylene produced is used in the manufacture of polyethylene; ethylene dichloride and vinyl chloride production uses about 20%, synthesis of ethylene oxide and derivatives account for about 12%, and styrene production consumes about 8% of the ethylene. Other important derivatives are ethanol, vinyl acetate, and acetaldehyde. Thermal cracking of hydrocarbons in the presence of steam is the most widely used process for producing ethylene.

Polyethylene, the most important derivative of ethylene, is produced by both high- and low-pressure processes to make high- and low-density, high-molecular-weight thermoplastic polymers. Aluminum alkyl catalysts (Ziegler polymerization) are used to polymerize ethylene to relatively low-molecular-weight, straight-chain hydrocarbon derivatives which are convertible to even-numbered-carbon linear olefins, alcohols, and acids. Commercial processes use palladium-catalyzed oxidation of ethylene to produce acetaldehyde, or if acetic acid is used as the solvent, vinyl acetate. Chlorination and oxychlorination processes are used to make vinyl chloride. Ethylene oxide is produced by silver-catalyzed oxidation of ethylene. Acid-catalyzed hydration of ethylene produces ethanol competitively with fermentation processes. *See* ACETYLENE; ALKENE; ETHYL ALCOHOL; ETHYLENE OXIDE; POLYMER. [R.K.Ba.]

**Ethylene glycol** A colorless, nearly odorless, sweet-tasting, hygroscopic liquid, formula $HOCH_2CH_2OH$. It is relatively nonvolatile and viscous and is the simplest member of the glycol family. Ethylene glycol freezes at −13°C (8.6°F), boils at 197.6°C (387.7°F), and is completely soluble in water, common alcohols, and phenol. Low molecular weight, low volatility, water solubility, and low solvent action on automobile finishes make ethylene glycol ideal as a radiator antifreeze and coolant. Other uses for this commodity chemical are in polyester resins, explosives, brake and shock-absorber fluids, and alkyl-type resins.

Dibasic acids or anhydrides react with ethylene glycol to form polyester condensation polymers. Reaction with terephthalic acid or its esters produces polyester resins which can be spun to fibers that find wide use in clothing and general fabrics applications. The polymer also has important film applications. *See* ETHYLENE OXIDE. [R.K.Ba.]

**Ethylene oxide** The simplest cyclic ether or epoxide, with the formula $C_2H_4O$, and the structure shown. It is also called epoxyethane and oxirane. Commercial

$$\overset{\displaystyle O}{\diagup\;\;\diagdown}$$
$$CH_2 — CH_2$$

processes use either air or oxygen to oxidize ethylene to ethylene oxide. *See* ETHER; ETHYLENE; HETEROCYCLIC COMPOUNDS.

Ethylene oxide is a colorless gas boiling at 10.4°C (50.7°F) and melting at −112°C (−169.6°F). Its vapors are flammable and explosive, and it is considered a relatively toxic liquid and gas. It is miscible in all proportions with water, alcohols, ethers, and other organic solvents.

About 50% of the ethylene oxide produced is converted to ethylene glycol; about 30% is used in the manufacture of non-ionic surfactants, glycol ethers, and ethanolamines. Gaseous ethylene oxide in $CO_2$ or difluoromethane is used as a fumigant and a sterilizing agent for medical equipment. *See* ETHYLENE GLYCOL. [R.K.Ba.]

**Ethylenediaminetetraacetic acid** A chelating agent for metallic ions, abbreviated EDTA. The structural formula is shown below. Tetrasodium EDTA is the

$$(HOOCH_2C)_2NCH_2CH_2N(CH_2COOH)_2$$

most common form in commerce, but other metallic chelates are marketed, for example, iron, zinc, and calcium. Tetrasodium EDTA is a white solid, very soluble in water and forming a basic solution. Prepared from ethylenediamine, formaldehyde, and sodium cyanide in basic solution, or from ethylenediamine and sodium chloroacetate, EDTA is a strong complexing and chelating agent. It reacts with many metallic ions to form soluble chelates. As such, it is widely used in analysis to retain alkaline earths and heavy metals in solution. The iron chelate is useful in lawn management and gardening as a replacement for ferrous sulfate (copperas). Calcium EDTA is used to control the deterioration of natural seawater in aquariums. Calcium disodium salt of EDTA is used in pharmaceuticals to prevent calcium depletion of the body during therapy. *See* CHELATION; COORDINATION COMPLEXES. [F.W.]

**Europium** A chemical element, Eu, atomic number 63, atomic weight 151.96, a member of the rare-earth group. The stable isotopes, $^{151}Eu$ and $^{153}Eu$, make up the naturally occurring element. The metal is the second most volatile of the rare earths and has a considerable vapor pressure at its melting point. It is very soft, is rapidly attacked by air, and really belongs more to the calcium-strontium-barium series than to the rare-earth series.

The element is attractive to the atomic industry, since the elements can be used in control rods and as nuclear poisons. These poisons are materials added to a nuclear reactor to balance the excess reactivity at start-up, and are so chosen that the poisons burn out at the same rate as the excess activity decreases. The television industry uses considerable quantities of phosphors, such as europium-activated yttrium orthovanadates, and other europium-activated yttrium phosphors have been patented. These phosphors give a brilliant red color and are used in the manufacture of television screens. *See* PERIODIC TABLE; RARE-EARTH ELEMENTS. [F.H.Sp.]

**Evaporation** The process by which a liquid is converted into a vapor. In the liquid phase, the substance is held together by intermolecular forces. As the temperature is raised, the molecules move more vigorously, and in increasingly high proportion have sufficient energy to escape from their neighbors. Evaporation is therefore slow at low temperatures but faster at higher temperatures. In an open vessel, the molecules escape from the vicinity of the liquid, and there is a net migration from the liquid to the atmosphere. In a closed vessel, net evaporation continues until the number of molecules in the vapor has risen to the stage at which the rate of return from the vapor to the liquid is equal to the rate of evaporation. At this stage there is a dynamic equilibrium between the liquid and its vapor, with evaporation and its reverse, condensation, occurring at the same rate. The pressure of the vapor in the closed vessel is called the vapor pressure of the substance; its value depends on the temperature. Boiling occurs in an open vessel (but not in a closed vessel) when the vapor pressure is equal to the ambient pressure. *See* BOILING POINT; VAPOR PRESSURE.

Evaporation is an endothermic (heat-absorbing) process because molecules must be supplied with energy to overcome the intermolecular forces. The enthalpy of vaporization, $\Delta_{vap}H$ (formerly, the latent heat of vaporization) is the heat required at constant pressure per mole of substance for vaporization. The entropy of vaporization, $\Delta_{vap}S$, at the boiling point, $T_b$, is equal to $\Delta_{vap}H/T_b$. According to Trouton's rule, for many liquids the entropy of vaporization is close to 85 J/K · mol. This value reflects the similar change in disorder that occurs when a liquid is converted into a gas. However, certain liquids (water and mercury among them) are more structured than others, and have a bigger entropy of vaporization than Trouton's rule suggests. *See* ENTHALPY; ENTROPY; THERMODYNAMIC PRINCIPLES.

Volatile liquids evaporate more rapidly than others at the same temperature. Such liquids have relatively weak intermolecular forces. In general, the rate of evaporation depends on the strengths of the intermolecular forces and the rate at which heat is supplied to the liquid. *See* INTERMOLECULAR FORCES; LIQUID. [P.W.A.]

**Evaporator** A device used to vaporize part or all of the solvent from a solution. The valuable product is usually either a solid or a concentrated solution of the solute. If a solid, the heat required for evaporation of the solvent must have been supplied to a suspension of the solid in the solution, otherwise the device would be classed as a drier. The vaporized solvent may be made up of several volatile components, but if any separation of these components is effected, the device is properly classed as a still or distillation column. When the valuable product is the vaporized solvent, an evaporator is sometimes mislabeled a still, such as water still, and sometimes is properly labeled, such as boiler-feedwater evaporator. In the great majority of evaporator installations, water is the solvent that is removed. *See* DISTILLATION.

Evaporators are used primarily in the chemical industry. For example, common salt is made by boiling a saturated brine in an evaporator. The salt precipitates as a solid in suspension in the brine. This slurry is pumped continuously to a filter, from which the solids are recovered and the liquid portion returned for further evaporation. Evaporators are widely used in the food industry, usually as a means of reducing volume to permit easier storage and shipment. Evaporators are also the most commonly used means of producing potable water from sea water or other contaminated sources.

The vaporization of solvent requires large amounts of heat. Provisions for transferring this heat to the solution constitute the largest element of evaporator cost and the principal means of distinguishing between types of evaporators. Practically all evaporators fall into one of the following categories:

1. Submerged-combustion evaporators: those heated by a flame that burns below the liquid surface, and in which the hot combustion gases are bubbled through the liquid.

2. Direct-fired evaporators: those in which the flame and combustion gases are separated from the boiling liquid by a metal wall, or heating surface.

3. Stem-heated evaporators: those in which steam or other condensable vapor is the source of heat, and in which the steam condenses on one side of the heating surface and the heat is transmitted through the wall to the boiling liquid. *See* EVAPORATION.

[F.C.S.]

**Excitation potential** The difference in potential between an excited atomic or molecular state and the ground state. The term is most generally used in connection with electron excitation, but it can be applied to excited molecular vibrational and rotational states.

A closely related term is excitation energy. If the unit of potential is taken as the volt and the unit of energy as the electron volt, then the two are numerically equal. According to the Bohr theory, there is a relationship between the wavelength of the photon associated with the transition and the excitation energies of the two states. Thus the basic equation for the emission or absorption of energy is as shown below, where

$$\frac{hc}{\lambda} = E_i - E_f$$

$h$ is Planck's constant, $c$ the velocity of light, $\lambda$ the wavelength of the photon, and $E_f$ and $E_i$ the energies of the final and initial states, respectively. *See* EXCITED STATE; IONIZATION POTENTIAL.

[G.H.M.]

**Excited state** In quantum mechanics, a stationary state of higher energy than the lowest stationary state or ground state of a particle or a system of particles. Customarily, only bound stationary states, which generally are at most denumerably infinite in number, are spoken of as excited, although the formal quantum theory often treats the noncountable unbound stationary states on an equal footing with the bound states.

[E.G.]

**Extended x-ray absorption fine structure (EXAFS)** The structured absorption on the high-energy side of an x-ray absorption edge. The absorption edges for an element are abrupt increases in x-ray absorption that occur when the energy of the incident x-ray matches the binding energy of a core electron (typically a 1*s* or a 2*p* electron). For x-ray energies above the edge energy, a core electron is ejected from the atom. The ejected core electron can be thought of as a spherical wave propagating outward from the absorbing atom. The photoelectron wavelength is determined by its kinetic energy, which is in turn determined by the difference between the incident x-ray energy and the core-electron binding energy. As the x-ray energy increases, the kinetic energy of the photoelectron increases, and thus its wavelength decreases.

The x-ray-excited photoelectron will be scattered by the neighboring atoms surrounding the absorbing atom. The portion of the photoelectron wave that is scattered back in the direction of the absorbing atom is responsible for the EXAFS oscillations. If the outgoing and backscattered photoelectron waves are out of phase and thus interfere destructively, there is a local minimum in the x-ray absorption cross section. At a higher x-ray energy (shorter photoelectron wavelength), constructive interference

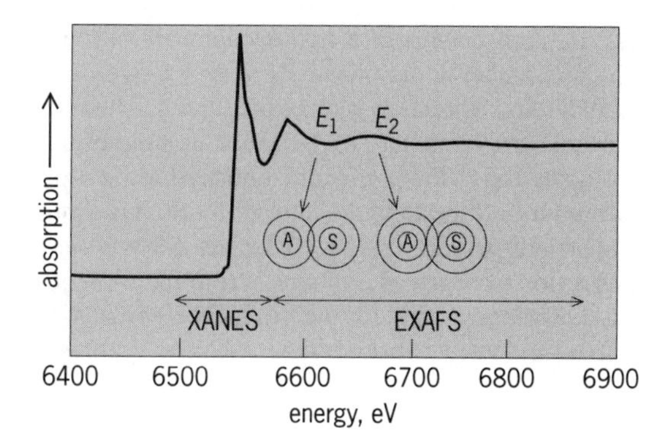


**X-ray absorption spectrum for manganese, showing XANES and EXAFS regions. As the x-ray energy increases from $E_1$ to $E_2$, the interference of the outgoing and backscattered photoelectron wave (shown schematically by concentric circles around the absorbing, *A*, and scattering, *S*, atoms) changes from destructive to constructive.**

leads to a local maximum in x-ray absorption (see illustration). EXAFS thus arises from photoelectron scattering, making it a spectroscopically detected scattering method. *See* SCATTERING OF ELECTROMAGNETIC RADIATION.

EXAFS typically refers to structured absorption from approximately 50 to 1000 eV or more above the absorption edge. X-ray absorption near edge structure (XANES) is often used to refer to the structure in the near (around 50 eV) region of the edge. X-ray absorption fine structure (XAFS) has gained some currency as a reference to the entire structured absorption region (XANES+EXAFS).

EXAFS spectra contain structural information comparable to that obtained from single-crystal x-ray diffraction. The principal advantage of EXAFS in comparison with crystallography is that EXAFS is a local structure probe and does not require the presence of long-range order. This means that EXAFS can be used to determine the local structure in noncrystalline samples. [J.P.Ha.]

**Extraction** A method of separating the constituents of a mixture utilizing preferential solubility of one or more components in a second phase. Commonly, this added second phase is a liquid, while the mixture to be separated may be either solid or liquid. If the starting mixture is a liquid, then the added solvent must be immiscible or only partially miscible with the original and of such a nature that the components to be separated have different relative solubilities in the two liquid phases.

Solvent extraction processes can be divided into two broad categories according to the origins of the differential solubility. On the one hand, it arises from purely physical differences between the two solutes, such as polarity, while in other cases it can be traced to definite chemical interaction between solute and solvent. Categories of major importance for the latter cases are ion-association systems and chelate compounds.

Liquid/solid extraction may be considered as the dissolving of one or more components in a solid matrix by simple solution, or by the formation of a soluble form by chemical reaction. The largest use of liquid/solid extraction is in the extractive metallurgical, vegetable oil, and sugar industries. The field may be subdivided into the following categories: leaching, washing extraction, and diffusional extraction. Leaching involves the contacting of a liquid and a solid (usually an ore) and the imposing of a chemical reaction upon one or more substances in the solid matrix so as to render them soluble. In washing extraction the solid is crushed to break the cell walls, permitting the valuable soluble product to be washed from the matrix. In diffusional extraction the soluble product diffuses across the denatured cell walls (no crushing involved) and is washed out of the solid.

Liquid/liquid extraction separates the components of a homogeneous liquid mixture on the basis of differing solubility in another liquid phase. Because it depends on differences in chemical potential, liquid/liquid extraction is more sensitive to chemical type than to molecular size. This makes it complementary to distillation as a separation technique. One of the first large-scale uses was in the petroleum industry for the separation of aromatic from aliphatic compounds. Liquid/liquid extraction also has found application for many years in the coal tar industry. On a smaller scale, extraction is a key process in the pharmaceutical industry for recovery of antibiotics from fermentation broths, in the recovery and separation of vitamins, and for the production of alkaloids from natural products. *See* CHEMICAL SEPARATION TECHNIQUES. [B.M.S.]

# F

**Fat and oil** Naturally occurring esters of glycerol and fatty acids that have commercial uses. Since fats and oils are triesters, they are commonly called triglycerides or simply glycerides. A glyceride may be designated a fat or oil, depending on its melting point. A fat is solid and an oil is liquid at room temperature. Some liquid waxes are incorrectly referred to as oils. *See* ESTER; TRIGLYCERIDE.

The structure of triglycerides is shown below, where $R_1$, $R_2$, and $R_3$ represent the alkyl chain of the fatty acid.

$$\begin{array}{c} H_2COC(O)R_1 \\ | \\ HCOC(O)R_2 \\ | \\ H_2COC(O)R_3 \end{array}$$

The physical and chemical properties of fats and oils are determined to a large extent by the types of fatty acids in the glyceride. It is possible for all the acids to be identical, but this is rare. Usually there are two or even three different acids esterified to each glycerol molecule.

In all commercially important glycerides, the fatty acids are straight-chain, and nearly all contain an even number of carbon atoms. Most fats and oils are based on $C_{16}$ and $C_{18}$ acids with zero to three ethylenic bonds. There are exceptions, such as coconut oil, which is rich in shorter-chain acids, and some marine oils, which contain acids with as many as 22 or more carbons and six or more ethylenic linkages.

The majority of fats and oils come from only a few sources. Plant sources are nuts or seeds, and nearly all terrestrial animal fats are from adipose tissue. Marine oils come principally from the whole body, although a small amount comes from trimmings. Plant fats and oils are obtained by crushing and solvent extraction. Animal and marine oils are nearly all recovered by rendering. This is a process of heating fatty tissue with steam or hot water to melt and free the glyceride, followed by separating the oil or fat from the aqueous layer. *See* SOLVENT EXTRACTION.

There are some uses for fats and oils in their native state, but ordinarily they are converted to more valuable products. The most important changes are hydrolysis and hydrogenation. Hydrolysis is commonly called splitting. The purpose is to hydrolyze the ester into its constituent glycerol and fatty acids, which are valuable intermediates for many compounds. Hydrogenation is the catalytic addition of hydrogen to ethylenic bonds, and is applicable to both acids and glycerides. The purpose of hydrogenation is usually to raise the melting point or to increase the resistance to oxidation. *See* HYDROGENATION; HYDROLYTIC PROCESSES. [H.M.H.]

**Fermium** A chemical element, Fm, atomic number 100, the eleventh element in the actinide series. Fermium does not occur in nature; its discovery and production have been accomplished by artificial nuclear transmutation of lighter elements. Radioactive isotopes of mass number 244–259 have been discovered. The total weight of fermium which has been synthesized is much less than one-millionth of a gram. *See* ACTINIDE ELEMENTS; PERIODIC TABLE.

Spontaneous fission is the major mode of decay for $^{244}Fm$, $^{256}Fm$, and $^{258}Fm$. The longest-lived isotope is $^{257}Fm$, which has a half-life of about 100 days. Fermium-258 decays by spontaneous fission with a half-life of 0.38 millisecond. This suggests the existence of an abnormality at this point in the nuclear periodic table. *See* NUCLEAR CHEMISTRY; TRANSURANIUM ELEMENTS. [G.T.S.]

**Ferricyanide** The common name for hexacyanoferrate(III), a compound containing the complex ion $[Fe(CN)_6]^{3-}$.

The $[Fe(CN)_6]^{3-}$ ion is kinetically unstable, and it dissociates to give the free cyanide anion, $CN^-$. It is therefore quite toxic. In contrast, the ferrocyanide ion, $[Fe(CN)_6]^{4-}$, is stable.

The sodium $[Na_3Fe(CN)_6]$ and potassium $[K_3Fe(CN)_6]$ salts have been isolated as ruby-red crystals and are photosensitive. The potassium salt reacts with metallic silver to produce silver ferrocyanide, and it is used in photographic processes. In addition, the $[Fe(CN)_6]^{3-}$ ion is used in blueprint materials, wood stains, and electroplating process, and as a mild oxidizing agent in organic synthesis.

The addition of Fe II to ferricyanide produces Prussian blue ($Fe_4$ III) [Fe II $(CN)_6]_3 \cdot xH_2O$, where $x = \sim 14$–16), a pigment discovered nearly 300 years ago. The structure of this mixed-valence complex has been determined by x-ray analysis and powder neutron diffraction studies. *See* COORDINATION CHEMISTRY; COORDINATION COMPLEXES; CYANIDE; IRON. [T.J.Me.]

**Film (chemistry)** A material in which one spatial dimension, thickness, is much smaller than the other two. Films can be conveniently classified as those that support themselves and those that exist only as layers on top of a supporting substrate. The latter are known as thin films and have their own specialized science and technology.

Thin films, from one to several hundred molecular layers, are generally defined as those that lie on a substrate, either liquid or solid. Monomolecular films on the surface of water can be made by adding appropriate materials in extremely small quantity to the surface. These materials are inappreciably soluble in water, rendered so by a large hydrocarbon or fluorocarbon functional group on the molecule or by other means; they are, however, able to dissolve in the surface of water by virtue of a hydrogen-bonding group or groups on the molecule.

Thin films can also be deposited directly onto solid substrates by evaporation of material in a vacuum, by sputtering, by chemical reaction (chemical vapor deposition), by ion plating, or by electroplating. *See* MONOMOLECULAR FILM.

Thin films play an extensive role in both traditional and emerging technologies. Examples are foams and emulsions, the active layers in semiconductors, the luminescent and protective layers in electroluminescent thin-film displays, and imaging and photoelectric devices. *See* EMULSION; FOAM.

Self-supported films have a nominal thickness not larger than 250 micrometers. Films of greater thickness are classified as sheets or foils. Self-supported films are commonly composed of organic polymers, either thermoplastic resins or cellulose-based materials. *See* POLYMER. [S.Ro.]

**Filtration** The separation of solid particles from a fluidsolids suspension of which they are a part by passage of most of the fluid through a septum or membrane that retains most of the solids on or within itself. The septum is called a filter medium, and the equipment assembly that holds the medium and provides space for the accumulated solids is called a filter. The fluid may be a gas or a liquid. The solid particles may be coarse or very fine, and their concentration in the suspension may be extremely low (a few parts per million) or quite high (>50%).

The object of filtration may be to purify the fluid by clarification or to recover clean, fluid-free particles, or both. In most filtrations the solids–fluid separation is not perfect. In general, the closer the approach to perfection, the more costly the filtration; thus the operator of the process cannot justify a more thorough separation than is required.

Gas filtration involves removal of solids (called dust) from a gas-solids mixture because: (1) the dust is a contaminant rendering the gas unsafe or unfit for its intended use; (2) the dust particles will ultimately separate themselves from the suspension and create a nuisance; or (3) the solids are themselves a valuable product that in the course of its manufacture has been mixed with the gas.

Three kinds of gas filters are in common use. Granular-bed separators consist of beds of sand, carbon, or other particles which will trap the solids in a gas suspension that is passed through the bed. Bag fitters are bags of woven fabric, felt, or paper through which the gas is forced; the solids are deposited on the wall of the bag. Air filters are light webs of fibers, often coated with a viscous liquid, through which air containing a low concentration of dust can be passed to cause entrapment of the dust particles.

Liquid filtration is used for liquid-solids separations in the manufacture of chemicals, polymer products, medicinals, beverages, and foods; in mineral processing; in water purification; in sewage disposal; in the chemistry laboratory; and in the operation of machines such as internal combustion engines.

Liquid filters are of two major classes, cake filters and clarifying filters. The former are so called because they separate slurries carrying relatively large amounts of solids. They build up on the filter medium as a visible, removable cake which normally is discharged "dry" (that is, as a moist mass), frequently after being washed in the filter. It is on the surface of this cake that filtration takes place after the first layer is formed on the medium. The feed to cake filters normally contains at least 1% solids. Clarifying filters, on the other hand, normally receive suspensions containing less than 0.1% solids, which they remove by entrapment on or within the filter medium without any visible formation of cake. The solids are normally discharged by backwash or by being discarded with the medium when it is replaced. *See* CLARIFICATION. [S.A.M.]

**Fischer-Tropsch process** The synthesis of hydrocarbons and, to a lesser extent, of aliphatic oxygenated compounds by the catalytic hydrogenation of carbon monoxide. The synthesis was discovered in 1923 by F. Fischer and H. Tropsch at the Kaiser Wilhelm Institute for Coal Research in Mülheim, Germany. The reaction is highly exothermic, and the reactor must be designed for adequate heat removal to control the temperature and avoid catalyst deterioration and carbon formation. The sulfur content of the synthesis gas must be extremely low to avoid poisoning the catalyst. *See* COAL GASIFICATION. [J.H.Fi]

**Flame** An exothermic reaction front or wave in a gaseous medium. Consider a uniform body of gas in which an exothermic chemical reaction (that liberates heat) is initiated by raising the temperature to a sufficiently high level; the reaction is started by a localized release of heat, as by a sufficiently energetic spark, and then spreads from the point of initiation. If the reaction is relatively slow, the whole gas will be involved before the initial region has finished reacting. If the reaction is relatively fast, the reaction zone will develop as a thin front or wave propagating into the unreacted gas, leaving fully reacted gas behind. If the front, in addition, shows luminosity (emission of light), the flame may be considered a classical example. However, perceptible emission of visible radiation is not essential to the definition.

Sufficient reaction rates may also be attained under special conditions (when the gas is very slowly heated inside a closed vessel) without very high temperatures if free radicals are generated in good concentration; this gives so-called cool flames.

The most common flame-producing reaction is combustion, which is broadly defined as a reaction between fuel and an oxidizer. The oxidizer is typically oxygen (usually in air), but a variety of other substances (for example, bromine with hydrogen) can play the same role in combination with the right fuel. While the overall theoretical reaction in a combustion flame—namely, fuel and oxidizer making fully oxidized products such as carbon dioxide and water vapor—is invariably simple, the actual reaction mechanism is typically very complex, involving many intermediate steps and compounds. Free radicals are generally present and figure prominently in the mechanism. *See* FREE RADICAL.

An overall reaction involving just one reactant is chemical decomposition, for example, ozone decomposing into oxygen. Decomposition flames are usually simpler chemically than combustion flames.

Combustion flames are broadly divided into premixed flames and diffusion flames. Premixed flames occur when fuel and oxidizer are mixed before they burn. Diffusion flames occur when fuel and oxidizer mix and burn simultaneously. The intermediate case, with partial premixing, has been of relatively low theoretical and practical interest. Flames are further categorized on the basis of shape, time behavior (stationary or moving), flow regime (laminar or turbulent), buoyancy regime (forced convection or natural convection), presence or absence of confinement (as by combustion chamber walls), and flow complications (such as swirling flow and crosswind). [H.A.Bec.]

**Fluidization** The processing technique employing a suspension or fluidization of small solid particles in a vertically rising stream of fluid—usually gas—so that fluid and solid come into intimate contact. This is a tool with many applications in the petroleum and chemical process industries. Suspensions of solid particles by vertically rising liquid streams are of lesser interest in modern processing, but have been shown to be of use, particularly in liquid contacting of ion-exchange resins. However, they come in this same classification and their use involves techniques of liquid settling, both free and hindered (sedimentation), classification, and density flotation. *See* ION EXCHANGE.

The interrelations of hydromechanics, heat transfer, and mass transfer in the gas-fluidized bed involve a very large number of factors. Because of the excellent contacting under these conditions, numerous chemical reactions are also possible—either between solid and gas, two fluidized solids with each other or with the gas, or most important, one or more gases in a mixture with the solid as a catalyst. In the usual case, the practical applications in plants have far outrun the exact understanding of the physical, and often chemical, interplay of variables within the minute ranges of each of the small particles and the surrounding gas phase.

With such excellent opportunities for heat and mass transfer to or from solids and fluids, fluidization has become a major tool in such fields as drying, roasting, and other processes involving chemical decomposition of solid particles by heat. An important application has been in the catalysis of gas reactions, wherein the excellent opportunity of heat transfer and mass transfer between the catalytic surface and the gas stream gives performance unequaled by any other system. *See* CATALYSIS; CRACKING; GAS ABSORPTION OPERATIONS; UNIT OPERATIONS. [D.F.O.]

**Fluorine** A chemical element, F, atomic number 9, the member of the halogen family that has the lowest atomic number and atomic weight. Although only the isotope with atomic weight 19 is stable, the other, radioactive isotopes between atomic weight 17 and 22 have been artificially prepared. Fluorine is the most electronegative element, and by a substantial margin the most chemically energetic of the nonmetallic elements. *See* PERIODIC TABLE.

**Properties.** The element fluorine is a pale yellow gas at ordinary temperatures. The odor of the element is somewhat in doubt. Some physical properties are listed in the table. The reactivity of the element is so great that it will react readily at ordinary temperatures with many other elementary substances, such as sulfur, iodine, phosphorus, bromine, and most metals. Since the products of the reactions with the nonmetals are in the liquid or gaseous state, the reactions continue to the complete consumption of the fluorine, frequently with the evolution of considerable heat and light. Reactions with the metals usually form a protective metallic fluoride which blocks further reaction, unless the temperature is raised. Aluminum, nickel, magnesium, and copper form such protective fluoride coatings.

Fluorine reacts with considerable violence with most hydrogen-containing compounds, such as water, ammonia, and all organic chemical substances whether liquids, solids, or gases. The reaction of fluorine with water is very complex, yielding mainly hydrogen fluoride and oxygen with less amounts of hydrogen peroxide, oxygen difluoride, and ozone. Fluorine displaces other nonmetallic elements from their compounds, even those nearest fluorine in chemical activity. It displaces chlorine from sodium chloride, and oxygen from silica, glass, and some ceramic materials. In the absence of hydrofluoric acid, however, fluorine does not significantly etch quartz or glass even after several hours at temperatures as high as 390°F (200°C).

Fluorine is a very toxic and reactive element. Many of its compounds, especially inorganic, are also toxic and can cause severe and deep burns. Care must be taken to prevent liquids or vapors from coming in contact with the skin or eyes.

**Natural occurrence.** At an estimated 0.065% of the Earth's crust, fluorine is roughly as plentiful as carbon, nitrogen, or chlorine, and much more plentiful than copper or lead, though much less abundant than iron, aluminum, or magnesium. Compounds whose molecules contain atoms of fluorine are widely distributed in nature. Many minerals contain small amounts of the element, and it is found in both sedimentary and igneous rocks.

**Physical properties of fluorine**

| Property | Value |
|---|---|
| Atomic weight | 18.998403 |
| Boiling point, °C | −188.13 |
| Freezing point, °C | −219.61 |
| Critical temperature, °C | −129.2 |
| Critical pressure, atm* | 55 |
| Density of liquid at b.p., g/ml | 1.505 |
| Density of gas at 0°C + 1 atm*, g/liter | 1.696 |
| Dissociation energy, kcal/mol | 36.8 |
| Heat of vaporization, cal/mol | 1510 |
| Heat of fusion, cal/mol | 121.98 |
| Transition temperature (solid), °C | −227.61 |

*1 atm = 101.325 kilopascals.

**Uses.** Fluorine-containing compounds are used to increase the fluidity of melts and slags in the glass and ceramic industries. Fluorspar (calcium fluoride) is introduced into the blast furnace to reduce the viscosity of the slag in the metallurgy of iron. Cryolite, $Na_2AlF_6$, is used to form the electrolyte in the metallurgy of aluminum. Aluminum oxide is dissolved in this electrolyte, and the metal is reduced electrically from the melt. The use of halocarbons containing fluorine as refrigerants was patented in 1930, and these volatile and stable compounds found a market in aerosol propellants as well as in refrigeration and air-conditioning systems. However, use of fluorocarbons as propellants has declined sharply because of concern over their possible damage to the ozone layer of the atmosphere. A use for fluorine that became prominent during World War II is in the enrichment of the fissionable isotope $^{235}U$; the most important process employed uranium hexafluoride. This stable, volatile compound was by far the most suitable material for isotope separation by gaseous diffusion.

While consumers are mostly unaware of the fluorine compounds used in industry, some compounds have become familiar to the general public through minor but important uses, such as additives to toothpaste and nonsticking fluoropolymer surfaces on frying pans and razor blades (for example Teflon).

**Compounds.** In all fluorine compounds the high electronegativity of this element suggests that the fluorine atom has an excess of negative charge. It is convenient, however, to divide the inorganic binary fluorides into saltlike (ionic lattice) nonvolatile metallic fluorides and volatile fluorides, mostly of the nonmetals. Some metal hexafluorides and the noble-gas fluorides show volatility that is frequently associated with a molecular compound. Volatility is often associated with a high oxidation number for the positive element.

The metals characteristically form nonvolatile ionic fluorides where electron transfer is substantial and the crystal lattice is determined by ionic size and the predictable electrostatic interactions. When the coordination number and valence are the same, for example, $BF_3$, $SiF_4$, and $WF_6$, the binding between metal and fluoride is not unusual, but the resulting compounds are very volatile, and the solids show molecular lattices rather than ionic lattice structures. For higher oxidation numbers, simple ionic lattices are less common and, while the bond between the central atom and fluorine usually still involves transfer of some charge to the fluorine, molecular structures are identifiable in the condensed phases.

In addition to the binary fluorides, a very large number of complex fluorides have been isolated, often with a fluoroanion containing a central atom of high oxidation number. The binary saltlike fluorides show a great tendency to combine with other binary fluorides to form a large number of complex or double salts.

The fluorine-containing compounds of carbon can be divided into fluorine-containing hydrocarbons and hydrocarbon derivatives (organic fluorine compounds) and the fluorocarbons and their derivatives. The fluorine atom attached to the aromatic ring, as in fluorobenzene, is quite unreactive. In addition, it reduces the reactivity of the molecule as a whole. Dyes, for example, that contain fluorine attached to the aromatic ring are more resistant to oxidation and are more light-fast than dyes that do not contain fluorine. Most aliphatic compounds, such as the alkyl fluorides, are unstable and lose hydrogen fluoride readily. These compounds are difficult to make and to keep and are not likely to become very important. *See* FLUOROCARBON; HALOGEN ELEMENTS.

[I.S.]

**Organic compounds.** The carbon compounds containing fluorine belong to several classes, depending on what other substituents besides fluorine are present. The physical properties and chemical reactivity of organic molecules containing fluorine are quite different when compared to the same molecules containing other halogen atoms, such as chlorine. This is due, in part, to a unique combination of the properties of fluorine, which include its small atomic size and high electronegativity. Stepwise replacement of several or all of the hydrogen atoms or other substituents attached to carbon is possible.

Many methods are available for creating a carbon-to-fluorine bond. A widely used method is to exchange a chlorine attached to carbon by reacting the compound with hydrofluoric acid. Elemental fluorine, which is very highly reactive, has also been used to prepare fluorine-containing compounds from a wide variety of organic compounds. The unusual property imparted to an organic molecule by fluorine substitution has led to the development of compounds that fulfill specific needs in refrigeration, medicine, agriculture, plastics, textiles, and other areas.

**Fluoroolefins.** These are a class of unsaturated carbon compounds containing fluorine; that is, they have a C=C in addition to other substituents. A typical fluoroolefin is tetrafluoroethylene ($F_2C{=}CF_2$). It is prepared from chlorodifluoromethane ($CHClF_2$), which loses HCl upon heating to produce $F_2C{=}CF_2$.

Many fluoroolefins combine with themselves or other olefins by the process of polymerization. Thus, polymerization of $F_2C{=}CF_2$ yields the polymer polytetrafluoroethylene (PTFE). This remarkable solid substance has outstanding physical and chemical properties. Nonstick polytetrafluoroethylene surfaces are used in kitchen utensils, bearings, skis, and many other applications. Since polytetrafluoroethylene is very viscous above its melting point, special methods have to be used for fabrication. For this reason, copolymers of tetrafluoroethylene with such olefins as ethylene have been developed. The chemical resistance of these copolymers is less than that of perfluorinated polymers. To obtain polymers with desired properties, the chemical processes to make them are carried out under rigorously controlled conditions. *See* COPOLYMER; POLYFLUOROOLEFIN RESINS; POLYMER; POLYMERIZATION.

There are many oxygen-containing fluorocarbons such as ethers, acids, ketones, and alcohols. Simple, fluorinated ethers are compounds of the type R-O-R, where R is a fluorinated alkyl group. The simple compound perfluoro ether ($F_3COCF_3$) is an analog of dimethyl ether. *See* ETHER.

Organofluorine chemicals offer some unique properties and solutions. In addition to the applications mentioned above, they are used in dyes, surfactants, pesticides, blood substitutes, textile chemicals, and biologically active compounds. *See* FLUOROCARBON; HALOGENATED HYDROCARBON. [V.N.M.R.]

**Fluorocarbon** Any of the organic compounds in which all of the hydrogen atoms attached to a carbon atom have been replaced by fluorine; also referred to as a perfluorocarbon. Fluorocarbons are usually gases or liquids at room temperature, depending on the number of carbon atoms in the molecule. A major use of gaseous fluorocarbons is in radiation-induced etching processes for the microelectronics industry; the most common one is tetrafluoromethane. Liquid fluorocarbons possess a unique combination of properties that has led to their use as inert fluids for cooling of electronic devices and soldering. Solubility of gases in fluorocarbons has also been used to advantage. For example, they have been used in biological cultures requiring oxygen, and as liquid barrier filters for purifying air. *See* HALOGENATED HYDROCARBON. [V.N.M.R.]

**Fluxional compounds** Molecules that undergo rapid intramolecular rearrangements among equivalent structures in which the component atoms are interchanged. The rearrangement process is usually detected by nuclear magnetic resonance (NMR) spectroscopy. With sufficiently rapid rates, a single resonance is observed in the

NMR spectrum for a molecule that might be expected to have several nonequivalent nuclei on the basis of its instantaneous structure.

Within organic chemistry, degenerate Cope rearrangements represented some of the first examples of interconversions between equivalent structures, but these were relatively slow. The rate of this rearrangement is rapid in more complex molecules. The epitome of degeneracy is reached in bullvalene, which has more than 1,200,000 equivalent structures and rapidly interconverts among them.

Fluxional molecules are frequently encountered in organometallic chemistry, and rapid rearrangements which involve migrations about unsaturated organic rings are commonly observed. The best known (called ring-whizzers) are cyclopentadienyl and cyclooctatetraene complexes of iron.

Inorganic structures also exhibit fluxional phenomena, and five-coordinate complexes provide the greatest number of well-known examples, one being phosphorus pentafluoride ($PF_5$).

The rearrangement of $PF_5$ involves interconversion of the trigonal bipyramidal molecule to a square pyramidal configuration and back. If two such nonequivalent structures are present in observable concentrations but interconvert rapidly to cause averaging in the NMR experiments, they are said to be stereochemically nonrigid. This term is generally taken to embrace all compounds that undergo rapid reversible intramolecular rearrangements. Thus, fluxional compounds are a subset of non-rigid compounds with equivalent structures. Nonequivalent structures, that is, tautomers, might be stereochemically nonrigid if they rearranged rapidly, but would not be considered fluxional.

Some workers prefer to reserve the term fluxional for molecules in which bonds are broken and reformed in the rearrangement process. Hence, of the examples above, only bullvalene and the iron complexes would be termed fluxional, whereas all would be considered stereochemically nonrigid. *See* TAUTOMERISM. [J.W.F.]

**Foam** A material made up of gas bubbles separated from one another by films of liquid. The bubbles are spherical when the liquid films separating them are thick (approximately 0.01 mm). Pure liquids do not foam; that is to say, they cannot produce liquid films of any permanence. Relatively permanent films are created only when a substance is present that is adsorbed at the surface of the liquid. Substances capable of being so adsorbed may be in true solution in the liquid or may be particles of a finely divided solid, which, because of poor wetting by the liquid, remain at the surface. In both cases, surface layers of the added substance are produced. The reluctance of the adsorbed substance to enter the bulk of the liquid preserves the surface and, hence, the thermodynamic stability of the foam. *See* SURFACTANT.

Although thermodynamically stable, a foam is mechanically fragile. Offsetting this fragility to some extent are mechanisms that provide the liquid films with resiliency and plasticity.

Although foams of exceptional stability are desired in some commercial applications, foam is a nuisance in many situations. A common recourse is the addition of chemical antifoams, which are usually insoluble liquids of very low surface tension. When a droplet of such a liquid is sprayed onto the foam or is carried into it by mechanical agitation, it spreads spontaneously and rapidly at the surface of the film, virtually sweeping the film away as it does so. *See* ADSORPTION; INTERFACE OF PHASES. [S.Ro.]

**Forensic chemistry** The application of chemistry to the study of physical materials or theoretical problems, the results of which may be entered into court as technical evidence. Boundaries are not sharply defined for forensic chemistry, and it includes topics that are not entirely chemical in nature.

Some of the items most often encountered in crime laboratories, and the information sought in regard to them, are: (1) body fluids and viscera to be analyzed for poisons, drugs, or alcohol, quantitation of which may assist in determining the dosage taken or the person's behavior prior to death; (2) licit and illicit pills, vegetable matter, and pipe residues for the presence of controlled substances; (3) blood, saliva, and seminal stains, usually in dried form, to be checked for species, type, and genetic data; (4) hairs, to determine if animal or human; if human, the race, body area of origin, and general characteristics; (5) fibers, to determine type (animal, vegetable, mineral, or synthetic), composition, dyes used, and processing marks; (6) liquor, for alcoholic proof, trace alcohols, sugars, colorants, and other signs of adulteration; (7) paint, glass, plastics, and metals, usually in millimeter-sized chips, to classify and compare to known materials; (8) inks on documents, to determine type, dye content, or possible age; also chemical obliterations and restoration of chaffed papers; (9) swabs from the hands of suspects, to be checked for the presence of gunshot residue; (10) debris from a fire or explosion scene, for the remains of the accelerant or explosive used. *See* ANALYTICAL CHEMISTRY. [M.J.C.]

**Formaldehyde** The simplest aldehyde, formula HCH=O. Because of its extreme reactivity, even with itself, it cannot be readily isolated or handled in the pure state. Therefore, it is produced and marketed as an aqueous solution (usually 37–50% formaldehyde by weight), sometimes known as Formalin. It is also sold as the solid hydrated polymer known as paraformaldehyde or paraform.

Formaldehyde is used principally to produce synthetic resins and adhesives by reaction with phenols, urea, and melamine. Other uses are in the manufacture of textiles, dyes, drugs, paper, leather, photographic materials, embalming agents, disinfectants, and insecticides. *See* ALDEHYDE; PHENOLIC RESIN. [L.M.]

**Francium** A chemical element, Fr, atomic number 87, an alkali metal element falling below cesium in group I of the periodic table. Distinguished by nuclear instability, francium exists only in short-lived radioactive forms, the most durable of which has a half-life of 21 min. The chief isotope of francium is actinium-K, an isotope of mass 223, which arises from the radioactive decay of the element actinium. From the properties of the known isotopes, it is reasonably certain that no long-lived form of element 87 will ever be found in nature or synthesized artificially.

The chemical properties of francium can be studied only on the tracer scale. The element has all the properties expected of the heaviest alkali element. With few exceptions, all the salts of francium are water-soluble. *See* ACTINIUM; ALKALI METALS; PERIODIC TABLE. [E.K.H.]

**Free energy** A term in thermodynamics which in different treatments may designate either of two functions defined in terms of the internal energy $E$ or enthalpy $H$, and the temperature-entropy product $TS$.

The function $(E - TS)$ is the Helmholtz free energy and is the function ordinarily meant by free energy in European references. The Gibbs free energy is the function $(H - TS)$. For the Lewis and Randall school of American chemical thermodynamics, this is the function meant by the free energy $F$. To avoid confusion with the symbol $F$ as applied elsewhere to the Helmholtz free energy, the symbol $G$ has also been used. Another development was the introduction of the name free enthalpy, with symbol $G$, for the Gibbs function.

For a closed system (no transfer of matter across its boundaries), the work which can be done in a reversible isothermal process is given by the series shown in Eq. (1).

$$W_{rev} = -\Delta A = -\Delta(E - TS) = -(\Delta E - T\Delta S) \quad (1)$$

For these conditions, $T\Delta S$ represents the heat given up to the surroundings. Should the process be exothermal, $T\Delta S < 0$, then actual work done on the surroundings is less than the decrease in the internal energy of the system. The quantity ($\Delta E - T\Delta S$) can then be thought of as a change in free energy, that is, as that part of the internal energy change which can be converted into work under the specified conditions. This then is the origin of the name free energy. Such an interpretation of thermodynamic quantities can be misleading, however; for the case in which $T\Delta S$ is positive, Eq. (1) shows that the decrease in "free" energy is greater than the decrease in internal energy. *See* CHEMICAL THERMODYNAMICS.

For constant temperature and pressure in a reversible process the decrease in the Gibbs function $G$ for the system again corresponds to a free-energy change in the above sense, since it is equal to the work which can be done by the closed system other than that associated with its change in volume $\Delta V$ under the given constant pressure $P$. The relations shown in Eq. (2) can be formed since $\Delta H = \Delta E + P\Delta V$.

$$\Delta G = -(\Delta H - T\Delta S) = W_{net} = W_{rev} - P\Delta V \quad (2)$$

Each of these free-energy functions is an extensive property of the state of the thermodynamic system. For a specified change in state, both $\Delta A$ and $\Delta G$ are independent of the path by which the change is accomplished. Only changes in these functions can be measured, not values for a single state.

The thermodynamic criteria for reversibility, irreversibility, and equilibrium for processes in closed systems at constant temperature and pressure are expressed naturally in terms of the function $G$. For any infinitesimal process at constant temperature and pressure, $-dG \geqq \delta w_{net}$. If $\delta w_{net}$ is never negative, that is, if the surroundings do no net work on the system, then the change $dG$ must be negative or zero. For a reversible differential process, $-dG > \delta w_{net}$; for an irreversible process, $-dG > \delta w_{net}$. The free energy $G$ thus decreases to a minimum value characteristic of the equilibrium state at the given temperature and pressure. At equilibrium, $dG = 0$ for any differential process taking place, for example, an infinitesimal change in the degree of completion of a chemical reaction. A parallel role is played by the work function $A$ for conditions of constant temperature and volume. Because temperature and pressure constitute more convenient working variables than temperature and volume, it is the Gibbs free energy which is the more commonly used in thermodynamics. *See* ENTROPY; THERMODYNAMIC PRINCIPLES. [P.J.B.]

**Free radical** Any molecule or atom which possesses one unpaired electron. There are some molecules which contain more than one unpaired electron (for example, oxygen); they normally are not considered as free radicals. Free radicals can be chemically very reactive (for example, the methyl radical) or they can be very stable entities (for example, nitric oxide).

Free radicals can be grouped into three major classes: atoms (for example, H, F, and Cl), inorganic radicals (for example, OH, CN, $NO_2$, and $ClO_3$), and organic radicals (for example, $CH_3$, $CH_3CH_2$, and $C_6H_6^-$). Such radicals are of great importance since they often appear as intermediates in thermal and photochemical reactions. Radicals are also known to initiate and propagate polymerization and combustion reactions.

In general, free radicals are formed by the rupture of a bond in a stable molecule with the production of two fragments, each with an unpaired electron. The resulting free radicals may participate in further reactions or may combine to reform the original compound.

There are many ways in which radicals can be generated—among these are thermal decomposition, electric discharge photochemical reactions, electrolysis at an electrode such as mercury or platinum, rapid mixing of two reactants, and gamma- or x-ray irradiation. [J.R.Bo.]

**Friedel-Crafts reaction** A substitution reaction, catalyzed by aluminum chloride, in which an alkyl (R—) group or an acyl (RCO—) group replaces a hydrogen atom of an aromatic nucleus. This general reaction is the most important member of a larger group of aromatic substitution reactions known to be catalyzed by conventional or Lewis acids.

In the classical alkylation reaction, an alkyl halide (RX) serves as the alkylating agent. Alkenes may be substituted for alkyl halides. For acylation of aromatic hydrocarbons, acyl halides have proved most valuable although acid anhydrides have also been used. *See* AROMATIC HYDROCARBON; SUBSTITUTION REACTION. [C.K.B.]

**Fuel cell** An electrical cell that converts the intrinsic chemical free energy of a fuel directly into direct-current electrical energy in a continuous catalytic process. As in the classical definition of catalysis, the fuel cell should not itself undergo change; that is, unlike the electrodes of a battery, its electrodes ideally remain invariant. For most fuel-oxidant combinations, the available free energy of combustion is somewhat less than the heat of combustion. In a typical thermal power conversion process, the heat of combustion of the fuel is turned into electrical work via a Carnot heat-engine cycle coupled with a rotating electrical generator. Since the Carnot conversion rarely proceeds at an efficiency exceeding 40% because of heat source and sink temperature limitations, the efficiency of conversion in a fuel cell can be greater than in a heat engine, especially in small devices. *See* CATALYSIS.

The fuel cell reaction usually involves the combination of hydrogen (H) with oxygen (O) [reaction (1)], as shown in the illustration. Under standard conditions of temperature

$$H_2(g) + {}^1/_2O_2(g) \rightarrow H_2O(l) \quad (1)$$

and pressure, 25°C (77°F) and 1 atm (100 kilopascals), the reaction takes place with a free-energy change $\Delta G = -56.69$ kcal (237 kilojoules) per mole of water. Since the formation of water involves two electrons, this value corresponds to $-1.23$ electronvolts (1 eV = 23.06 kcal/equivalent). Thus, at thermodynamic equilibrium (zero current),

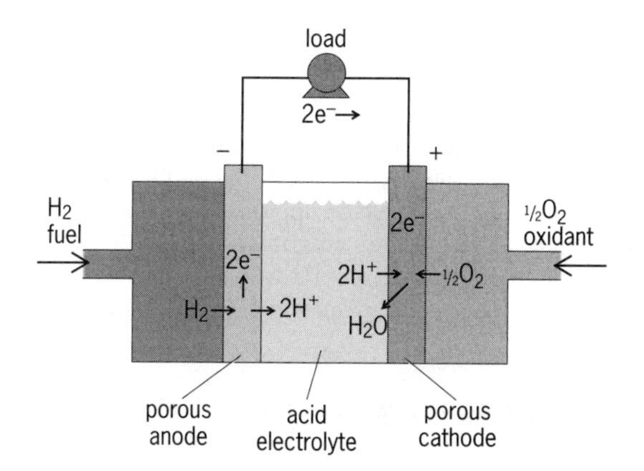


**Diagram showing the principle of operation of a fuel cell. (*After A. J. Appleby and F. R. Foulkes, Fuel Cell Handbook, Van Nostrand Reinhold, 1989*)**

the cell voltage should be 1.23 V, yielding a theoretical efficiency based on the heat of combustion [$\Delta H$ for $H_2O(l) = -1.48$ eV] of 83.1%. *See* FREE ENERGY.

At a net (nonzero) current, all cells show losses in cell voltage ($V$). In low-temperature fuel cells, these are due largely to the kinetic slowness (irreversibility) of the oxygen reduction reaction, which requires the breaking of a double bond with transfer of four electrons per molecule in a complex sequence of reactions. In high-temperature systems, oxygen reduction losses are less significant, since the reaction rate increases with temperature. However, the available free energy then decreases, falling to a value corresponding to about 1.0 V at 1000°C (1832°F). A further thermodynamic loss results from high cell fuel (or oxidant) conversion to avoid waste, so that the effective reversible potential is displaced from the standard state. Thus, at high temperature, the major loss is thermodynamic, which tends to compensate for the irreversible oxygen electrode losses at low temperature. As a result, cell voltages under typical loads vary from about 0.6 V for simple terrestrial cells to 1.0 V for aerospace cells. Cell voltage falls with increasing current per unit area. Since thermal efficiency is given by $V/1.48$, cell performance is a compromise between relative cost (that is, kilowatts available per unit area) and fuel efficiency, to give the lowest cost of electricity for a given application. *See* OXIDATION-REDUCTION.

**Fuels.** While any chemically suitable fuel, including metals such as lithium (Li), sodium (Na), aluminum (Al), and zinc (Zn), may be used in a fuel cell, hydrocarbons (for example, natural gas) will not react at a significant rate in low-temperature fuel cells. They will crack thermally before reacting electrochemically if injected directly into high-temperature fuel cells. Simple low-power units operating directly on methanol at ambient temperature do find some use, and liquid-fueled hydrazine cells have also found specialized applications. However, the high manufacturing energy requirement for hydrazine, together with its high cost and hazardous nature, leaves hydrogen the only suitable general high-performance fuel candidate. *See* HYDRAZINE; HYDROCARBON; HYDROGEN; METHANOL.

For practical fuel cells, hydrogen can be produced from readily available fuels such as clean light distillate (for example, naphtha), usually by steam reforming, or from coal via gasification at high temperature (the direct use of coal or carbon has been abandoned). In the high-temperature cells under certain conditions, internal steam reforming of simple hydrocarbons and alcohols (for example methane and methanol) can take place by the injection of the fuel with steam, which avoids cracking. Since methanol fuel reacts only slowly at low temperature, it is also steam-reformed to hydrogen. Methanol reforming takes place at only about 250°C (480°F), giving mixtures of hydrogen and carbon dioxide ($CO_2$) with a small amount of carbon monoxide (CO). In contrast, steam reforming of higher-molecular-weight alcohols or clean light distillates requires temperatures in excess of 700°C (1290°F). This favors mixtures of hydrogen and carbon monoxide, as in coal synthesis gas. [A.J.A.]

**Fugacity** A function introduced by G. N. Lewis to facilitate the application of thermodynamics to real systems. Thus, when fugacities are substituted for partial pressures in the mass action equilibrium constant expression, which applies strictly only to the ideal case, a true equilibrium constant results for real systems as well.

The fugacity $f_i$ of a constituent $i$ of a thermodynamic system is defined by the following equation (where $\mu_i$ is the chemical potential and $\mu_i^*$ is a function of temperature only),

$$\mu_i = \mu_i^* + RT \ln f_i$$

in combination with the requirement that the fugacity approach the partial pressure as the total pressure of the gas phase approaches zero. At a given temperature, this is possible only for a particular value for $\mu_i^*$, which may be shown to correspond to the chemical potential the constituent would have as the pure gas in the ideal gas state at 1 atm pressure. This definition makes the fugacity identical to the partial pressure in the ideal gas case. For real gases, the ratio of fugacity to partial pressure, called the fugacity coefficient, will be close to unity for moderate temperatures and pressures. At low temperatures and appropriate pressures, it may be as small as 0.2 or less, whereas at high pressures at any temperature it can become very large. *See* CHEMICAL EQUILIBRIUM; CHEMICAL THERMODYNAMICS; GAS. [P.J.B.]

**Fullerene** A hollow, pure carbon molecule in which the atoms lie at the vertices of a polyhedron with 12 pentagonal faces and any number (other than one) of hexagonal faces. The molecule was named after R. Buckminster Fuller, the inventor of geodesic domes, which conform to the same underlying structural formula.

Buckminsterfullerene ($C_{60}$ or fullerene-60; Fig. 1) is the archetypal member of the fullerenes. Other stable members of the fullerene family have similar structures (Fig. 2). The fullerenes can be considered, after graphite and diamond, to be the third well-defined allotrope of carbon. Macroscopic amounts of various fullerenes were first isolated in 1990, and since that time it has been discovered that members of this class of spheroidal organic molecules have numerous novel physical and chemical properties. The fullerenes promise to have synthetic, pharmaceutical, and industrial applications. Derivatives have been found to exhibit fascinating electrical and magnetic behavior, in particular superconductivity and ferro-magnetism. *See* CARBON; GRAPHITE.

**Structures and properties.** In the fullerene molecule an even number of carbon atoms are arrayed over the surface of a closed hollow cage. Each atom is trigonally linked to its three near neighbors by bonds that delineate a polyhedral network, consisting of 12 pentagons and $n$ hexagons. (Such structures conform to Euler's theorem for polyhedrons in that $n$ may be any number other than one including zero.) All 60 atoms in fullerene-60 are equivalent and lie on the surface of a sphere distributed with the symmetry of a truncated icosahedron. The 12 pentagons are isolated and interspersed symmetrically among 20 linked hexagons.

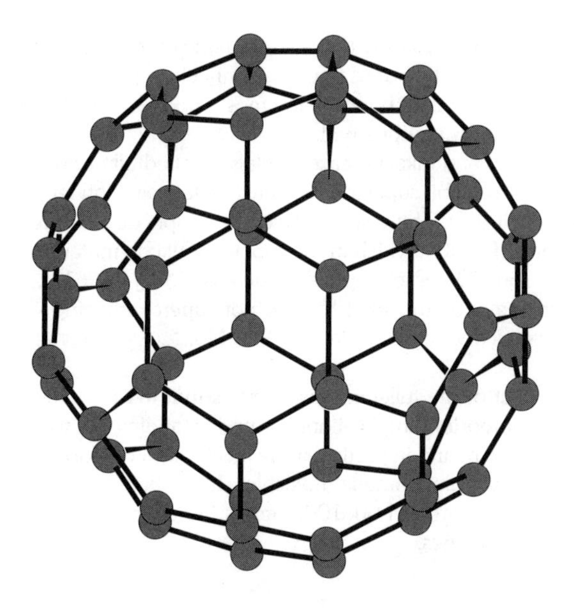

**Fig. 1. Structure of fullerene-60 ($C_{60}$).**

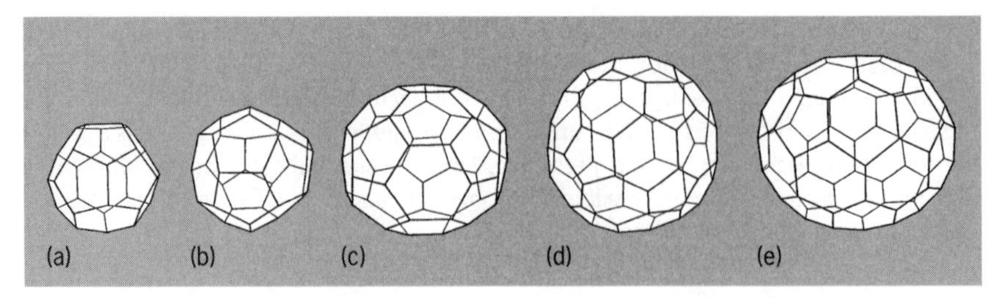


**Fig. 2. Some of the more stable members of the fullerene family. (*a*) $C_{28}$. (*b*) $C_{32}$. (*c*) $C_{50}$. (*d*) $C_{60}$. (*e*) $C_{70}$.**

In benzene solution, fullerene-60 is magenta and fullerene-70 red. Fullerene-60 forms translucent magenta face-centered cubic (fcc) crystals that sublime. The ionization energy is 7.61 eV and the electron affinity is 2.6–2.8 eV. The strongest absorption bands lie at 213, 257, and 329 nanometers. Studies with nuclear magnetic resonance spectroscopy yield a chemical shift of 142.7 parts per million; this result is commensurate with an aromatic system.

Solid $C_{60}$ exhibits interesting dynamic behavior in that at room temperature the individual round molecules in the face-centered cubic crystals are rotating isotropically (that is, freely) at around $10^8$ Hz. At around 260 K (8.3°F) there is a phase transition to a simple cubic (sc) lattice accompanied by an abrupt lattice contraction. Rotation is no longer free, and the individual molecules make rotational jumps between two favored (relative) orientational configurations—in the lower-energy one a double bond lies over a pentagon, and in the other it lies over a hexagon. At 90 K (−300°F) the individual molecules stop rotating altogether, freezing into an orientationally disordered crystal involving a mix of the two configurations.

**Chemistry and formation.** Fullerene-60 behaves as a soft electrophile, a molecule that readily accepts electrons during a primary reaction step. It can readily accept three electrons and perhaps even more. The molecule can be multiply hydrogenated, methylated, ammonated, and fluorinated. It forms exohedral complexes in which an atom (or group) is attached to the outside of the cage, as well as endohedral complexes in which an atom is trapped inside the cage.

The $C_{60}$ molecule behaves as though it has only a single resonance form—one in which the 30 double bonds are localized in the bonds that interconnect the pentagons. This is a key factor, as addition to these double bonds is the most important reaction as far as the application of $C_{60}$ in synthesis is concerned.

On exposure of $C_{60}$ to certain alkali and alkaline earth metals, exohedrally doped crystalline materials are produced that exhibit superconductivity at relatively high temperatures (10 to 33 K or −440 to −400°F). The $C_{60}$ molecule has a triply degenerate lowest unoccupied molecular orbital (LUMO), which in the superconducting materials is half filled, containing three electrons. Other ionic phases, such as $M_nC_{60}$ ($n = 1$, 2, 4, 6, where M is the intercalated metal atom), exist but are not superconducting—they appear to be metallic or semiconductor/insulators. *See* MOLECULAR ORBITAL THEORY.

Perhaps the most important aspect of the fullerene discovery is that the molecule forms spontaneously. This fact has important implications for understanding the way in which extended carbon materials form, and in particular the mechanism of graphite growth and the synthesis of large polycyclic aromatic molecules. It has become clear that as far as pure carbon aggregates of around 60–1000 atoms are concerned, the most stable species are closed-cage fullerenes. [J.Hare; H.W.Kr.]

**Fulminate** A compound containing the —ONC group and derived from fulminic acid, HONC. Fulminates are isomeric with cyanates; that is, cyanates have the same atoms but in different arrangement, —OCN.

The fulminates have been commercially important because of the use of mercury fulminate, $Hg(ONC)_2$, in priming compositions and as an initial detonating agent; lead azide is replacing mercury fulminate as a detonating agent. *See* CYANATE. [E.E.W.]

**Fundamental constants** That group of physical constants which play a fundamental role in the basic theories of physics. These constants include the speed of light in vacuum, $c$; the magnitude of the charge on the electron, $e$, which is the fundamental unit of electric charge; the mass of the electron, $m_e$; Planck's constant, $h$; and the fine-structure constant, $\alpha$.

These five quantities typify the different origins of the fundamental constants: $c$ and $h$ are examples of quantities which appear naturally in the mathematical formulation of certain physical theories—Einstein's theories of relativity, and quantum theory, respectively; $e$ and $m_e$ are examples of quantities which characterize the elementary particles of which all matter is constituted; and $\alpha$, the fundamental constant of quantum electrodynamics (QED), is an example of quantities which are combinations of other fundamental constants, but are actually constants in their own right since the same combination always appears together in the basic equations of physics.

**Recommended values (1986) of selected fundamental physical constants**

| Quality | Symbol | Numerical value* | Units† | Relative uncertainty, ppm |
|---|---|---|---|---|
| Speed of light in vacuum | $c$ | 299792458 | m/s | (defined) |
| Constant of gravitation | $G$ | 6.67259(85) | $10^{-11}$ m$^3$/(kg · s$^2$) | 128 |
| Planck constant | $h$ | 6.6260755(40) | $10^{-34}$ J · s | 0.60 |
| Elementary charge | $e$ | 1.60217733(49) | $10^{-19}$ C | 0.30 |
| Magnetic flux quantum, $h/(2e)$ | $\Phi_0$ | 2.06783461(61) | $10^{-15}$ Wb | 0.30 |
| Fine-structure constant, | $\alpha$ | 7.29735308(33) | $10^{-3}$ | 0.045 |
| $\mu_0 ce^2/(2h)$ | $\alpha^{-1}$ | 137.0359895(61) | | 0.045 |
| Electron mass | $m_e$ | 9.1093897(54) | $10^{-31}$ kg | 0.59 |
| Proton mass | $m_p$ | 1.6726231(10) | $10^{-27}$ kg | 0.59 |
| Neutron mass | $m_n$ | 1.6749286(10) | $10^{-27}$ kg | 0.59 |
| Proton-electron mass ratio | $m_p/m_e$ | 1836.152701(37) | | 0.020 |
| Rydberg constant, $m_e c\alpha^2/(2h)$ | $R_\infty$ | 1.0973731534(13) | m$^{-1}$ | 0.0012 |
| Bohr radius, $\alpha/(4\pi R_\infty)$ | $a_0$ | 5.29177249(24) | $10^{-11}$ m | 0.045 |
| Compton wavelength of the electron, $h/(m_e c) = \alpha^2/(2R_\infty)$ | $\lambda_c$ | 2.42631058(22) | $10^{-12}$ m | 0.089 |
| Classical electron radius, $\mu_0 e^2/(4\pi m_e) = \alpha^3/(4\pi R_\infty)$ | $r_e$ | 2.81794092(38) | $10^{-15}$ m | 0.13 |
| Bohr magneton, $eh/(4\pi m_e)$ | $\mu_B$ | 9.2740154(31) | $10^{-24}$ J/T | 0.34 |
| Electron magnetic moment in Bohr magnetons | $\mu_e/\mu_B$ | 1.001159652193(10) | | $10^{-5}$ |
| Nuclear magneton, $eh/(4\pi m_p)$ | $\mu_N$ | 5.0507866(17) | $10^{-27}$ J/T | 0.34 |
| Proton magnetic moment in nuclear magnetons | $\mu_p/\mu_N$ | 2.792847386(63) | | 0.023 |
| Boltzmann constant | $k$ | 1.380658(12) | $10^{-23}$ J/K | 8.5 |
| Avogadro constant | $N_A$ | 6.0221367(36) | $10^{-23}$/mol | 0.59 |
| Fataday constant, $N_A e$ | $F$ | 96485.309(29) | C/mol | 0.30 |
| Molar gas constant, $N_A k$ | $R$ | 8.314510(70) | J/(mol-K) | 8.4 |

*The digits in parentheses represent the one-standard-deviation uncertainties in the last digits of the quoted value.

†C = coulomb, J = joule, kg = kilogram, m = meter, mol = mole, s = second, T = tesla, Wb = weber.

Reliable numerical values for the fundamental physical constants are required for two main reasons. First, they are necessary if quantitative predictions from physical theory are to be obtained. Second, and even more important, the self-consistency of the basic theories of physics can be critically tested by a careful intercomparison of the numerical values of fundamental constants obtained from experiments in the different fields of physics. In general, the accuracy of fundamental constants determinations has continually improved over the years. Whereas in the past, 100 ppm (0.01%) and even 1000 ppm (0.1%) measurements were commonplace, today 0.01 ppm and better determinations are not unusual (ppm = parts per million).

Complex relationships can exist among groups of constants and conversion factors, and a particular constant may be determined either directly by measurement or indirectly by an appropriate combination of other directly measured constants. If the direct and indirect values have comparable accuracy, then both must be taken into account in order to arrive at a best value for that quantity. (By best value is meant that value believed to be closest to the true but unknown value.) Generally, each of the several routes which can be followed to a particular constant, both direct and indirect, will give a slightly different numerical value. Such a situation may be satisfactorily handled by the mathematical method known as least-squares. This technique provides a self-consistent procedure for calculating best "compromise" values of the constants from all of the available data. It automatically takes into account all possible routes and determines a single final value for each constant being calculated. It does this by weighting the different routes according to their relative uncertainties. The appropriate weights follow from the uncertainties assigned the individual measurements constituting the original set of data.

The 1986 least-squares adjustment, carried out under the auspices of the CODATA Task Group on Fundamental Constants, succeeded a CODATA adjustment in 1973 by E. R. Cohen and B. N. Taylor; CODATA, the Committee on Data for Science and Technology, is an interdisciplinary committee of the International Council of Scientific Unions. Recommended values are shown in the table. [E.R.Co.]

**Furan** One of a group of organic heterocyclic compounds containing a diunsaturated ring of four carbon atoms and one oxygen atom. Furan (**1**) is a typical member of the group. Furfural (**2**) and some of its close relatives, such as furfuryl alcohol, tetrahydrofurfuryl alcohol, and tetrahydrofuran, are important chemicals of commerce. *See* HETEROCYCLIC COMPOUNDS.

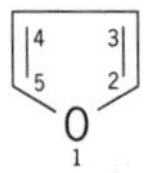

4 3
5 2
O
1
(**1**)

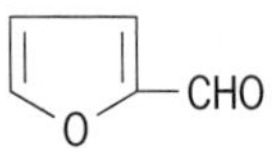

—CHO
O
(**2**)

Furan is a colorless, volatile liquid, bp 31.4°C (88.5°F), which is stable to alkali but not to mineral acid. Its water solubility is approximately 1% at room temperature. On exposure to air, furan decomposes very slowly by autoxidation. Substituted furans, particularly negatively substituted furans, are much less sensitive. The furan system is aromatic. Nitration, halogenation, acylation, mercuration, and sulfonation reactions occur with relative ease. [W.J.Ge.]

**Fused-salt phase equilibria** Conditions in which two or more phases of fused-salt mixtures can coexist in thermodynamic equilibrium. Phase diagrams of these equilibrium conditions summarize basic knowledge about fused salts. Numerous advances in the technologies which are based on high-temperature chemistry have become possible through the increase in knowledge about fused salts. The increasingly significant role of fused salts in industrial processes is evident in the widening application of these materials as heat-transfer media, in extractive metallurgy, in nonaqueous reprocessing of nuclear reactor fuels, and in the development of nuclear reactors which create more fuel than they consume (breeder reactors). *See* PHASE EQUILIBRIUM.

Fused-salt mixtures find application in technology when the need arises for liquids which are stable at high temperatures. For most applications, suitably low melting temperatures and low vapor pressures are primary considerations. To some extent these requirements are conflicting, because salts which are useful in obtaining low freezing temperatures often tend to have appreciable covalent character and therefore to exhibit unfavorably high vapor pressures.

As a special class of liquids, one which is composed entirely of positively and negatively charged ions undiluted by weak-electrolyte supporting media, fused salts are used in many different types of research. For example, advances in solution theory, thermodynamics, and crystal chemistry have come about through studies of fused-salt systems. *See* FUSED-SALT SOLUTION.

A close connection between fused-salt phase diagrams and geochemistry stems from the model principle developed by V. M. Goldschmidt, who noted that isomorphic structures are assumed by ions of the same proportionate size and stoichiometric relations but of different charge. Thus the fluorides of beryllium, calcium, and magnesium, for example, are structural models for silicon dioxide ($SiO_2$), titanium dioxide ($TiO_2$), and zirconium dioxide ($ZrO_2$). The fluoride structures are referred to as weakened models because of the smaller electrostatic forces resulting from smaller ionic charges; they have been useful for comparisons with oxide and silicate systems. According to Goldschmidt's interpretation, saltlike materials were derived from components such as water ($H_2O$), carbon dioxide ($CO_2$), sulfur trioxide ($SO_3$), chlorine ($Cl_2$), and fluorine ($F_2$), which were volatilized from molten magmas as they crystallized. Crystallization equilibria in fused-salt systems therefore provide a convenient way to study the mechanisms occurring in the formation of igneous rocks. [R.E.Th.]

**Fused-salt solution** A nonaqueous solvent system particularly useful in coordination chemistry. Fused salts are a large class of liquids which are composed largely of ions. Many simple inorganic salts melt at rather high temperatures (greater than 600°C or 1000°F), forming liquids which have high specific conductivities, 1–6 $ohm^{-1} \cdot cm^{-1}$. There are, however, a number of exceptions to this generalization; for example, the electrical conductivity of $AlCl_3$ decreases sharply upon melting due to the formation of molecular liquid ($Al_2Cl_6$). *See* HIGH-TEMPERATURE CHEMISTRY; SOLVENT.

The use of binary or ternary melt compositions results in liquids which typically have much lower melting temperatures and somewhat lower specific conductivities than do pure salts. The choice of melts for use as solvents is frequently based on such considerations as the availability and cost, the lowest melting temperature attainable, and the ease of purification of the solvent, as well as the width of the electrochemical span and the spectroscopic transparency of the melt. Several molten-salt solvents, such as the LiCl–KCl eutectic and the equimolar $NaNO_3$–$KNO_3$ melt, have been extensively studied, and many of their physical and chemical properties are well known. Many melt systems, particularly ternary and more complex compositions, have been only partially characterized; their physical properties are estimated by using the data available for less complex systems, such as the binary component melts.

One use of fused salts is as media for organic reactions. Not only does the fused-salt environment provide for a better thermal control of the reaction (heat dissipation is readily possible), but the fused salt may serve as a catalyst. For example, molten $SbCl_3$

and $ZnCl_2$ have been found to be effective hydrocracking catalysts for coal. It has been found that polycyclic hydrocarbons, such as anthracene, undergo several types of reactions in $SbCl_3$ melts, including formation of radical cations and protonated species which react further to form condensed systems, such as anthra[2.1-*a*]aceanthrylene. *See* CATALYSIS.

Another technological area of great interest in which molten salts play a key role is that of advanced batteries and fuel cells. Thus, LiCl–KCl eutectic is the solvent in the rechargeable Li(Al)–FeS (or $FeS_2$) battery, and sodium polysulfide melts are employed in the sodium/sulfur battery which operates at about 350°C (660°F). The rechargeable cell Na/$Na^+$ conductor/S(IV) in $AlCl_3$–NaCl has an open circuit voltage of 4.2 V at 200–250°C (390–480°F) as well as high energy densities.

Molten carbonates, such as the ternary $Li_2CO_3$–$Na_2CO_3$–$K_2CO_3$ melt, are used in fuel cells which employ $H_2$/CO mixtures and oxygen as electrochemical fuels. *See* FUEL CELL.

[G.Ma.]

# G

**Gadolinium** A metallic chemical element, Gd, atomic number 64 and atomic weight 157.25, belonging to the rare-earth group. The naturally occurring element is composed of eight isotopes. It is named in honor of the Swedish scientist J. Gadolin. The oxide, $Gd_2O_3$, in powdered form is white, and solutions of the salt are colorless. Gadolinium metal is paramagnetic and becomes strongly ferromagnetic below room temperatures. The Curie point, where this transition occurs, is about 16 K. *See* PERIODIC TABLE; RARE-EARTH ELEMENTS. [F.H.S.]

**Gallium** A chemical element, Ga, atomic number 31, atomic weight 69.72. Gallium is a member of group 13 and the fourth period of the periodic table (IUPAC). *See* PERIODIC TABLE.

The major commercial sources of gallium are bauxite, containing gallite ($CuGaS_2$), and zinc and germanium sulfides. Normal ore-grade deposits usually contain substantially less than 0.1% gallium. In the United States the bauxite deposits in Arkansas and the zinc deposits in Oklahoma are the main sources of domestic production. Much of the gallium used in the United States is imported from Switzerland and Germany, with lesser amounts from Canada and France.

Gallium is a unique element in that it possesses the largest liquid range of any element. Its normal freezing point of 29.78°C (85.60°F) is lower than any metal except mercury and cesium. Its boiling point is in the vicinity of 2420°C (4388°F), although there is some uncertainty owing to the reactivity of gallium with the container material at this temperature.

The valence-electron notation of gallium corresponding to its ground-state term is [Ar, $3d^{10}4s^24p^1$], which accounts for the maximum oxidation state of III in its chemistry. Compounds of formal oxidation state II and I are also known.

Approximately 95% of the gallium consumed in the United States and presumably in the world is used in the electronics industry. Minor quantities have been used or studied for use in thermometers, low-melting solders, as a heat-transfer fluid, in arc lamps, batteries, vanadium-gallium superconductors, and in catalyst mixtures.

The most important gallium semiconductors are gallium arsenide (GaAs) and gallium phosphide (GaP). The magnitude of the energy gap in GaAs favors its use in transistors. The electron mobility in GaAs is very much higher than the hole mobility; in contrast, the electron and hole mobility in GaP are of similar magnitude and very much lower than in GaAs. By doping with the appropriate elements, these properties can be altered. Electron transport (*n*-type) GaP semiconductors are used in rectifiers, hole transport (*p*-type) in light sources and photocells. *n*-Type GaAs semiconductors are used in injection lasers and *p*-type GaAs in electroluminescent transistors. *See* SEMICONDUCTOR.

GaN is prepared by the reaction of metallic gallium or $Ga_2O_3$ at elevated temperature with ammonia, and the other semiconductors by direct reaction with the elements or $Ga_2O$ at high temperature. *See* ALUMINUM; INDIUM; THALLIUM. [E.M.L.]

**Gas** A phase of matter characterized by relatively low density, high fluidity, and lack of rigidity. A gas expands readily to fill any containing vessel. Usually a small change of pressure or temperature produces a large change in the volume of the gas. The equation of state describes the relation between the pressure, volume, and temperature of the gas. In contrast to a crystal, the molecules in a gas have no long-range order.

At sufficiently high temperatures and sufficiently low pressures, all substances obey the ideal-gas, or perfect-gas, equation of state below, where $p$ is the pressure, $T$ is the

$$p\overline{V} = RT$$

absolute temperature, $\overline{V}$ is the molar volume, and $R$ is the gas constant. Absolute temperature $T$ is expressed on the Kelvin scale. The gas constant is 8.314 joules/(mole K). The molar volume is the molecular weight divided by the gas density.

At lower temperatures and higher pressures, the equation of state of a real gas deviates from that of a perfect gas. Various empirical relations have been proposed to explain the behavior of real gases. [C.F.C.; J.O.H.]

**Gas absorption operations** The separation of solute gases from gaseous mixtures of noncondensables by transfer into a liquid solvent. This recovery is achieved by contacting the gas stream with a liquid that offers specific or selective solubility for the solute gas or gases to be recovered. The operation of absorption is applied in industry to purify process streams or recover valuable components of the stream. It is used extensively to remove toxic or noxious components (pollutants) from effluent gas streams. *See* ABSORPTION.

The absorption process requires the following steps: (1) diffusion of the solute gas molecules through the host gas to the liquid boundary layer based on a concentration gradient, (2) solvation of the solute gas in the host liquid based on gas-liquid solubility, and (3) diffusion of the solute gas based on concentration gradient, thus depleting the liquid boundary layer and permitting further solvation. The removal of the solute gas from the boundary layer is often accomplished by adding neutralizing agents to the host liquid to change the molecular form of the solute gas. This process is called absorption accompanied by chemical reaction. *See* DISTILLATION. [A.J.T.]

**Gas and atmosphere analysis** Qualitative identifications and quantitative determinations of substances essential for the evaluation of the air quality in the ambient air and in the industrial workplace.

**Qualitative identification.** The qualitative identification of air pollutants may require the use of several instruments which provide complementary information about composition and structure. Since the entire sample is often limited to milligram or microgram quantities, the classical identification methods, such as boiling point and refractive index determinations, functional group tests, combustion analyses, and derivative preparations, have been largely replaced by instrumental methods. Information for identification purposes is now generally obtained from instruments such as mass, nuclear magnetic resonance, infrared, and ultraviolet spectrometers that rely upon the response of a molecule to an energy probe.

Mass spectroscopy is probably the single most powerful technique for the qualitative identification of volatile organic compounds, and has been particularly useful in the identification of many environmental contaminants. When a sample is introduced into the mass spectrometer, electron bombardment causes the parent molecule to lose an electron and form a positive ion. Some of the parent ions are also fragmented into characteristic daughter ions, while other ions remain intact. All of the ions are accelerated, separated, and focused on an ion detector by means of either a magnetic field or a quadrupole mass analyzer. Using microgram quantities of pure materials, the mass spectrometer yields information about the molecular weight and the presence of other atoms, such as nitrogen, oxygen, and halogens, within the molecule. In addition, the fragmentation pattern often provides a unique so-called fingerprint of a molecule, allowing positive identification. If the gas is a mixture, interpretation of the mass spectral data is difficult since the fragmentation patterns are superimposed. However, interfacing the mass spectrometer to a gas chromatograph provides an elegant solution to this problem. *See* MASS SPECTROMETRY.

A gas chromatograph is essentially a highly efficient apparatus for separating a complex mixture into individual components. When a mixture of components is injected into a gas chromatograph equipped with an appropriate column, the components travel down the column at different rates and therefore reach the end of the column at different times. The mass spectrometer located at the end of the column can then analyze each component separately as it leaves the column. In essence, the gas chromatograph allows the mass spectrometer to analyze a complex mixture as a series of pure components. More than 100 compounds have been identified and quantified in automobile exhaust by using a gas chromatograph–mass spectrometer combination. *See* GAS CHROMATOGRAPHY.

**Quantitative analysis.** The methods employed chiefly for quantification can be classified for convenience into direct and indirect procedures. Direct-reading instruments are generally portable and may analyze and display their results in a few seconds or minutes, and can operate in a continuous or semicontinuous mode. Indirect methods are those involving collection and storage of a sample for subsequent analysis.

Direct methods utilize colorimetric indicating devices and instrumental methods. Three types of direct-reading colorimetric indicators have been utilized: liquid reagents, chemically treated papers, and glass tubes containing solid chemicals (detector tubes). The simplest of these methods is the detector tube. Detector tubes are constructed by filling a glass tube with silica gel coated with color-forming chemicals. For use, the ends of the sealed tube are broken and a specific volume of air, typically 6 $in.^3$ (100 $cm^3$), is drawn through the tube at a controlled rate. Detector tubes for analyzing approximately 400 different gases are commercially available. Accuracy is sometimes low, and detector tubes for only 25 gases meet the National Institute for Occupational Safety and Health (NIOSH) accuracy requirement of $\pm25\%$. For some gases, semicontinuous analyzers have been developed that operate by pulling a fixed volume of air through a paper tape impregnated with a color-forming reagent. The intensity of the color is then measured for quantification. Phosgene, arsine, hydrogen sulfide, nitric oxide, chlorine, and toluene diisocyanate have been analyzed by indicating tapes. *See* COLORIMETRY.

With the availability of stable and sensitive electronics, direct-reading instruments capable of measuring gases directly at the parts-per-billion range were developed. Most direct-reading instruments contain a sampling system, electronics for processing signals, a portable power supply, a display system, and a detector. The detector or sensor is a component that is capable of converting some characteristic property of the gas into an electrical signal. While there are dozens of properties for the bases of operation of these detectors, the most sensitive and popular detectors are based on electrical or thermal conductivity, ultraviolet or infrared absorption, mass spectrometry, electron capture, flame ionization, flame photometry, heat of combustion, and chemiluminescence.

Many of these detectors respond to the presence of $10^{-9}$ g quantities, and even to $10^{-12}$ g levels. In addition to improved accuracy, precision, and analysis time, another advantage is that most instruments produce an electrical signal which can be fed into a computer for process control, averaging, and record keeping. Rapid fluctuations and hourly, daily, and yearly averages are readily obtained.

For indirect methods of quantification, the main collection devices are freeze traps, bubblers, evacuated bulbs, plastic bags, and solid sorbents. Because of their convenience, solid sorbents dominate collection procedures. NIOSH developed a versatile method for industrially important vapors, based on the sorption of the vapors on activated charcoal or, to a lesser extent, on other solid sorbents such as silica gel and porous polymers. Typically, in this technique a few liters of air are pulled through a glass tube containing about 0.004 oz (100 mg) of charcoal. The charcoal tube is only 7 cm × 6 mm (3 in. × 0.2 in.), and has the advantage that it can be placed on the worker's lapel. A battery-operated pump small enough to fit into a shirt pocket is connected by a plastic tube to the collecting device, so that the contaminants are continuously collected from the breathing zone of the worker. Many solvent vapors and gases are efficiently trapped and held on the charcoal. The ends of the sample tube are then capped, and the tube is returned to a laboratory for analysis. In the laboratory the tube is broken open, and the charcoal is poured into carbon disulfide to desorb the trapped vapors. Following desorption, a sample of the solution is injected into a gas chromatograph for quantification.

This technique has been highly successful for several classes of compounds, such as aromatics, aliphatics, alcohols, esters, aldehydes, and chlorinated compounds. Sulfur- and nitrogen-containing compounds can also be analyzed by using a gas chromatograph which is equipped with a sulfur- or nitrogen-sensitive detector.

[W.R.Bu.; M.Gl.; L.S.Be.]

**Gas chromatography** A method for the separation and analysis of complex mixtures of volatile organic and inorganic compounds. Most compounds with boiling points less than about 250°C (480°F) can be readily analyzed by this technique. A complex mixture is separated into its components by eluting the components from a heated column packed with sorbent by means of a moving-gas phase. *See* CHROMATOGRAPHY.

Gas chromatography may be classified into two major divisions: gas-liquid chromatography, where the sorbent is a nonvolatile liquid called the stationary-liquid phase, coated as a thin layer on an inert, granular solid support, and gas-solid chromatography, where the sorbent is a granular solid of large surface area. The moving-gas phase, called the carrier gas, is an inert gas such as nitrogen or helium which flows through the chromatographic column packed with the sorbent. The solute partitions, or divides, itself between the moving-gas phase and the sorbent and moves through the column at a rate dependent upon its partition coefficient, or solubility, in the liquid phase (gas-liquid chromatography) or upon its adsorption coefficient on the packing (gas-solid chromatography) and the carrier-gas flow rate.

The apparatus used in gas chromatography consists of four basic components: a carrier-gas supply and flow controller, a sample inlet system providing a means for introduction of the sample, the chromatographic column and associated column oven, and the detector system.

Qualitative and quantitative information is obtained from analyzing the peaks appearing on a chromatogram. Combination of gas chromatography with mass spectrometry provides the ultimate in qualitative information and has been used extensively in research. *See* MASS SPECTROMETRY; QUALITATIVE CHEMICAL ANALYSIS. [R.S.J.]

**Gas constant** The universal constant $R$ that appears in the ideal gas law, Eq. (1), where $P$ is the pressure, $V$ the volume, $n$ the amount of substance, and $T$

$$PV = nRT \tag{1}$$

the thermodynamic (absolute) temperature. The gas constant is universal in that it applies to all gases, providing they are behaving ideally (in the limit of zero pressure). The gas constant is related to the more fundamental Boltzmann constant, $k$, by Eq. (2),

$$R = N_A k \tag{2}$$

where $N_A$ is the Avogadro constant (the number of entities per mole). The best modern value in SI units is $R = 8.314\ 472\ (15)$ J/K · mol, where the number in parentheses represents the uncertainty in the last two digits. *See* GAS.

According to the equipartition principle, at a temperature $T$, the average molar energy of each quadratic term in the expression for the energy is $(1/2)RT$; as a consequence, the translational contribution to the molar heat capacity of a gas at constant volume is $(3/2)R$; the rotational contribution of a linear molecule is $R$.

Largely because $R$ is related to the Boltzmann constant, it appears in a wide variety of contexts, including properties unrelated to gases. Thus, it occurs in Boltzmann's formula for the molar entropy of any substance, Eq. (3), where $W$ is the number of

$$S = R \ln W \tag{3}$$

arrangements of the system that are consistent with the same energy; and in the Nernst equation for the potential of an electrochemical cell, Eq. (4), where $E^\circ$ is a standard

$$E = E^\circ - (RT/nF) \ln Q \tag{4}$$

potential, $F$ is the Faraday constant, and $Q$ is a function of the composition of the cell. The gas constant also appears in the Boltzmann distribution for the population of energy levels when the energy of a level is expressed as a molar quantity. *See* ELECTRODE POTENTIAL; ENTROPY. [P.W.A.]

**Gel** A continuous solid network enveloped in a continuous liquid phase; the solid phase typically occupies less than 10 vol % of the gel. Gels can be classified in terms of the network structure. The network may consist of agglomerated particles (formed, for example, by destabilization of a colloidal suspension; illus. *a*); a "house of cards" consisting of plates (as in a clay) or fibers (illus. *b*); polymers joined by small crystalline regions (illus. *c*); or polymers linked by covalent bonds (illus. *d*).

In a gel the liquid phase does not consist of isolated pockets, but is continuous. Consequently, salts can diffuse into the gel almost as fast as they disperse in a dish of free liquid. Thus, the gel seems to resemble a saturated household sponge, but it is distinguished by its colloidal size scale: the dimensions of the open spaces and of the solid objects constituting the network are smaller (usually much smaller) than a micrometer. This means that the interface joining the solid and liquid phases has an area on the order of 1000 $m^2$ per gram of solid. As a result, the properties of a gel are controlled by interfacial and short-range forces, such as van der Waals, electrostatic, and hydrogen-bonding. Factors that influence these forces, such as introduction of salts or another solvent, application of an electric field, or changes in pH or temperature, affect the interaction between the solid and liquid phases. Variations in these parameters can induce huge changes in volume as the gel imbibes or expels liquid, and this phenomenon is exploited to make mechanical actuators or hosts for controlled release of drugs from gels. For example, a polyacrylamide gel (a polymer linked by covalent bonds) shrinks dramatically when it is transferred from a dish of water (a good solvent)

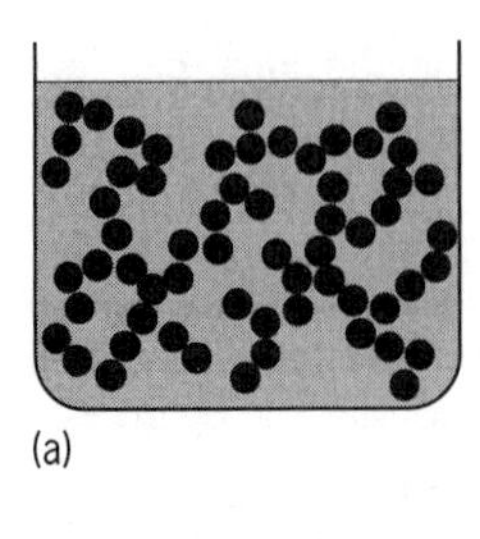


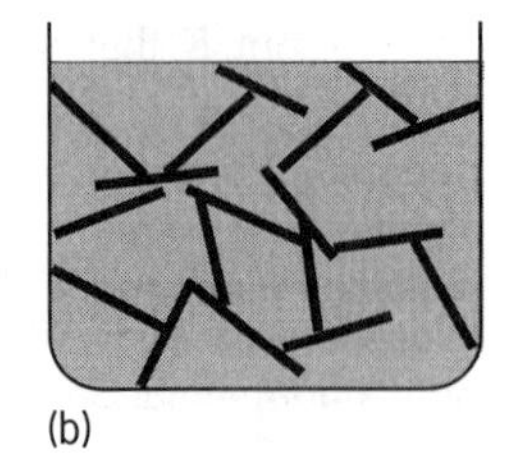


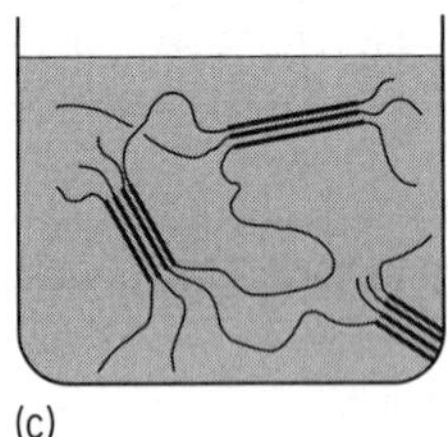


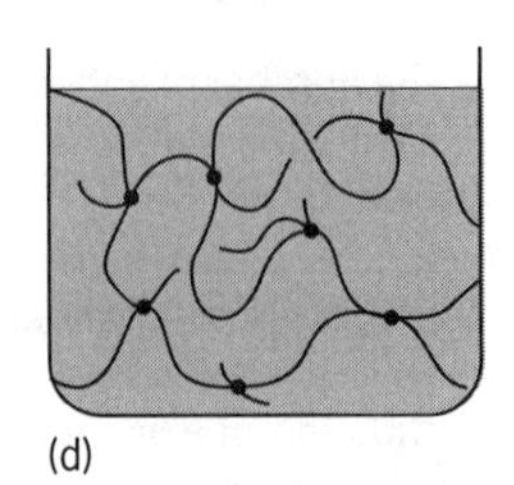


**Gel structures. (*a*) Agglomerated particles. (*b*) Framework of fibers or plates. (*c*) Polymers linked by crystalline junctions. (*d*) Polymers linked by covalent bonds. (*After M. Djabourov, Architecture of gelatin gels, Contemp. Phys., 29(3):273–297, 1988*)**

to a dish of acetone (a poor solvent), because the polymer chains tend to favor contact with one another rather than with acetone, so the network collapses onto itself. Conversely, the reason that water cannot be gently squeezed out of such a gel is that the network has a strong affinity for the liquid, and virtually all of the molecules of the liquid are close enough to the solid-liquid interface to be influenced by those attractive forces. *See* HYDROGEN BOND; INTERMOLECULAR FORCES.

The most striking feature of a gel is its elasticity: if the surface of a gel is displaced slightly, it springs back to its original position. If the displacement is too large, gels, except those with polymers linked by covalent bonds, may suffer some permanent plastic deformation, because the network is weak. The process of gelation, which transforms a liquid into an elastic gel, may begin with a change in pH that removes repulsive forces between the particles in a colloidal suspension, or a decrease in temperature that favors crystallization of a solution of polymers or the initiation of a chemical reaction that creates or links polymers. *See* PH.

Many inorganic gels can be made from solutions of salts or metallorganic compounds, and this offers several advantages in ceramics processing: the reactants are readily purified; the components can be intimately mixed in the solution or sol stage; the sols can be applied as coatings, drawn into fibers, emulsified or spray-dried to make particles, or molded and gelled into shapes. Hybrid materials can be made by combining organic and inorganic components in the gel. Many hybrids have such compliant networks that they collapse completely during drying, leaving a dense solid; therefore, they can be used as protective coatings (on plastic eyeglass lenses, for example) without heat treatment. Hybrid gels show great promise for active and integrated optics, because optically active organic molecules retain their activity while encapsulated in the gel matrix. [G.W.Sc.]

**Gel permeation chromatography** A separation technique involving the transport of a liquid mobile phase through a column containing the separation medium, a porous material. Gel permeation chromatography (GPC), also called size exclusion chromatography and gel filtration, affords a rapid method for the separation of oligomeric and polymeric species. The separation is based on differences in molecular size in solution. It is of particular importance for research in biological systems

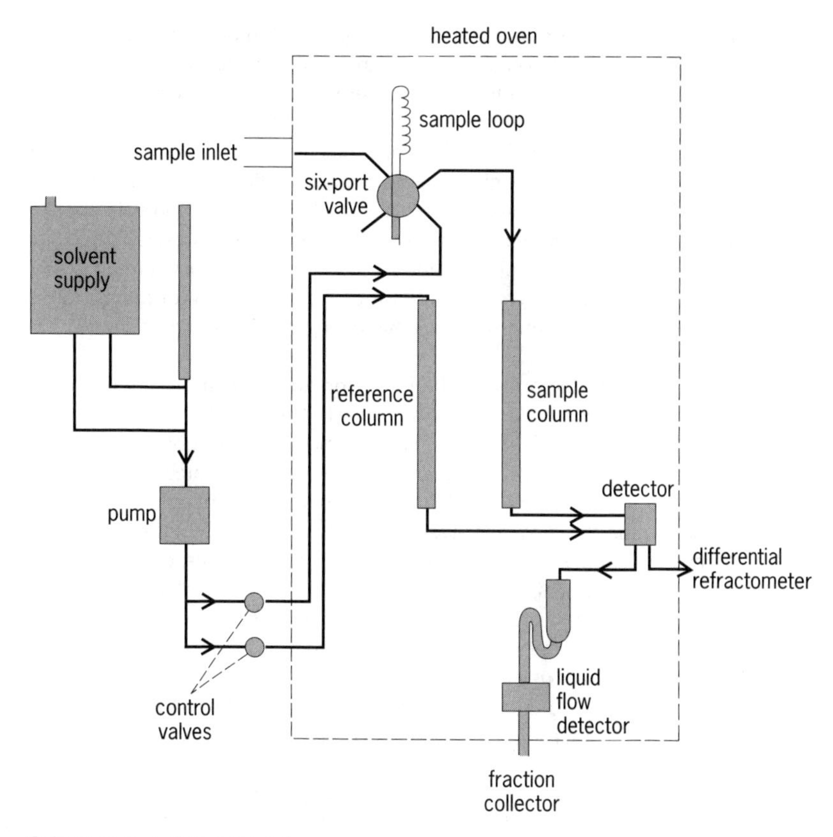


**Gel permeation chromatograph.**

and is the method of choice for determining molecular weight distribution of synthetic polymers.

The separation medium is a porous solid, such as glass or silica, or a cross-linked gel which contains pores of appropriate dimensions to effect the separation desired. The liquid mobile phase is usually water or a buffer for biological separations, and an organic solvent that is appropriate for the sample and is compatible with the column packing for synthetic polymer characterization. Solvent flow may be driven by gravity, or by a high-pressure pump to achieve the desired flow rate through the column. The sample to be separated is introduced at the head of the column (see illustration). As it progresses through the column, small molecules can enter all pores larger than the molecule, while larger molecules can fit into a smaller number of pores, again only those larger than the molecule. Thus, the larger the molecule, the smaller is the amount of pore volume available into which it can enter. The sample emerges from the column in the inverse order of molecular size; that is, the largest molecules emerge first followed by progressively smaller molecules. In order to determine the amount of sample emerging, a concentration detector is located at the end of the column. Additionally, detectors may be used to continuously determine the molecular weight of species eluting from the column. The volume of solvent flow is also monitored to provide a means of characterizing the molecular size of the eluting species.

As the sample emerges from the column, a concentration detector signal increases above the baseline, passes through a maximum, and returns to the baseline. This signal provides a relative concentration of emerging species and is recorded as a function of elution volume. Molecular-weight detectors based on light scattering or viscometry produce a signal, independent of molecular weight, that increases with increasing molecular weight for a given sample concentration. The molecular-weight chromatograms therefore are skewed with respect to the concentration detector signal. Typically, constant-volume pumps are used, and the time axis is transformed into a volume axis. Devices such as siphons that empty after a known volume of solvent has collected, or electronic circuitry that measures the transit time of a thermal pulse between two points in the flowing solvent, have been used to provide a measurement of the volumetric flow rate.

Gel permeation chromatography has had widespread applications. For determining the molecular weight of synthetic polymers, at least 50 types of polymers have been characterized. These include alkyd resins, natural and synthetic rubbers, cellulose esters, polyolefins, polyamides, polyesters, polystyrenes, polyacrylates, uncured epoxy, urethane and phenolic resins, and a wide variety of oligomeric materials. Additionally, the ability to determine the molecular-weight distribution and changes in distribution has led to many applications in areas such as blending distributions, chain-length studies in semicrystalline polymers, interactions in solution, radiation studies, mechanical degradation studies, mechanisms of polymerization research, polymerization reactor control, and evaluation of the processing of polymers. In the field of natural and biological polymers, numerous systems have been separated and analyzed. Among these are acid phosphatases, adrenalin, albumin, amino acids and their derivatives, enzymes, blood group antibodies, collagen and related compounds, peptides, and proteins.

Additionally, gel permeation chromatography is capable of making separations of low-molecular-weight compounds. This is particularly important when both low- and high-molecular-weight species are present in the same sample. *See* CHROMATOGRAPHY; POLYMER. [A.R.C.]

**Gelatin** A protein extracted after partial hydrolysis of collagenous raw material from the skin, white connective tissue, and bone of animals. It is a linear polymer of amino acids, most often with repeating glycine-proline-proline and glycine-proline-hydroxyproline sequences in the polypeptide linkages. *See* PROTEIN.

The unique characteristics of gelatin are: reversible sol-to-gel formation, amphoteric properties, swelling in cold water, film-forming properties, viscosity-modifying properties, and protective colloid properties. Gelatin contains 26.4–30.5% glycine, 14.8–18% proline, 13.3–14.5% hydroxyproline, 11.1–11.7% glutamic acid, 8.6–11.3% alanine, and in decreasing order arginine, aspartic acid, lysine, serine, leucine, valine, phenylalanine, threonine, isoleucine, hydroxylysine, histidine, methionine, and tyrosine. Absence of only two essential amino acids—tryptophan and methionine—makes gelatin a good dietary food supplement.

The principal uses of gelatin are in foods, pharmaceuticals, and photographic industries. Other uses of industrial gelatin are in the field of microencapsulation, health and cosmetics, and plastics. [F.V.]

**Germanium** A brittle, silvery-gray, metallic chemical element, Ge, atomic number 32, atomic weight 72.59, melting point 937.4°C (1719°F), and boiling point 2830°C

(5130°F), with properties between silicon and tin. Germanium is distributed widely in the Earth's crust in an abundance of 6.7 parts per million (ppm). Germanium is found as the sulfide or is associated with sulfide ores of other elements, particularly those of copper, zinc, lead, tin, and antimony. *See* PERIODIC TABLE.

Germanium has a metallic appearance but exhibits the physical and chemical properties of a metal only under special conditions since it is located in the periodic table where the transition from nonmetal to metal occurs. At room temperature there is little indication of plastic flow and consequently it behaves like a brittle material.

As it exists in compounds, germanium is either divalent or tetravalent. The divalent compounds (oxide, sulfide, and all four halides) are easily reduced or oxidized. The tetravalent compounds are more stable. Organogermanium compounds are many in number and, in this respect, germanium resembles silicon. Interest in organogermanium compounds has centered around their biological action. Germanium in its derivatives appears to have a lower mammalian toxicity than tin or lead compounds.

The properties of germanium are such that there are several important applications for this element, especially in the semiconductor industry. The first solid-state device, the transistor, was made of germanium. Single-crystal germanium is used as a substrate for vapor-phase growth of GaAs and GaAsP thin films in some light-emitting diodes. Germanium lenses and filters are used in instruments operating in the infrared region of the spectrum. Mercury-doped and copper-doped germanium are used as infrared detectors; synthetic garnets with magnetic properties may have applications for high-power microwave devices and magnetic bubble memories; and germanium additives increase usable ampere-hours in storage batteries. [P.S.G.]

**Glucose** A monosaccharide also known as D-glucose, D-glucopyranose, grape sugar, corn sugar, dextrose, and cerelose. The structure of $\alpha$-D-glucose is shown below.

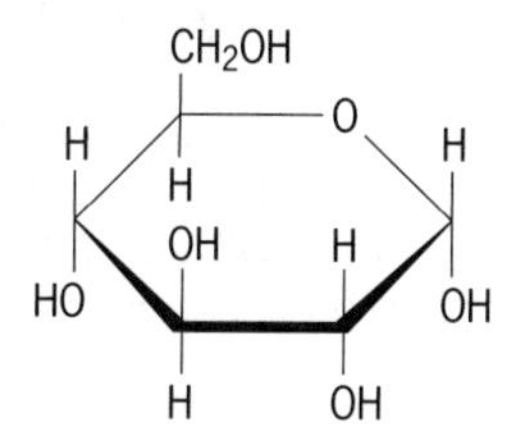


Glucose in free or combined form is not only the most common of the sugars but is probably the most abundant organic compound in nature. It occurs in free state in practically all higher plants. It is found in considerable concentrations in grapes, figs, and other sweet fruits and in honey. In lesser concentrations, it occurs in the animal body fluids, for example, in blood and lymph. Urine of diabetic patents usually contains 3–5%.

Cellulose, starch, and glycogen are composed entirely of glucose units. Glucose is also a major constituent of many oligosaccharides, notably sucrose, and of many glycosides. It is produced commercially from cornstarch by hydrolysis with dilute mineral acid. The commercial glucose so obtained is used largely in the manufacture of confections and in the wine and canning industries.

D-Glucose is the principal carbohydrate metabolite in animal nutrition; it is utilized by the tissues, and it is absorbed from the alimentary tract in greater amounts than any other monosaccharide. Glucose could serve satisfactorily in meeting at least 50% of the entire energy needs of humans and various animals.

Glucose enters the bloodstream by absorption from the small intestine. It is carried via the portal vein to the liver, where part is stored as glycogen, the remainder reentering the circulatory system. Another site of glycogen storage is muscle tissue.

Glucose is readily fermented by yeast, producing ethyl alcohol and carbon dioxide. It is also metabolized by many bacteria, resulting in the formation of various degradation products, such as hydrogen, acetic and butyric acids, butyl alcohol, acetone, and many others. *See* CARBOHYDRATE.

**Glycerol** The simplest trihydric alcohol, with the formula $CH_2OHCHOHCH_2OH$. The name glycerol is preferred for the pure chemical, but the commercial product is usually called glycerin. It is widely distributed in nature in the form of its esters, called glycerides. The glycerides are the principal constituents of the class of natural products known as fats and oils.

When pure, glycerin is a colorless, odorless, viscous liquid with a sweet taste. It is completely soluble in water and alcohol but is only slightly soluble in many common solvents, such as ether, ethyl acetate, and dioxane. Glycerin is insoluble in hydrocarbons. It boils at 290°C (554°F) at atmospheric pressure and melts at 17.9°C. Its specific gravity is 1.262 at 25°C (77°F) referred to water at 25°C, and its molecular weight is 92.09. It has a very low mammalian toxicity.

Glycerin is used in nearly every industry. With dibasic acids, such as phthalic acid, it reacts to make the important class of products known as alkyd resins, which are widely used as coating and in paints. It is used in innumerable pharmaceutical and cosmetic preparations; it is an ingredient of many tinctures, elixirs, cough medicines, and anesthetics; and it is a basic medium for toothpaste. In foods, it is an important moistening agent for baked goods and is added to candies and icings to prevent crystallization. It is used as a solvent and carrier for extracts and flavoring agents and as a solvent for food colors. Many specialized lubrication problems have been solved by using glycerin or glycerin mixtures. Many millions of pounds are used each year to plasticize various materials.

Several grades of glycerin are marketed, including high gravity, dynamite, yellow distilled, USP (U.S. Pharmacopoeia), and CP (chemically pure). USP grade is water-white and suitable for use in foods, pharmaceuticals, and cosmetics, or for any purpose where the product is designed for human consumption. *See* ALCOHOL; FAT AND OIL; POLYOL. [P.H.C.]

**Gold** A chemical element, Au, atomic number 79 and atomic weight 196.967, a deep yellow, soft, and very dense metal. Gold is classed as a heavy metal and as a noble metal; commercially, it is the most familiar of the precious metals. Copper, silver, and gold are in the same group of the periodic table of elements. The Latin name for gold, *aurum* (glowing dawn), is the source of the chemical symbol Au. There is only one stable isotope of gold, that of mass number 197. *See* PERIODIC TABLE.

**Uses.** Consumption of gold in jewelry accounts for about three-fourths of the world's production of gold. Industrial applications, especially electronic, consume another 10–15%. The remainder is divided among medical and dental uses, coinage, and bar stock for governmental and private holdings. Gold coins and most decorative gold objects are actually gold alloys, because the metal itself is too soft (2.5–3 on Mohs scale) to be useful with frequent handling.

Radioactive $^{198}Au$ is used in medical irradiation, in diagnosis, and in a number of industrial applications as a tracer. Another tracer use is in the study of movement of sediment on the ocean floor in and around harbors. The properties of gold toward radiant energy have led to development of efficient energy reflectors for infrared heaters and cookers and for focusing and retention of heat in industrial processes.

**Occurrence.** Gold occurs widely throughout the world, but usually very sparsely, so that it is quite a rare element. Sea water contains low concentrations of gold, on the order of 10 $\mu$g per ton (10 parts of gold per trillion parts of water). Somewhat higher concentrations accumulate on plankton or on the ocean bottom. At present, no economically feasible process is visualized for extracting gold from the sea. Native, or metallic, gold and various telluride minerals are the only forms of gold found on land. Native gold may occur in veins among rocks and ores of other metals, especially quartz or pyrite, or it may be scattered in sands and gravel (alluvial gold).

**Properties.** The density of gold is 19.3 times that of water at 20°C (68°F), so that 1 $ft^3$ of gold weighs about 1200 lb (1 $m^3$, about 19,000 kg). Masses of gold, like those of other precious metals, are measured on the troy scale, which counts 12 oz to the pound. Gold melts at 1064.43°C (1947.97°F) and boils at 2860°C (5180°F). It is somewhat volatile well below its boiling point. Gold is a good conductor of heat and electricity. It is the most malleable and ductile metal. It can easily be made into translucent sheets 0.0000039 in. (0.00001 mm) thick or drawn into wire weighing only 0.00005 oz/ft (0.5 mg/m). The quality of gold is expressed on the fineness scale as parts of pure gold per thousand parts of total metal, or on the karat scale as parts of pure gold per 24 parts of total metal. Gold readily dissolves in mercury to form amalgams. Gold is one of the least active metals chemically. It does not tarnish or burn in air. It is inert to strong alkaline solutions and to all pure acids except selenic acid.

**Compounds.** Gold may be either unipositive or tripositive in its compounds. So strong is the tendency for gold to form complexes that all the compounds of the 3+ oxidation state are complex. The compounds of the 1+ oxidation state are not very stable and tend to be oxidized to the 3+ state or reduced to metallic gold. All compounds of either oxidation state are easy to reduce to the metal.

In its complex compounds gold forms bonds most readily and stably with halogens and sulfur, less stably with oxygen and phosphorus, and only weakly with nitrogen. Bonds between gold and carbon are fairly stable, as in the cyanide complexes and a variety of organogold compounds. [W.E.C.]

**Gram-equivalent weight** A quantity of a substance that contains the same number (known as the Avogadro number) of molecules as the number of atoms contained in exactly 12.000 g of carbon-12 ($^{12}C$). This convention stems from the concept that the central principle guiding chemical calculations is the relation of quantities of reacting substances to the numbers of molecules involved.

An added convenience in stoichiometric calculations is to incorporate the combining capacity ($n$), as well as the number of molecules, so that an equivalence of reacting substances is implicit without the need to examine balanced equations each time. Thus, the gram-equivalent weight of a substance is its gram-molecular weight divided by $n$. In acid-base reactions, $n$ of the acid or base is given by the number of protons released or consumed in the reaction. For hydrochloric acid (HCl), ammonia ($NH_3$), acetic acid ($CH_3COOH$), and the acetate ion [$CH_3COO^-$; as in sodium acetate ($NaOOCCH_3$)], $n = 1$. For carbonic acid ($H_2CO_3$), sodium carbonate ($Na_2CO_3$), and ethylenediamine ($H_2NCH_2CH_2NH_2$), $n = 2$.

In precipitation and other metathetical reactions, the charge (valence) of the ion involved governs the value of $n$. Thus, $n = 3$ for ferric chloride ($FeCl_3$) and $n = 6$ for ferric sulfate [$Fe_2(SO_4)_3$] when precipitation yields either ferric hydroxide [$Fe(OH)_3$], barium sulfate ($BaSO_4$), or silver chloride (AgCl). When oxidation-reduction is involved, the change of valence, rather than the valence itself, defines *n*. When the ferric ion ($Fe^{3+}$) acts as an oxidant [with the ferrous ion ($Fe^{2+}$) as product], $n = 1$ for ferric chloride ($FeCl_3$) and $n = 2$ for ferric sulfate. When potassium permanganate ($KMnO_4$) reacts in acid medium, $n = 5$, but in neutral or basic solution $n = 3$.

Further changes in the definition arise in dealing with metal complex formation, such as $FeF_6^{3-}$, $FeCl_4^-$, $Zn(EDTA)^{2+}$, and $Bi(EDTA)^-$; EDTA is the polydentate metal chelating agent ethylenediaminetetraacetate ion. The combining capacity of metals in complex formation depends on their coordination number, which, as these examples demonstrate, differs with different complexing agents, or ligands. *See* ELECTROCHEMICAL EQUIVALENT; EQUIVALENT WEIGHT; ETHYLENEDIAMINETETRAACETIC ACID; MOLE; STOICHIOMETRY; VALENCE. [H.Frei.]

**Gram-molecular weight** The molecular weight of an element or compound expressed in grams (g), that is, the molecular weight on a scale on which the atomic weight of the $^{12}C$ isotope of carbon is taken as 12 exactly. This replaces the earlier scale on which the atomic weight of oxygen was taken as 16.00 g. In the International System of Units, gram-molecular weight is replaced by the mole.

The ratio of the gram-molecular weights of any two elements or compounds must be identical with the ratio of the absolute weights of their individual molecules. Therefore, the gram-molecular weights of all elements or compounds contain the same number of molecules. This number, called the Avogadro number, *N*, is $6.022 \times 10^{23}$. *See* MOLE; MOLECULAR WEIGHT; RELATIVE MOLECULAR MASS. [T.C.W.]

**Graphite** A low-pressure polymorph of carbon (the common high-pressure polymorph being diamond). Graphite is metallic in appearance and very soft. Crystals of graphite are infrequently encountered since the mineral usually occurs as earthy, foliated, or columnar aggregates often mixed with iron oxide, quartz, and other minerals. *See* CARBON.

The sheetlike character of the graphite atomic arrangement results in distinctive physical properties. The mineral is very soft, with hardness $1^1/_2$; it soils the fingers and leaves a black streak on paper, hence its use in pencils. The specific gravity is 2.23, often less because of the presence of pore spaces and impurities. The color is black in earthy material to steel-gray in plates, and thin flakes are deep blue in transmitted light. Graphite is a conductor of electricity.

The major sources of graphite are in gneisses and schists, where the mineral occurs in foliated masses mixed with quartz, mica, and so on. Noteworthy localities include the Adirondack region of New York, Korea, and Ceylon. In Sonora, Mexico, graphite occurs as a product of metamorphosed coal beds. Graphite is also observed in meteorites. [P.B.M.]

**Synthetic graphite.** Commercially produced synthetic graphite is a mixture of crystalline graphite and cross-linking intercrystalline carbon. Its physical properties are the result of contributions from both sources. Thus, among engineering materials, synthetic graphite is unusual because a wide variation in measurable properties can occur without significant change in chemical composition.

At room temperature the thermal conductivity of synthetic graphite is comparable to that of aluminum or brass. An unusual property of graphite is its increased strength at high temperature. Graphite is resistant to thermal shock because of its high thermal conductivity and low elastic modulus. It is one of the most inert materials with respect to chemical reaction with other elements and compounds. It is subject to oxidation, and reaction with and solution in some metals.

Graphite has many uses in the electrical, chemical, metallurgical, nuclear, and rocket fields: electrodes in electric furnaces producing carbon steel, alloy steel, and ferroalloys; anodes for the electrolytic production of chlorine, chlorates, magnesium, and sodium; motor and generator brushes; sleeve-type bearings and seal rings; rocket motor nozzles; missile nose cones; metallurgical molds and crucibles; linings for chemical reaction vessels; and, in a resin-impregnated impervious form, for heat exchangers, pumps, pipings, valves, and other process equipment.

**Graphite (carbon) fibers.** Carbon fibers are filamentary forms of carbon, with a fiber diameter normally in the 6–10-micrometer range. The product is offered in the form of yarns or tows containing from 1000 to 500,000 filaments per strand. The fibers offer a unique combination of properties. They are flexible, lightweight, thermally and to a large extent chemically inert, and are good thermal and electrical conductors. In their high-performance varieties, carbon fibers are very strong and can be extremely stiff.

The principal use of high-performance carbon fibers is as the reinforcing component in structural epoxy matrix composites. Due to initially high cost, the original applications were almost exclusively for lightweight, high-stiffness, and high-strength composites for the aerospace industry. The second major usage of high-performance carbon fibers is in sporting goods, such as golf club shafts, tennis rackets, fishing rods, and sailboat structures. The major matrix material for both applications is epoxy. [H.F.V.]

**Gravimetric analysis** That branch of quantitative chemical analysis in which a desired constituent is converted (usually by precipitation) to a pure compound or element of definite, known composition, and is weighed. In a few cases, a compound or element is formed which does not contain the constituent but bears a definite mathematical relationship to it. In either case, the amount of desired constituent can be determined from the weight and composition of the precipitate.

At least two weighings are required for each analysis—the original sample, and the dried or ignited residue. From these weights, the percentage or proportion of the desired constituent may be calculated from the equation below.

$$\%A = \frac{\text{wt of residue} \times \text{factor} \times 100}{\text{wt of sample}}$$

The factor is determined from a knowledge of the chemical relationships between the weight of substance A contained in, or equivalent to, a fixed weight of residue of known composition. *See* ANALYTICAL CHEMISTRY ; QUANTITATIVE CHEMICAL ANALYSIS; STOICHIOMETRY. [S.G.S.]

**Grignard reaction** A reaction between an alkyl or aryl halide and magnesium metal in a suitable solvent, usually absolute ether. The organomagnesium halides produced by this reaction are known as Grignard reagents and are useful in many chemical syntheses. The structure of a Gignard reagent is usually written RMgX, where X represents a halogen.

The scope of the Grignard reaction is extremely broad, and Grignard reagents have been prepared from many kinds of alkyl and aryl halides. In general, alkyl chlorides, bromides, and iodides and aryl bromides and iodides react readily. A few halides, such as aryl chlorides, react very sluggishly. *See* ORGANOMETALLIC COMPOUND. [P.E.F.]

**Gum** A class of high-molecular-weight molecules, usually with colloidal properties, which in an appropriate solvent or swelling agent are able to produce gels at low dry-substance content. The molecules are either hydrophilic or hydrophobic. The term gum is applied to a wide variety of substances of gummy characteristics, and therefore cannot be precisely defined. *See* GEL.

Various rubbers are considered to be gums, as are many synthetic polymers, high-molecular-weight hydrocarbons, or other petroleum products. Chicle for chewing gum is an example of a hydrophobic polymer which is termed a gum but is not frequently classified among the gums. Quite often listed among the gums are the hydrophobic resinous saps that often exude from plants and are commercially tapped in balsam (gum balsam) and other evergreen trees (gum resin). Incense gums such as myrrh and frankincense are likewise fragrant plant exudates.

Usually, however, the term gum, as technically employed in industry, refers to plant polysaccharides or their derivatives. Modern usage of the term includes water-soluble derivatives of cellulose and derivatives and modifications of other polysaccharides which in the natural form are insoluble. Usage, therefore, also includes with gums the ill-defined group of plant slimes called mucilages. *See* COLLOID; POLYSACCHARIDE.

Gums are used in foods as stabilizers and thickeners. They form viscous solutions which prevent aggregation of the small particles of the dispersed phase. In this way they aid in keeping solids dispersed in chocolate milk, air in whipping cream, and fats in salad dressings. Gum solutions also retard crystal growth in ice cream (ice crystals) and in confections (sugar crystals). Their thickening and stabilizing properties make them useful in water-base paints, printing inks, and drilling muds. Because of these properties they also are used in cosmetics and pharma-ceuticals as emulsifiers or bases for ointments, greaseless creams, toothpastes, lotions, demulcents, and emollients. The adhesive properties of gums make them useful in the production of cardboard, postage stamps, gummed envelopes, and as pill binders. Other applications include the production of dental impression molds, fibers (alginate rayon), soluble surgery films and gauze, blood anticoagulants, plasma extenders, beverage-clarifying agents, bacteriological culture media, half-cell bridges, and tungsten-wire-drawing lubricants. [R.L.Wh.]

**Hafnium** A metallic element, symbol Hf, atomic number 72, and atomic weight 178.49. There are five naturally occurring isotopes. It is one of the less abundant elements in the Earth's crust. *See* PERIODIC TABLE.

Hafnium is a lustrous, silvery metal that melts at about 2222°C (4032°F). Reported values of the boiling point vary greatly, from about 2500 to about 5100°C (4530 to 9200°F). There are virtually no uses of the metal other than in control rods for nuclear reactors.

The chemistry of hafnium is almost identical with that of zirconium. The similarity of hafnium to zirconium is a consequence of the lanthanide contraction, which brings the ionic radii to very nearly identical values. Before (and since) the discovery of hafnium, this element was extracted with zirconium from its ores and passed with zirconium into all derivatives. Since the chemical properties are so similar, there has been no incentive to separate the hafnium except for making nuclear studies and components of nuclear reactors. *See* ZIRCONIUM. [W.B.B.]

**Half-life** The time required for one-half of a given material to undergo chemical reactions; also, the average time interval required for one-half of any quantity of identical radioactive atoms to undergo radioactive decay.

The concept of the time required for all of the material to react is meaningless, because the reaction goes very slowly when only a small amount of the reacting material is left and theoretically an infinite time would be required. The time for half completion of the reaction is a definite and useful way of describing the rate of a reaction.

The specific rate constant $k$ provides another way of describing the rate of a chemical reaction. This is shown in a first-order reaction, Eq. (1), where $c_0$ is the initial

$$k = \frac{2.303}{t} \log \frac{c_0}{c} \quad (1)$$

concentration and $c$ is the concentration at time $t$. The relation between specific rate constant and period of half-life, $t_{1/2}$, in a first-order reaction is given by Eq. (2).

$$t_{1/2} = \frac{2.303}{k} \log \frac{1}{1/2} = \frac{0.693}{k} \quad (2)$$

*See* CHEMICAL DYNAMICS. [F.D.]

The activity of a source of any single radioactive substance decreases to one-half in 1 half-period, because the activity is always proportional to the number of radioactive atoms present. For example, the half-period of $^{60}$Co (cobalt-60) is $t_{1/2}$, = 5.3 years. Then a $^{60}$Co source whose initial activity was 100 curies will decrease to 50 curies in 5.3 years. In 1 additional half-period this activity will be further reduced by the factor $^1/_2$. Thus, the fraction of the initial activity which remains is $^1/_2$ after one half-period, $^1/_4$ after two half-periods, $^1/_8$ after three half-periods, $^1/_{16}$, after four half-periods, and so on.

The half-period is sometimes also called the half-value time or, with less justification, the half-life. [R.D.E.]

**Halide** A compound containing one of the halogens [fluorine (F), chlorine (Cl), bromine (Br), iodine (I)] and another element or organic group. Halides have the general formula $M_xX_y$, where M is a metal or organic group and X is a halogen. Halides are composed of almost every element in the periodic table, and they are referred to as fluorides, chlorides, bromides, or iodides. *See* HALOGEN ELEMENTS; PERIODIC TABLE.

The halides are divided into classes that reflect the nature of bonding between the halogen and metal or organic species. The bonding of halides ranges from purely ionic to essentially covalent. The classes include ionic halides, molecular halides, halides and halogens that behave as ligands in coordination complexes, and organic halides.

Ionic halides such as sodium chloride (NaCl) and potassium chloride (KCl) are prepared from the vigorous reaction of the alkali and alkaline-earth metals with the halogens. These compounds possess high melting and boiling points and are soluble in very polar solvents. Ionic halides are extremely important to the chemical industry, where they are used to produce commodity chemicals such as sodium hydroxide (NaOH), hydrochloric acid (HCl, a hydrogen halide), and potassium nitrate ($KNO_3$).

The organic halides are divided into the alkyl halides (haloalkanes) and the aryl halides. The alkyl halides have the general formula RX, where R is any alkyl group and X is one of the halogens; for example, 1-chlorobutane ($CH_3CH_2CH_2CH_2Cl$). Halides are good leaving groups in nucleophilic substitution reactions and are good nucleophiles. The aryl halides are compounds where the halogen is attached directly to an aromatic ring and have the general formula ArX, where Ar is an aromatic group. *See* COORDINATION CHEMISTRY; COORDINATION COMPLEXES; ELECTROPHILIC AND NUCLEOPHILIC REAGENTS; HALOGENATED HYDROCARBON; HALOGENATION. [T.J.Me.]

**Halogen elements** The halogen family consists of the elements fluorine, F; chlorine, Cl; bromine, Br; iodine, I; and astatine, At. All the halogen elements except astatine exist in the Earth's crust and atmosphere.

The halogens are the best-defined family of elements. They have an almost perfect gradation of physical properties. The increase in atomic weight from fluorine through iodine is paralleled by increases in density, melting and boiling points, critical temperature and pressure, heats of fusion and vaporization, and even in progressively deeper color (fluorine is pale yellow; chlorine, yellow-green; bromine, dark red; and iodine, deep violet).

Although all halogens generally undergo the same types of reactions, the extent and ease with which these reactions occur vary markedly. Fluorine in particular has the usual tendency of the lightest member of a family of elements to exhibit reactions not comparable to the other members. Each halogen must be considered individually, both in its preparation and in its reaction. *See* ASTATINE; BROMINE; CHLORINE; HALIDE; HALOGENATION; IODINE. [R.J.C.; A.A.G.]

**Halogenated hydrocarbon** An aliphatic or aromatic hydrocarbon in which one or more hydrogen atoms are substituted by halogen. *See* HALOGEN ELEMENTS; HALOGENATION.

Alkyl halides are compounds in which one hydrogen of an alkane has been replaced by halogen [fluorine (F), chlorine (Cl), bromine (Br), or iodine (I)], for example, bromoethane (ethyl bromide; $CH_3CH_2Br$). Many alkyl halides have been prepared; the chlorides and bromides are most useful and most common. Alkyl halides are important starting materials for the preparation of many other functionally substituted compounds. A general reaction for chlorides, bromides, and iodides is nucleophilic substitution, in which an ion or molecule with an available electron pair (a nucleophile) displaces a halide ion. *See* QUATERNARY AMMONIUM SALTS.

Compounds with halogen bonded directly to a benzene or other aromatic ring are called aryl halides. Halogen is introduced by electrophilic substitution, with a Lewis acid catalyst such as $FeCl_3$ or $FeBr_3$ to enhance the positive character of the halogen.

In fluorocarbons, every hydrogen atom is replaced by fluorine. Fluorocarbons can be named simply by using the prefix perfluoro- with the parent name. Because of the small atomic radius and high electronegativity of fluorine, these compounds are chemically inert and have properties quite unlike those of other halogenated organic compounds. *See* FLUOROCARBON.

Hydrofluorocarbons contain combinations of fluorine and hydrogen to satisfy the valency requirement of carbon. Hydrofluorocarbons are also known as HFCs. The development of specific molecules for particular applications, previously satisfied by chlorofluorocarbons, has been international in scope. Because they contain hydrogen, hydrofluorocarbons are more likely to be degraded in the lower regions of the atmosphere. Since they do not contain chlorine, these compounds do not contribute to ozone depletion.

Perfluorocarbons contain only carbon and fluorine. They are named by using the prefix perfluoro- along with the name of the equivalent hydrocarbon. Perfluorocarbons are chemically very inert and also have excellent thermal stability. This inertness and the resulting long atmospheric lifetime is reflected in higher global warming potentials compared to hydrofluorocarbons, hydrochlorofluorocarbons, and chlorofluorocarbons. This is due, partly, to the high strength of the C-F bond. The electronegativity of fluorine shields the carbon backbone from chemical attack. Under normal conditions, perfluorocarbons are unaffected by strong acids or bases and by oxidizing or reducing agents. *See* ELECTRONEGATIVITY.

A number of compounds with two or more halogen atoms are of special importance. Methane ($CH_4$) can be substituted with as many as four halogen atoms to give compounds such as $CH_2Cl_2$, $CHI_3$, and $CF_3Br$. Several polyhalomethane, -ethane, and -ethylene derivatives have major uses, and they are industrial chemicals produced in large quantities.

Chlorination of methane leads to mixtures of mono-, di-, tri-, and tetrachloro products. The relative amounts can be controlled by adjusting the ratio of starting materials. Methyl chloride is manufactured by this chlorination process and also by reaction of methanol and HCl. Methylene chloride, $CH_2Cl_2$, is the major product from methane, and is utilized primarily as a cleaning solvent or as a blowing agent for plastic foam. Methylene chloride is more volatile and much less toxic than $CHCl_3$ or $CCl_4$.

Chlorofluorocarbons are methane and ethane derivatives with all hydrogen atoms replaced by combinations of chlorine and fluorine. Chlorofluorocarbons are known collectively as CFCs. The first of these compounds, dichlorofluoromethane ($CCl_2F_2$), was introduced as a nontoxic, nonflammable working fluid in refrigeration equipment to replace ammonia and sulfur dioxide. Other compounds were developed to meet the requirements of specific uses such as air-conditioning equipment in buildings and vehicles, and propellants for aerosols.

Hydrochlorofluorocarbons contain combinations of hydrogen, chlorine, fluorine, and carbon to satisfy the valency requirement of carbon. Also known as HCFs, the hydrochlorofluorocarbons have been developed as interim substitutes for chlorofluorocarbons. Since they have at least one hydrogen atom in the molecule, they are more likely to be degraded in the troposphere by reaction with hydroxyl (OH) radicals. Thus, the potential that hydrochlorofluorocarbons have to deplete ozone by migration to the stratosphere is reduced. [J.A.Mo.; V.N.M.R.]

**Halogenation** A chemical reaction or process which results in the formation of a chemical bond between a halogen atom and another atom. Reactions resulting in the formation of halogen-carbon bonds are especially important. The halogenated compounds produced are employed in many ways, for example, as solvents, intermediates for numerous chemicals, plastic and polymer intermediates, insecticides, fumigants, sterilants, refrigerants, additives for gasoline, and materials used in fire extinguishers. *See* HALOGEN ELEMENTS.

Halogenation reactions can be subdivided in several ways, for example, according to the type of halogen (fluorine, chlorine, bromine, or iodine), type of material to be halogenated (paraffin, olefin, aromatic, hydrogen, and so on), and operating conditions and methods of catalyzing or initiating the reaction.

Halogenation reactions with elemental chlorine, bromine, and iodine are of considerable importance. Because of high exothermocities, fluorinations with elemental fluorine tend to have high levels of side reactions. Consequently, elemental fluorine is generally not suitable for direct fluorination. Two types of reactions are possible with these halogen elements, substitution and addition.

Substitution halogenation is characterized by the substitution of a halogen atom for another atom (often a hydrogen atom) or group of atoms (or functional group) on paraffinic, olefinic, aromatic, and other hydrocarbons. A chlorination reaction of importance that involves substitution is that between methane and chlorine. *See* SUBSTITUTION REACTION.

Addition halogenation involves a halogen reacting with an unsaturated hydrocarbon. Chlorine, bromine, and iodine react readily with most olefins; the reaction between ethylene and chlorine to form 1,2-dichloroethane is of considerable commercial importance, since it is used in the manufacture of vinyl chloride.

Addition reactions with bromine or iodine are frequently used to measure quantitatively the number of —CH=CH— (or ethylenic-type) bonds in organic compounds. Bromine numbers or iodine values are measures of the degree of unsaturation of the hydrocarbons.

Substitution halogenation on the aromatic ring can be made to occur via ionic reactions. The chlorination reactions with elemental chlorine are similar to those used for addition chlorination of olefins. *See* HALOGENATED HYDROCARBON. [L.F.A.]

**Hassium** A chemical element, symbol Hs, atomic number 108. It was synthesized and identified in 1984 by using the Universal Linear Accelerator (UNILAC) at Darmstadt, West Germany, by the same team (led by P. Armbruster and G. Müzenberg) which first identified bohrium and meitnerium. The isotope $^{265}$Hs was produced in a fusion reaction by bombarding a $^{208}$Pb target with a beam of $^{58}$Fe projectiles.

The discovery of bohrium and meitnerium was made by detection of isotopes with odd proton and neutron numbers. In this region, odd-odd nuclei show the highest stability against fission. Elements with an even atomic number are intrinsically less stable against spontaneous fission. The isotopes of hassium were expected to decay by spontaneous fission—which explains why meitnerium was synthesized before hassium.

As in the case of bohrium and meitnerium, the isotope was produced by fusion in a one-neutron deexcitation channel; in this case the compound system was $^{266}$Hs. The reaction mechanism again was cold fusion. The isotope $^{265}$Hs has a half-life of about 2 ms and decays by emission of an alpha particle of 10.36 MeV.

The elements bohrium, hassium, and meitnerium are stabilized by shell effects against spontaneous fission; this special stability may occur because these nuclei may prefer a sausage shape which is predicted to be energetically most favorable for them. *See* BOHRIUM; MEITNERIUM; PERIODIC TABLE; TRANSURANIUM ELEMENTS. [P.Ar.]

**Hazardous waste** Any solid, liquid, or gaseous waste materials that, if improperly managed or disposed of, may pose substantial hazards to human health and the environment. Every industrial country in the world has had problems with managing hazardous wastes. Improper disposal of these waste streams in the past has created a need for very expensive cleanup operations. Efforts are under way internationally to remedy old problems caused by hazardous waste and to prevent the occurrence of other problems in the future.

A waste is considered hazardous if it exhibits one or more of the following characteristics: ignitability, corrosivity, reactivity, and toxicity. Ignitable wastes can create fires under certain conditions; examples include liquids, such as solvents, that readily catch fire, and friction-sensitive substances. Corrosive wastes include those that are acidic and those that are capable of corroding metal (such as tanks, containers, drums, and barrels). Reactive wastes are unstable under normal conditions. They can create explosions, toxic fumes, gases, or vapors when mixed with water. Toxic wastes are harmful or fatal when ingested or absorbed. When they are disposed of on land, contaminated liquid may drain (leach) from the waste and pollute groundwater.

Hazardous wastes may arise as by-products of industrial processes. They may also be generated by households when commercial products are discarded. These include drain openers, oven cleaners, wood and metal cleaners and polishes, pharmaceuticals, oil and fuel additives, grease and rust solvents, herbicides and pesticides, and paint thinners.

The predominant waste streams generated by industries in the United States are corrosive wastes, spent acids, and alkaline materials used in the chemical, metal-finishing, and petroleum-refining industries. Many of these waste streams contain heavy metals, rendering them toxic. Solvent wastes are generated in large volumes both by manufacturing industries and by a wide range of equipment maintenance industries that generate spent cleaning and degreasing solutions. Reactive wastes come primarily from the chemical industries and the metal-finishing industries. The chemical and primary-metals industries are the major sources of hazardous wastes.

There is a growing acceptance throughout the world of the desirability of using waste management hierarchies for solutions to problems of hazardous waste. A typical sequence involves source reduction, recycling, treatment, and disposal. Source reduction comprises the reduction or elimination of hazardous waste at the source, usually within a process. Recycling is the use or reuse of hazardous waste as an effective substitute for a commercial product or as an ingredient or feedstock in an industrial process.

Treatment is any method, technique, or process that changes the physical, chemical, or biological character of any hazardous waste so as to neutralize such waste; to recover energy or material resources from the waste; or to render such waste nonhazardous, less hazardous, safer to manage, amenable for recovery, amenable for storage, or reduced in volume. Disposal is the discharge, deposit, injection, dumping, spilling, leaking, or placing of hazardous waste into or on any land or body of water so that the waste or any constituents may enter the air or be discharged into any waters, including groundwater.

There are various alternative waste treatment technologies, for example, physical treatment, chemical treatment, biological treatment, incineration, and solidification or stabilization treatment. These processes are used to recycle and reuse waste materials, reduce the volume and toxicity of a waste stream, or produce a final residual material that is suitable for disposal. The selection of the most effective technology depends upon the wastes being treated.

There are abandoned disposal sites in many countries where hazardous waste has been disposed of improperly in the past and where cleanup operations are needed to restore the sites to their original state. Cleaning up such sites involves isolating and containing contaminated material, removal and redeposit of contaminated sediments, and in-place and direct treatment of the hazardous wastes involved. As the state of the art for remedial technology improves, there is a clear preference for processes that result in the permanent destruction of contaminants rather than the removal and storage of the contaminating materials. [H.M.F.]

**Heat capacity** The quantity of heat required to raise a unit mass of homogeneous material one unit in temperature along a specified path, provided that during the process no phase or chemical changes occur, is known as the heat capacity of the material. Moreover, the path is so restricted that the only work effects are those necessarily done on the surroundings to cause the change to conform to the specified path. The path is usually at either constant pressure or constant volume.

In accordance with the first law of thermodynamics, heat capacity at constant pressure $C_p$ is equal to the rate of change of enthalpy with temperature at constant pressure $(\partial H/\partial T)_p$. Heat capacity at constant volume $C_v$ is the rate of change of internal energy with temperature at constant volume $(\partial U/\partial T)_v$. Moreover, for any material, the first law yields the relation shown in the equation below.

$$C_p - C_v = \left[P + \left(\frac{\partial U}{\partial V}\right)\right]_T \left(\frac{\partial U}{\partial T}\right)_P$$

*See* ENTHALPY; INTERNAL ENERGY; THERMODYNAMIC PRINCIPLES. [H.C.W.]

**Heavy water** A form of water in which the hydrogen atoms of mass 1 ($^1H$) ordinarily present in water are replaced by deuterium (D or $^2H$), the heavy stable isotope of hydrogen of mass 2. The molecular formula of heavy water is $D_2O$ (or $^2H_2O$). *See* DEUTERIUM.

Because the mass difference between $^1H$ and $^2H$ is the largest for any pair of stable (nonradioactive) isotopes in the periodic table, many of the physical and chemical properties of the pure isotopic species and their respective compounds differ to a significant extent. Selected physical properties of $^1H_2O$ and $^2H_2O$ are compared in the table.

**Physical properties of ordinary and heavy water**

| Property | $^1H_2O$ | $^2H_2O$ ($D_2O$) |
|---|---|---|
| Molecular weight, $^{12}C$ scale | 18.015 | 20.028 |
| Melting point, °C | 0.00 | 3.81 |
| Normal boiling point, °C | 100.00 | 101.42 |
| Temperature of maximum density, °C | 3.98 | 11.23 |
| Density at 25°C, g/cm$^3$ | 0.99701 | 1.1044 |
| Critical constants | | |
| Temperature, °C | 374.1 | 371.1 |
| Pressure, mPa | 22.12 | 21.88 |
| Volume, cm$^3$/mol | 55.3 | 55.0 |
| Viscosity at 55°C, mPa · s | 0.8903 | 1.107 |
| Refractive index, $n_D^{20}$ | 1.3330 | 1.3283 |

Heavy water, judging from its higher melting and boiling points, its higher viscosity, and its surprisingly high temperature of maximum density, is a distinctly more structured liquid than is ordinary water. Heavy water is more extensively hydrogen-bonded, and the hydrogen bonds formed by $^{2}H$ are somewhat stronger than are those of $^{1}H$.

The only large-scale use of heavy water in industry is as a moderator in nuclear reactors. Small amounts of heavy water are used to grow fully deuterated organisms, which serve as a source of fully deuterated compounds of biological importance. These are finding increasing use in research techniques such as small-angle neutron scattering, in high-resolution nuclear magnetic resonance spectroscopy of immobilized samples, and in the study of isotope effects. *See* WATER. [J.J.K.]

**Helium** A gaseous chemical element, He, atomic number 2 and atomic weight 4.0026. Helium is one of the noble gases in group 18 of the periodic table. It is the second lightest element. The world's chief source of helium is a group of natural gas fields in the United States. *See* INERT GASES; PERIODIC TABLE.

Helium is a colorless, odorless, and tasteless gas. It has the lowest solubility in water of any known gas. It is the least reactive element and forms essentially no chemical compounds. The density and the viscosity of helium vapor is very low. Thermal conductivity and heat content are exceptionally high. Helium can be liquefied, but its condensation temperature is the lowest of any known substance. The properties of helium are given in the table.

**Properties of helium**

| Property | Value |
|---|---|
| Atomic number | 2 |
| Atomic weight | 4.0026 |
| Melting point* at 25.2 atm pressure | −272.1°C (1.1 K) |
| Triple point (solid, helium I, helium II) | −271.37°C (1.78 K) |
| Triple point = λ-point (helium gas, helium I, helium II) | −270.96°C (2.19 K) |
| Boiling point at 1 atm pressure | −268.94°C (4.22 K) |
| Gas density at 0°C and 1 atm pressure, g/liter | 0.17847 |
| Liquid density at its boiling point, g/ml | 0.1249 |
| Solubility in water at 20°C, ml helium (STP)/1000 g water at 1 atm partial pressure of helium | 8.61 |

*The melting point varies with the pressure.

Helium was first used as a lifting gas in balloons and dirigibles. This use continues for high-altitude research and for weather balloons. The principal use of helium is in inert gas–shielded arc welding. The greatest potential for helium use continues to emerge from extreme-low-temperature applications. Helium is the only refrigerant capable of reaching temperatures below 14 K (−434°F). The chief value of ultralow temperature is the development of the state of superconductivity, in which there is virtually zero resistance to the flow of electricity. Other helium applications include use as a pressurizing gas in liquid-fueled rockets, in helium-oxygen breathing mixtures for divers, as a working fluid in gas-cooled nuclear reactors, and as a carrier gas for chemical analysis by gas chromatography.

Terrestrial helium is believed to be formed in natural radioactive decay of heavy elements. Most of this helium migrates to the surface and enters the atmosphere. The atmospheric concentration of helium (5.25 parts per million at sea level) could be expected to be higher. However, its low molecular weight permits helium to escape into space from the upper atmosphere at a rate roughly equal to its formation. Natural gases contain helium at concentrations higher than in the atmosphere. [A.W.F.]

Helium is an element with a closed electronic shell, a large ionization potential, and a low polarizability, which makes it a very unlikely candidate to form chemical bonds. However, solid helium compounds have been found to form at high pressure, one with nitrogen [$He(N_2)_{11}$] and one with neon [$Ne(He)_2$]. These compounds belong to a class known as van der Waals compounds. *See* INTERMOLECULAR FORCES.

Other helium compounds have also been observed in a clathrate hydrate, $He(H_2O)_{6+\delta}$, and helium has been detected inside the carbon molecule buckminsterfullerene ($C_{60}$), forming $HeC_{60}$. Mixtures of helium and other components prevail under conditions of high pressure in the outer planets of the solar system and their satellites. Therefore, it is believed that helium compounds are important in the modeling of the interiors of such celestial bodies. The formation of helium compounds at high pressures illustrates that under such conditions different chemical behavior occurs compared to that observed under ambient conditions. *See* CHEMICAL BONDING; CLATHRATE COMPOUNDS; FULLERENE. [W.L.V.]

Helium-3 is a rare stable isotope of helium was discovered by L. W. Alvarez and R. Cornog in 1939. Its concentration in nature is so low, approximately one part per hundred million in well helium, that it was 1951 before sufficient quantities of pure gas became available for experimentation. The gas was then, and continues to be, obtained as a by-product from the decay of tritium, the heavy isotope of hydrogen. Tritium is produced in a nuclear reactor from the reaction between lithium and a neutron.

The $^3He$ nucleus is composed of two protons and one neutron, one fewer than for $^4He$; as a consequence, $^3He$ is a fermion whereas $^4He$ is a boson. The two isotopes are the exemplars of Fermi-Dirac and Bose-Einstein systems, respectively. It is principally for this reason that helium, an apparently featureless chemical element, has been studied intensively. *See* BOSE-EINSTEIN STATISTICS; TRITIUM. [B.M.A.]

**Heparin** A highly sulfated mucopolysaccharide with blood anticoagulant activity, isolated from mammalian (chiefly beef) tissues. Heparin was first found in abundance in the liver, hence the name, but it is present in substantial amounts in the spleen, muscle, and lung as well. In the blood of most mammals, heparin is an antagonist to thrombin, prothrombin, and thromboplastin. It lessens the tendency of platelets to agglutinate. It is used in the treatment of venous thrombosis, embolism, myocardial infarction, and certain types of cerebral thrombosis. *See* POLYSACCHARIDE. [F.W.]

**Heterocyclic compounds** Cyclic compounds in which the rings include at least one atom of an element different from the rest. Most types of heterocyclic compounds studied to date are organic compounds. An example of an organic heterocyclic compound is oxazoline (**1**); an example of an inorganic heterocyclic compound is the phosphonitrilic chloride (**2**). The smallest possible ring is three-membered, for example, ethylene oxide (**3**), but very large rings are possible, as in the crown ethers, for example,

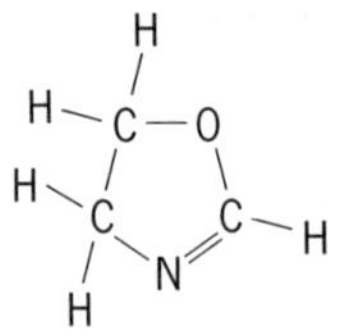

(1) 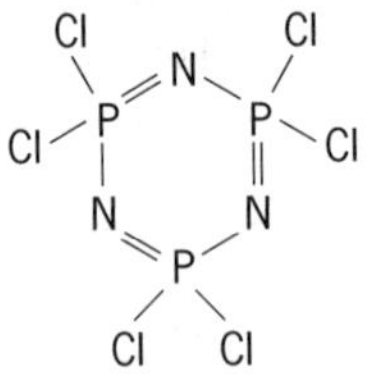 (2) 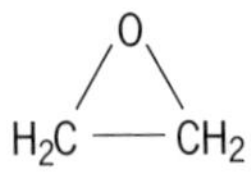 (3)

18-crown-6 (**4**). The cycle may contain only single bonds and is thus saturated; it may

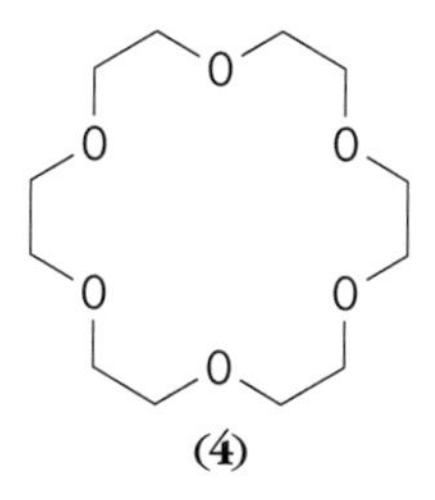

(4)

include one or more double bonds; or it may possess aromatic unsaturation characteristics of benzene, that is, it is heteroaromatic. Heterocyclic compounds can contain more than one ring, either heterocyclic or homocyclic.

Naturally occurring heterocyclic compounds are extremely common as, for example, most alkaloids, sugars, vitamins, DNA and RNA, enzymic cofactors, plant pigments, many of the components of coal tar, many natural pigments (such as indigo, chlorophyll, hemoglobin, and the anthocyanins), antibiotics (such as penicillin and streptomycin), and some of the essential amino acids (for example, tryptophan), and many of the peptides (such as oxytocin). Some of the most important naturally occurring high polymers are heterocyclic, including starch and cellulose. The major groups of natural products that are not mainly heterocyclic are the fats and most of the terpenes, steroids, and essential $\alpha$-amino acids, though exceptions do exist.

Heterocyclic compounds may be named systematically. Many heterocycles, however, have nonsystematic names that are usually preferred by practicing chemists over the systematic ones. In the systematic approach to nomenclature the ring size is denoted by the appropriate stem. For example, three-membered saturated rings without nitrogen would have a name ending in -irane. The nature of the heteroatom is denoted by such prefixes as oxa-, thia-, or aza-, for oxygen, sulfur, or nitrogen, respectively. Thus, ethylene oxide (**3**) becomes oxirane. A five-membered unsaturated ring would have a name ending in -ole. A six-membered unsaturated ring containing nitrogen would have a name ending in -ine according to this scheme. Actually, the trivial names for many systems are commonly accepted, and the systematic names are not often used.

For details about specific heterocyclic systems *see* FURAN; HETEROCYCLIC POLYMER; INDOLE; PYRIDINE; PYRIMIDINE; PYRROLE; THIOPHENE. [R.A.A.]

**Heterocyclic polymer** Essentially, linear high polymers comprising heterocyclic rings, or groups of rings, linked together by one or more covalent bonds. As the search has continued for polymeric materials having useful properties at high temperatures (500°C or 930°F, or higher), much attention has been given to heterocyclic polymers. As a group such polymers are often both mechanically rigid and inherently resistant to thermal degradation. *See* POLYMER.

Some of these polymers form molecules in which the rings are fused together, as shown symbolically in the illustration (ladder polymers), and some form molecules in which fused rings are joined by single bonds (stepladder polymers). Similar considerations hold for simple aromatic systems, for example, linear polymers of benzene, but the heterocyclic systems have, in general, been more useful in application.

Three heterocyclic polymers have been developed to the point of commercial availability; polyimides, polybenzimidazoles, and polybenzothiazoles. At least the first two appear to have established specialized markets.

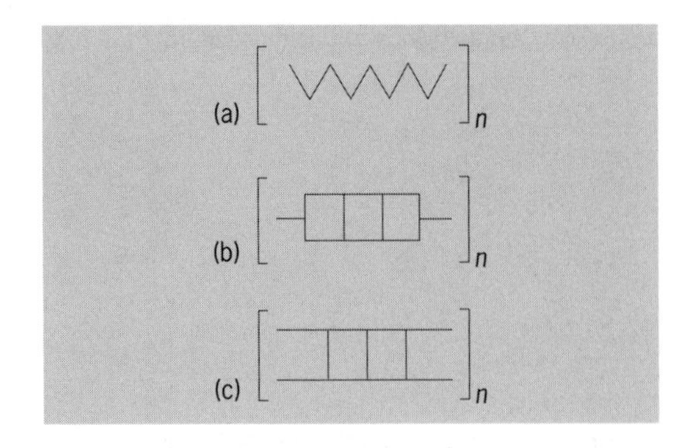


**Structural units in linear polymers. (*a*) Simple linear polymer. (*b*) Stepladder polymer. (*c*) Ladder polymer.**

Major applications for these rather expensive polymers are as metal-to-metal adhesives and as laminating resins for fibrous composites for structural applications in the aerospace industry. Other applications requiring both strength and resistance to oxidation at elevated temperatures have developed, including valve seats, bearings, and turbine blades. *See* POLYMERIC COMPOSITE. [J.A.M.]

**Heterogeneous catalysis** A chemical process in which the catalyst is present in a separate phase. In the usual case, the catalyst is a solid and the reactants and products are in gaseous or liquid phases. *See* CATALYSIS.

Heterogeneous catalysis proceeds by the formation and subsequent reaction of chemisorbed complexes which can be considered to be surface chemical compounds. A very simple case, where $A \rightarrow B$ is slow in the absence of catalyst, is shown in the following reaction:

$$\left.\begin{array}{l} {}^{*} + A \rightarrow {}^{*}A \\ A^{*} \rightarrow B^{*} \\ B^{*} \rightarrow B + {}^{*} \\ A \rightarrow B \end{array}\right\} \text{chain propagating steps}$$

Reaction $A \rightarrow B$ is fast if the three preceding steps are fast. Here, * represents a catalytic site on the surface of the catalyst, $A^{*} \rightarrow B^{*}$ is called a surface reaction, ${}^{*} + A \rightarrow {}^{*}A$ represents the chemisorption of A, and $B^{*} \rightarrow B + {}^{*}$ represents desorption of B. *See* ADSORPTION.

With most sets of reactants, more than one reaction will be thermodynamically possible. The degree to which a given catalyst favors one reaction compared with other possible reactions is called the selectivity of the catalyst for the reaction. Two aspects of a catalyst are of particular importance: its selectivity, and its activity, which can be taken as the rate of conversion of reactants by a given amount of catalyst under specified conditions. Ideally, the rate will be proportional to the amount of catalyst.

The first important heterogeneous catalytic process to be used in the chemical industry was the manufacture of sulfuric acid from sulfur trioxide by the contact process in 1875. By the 1950s, heterogeneous catalytic processes had come to dominate the petroleum, petrochemical, and chemical industries. Today, about 70% of the crude oil refined in the United States is exposed to at least one heterogeneous catalytic process. Heterogeneous catalysis is a critical feature in energy conservation and interconversion, and is a key feature in the production of synthetic fuels from coal and oil shale. *See* COAL GASIFICATION; FISCHER-TROPSCH PROCESS.

Since catalytic activity will ordinarily be proportional to surface area, most catalysts are used in forms with large specific areas. Higher-area metal powders are often used for liquid-phase reactions in batch reactors. For example, finely divided nickel is used for the hydrogenation of unsaturated glycerides in the manufacture of margarine from vegetable oils. Supported catalysts are widely used. In these, the catalytic ingredient is dispersed in the internal porosity of such supports as silica gel, $\gamma$-alumina, and charcoals. These supports have large areas in the internal porosity, and their average pore diameters are 2–20 nm. Supported catalysts have the advantage that the area of the catalytic ingredient can be very large.

One important type of catalyst exposes strongly acidic sites in its internal porosity. Such catalysts are used to crack larger molecules of hydrocarbon into smaller ones in petroleum refining. Other catalysts, called dual-functional catalysts, have a hydrogenating catalytic ingredient on an acidic support. These are also of major importance in processing petroleum. *See* CRACKING; HYDROCRACKING. [R.L.Bu.; G.L.H.]

**High-pressure chemistry** Chemistry at very high pressures, arbitrarily chosen to be above $10^4$ bars (1 gigapascal), and mainly concerned with solid and liquid states. At 25°C (77°F) and $10^4$ bars (1 GPa), nearly all ordinary gases are liquid or solid, and only a few liquids are not frozen; thus most high-pressure chemistry involves either higher temperatures, at which chemical reactions can occur at appreciable rates, or studies of internal arrangements in solids.

From 1 bar ($10^2$ kilopascals) to about $10^5$ bars (10 GPa), normal low-pressure chemical behavior prevails, and only minor departures from the usual valence and coordination rules are found. However, many interesting changes in materials can be effected in this pressure range as atoms are forced into new bonding arrangements. From $10^5$ to $10^9$ bars (10 to $10^5$ GPa), the energy added by compression becomes comparable with chemical bond energies, so that outer-shell electronic orbits are distorted and atoms and molecules change in character. A general tendency toward more metallic behavior is observed as the electrons become less strongly fixed to particular atoms, and chemical bonds may be broken. Upward of about $10^9$ bars ($10^5$ GPa), the delocalization of electrons is extensive, and the material consists of a mixture of ions and electrons, so that chemical bonds are of little importance. The boundaries on these three pressure ranges are, of course, only approximate, and show some variation according to the temperature and the atoms involved.

The simplest effect of high pressure is the closer compression of atoms. The noble gases and alkali metals are quite compressible, whereas most oxides and the stronger metals are considerably stiffer. However, at a pressure exceeding about $10^5$ bars (10 GPa), most of the easily compressed electronic clouds are tightened up, and the compressibilities of most substances approach each other.

Substances which consist of large molecules are easily stiffened or frozen by high pressures. The mobility of the molecules is sharply decreased by a sort of interlocking and tangling effect; thus for the substance to be sheared, chemical bonds must be broken, a process which requires considerable energy. This stiffening phenomenon limits the study of most reactions of organic molecules to low pressures because they are rather large and "freeze" easily, but yet are usually not stable enough to withstand the temperatures necessary for liquefaction or intermolecular reactions. [R.H.W.]

**High-pressure processes** Changes in the chemical or physical state of matter subjected to high pressure. The earliest high-pressure chemical process of commercial importance were the Haber synthesis of ammonia from hydrogen and nitrogen and the synthesis of diamonds from graphite. Raising the pressure on a system may result in several kinds of change. It causes a gas or vapor to become a liquid, a liquid to become a solid, a solid to change from one molecular arrangement to another, and a gas to dissolve to a greater extent in a liquid or solid. These are physical changes. A chemical reaction under pressure may proceed in such a fashion that at equilibrium more of the product forms than at atmospheric pressure; it may also take place more rapidly under pressure; and it may proceed selectively, forming more of the desired product among multiple possible products.

Pressures higher than that of the atmosphere are expressed in bars and kilobars as well as in other units. A bar is $10^5$ pascals, or $10^5$ newtons per square meter, which are the units for pressure in the International System of units. These units are too small for convenient use in high-pressure processes, hence the bar is used. The bar equals 0.9869 standard atmosphere, 760 mmHg.

Increasing the pressure on a gas or vapor compresses it to a higher density and so to a smaller volume. If the pressure exceeds the vapor pressure, the vapor will condense to a liquid which occupies a still smaller volume. A vapor may be condensed at a higher temperature when it is under pressure; this permits the use of cooling water to remove the latent heat instead of more costly refrigeration.

Solids also change from a less dense phase to a more dense phase under the influence of increases in pressure. The density of diamond is about 1.6 times greater than that of graphite because of a change in the spatial arrangement of the carbon atoms. The temperatures and pressures used in the commercial synthesis of diamond range up to 5000°F (3000 K) and 100,000 atm. A molten metal is required as a catalyst to permit the atomic rearrangement to take place at economical rates of conversion. Metals such as tantalum, chromium, and iron form a film between graphite and diamond.

In a manner similar to its effect during a physical change in which the volume of a system decreases, pressure also favors a chemical change where the volume of the products is less than the volume of the reactants. This is Le Chatelier's principle, which applies to systems in equilibrium. This general principle may be derived more precisely by thermodynamic reasoning, and thermodynamics is used to predict the effect of pressure on physical and chemical changes which lead to an equilibrium state.

Ammonia is formed according to the reaction shown below. At 1 atm only a fraction

$$N_2 + 3H_2 \rightleftharpoons 2NH_3$$

of 1% ammonia is formed. The ammonia content increases greatly when the pressure is raised. At 100 atm and 392°F (200°C) there would be about 80% ammonia at equilibrium. However, a very long time is required to form ammonia under these conditions, and consequently commercial processes operate at higher temperatures and pressures and use a catalyst to obtain higher rates of reaction. Many combinations of pressure and temperature have been used. The largest number of plants now operate in the region of 300 atm and 840–930°F (450–500°C). A higher-pressure process is carried out at about 1000 atm and 930–1200°F (500–650°C).

Methanol is synthesized from hydrogen and carbon monoxide at 200 atm and 600°F (315°C) in a similar manner. The catalyst contains aluminum oxide, zinc oxide, chromium oxide, and copper. Higher alcohols are produced at pressures of 200–1000 atm and temperatures up to 1000°F (538°C) with a similar catalyst to which potassium carbonate or chromate has been added.

Polyethylene has been produced at pressures in the ranges 3–4, 20–30, 40–60, and 1000–3000 atm. The last is probably the highest pressure yet used in the commercial synthesis of an organic chemical product. The ethylene is polymerized in a stainless steel tubular reactor at 375°F (191°C) with small amounts of oxygen as a catalyst.

Phenol can be formed from chlorobenzene mixed with 18% sodium hydroxide solution at a pressure of 330 atm. Pressure is employed in this instance to maintain the mixture in the liquid phase at a temperature high enough for the hydrolysis reaction to proceed at an acceptable rate.

Hydrocracking and hydrodesulfurization in the refining of gasoline and fuel oils are carried out at pressures up to 200 atm and temperatures of 800°F (427°C) and higher. *See* HYDROCRACKING; HYDROGENATION. [E.W.C.]

**High-temperature chemistry** The study of chemical phenomena occurring above 500 K (227°C or 440°F). High temperatures represent one of the important variables available to scientists for increasing the variety of possible chemical reactions over that expected for classical ground-state atoms and molecules. The relative population of excited rotational, vibrational, and electronic states can be enhanced by increasing the temperature and thus can effectively create new species and new mechanisms for reaction. The potentialities of this approach are well illustrated by the three laws of high-temperature chemisty: (1) At high temperatures everything reacts with everything. (2) The higher the temperature, the faster the reaction. (3) The products may be anything.

High temperatures also provide a common tie among the various options for energy production, conversion, or storage. For maximum thermodynamic efficiency, an energy production cycle should operate with a working fluid at as high a temperature as possible, and exhaust the spent fluid at as low a temperature as possible. Thus, in the combustion of coal to produce electric power or in the combustion of gasoline or diesel fuel to propel a car or an airplane, there is a need for materials of construction which allow operation of such devices at higher temperatures.

It is convenient to discuss temperatures in terms of energy and to note that 11,500 K (20,200°F) corresponds to 1 electronvolt. In this sense, the particles emitted by radioactive nuclei or accelerated in cyclotrons and synchrotrons, which have energies in the keV, MeV, and BeV ranges, are effectively at temperatures of $\sim 10^7$ K, $\sim 10^{10}$ K, and $\sim 10^{13}$ K, respectively, and "high-energy physics" is synonymous with "ultra-high-temperature chemistry."

Traditional high-temperature chemistry in the last several decades has been mainly concerned with phenomena in the range of 500–3000 K, although exotic flames can produce temperatures up to ~6000 K, shock waves can generate temperatures up to ~25,000 K, electric arcs can be operated in constricted modes to produce temperatures of ~50,000 K, and nuclear processes begin to occur at temperatures in the millions-of-degrees range. Laser excitation of selected energy states can produce species with effective temperatures in the range of $10^8$ K. [J.L.M.]

**Holmium** A chemical element, Ho, atomic number 67, atomic weight 164.93, a metallic element belonging to the rare-earth group. The stable isotope $^{165}$Ho makes up 100% of the naturally occurring element. The metal is paramagnetic, but as the temperature is lowered, it changes to antiferromagnetic and then to the ferromagnetic system. *See* PERIODIC TABLE; RARE-EARTH ELEMENTS. [F.H.Sp.]

**Homogeneous catalysis** A process in which a catalyst is in the same phase as the reactant. A homogeneous catalyst is molecularly dispersed (dissolved) in the reactants, which are most commonly in the liquid state. Catalysis of the transformation of organic molecules by acids or bases represents one of the most widespread types of homogeneous catalysis. In addition, the catalysis of organic reactions by metal complexes in solution has grown rapidly in both scientific and industrial importance. *See* CATALYSIS. [D.F.]

**Hydrate** A particular form of a solid compound which has water in the form of $H_2O$ molecules associated with it. For example, anhydrous copper sulfate is a white solid with the formula $CuSO_4$. When crystallized from water, a blue crystalline solid which contains water molecules as part of the crystals is formed. Analysis shows that the water is present in a definite amount, and the hydrate may be given the formula $CuSO_4 \cdot 5H_2O$. Four of the water molecules are attached to the copper ion in the manner of coordination complexes, and the fifth water molecule is related to the sulfate and presumably held by hydrogen bonding. *See* HYDROGEN BOND.

Water can also be present in definite proportions in the crystal without being associated directly with the anion or cation. The water occupies a definite place in the crystal lattice. Alums, with their 12 molecules of water, are examples of this. *See* ALUM.

Gas hydrates (gas clathrates) are crystalline compounds in which an isometric (cubic) ice ($H_2O$) lattice contains cages that incorporate small guest gas molecules. They are stable at moderate to high pressures and low temperatures, above and below the ice point. These ice lattices are stable only when the cages contain a gas molecule. The pressure and temperature constraints restrict them to oceanic continental margins in the uppermost few hundred meters of slope and rise sediments where water depths exceed 300–500 m (1000–1600 ft), and to permafrost in polar regions. Under the ocean, the amount of gas hydrates is at least an order of magnitude higher than in permafrost.

Methane ($CH_4$) hydrate is the dominant natural gas hydrate on Earth. One cubic meter of methane hydrate when dissociated can contain 165–180 $m^3$ of methane gas. The total amount of methane in gas hydrates is estimated to be very large; about $10^{19}$ g of methane carbon is stored in them, approximately twice that in fossil fuels.

Recent interest in natural gas hydrates, most of which are methane hydrates, has resulted from the recognition that global warming may destabilize the enormous quantities of methane hydrate in shallow marine slope sediments and permafrost. The environmental impact of releasing large quantities of methane into the ocean and atmosphere could have important consequences. The fossil fuel resource potential of the enormous quantities of marine methane hydrates is being evaluated. *See* METHANE. [F.W.; M.Ka.]

**Hydration** The incorporation of molecular water into a complex with the molecules or units of another species. The complex may be held together by relatively weak forces or may exist as a definite compound. Many salts form solid hydrates when exposed to water vapor under certain conditions of temperature and pressure. Water is lost from these compounds when they are heated or when the water vapor pressure falls below a minimum value. Solids forming hydrates at low pressures are used as drying agents. *See* DELIQUESCENCE; DESICCANT; EFFLORESCENCE; HYDRATE; SOLUTION; SOLVATION. [F.J.J.]

**Hydrazine** A colorless liquid, $H_2NNH_2$ (boiling point 114°C or 237°F), with a musty, ammonialike odor. Physically it is similar to water, but chemically it is reducing, decomposable, basic, and bifunctional. Its derivatives range from simple salts to ring compounds, polymers, and coordination complexes. Major uses of hydrazine include such diverse applications as rocket fuels, corrosion inhibition in boilers, syntheses of biologically active materials and in rubber curing and foam-rubber production. [T.H.D.]

**Hydride** The isolated atomic hydrogen anion, $H^-$. It consists of a singly charged positive nucleus and two electrons. The electron-electron repulsion almost overwhelms the nuclear-electron attraction. Thus, the "extra" electron is held weakly and is readily donated. Ionic salts containing this large and easily polarized ion are highly reactive, strongly basic, and powerfully reducing. This makes them important reagents despite the fact that they are readily destroyed by the presence of the relatively acidic compound water ($H_2O$) or by exposure to the relatively oxidizing dioxygen ($O_2$) as found in air. *See* ELECTRON CONFIGURATION.

The term hydride also refers to salts containing the $H^-$ anion and a highly electropositive alkali or alkaline-earth metal as the cation. The salt names reflect this high ionic character, for example, sodium hydride (NaH). In such salts the ionic radius of $H^-$ is comparable to that of $Cl^-$.

There are complex metal hydrides that are formed from the formal reaction of $H^-$ salts with some more covalent metal or metalloid hydrogen compound. Among the earliest to be investigated were lithium aluminum hydride ($LiAlH_4$) and sodium borohydride ($NaBH_4$).

There are compounds with metal-hydrogen bonds that are also referred to as hydrides. For example, hydridocarbonyl [$HCo(CO)_4$] and dihydridotetracarbonyl iron [$H_2Fe(CO)_4$] are often named cobalt tetracarbonyl hydride and iron tetracarbonyl dihydride by analogy to corresponding halides. *See* COORDINATION COMPLEXES; METAL HYDRIDES.

Ideally, the term hydride should be reserved for those species that contain $H^-$ or that at least formally transfer this ion to another substance in a so-called hydride transfer reaction. Such reactions are found in the industrial synthesis of 2,2,4-trimethylpentane (isooctane) from isobutylene and isobutane, as well as in many of the classical organic chemistry named reactions. Hydride transfer is also important in most of the biologically important oxidation-reduction reactions of the vitamin niacin (nicotinamide) as found in the forms of nicotinamide adenine dinucleotide/hydrogenated nicotinamide adenine dinucleotide ($NAD^+$/NADH) and nicotinamide adenine dinucleotide phosphate ($NADP^+$)/NADPH. [J.F.Li.]

**Hydrido complexes** Complex hydrides containing a hydride ligand bonded to a central atom. The prefix hydro instead of hydrido is sometimes used. All are soluble in aromatic hydrocarbons. They are rather expensive, but their specific reducing powers make them attractive for synthesizing high-value products, such as pharmaceuticals, flavorings, fragrances, dyes, and insecticides. *See* COORDINATION COMPLEXES; METAL HYDRIDES. [J.C.W.]

**Hydroboration** The process of producing organoboranes by the addition of diborane to unsaturated organic compounds. In ether solvents the addition of diborane to such molecules is exceedingly rapid and essentially quantitative. This reaction therefore makes the organoboranes readily available. Such organoboranes are finding increasing application as intermediates for organic synthesis.

Diborane is highly soluble in tetrahydrofuran, where it exists as the addition compound tetrahydrofuran-borane. Such solutions are often used for hydroboration, and merely involve bringing the two reactants together as indicated by reaction (1).

$$3RCH{=}CH_2 + C_4H_8O{:}BH_3 \rightarrow (RCH_2CH_2)_3B + C_4H_8O \qquad (1)$$

Alternatively, sodium borohydride may be utilized to achieve hydroboration by the addition of boron trifluoride etherate. This is shown by reaction (2). Usually the

$$12RCH{=}CH_2 + 3NaBH_4 + 4(C_2H_5)_2O{:}BF_3 \rightarrow 4(RCH_2CH_2)_3B + 3NaBF_4 + 4(C_2H_5)_2O \quad (2)$$

organoborane is not isolated but is utilized in place, similar to applications of the Grignard reagent in synthesis.

Organoboranes are among the most versatile synthetic intermediates that are available to the organic chemist. The hydroboration reaction has made these intermediates readily available. *See* BORANE. [H.C.B.]

**Hydrocracking** A catalytic, high-pressure process flexible enough to produce either of the two major light fuels—high octane gasoline or aviation jet fuel. It proceeds by two main reactions: adding hydrogen to molecules too massive and complex for gasoline and then cracking them to the required fuels. The process is carried out by passing oil feed together with hydrogen at high pressure (1000–2500 pounds per square inch gage or 7–17 megapascals) and moderate temperatures (500–750°F or 260–400°C) into contact with a bifunctional catalyst, comprising an acidic solid and a hydrogenating metal component. Gasoline of high octane number is produced, both directly and through a subsequent step such as catalytic reforming; jet fuels may also be manufactured simply by changing conditions with the same catalysts.

Generally, the process is used as an adjunct to catalytic cracking. Oils, which are difficult to convert in the catalytic process because they are highly aromatic and cause rapid catalyst decline, can be easily handled in hydrocracking, because of the low cracking temperature and the high hydrogen pressure, which decreases catalyst fouling. However, the most important components in any feed are the nitrogen-containing compounds, because these are severe poisons for hydrocracking catalysts and must be almost completely removed.

The products from hydrocracking are composed of either saturated or aromatic compounds; no olefins are found. In making gasoline, the lower paraffins formed have high octane numbers. The remaining gasoline has excellent properties as a feed to catalytic reforming, producing a highly aromatic gasoline which, with added lead, easily attains 100 octane number. Another attractive feature of hydrocracking is the low yield of gaseous components, such as methane, ethane, and propane, which are less desirable than gasoline.

The hydrocracking process is being applied in other areas, notably, to produce lubricating oils and to convert very asphaltic and high-boiling residues to lower-boiling fuels. *See* AROMATIZATION; CRACKING; HYDROGENATION. [C.P.B.]

**Hydroformylation** An aldehyde synthesis process that falls under the general classification of a Fischer-Tropsch reaction but is distinguished by the addition of an olefin feed along with the characteristic carbon monoxide and hydrogen. In the oxo process for alcohol manufacture, hydroformylation of olefins to aldehydes is the first step. The second step is the hydrogenation of the aldehydes to alcohols. At times the term "oxo process" is used in reference to the hydroformylation step alone. In the hydroformylation step, olefin, carbon monoxide, and hydrogen are reacted over a cobalt catalyst to produce an aldehyde which has one more carbon atom than the feed

$$R{-}CH{=}CH_2 + CO + H_2 \xrightarrow{Co} R{-}CH_2{-}CH_2{-}CHO$$

$$R{-}CH{=}CH_2 + CO + H_2 \xrightarrow{Co} R{-}CH(CHO){-}CH_3$$

olefin. As in the reaction above, the olefin conversion takes place by the addition of a formyl group (CHO) and a hydrogen atom across the double bond. *See* FISCHER-TROPSCH PROCESS.

The aldehyde is then treated with hydrogen to form the alcohol. In commercial operations, the hydrogenation step is usually performed immediately after the hydroformylation step in an integrated system.

A wide range of carbon-number olefins, $C_2$—$C_{16}$, have been used as feeds. Propylene, heptene, and nonene are frequently used as feedstocks to produce normal and isobutyl alcohol, isooctyl alcohol, and primary decyl alcohol, respectively. Feed streams to oxo units may be single-carbon-number or mixed-carbon-number olefins.

The lower-carbon-number alcohols such as butanols are used primarily as solvents, while the higher-carbon-number alcohols go into the manufacture of plasticizers, detergents (surfactants), and lubricants. [D.L.H.]

**Hydrogen** The first chemical element in the periodic system. Under ordinary conditions it is a colorless, odorless, tasteless gas composed of diatomic molecules, $H_2$. The hydrogen atom, symbol H, consists of a nucleus of unit positive charge and a single electron. It has atomic number 1 and an atomic weight of 1.00797. The element is a major constituent of water and all organic matter, and is widely distributed not only on the Earth but throughout the universe. There are three isotopes of hydrogen: protium, mass 1, makes up 99.98% of the natural element; deuterium, mass 2, makes up about 0.02%; and tritium, mass 3, occurs in extremely small amounts in nature but may be produced artificially by various nuclear reactions. *See* DEUTERIUM; PERIODIC TABLE; TRITIUM.

**Physical properties.** Ordinary hydrogen has an atomic weight of 1.00797, and a molecular weight of 2.01594. The gas has a density at 0°C (32°F) and 1 atm ($10^5$ pascals) of 0.08987 g/liter ($5.610 \times 10^{-3}$ lb/ft$^3$). Its specific gravity, compared to air, is 0.0695. The lightest substance known, it has a buoyancy in air of 1.203 g/liter ($7.510 \times 10^{-2}$ lb/ft$^3$) [see table].

Hydrogen dissolves in water to the extent of 0.0214 volume per volume of water at 0°C (32°F), 0.018 volume at 20°C (68°F), and 0.016 volume at 50°C (122°F). It is

**Properties of hydrogen***

| Property | Value |
|---|---|
| Melting point | −259.34°C |
| Boiling point at 1 atm | −252.87°C |
| Density of solid at −259.34°C | 0.0858 g/cm$^3$ |
| Density of liquid at −252.87°C | 0.0708 g/cm$^3$ |
| Critical temperature | −240.17°C |
| Critical pressure | 12.8 atm |
| Critical density | 0.0312 g/cm$^3$ |
| Specific heat at constant pressure | |
| Gas at 25°C | 3.42 cal/(g)(°C) |
| Liquid at −256°C | 1.93 cal/(g)(°C) |
| Solid at −259.8°C | 0.63 cal/(g)(°C) |
| Heat of fusion at −259.34°C | 13.9 cal/g |
| Heat of vaporization at −252.87°C | 107 cal/g |
| Thermal conductivity at 25°C | 0.000444 cal/(cm)(s)(°C) |
| Viscosity at 25°C | 0.00892 centipoise |

*The low-temperature properties all refer to hydrogen which has the parahydrogen concentration corresponding to its low-temperature equilibrium value.

somewhat more soluble in organic solvents, and 0.078 volume dissolves in 1 volume of ethanol at 25°C (77°F). Many metals absorb hydrogen. Palladium is particularly notable in this respect, and dissolves about 1000 times its volume of the gas. The adsorption of hydrogen in steel may cause "hydrogen embrittlement," which sometimes leads to the failure of chemical processing equipment.

The hydrogen atom has an ionization potential of 13.54 volts. The hydrogen nucleus (proton, mass 1) has a spin of $\frac{1}{2}\hbar$ and a magnetic moment of 2.79270 nuclear magnetons. Its absorption cross section for thermal neutrons is $0.332 \times 10^{-24}$ cm$^2$.

**Chemical properties.** At ordinary temperatures hydrogen is a comparatively unreactive substance unless it has been activated in some manner, for example, by a suitable catalyst. At elevated temperatures it is highly reactive.

Although ordinarily diatomic, molecular hydrogen dissociates at high temperatures into free atoms. Atomic hydrogen is a powerful reducing agent, even at room temperature. It reacts with the oxides and chlorides of many metals, including silver, copper, lead, bismuth, and mercury, to produce the free metals. It reduces some salts, such as nitrates, nitrites, and cyanides of sodium and potassium, to the metallic state. It reacts with a number of elements, both metals and nonmetals, to yield hydrides such as sodium hydride (NaH), potassium hydride (KH), and phosphorus hydride ($PH_3$). Sulfur forms a number of hydrides; the simplest is $H_2S$. With oxygen atomic hydrogen yields hydrogen peroxide, $H_2O_2$. With organic compounds atomic hydrogen reacts to produce a complex mixture of products. With ethylene, $C_2H_4$, for example, the products include ethane, $C_2H_6$, and butane $C_4H_{10}$. The heat liberated when hydrogen atoms recombine to form hydrogen molecules is used to obtain very high temperatures in atomic hydrogen welding.

Hydrogen reacts with oxygen to form water. At room temperature this reaction is immeasurably slow, but is accelerated by catalysts, such as platinum, or by an electric spark.

With nitrogen, hydrogen reacts to give ammonia. Hydrogen reacts at elevated temperatures with a number of metals, including lithium, sodium, potassium, calcium, strontium, and barium, to give hydrides. *See* METAL HYDRIDES.

**Principal compounds.** Hydrogen is a constituent of a very large number of compounds containing one or more other elements. Such compounds include water, acids, bases, most organic compounds, and many minerals. Compounds in which hydrogen is combined with a single other element are commonly referred to as hydrides. These may be divided into three general classes: the ionic or saltlike hydrides, the covalent or molecular hydrides, and the transition metal hydrides. *See* HYDRIDE.

**Preparation.** A large number of methods may be used to prepare hydrogen gas. The choice of method is determined by such factors as the quantity of hydrogen desired, the purity required, and the availability and cost of raw materials. Among the processes frequently used are the reactions of metals with water or acides, the electrolysis of water, the reaction of steam with hydrocarbons or other organic materials, and the thermal decomposition of hydrocarbons.

**Uses.** The largest single use of hydrogen is in the synthesis of ammonia. Ammonia plants are often built adjacent to petroleum refineries or coking plants to utilize by-product hydrogen that might otherwise be wasted. An important use for hydrogen is in petroleum-refining operations, such as hydrocracking and hydrogen treatment for removal of sulfur. Large quantities of hydrogen are consumed in the catalytic hydrogenation of unsaturated liquid vegetable oils to make solid fats. Hydrogenation is used in the manufacture of organic chemicals, such as alcohols from esters and glycerides, amines from nitriles, and cycloparaffins from aromatic hydrocarbons. Methanol is produced commercially by reaction of hydrogen with carbon monoxide. Reaction of hydrogen with chlorine is a major source of hydrochloric acid. [L.K.]

**Liquid hydrogen.** Liquid hydrogen, a clear colorless fluid which boils at −252.87°C (−423.17°F) and has the smallest boiling point density of any known liquid, was first produced in 1898. James Dewar was successful with this experiment because his development of the glass vacuum flask enabled him to retain this low-boiling-point liquid. For many years, hydrogen was liquefied by using high-pressure processes that are dangerous, since small air (oxygen) impurities in the hydrogen can result in an explosion. Modern liquefiers, however, generally use closed-circuit helium refrigeration cycles which condense the hydrogen at low pressure, and the hazards are considerably reduced.

The density of liquid hydrogen is roughly 790 times greater than that of the gas under normal conditions, so quantities of hydrogen in liquid form are transported conveniently and economically in relatively light-weight vacuum-jacketed containers that are open to the atmosphere. The alternative, to compress the gas at room temperature into heavy-walled steel containers, involves much less shipping efficiency. The space program uses large amounts of hydrogen for rocket fuels, in conjunction with oxygen or fluorine, and here the ability to transport hydrogen in liquid form with minimal insulation has great advantages. Liquid (and solid) hydrogen (and its heavier isotopes, deuterium and tritium) can be used for the production of electrical power by nuclear fusion. [C.A.Sw.]

**Spin-polarized atomic hydrogen.** Spin-polarized atomic hydrogen is a special preparation of the element which is expected to remain a gas down to the absolute zero of temperature. It is created by dissociating the hydrogen molecule ($H_2$), exposing the atoms to a high magnetic field which aligns the electronic spins, and storing them at temperatures so low that thermal effects are unlikely to disturb the spin alignment. Typically, spin-polarized hydrogen is stored at 0.3 K in a magnetic field of 8 tesla.

Spin-polarized hydrogen has been used to study the quantum theory of weakly interacting gases and has had applications in the areas of polarized proton sources for high-energy physics and cryogenic hydrogen masers. The ultimate goal is to observe the Bose-Einstein phase transition to a state in which a finite fraction of the atoms in the gas essentially come to rest. The gas in this state is also expected to be a superfluid. The transition occurs when the temperature of the gas is so slow that the quantum-mechanical wavelength associated with the particles becomes comparable to the interparticle spacing. [T.J.Gr.]

**Hydrogen bond** The interaction which occurs when a hydrogen atom, covalently bonded to an electronegative atom (as in A—H), interacts with another atom to form the aggregate A—H ··· Y. The shortest and strongest bond is indicated as A—H, while the secondary and weaker interaction is written as H ··· Y. Thus A—H is a proton donor, while (Y) is a proton acceptor which often contains lone pair electrons and can act as a base. The strongest hydrogen bonds are formed between the most electronegative (A) atoms such as fluorine, nitrogen, and oxygen which interact with (Y) atoms having electronegativity greater than that of hydrogen (C, N, 0, S, Se, F, Cl, Br, I). The weakest of hydrogen bonds are formed by acidic protons of C—H groups, as in chloroform and acetylene, and by olefinic and aromatic $\pi$-electrons acting as (Y).

The weaker the hydrogen bond, the shorter the lifetime of the complex it forms. An important aspect of weak hydrogen bond formation is that the different molecular aggregates which do form can be easily and reversibly transformed. Thus the small energy changes resulting in the rapid making and breaking of hydrogen bonds in biological systems are of great importance; for example, hydrogen bonding determines the configuration of the famous α-helix of DNA, and the structures of most proteins, thereby serving an important function in determining the nature of all living things. *See* CHEMICAL BONDING; PROTEIN. [J.M.Wi.]

**Hydrogen fluoride** The hydride of fluorine and the first member of the family of halogen acids. Anhydrous hydrogen fluoride is a mobile, colorless liquid that fumes strongly in air. It has the empirical formula HF, melts at −83°C, and boils at 19.8°C. The vapor is highly aggregated, and gaseous hydrogen fluoride deviates from perfect gas behavior to a greater extent than any other gaseous substance known. Aggregate formation in both the vapor and liquid phase arises from unusually strong hydrogen-bond interactions. *See* HYDROGEN BOND.

Anhydrous hydrogen fluoride is an extremely powerful acid, exceeded in this respect only by 100% sulfuric acid. Because anhydrous hydrogen fluoride is a superacid, many organic solutes dissolve in it to form stable carbonium ions. Alkali metal fluorides and silver fluoride dissolve readily in hydrogen fluoride to form conducting solutions. Anhydrous hydrogen fluoride dissolves a wide variety of organic compounds. Aqueous solutions of hydrogen fluoride (hydrofluoric acid) are relatively weakly acidic as compared to hydrochloric acid.

Hydrogen fluoride is a widely used industrial chemical. The largest use is in making fluorine-containing refrigerants (Freons, Genetrons). An increasingly important use of hydrogen fluoride is in the preparation of organic fluorocarbon compounds.

Both hydrogen fluoride and hydrofluoric acid cause unusually severe burns; appropriate precautions must be taken to prevent any contact of the skin or eyes with either the liquid or the vapor. *See* HALOGENATED HYDROCARBON. [J.J.K.]

**Hydrogen ion** A proton combined with a number of water molecules. It is often written as $H_3O^+$ and called the hydronium ion. However, this species is best considered as an excess proton on a tetrahedral group of four water molecules and so would be designated as $H_9O_4{}^+$. For simplicity, it is most commonly written as $H^+(aq)$.

Since it is formed by the self-ionization of water, the hydrogen ion is present in all aqueous solutions. This formation also means that $H^+(aq)$ is always found in the company of the hydroxide ion, $OH^-(aq)$. The equilibrium relationship between the concentrations of these two species is a very important property of water. *See* IONIC EQUILIBRIUM.

The $H^+(aq)$ and $OH^-(aq)$ concentrations in pure water are equal to each other with a value of $10^{-7}$ mole/liter. Any aqueous solution with this concentration of $H^+(aq)$ is called a neutral solution. If the $H^+(aq)$ concentration is greater than $10^{-7}$ mole/liter, the solution is called acidic. Basic solutions are those in which the $H^+(aq)$ concentration is less than $10^{-7}$ mole/liter. As the $H^+(aq)$ concentration increases the $OH^-(aq)$ concentration must decrease, and vice versa. In the most straightforward system, acids are substances that can donate an $H^+(aq)$, and bases are substances that can accept one. *See* ACID AND BASE.

Hydrogen ion concentration determines the course of many chemical reactions that occur in living organisms and in the chemical industry. The control of hydrogen ion concentration is achieved in living organisms and in the laboratory by buffer systems. These are chemical mixtures designed to resist change in hydrogen ion concentration. *See* BUFFERS (CHEMISTRY).

Another property of the $H^+(aq)$ is important in both theoretical and practical ways. $H^+(aq)$ is the best conductor of electricity of any ion in aqueous solution. Its conductance at 77°F (25°C) is almost five times as large as the next-most-conducting ion. *See* ELECTROLYTIC CONDUCTANCE.

The hydrogen ion concentration can vary over fourteen powers of 10. To avoid dealing with such exponentials, the concept of pH is used. Since all aqueous solutions contain both hydrogen ion and hydroxide ion, it is possible to define all degrees of acidity and basicity on the pH scale. *See* PH.

Two general methods are used for the determination of hydrogen ion concentrations. For relatively crude work, colorimetric methods are commonly used. These methods depend on the fact that certain natural and synthetic dyes have colors that depend on the hydrogen ion concentration. At times, paper is impregnated with such an indicator. In most precise work, a potentiometric method is used for the determination of hydrogen ion concentration. This method depends on an electrode whose potential is sensitive to hydrogen ion concentration. The only electrode commonly in use for practical pH measurements is the glass electrode. *See* ACID-BASE INDICATOR; COLORIMETRY; TITRATION. [G.A.]

**Hydrogen peroxide** A binary compound of hydrogen and oxygen, empirical formula $H_2O_2$, used mostly in dilute aqueous solutions as an oxidizing agent. Its most remarkable feature is its tendency to decompose readily into water and oxygen.

Anhydrous hydrogen peroxide is a clear, colorless liquid, of nearly the same viscosity and dielectric constant as water, but of greater density. Like water, it is strongly associated through hydrogen bonds. It boils at 150°C (300°F) with violent, sometimes explosive decomposition. Decomposition by light begins only in the near ultraviolet. As a solvent, hydrogen peroxide resembles water, except that acids and bases show much lower electrical conductivity. Although a fairly strong oxidant, it can act as a mild reducing agent, for example, with permanganates and perchromates.

Hydrogen peroxide is used mainly for bleaching cotton and other fibers, natural or synthetic. Increasing amounts are used in the pulp and paper industry. Its well-known cosmetic use as hair bleach consumes relatively little of the commercial 10% (30 volume) solution. In medicine it is useful for cleansing wounds and cuts, although its antiseptic action is rather slow. A limited but important use of the concentrated peroxide is for energy production in rockets, submarines (during submersion), airplanes (at takeoff), and the steering of space vessels. *See* CHEMICAL FUEL.

Hydrogen peroxide, especially when concentrated, requires great care in handling and storing. When dropped on paper or wood, it can start a fire. Contact with the skin causes blotches that can be painful, but they disappear after a few hours without leaving traces. *See* HYDROGEN; OXYGEN; PEROXIDE. [P.A.G.]

**Hydrogenation** The chemical reaction of hydrogen with another substance, generally an unsaturated organic compound, and usually under the influence of temperature, pressure, and catalysts. There are several types of hydrogenation reactions. They include: (1) the addition of hydrogen to reactive molecules; (2) the incorporation of hydrogen accompanied by cleavage of the starting molecules (hydrogenolysis); and (3) reactions in which isomerization, cyclization, and so on, result.

Hydrogenation is synonymous with reduction in which oxygen or some other element (most commonly nitrogen, sulfur, carbon, or halogen) is withdrawn from, or hydrogen is added to, a molecule. When hydrogenation is capable of producing the desired reduction product, it is generally the simplest and most efficient procedure.

Hydrogenation is used extensively in industrial processes. Important examples are the synthesis of methanol, liquid fuels, hydrogenated vegetable oils, fatty alcohols from the corresponding carboxylic acids, alcohols from aldehydes prepared by the aldol reaction, cyclohexanol and cyclohexane from phenol and benzene, respectively, and hexamethylenediamine for the synthesis of nylon from adiponitrile. *See* DEHYDROGENATION; FISCHER-TROPSCH PROCESS; HYDROFORMYLATION; HYDROGEN; OXIDATION-REDUCTION.

[R.Le.]

**Hydrolysis** A chemical reaction in which splitting of a molecule by water occurs. Hydrolysis as applied to organic molecules can be considered a reversal of such reactions as esterification and amide formation. The hydrolysis of esters [reaction (1)] and of amides [reaction (2)] is shown. Other classes of organic compounds that are

$$\underset{\text{Ethyl acetate}}{CH_3COOC_2H_5} + H_2O \rightarrow \underset{\text{Acetic acid}}{CH_3COOH} + \underset{\text{Ethanol}}{C_2H_5OH} \qquad (1)$$

$$\underset{\text{Acetamide}}{CH_3CONH_2} + H_2O \rightarrow \underset{\text{Acetic acid}}{CH_3COOH} + \underset{\text{Ammonia}}{NH_3} \qquad (2)$$

subject to hydrolysis include acetals, acyl and alkyl halides, ketals, and peptides. While the overall hydrolysis reaction [as in reactions (1) and (2)] appears to involve the addition of the water molecule ($H_2O$), the reaction is in fact more complicated. There are several reaction steps, such as the formation of a complex with either a proton ($H^+$; acid-catalyzed hydrolysis) or a hydroxyl ion ($OH^-$; base-catalyzed hydrolysis), followed by elimination of these ions to give the overall equation. The hydrolysis reaction is frequently encountered in biological systems. The kinetics of these reactions are greatly enhanced by the action of enzymes (biological catalysts) such as the esterases and the peptidases. *See* ENZYME; HYDROGEN ION.

In inorganic chemistry, hydrolysis, also called aquation, represents a class of reactions involving metal coordination complexes in which one of the coordinated ligands is displaced by either $H_2O$ or $OH^-$. Hydrolysis is a special case of the class of reactions termed ligand displacement reactions [reaction (3), where M is a metal, L and X are

$$L_nMX + Y \rightarrow L_nMY + X \qquad (3)$$

ligands, and Y is the displacing ligand ($H_2O$ in hydrolytic reactions)]. *See* COORDINATION COMPLEXES; LIGAND.

[H.Frei.]

**Hydrolytic processes** Reactions of both organic and inorganic chemistry wherein water effects a double decomposition with another compound, hydrogen going to one component, hydroxyl to another, as in reactions (1)–(3). Although the word

$$XY + H_2O \rightarrow HY + XOH \qquad (1)$$

$$KCN + H_2O \rightarrow HCN + KOH \qquad (2)$$

$$C_5H_{11}Cl + H_2O \rightarrow HCl + C_5H_{11}OH \qquad (3)$$

"hydrolysis" means decomposition by water, cases in which water brings about effective hydrolysis unaided are rare, and high temperatures and pressures are usually necessary. *See* HYDROLYSIS.

Hydrolytic reactions may be classified as follows: (1) hydrolysis with water alone; (2) hydrolysis with dilute or concentrated acid; (3) hydrolysis with dilute or concentrated alkali; (4) hydrolysis with fused alkali with little or no water at high temperature.

In the field of organic chemistry, the term "hydrolysis" has been extended to cover the numerous reactions in which alkali or acid is added to water. An example of an alkaline-condition hydrolytic process is the hydrolysis of esters, reaction (4), to produce

$$CH_3COOC_2H_5 + NaOH \rightarrow CH_3COONa + C_2H_5OH \qquad (4)$$

alcohol. An example of an acidic-condition process is the hydrolysis of olefin to alcohol in the presence of phosphoric acid, reaction (5). The addition of acids or alkalies hastens

$$C_2H_4 + H_2O \xrightarrow{H_3PO_4} C_2H_5OH \qquad (5)$$

such reactions even it it does not initiate the reaction.

Perhaps some of the oldest and largest-volume hydrolysis technology is involved in soap manufacture. In the first step, glyceryl stearate acid, a fat, is hydrolyzed with water to yield stearic acid and glycerin. In the second step, the stearic acid is neutralized with caustic soda to give sodium stearate, the soap, and water. *See* FAT AND OIL.

Hydrolytic processes account for a huge product volume. Conversion of starch such as corn starch into maltose and glucose (sugar syrups) by treatment with hydrochloric acid is a major industry. Similarly, the production of furfural from pen-tosans of oat hulls or other cereal by-products such as corn cobs, rice hulls, or cottonseed bran is another commercial hydrolytic process. [D.L.H.]

**Hydroxide** A compound containing the hydroxide ion ($OH^-$) and having the general formula $M(OH)_n$, where M represents a metal. Hydroxides are a subset of compounds containing the hydroxyl group (—OH) and range in chemical character from strongly basic, to amphoteric (having both acidic and basic characteristics), to essentially acidic. The hydroxide ion has a closed-shell electronic structure with a singlet ground state.

In the Lewis acid-base scheme, where a base is defined as an electron pair donor and an acid as an electron pair acceptor, a typical hydroxide decreases in base strength as attraction for electrons of the cation increases. For example, the hydroxides of electropositive elements such as the alkali metals and alkaline earths tend to be bases. When these ionic compounds are dissolved in water, they form metal ions and hydroxide ions, as in reaction (1), where S represents a solid and aq represents a water

$$M^+(OH^-)_n(S) + nH_2O \longrightarrow M^{n+}(aq) + nOH^-(aq) \qquad (1)$$

solution. The hydroxides of nonmetals (X) such as boron hydroxide [$B(OH)_3$], where the X-O bonds are covalent, are generally acidic, as shown in reaction (2), where $H_3O^+$

$$X(OH)_n + nH_2O \longrightarrow XO_n^{n-}(aq) + nH_3O^+(aq) \qquad (2)$$

is the hydronium ion. The amphoteric hydroxides may dissociate by either mechanism, depending on the presence of strong acids or bases. *See* ACID AND BASE; HYDROGEN ION.

The alkali metal hydroxides such as sodium hydroxide (NaOH) are extremely important as reagents in metallurgy and photography and in the manufacture of soaps and detergents. Calcium hydroxide [$Ca(OH)_2$], known as slaked lime, is used in the preparation of mortar for brick laying. Minerals such as brucite [$Mg(OH)_2$] and pyrochroite [$Mn(OH)_2$] are naturally occurring hydroxides. [T.J.Me.]

**Hydroxyl** A chemical group in which oxygen and hydrogen are bonded and act as a single entity. In inorganic chemistry the hydroxyl group is known as the hydroxide ion ($OH^-$), and it is frequently bonded to metal cations, for example, sodium hydroxide (NaOH). In organic chemistry it frequently acts as a functional group, for example, in an alcohol (ROH, where R represents an alkyl group). *See* ACID AND BASE.

Many of the intermediate redox forms of dioxygen are toxic and damage important biomolecules. Much of this toxicity is thought to involve the generation and reactivity of hydroxyl ($\cdot OH$), which is sometimes called the hydroxy radical. The most common means for producing hydroxyl is the reaction of a reducing agent with hydrogen peroxide ($H_2O_2$). Transition-metal ions, such as ferrous ion ($Fe^{2+}$), are the most common reducing agents for generating $\cdot OH$.

Once generated, hydroxyl is a potent one-electron oxidant that forms the very stable $OH^-$ ion, and it abstracts hydrogen atoms from organic molecules that contain C—H bonds to form the stronger O—H bond in water. The reaction of a radical, which contains an uneven number of electrons, with a molecule, which contains an even number of electrons paired in bonds, must generate a radical, because the number of electrons cannot change during the reaction. Thus, most reactions of radicals generate new radicals in processes called radical chain reactions. Reactions of radicals with molecules will continue to produce new radicals until other odd-electron species (such as transition-metal ions or other radicals) react with the radicals to produce even-electron molecules via termination reactions. *See* Chain reaction (chemistry); Transition elements. [H.H.T.]

**Hypervalent compounds** Group 1, 2, and 13–18 compounds which contain a number ($N$) of formally assignable electrons of more than eight (octet) in a valence shell directly associated with the central atom (X) in direct bonding with a number of ligands ($L$). The designation $N$-X-$L$ is conveniently used to describe hypervalent molecules. *See* Electron configuration; Ligand; Valence.

In the periodic table, compounds of main group elements in the second row (such as carbon, nitrogen, and oxygen) have eight valence electrons. As such, the fundamental shapes of their atoms are linear (such as acetylene, sp orbital), triangular, (such as ethylene, $sp^2$ orbital), and tetrahedral (such as methane, $sp^3$ orbital). In contrast, main group elements in the third row of the periodic table (such as silicon, phosphorus, and sulfur) may contain more than eight electrons in a valence shell. These are called hypervalent compounds. Fundamental shapes of *10*-X-*5* molecules (including *10*-X-*4* molecules bearing a pair of unshared electrons) are trigonal bipyramid (TBP) or square pyramid (SP), and those of *12*-X-*6* (including *12*-X-*5* bearing a pair of unshared electrons) are octahedral. Hence, there is an apparent similarity in shape between hypervalent compounds and organotransition metal compounds. *See* Molecular orbital theory.

The hydrogen bond is one of the best examples of a hypervalent bond. The covalent nature of the hydrogen bond of the [N—H···N] system has recently been established experimentally and theoretically. The structure of $(HF)_5$, shown below, is pentagonal,

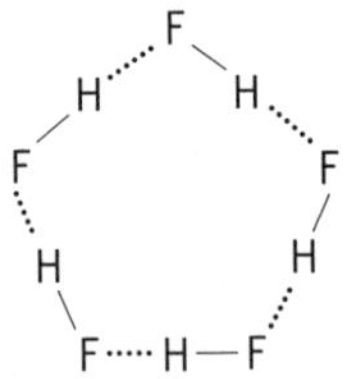

with a bond length for F—H of 1 Å and H···F of 1.5 Å. The hydrogen atom shifts rapidly between the two fluorines in the range of 0.5 Å. However, $[F—H—F]^-K^+$, (*4*-H-*2*), is linear, and the two F—H bond lengths are equal to 1.13 Å, which is typical for a hypervalent bond. *See* Chemical bonding; Hydrogen bond.

In order to accept extra electrons in a valence shell, an electron-rich and polarized sigma bond contains an apical bond (three-center, four-electron bond, which is defined as a hypervalent bond by molecular orbital theory) on the central atom. One of the most unstable hypervalent molecules is $[F—F—F]^-Li^+$, (*10*-F-*2*). It is linear and the F—F bond is calculated to be 1.701 Å, which is elongated from that of F—F

(1.412 Å) by accepting a fluoride ion. This is essentially the same as that of $[F—H—F]^{-}Li^{+}$, (4-F-2). [K.Ak.]

**Hypohalous acid** An oxyacid of a halogen [fluorine (F), chlorine (Cl), bromine (Br), iodine (I), or astatine (At)] possessing the general chemical formula HOX, where X is the halogen atom. The chemical behavior of hypofluorous acid (HOF) is dramatically different from the heavier hypohalous acids which, as a group, exhibit similar properties. These differences are attributed primarily to the high electronegativity and small size of the fluorine atom, which cause HOF to be an extremely strong oxidant with an anomalously weak O-F bond. Thus, the molecule is highly reactive and relatively unstable. (Gaseous HOF decomposes to HF and $O_2$ at room temperature with a half-life of about 1 h, and the liquid has a tendency to explode.) Because the most electronegative element in HOF is fluorine, whereas the other halogen atoms are less electronegative than oxygen, the O-X bond polarities are reversed in HOF and the heavier congeners. HOF therefore acts primarily as an oxygenating and hydroxylating agent, whereas the other hypohalous acids are electrophilic halogenating agents. For example, HOF hydroxylates aromatic compounds to form phenols and reacts instantaneously with water to give hydrogen peroxide ($H_2O_2$), whereas hypochlorous acid (HOCl) chlorinates aromatic compounds and is unreactive toward water. *See* ASTATINE; HALOGEN ELEMENTS.

Hypochlorous, hypobromous, and hypoiodous acid solutions are formed by disproportionation of the corresponding halogen, as in the reaction below.

$$X_2 + H_2O \rightleftharpoons H^+ + X^- + HOX$$

Hypohalous acids are weak acids, that is, $HOX \rightleftharpoons H^+ + OX^-$ acids whose dissociation constants increase in the order HOI < HOBr < HOCl. The decomposition mechanisms are complex with rates that vary widely, increasing in the order $X = Cl^- < Br^- < I^-$. Hypochlorite solutions (for example, commercial bleach) are stable at room temperature and below for months but decompose at elevated temperatures; hypobromite solutions are stable only at reduced temperatures; and hypoiodite generally decays within minutes of its formation.

HOCl and HOBr and their anions are powerful oxidants that react rapidly with a wide range of organic and inorganic reductants. In addition, both acids halogenate aromatic compounds and form halohydrins with unsaturated organic compounds, and N-chloro compound with nitrogen bases. In general, HOCl reacts much more rapidly than $OCl^-$.

HOCl and HOBr are potent cytotoxins. Hypochlorite was first used as a disinfectant around the beginning of the nineteenth century, and has subsequently been widely applied to problems in public sanitation. Recent studies with bacteria indicate that death is accompanied by disruption of metabolic functions associated with the plasma membrane, including inactivation of the adenosine triphosphate synthase and proteins involved with active transport of metabolites, and (in aerobes) inhibition of respiration. This pattern of reactivity suggests that toxicity arises from disruption of the energy-transducing capabilities of the cell. [J.K.Hu.]

# I

**Indium** A chemical element, In, atomic number 49, a member of group 13 and the fifth period of the periodic table. Indium has a relative atomic weight of 114.82.

Indium occurs in the Earth's crust to the extent of about 0.000001% and is normally found in concentrations of 0.1% or less. It is widely distributed in many ores and minerals but is largely recovered from the flue dusts and residues of zinc-processing operations.

Indium is used in soldering lead wires to germanium transistors and as a component of the intermetallic semiconductor used for germanium transistors. Indium arsenide, antimonide, and phosphide are semiconductors with unique properties. Other uses of indium are sleeve-type bearings to reduce corrosion and wear, glass-sealing alloys, and dental alloys. *See* GERMANIUM; PERIODIC TABLE. [E.M.L.]

**Indole** The parent compound of a group of organic heterocyclic compounds containing the indole nucleus, which is a benzene ring fused to a pyrrole ring as in indole itself (**1**). The importance of the indole ring lies in its presence in a large number of naturally occurring compounds. *See* PYRROLE.

Indole can exist in two tautomeric forms, the more stable enamine form (**1**) and the 3-H-indole or imine form (**2**). Unsubstituted 3-H-indoles (sometimes called indolenines) and a structural isomer of indole, isoindole (**3**), are not stable, but have been shown to be reaction intermediates. They are isolable when properly substituted.

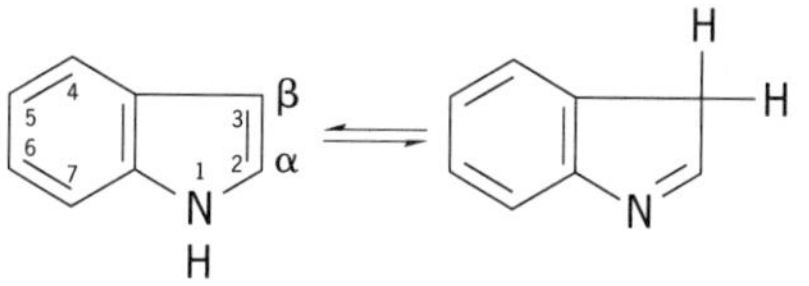

(1) (2)

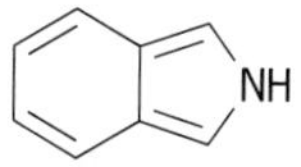

(3)

Indole (**1**) is a steam-volatile, colorless solid, melting point 52.5°C (126.5°F), boiling point 253°C (487°F). It is found in small amounts in coal tar, feces, and flower oils. Despite the presence of nitrogen, indole is not basic in the sense that it dissolves in aqueous acid or turns litmus blue. In fact, the hydrogen on the nitrogen is about as acidic as an aliphatic alcohol hydrogen. Indole is an aromatic compound and undergoes electrophilic substitutions much like benzene, although it is much more reactive than benzene. Its reactivity is comparable to that of phenol, and it undergoes a number of reactions similar to those of phenol. Indole itself reacts slowly with air and rapidly with most oxidizing agents to give intractable polymeric tars.

The indole alkaloids are a large group of substances containing the indole nucleus and can be isolated from plants. They contain a number of physiologically active materials, such as strychnine, reserpine, some forms of curare poison, and the rye-fungus drug, ergot. Lysergic acid diethylamide (LSD) is a synthetic, and not a naturally occurring, substance. There are many hundreds of other indole alkaloids. *See* ALKALOID; HETEROCYCLIC COMPOUNDS. [J.M.Bo.]

**Inert gases** The inert gases, listed in the table, constitute group 18 of the periodic table of the elements. They are now better known as the noble gases, since stable compounds of xenon have been prepared. The noble gases are all monatomic.

**The inert gases**

| Name | Symbol | Atomic number | Atomic weight |
|---|---|---|---|
| Helium | He | 2 | 4.0026 |
| Neon | Ne | 10 | 20.183 |
| Argon | Ar | 18 | 39.948 |
| Krypton | Kr | 36 | 83.80 |
| Xenon | Xe | 54 | 131.30 |
| Radon | Rn | 86 | (222) |

All these gases occur to some extent in the Earth's atmosphere, but the concentrations of all but argon are exceedingly low. Argon is plentiful, constituting almost 1% of the air.

All the gases are colorless, odorless, and tasteless. They are all slightly soluble in water, the solubility increasing with increasing molecular weight. They can be liquefied at low temperatures, the boiling point being proportional to the atomic weight. All but helium can be solidified by reducing the temperature sufficiently, and helium can be solidified at temperatures of less than 2°F above absolute zero (0–1 K) by the application of an external pressure of 25 atm (2.5 megapascals) or more. *See* ARGON; HELIUM; KRYPTON; NEON; RADON; XENON. [A.W.F.]

**Inhibitor (chemistry)** A substance which is capable of stopping or retarding a chemical reaction. To be technically useful, such compounds must be effective in low concentrations, usually under 1%. The type of reaction which is most easily inhibited is the free-radical chain reaction. Vinyl polymerization and autoxidation are two important examples of the class. Another reaction type for which inhibitors have been found is corrosion, particularly in aqueous systems. *See* ANTIOXIDANT; FREE RADICAL; POLYMERIZATION. [L.R.M.]

**Inorganic chemistry** The chemical reactions and properties of all the elements in the periodic table and their compounds, with the exception of the element carbon. The chemistry of carbon and its compounds falls in the domain of organic chemistry. The boundaries of inorganic chemistry with the other major areas of chemistry are not precisely defined, and it is often a matter of taste as to whether a particular topic is to be included in the field of inorganic chemistry or is to be considered physical or even organic chemistry. Investigations into theoretical inorganic chemistry or the study of problems in inorganic chemistry by quantitative and sophisticated physical methods may be considered either inorganic or physical chemistry quite arbitrarily. In similar fashion, organometallic compounds may be considered to be in the sphere of either inorganic or organic chemistry. To an increasing extent, the inorganic chemist is concerned with problems that once were considered the prerogative of physical chemists, organic chemists, or even biochemists. *See* BIOINORGANIC CHEMISTRY; CHEMICAL DYNAMICS; COORDINATION CHEMISTRY; NUCLEAR CHEMISTRY; ORGANOMETALLIC COMPOUND; PHYSICAL CHEMISTRY; SOLID-STATE CHEMISTRY. [J.J.K.]

**Inorganic photochemistry** The study of the light-induced behavior of various metal compounds. The physical and chemical properties of substances are generally altered by the absorption of light. Typical metal compounds have a characteristic number (coordination number) of molecules or ions (ligands) directly bonded to the metal center. For example a six-coordinate compound has the general formula $ML_6^{n+}$. Many of these compounds are colored, and much interest has been aroused by speculation that some metal compounds could mediate the transformation of solar radiation into useful chemical or electrical energy.

The photochemistry of metal compounds has grown in concert with modern theories of the electronic structure of molecules and of chemical bonding in molecules. Photochemical studies are often designed to probe and test these theories. The range of pertinent studies spans most of the subdisciplines of chemistry and includes or bears on such topics as photo-physics, the development of laser materials, catalysis, photo-synthesis, oxidation-reduction chemistry, acid-base chemistry, organometallic chemistry, metalloenzyme chemistry, solid-state chemistry, and surface chemistry. *See* CHEMICAL BONDING; CHEMICAL DYNAMICS; COORDINATION CHEMISTRY; LASER PHOTOCHEMISTRY; PHOTOCHEMISTRY. [J.F.E.]

**Inorganic polymer** A giant molecule linked by covalent bonds but with an absence or near-absence of hydrocarbon units in the main molecular backbone; these may be included as pendant side chains. Carbon fibers, graphite, and so forth are considered inorganic polymers.

Some special characteristics of many inorganic polymers are a higher Young's modulus and a lower failure strain compared with organic polymers. Relatively few inorganic polymers dissolve in the true sense, or alternatively, if they swell, few can revert. Crystallinity and high glass transition temperatures are also much more common than in organic polymers. In highly cross-linked inorganic polymers, stress relaxation frequently involves bond interchange.

Inorganic polymers can be classified in a number of ways. Some are based on the composition of the backbone, such as the silicones (Si—O), the phosphazenes (P—N), and polymeric sulfur (S—S). Others are based on their connectivity, that is, the number of network bonds linking the repeating unit into the network. Thus the silicones based on $R_2SiO$, the phosphazenes based on $NPX_2$, and polymeric sulfur each have a connectivity of two, while boric oxide based on $B_2O_3$ has a connectivity of three, and amorphous silica based on $SiO_2$ has one of four (see illustration).

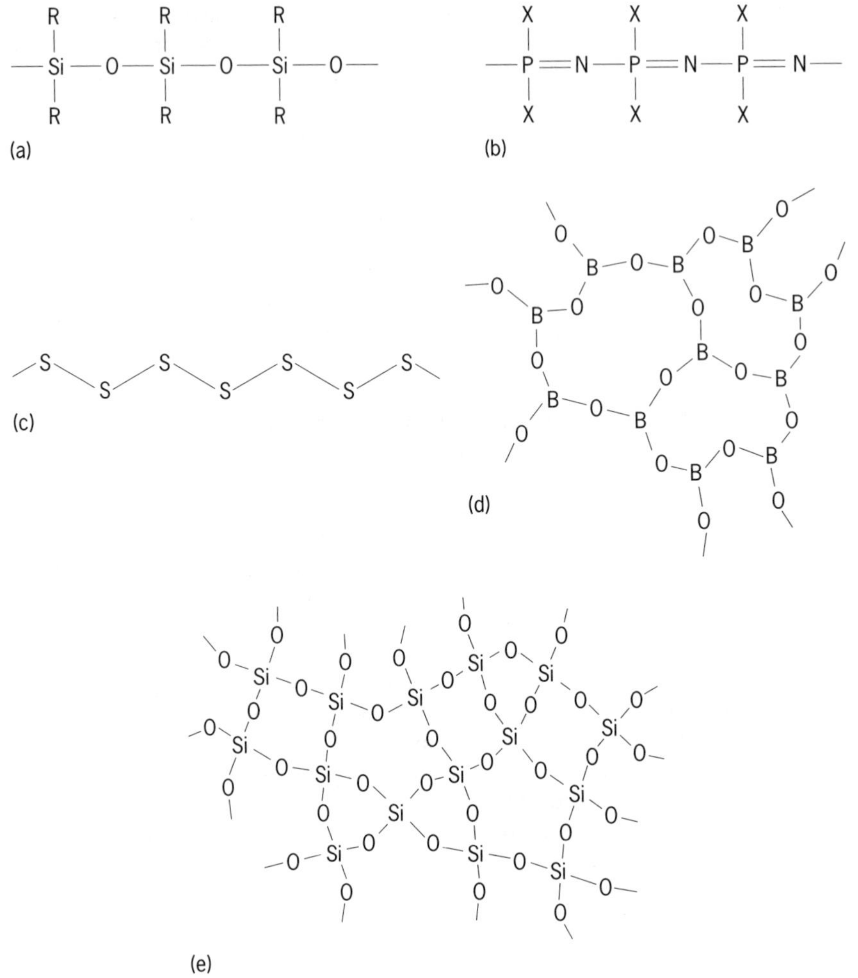

**Polymers with varying connectivities: (*a*) siloxanes, two; (*b*) phosphazenes, two; (*c*) sulfur, two; (*d*) boric oxide, three; (*e*) amorphous silica, four. (*After N. H. Ray, Inorganic Polymers, Academic Press. 1978*).**

Among the well-known inorganic polymers are silicones, chalcogenide glasses, graphite, boron polymers, and silicate polymers. *See* GRAPHITE; SILICONE RESINS. [R.A.S.]

**Intercalation compounds** Crystalline or partially crystalline solids consisting of a host lattice containing voids into which guest atoms or molecules are inserted. Candidate hosts for intercalation reactions may be classified by the number of directions (0 to 3) along which the lattice is strongly bonded and thus unaffected by the intercalation reaction. Isotropic, three-dimensional lattices (including many oxides and zeolites) contain large voids that can accept multiple guest atoms or molecules. Layer-type, two-dimensional lattices (graphite and clays) swell up perpendicular to the layers when the guest atoms enter. The chains in one-dimensional structures (polymers such as polyacetylene) rotate cooperatively about their axes during the intercalation reaction to form channels that are occupied by the guest atoms. In the intercalation family based on solid $C_{60}$ (buckminsterfullerene), the zero-dimensional host lattice consists of 60-atom carbon clusters with strong internal bonding but weak intercluster bonding. These clusters pack together like hard 1-nm-diameter spheres, creating interstitial voids which are large enough to accept most elements in the periodic table. The proportions of guest and host atoms may be varied continuously in many of these materials, which are therefore not true compounds. Many ternary and quaternary substances, containing two or three distinct guest species, are known. The guest may be an atom or inorganic molecule (such as an alkali metal, halogen, or metal halide), an organic molecule (for example, an aromatic such as benzene, pyridine, or ammonia), or both. *See* FULLERENE; GRAPHITE; POLYMER.

Many applications of intercalation compounds derive from the reversibility of the intercalation reaction. The best-known example is pottery: Water intercalated between the silicate sheets makes wet clay plastic, while driving the water out during firing results in a dense, hard, durable material. Many intercalation compounds are good ionic conductors and are thus useful as electrodes in batteries and fuel cells. A technology for lightweight rechargeable batteries employs lithium ions which shuttle back and forth between two different intercalation electrodes as the battery is charged and discharged: vanadium oxide (three-dimensional) and graphite (two-dimensional). Zeolites containing metal atoms remain sufficiently porous to serve as catalysts for gas-phase reactions. Many compounds can be used as convenient storage media, releasing the guest molecules in a controlled manner by mild heating. *See* CLAY, COMMERCIAL; FUEL CELL; SOLID-STATE BATTERY. [J.E.Fi.]

**Interface of phases** The boundary between any two phases. Among the three phases—gas, liquid, and solid—five types of interfaces are possible: gas-liquid, gas-solid, liquid-liquid, liquid-solid, and solid-solid. The abrupt transition from one phase to another at these boundaries, even though subject to the kinetic effects of molecular motion, is statistically a surface only one or two molecules thick.

A unique property of the surfaces of the phases that adjoin at an interface is the surface energy which is the result of unbalanced molecular fields existing at the surfaces of the two phases. Within the bulk of a given phase, the intermolecular forces are uniform because each molecule enjoys a statistically homogeneous field produced by neighboring molecules of the same substance. Molecules in the surface of a phase, however, are bounded on one side by an entirely different environment, with the result that there are intermolecular forces that then tend to pull these surface molecules toward the bulk of the phase. A drop of water, as a result, tends to assume a spherical shape in order to reduce the surface area of the droplet to a minimum.

At an interface, there will be a difference in the tendencies for each phase to attract its own molecules. Consequently, there is always a minimum in the free energy of the surfaces at an interface, the net amount of which is called the interfacial energy in units of joules/$cm^2$. The interfacial energy can also be expressed as surface tension in units of millinewtons per meter.

The surface energy at an interface may be altered by the addition of solutes that migrate to the surface and modify the molecular forces there, or the surface energy may be changed by converting the planar interfacial boundary to a curved surface.

At liquid-solid interfaces, where the confluence of the two phases is usually termed wetting, a critical factor called the contact angle is involved. A drop of water placed on a paraffin surface, for example, retains a globular shape, whereas the same drop of water placed on a clean glass surface spreads out into a thin layer. In the first instance,

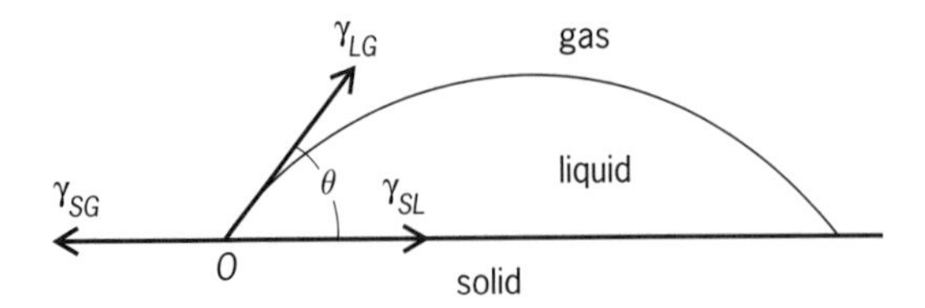


**Contact angle at interface of three phases.**

the contact angle is practically 180°, and in the second instance, it is practically 0°. The study of contact angles reveals the interplay of interfacial energies at three boundaries. The illustration is a schematic representation of the cross section of a drop of liquid on a solid. There are solid-liquid, solid-gas, and liquid-gas interfaces that meet in a linear zone at *O*.

The measurement of interfacial energies is made directly only upon liquid-gas and liquid-liquid interfaces. In measuring the liquid-gas interfacial energy (surface tension), the methods of capillary rise, drop weight on pendant drop, bubble pressure, sessile drops, Du Nuoy ring, vibrating jets, and ultrasonic action are among those used. There is a small but appreciable temperature effect upon surface tension, and this property is used to determine small differences in the surface tension of a liquid by placing the two ends of a liquid column in a capillary tube whose two ends are at different temperatures. The determination of interfacial energies at other types of interfaces can be inferred only by indirect methods. *See* FLOTATION; FOAM; FREE ENERGY; PHASE EQUILIBRIUM; SURFACE TENSION. [W.H.S.]

**Interhalogen compounds** The elements of the halogen family (fluorine, chlorine, bromine, and iodine) possess an ability to react with each other to form a series of binary interhalogen compounds (halogen halides) of general composition given

**Known interhalogen compounds**

| | XY | $XY_3$ | $XY_5$ | $XY_7$ |
|---|---|---|---|---|
| | ClF | $ClF_3$ | $ClF_5$ | $IF_7$ |
| mp | −154°C | −76°C | −103°C | |
| bp | −101°C | 12°C | −14°C | 4.77 (sublimes) |
| | BrF | $BrF_3$ | $BrF_5$ | |
| mp | ≈−33° | 8.77°C | −62.5°C | |
| bp | ≈20°C | 125°C | 40.3°C | |
| | IF | $IF_3$ | $IF_5$ | |
| mp | – | −28°C | 10°C | |
| bp | – | – | 101°C | |
| | BeCl | $ICl_3$* | | |
| mp | ≈−54°C | 101°C | | |
| bp | – | – | | |
| | ICl† | | | |
| mp | 27.2°C($\alpha$) | | | |
| bp | ≈100°C | | | |
| | IBr | | | |
| mp | 40°C | | | |
| bp | 119°C | | | |

* In the solid state the compound forms a dimer.
† Unstable $\beta$-modification exists, mp 14°C.

by $XY_n$, where $n$ can have the values 1, 3, 5, and 7, and where X is the heavier (less electronegative) of the two elements. All possible diatomic compounds of the first four halogens have been prepared. In other groups a varying number of possible combinations is absent. Although attempts have been made to prepare ternary interhalogens, they have been unsuccessful; there is considerable doubt that such compounds can exist. A list of known interhalogen compounds and some of their physical properties is given in the table. *See* HALOGEN ELEMENTS.

The reactivity of the polyhalides reflects the reactivity of the halogens they contain. In general, they behave as strong oxidizing and halogenating agents. Most halogen halides (especially halogen fluorides) readily attack metals, yielding the corresponding halide of the more electronegative halogen. All halogen halides readily react with water. Such reactions can be quite violent and, with halogen fluorides, they may be explosive. They readily react with aliphatic and aromatic hydrocarbons and with oxygen- or nitrogen-containing compounds. [T.Su.]

**Intermolecular forces** Attractive or repulsive interactions that occur between all atoms and molecules. Intermolecular forces become significant at molecular separations of about 1 nanometer or less, but are much weaker than the forces associated with chemical bonding. They are important, however, because they are responsible for many of the physical properties of solids, liquids, and gases. These forces are also largely responsible for the three-dimensional arrangements of biological molecules and polymers.

Intermolecular forces can be classified into several types, of which two are universal. The attractive force known as dispersion arises from the quantum-mechanical fluctuation of the electron density around the nucleus of each atom. At distances greater than 1 nm or so, the electrons of each atom move independently of the other, and the charge distribution is spherically symmetric. At shorter distances, an instantaneous fluctuation of the charge density in one atom can affect the other. If the electrons of one atom move briefly to the side nearer the other, the electrons of the other atom are repelled to the far side. In this configuration, both atoms have a small dipole moment, and they attract each other electrostatically. At another moment, the electrons may move the other way, but their motions are correlated so that an attractive force is maintained on average. Molecular orbital theory shows that the electrons of each atom are slightly more likely to be on the side nearer to the other atom, so that each atomic nucleus is attracted by its own electrons in the direction of the other atom.

At small separations the electron clouds can overlap, and repulsive forces arise. These forces are described as exchange-repulsion, and are a consequence of the Pauli exclusion principle, a quantum-mechanical effect which prevents electrons from occupying the same region of space simultaneously. To accommodate it, electrons are squeezed out from the region between the nuclei, which repel each other as a result. Each element can be assigned, approximately, a characteristic van der Waals radius; that is, when atoms in different molecules approach more closely than the sum of their radii, the repulsion ennergy increases sharply. It is this effect that gives molecules their characteristic shape, leading to steric effects in chemical reactions. *See* STERIC EFFECT (CHEMISTRY).

The other important source of intermolecular forces is the electrostatic interaction. When molecules are formed from atoms, electrons flow from electropositive atoms to electronegative ones, so that the atoms become somewhat positively or negatively charged. In addition, the charge distribution of each atom may be distorted by the process of bond formation, leading to atomic dipole and quadrupole moments. The electrostatic interaction between these is an important source of intermolecular forces, especially in polar molecules, but also in molecules that are not normally thought of as highly polar. The electrostatic field of a molecule may cause polarization of its neighbors, and this leads to a further induction contribution to the intermolecular interaction. An induction interaction can often polarize both molecules in such a way as to favor interactions with further molecules, leading to a cooperative network of intermolecular attractions. This effect is important in the network structure of water and ice. *See* WATER.

Intermolecular forces are responsible for many of the bulk properties of matter in all its phases. A realistic description of the relationship between pressure, volume, and temperature of a gas must include the effects of attractive and repulsive forces between molecules. The viscosity, diffusion, and surface tension of liquids are examples of physical properties which depend strongly on intermolecular forces. Intermolecular forces are also responsible for the ordered arrangement of molecules in solids, and account for their elasticity and properties (such as the velocity of sound in materials). [A.J.S.]

**Internal energy** A characteristic property of the state of a thermodynamic system, introduced in the first law of thermodynamics. For a static, closed system (no bulk motion, no transfer of matter across its boundaries), the change in internal energy for a process is equal to the heat absorbed by the system from its surroundings minus the work done by the system on its surroundings. Only a change in internal energy can be measured, not its value for any single state. For a given process, the change in internal energy is fixed by the initial and final states and is independent of the path by which the change in state is accomplished. *See* THERMODYNAMIC PRINCIPLES. [P.J.B.]

**Iodine** A nonmetallic element, symbol I, atomic number 53, relative atomic mass 126.9045, the heaviest of the naturally occurring halogens. Under normal conditions iodine is a black, lustrous, volatile solid; it is named after its violet vapor. *See* HALOGEN ELEMENTS; PERIODIC TABLE.

The chemistry of iodine, like that of the other halogens, is dominated by the facility with which the atom acquires an electron to form either the iodide ion $I^-$ or a single covalent bond —I, and by the formation, with more electronegative elements, of compounds in which the formal oxidation state of iodine is +1, +3, +5, or +7. Iodine is more electropositive than the other halogens, and its properties are modulated by: the relative weakness of covalent bonds between iodine and more electropositive elements; the large sizes of the iodine atom and iodide ion, which reduce lattice and solvation enthalpies for iodides while increasing the importance of van der Waals forces in iodine compounds; and the relative ease with which iodine is oxidized, Some properties of iodine are listed in the table. *See* ASTATINE; BROMINE; CHEMICAL BONDING; CHLORINE.

Iodine occurs widely, although rarely in high concentration and never in elemental form. Despite the low concentration of iodine in sea water, certain species of seaweed can extract and accumulate the element. In the form of calcium iodate, iodine is found in the caliche beds in Chile. Iodine also occurs as iodide ion in some oil well brines in California, Michigan, and Japan.

The sole stable isotope of iodine is $^{127}I$ (53 protons, 74 neutrons). Of the 22 artificial isotopes (masses between 117 and 139), the most important is $^{131}I$, with a half-life of 8 days. It is widely used in radioactive tracer work and certain radiotherapy procedures.

Iodine exists as diatomic $I_2$ molecules in solid, liquid, and vapor phases, although at elevated temperatures (>200°C or 390°F) dissociation into atoms is appreciable. Short intermolecular I . . . I distances in the crystalline solid indicate strong inter-molecular van der Waals forces. Iodine is moderately soluble in nonpolar liquids, and the violet color of the solutions suggests that $I_2$ molecules are present, as in iodine vapor.

**Some important properties of iodine**

| Property | Value |
|---|---|
| Electronic configuration | $[Kr]4d^{10}5s^25p^5$ |
| Relative atomic mass | 126.9045 |
| Electronegativity (Pauling scale) | 2.66 |
| Electron affinity, eV | 3.13 |
| Ionization potential, eV | 10.451 |
| Covalent radius, —I, nm | 0.133 |
| Ionic radius, $I^-$, nm | 0.212 |
| Boiling point, °C | 184.35 |
| Melting point, °C | 113.5 |
| Specific gravity (20/4) | 4.940 |

Although it is usually less vigorous in its reactions than the other halogens, iodine combines directly with most elements. Important exceptions are the noble gases, carbon, nitrogen, and some noble metals. The inorganic derivatives of iodine may be grouped into three classes of compounds: those with more electropositive elements, that is, iodides; those with other halogens; and those with oxygen. Organoiodine compounds fall into two categories: the iodides; and the derivatives in which iodine is in a formal positive oxidation state by virtue of bonding to another, more electronegative element. *See* GRIGNARD REACTION; HALOGENATED HYDROCARBON; HALOGENATION.

Iodine appears to be a trace element essential to animal and vegetable life. Iodide and iodate in sea water enter into the metabolic cycle of most marine flora and fauna, while in the higher mammals iodine is concentrated in the thyroid gland, being converted there to iodinated amino acids (chiefly thyroxine and iodotyrosines). They are stored in the thyroid as thyroglobulin, and thyroxine is apparently secreted by the gland. Iodine deficiency in mammals leads to goiter, a condition in which the thyroid gland becomes enlarged.

The bactericidal properties of iodine and its compounds bolster their major uses, whether for treatment of wounds or sterilization of drinking water. Also, iodine compounds are used to treat certain thyroid and heart conditions, as a dietary supplement (in the form of iodized salt), and for x-ray contrast media.

Major industrial uses are in photography, where silver iodide is a constituent of fast photographic film emulsions, and in the dye industry, where iodine-containing dyes are produced for food processing and for color photography. *See* DYE. [C.A.]

**Ion** An atom or group of atoms that bears an electric charge. Positively charged ions are called cations, and negatively charged ions are called anions. When a single atom gains or loses an electron, monoatomic ions are formed. For example, reaction of the element sodium (Na) with the element chlorine (Cl) leads to the transfer of electrons from Na to Cl to form $Na^+$ cations and $Cl^-$ anions. In general, atoms of metallic elements (on the left side of the periodic table) lose electrons to form cations, while atoms of nonmetallic atoms (on the right side of the periodic table) gain electrons to form anions. Ions can bear multiple charges, as in the magnesium ion ($Mg^{2+}$) or the nitride ion ($N^{3-}$). The charge on monoatomic ions is usually the same for elements in the same column of the periodic table; for example, hydrogen (H), sodium, lithium (Li), potassium (K), rubidium (Rb), and cesium (Cs) all form +1 ions. *See* PERIODIC TABLE.

Ions can also comprise more than one atom and are then called polyatomic ions. For example, the ammonium ion ($NH_4^+$) carries a positive charge and is composed of one nitrogen atom and four hydrogen atoms. The nitrate ion ($NO_3^-$) is composed of one nitrogen atom and three oxygen atoms and carries a single negative charge. Polyatomic ions are usually depicted inside brackets with superscripted charges, as shown in the structure below.

$$\left[ \begin{array}{c} O \\ | \\ N \\ O \quad O \end{array} \right]^-$$

Nitrate ion

Anions and cations can combine to form solid materials called salts, which are named by the cation name followed by the anion name. For a salt composed of the polyatomic ions ammonium and nitrate, the formula is $NH_4NO_3$ and the name is ammonium nitrate. For monoatomic ions, the cation name is the same as the element and the anion name is the element name with the ending -ide. Thus, common table salt, NaCl, is called sodium chloride. The ratio of anions to cations must always be such that an electrically neutral material is produced. Thus, magnesium nitrate must contain one magnesium for every two nitrates, giving the formula $Mg(NO_3)_2$. *See* SALT (CHEMISTRY).

[H.H.T.]

**Ion exchange** The reversible exchange of ions of the same charge between a solution and an insoluble solid in contact with it; or between two immiscible solvents, one of which contains a soluble material with immobilized ionic groups. Ions are atoms or molecules containing charge-bearing groups. Their interactions are dominated by the electrostatic forces between charge centers. These interactions are attractive when the ions are of opposite charge, or repulsive when the ions have the same charge. Ions with a net negative charge are called anions, and those with a net positive charge are cations.

A unique property of ions is their capacity to render gases and liquids conducting, and conductivity is a universal method of detecting ions. Ions in solution are in rapid motion and have no distinct partners. Ions in an electric field migrate to the electrode of opposite charge with a velocity roughly proportional to their charge-to-size ratio. This process is known as electrophoresis, and it is one method used to separate and identify ions. *See* ELECTROPHORESIS.

Ions can also be separated on the basis of their equilibrium with a system containing immobilized ions of opposite charge. Ions can be immobilized by virtue of their location in a rigid matrix. Associated with these fixed ionic sites are mobile counterions of opposite charge. Solution ions with a higher affinity than the counterions for the fixed sites will displace them from the fixed sites and remain localized in the vicinity of the fixed sites. Simultaneously the solution is enriched in the counterions originally localized at the fixed sites. This exchange process for ions of the same charge type is called ion exchange. In a column containing the immobilized ions as part of the stationary phase and the solution of competing ions as the mobile phase, the sample ions can be separated by the repeated equilibrium steps involved as they are transported through the column until they exit it, and are detected. This is an example of ion-exchange chromatography, an important method of separating and identifying ions.

**Ion-exchange materials.** Ion-exchange polymers are based on styrene and divinylbenzene and, to a lesser extent, polymers prepared from divinylbenzene, or a similar cross-linking agent, and acrylic, methacrylic, or hydroxyalkyl methacrylic acids and esters. These are usually prepared in bead form.

Ion exchangers prepared for the isolation or separation of cations must have negatively charged functional groups incorporated into the polymer backbone. The most common groups are sulfonic and carboxylic acids. Sulfonic acid groups are introduced by reacting the polymer beads with fuming sulfuric acid or a similar reagent. Similarly, carboxylic acid groups can be introduced by a number of common chemical reactions or by hydrolysis of the ester group or oxidation of hydroxyalkyl groups in methyl methacrylate or hydroxyalkyl methacrylate polymers, respectively. Other common functional groups used in cation exchangers include phosphoric acid and phenol and, to a lesser extent, phosphinic, arsonic, and selenonic acids.

A common approach for the preparation of anion exchangers is to react the styrene-divinylbenzene polymer with chloromethylmethyl ether in the presence of a catalyst, which adds the side chain, $—CH_2Cl$; then this chloromethylated product is treated with an amine to introduce the charged functional group. A tertiary amine produces a quaternary ammonium group, while primary and secondary amines give products that are charged only in contact with solutions of low pH. As well as simple alkyl and benzyl amines, hydroxyalkyl amines are used to introduce functional groups of the type $[—CH_2N(CH_3)_2C_2H_4OH]^+$. *See* QUATERNARY AMMONIUM SALTS.

Silica-based materials are used primarily in chromatography because of the favorable mechanical and physical properties of the silica ($SiO_2$) gel support matrix. Ion-exchange groups are introduced by reacting the surface silanol groups of the porous silica particles with silanizing reagents containing the desired functional group (R).

Hydrous oxides of elements of groups 14, 15, and 16 of the periodic table can be used as selective ion exchangers. The most important hydrous oxides used for the separation of organic and inorganic ions are alumina ($Al_2O_3 \cdot nH_2O$), silica ($SiO_2 \cdot nH_2O$), and zirconia ($ZrO_2 \cdot nH_2O$). Silica, by virtue of the presence of surface silanol groups, is used as a cation exchanger at pH $> 2$. Alumina is amphoteric and can be used as an anion exchanger at low pH and a cation exchanger at high pH. Alumina has the advantage over silica of being chemically stable over a wide pH range. The ion-exchange capacity of silica and alumina is controlled by the pH of the solution in contact with the oxides, since this controls the number of ionized surface functional groups. Alumina is used to isolate nitrogen-containing drugs and biochemically active substances from biological fluids, thus minimizing matrix interferences in their subsequent chromatographic analysis.

**Applications.** Ion exchange has numerous applications for industry and for laboratory research. By the quantity of materials used, water conditioning is the most important. Ion exchange is one of the primary analytical methods used to identify and quantify the concentration of ions in a wide range of environmental, biological, and industrial samples.

Natural water from rivers and wells is never pure; it is usually hard, that is, it contains calcium and magnesium salts that form curds with soap and leave hard crusts in pipes and boilers. Hard water is softened by passage through a cartridge or bed of cation exchanger in the sodium form (the mobile counterions are sodium in this case).

Many industrial and laboratory processes require a supply of pure water with a very low concentration of salts. This can be achieved by passing water through a bed of mixed strong cation exchanger in the hydrogen form and a strong anion exchanger in the hydroxide form. The cation exchanger removes all the cations from the water by replacing them by hydrogen ions. The anions are removed by the anion exchanger and replaced by hydroxide ions. The hydrogen and hydroxide ions combine to form water.

Toxic ions such as mercury ($Hg^{2+}$), lead ($Pb^{2+}$), chromate ($CrO_4^{2-}$), and ferrocyanide [$Fe(CN)_6^{4-}$] are removed by ion exchange from industrial wastewaters prior to their discharge into the environment. Ion exchangers are used to recover precious metals such as gold ($Au^+$), platinum ($Pt^+$), and silver ($Ag^+$) in a useful form from mine workings and metalworking factories. Ion exchange is frequently used to decontaminate waste and concentrate radioactive elements from the nuclear industry.

Ion exchange is used on the laboratory scale for isolation and preconcentration of ions prior to instrumental analysis and to obtain preparative scale quantities of material for use in laboratory studies. Ion exchange is often employed in conjunction with activation analysis to isolate individual elements for quantification by radiochemical detection. Modern chromatographic techniques employ ion exchangers of small particle size and favorable mass-transfer characteristics, and operate at high pressures, providing better resolution of mixtures in a shorter time than with conventional gravity-flow-controlled separations.

Biotechnology requires reliable, efficient methods to purify commercial-scale quantities of proteins, peptides, and nucleic acids for use in the pharmaceutical, agricultural, and food industries. Ion exchange is widely used in the isolation and purification of these materials. Typical applications include the removal of ionic compounds used in the production process, the elimination of endotoxins and viruses, the removal of host-cell proteins and deoxyribonucleic acid (DNA), and the removal of potentially hazardous variants of the main product. [C.F.P.]

**Membranes.** Ion-exchange membranes are a class of membranes that bear ionic groups and therefore have the ability to selectively permit the transport of ions through themselves. In biological systems, cell membranes and many other biological membranes contain ionic groups, and the conduction of ions is essential to their function. Synthetic ion-exchange membranes are used in fuel cells, electrochemical processes for chlorine manufacture and desalination, membrane electrodes, and separation processes. Ion-exchange membranes typically consist of a thin-film phase, usually polymeric, to which have been attached ionizable groups. Numerous polymers have been used, including polystyrene, polyethylene, polysulfone, and fluorinated polymers. Ionic groups attached to the polymer include sulfonate ($—SO_3^-$), carboxylate ($—COO^-$), tetralkylammonium ($—N(CH_3)_4^+$), phosphonate ($—PO_3H^-$), and many others. [N.N.Li; S.F.Y.]

**Ion-selective membranes and electrodes** Membrane-based devices, involving permselective, ion-conducting materials, used for measurement of activities of species in liquids or partial pressures in the gas phase. Permselective means that ions of one sign may enter and pass through a membrane.

Ion-selective electrodes are classified mainly according to the physical state of the ion-responsive membrane material, and not with respect to the ions sensed.

Glass membrane electrodes are mainly used for hydrogen ion activity measurements. They predate the wider variety of membrane electrodes developed after 1960. Electrodes based on water-insoluble inorganic salts include sensors for $F^-$, $Cl^-$, $Br^-$, $I^-$, $CN^-$, $SCN^-$, $S^{2-}$, $Ag^+$, $Cu^{2+}$, $Cd^{2+}$, and $Pb^{2+}$. The compounds used are silver salts, mercury salts, sulfides of Cu, Pb, and Cd, and rare-earth salts. All of these are so-called white metals whose aqueous cations (except $La^{3+}$) are labile. Electrodes using liquid-ion exchangers are supported in the voids of inert polymers such as cellulose acetate, or in transparent films of polyvinyl chloride, and provide extensive examples of devices for sensing. Electrodes with chemical reactions interposed between the sample and the sensor surface permit a new degree of freedom in design of sensors for species which do not directly respond at an electrode surface. Two primary examples are the categories of gas sensors and of electrodes which use enzyme-catalyzed reactions. *See* ELECTRODE; ION EXCHANGE.

Electrodes for many species are, for the most part, commercially available. Applications may be batch or continuous. Important batch examples are potentiometric titrations with ion-selective electrode end-point detection, determination of stability constants of complexes and speciation identity, solubility and activity coefficient determinations, and monitoring of reaction kinetics, especially for oscillating reactions. Ion, selective electrodes serve as liquid chromatography detectors and as quality-control monitors in drug manufacture. Applications occur in air and water quality (soil, clay, ore, natural-water, water-treatment, sea-water, and pesticide analyses); medical and clinical laboratories (serum, urine, sweat, gastric-juices, extra-cellular-fluid, dental-enamel, and milk analyses); and industrial laboratories (heavy-chemical, metallurgical, glass, beverage, and household-product analyses). *See* ANALYTICAL CHEMISTRY; CHROMATOGRAPHY; TITRATION. [R.P.B.]

**Ion-solid interactions** Physical processes resulting from the collision of energetic ions, atoms, or molecules with condensed matter. These include elastic and inelastic backscattering of the projectile, penetration of the solid by the projectile, emission of electrons and photons from the surface, sputtering of neutral atoms and ions, production of defects in crystals, creation of nuclear tracks in insulating solids, and electrical, chemical, and physical changes to the irradiated matter resulting from the passage or implantation of the projectile.

When an energetic ion impinges upon the surface of condensed matter, it experiences a series of elastic and inelastic collisions with the atoms which lie in its path. These collisions occur because of the electrical forces between the nucleus and electrons of the projectile and those of the atoms which constitute the solid target. They result in the transformation of the kinetic energy of the projectile into internal excitation of the solid.

One of the most simple interactions occurs when the projectile collides with a surface atom and bounces back in generally the opposite direction from which it came. This process is known as backscattering. Its observation in 1911 led Ernest Rutherford to conclude that most of the matter in atoms is concentrated in a small nucleus. Now it is used as an analytical technique to measure the masses and locations of atoms on and near a surface. This technique for surface characterization is appropriately named Rutherford backscattering analysis, and is most commonly performed with alpha particles of about 2 MeV. Another backscattering technique, known as ion-scattering spectrometry, uses projectiles with energies of perhaps 2 keV.

Although backscattering events are well enough understood to be used as analytical tools, they are relatively rare because they represent nearly head-on collisions between two nuclei. Far more commonly, a collision simply deflects the projectile a few degrees from its original direction and slows it somewhat, transferring some of its kinetic energy to the atom that is struck. Thus, the projectile does not rebound from the surface but penetrates deep within the solid, dissipating its kinetic energy in a series of grazing collisions.

The capacity of a solid to slow a projectile is called the stopping power, and is defined as the amount of energy lost by the projectile per unit length of trajectory in the solid. Stopping power is of central importance for many phenomena because it measures the capacity of a projectile to deposit energy within a thin layer of the solid and this energy drives secondary processes associated with penetration. In many insulating solids (including mica, glasses, and some plastics) the passage of an ion with a large electronic stopping power creates a unique form of radiation damage known as a nuclear track. When the substance is chemically etched, conical pits visible under an ordinary microscope are produced where ionizing particles have penetrated. The passage of single projectiles may thereby be observed.

In the nuclear stopping region it is relatively likely that the projectile will transfer significant amounts of energy to individual target atoms. These atoms will subsequently strike others, and eventually a large number of atoms within the solid will be set in motion. This disturbance is known as a collision cascade. Collision cascades may cause permanent damage to materials, induce mixing of layers in the vicinity of interfaces, or cause sputtering if they occur near surfaces.

Ion implantation is used in the manufacture of integrated circuits and in the improvement of surface properties of metals. Ion-solid processes permit highly sensitive analyses for trace elements, the characterization of materials and surfaces, and the detection of ionizing radiation. Techniques employing them include secondary ion mass spectrometry (SIMS) for elemental analysis and imaging of surfaces, proton-induced x-ray emission (PIXE), ion-scattering spectrometry (ISS), and Rutherford backscattering analysis (RBS). They are also fundamental to the operation of silicon surface-barrier detectors which are used for the measurement of particle radiation, and of nuclear track detectors which are used in research as diverse as the dating of meteorites and the search for magnetic monopoles. *See* ACTIVATION ANALYSIS; PROTON-INDUCED X-RAY EMISSION (PIXE); SECONDARY ION MASS SPECTROMETRY (SIMS). [R.A.We.]

**Ionic equilibrium** An equilibrium in a chemical reaction in which at least one ionic species is produced, consumed, or changed from one medium to another.

The wide variety of types of ionic equilibrium possible include: dissolution of an un-ionized substance, for example, the dissolution of hydrogen chloride (a gas) in water (an ionizing solvent), reactions (1)–(3); dissolution of a crystal in water, such

$$HCl(g) \rightleftharpoons H^+ + Cl^- \quad (1)$$

$$HCl(g) + H_2O \rightleftharpoons H_3O^+ + Cl^- \quad (2)$$

$$HCl(g) + 4H_2O \rightleftharpoons H_9O_4{}^+ + Cl^- \quad (3)$$

as the dissociation of solid silver chloride, reaction (4); dissociation of a strong acid, for example, nitric acid, $HNO_3$, dissociates as it dissolves in water, as in reaction (5); dissociation of an ion in water, for example, the bisulfate ion, $HSO_4{}^-$, dissociates in water, as in reaction (6); and dissociation of water itself, is represented by reaction (7).

$$AgCl(Crystal) \rightleftharpoons Ag^+ + Cl^- \quad (4)$$

$$HNO_3 + H_2O \rightleftharpoons H_3O^+ + NO_3{}^- \quad (5)$$

$$HSO_4{}^- + H_2O \rightleftharpoons H_3O^- + SO_4{}^{2+} \quad (6)$$

$$2H_2O \rightleftharpoons H_3O^+ + OH^- \quad (7)$$

*See* ACID AND BASE; CHEMICAL EQUILIBRIUM; HYDROLYSIS. [T.F.Y.]

**Ionization** The process by which an electron is removed from an atom, molecule, or ion. It is of basic importance to electrical conduction in gases and liquids. In the simplest case, ionization may be thought of as a transition between an initial state consisting of a neutral atom and a final state consisting of a positive ion and a free electron. In more complicated cases, a molecule may be converted to a heavy positive ion and a heavy negative ion which are separated. [G.H.M.]

**Ionization potential** The potential difference through which a bound electron must be raised to free it from the atom or molecule to which it is attached. In particular, the ionization potential is the difference in potential between the initial state, in which the electron is bound, and the final state, in which it is at rest at infinity.

The ionization potential for the removal of an electron from a neutral atom other than hydrogen is more correctly designated as the first ionization potential. The potential associated with the removal of a second electron from a singly ionized atom or molecule is then the second ionization potential, and so on. [G.H.M.]

**Ionophore** A substance that can transfer ions from a hydrophilic medium, such as water, into a hydrophobic medium, such as hexane or a biological membrane, where the ions typically would not be soluble; also known as an ion carrier. The ions transferred are usually metal ions, for example, lithium ($Li^+$), sodium ($Na^+$), potassium ($K^+$), magnesium ($Mg^{2+}$), and calcium ($Ca^{2+}$); but there are ionophores that promote the transfer of other ions, such as ammonium ion ($NH_4^+$) or amines of biological interest. *See* ION.

There are two mechanisms by which ionophores promote the transfer of ions across hydrophobic barriers: ion-ionophore complex formation and ion channel formation. In complex formation, the ion forms a coordination complex with the ionophore in which there is a well-defined ratio (typically 1:1) of ion to ionophore. The ionophore wraps around the ion so that the ion exists in the polar interior of the complex, while the exterior is predominantly hydrophobic in character and as such is soluble in nonpolar media. The ion is coordinated by oxygen atoms present in the ionophore molecule through iondipole interactions. The ionophore molecule essentially acts as the solvent for the ion, replacing the aqueous solvation shell that normally surrounds the ion. *See* COORDINATION COMPLEXES.

Ionophores that act via ion channel formation are found in biological environments. The molecule forms a polar channel in an otherwise nonpolar cell membrane, allowing passage of small ions either into or out of the cell.

Naturally occurring ionophores fall into four classes, each of which has antibiotic activity: peptide, cyclic depsipeptide, macrotetrolide, and polyether ionophores. The biological activity of ionophore antibiotics is due to their ability to disrupt the flow of ions either into or out of cells. Under normal conditions, cells have a high internal concentration of potassium ions but a low concentration of sodium ions. The concentration of ions in the extracellular medium is just the reverse, high in sodium ions but low in potassium ions. This imbalance, which is necessary for normal cell function, is maintained by a specific transport protein (sodium-potassium adenosine triphosphatase) in the cell membrane that pumps sodium ions out of the cell in exchange for potassium ions. Ionophore antibiotics can disrupt this ionic imbalance by allowing ions to penetrate the cell membrane as ion-ionophore complexes or via the formation of ion channels. Gram-positive bacteria appear to be particularly sensitive to the effect of ionophores perturbing normal ion transport. [C.A.V.]

**Iridium** A chemical element, Ir, atomic number 77, relative atomic weight 192.22. Iridium is a transition metal and shares similarities with rhodium as well as the other platinum metals, including palladium, platinum, ruthenium, and osmium. The atom in the gas phase has the electronic configuration $1s^2$, $2s^2$, $2p^6$, $3s^2$, $3p^6$, $3d^{10}$, $4s^2$, $4p^6$ $4d^{10}$, $4f^{14}$, $5s^2$, $5p^6$, $5d^7$, $6s^2$. The ionic radius for $Ir^{3+}$ is 0.068 nanometer and its metallic radius is 0.1357 nm. Metallic iridium is slightly less dense than osmium, which is the densest of all the elements. *See* PERIODIC TABLE.

**Physical properties of iridium metal**

| Property | Value |
|---|---|
| Crystal structure | Face-centered cubic |
| Lattice constant $a$ at 25°C, nm | 0.38394 |
| Thermal neutron capture cross section, barns | 440 |
| Density at 25°C, g/cm$^3$ | 22.560 |
| Melting point | 2443°C (4429°F) |
| Boiling point | 4500°C (8130°F) |
| Specific heat at 0°C, cal/g | 0.0307 |
| Thermal conductivity 0–100°C, cal cm/cm$^2$ s °C | 0.35 |
| Linear coefficient of thermal expansion | |
| 20–100°C, $\mu$in./in./°C | 6.8 |
| Electrical resistivity at 0°C, microhm-cm | 4.71 |
| Temperature coefficient of electrical resistance | |
| 0–100°C/°C | 0.00427 |
| Tensile strength (1000 lb/in.$^2$) | |
| Soft | 160–180 |
| Hard | 300–360 |
| Young's modulus at 20°C | |
| lb/in.$^2$, static | $75.0 \times 10^6$ |
| lb/in.$^2$, dynamic | $76.5 \times 10^6$ |
| Hardness, diamond pyramid number | |
| Soft | 200–240 |
| Hard | 600–700 |
| $\Delta H_{fusion}$, kJ/mol | 26.4 |
| $\Delta H_{vaporization}$, kJ/mol | 612 |
| $\Delta H_f$ monoatomic gas, kJ/mol | 669 |
| Electronegativity | 2.2 |

The abundance of iridium in the Earth's crust is very low, 0.001 ppm. For mining purposes, it is generally found alloyed with osmium in materials known as osmiridium and iridiosmium, with iridium contents ranging from 25 to 75%.

Solid iridium is a silvery metal with considerable resistance to chemical attack. Upon atmospheric exposure the surface of the metal is covered with a relatively thick layer of iridium dioxide ($IrO_2$). Important physical properties of metallic iridium are given in the table.

Because of its scarcity and high cost, applications of iridium are severely limited. Although iridium metal and many of its complex compounds are good catalysts, no large-scale commercial application for these has been developed. In general, other platinum metals have superior catalytic properties. The high degree of thermal stability of elemental iridium and the stability it imparts to its alloys does give rise to those applications where it has found success. Particularly relevant are its high melting point (2443°C or 4429°F), its oxidation resistance, and the fact that it is the only metal with good mechanical properties that survives atmospheric exposure above 1600°C (2910°F). Iridium is alloyed with platinum to increase tensile strength, hardness, and corrosion resistance. However, the workability of these alloys is decreased. These alloys find use as electrodes for anodic oxidation, for containing and manipulating corrosive chemicals, for electrical contacts that are exposed to corrosive chemicals, and as primary standards for weight and length. Platinum-iridium alloys are used for electrodes in spark plugs that are unusually resistant to fouling by antiknock lead additives. Iridium-rhodium thermocouples are used for high-temperature applications, where they have unique stability. Very pure iridium crucibles are used for growing single crystals of gadolinium gallium garnet for computer memory devices and of yttrium aluminum garnet for solid-state lasers. The radioactive isotope, $^{192}Ir$, which is obtained synthetically from $^{191}Ir$ by irradiation of natural sources, has been used as a portable gamma source for radiographic studies in industry and medicine. *See* PLATINUM. [A.L.Ba.]

**Iron** A chemical element, Fe, atomic number 26, and atomic weight 55.847. Iron is the fourth most abundant element in the crust of the Earth (5%). It is a malleable, tough, silver-gray, magnetic metal. It melts at 1540°C, boils at 2800°C, and has a density of 7.86 g/cm$^3$. The four stable, naturally occurring isotopes have masses of 54, 56, 57, and 58. The two main ores are hematite, $Fe_2O_3$, and limonite, $Fe_2O_3 \cdot 3H_2O$. Pyrites, $FeS_2$, and chromite, $Fe(CrO_2)_2$, are mined as ores for sulfur and chromium, respectively. Iron is found in many other minerals, and it occurs in groundwaters and in the red hemoglobin of blood. *See* PERIODIC TABLE.

The greatest use of iron is for structural steels; cast iron and wrought iron are made in quantity, also. Magnets, dyes (inks, blueprint paper, rouge pigments), and abrasives (rouge) are among the other uses of iron and iron compounds.

There are several allotropic forms of iron. Ferrite or $\alpha$-iron is stable up to 760°C (1400°F). The change of $\beta$-iron involves primarily a loss of magnetic permeability because the lattice structure (body-centered cubic) is unchanged. The allotrope called $\gamma$-iron has the cubic close-packed arrangements of atoms and is stable from 910 to 1400°C (1670 to 2600°F). Little is known about $\delta$-iron except that it is stable above 1400°C (2600°F) and has a lattice similar to that of $\alpha$-iron.

The metal is a good reducing agent and, depending on conditions, can be oxidized to the 2+, 3+, or 6+ state. In most iron compounds, the ferrous ion, iron(II), or ferric ion, iron(III), is present as a distinct unit. Ferrous compounds are usually light yellow to dark green-brown in color; the hydrated ion, $Fe(H_2O)_6{}^{2+}$, which is found in many compounds and in solution, is light green. This ion has little tendency to form coordination complexes except with strong reagents such as cyanide ion, polyamines, and porphyrins. The ferric ion, because of its high charge (3+) and its small size, has a strong tendency to hold anions. The hydrated ion, $Fe(H_2O)_6{}^{3+}$, which is found in solution, combines with $OH^-$, $F^-$, $Cl^-$, $CN^-$, $SCN^-$, $N_3{}^-$, $C_2O_4{}^{2-}$, and other anions to form coordination complexes. *See* COORDINATION CHEMISTRY.

An interesting aspect of iron chemistry is the array of compounds with bonds to carbon. Cementite, $Fe_3C$, is a component of steel. The cyanide complexes of both ferrous and ferric iron are very stable and are not strongly magnetic in contradistinction to most iron coordination complexes. The cyanide complexes form colored salts. *See* TRANSITION ELEMENTS. [J.O.E.]

**Isocyanate** A derivative of isocyanic acid. Isocyanates are represented by the general formula R—N=C=O, where R is predominantly alkyl or aryl; however, stable isocyanates in which the N=C=O group is linked to elements such as sulfur, silicon, phosphorus, nitrogen, or the halogens have also been prepared. Most members of this class of compounds are liquids that are sensitive to hydrolysis and are strong lacrimators. Isocyanates are extremely reactive, especially toward substrates containing active hydrogen. They have found wide use in the commercial manufacture of polyurethanes, which are used in making rigid and flexible foam, elastomers, coatings, and adhesives.

Diisocyanates react with difunctional reagents, such as diols, to form addition polymers with a wide variety of properties. The flexibility in the choice of starting materials (diisocyanate, diol, diamine, diacid, and so forth) and consequently in the multitude of possible adducts makes this product group unique in the field of polymeric materials.

Two aromatic diisocyantes, tolylene diisocyanate [TDI; structure (1)] and di(4-isocyanatophenyl)methane [MDI; structure (2)], have become the major starting

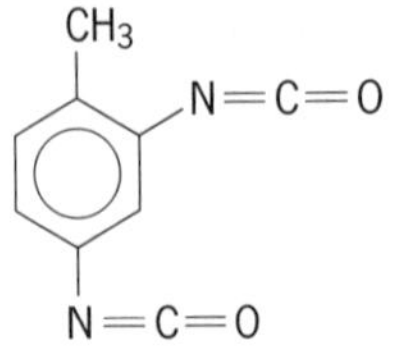

$CH_3$ N=C=O N=C=O

**(1)**

O=C=N N=C=O

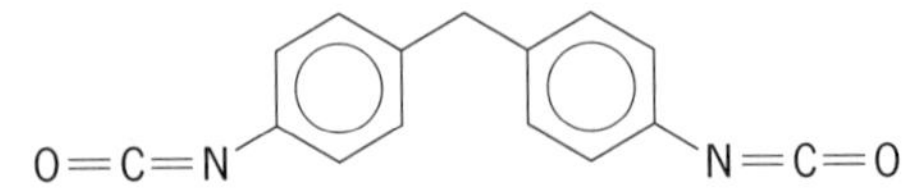

**(2)**

materials for a family of polymeric products, such as flexible and rigid polyurethane foams used in construction and appliance insulation, automotive seating, and furniture. Elastomers based on MDI, polyols, and polyamines are widely used in the automotive industry, where reaction injection molding technology is used for the manufacture of exterior parts such as body panels and bumpers. *See* POLYOL.

Thermoplastic polyurethane elastomers (TPU) are used in the molding and extrusion of many industrial and consumer products with superior abrasion resistance and toughness. *See* POLYURETHANE RESINS.

The trimerization to polyisocyanurates and the formation of polyamides from dicarboxylic acids have been used to synthesize polymers with excellent thermal properties. Aliphatic diisocyanates, notably 1,6-diisocyanatohexane (HDI), fully hydrogenated MDI ($H_{12}$MDI), and isophorone diisocyanate (IPDI) have become building blocks for color-stable polyurethane coatings and elastomers with high abrasion resistance. *See* POLYAMIDE RESINS; POLYMER. [R.H.Ri.]

**Isoelectric point** The pH of a dispersion medium of a colloidal suspension or an ampholyte at which the solute does not move in an electrophoretic field. The term isoelectric point is abbreviated pI.

Ampholytes are molecules with acid as well as basic functional groups. When dissolved in a suitable medium, ampholytes may acquire positive and negative charges by dissociation or by accepting or losing protons, thereby becoming bipolar ions (zwitterions). Ampholytes may be as small as glycine and carry just one chargeable group each; or as large as polyampholytes (polyelectrolytes that carry positive charges, negative charges, or both). They may possess molecular weights in the hundreds of thousands like proteins or in the millions like nucleic acids, and carry many hundreds of chargeable groups. *See* ION; NUCLEIC ACID; PH; PROTEIN.

An example of establishing the isoelectric point is shown by the course of the pH changes during the titration of alanine [$NH_2CH(CH_3)COOH$], a 1:1 ampholyte, meaning a molecule that carries one positively and one negatively ionizable group. Starting from acid solution (see illustration), relatively small pH changes (with alkali as the titrant) are observed between pH 2 and 3 (acidic), and again between pH 9.5 and 10.5 (alkaline), caused by the buffering capacity of the carboxyl (—COOH) and amine (—$NH_2$) groups as weak electrolytes. The pH for $^1/_2$ equivalence corresponds to the pK of the acid function (a value related to the equilibrium constant), where one-half of the alanine molecules still carry only a positive charge ($-NH_3^+$), while the other half are also negatively charged ($-COO^-$). Thus alanine exists in the form of zwitterions. *See* PK.

For molecules that carry four or more chargeable groups, that is, for polyelectrolytes, the courses of the overall titration curves may no longer reflect the individual dissociation steps clearly, as the dissociation areas usually overlap. The isoelectric point then becomes an isoelectric range, such as for pigskin (parent) gelatin, a protein that exhibits an electrically neutral isoelectric range from pH 7 to pH 9.

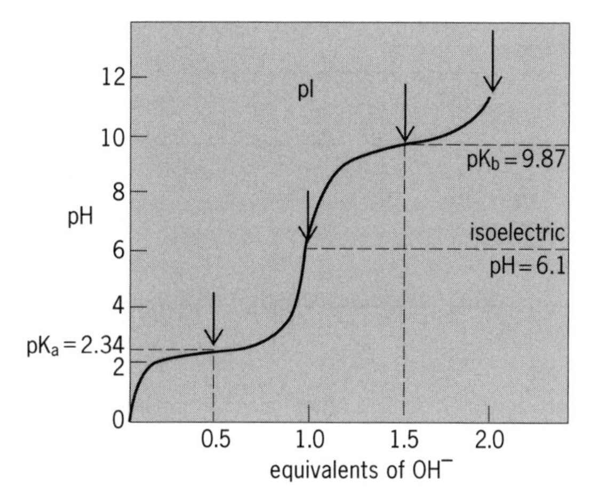


**Titration of alanine with sodium hydroxide (NaOH), showing the course of pH with added fractional equivalents. The four arrows show, from left to right, the p$K_a$ [$^1/_2$ cations: $NH_3^+(CH_3)COOH$, and $^1/_2$ zwitterions: $NH_3^+(CH_3)CHCOO^-$]; the pI (all zwitterions); the p$K_b$ ($^1/_2$ zwitterions, $^1/_2$ anions); and the end of titration, when all alanine molecules are in the anionic form: $NH_2(CH_3)CHOO^-$.**

Since ampholytes in an electric field migrate according to their pI with a specific velocity to the cathode or anode, the blood proteins, for example, can be separated by the techniques of gel or capillary-zone electrophoresis. *See* ELECTROPHORESIS; TITRATION.

The notion that some ampholytes may pass with changing pH through a state of zero charge (zero zeta potential) on their way from the positively to the negatively charged state has become so useful for specifying and handling polyampholytes that it was extended to all kinds of colloids, and to solid surfaces that are chargeable in contact with aqueous solutions. Practically all metal oxides, hydroxides, or hydroxyoxides become charged by the adsorption of hydrogen ions ($H^+$) or hydroxide ions ($OH^-$), while remaining neutral at a specific pH. Strictly speaking, the isoelectric point of electrophoretically moving entities is given by the pH at which the zeta potential at the shear plane of the moving particles becomes zero. The point of zero charge at the particle (solid or surface) is somewhat different but often is not distinguished from the isoelectric point. It is determined by solubility minima or, for solid surfaces, is found by the rate of slowest adsorption of colloids (for example, latexes) of well-defined charge.

The important separation technique of ion-exchange chromatography is based on the selective adsorption of ampholytes on the resins with which the column is filled, at a given pH. For example, the larger the net positive charge of an ampholyte, the more strongly will it be bound to a negative ion-exchange resin and the slower will it move through the column. By rinsing with solutions of gradually increasing pH, the ampholytes of a mixture can be eluted and made to emerge separately from the column and be collected. Automated amino acid analyzers are built on this principle. *See* AMINO ACIDS; COLLOID; ELECTROKINETIC PHENOMENA; ION EXCHANGE; ION-SELECTIVE MEMBRANES AND ELECTRODES. [F.R.E.]

**Isoelectronic sequence** A term used in spectroscopy to designate the set of spectra produced by different chemical elements ionized in such a way that their atoms or ions contain the same number of electrons. The sequence in the table is an example. Since the neutral atoms of these elements each contain *Z* electrons, removal of one electron from scandium, two from titanium, and so forth, yields a series of ions all of which have 20 electrons. Isoelectronic sequences are useful in predicting unknown

**Example of isoelectronic sequence**

| Designation of spectrum | Emitting atom or ion | Atomic number, $Z$ |
|---|---|---|
| CaI | Ca | 20 |
| ScII | $Sc^{+}$ | 21 |
| TiIII | $Ti^{2+}$ | 22 |
| VIV | $V^{3+}$ | 23 |
| CrV | $Cr^{4+}$ | 24 |
| MnVI | $Mn^{5+}$ | 25 |

spectra of ions belonging to a sequence in which other spectra are known. *See* ATOMIC STRUCTURE AND SPECTRA. [F.A.J./W.W.W.]

**Isomerization** Rearrangement of the atoms within hydrocarbon molecules. Isomerization processes of practical significance in petroleum chemistry are (1) migration of alkyl groups, (2) shift of a single-carbon bond in naphthenes, and (3) double-bond shift in olefins. *See* AROMATIZATION; CRACKING; MOLECULAR ISOMERISM. [G.E.L.]

**Isotope dilution techniques** Analytical techniques that involve the addition to a sample of an isotopically labeled compound. Soon after the discovery of the stable heavy isotopes of hydrogen, carbon, nitrogen, and oxygen, their value in analytical chemistry was recognized. Stable isotopes were particularly useful in the analysis of complex mixtures of organic compounds where the isolation of the desired compound with satisfactory purity was difficult and effected only with low or uncertain yields. The addition of a known concentration of an isotopically labeled compound to a sample immediately produces isotope dilution if the particular compound is present in the sample. After thorough mixing of the isotopically labeled compound with the sample, any technique that determines the extent of the isotopic dilution suffices to establish the original concentration of the compound in the mixture. Isotope dilution techniques exploit the difficulty in the separation of isotopes, with the isotopically labeled "spike" following the analytically targeted compound through a variety of separation procedures prior to isotopic analysis.

The technique depends on the availability of a stable or radioisotope diluent with isotope abundance ratios differing markedly from those of the naturally occurring elements. With monoisotopic elements, such as sodium or cesium, radioactive elements of sufficiently long life can be used in isotope dilution techniques.

The original applications of isotope dilution were by biochemists interested in complex mixtures of organic compounds. In these studies care had to be taken to ensure the stability of the labeled compound and its resistance to isotopic exchange reactions. Nitrogen-15–labeled glycine for example, could be used to determine glycine in a mixture of amino acids obtained from a protein. Deuterium-labeled glycine could not be used reliably if the deuterium isotopes were attached to the glycine amino or carboxyl group, because in these locations deuterium is known to undergo exchange reactions with hydrogens in the solvent or in other amino acids. Deuterium is very useful in elemental isotopic analysis where total hydrogen or exchangeable hydrogen concentrations are desired. *See* BIOCHEMISTRY; DEUTERIUM.

Applications of isotope dilution techniques have also been found in geology, nuclear science, and materials science. These applications generally focus on the very high sensitivity attainable with these techniques. Isotopes of argon, uranium, lead, thorium, strontium, and rubidium have been used in geologic age determinations of minerals and meteorites. Taking the estimated error as a measurement of sensitivity, isotopic dilution analyses of uranium have been done down to 4 parts in $10^{12}$ and on thorium to 8 parts in $10^{9}$. Studies in geology and nuclear science require the determination of trace amounts of radiogenic products. If the half-life and decay scheme of the parent nuclide is known, then isotopic dilution determinations of parent and daughter isotopes provide a basis for the calculation of the age of the sample. If the age or history of the sample is known, then determination of the trace concentrations of isotopes provides information on pathways of nuclear reactions. [L.Fr.]

# K

**Ketene** A member of a class of organic compounds with the C=C=O group as a common structural element. Ketenes are derivatives of carboxylic acids, from which they are (hypothetically) formed by abstraction of water; they can therefore be considered to be inner anhydrides of acids, as opposed to the common carboxylic acid anhydrides formed from two molecules of a carboxylic acid.

Like real anhydrides, ketenes are acylating agents which readily undergo reactions with many compounds containing active hydrogens. They are relatively labile compounds, and only a limited number have been prepared and isolated. Many ketenes have been prepared in place and reacted immediately. *See* ACID ANHYDRIDE. [R.H.Ri.]

**Ketone** One of a class of chemical compounds of the general formula

$$\begin{matrix} R \diagdown \\ \quad C{=}O \\ R' \diagup \end{matrix}$$

R and R′ are alkyl, aryl, or heterocyclic radicals. The groups R and R′ may be the same or different or incorporated into a ring as in cyclopentanone:

$$\underbrace{CH_2CH_2CH_2CH_2C}{=}O$$

The ketones acetone and methyl ethyl ketone are used as solvents. Ketones are important intermediates in the syntheses of organic compounds.

By common nomenclature rules, the R and R′ groups are named, followed by the word ketone—for example, $CH_3CH_2COCH_2CH_3$ (diethyl ketone), $CH_3COCH(CH_3)_2$ (methyl isopropyl ketone), and $C_6H_5COC_6H_5$ (diphenyl ketone). The nomenclature of the International Union of Pure and Applied Chemistry uses the hydrocarbon name corresponding to the maximum number of carbon atoms in a continuous chain in the ketone molecule, followed by "-one," and preceded by a number designating the position of the carbonyl group in the carbon chain. The first two ketones above are named 3-pentanone and 3-methyl-3-butanone.

The lower-molecular-weight ketones are colorless liquids. Acetone and methyl ethyl ketone are miscible with water; the water solubility of the higher homologs decreases with increasing number of carbon atoms. Because of their characteristic odors, various ketones are of use in the flavoring and perfumery industry.

Addition to the carbonyl group is the most important type of ketone reaction. Ketones are generally less reactive than aldehydes in addition reactions. Methyl ketones are more reactive than the higher ketones because of steric group effects. [P.E.F.]

**Kinetic methods of analysis** The measurement of reaction rates for the analytical determination of the initial concentrations of the species taking part in chemical reactions. This technique can be used since, in most cases, the rates or velocities of chemical reactions are directly proportional to the concentrations of the species taking part in the reactions.

The rate of a chemical reaction is measured by experimentally following the concentration of some reactant or product as a function of time as the mixture proceeds from a nonequilibrium to an equilibrium or static state (steady state). Kinetic techniques of analysis have the inherent problem of the difficulty of making measurements on a dynamic system. However, kinetic methods often have advantages over equilibrium techniques in spite of the increased experimental difficulty. For example, the equilibrium differentiations or distinctions attainable for the reactions of very closely related compounds are often very small and not sufficiently separated to resolve the individual concentrations of a mixture without prior separation. But the kinetic differentiations or distinctions obtained when such compounds are reacted with a common reagent are often quite large and permit simultaneous analysis. *See* CHEMICAL EQUILIBRIUM.

A further advantage of kinetic methods is that they permit a larger number of chemical reactions to be used analytically. Many reactions, both inorganic and organic, are not sufficiently well behaved to be employed analytically by equilibrium or thermodynamic techniques. Many reactions attain equilibrium too slowly; side reactions occur as the reactions proceed to completion, or the reactions are not sufficiently quantitative (do not go to completion) to be applicable. However, a kinetic-based technique can often be employed in these cases simply by measuring the reaction rate of these reactions during the early or initial portion of the reaction period. Also, the measurement of the rates of catalyzed reactions generally is a considerably more sensitive analytical method for the determination of trace amounts of a large number of species than equilibrium methods. *See* CATALYSIS. [H.B.Ma.]

**Krypton** A gaseous chemical element, Kr, atomic number 36, and atomic weight 83.80. Krypton is one of the noble gases in group 18 of the periodic table. Krypton is a colorless, odorless, and tasteless gas. The table gives some physical properties of krypton. The principal use for krypton is in filling electric lamps and electronic devices of various types. Krypton-argon mixtures are widely used to fill fluorescent lamps. *See* INERT GASES; PERIODIC TABLE.

**Physical properties of krypton**

| Property | Value |
|---|---|
| Atomic number | 36 |
| Atomic weight (atmospheric krypton only) | 83.80 |
| Melting point, triple point °C | −157.20 |
| Boiling point at 1 atm pressure, °C | −153.35 |
| Gas density at 0°C and 1 atm pressure, g/liter | 3.749 |
| Liquid density at its boiling point, g/mi | 2.413 |
| Solubility in water at 20°C, ml krypton (STP) per 1000 g water at 1 atm partial pressure krypton | 59.4 |

The only commercial source of stable krypton is the air, although traces of krypton are found in minerals and meteorites.

A mixture of stable and radioactive isotopes of krypton is produced in nuclear reactors by the slow-neutron fission of uranium. It is estimated that about $2 \times 10^{-8}\%$ of the weight of the Earth is krypton. Krypton also occurs outside the Earth. [A.W.F.]

# L

**Lactam** A cyclic amide that is the nitrogen analog of a lactone. For example, a $\gamma$-aminobutyric acid readily forms $\gamma$-butyrolactam (also known as 2-pyrrolidinone) upon heating, as in the reaction below. The tautomeric enol form of a lactam is known as a lactim.

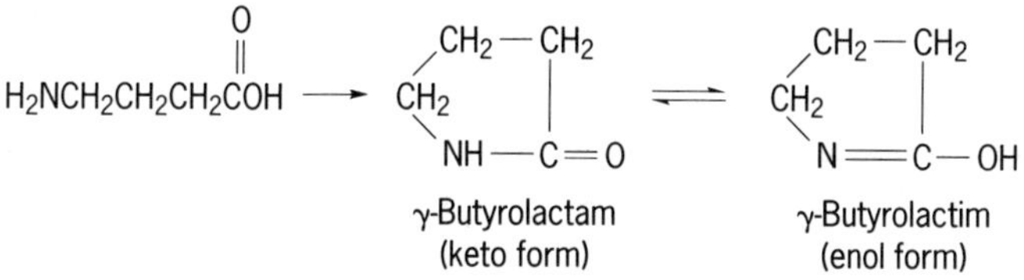

$$H_2NCH_2CH_2CH_2\overset{\overset{O}{\|}}{C}OH \longrightarrow \underset{\gamma\text{-Butyrolactam (keto form)}}{\begin{array}{c} CH_2-CH_2 \\ CH_2 \quad\quad | \\ NH-C{=}O \end{array}} \rightleftharpoons \underset{\gamma\text{-Butyrolactim (enol form)}}{\begin{array}{c} CH_2-CH_2 \\ CH_2 \quad\quad | \\ N{=}C-OH \end{array}}$$

γ-Butyrolactam
(keto form)

γ-Butyrolactim
(enol form)

The $\delta$-amino acids similarly form $\delta$ (six-membered-ring) lactams upon heating, but larger- and smaller-ring lactams must be made by indirect methods.

Several lactams are of considerable industrial importance. 2-Pyrrolidinone and 1-methyl-2-pyrrolidinone are made by heating $\gamma$-butyrolactone with ammonia and methylamine, respectively. They are useful specialty solvents. Vinylation of 2-pyrrolidinone with acetylene gives 1-vinyl-2-pyrrolidinone, which is polymerized to a substance commonly used in aerosol hair sprays.

The ß-lactam antibiotics comprise two groups of clinically important therapeutic agents, the penicillins and the cephalosporins. In both cases they contain a four-membered or ß-lactam ring which has its nitrogen atom and a carbon atom in common with another ring. Such substances are derived commercially from fermentation processes, followed usually by chemical manipulation of the functional groups. *See* AMINO ACIDS ; BIOCHEMICAL ENGINEERING; LACTONE. [P.E.F.]

**Lactate** A salt or ester of lactic acid ($CH_3CHOHCOOH$). In lactates, the acidic hydrogen of the carboxyl group has been replaced by a metal or an organic radical. Lactates are optically active, with a chiral center at carbon 2. Commercial fermentation produces either the dextrorotatory (*R*) or the levorotatory (*S*) form, depending on the organism involved. *See* OPTICAL ACTIVITY.

The *R* form of lactate occurs in blood and muscle as a product of glycolysis. Lack of sufficient oxygen during strenuous exercise causes enzymatic (lactate dehydrogenase) reduction of pyruvic acid to lactate, which causes tiredness, sore muscles, and even muscle cramps. During renewed oxygen supply (rest) the lactate is reoxidized to pyruvic acid and the fragments enter the Krebs (citric acid) cycle. The plasma membranes of muscle and liver are permeable to pyruvates and lactates, permitting the blood to transport them to the liver (Cori cycle). Lactates also increase during fasting and in diabetics.

Lactates are found in certain foods (sauerkraut), and may be used for flour conditioning and in food emulsification. Alkali-metal salts act as blood coagulants and are used in calcium therapy, while esters are used as plasticizers and as solvents for lacquers. *See* ESTER; SALT (CHEMISTRY). [E.H.H.]

**Lactone** A cyclic, intramolecular ester derived from a hydroxy acid. Simple lactones are designated $\alpha$, $\beta$, $\gamma$, $\delta$ and so forth. Five- and six-membered lactones are very readily obtained by cyclization of a hydroxy acid or precursor as shown in reactions (1)–(3). Lactones with three- and four-membered rings are also known. *See* ESTER.

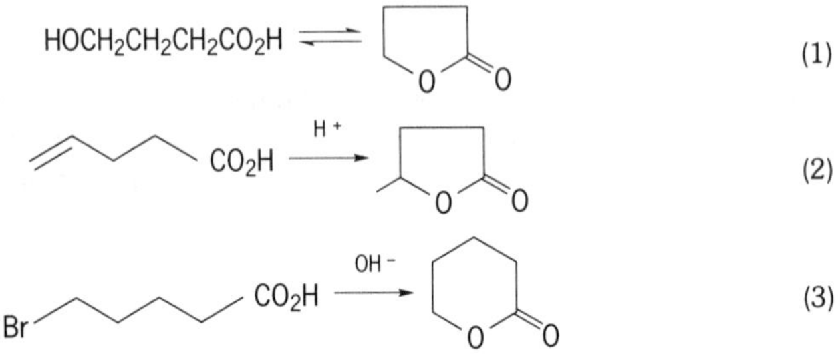

$HOCH_2CH_2CH_2CO_2H \rightleftharpoons$ (1)

$CO_2H \xrightarrow{H^+}$ (2)

Br $CO_2H \xrightarrow{OH^-}$ (3)

Lactones in various forms are found in numerous naturally occurring compounds. Unsaturated $\gamma$-lactone rings (**1** and **2**) are present in many components of essential oils. Ascorbic acid [vitamin C; (**3**)] is a carbohydrate lactone. *See* ASCORBIC ACID.

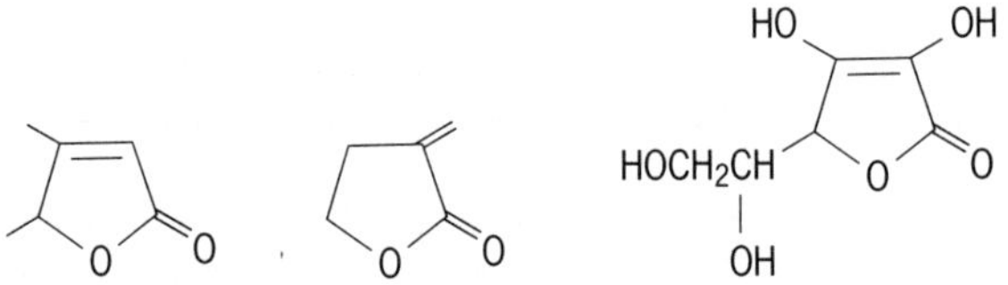

(1) (2) (3) HO OH $HOCH_2CH$ OH

A major development in the chemistry of naturally occurring compounds has been the isolation from microorganisms of a number of macrocyclic lactones with rings containing from 12 to more than 30 atoms. These substances include immunosuppressive agents and antibiotics such as erythromycin. Although they represent some of the most complex organic compounds known, several of these macrocyclic lactones have been synthesized. *See* ORGANIC CHEMISTRY. [P.E.F.; J.A.Mo.]

**Lanthanide contraction** The name given to an unusual phenomenon encountered in the rare-earth series of elements. The radii of the atoms of the members of this series decrease slightly as the atomic number increases. As the charge on the nucleus increases across the rare-earth series, all electrons are pulled in closer to the nucleus so that the radii of the rare-earth ions decrease slightly as the compounds go across the rare-earth series. Any given compound of the rare earths is very likely to crystallize with the same structure as any other rare earth. However, the lattice parameters become smaller and the crystal denser as the compounds proceed across the series. This contraction of the lattice parameters is known as the lanthanide contraction. *See* RARE-EARTH ELEMENTS. [F.H.Sp.]

**Lanthanum** A chemical element, La, atomic number 57, atomic weight 138.91. Lanthanum, the second most abundant element in the rare-earth group, is a metal. The naturally occurring element is made up of the isotopes $^{138}La$, 0.089%, and $^{139}La$,

99.91%. $^{138}La$ is a radioactive positron emitter with a half-life of $1.1 \times 10^{11}$ years. The element was discovered in 1839 by C. G. Mosander and occurs associated with other rare earths in monazite, bastnasite, and other minerals. It is one of the radioactive products of the fission of uranium, thorium, or plutonium. Lanthanum is the most basic of the rare earths and can be separated rapidly from other members of the rare-earth series by fractional crystallization. Considerable quantities of it are separated commercially, since it is an important ingredient in glass manufacture. Lanthanum imparts a high refractive index to the glass and is used in the manufacture of expensive lenses. The metal is readily attacked in air and is rapidly converted to a white powder. Lanthanum becomes a superconductor below about 6 K (−449°F) in both the hexagonal and face-centered crystal forms. *See* PERIODIC TABLE; RARE-EARTH ELEMENTS.

[F.J.Sp.]

**Laser photochemistry** A branch of chemistry in which reactions are induced or altered by laser light. The initial part of any photochemical reaction involves an optical transition to some excited state of molecule. These excited states could involve electronic, vibrational, and rotational excitation. The particular photochemical product that results from the absorption of light depends on the specific excited state species created during the irradiation. Thus the properties of the light source often determine the photochemical product. Lasers have had an immense impact on the field of photochemistry by providing scientists with an intense, polarized, and nearly monochromatic source of light. There are lasers that extend from wavelengths of less than 110 nanometers (vacuum ultraviolet) to more than 100,000 nm (far-infrared); for comparison, the entire visible spectrum extends from only 400 nm (violet) to 700 nm (red).

Use of lasers in photochemical applications provides three main advantages over conventional light sources such as discharge or arc lamps. First, lasers are generally more powerful than conventional light sources. A continuous-wave argon ion laser can produce 10 W at 514.5 nm. Pulsed lasers, which compress the light energy into very short time periods ($10^{-6}$ to $10^{-12}$ s), can generate correspondingly higher peak powers, typically from $10^3$ to $10^9$ W. These very short, intense pulses of light are used by the photochemist to monitor the time evolution of excited-state species or the appearance rate of products in a photochemical reaction.

Second, laser light can be collimated into a beam with a very small divergence angle, routinely less than 1/100 of a degree. The high degree of collimation of laser light permits efficient illumination at a chosen point within the sample, which could be far from the light source. This collimation permits the photochemist to confine the photochemical activation to some very small and precisely located area, such as in the fabrication and repair of microelectronic devices. In addition, the photochemical event, which has a very low absorption cross section, can be induced by photolyzing a very small region of the chemical sample with the total laser output power.

Third, laser light is exceptionally pure in color. The spectral purity of light can be described by a band width measured in wavenumbers ($cm^{-1}$), defined as the frequency width divided by the speed of light. While the full visible spectrum is 10,000 $cm^{-1}$ wide, a typical laser may be as narrow as a few $cm^{-1}$. When conventional sources are used to irradiate molecules that have many possible excited-state transitions, a distribution of excited molecules results. These molecules may have been excited to several different electronic states with many possible vibrational or rotational energies. Under such conditions, it is often impossible to identify the excited electronic states that produced the various photochemical products. The high spectral purity of the laser obviates this problem, and the identity of the excited electronic state is almost always known to the laser photochemist.

The two general areas of laser-induced photochemistry are vibrational photochemistry and electronic photochemistry. In vibrational photochemistry the chemical reaction occurs entirely on the ground electronic state of the molecule, whereas electronic photochemistry occurs via some excited electronic state, usually involving the first excited singlet state (all electrons spin-paired) and possibly the lowest-lying triplet state (two electrons spin-unpaired) for most polyatomic molecules. Although electronic photochemistry has been profoundly affected by the advent of lasers, vibrational photochemistry, which requires intense light sources, is not even possible without lasers. *See* ELECTRON SPIN; TRIPLET STATE.

Because of the advantages of lasers, there is a constant demand for the development of lasers that will serve in new regions of the spectrum. In spite of the advantages of lasers, conventional light sources are still widely used in most photochemical investigations and commercial processes because of their simplicity and low cost. [D.Sn.]

**Laser spectroscopy** Spectroscopy with laser light or, more generally, studies of the interaction between laser radiation and matter. Lasers have led to a rejuvenescence of classical spectroscopy, because laser light can far surpass the light from other sources in brightness, spectral purity, and directionality, and if required, laser light can be produced in extremely intense and short pulses. The use of lasers can greatly increase the resolution and sensitivity of conventional spectroscopic techniques, such as absorption spectroscopy, fluorescence spectroscopy, or Raman spectroscopy. Moreover, interesting new phenomena have become observable in the resonant interaction of intense coherent laser light with matter. Laser spectroscopy has become a wide and diverse field, with applications in numerous areas of physics, chemistry, and biology. *See* SPECTROSCOPY. [T.W.Ha.]

**Lawrencium** A chemical element, symbol Lr, atomic number 103. Lawrencium, named after E. O. Lawrence, is the eleventh transuranium element; it completes the actinide series of elements. *See* ACTINIDE ELEMENTS; PERIODIC TABLE; TRANSURANIUM ELEMENTS.

The nuclear properties of all the isotopes of lawrencium from mass 255 to mass 260 have been established. $^{260}$Lr is an alpha emitter with a half-life of 3 min and consequently is the longest-lived isotope known. [A.Gh.]

**Lead** A chemical element, Pb, atomic number 82 and atomic weight 207.19. Lead is a heavy metal (specific gravity 11.34 at 16°C or 61°F), of bluish color, which tarnishes to dull gray. It is pliable, inelastic, easily fusible, melts at 327.4°C (621.3°F), and boils at 1740°C (3164°F). The normal chemical valences are 2 and 4. It is relatively resistant to attack by sulfuric and hydrochloric acids but dissolves slowly in nitric acid. Lead is amphoteric, forming lead salts of acids as well as metal salts of plumbic acid. Lead forms many salts, oxides, and organometallic compounds. *See* PERIODIC TABLE.

Industrially, the most important lead compounds are the lead oxides and tetraethyllead. Lead forms alloys with many metals and is generally employed in the form of alloys in most applications. Alloys formed with tin, copper, arsenic, antimony, bismuth, cadmium, and sodium are all of industrial importance.

Lead compounds are toxic and have resulted in poisoning of workers from misuse and overexposure. However, lead poisoning is presently rare because of the industrial application of modern hygienic and engineering controls. The greatest hazard arises from the inhalation of vapor or dust. In the case of organolead compounds, absorption through the skin may become significant. Some of the symptoms of lead poisoning are headaches, dizziness, and insomnia. In acute cases there is usually stupor, which progresses to coma and terminates in death. The medical control of employees engaged in lead usage involves precise clinical tests of lead levels in blood and urine. With such control and the proper application of engineering control, industrial lead poisoning may be entirely prevented.

Lead rarely occurs in its elemental state. The most common ore is the sulfide, galena. The other minerals of commercial importance are the carbonate, cerussite, and the sulfate, anglesite, which are much more rare. Lead also occurs in various uranium and thorium minerals, arising directly from radioactive decay. Commercial lead ores may contain as little as 3% lead, but a lead content of about 10% is most common. The ores are concentrated to 40% or greater lead content before smelting.

The largest single use of lead is for the manufacture of storage batteries. Other important applications are for the manufacture of tetraethyllead, cable covering, construction, pigments, solder, and ammunition.

Organolead compounds are being developed for applications such as catalysts for polyurethane foams, marine antifouling paint toxicants, biocidal agents against gram-positive bacteria, protection of wood against marine borers and fungal attack, preservatives for cotton against rot and mildew, molluscicidal agents, anthelmintic agents, wear-reducing agents in lubricants, and corrosion inhibitors for steel.

Because of its excellent resistance to corrosion, lead finds extensive use in construction, particularly in the chemical industry. It is resistant to attack by many acids because it forms its own protective oxide coating. Because of this advantageous characteristic, lead is used widely in the manufacture and handling of sulfuric acid.

Lead has long been used as protective shielding for x-ray machines. Because of the expanded applications of atomic energy, radiation-shielding applications of lead have become increasingly important.

Lead sheathing for telephone and television cables continues to be a sizable outlet for lead. The unique ductility of lead makes it particularly suitable for this application because it can be extruded in a continuous sheath around the internal conductors.

The use of lead in pigments has been a major outlet for lead but is decreasing in volume. White lead, $2PbCO_3 \cdot Pb(OH)_2$, is the most extensively used lead pigment. Other lead pigments of importance are basic lead sulfate and lead chromates.

A considerable variety of lead compounds, such as silicates, carbonates, and salts of organic acids, are used as heat and light stabilizers for polyvinyl chloride plastics. Lead silicates are used for the manufacture of glass and ceramic frits, which are useful in introducing lead into glass and ceramic finishes. Lead azide, $Pb(N_3)_2$ is the standard detonator for explosives. Lead arsenates are used in large quantities as insecticides for crop protection. Litharge (lead oxide) is widely employed to improve the magnetic properties of barium ferrite ceramic magnets. Also, a calcined mixture of lead zirconate and lead titanate, known as PZT, is finding increasing markets as a piezoelectric material.

[H.S.; J.D.J.]

**Le Chatelier's principle** A description of the response of a system in equilibrium to a change in one of the variables determining the equilibrium.

For any chemical reaction equilibrium or phase equilibrium, an increase in temperature at constant pressure shifts the equilibrium in the direction in which heat is absorbed by the system. *See* CHEMICAL EQUILIBRIUM; PHASE EQUILIBRIUM.

For any reaction equilibrium or phase equilibrium, an increase in pressure at constant temperature shifts the equilibrium in the direction in which the volume of the system decreases.

For a reaction equilibrium in a dilute solution, addition of a small amount of a solute species that participates in the reaction will shift the equilibrium in the direction that uses up some of the added solute.

For an ideal-gas reaction equilibrium, addition at constant temperature and volume of a species that participates in the reaction will shift the equilibrium in the direction that consumes some of the added species. For an ideal-gas reaction equilibrium, addition at constant temperature and pressure of a species that participates in the reaction might shift the equilibrium to produce more of the added species or might shift the equilibrium to use up some of the added species; the direction of the shift depends on the reaction, on which species is added, and on the initial composition of the equilibrium mixture. [I.N.L.]

**Ligand** A molecule with an affinity to bind to a second atom or molecule. This affinity can be described in terms of noncovalent interactions, such as the type of binding that occurs in enzymes that are specific for certain substrates; or of a mode of binding where an atom or groups of atoms are covalently bound to a central atom, as in the case of coordination complexes and organometallic compounds. Ligands of the latter type can be further distinguished by the nature of the orbitals used in bond formation. *See* ENZYME.

When a protein binds to another molecule, that molecule may be referred to as a ligand. The site where the ligand is bound is known as the binding or active site of the protein. In order for a molecule to be classified as a ligand for a protein, several weak interactions such as hydrophobic, van der Waals, and hydrogen bonding must take place simultaneously. Therefore, the binding of a ligand by a protein is generally quite specific. *See* CHEMICAL BONDING; COORDINATION CHEMISTRY. [T.J.Me.]

**Ligand field theory** An essentially ionic approach to chemical bonding which is often used with coordination compounds. These compounds consist of a central transition-metal ion that is surrounded by a regular array of coordinated atoms or ligands. Accordingly, the ligands are assumed to be sources of negative charge which perturb the energy levels of the central metal ion. In this respect the ligands subject the metal ion to an electric field which is analogous to the electric or crystal field produced by the regular distribution of nearest neighbors within an ionic crystalline lattice. For example, the crystal field produced by the Cl ion ligand in octahedral $TiCl_6^{3-}$ is considered to be similar to that produced by the octahedral array of the six Cl ions about each Na ion in NaCl. The Na ion with its rare-gas configuration has an electronic charge distribution which is spherically symmetric both within and without the crystal field. The paramagnetic Ti(III) ion, which possesses one 3*d* electron ($d^1$), has a spherically symmetric charge distribution only in the absence of the crystal field produced by the ligands. The presence of the ligands destroys the spherical symmetry and produces a more complex set of energy levels within the central metal ion. The crystal field theory allows the energy levels to be calculated and related to experimental observation. [R.A.D.W.]

**Light-scattering photometry** Optical methods used to measure the extent of scattering of light by particles suspended in fluids or by macromolecules in solution. Two different approaches are employed. The photometric measurement of the extent of attenuation of an incident light beam as it passes through the scattering medium is known as turbidimetry. The measurement of the intensity of light scattered to a detector which is not in the path of the incident light (often at right angles to it) is known as nephelometry—literally, the measurement of cloudiness.

Nephelometry and turbidimetry are also used for quantitative analytical chemical measurements. These methods were formerly considered relatively nonprecise and were used only to obtain approximate concentration information. With the advent of microprocessor-based instrumentation, however, it has been possible to overcome such prior limitations as nonlinearity of response with concentration, and these methods have become increasingly popular, particularly in the field of clinical chemistry.

Turbidimetric analysis involves measurement of the intensity of light that is transmitted through a solution or suspension. For strongly turbid samples containing many particles or particles that are large compared with the wavelength of visible light, turbidimetry is the method of choice, and it is most often performed on a colorimeter or spectrophotometer. For a limited range of particle concentrations in suspensions, this method of measurement gives fairly good precision. *See* COLORIMETRY; SPECTROSCOPY.

The relative ease of discriminating the presence of scattered light from the dark background present in the absence of scattering makes nephelometric measurement an extremely sensitive tool for analytical purposes. For a relatively clear solution, where the light is only weakly scattered because of a low concentration of particles or the presence of particles which are extremely small compared with the wavelength of the incident light, nephelometry is generally the tool best suited to measurement. In principle, nephelometric measurements can be made at the detection angle and with the wavelength of light most suitable for a particular application. The most notable analytical application of nephelometry is in the quantitative analysis of specific human serum proteins. [J.C.St.; A.F.-T.C.]

**Lignin** A polymer found extensively in the cell walls of all woody plants, Lignin, one of the most abundant natural polymers, constitutes one-fourth to one-third of the total dry weight of trees. It combines with hemicellulose materials to help bind the cells together and direct water flow. *See* POLYMER.

Several methods have been devised for isolating lignin from wood. Some isolation methods are based on acid treatments in which the carbohydrate components (cellulose and hemicelluloses) are hydrolyzed to water-soluble materials. However, with such procedures, serious doubts exist as to whether the isolated lignin is representative of the "native" lignin. Enzymatic digestion of the carbohydrate in wood meal is a lengthy, tedious procedure but offers the greatest promise of leaving lignin unaltered during isolation.

Structural studies on lignin have been hampered by the random, cross-linked nature of the polymer. The relative proportions of the monomers which make up lignin (**1** and **2**) vary with the plant species. Lignin is formed in the plant by an enzymatic

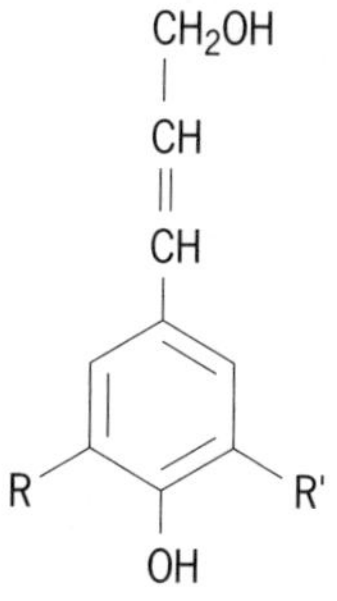

$CH_2OH$, CH, CH, R, R', OH

(1)

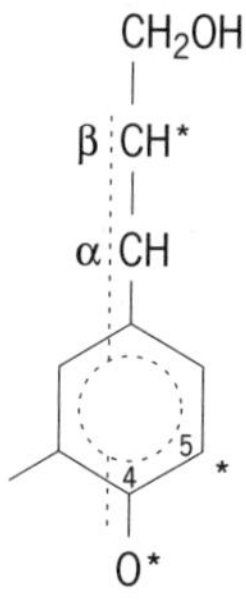

$CH_2OH$, β CH*, α CH, 5, 4, *, O*

(2)

dehydrogenation of the monomers. The * designations shown on the structure (**2**) indicate the principal sites for coupling monomers.

In general, the markets for lignin products are not large or attractive enough to compensate for the cost of isolation and the energy derived from its burning. An exception is lignosulfonate, which is obtained during paper production either directly from sulfite pulping liquors or by sulfonation of acid-precipitated kraft lignin; its markets include a dispersant in carbon black slurries, clay products, dyes, cement, oil drilling muds, an asphalt emulsifier, a binder for animal feed pellets, a conditioner for boiler water or cooling water, and an additive to lead-acid storage battery plate expanders. *See* WOOD CHEMICALS. [D.R.D.]

**Liquid** A state of matter intermediate between that of crystalline solids and gases. Macroscopically, liquids are distinguished from crystalline solids in their capacity to flow under the action of extremely small shear stresses and to conform to the shape of a confining vessel. Liquids differ from gases in possessing a free surface and in lacking the capacity to expand without limit. On the scale of molecular dimensions liquids lack the long-range order that characterizes the crystalline state, but nevertheless they possess a degree of structural regularity that extends over distances of a few molecular diameters. In this respect, liquids are wholly unlike gases, whose molecular organization is completely random.

Liquids possess important transport properties, notably their capacity to transmit heat (thermal conductivity), to transfer momentum under shear stresses (viscosity), and to attain a state of homogeneous composition when mixed with other miscible liquids (diffusion). These nonequilibrium properties of liquids are well understood in macroscopic terms and are exploited in large-scale engineering and chemical-process operations. *See* GAS. [N.H.N.]

**Liquid chromatography** A method of chemical separation that involves passage of a liquid phase through a solid phase and relies on subtle chemical interactions to resolve complex mixtures into pure compounds. A small amount of the sample to be separated is injected onto the top of a column that is densely packed with spherical particles of small diameter, that is, the stationary phase. A liquid solvent, the mobile phase, flows through the column continuously to carry the sample from the top to the bottom of the column. During passage through the column, the components of the sample are transferred back and forth continuously between the two phases, and small thermodynamic differences in the chemical interactions of the various sample components with the mobile and stationary phases slow the passage of some solutes more than others and lead to their separation. The technique can be performed on very small scales for chemical analysis, dealing with micrograms or even nanograms of sample, or it can be performed on an industrial scale for purification of commercial products. The technique has great resolving power.

In the late 1960s, workers realized that to achieve maximum performance they needed to make the stationary-phase particles very small. As the stationary-phase particles became smaller, they packed too densely to permit gravity-driven flow of the mobile phase. It was necessary to force the mobile phase through the column under high pressure, and the technique was named high-pressure liquid chromatography (HPLC). The meaning of the acronym has been changed to high-performance liquid chromatography, for the elegant separations that are possible. Typical stationary-phase particles are monodisperse, macroporous silica particles either 3 or 5 micrometers in diameter, and the column lengths for analytical scale separations are on the order of 2–10 in. (5–25 cm), with inside diameters of about 0.16 in. (4 mm). The columns are made of stainless steel, which is relatively inert chemically and able to withstand the high pressures applied to the top of the column. Since these columns require pressures of a few hundred to a few thousand pounds per square inch, depending on the mobile-phase flow velocity desired, a high-pressure metering pump is an integral part of a modern liquid chromatograph. Most instruments include a means of performing gradient elution, that is, making a continuous change in the composition of the liquid mobile phase during the separation process. Gradient elution can be performed by using a separate pump for each solvent and changing the relative proportions during the separation, or by using a proportioning valve between the solvent reservoirs and the pump. An injector valve is used to introduce a small volume of sample, typically 5–100 microliters, onto the top of the column without interrupting the mobile-phase flow. These valves can be operated manually, or they can be programmed to perform injections from a tray of samples for routine analyses. After the sample traverses the column, a flow-through detector is employed to generate the chromatogram, which is the visual representation of the separation. Detectors can provide both quantitative and qualitative information about the separated components. Temperature control of the column and detector is important; they are generally operated at or near room temperature, and temperature fluctuations can adversely affect the reproducibility of the separation and detection steps.

Liquid chromatography very much depends on the highly selective chemical interactions that occur in both the mobile and stationary phases. Rapid separations have become possible for compounds whose difference in free energy of transfer between the two phases is only a few calories per mole. Columns exhibiting virtually every type of possible selectivity exist, including selectivity by shape, charge, size, and optical activity. Additional selectivity can be generated through manipulations of the mobile phase; additives that interact with the solute in the mobile phase can create unique selectivities in columns that do not normally show that type of selectivity.

In addition to facilitating chemical analysis, liquid chromatography can be used to obtain physicochemical information. Diffusion coefficients, kinetic parameters, critical micelle concentrations of surfactants, and other information have been estimated from chromatographic data. The most common application is the estimation of hydrophobic parameters, especially as models of biological or environmental partitioning processes (most frequently, of octanol-water partitioning). Bioavailability, bioaccumulation, soil sorption, and various other factors are estimated based on linear free-energy relationships. *See* CHROMATOGRAPHY. [J.G.D.]

**Liquid crystals** A state of matter that mixes the properties of both the liquid and solid states. Liquid crystals may be described as condensed fluid states with spontaneous anisotropy. They are categorized in two ways: thermotropic liquid crystals, prepared by heating the substance, and lyotropic liquid crystals, prepared by mixing two or more components, one of which is rather polar in character, for example, water. Thermotropic liquid crystals are divided, according to structural characteristics, into two classes, nematic and smectic.

**Nematic structure.** Nematic liquid crystals are subdivided into the ordinary nematic and the chiral-nematic (or cholesteric). The molecules in the ordinary nematic structure maintain a parallel or nearly parallel arrangement to each other along the long molecular axes (illus. *a*). They are mobile in three directions and can rotate about one axis. This structure is one-dimensional.

When the nematic structure is heated, it is generally transformed into the isotropic liquid where the completely disordered motion of the molecules produces a phase in which all directions are equivalent. The nematic structure is the highest-temperature mesophase in thermotropic liquid crystals. The energy required to deform a nematic

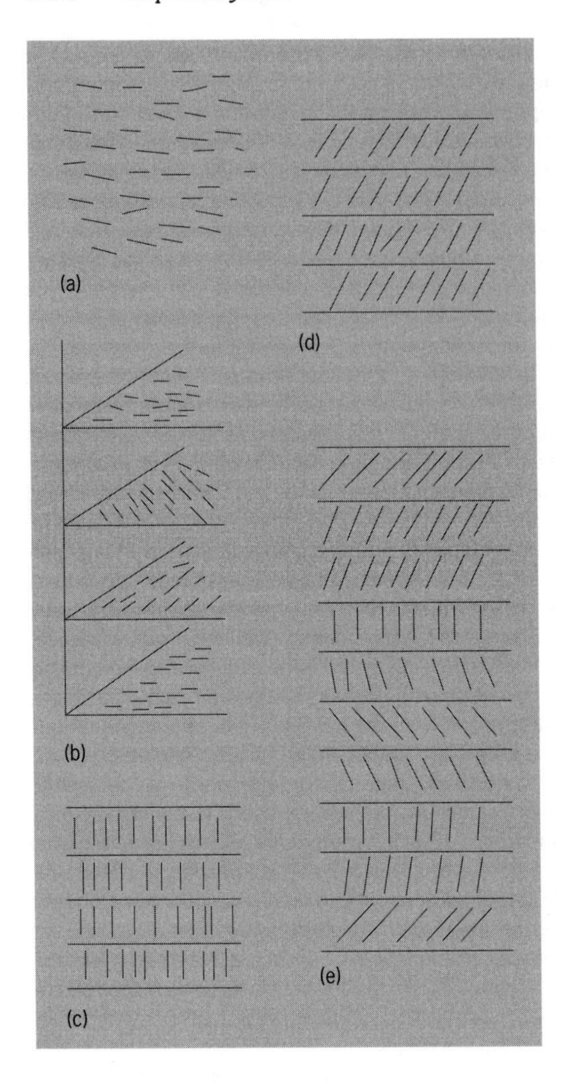


**Molecular arrangement in liquid crystals. (*a*) Nematic. (*b*) Chiral-nematic. (*c*) Smectic A. (*d*) Smectic C. (*e*) Smectic C* ($S_c^+$). (*After R. Bline, Solid and liquid crystalline ferroelectrics and antiferroelectrics, Condensed Matter News 1(1):17–23, 1991*)**

liquid crystal is so small that even the slightest perturbation caused by a dust particle can distort the structure considerably.

In the chiral-nematic structure (illus. *b*), the direction of the long axis of the molecule in a given layer is slightly displaced from the direction of the molecular axes of the molecules in an adjacent layer. If a twist is applied to this molecular packing, a helical structure is formed. The helix has a pitch that is temperature-sensitive. The helical structure serves as a diffraction grating for visible light.

**Smectic structure.** The term smectic includes all thermotropic liquid crystals that are not nematics. In the smectic phase, not only is the small amount of orientational order of nematic liquid crystals present, but there is also a small amount of positional order. In most smectic structures, the molecules are free to bounce around randomly; but they tend to point along a specific direction and arrange themselves in layers, either in neat rows or randomly distributed. The molecules can move in two directions in the plane and rotate about one axis.

Smectic liquid crystals may have structured or unstructured strata. Structured smectic liquid crystals have long-range order in the arrangement of molecules in layers to form a regular two-dimensional lattice. The most common of the structured liquid crystals is smectic B. Molecular layers are in well-defined order, and the arrangement of the molecules within the strata is also well ordered. The long axes of the molecules lie perpendicular to the plane of the layers. In the smectic A (illus. *c*) structure, molecules are also packed in strata, but the molecules in a stratum are randomly arranged. The long axes of the molecules in the smectic A structure lie perpendicular to the plane of the layers. Molecular packing in a smectic C (illus. *d*) is the same as that in smectic A, except the molecules in the stratum are tilted at an angle to the plane of the stratum.

**Applications.** Liquid crystals are a state of matter that combines a kind of long-range order (in the sense of a solid) with the ability to form droplets and to pour (in the sense of waterlike liquids). They also exhibit properties of their own such as the ability to form monocrystals with the application of a normal magnetic or electric field; an optical activity of a magnitude without parallel in either solids or liquids; and a temperature sensitivity that results in a color change in certain liquid crystals. As such, liquid crystals have many applications. They are used as displays in digital wristwatches, calculators, and panel meters. They can be used to record, store, and display images which can be projected onto a large screen. Direct and active-matrix liquid-crystal displays (LCDs) are used in several areas from laptop computers to communication equipment such as television teleconferencing systems, portable and high-definition television (HDTV), and video games. [G.H.B.; J.W.D.; D.Fi.]

**Lithium** A chemical element, Li, atomic number 3, and atomic weight 6.939. Lithium heads the alkali metal family in the periodic table. In nature it is a mixture of the isotopes $^{6}Li$ and $^{7}Li$. Lithium, the lightest solid element, is a soft, low-melting, reactive metal. In many physical and chemical properties it resembles the alkaline-earth metals as much as, or more than, it does the alkali metals. *See* ALKALINE-EARTH METALS; PERIODIC TABLE.

The major industrial use of lithium is in the form of lithium stearate as a thickener for lubricating greases. Other important uses of lithium compounds are in ceramics, specifically in porcelain enamel formulation; as an additive to give longer life and higher output in alkaline storage batteries; and in welding and brazing fluxes.

Lithium is a moderately abundant element and is present in the Earth's crust to the extent of 65 parts per million (ppm). This places lithium a little below nickel, copper, and tungsten, and a little above cerium and tin in abundance.

Noteworthy among lithium's physical properties are the high specific heat (heat capacity), large temperature range of the liquid phase, high thermal conductivity, low viscosity, and very low density. Lithium metal is soluble in liquid ammonia and is slightly soluble in the lower aliphatic amines, such as ethyl-amine. It is insoluble in hydrocarbons.

Lithium undergoes a large number of reactions with both organic, and inorganic, reagents. It reacts with oxygen to form the monoxide, $Li_2O$, and the peroxide, $Li_2O_2$. Lithium is the only alkali metal that reacts with nitrogen at room temperature to form a nitride, $Li_3N$, which is black. Lithium reacts readily with hydrogen at about 930°F (500°C) to form lithium hydride, LiH. The reaction of lithium metal with water is exceedingly vigorous. Lithium reacts directly with carbon to form the carbide, $Li_2C_2$. Lithium combines readily with the halogens, forming halides with the emission of light. While lithium does not react with paraffin hydrocarbons, it does undergo addition reactions with arylated alkenes and with dienes. Lithium also reacts with acetylenic compounds, forming lithium acetylides, which are important in the synthesis of vitamin A.

The most important lithium compound is lithium hydroxide. It is a white powder, and the material of commerce is actually lithium hydroxide monohydrate, $LiOH \cdot H_2O$. Lithium carbonate, $LiCO_3$, finds application in the ceramic industries and in medicine as an antidepressant. Both lithium halides, lithium chloride and lithium bromide, form concentrated brines with ability to absorb moisture over a wide temperature range; these brines are used in commercial air conditioning systems. [M.Si.]

**Luminescence analysis** Methods of chemical analysis in which analyte concentration is related to luminescence intensity or some other property of luminescence. Photoluminescence, particularly fluorescence, is the most widely used type of luminescence for chemical analysis. However, there are also important analytical applications of both chemiluminescence and bioluminescence.

Luminescence is light that accompanies the transition from an electronically excited atom or molecule to a lower energy state. The forms of luminescence are distinguished by the method used to produce the electronically excited species. When produced by absorption of incident radiation, the light emission is known as photoluminescence. Photoluminescence that is short-lived ($10^{-8}$ s or less between excitation and emission) is known as fluorescence. Photoluminescence that is longer-lived (from $10^{-6}$ s all the way up to seconds) is known as phosphorescence. The reason for the difference in lifetime is that fluorescence involves an allowed, high-probability, transition while phosphorescence involves a forbidden, low-probability, transition.

Chemiluminescence is observed when the electronically excited atom or molecule is formed as the product of a chemical reaction. If the light-producing reaction occurs in nature, such as the light emitted by fireflies, it is known as bioluminescence. *See* CHEMILUMINESCENCE.

Photoluminescence excitation spectra are determined by measuring emission intensity at a fixed wavelength while varying the wavelength of the incident light used to produce the electronically excited species responsible for emission. The excitation spectrum is a measure of the efficiency of electronic excitation as a function of excitation wavelength. Photoluminescence emission spectra are determined by exciting at a fixed wavelength and varying the wavelength at which emission is observed. The

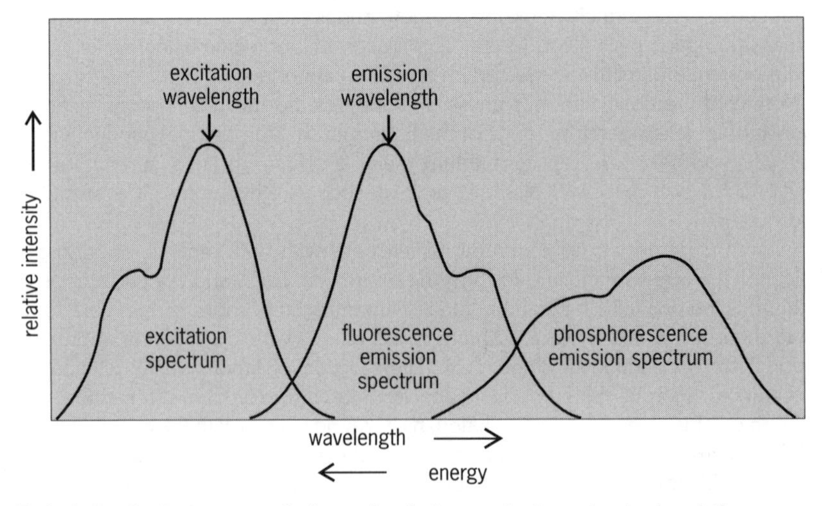


**Typical photoluminescence excitation and emission spectra for molecules in solution.**

illustration shows typical excitation and fluorescence and phosphorescence emission spectra for a molecule in solution. Between excitation and emission, electronically excited molecules normally lose some of their energy because of relaxation processes. As a consequence, the emission spectrum is at longer wavelengths, that is, at lower energy, than the excitation spectrum. Because the magnitude of the energy loss due to relaxation processes is greater for phosphorescence than for fluorescence, phosphorescence occurs at longer wavelengths than fluorescence. The spectra extend over a range of wavelengths, because in solution, molecules exist in a continuous distribution of vibrational and rotational energy levels.

Observed luminescence intensities depend on three factors: (1) the number of electronically excited molecules or atoms produced by the excitation process; (2) the fraction of electronically excited molecules that emit light as they relax to a lower energy state; and (3) the fraction of the emitted luminescence that impinges on the detector and is measured. In the case of photoluminescence, the number of excited molecules is proportional to incident excitation intensity, the concentration of the luminescent species, and the efficiency with which the luminescence species absorbs the incident radiation. In the case of chemiluminescence and bioluminescence, the number of excited molecules depends on reactant concentrations and the efficiency with which the reaction pathway leads to production of the excited state. The dependence of intensity on concentration is the basis for chemical analysis based on luminescence.

By far the most important advantage of luminescence methods of chemical analysis is their ability to measure extremely low concentrations. This advantage arises because luminescence is measured relative to weak background signals. In contrast, methods based on the absorption of light require that a small difference between two large signals be measured. [W.R.Se.]

**Lutetium** A chemical element, Lu, atomic number 71, atomic weight 174.97, a very rare metal and the heaviest member of the rare-earth group. The naturally occurring element is made up of the stable isotope $^{175}Lu$, 97.41%, and the long-life $\beta$-emitter $^{176}Lu$ with a half-life of $2.1 \times 10^{10}$ years. *See* PERIODIC TABLE.

Lutetium, along with yttrium and lanthanum, is of interest to scientists studying magnetism. All of these elements form trivalent ions with only subshells which have been completed, so they have no unpaired electrons to contribute to the magnetism. Their radii with regard to the other rare-earth ions or metals are very similar so they form at almost all compositions either solid solutions or mixed crystals with the strongly magnetic rare-earth elements. Therefore, the scientist can dilute the magnetically active rare earths in a continuous manner without changing appreciably the crystal environment. *See* MAGNETOCHEMISTRY; RARE-EARTH ELEMENTS. [F.H.Sp.]

# M

**Macrocyclic compound** An organic compound that contains a large ring. In the organic chemistry of alicyclic compounds, a closed chain of 12 carbon (C) atoms is usually regarded as the minimum size for a large ring; crown ethers are similarly defined. Macrocyclic compounds may be a single, continuous thread of atoms, as in cyclododecane [$(CH_2)_{12}$], or they may incorporate more than one strand or other ring systems (subcyclic units) within the macrocycle or macroring. In addition, macrocycles may be composed of aromatic rings that confer considerable rigidity upon the cyclic system. These aromatic rings may be joined together or coupled by spacer units consisting of one or more carbon atoms. *See* AROMATIC HYDROCARBON.

**Classes of macrocyclic polyethers.** Crown ethers are generally composed of repeating ethylene ($CH_2CH_2$) units separated by noncarbon atoms such as oxygen (O), nitrogen (N), sulfur (S), phosphorus (P), or silicon (Si). By far, the most common heteroatom present in the macrorings of crowns [X in $(XCH_2CH_2)_n$] is oxygen; but as more intricate structures are prepared, nitrogen, sulfur, phosphorus, silicon, or siloxy residues are becoming much more common.

By adding a third strand to the simple macrocyclic polyethers, three-dimensional compounds based on the crown framework are formed. Typically, two of the oxygen atoms across the ring from each other are replaced by nitrogens, and a third ethyleneoxy chain is attached to them. Known as cryptands, these structures completely encapsulate cations smaller than their internal cavities and strongly bind the most similar in size.

Two crown ether rings may be held together by a crown-ether-like strand to give a bicyclic cryptand. These have sometimes been referred to as ditopic receptors because they possess two distinct binding sites.

Lariat ethers, spherands, calixarenes, cavitands, and carcerands are other types of macrocyclic compounds, all of which are capable of encapsulating "guest" molecules in their interior cavities.

Cyclophane is the name given to macrocyclic compounds that contain organic (usually aromatic) rings as part of a cavity-containing structure. The first such compound was [2.2]-paracyclophane. In it, two benzene rings are joined by ethylene ($CH_2CH_2$) chains in their para positions. *See* CYCLOPHANE.

**Complexation phenomena.** It is the ability of these macrocyclic host compounds to complex a variety of guest species that makes these structures interesting. A crown ether can be described as a doughnut with an electron-rich and polar hole and a greasy or lipophilic (hydrophobic) exterior. As a result, these compounds are usually quite soluble in organic solvents but accommodate positively charged species in their holes.

A variety of organic cations have been found to complex with crown ethers and related hosts. It has been suggested that for a host-guest interaction to occur, the host must have convergent binding sites and the guest must have divergent sites. This is illustrated by the interaction between optically active dibinaphtho-22-crown-6 and optically active phenethylammonium chloride. The crown ether oxygen atoms converge to the center of a hole and the ammonium hydrogens diverge from nitrogen. Three complementary O—H—N hydrogen bonds stabilize the complex. In this particular case, different steric interactions between the optically active crown and the enantiomers of the complex permit resolution of the salt.

Other organic cations have also been complexed, either by insertion of the charged function in the crown's polar hole or by less distinct interactions observed in the solid state. *See* COORDINATION CHEMISTRY; COORDINATION COMPLEXES.

**Applications.** The striking ability of neutral macrocyclic polyethers to complex with alkali and alkaline-earth cations as well as a variety of other species has proved of considerable interest to the chemistry community. Crown ethers may complex the cation associated with an organic salt and cause separation of the ions. In the absence of cations to neutralize them, many anions show considerably enhanced reactivity. *See* ORGANIC REACTION MECHANISM.

One of the important modern developments in synthetic chemistry was the use of the phase-transfer technique. Nucleophiles such as cyanide are often insoluble in media that dissolve organic compounds with which they react. Thus 1-bromooctane may be heated in the presence of sodium cyanide for days with no product formation. When a crown ether is added, two things change. First, solubility is enhanced because the crown wraps about the cation, making it more lipophilic. This, in turn, makes the entire salt more lipophilic. Second, by solvating the cation, the association between cation and anion and the interactions with solvent are weakened, thus activating the anion for reaction. This approach has been used to assist the dissolution of potassium permanganate ($KMnO_4$) in benzene in which solvent permanganate is a powerful oxidizing agent. One striking example of solubilization is the displacement of chloride ($Cl^-$) by fluoride ($F^-$) in dimethyl 2-chloroethylene-1,1-dicarboxylate by using the KF complex of dicyclohexano-18-crown-6. In this reaction, a crown provides solubility for an otherwise insoluble or marginally soluble salt. Use of crowns to transfer a salt from the solid phase into an organic phase is often referred to as solid-liquid phase-transfer catalysis. *See* CATALYSIS; PHASE-TRANSFER CATALYSIS.

Since crown ethers and related species complex cations selectively, they can be used as sensors. Crowns have been incorporated into electrodes for this purpose, and crowns having various appended chromophores have been prepared. When a cation is bound within the macroring, a change in electron density is felt in the chromophore. The chromophores are often nitroaromatic residues and therefore highly colored. The color change that accompanies complexation can be easily detected and quantitated. *See* ION-SELECTIVE MEMBRANES AND ELECTRODES. [G.W.G.]

**Magnesium** A metallic chemical element, Mg, in group 2 of the periodic system, atomic number 12, atomic weight 24.312. Magnesium is silvery white and extremely light in weight. The specific gravity is 1.74, and the density is 1740 kg/m$^3$ (0.063 lb/in.$^3$ or 108.6 lb/ft$^3$). Because of this lightness combined with alloy strength suitable for many structural uses, magnesium has long been known as industry's lightest structural metal. *See* PERIODIC TABLE.

With a density only two-thirds that of aluminum, magnesium is used in countless applications where weight saving is an important consideration. The metal also has, however, many desirable chemical and metallurgical properties which account for its extensive use in a variety of nonstructural applications.

Magnesium is very abundant in nature, occurring in substantial amounts in many rock-forming minerals such as dolomite, magnesite, olivine, and serpentine. In addition,

**Table 1. Physical properties of primary magnesium (99.9% pure)**

| Property | Value |
|---|---|
| Atomic number | 12 |
| Atomic weight | 24.312 |
| Atomic volume, $cm^3$/g-atom | 14.0 |
| Crystal structure | Close-packed hexagonal |
| Electron arrangement in free atoms | (2) (8) 2 |
| Mass numbers of the isotopes | 24, 25, 26 |
| Percent relative abundances of $^{24}Mg$, $^{25}Mg$, $^{26}Mg$ | 77, 11.5, 11.5 |
| Density, g/$cm^3$ at 20°C | 1.738 |
| Specific heat, cal/g/°C at 20°C (1 cal = 4.2 joules) | 0.245 |
| Melting point, °C | 650 |
| Boiling point, °C | 1110±10 |

magnesium is also found in sea water, subterranean brines, and salt beds. It is the third most abundant structural metal in the Earth's crust, exceeded only by aluminum and iron.

Some of the properties of magnesium in metallic form are listed in Table 1. Magnesium is very active chemically. It will actually displace hydrogen from boiling water, and a large number of metals can be prepared by thermal reduction of their salts and oxides with magnesium. The metal will combine with most nonmetals and with practically all acids. Magnesium reacts only slightly or not at all with most alkalies and many organic chemicals, including hydrocarbons, aldehydes, alcohols, phenols, amines, esters, and most oils. As a catalyst, magnesium is useful for promoting organic condensation, reduction, addition, and dehalogenation reactions. It has long been used for the synthesis of complex and special organic compounds by the well-known Grignard reaction. Principal alloying ingredients include aluminum, manganese, zirconium, zinc, rare-earth metals, and thorium.

**Table 2. Principal magnesium compounds and uses**

| Compound | Uses |
|---|---|
| Magnesium carbonate | Refractories, production of other magnesium compounds, water treatment, fertilizers |
| Magnesium chloride | Cell feed for production of metallic magnesium, oxychloride cements, refrigerating brines, catalyst in organic chemistry, production of other magnesium compounds, flocculating agent, treatment of foliage to prevent fire and resist fire, magnesium melting and welding fluxes |
| Magnesium hydroxide | Chemical intermediate, alkali, medicinal |
| Magnesium oxide | Insulation, refractories, oxychloride and oxysulfate cements, fertilizers, rayon-textile processing, water treatment, papermaking, household cleaners, alkali, pharmaceuticals, rubber filler catalyst |
| Magnesium sulfate | Leather tanning, paper sizing, oxychloride and oxysulfate cements, rayon delustrant, textile dyeing and printing, medicinal, fertilizer ingredient, livestock-food additive, ceramics, explosives, match manufacture |

Magnesium compounds are used extensively in industry and agriculture. Table 2 lists the major magnesium compounds and indicates some of their more significant applications. [W.H.Gr.; S.C.E.]

**Magnetochemistry** The branch of chemistry which studies the interrelationship between a magnetic held and atomic and molecular structures.

A substance in a magnetic field acquires an intensity of magnetization which may be either smaller or larger than that induced in a vacuum by the same field. In the first case, the substance is said to be diamagnetic. In the second case, the substance may be paramagnetic, ferromagnetic, or antiferromagnetic.

Diamagnetism, a universal property of matter, is usually of the order of magnitude $10^{-6}$ to $10^{-5}$. Temperature-dependent paramagnetism, on the other hand, arises only when an atom, ion, or molecule possesses a permanent magnetic moment either in the ground state or in an excited state. A permanent magnetic moment is the result of the presence of one or more unpaired electrons. Paramagnetic susceptibilities are of the order of magnitude $10^{-4}$ to $10^{-3}$.

A substance composed of atoms with permanent magnetic moments which are very near to one another (for example, iron metal) may display ferromagnetism. This phenomenon occurs when large numbers of the atoms with permanent magnetic moments interact so that their individual moments align in a parallel fashion, giving rise to a large resultant moment.

On the other hand, a similar substance (for example, manganese metal) may display antiferromagnetism. Here, the magnetic moments align in an antiparallel fashion, thus largely canceling the individual magnetic moments of the atoms. Parallel versus antiparallel alignment depends, among other factors, upon interatomic distances. *See* ATOMIC STRUCTURE AND SPECTRA; ELECTRON PARAMAGNETIC RESONANCE (EPR) SPECTROSCOPY; MOLECULAR STRUCTURE AND SPECTRA. [D.M.Gr.]

**Maillard reaction** A nonenzymatic chemical reaction involving condensation of an amino group and a reducing group, resulting in the formation of intermediates which ultimately polymerize to form brown pigments (melanoidins). The reaction was named for the French biochemist Louis-Camille Maillard. It is of extreme importance to food chemistry, especially because of its ramifications in terms of food quality. *See* AMINE; REACTIVE INTERMEDIATES.

There are three major stages of the reaction. The first comprises glycosylamine formation and rearrangement *N*-substituted-1-amino-l-deoxy-2-ketose (Amadori compound). The second phase involves loss of the amine to form carbonyl intermediates, which upon dehydration or fission form highly reactive carbonyl compounds through several pathways. The third phase occurring upon subsequent heating involves the interaction of the carbonyl flavor compounds with other constituents to form brown nitrogen-containing pigments (melanoidins). These are highly desirous compounds in certain foods browned by heating in the presence of oxygen.

The Maillard reaction is considered undesirable in some biological and food systems. The interaction of carbonyl and amine compounds might damage the nutritional quality of proteins by reducing the availability of lysine and other essential amino acids and by forming inhibitory or antinutritional compounds. The reaction is also associated with undesirable flavors and colors in some foods, particularly dehydrated foods. *See* AMINO ACIDS; CARBONYL. [M.E.B.]

**Manganese** A metallic element, Mn, atomic number 25, and atomic weight 54.9380 g/mole. Manganese is one of the transition elements of the first long period

**Properties of manganese**

| Property | Value |
|---|---|
| Atomic number | 25 |
| Atomic weight, g/mole | 54.9380 |
| Naturally occurring isotope | $^{55}Mn$ (100%) |
| Electronic configuration | $[Ar]3d^54s^2$ |
| Electronegativity | 1.5 |
| Metal radius, picometers | 127 |
| Melting point, °C (°F) | 1244 ± 3 (2271 ± 5.4) |
| Boiling point. °C (°F) | 1962 (3563) |
| Density (25 °C or 77°F), g/cm$^3$ (oz/in.$^3$) | 7.43 (4.30) |
| Electrical resistivity, ohm·cm | $185 \times 10^{-6}$ |

of the periodic table, falling between chromium and iron. The principal properties of manganese are given in the table. It is the twelfth most abundant element in the Earth's crust (approximately 0.1%) and occurs naturally in several forms, primarily as the silicate ($MnSiO_3$) but also as the carbonate ($MnCO_3$) and a variety of oxides, including pyrolusite ($MnO_2$) and hausmannite ($Mn_3O_4$). Weathering of land deposits has led to large amounts of the oxide being washed out to sea, where they have aggregated into the so-called manganese nodules containing 15–30% Mn. Vast deposits, estimated at over $10^{12}$ metric tons, have been detected on the seabed, and a further $10^7$ metric tons is deposited every year. The nodules also contain smaller amounts of the oxides of other metals such as iron (Fe), cobalt (Co), nickel (Ni), and copper (Cu). The economic importance of the nodules as a source of these important metals is enormous. *See* PERIODIC TABLE.

Manganese is more electropositive than its near neighbors in the periodic table, and consequently more reactive. The bulk metal undergoes only surface oxidation when exposed to atmospheric oxygen, but finely divided metal is pyrophoric.

Manganese is a trace element essential to a variety of living systems, including bacteria, plants, and animals. In contrast to iron (Fe), its neighbor in the periodic table, the exact function of the manganese in many of these systems was determined only recently. The manganese superoxide dismutases have been isolated from bacteria, plants, and animals, and are relatively small enzymes with molecular weights of approximately 20,000. The function of the enzyme is believed to be protection of living tissue from the harmful effects of the superoxide ion ($O_2{}^-$), a radical formed from partial reduction of $O_2$ in the cells of respiring ($O_2$-utilizing) cells.

The most important biological role yet recognized for manganese is in the enzyme responsible for photosynthetic water oxidation to oxygen in plants and certain photosynthetic bacteria. This reaction represents the source of oxygen gas on the Earth and is therefore responsible for the development of the most common forms of life.

All steels contain some manganese, the major advantage being an increase in hardness, although it also serves as a scavenger of oxygen and sulfur impurities that would induce defects and consequent brittleness in the steel. Manganese even has some use in the electronics industry, where manganese dioxide, either natural or synthetic, is employed to produce manganese compounds possessing high electrical resistivity; among other applications, these are utilized as components in every television set. *See* ELECTROLYSIS; TRANSITION ELEMENTS. [G.Ch.]

**Mass number** The mass number *A* of an atom is the total number of its nuclear constituents, or nucleons, as the protons and neutrons are collectively called. The mass number is placed before and above the elemental symbol, thus $^{238}U$. The mass number gives a useful rough figure for the atomic mass; for example, $^{1}H = 1.00814$ atomic mass units (amu), $^{238}U = 238.124$ amu, and so on. *See* ATOMIC NUMBER. [H.E.D.]

**Mass spectrometry** An analytical technique for identification of chemical structures, determination of mixtures, and quantitative elemental analysis, based on application of the mass spectrometer. Determination of organic and inorganic molecular structure is based on the fragmentation pattern of the ion formed when the molecule is ionized; further, because such patterns are distinctive, reproducible, and additive, mixtures of known compounds may be analyzed quantitatively. Quantitative elemental analysis of organic compounds requires exact mass values from a high-resolution mass spectrometer; trace analysis of inorganic solids requires a measure of ion intensity as well. *See* MASS SPECTROSCOPE.

For analysis of organic compounds the principal methods are electron impact, chemical ionization, field ionization, field desorption, particle bombardment, laser desorption, and electrospray.

In electron impact, when a gaseous sample of a molecular compound is ionized with a beam of energetic (commonly 70-V) electrons, part of the energy is transferred to the ion formed by the collision, as shown in the reaction below.

$$\text{A—B—C} + e \longrightarrow \text{A—B—C}^{+} + e + e$$

For most molecules the production of cations is favored over the production of anions by a factor of about $10^4$, and the following discussion pertains to cations. The ion corresponding to the simple removal of the electron is commonly called the molecular ion and normally will be the ion of greatest $m/e$ ratio in the spectrum. In the ratio, $m$ is the mass of the ion in atomic mass units (daltons) and $e$ is the charge of the ion measured in terms of the number of electrons removed (or added) during ionization. Occasionally the ion is of vanishing intensity, and sometimes it collides with another molecule to abstract a hydrogen or another group. In these cases an incorrect assignment of the molecular ion may be made unless further tests are applied. Proper identification gives the molecular weight of the sample.

The remaining techniques were devised generally to circumvent the problem of the weak or vanishingly small intensity of a molecular ion.

In chemical ionization the ions to be analyzed are produced by transfer of a heavy particle ($H^+$, $H^-$, or heavier) to the sample from ions produced from a reactant gas.

Field ionization and field desorption is used for less volatile material. The sample is ionized when it is in a very high field gradient (several volts per angstrom) near an electrode surface. The molecular potential well is distorted so that an electron tunnels from the molecule to the anode. The ion thus formed is repelled by the anode. Typically, the lifetime of the ion in the mass spectrometer source is much less ($10^{-12}$ to $10^{-9}$ s) than in electron impact. Because little energy is transferred as internal energy and the ion is removed rapidly, little fragmentation occurs, and the molecular weight is more easily determined.

In electrohydrodynamic ionization, a high electric-field gradient induces ion emission from a droplet of a liquid solution, that is, the sample and a salt dissolved in a solvent of low volatility. An example is the sample plus sodium iodide (NaI) dissolved in glycerol. The spectra include peaks due to cationized molecules of $MNa^+$, $MNa(C_3H_8O_3)_n{}^+$, and $Na(C_3H_8O_3)_n{}^+$.

A sample heated very rapidly may vaporize before it pyrolyzes. Techniques for heating by raising the temperature of a source probe on which the sample is coated by 200 K (360°F)/s have been developed. Irradiation of an organic sample with laser radiation can move ions of mass up to 1500 daltons into the gas phase for analysis. This technique is the most compatible with analyzers that require particularly low pressures such as ion cyclotron resonance. Time-resolved spectra of surface ejecta are proving to be the most useful kinds of laser desorption spectra available.

A solid sample or a sample in a viscous solvent such as glycerol may have ions sputtered from its surface by bombardment with accelerated electrons, ions, or neutrals. Bombardment by electrons is achieved simply by inserting a probe with the sample directly into the electron beam of an electron impact source (in-beam electron ionization).

In laser desorption, the energy of laser photons may be used to remove an analyte from a surface and ionize it for mass-spectrometric analysis. In virtually all cases involving polar compounds of high molecular weight, the analyte is prepared in or on an organic matrix that is coated on the surface to be irradiated. The matrix material assists in the ionization process; the technique is known as matrix-assisted laser desorption ionization (MALDI). Numerous organic materials have been investigated. Molecular weights in excess of 700,000 have been measured with this technique, and the technique is particularly suited for the molecular-weight determination of large polar biological molecules; for example, enzymes and intact antibodies have been analyzed.

In electrospray ionization, a solution sprayed through a nozzle of very small diameter into a vacuum with an electric field having a gradient of several hundred to a thousand volts per centimeter produces gaseous ions from solutes effectively. Electrospray ionization is the only mass-spectrometric technique that produces a large fraction of multiply charged ions from an organic or biological analyte. Since mass spectrometers measure mass-to-charge ratios of ions, not simply their mass, electrospray ionization has the advantage of permitting ions of very high mass to be analyzed without special mass-analysis instrumentation; for example, an ion of mass 120,000 daltons carrying 60 positive charges appears at mass-to-charge 2000, within the range of many mass analyzers. This technique has been used to measure the masses of ions from molecules of masses up to about 200,000 daltons. Since the distribution of charged species reflects to some extent the degree of protonation in solution, a signal that reflects the folding of a protein, there is evidence that solution conformations of proteins can be studied by electrospray ionization mass spectrometry.

Measurement using a mass spectrometer takes advantage of the mass dependency in the equations of motion of an ion in an electric or magnetic field. Three common mass analyzers are magnetic-sector, quadrupole, and time-of-flight. An ion traversing the magnetic field of a sector instrument successfully reaches the detector when its mass, $m$, corresponds to $m/z = r^2B^2/2V$, where $z$ is the charge of the ion, $r$ is the radius of the flight tube, $B$ is the magnetic field strength, and $V$ is the acceleration voltage. A mass spectrum can be recorded by varying $B$ or $V$. A quadrupole is an array of four parallel rods electronically connected such that a radio frequency (RF) is applied to opposing rods, and the waveforms applied to adjacent rods are out of phase. In addition, a direct-current (DC) component is added to the rods such that one opposing pair has +DC and the other −DC. Ions entering the quadrupole follow complex sinusoidal paths. An ion successfully traversing the rods to the detector is mass-selected by the amplitude of the applied radio frequency and the amount of resolution from adjacent masses by the DC/RF ratio. A spectrum is detected by scanning the RF amplitude and the DC voltage in a fixed ratio. The time-of-flight analyzer is a tube with an acceleration field at one end and a detector at the other. Ions are pulsed and accelerated down the tube, starting a clock. Arrival times at the detector are converted to mass by solving the equation $m/z = 2V(t/L)^2$, where $V$ is the acceleration potential, $t$ is the arrival time, and $L$ is the flight tube length.

The mass analyzers are characterized by their mass resolving power (RP), scan or acquisition rate, and mass range. Sector instruments are usually assembled with an additional energy filter and can be operated up to 100,000 RP. Resolving power is defined as $M/\Delta M$, where $M$ is the mass number of the observed mass and $\Delta M$ is the mass difference in two signals with a 10% valley between them. Increased resolution in sectors comes at the sacrifice of sensitivity. To increase resolution, the slits are narrowed on the ion beam. It is typical to lose 90% of the signal going from 1000 to 10,000 RP. Magnetic instruments are available with mass ranges up to 10,000 atomic mass units (amu) for singly charged ions at full acceleration potential. To obtain good quality, data sectors are scanned relatively slowly—a practical rate is 5 s per decade (1000–100 amu). Quadrupoles are usually operated at unit resolution. For example, 100 is resolved from 101, or 1000 from 1001. Quadrupoles can be scanned at 2000 amu/s and are available with mass range up to 4000 amu. Fast scanning and high-gas-pressure tolerance make quadrupoles popular for mass spectrometer–chromatographic hybrid systems. Modern time-of-flight analyzers are capable of 10,000 RP, 50% valley definition of $\Delta M$. Acquisition rates can be 100 spectra per second. The time-of-flight mass range is theoretically unlimited, and singly charged ions over 100,000 amu have been observed. *See* ATOMIC NUMBER; MASS NUMBER. [T.D.Wi.]

The analysis of solid inorganic samples can be made either by vaporization in a Knudsen cell arrangement at very high temperatures or by volatilization of the sample surface so that particles are atomized and ionized with a high-energy spark (for example, 20,000 eV). The wide range of energies given to the particles requires a double-focusing mass spectrometer for analysis. Detection in such instruments is by photographic plate; exposures for different lengths of time are recorded sequentially, and the darkening of the lines on the plate is related empirically to quantitative composition by calibration charts. The method is useful for trace analysis (parts per billion) with accuracy ranging from 10% at higher concentration levels to 50% at trace levels. Methods for improving accuracy, including interruption and sampling of the ion beam, are under development.

Secondary ion mass spectrometry is most commonly used for surface analysis. A primary beam of ions accelerated through a few kilovolts is focused on a surface; ions are among the products sputtered from this surface, and they may be directly analyzed in a quadrupole filter. Sputtered material may also be analyzed by ionization of the neutrals in an inductively coupled plasma and subsequent mass analysis. This method produces ions with a lower energy spread than spark source mass spectrometry and, since detection then becomes less of a problem, has been supplanting spark source methods. *See* LASER SPECTROSCOPY; SECONDARY ION MASS SPECTROMETRY (SIMS); SPECTROSCOPY. [M.M.Bu.]

**Mass spectroscope** An instrument used for determining the masses of atoms or molecules found in a sample of gas, liquid, or solid. It is analogous to the optical spectroscope, in which a beam of light containing various colors (white light) is sent through a prism to separate it into the spectrum of colors present. In a mass spectroscope, a beam of ions (electrically charged atoms or molecules) is sent through a combination of electric and magnetic fields so arranged that a mass spectrum is produced (see illustration). If the ions fall on a photographic plate which after development shows the mass spectrum, the instrument is called a mass spectrograph; if the spectrum is allowed to sweep across a slit in front of an electrical detector which records the current, it is called a mass spectrometer.

Mass spectroscopes are used in both pure and applied science. Atomic masses can be measured very precisely. Because of the equivalence of mass and energy, knowledge of nuclear structure and binding energy of nuclei is thus gained. The relative

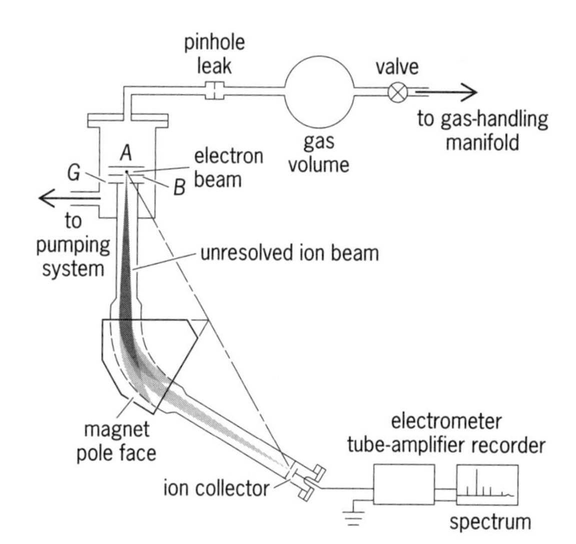


**Schematic drawing of mass spectrometer tube. Ion currents are in the range $10^{-10}$ to $10^{-15}$ ampere and require special electrometer tube amplifiers for their detection. In actual instruments the radius of curvature of ions in a magnetic field is 4–6 in. (10–15 cm).**

abundances of the isotopes in naturally occurring or artificially produced elements can be determined.

Empirical and theoretical studies have led to an understanding of the relation between molecular structure and the relative abundances of the fragments observed when a complex molecule, such as a heavy organic compound, is ionized. When a high-resolution instrument is employed, the masses of the molecular or fragment ions can be determined so accurately that identification of the ion can frequently be made from the mass alone. *See* MASS SPECTROMETRY.

Because chemical compounds may have mass spectra as unique as fingerprints, mass spectroscopes are widely used in industries such as oil refineries, where analyses of complex hydrocarbon mixtures are required. [A.O.N.]

**Matrix isolation** A technique for providing a means of maintaining molecules at low temperature for spectroscopic study. This method is particularly well suited for preserving reactive species in a solid, inert environment. Elusive molecular fragments, such as free radicals that may be postulated as important controlling intermediates for chemical transformations used in industrial reactions, high-temperature molecules that are in equilibrium with solids at very high temperatures, and molecular ions that are produced in plasma discharges or by high-energy radiation all can be examined by using absorption (infrared, visible, and ultraviolet), electron-spin resonance, and laser-excitation spectroscopes.

The experimental apparatus for matrix isolation experiments is designed with the method of generating the molecular transient and performing the spectroscopy in mind. The illustration shows the cross section of a vacuum vessel used for absorption spectroscopic measurements. The matrix sample is introduced through the spray-on line; argon is the most widely used matrix gas, although neon, krypton, xenon, and nitrogen are also used. The reactive species can be generated in a number of ways: mercury-arc photolysis of a trapped precursor molecule through the quartz window, evaporation from a Knudsen cell in the heater, chemical reaction of atoms evaporated from the Knudsen cell with molecules deposited through the spray-on line, and vacuum-ultraviolet photolysis of molecules deposited from the spray-on line by radiation from discharge-excited atoms flowing through the tube. For laser excitation studies,

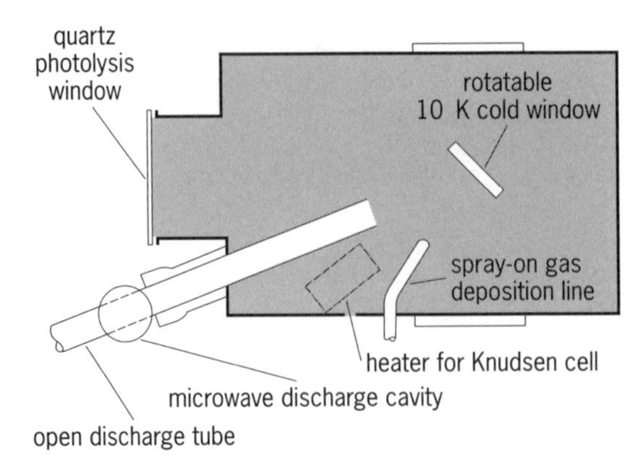


**Vacuum-vessel base cross section for matrix photoionization experiments.**

the sample is deposited on a tilted copper wedge which is grazed by the laser beam, and light emitted or scattered at approximately 90° is examined by a spectrograph. In electronspin resonance studies, the sample is condensed on a sapphire rod that can be lowered into the necessary waveguide and magnet.

The matrix isolation technique enables spectroscopic data to be obtained for reactive molecular fragments, many of which cannot be studied in the gas phase. [L.A.]

**Mechanical separation techniques** A group of laboratory and production operations whereby the components of a polyphase mixture are separated by mechanical methods into two or more fractions of different mechanical characteristics. The separated fractions may be homogeneous or heterogeneous, particulate or nonparticulate.

The techniques of mechanical separation are based on differences in phase density, in phase fluidity, and in such mechanical properties of particles-as size, shape, and density; and on such particle characteristics as wettability, surface charge, and magnetic susceptibility. Obviously, such techniques are applicable only to the separation

**Types of mechanical separator**

| Materials separated | Separators |
|---|---|
| Liquid from liquid | Settling tanks, liquid cyclones, centrifugal decanters, coalescers |
| Gas from liquid | Still tanks, deaerators, foam breakers |
| Liquid from gas | Settling chambers, cyclones, electrostatic precipitators, impingement separators |
| Solid from liquid | Filters, centrifugal filters, clarifiers, thickeners, sedimentation centrifuges, liquid cyclones, wet screens, magnetic separators |
| Liquid from solid | Presses, centrifugal extractors |
| Solid from gas | Settling chambers, air filters, bag filters, cyclones, impingement separators, electrostatic and high-tension precipitators |
| Solid from solid | |
| By size | Screens, air and wet classifiers, centrifugal classifiers |
| By other characteristics | Air and wet classifiers, centrifugal classifiers, jigs, tables, spiral concentrators, flotation cells, dense-medium separators, magnetic separators, electrostatic separators |

of phases in a heterogeneous mixture. They may be applied, however, to all kinds of mixtures containing two or more phases, whether they are liquid-liquid, liquid-gas, liquid-solid, gas-solid, solid-solid, or gas-liquid-solid.

Methods of mechanical separations fall into four general classes: (1) those employing a selective barrier such as a screen or filter cloth; (2) those depending on difference in phase density alone (hydrostatic separators); (3) those depending on fluid and particle mechanics; and (4) those depending on surface or electrical characteristics of particles. A wide variety of separation devices have been devised and are in use. The more important kinds of equipment are listed in the table, grouped according to the phases involved. *See* CENTRIFUGATION; CLARIFICATION; ELECTROSTATIC PRECIPITATOR; FILTRATION; SEDIMENTATION (INDUSTRY). [S.A.M.]

**Meitnerium** The seventeenth of the synthetic transuranium elements. Element 109 falls in column 9 of the periodic table under the elements cobalt, rhodium, and iridium. It is expected to have chemical properties similar to those of iridium. *See* IRIDIUM; PERIODIC TABLE; TRANSURANIUM ELEMENTS.

Element 109 was discovered in 1982 by a team under P. Armbruster and G. Münzenberg at the Gesellschaft für Schwerionenforschung (GSI) at Darmstadt, Germany. In a sequence of bombardments of bismuth-209 targets with beams of ions of titanium-50, chromium-54, and iron-58, the compound systems $^{259}105$, $^{263}107$, and $^{267}109$ were produced. The decay analysis of the isotopes produced showed in the case of elements 105 and 107 the production of $^{258}105$ and $^{262}107$ by reaction channels in which one neutron is emitted. These isotopes have odd neutron and proton numbers and possess a special stability against spontaneous fission. It was shown that alpha-particle decay dominated the decay chains. Spontaneous fission occurs through a 30% electron capture branch of $^{256}105$ in $^{258}104$. Three decay chains were observed for the three reactions ending by fission of $^{258}104$, and the decay of the first atom of element 109 was observed. *See* DUBNIUM; NUCLEAR REACTION; RUTHERFORDIUM.

The single atom of element 109 was produced at a bombarding energy of 299 MeV in the reaction between iron-58 and bismuth-209. A total dose of $7 \times 10^{17}$ ions was used to bombard thin layers of bismuth during a 250-h irradiation time. [P.Ar.]

**Melting point** The temperature at which a solid changes to a liquid. For pure substances, the melting or fusion process occurs at a single temperature, the temperature rise with addition of heat being arrested until melting is complete.

Melting points reported in the literature, unless specifically stated otherwise, have been measured under an applied pressure of 1 atm ($10^5$ pascals), usually 1 atm of air. (The solubility of air in the liquid is a complicating factor in precision measurements.) Upon melting, all substances absorb heat, and most substances expand; consequently an increase in pressure normally raises the melting point. A few substances, of which water is the most notable example, contract upon melting; thus, the application of pressure to ice at 32°F (0°C) causes it to melt. Large changes in pressure are required to produce significant shifts in the melting point.

For solutions of two or more components, the melting process normally occurs over a range of temperatures, and a distinction is made between the melting point, the temperature at which the first trace of liquid appears, and the freezing point, the higher temperature at which the last trace of solid disappears, or equivalently, if one is cooling rather than heating, the temperature at which the first trace of solid appears. *See* PHASE EQUILIBRIUM; SOLUTION; SUBLIMATION; TRIPLE POINT. [R.L.S.]

**Membrane distillation** A separation method in which a nonwetting, microporous membrane is used with a liquid feed phase on one side of the membrane and a condensing, permeate phase on the other side. Separation by membrane distillation is based on the relative volatility of various components in the feed solution. The driving force for transport is the partial pressure difference across the membrane. Separation occurs when vapor from components of higher volatility passes through the membrane pores by a convective or diffusive mechanism.

Membrane distillation shares some characteristics with another membrane-based separation known as pervaporation, but there also are some vital differences. Both methods involve direct contact of the membrane with a liquid feed and evaporation of the permeating components. However, while membrane distillation uses porous membranes, pervaporation uses nonporous membranes.

Membrane distillation systems can be classified broadly into two categories: direct-contact distillation and gas-gap distillation. These terms refer to the permeate or condensing side of the membrane; in both cases the feed is in direct contact with the membrane. In direct-contact membrane distillation, both sides of the membrane contact a liquid phase; the liquid on the permeate side is used as the condensing medium for the vapors leaving the hot feed solution. In gas-gap membrane distillation, the condensed permeate is not in direct contact with the membrane.

Potential advantages of membrane distillation over traditional evaporation processes include operation at ambient pressures and lower temperatures as well as ease of process scaleup. *See* CHEMICAL SEPARATION TECHNIQUES; MEMBRANE SEPARATIONS.

[S.S.K.; N.N.L.]

**Membrane mimetic chemistry** The study of processes and reactions whose developments have been inspired by the biological membrane. Faithful modeling of the biomembrane is not an objective of membrane mimetic chemistry. Rather, only the essential components of natural systems are recreated from relatively simple, synthesized molecules. (The term membrane mimetic is more restrictive than the term biomimetic. Biomimetic chemistry is directed at the mechanistic elucidation of biochemical reactions and at the development of new compounds modeled on specific biological systems.)

Various surfactant aggregate systems have been used in membrane mimetics.

Surfactants (detergents) contain distinct hydrophobic (apolar) and hydrophilic (polar) regions. Depending on the chemical structure of their hydrophilic polar head groups, surfactants can be neutral, positively charged, or negatively charged.

Aqueous micelles are spherical aggregates, 4–8 nanometers in diameter, formed dynamically from surfactants in water above a characteristic concentration, the critical micelle concentration. *See* MICELLE.

Monomolecular layers are formed by spreading naturally occurring lipids or synthetic surfactants, dissolved in volatile solvents, over water in a trough. The polar head groups of the surfactants are in contact with water, the subphase, while their hydrocarbon tails protrude above it. *See* MONOMOLECULAR FILM.

Other systems used in membrane mimetics are multilayer assemblies (Langmuir-Blodgett films), bilayer lipid membranes, and vesicles prepared by sonication from naturally occurring lipids. *See* SONOCHEMISTRY.

Membrane mimetic chemistry has become a versatile chemical tool. Applications of compartmentalization of reactants in membrane mimetic systems involve altered reaction rates, products, stereochemistries, and isotope distributions. Monolayers and organized multilayers can be employed profitably as molecular electronic devices. Opportunities also exist for using different surfactant aggregates with polymeric membranes for the control and regulation of reverse osmosis and ultrafiltration. *See* SURFACTANT; ULTRAFILTRATION. [J.H.Fe.]

**Membrane separations** Processes for separating mixtures by using thin barriers (membranes) between two miscible fluids. A suitable driving force across the membrane, for example concentration or pressure differential, leads to the preferential transport of one or more feed components.

Membrane separation processes are classified under different categories depending on the materials to be separated and the driving force applied: (1) In ultrafiltration, liquids and low-molecular-weight dissolved species pass through porous membranes while colloidal particles and macromolecules are rejected. The driving force is a pressure difference. (2) In dialysis, low-molecular-weight solutes and ions pass through while colloidal particles and solutes with molecular weights greater than 1000 are rejected under the conditions of a concentration difference across the membrane. (3) In electrodialysis, ions pass through the membrane in preference to all other species, due to a voltage difference. (4) In reverse osmosis, virtually all dissolved and suspended materials are rejected and the permeate is a liquid, typically water. (5) For gas and liquid separations, unequal rates of transport can be obtained through nonporous membranes by means of a solution and diffusion mechanism. Pervaporation is a special case of this separation where the feed is in the liquid phase while the permeate, typically drawn under subatmospheric conditions, is in the vapor phase. (6) In facilitated transport, separation is achieved by reversible chemical reaction in the membrane. High selectivity and permeation rate may be obtained because of the reaction scheme. Liquid membranes are used for this type of separation. *See* DIALYSIS; ION-SELECTIVE MEMBRANES AND ELECTRODES; OSMOSIS; TRANSPORT PROCESSES; ULTRAFILTRATION. [N.N.L; S.S.K.]

**Mendelevium** A chemical element, Md, atomic number 101, the twelfth member of the actinide series of elements. Mendelevium does not occur in nature; it was discovered and is prepared by artificial nuclear transmutation of a lighter element. Known isotopes of mendelevium have mass numbers from 248 to 258 and half-lives from a few seconds to about 55 days. They are all produced by charged-particle bombardments of more abundant isotopes. The amounts of mendelevium which are produced and used for studies of chemical and nuclear properties are usually less than about a million atoms; this is of the order of a million times less than a weighable amount. Studies of the chemical properties of mendelevium have been limited to a tracer scale. The behavior of mendelevium in ion-exchange chromatography shows that it exists in aqueous solution primarily in the 3+ oxidation state characteristic of the actinide elements. However, it also has a dipositive (2+) and a monopositive (1+) oxidation state. *See* ACTINIDE ELEMENTS; PERIODIC TABLE; TRANSURANIUM ELEMENTS. [G.T.S.]

**Mercaptan** One of a group of organosulfur compounds which are also called thiols or thio alcohols and which have the general structure RSH. Aromatic thiols are called thiophenols, and biochemists often refer to thiols as sulfhydryl compounds. The unpleasant odor of volatile thiols causes them to be classed as stenches, but the odors of many solid thiols are not unpleasant.

Mercaptans (1) form salts with bases, (2) are easily oxidized to disulfides and higher oxidation products such as sulfonic acids, (3) react with chlorine (or bromine) to form sulfenyl chlorides (or bromides), and (4) undergo additions to unsaturated compounds, such as olefins, acetylenes, aldehydes, and ketones. The insoluble mercury salts (mercaptides) are used to isolate and identify mercaptans. *See* ORGANOSULFUR COMPOUND. [N.K.]

**Mercury (element)** A chemical element, Hg, atomic number 80 and atomic weight 200.59. Mercury is a silver-white liquid at room temperature (melting point −38.89°C or −37.46°F); it boils at 357.25°C (675.05°F) under atmospheric pressure. It is a noble metal that is soluble only in oxidizing solutions. Solid mercury is as soft as lead. The metal and its compounds are very toxic. With some metals (gold, silver, platinum, uranium, copper, lead, sodium, and potassium, for example) mercury forms solutions called amalgams. *See* TRANSITION ELEMENTS.

In its compounds, mercury is found in the 2+, 1+, and lower oxidation states, for example, $HgCl_2$, $Hg_2Cl_2$, or $Hg_3(AsF_6)_2$. Often the mercury atoms are doubly covalently bonded, for example, Cl—Hg—Cl or Cl—Hg—Hg—Cl. Some mercury(II) salts, for example, $Hg(NO_3)_2$ or $Hg(ClO_4)_2$, are quite soluble in water and dissociate normally. The aqueous solutions of these salts react as strong acids because of hydrolysis. Other mercury(II) salts, for example, $HgCl_2$ or $Hg(CN)_2$, also dissolve in water, but exist in solution as only slightly dissociated molecules. There are compounds in which mercury atoms are bound directly to carbon or nitrogen atoms, for example, $H_3C$—Hg—$CH_3$ or $H_3C$—CO—NH—Hg—NH—CO—$CH_3$. In complex compounds, for example, $K_2(HgI_4)$, mercury often has three or four bonds.

Metallic mercury is used as a liquid contact material for electrical switches, in vacuum technology as the working fluid of diffusion pumps, for the manufacture of mercury-vapor rectifiers, thermometers, barometers, tachometers, and thermostats, and for the manufacture of mercury-vapor lamps. It finds application for the manufacture of silver amalgams for tooth fillings in dentistry. Of importance in electrochemistry are the standard calomel electrode, used as the reference electrode for the measurement of potentials and for potentiometric titrations, and the Weston standard cell.

Mercury is commonly found as the sulfide, HgS, frequently as the red cinnabar and less often as the black metacinnabar. A less common ore is the mercury(I) chloride. Occasionally the mercury ore contains small drops of metallic mercury.

The surface tension of liquid mercury is 484 dynes/cm, six times greater than that of water in contact with air. Hence, mercury does not wet surfaces with which it is in contact. In dry air metallic mercury is not oxidized. After long standing in moist air, however, the metal becomes coated with a thin layer of oxide. In air-free hydrochloric acid or in dilute sulfuric acid, the metal does not dissolve. Conversely, it is dissolved by oxidizing acids (nitric acid, concentrated sulfuric acid, and aqua regia). [K.B.]

**Meso-ionic compound** A member of a class of five-membered ring heterocycles (and their benzo derivatives) which possess a sextet of $\pi$ electrons in association with the atoms composing the ring but which cannot be represented satisfactorily by any one covalent or polar structure.

Two main types, depending formally upon the origin of the electrons in the $\pi$ system, have been identified; they are exemplified by compounds (**1**) and (**2**). In structure (**1**)

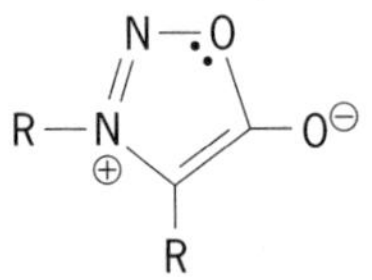

(1) 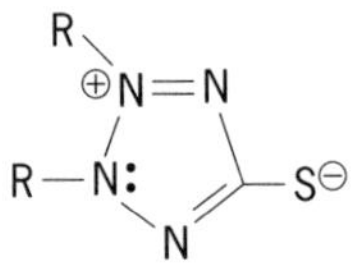 (2)

the nitrogen and oxygen atoms, 1,3 to each other, are shown as donating two electrons each to the total of eight electrons in the whole $\pi$ system, whereas in structure (**2**) the two middle nitrogen atoms, 1,2 to each other, are the two-electron donors.

The term "satisfactorily" in the definition refers to the fact that the charge in the ring cannot be associated exclusively with one ring atom. Thus, these compounds are in sharp contrast with other dipolar structures, such as ylides, and such compounds are not considered meso-ionic.

There has been considerable interest in the pharmacological activity of meso-ionic compounds, and derivatives have shown a variety of antibiotic, anthelminthic, antidepressant, and anti-inflammatory properties. [J.P.Fr.]

**Metal cluster compound** A compound in which two or more metal atoms are bonded to one another. Metal cluster compounds bridge the gap between the solid-state chemistry of the metals—or their lower-valent oxides, chalcogenides, and related salts—and the complexes of the metals in which each metal ion is completely surrounded by and bonded to a set of ligands or ions. The latter group comprises the classical coordination chemistry of metal ions. *See* COORDINATION CHEMISTRY; COORDINATION COMPLEXES; SOLID-STATE CHEMISTRY.

Interest in metal cluster compounds arises from unique features of their chemistry: (1) Cluster compounds provide models for studying fundamental reactions on surfaces. (2) There is a hope that cluster compounds may provide entry to new classes of catalysts that may be tailored to specific syntheses and may thus be more selective than existing processes. (3) The nature of the bonding in cluster compounds is an area wherein experiment and theory are continuously challenging each other. (4) The systematic synthesis of mixed metal clusters may provide for the development of new types of supported catalysts (the discrete clusters are deposited on supports such as alumina, silica, or zeolites). *See* ATOM CLUSTER; NANOSTRUCTURE. [M.H.Ch.]

**Metal hydrides** A compound in which hydrogen is bonded chemically to a metal or metalloid element. The compounds are classified generally as ionic, transition metal, and covalent hydrides. Covalent hydrides are of two subtypes, binary and complex. Certain hydrides have achieved a position of modest industrial importance, but most are of theoretical interest only.

Under extreme conditions such as in electric discharges, many metals form volatile, short-lived transient hydrides of the general formula type MH. Although some of these can be prepared experimentally, most are observed only by their spectra. They are important in studying molecular bonding. The action of atomic hydrogen at low temperatures forms surface films of unstable hydrides with many metals. *See* HYDRIDO COMPLEXES; HYDROGEN. [J.C.W.]

**Metallocene catalyst** A transition-metal atom sandwiched between ring structures having a well-defined single catalytic site and well-understood molecular structure used to produce uniform polyolefins with unique structures and physical properties. *See* CATALYSIS; COORDINATION CHEMISTRY; COORDINATION COMPLEXES; METALLOCENES; ORGANOMETALLIC COMPOUND.

In the early 1980s, W. Kaminsky discovered that an appropriate co-catalyst activated metallocene compounds of group 4 metals, that is, titanium, zirconium, and hafnium, for alpha-olefin polymerization, attracting industrial interest. This observation led to the synthesis of a great number of metallocene compounds for the production of polymers already made industrially, such as polyethylene and polypropylene, and new materials. Polymers produced with metallocene catalysts represent a small fraction of the entire polyolefin market, but experts agree that such a fraction will increase rapidly in the future. *See* POLYMER; POLYMERIZATION; POLYOLEFIN RESINS.

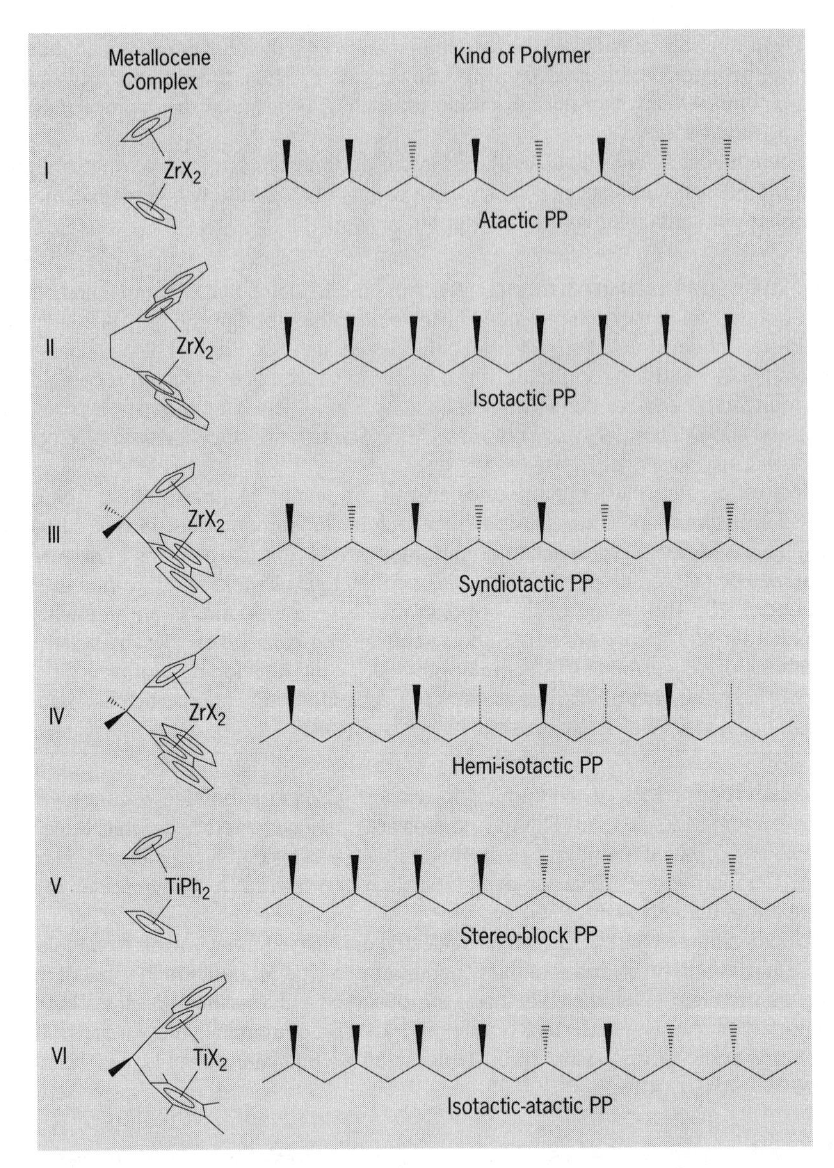


**Correlation between the metallocene structure and the obtained polymer microstructure. PP = polypropylene.**

The simplest metallocene precursor has the formula $Cp_2MX_2$, where M is one of the group 4 metals (mainly Zr and Ti) and X are halogen atoms (mainly chlorine, Cl). The latter are known as mobile ligands because during polymerization they are substituted or removed. A typical co-catalyst, in the absence of which the activity is very low, is methylaluminoxane (MAO), an oligomeric compound described by the formula $(CH_3AlO)_n$, whose structure is not yet fully understood. MAO plays several roles: it alkylates the metallocene precursor by replacing chlorine atoms with methyl groups; it produces the catalytic active ion pair $Cp_2MCH_3^+/MAO^-$, where the cationic moiety is considered responsible for polymerization and $MAO^-$ acts as weakly coordinating anion.

The simplest metallocene structures are easily modified by replacing the Cp ligands with other variously substituted derivatives. In this way, a great number of catalysts with different steric and electronic properties are generated. The catalysts contain two C5 ring derivatives, always lying on tilted planes, which can be bridged or unbridged. Some examples are shown in the illustration, where the influence of the metallocene structure on the microstructure of the polymer product is also shown.

Because activity, stereospecificity, regiospecificity, and relative reactivity toward different monomers depend on the catalysts' characteristics, the metallocene systems offer the advantage of controlling the product through modifications of their chemical structure. [R.F.; F.G.; L.Lo.]

**Metallocenes** Bis-cyclopentadienyl derivatives of transition metals whose bonding involves overlap of *ns*, $(n-1)d$, and *np* orbitals of the metal with molecular orbitals of appropriate symmetry of each cyclopentadienyl ring. The resulting complexes often possess two parallel rings (sandwich structure), but in some cases, for example those involving the titanium subgroup of metals, the rings are canted (see illustration). Metals in the periodic table commonly known to form metallocene complexes are titanium, zirconium, hafnium, vanadium, chromium, molybdenum, tungsten, manganese, iron, ruthenium, osmium, cobalt, rhodium, and nickel. *See* COORDINATION COMPLEXES.

The reactions of metallocenes can be divided into two classes: the first is typified by the iron triad, and comprises essentially the reactions of aromatic molecules; the second consists of the reactions of the other metallocenes where the 18-electron rare-gas configuration is not found. Reactions in these latter systems often lead to a product where the 18-electron rule is obeyed.

Ferrocene is a very electron-rich system, and undergoes electrophilic substitution with great rapidity. For example, acylation proceeds about $10^6$ times faster than that of benzene under similar conditions. Ferrocene also undergoes several other typical aromatic substitution reactions besides acylation, including sulfonation, dimethylaminomethylation (Mannich reaction), metalation, and the like. Bis substitution tends to factor a product where each ring is monosubstituted, although several cases are known where two substituents are introduced into one ring. Ferrocene is oxidized and deactivated under conditions for nitration and halogenation.

Uses of metallocenes include reaction of chromocene with alumina to make a polymerization catalyst for ethylene. Ferrocene and some alkyl-substituted ferrocenes have been used as moderators in high-temperature combustions such as occur in solid rocket fuels. A cyclopentadienyl complex, $CH_3C_5H_4Mn(CO)_3$, briefly replaced

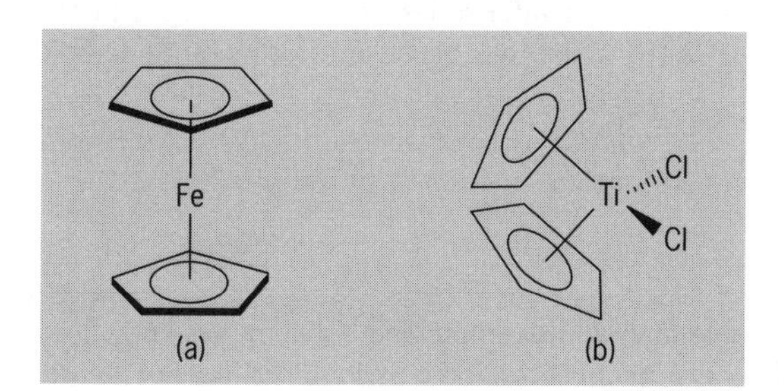


**Metallocene structures. (*a*) Staggered sandwich structure of ferrocene. (*b*) Canted cyclopentadienyl ring structure of titanocene dichloride. The distribution of the ligands about the Ti atom is tetrahedral.**

tetraethyllead as an octane booster and antiknock agent in liquid fuels. *See* ORGANOMETALLIC COMPOUND. [D.W.Sl.]

**Metallochaperones** A family of proteins that shuttle metal ions to specific sites within a cell. The target sites for metal delivery include a number of metalloenzymes, or proteins that bind metal ions, such as copper, zinc, or iron, and use these ions as cofactors to carry out essential biochemical reactions. Metallochaperones escort the ion to a specific intracellular location and facilitate incorporation of the metal into designated metalloenzymes. *See* BIOINORGANIC CHEMISTRY.

The bulk of current knowledge on metallochaperones is restricted to copper, although it is reasonable to assume that a distinct class of proteins is responsible for the incorporation of other metal ion cofactors into metalloenzymes. Among the metallochaperones that have been studied in detail are a family of three copper chaperones. These molecules operate in eukaryotic (nucleated) cells to direct copper to distinct intracellular locations: the mitochondria, the secretory pathway, and the cytosol. The first copper chaperone identified, COX17, is a small protein that specifically directs copper to the mitochondria. The copper delivered by COX17 is inserted into the metalloenzyme cytochrome oxidase, needed for respiration. A second copper chaperone identified was ATX1, which carries copper to the secretory pathway, a cellular compartment that functions to shuttle proteins toward the cell surface. The metal delivered by ATX1 is incorporated into copper enzymes destined for the cell surface or the extracellular milieu. The most recently identified copper chaperone is CCS, which specifically delivers copper to a single metalloenzyme, superoxide dismutase. This copper-requiring enzyme is located in the soluble cytosolic compartment of the cell and acts to detoxify harmful reactive oxygen species.

Intracellular copper is normally present at exquisitely low levels, and activation of copper enzymes is wholly dependent upon copper chaperones. Copper not only is an essential nutrient but also is quite toxic to living cells, and elaborate detoxification mechanisms prevent the free metal ion from accumulating to any substantial degree. The copper-requiring metalloenzymes cannot compete for these vanishingly low levels of available metal, explaining the requirement for the copper metallochaperones. [V.C.C.]

**Metalloid** An element which exhibits the external characteristics of a metal but behaves chemically both as a metal and as a nonmetal. Arsenic and antimony, for example, are hard crystalline solids that are definitely metallic in appearance. They may, however, undergo reactions that are characteristic of both metals and nonmetals. However, only when this dualistic chemical behavior is very marked and the external appearance metallic is the element commonly called a metalloid. *See* NONMETAL. [F.J.J.]

**Methane** A member of the alkane or paraffin series of hydrocarbons with the formula shown below. Methane is called marsh gas because it forms by anaerobic

```
    H
    |
H — C — H
    |
    H
```

bacterial decomposition of vegetable matter in swampy land. Coal miners know it as firedamp because mixtures with air are combustible. It is a major constituent of natural gas (50–90%) and of coal gas. It forms in large amounts in sewage disposal processes, especially in anaerobic digestion. As a liquid it freezes at −182.6°C (−296.7°F) and boils at −161.6°C (−258.9°F).

In addition to its use as a fuel, methane is important as a source of organic chemicals and of hydrogen. Its reaction with steam at high temperatures in the presence of catalysts yields carbon monoxide and hydrogen (synthesis gas), which can be catalytically converted to liquid alkanes (Fischer-Tropsch process) or to methanol and other alcohols. *See* ALKANE; FISCHER-TROPSCH PROCESS; HYDROFORMYLATION. [L.S.]

**Methanol** The first member of the homologous series of aliphatic alcohols, with the formula $CH_3OH$. It is produced commercially from a mixture of carbon monoxide (CO) and hydrogen ($H_2$). Methanol is a highly flammable liquid, boiling point 64.7°C (149°F), and is miscible with water and most organic liquids. It is a highly poisonous substance; sublethal amounts can cause permanent blindness. *See* ALCOHOL.

Methanol is one of the major industrial organic chemicals. Its major derivatives are methyl tertiary butyl ether (MTBE), formaldehyde, and acetic acid. Other derivatives and uses include chloromethanes, methyl methacrylate, methylamines, dimethyl terephthalate, solvents (such as glycol methyl ethers), antifreeze, and fuels. [J.A.Mo.]

**Micelle** A colloidal aggregate of a unique number (50→100) of amphipathic molecules, which occurs at a well-defined concentration called the critical micelle concentration. In polar media such as water, the hydrophobic part of the amphiphiles forming the micelle tends to locate away from the polar phase while the polar parts of the molecule (head groups) tend to locate at the polar micelle solvent interface. A micelle may take several forms, depending on the conditions and composition of the system, such as distorted spheres, disks, or rods (see illustration). Micelles are formed in nonpolar media such as benzene, where the amphiphiles cluster around small water droplets in the system, forming an assembly known as a reversed micelle.

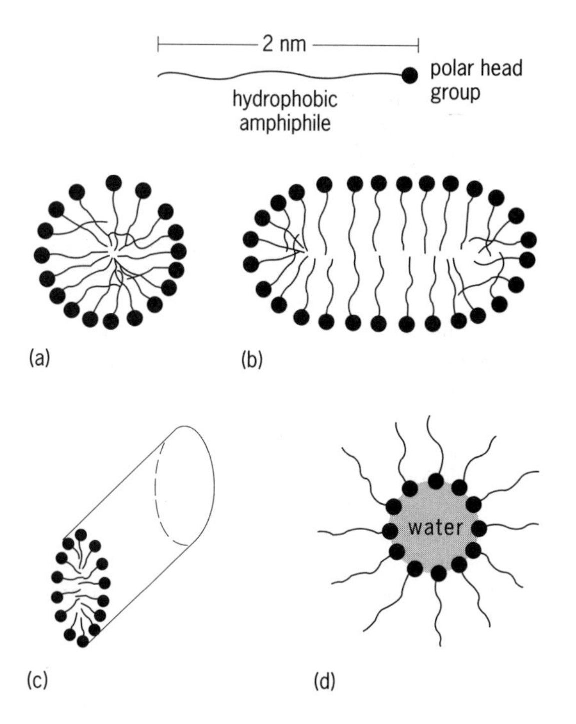


**Form of an amphiphile and several forms of micelle: *(a)* spherical, *(b)* disk, *(c)* rod, and *(d)* reversed.**

Micellar systems have the unique property of being able to solubilize both hydrophobic and hydrophilic compounds. They are used extensively in industry for detergency and as solubilizing agents. *See* SOAP. [J.K.T.]

**Microdialysis sampling** An approach for sampling the extracellular space of essentially any tissue or fluid compartment in the body. Continuous sampling can be performed for long periods with minimal perturbation to the experimental animal. Microdialysis provides a route for sampling the extracellular fluid without removing fluid, and administering compounds without adding fluid. The resulting sample is clean and amenable to direct analysis.

Microdialysis sampling is performed by implanting a short length of hollow-fiber dialysis membrane at the site of interest. The fiber is slowly perfused with a sampling solution (the perfusate) having an ionic composition and pH that closely matches the extracellular fluid of the tissue being sampled. Low-molecular-weight compounds in the extracellular fluid diffuse into the fiber and are swept to a collection vial for subsequent analysis. The system is analogous to an artificial blood vessel that can deliver compounds and remove the resulting metabolites. *See* DIALYSIS; MEMBRANE SEPARATIONS.

Microdialysis is a diffusion-controlled process. The perfusion rate through the probe is generally in the range of 0.5 to 5.0 ml/min. At this flow rate, there is no net flow of liquid across the dialysis membrane. The driving force for mass transport is the concentration gradient between the extracellular fluid and the fluid in the probe. *See* DIFFUSION; TRANSPORT PROCESSES.

The greatest use of microdialysis sampling has been in the neurosciences. Microdialysis probes can be implanted in specific brain regions of conscious animals in order to correlate neurochemical activity with behavior. Most studies have focused on determining dopamine or the other monoamine neurotransmitters.

Microdialysis probes have been implanted in the skin of experimental animals and humans to determine the transdermal delivery of drugs from ointments. Delivery of anticancer drugs to tumors has been studied using microdialysis.

Microdialysis sampling has also been used to study the metabolism of compounds in vivo. Metabolic organs such as the liver and kidneys have been studied by microdialysis sampling. By also sampling the bile by microdialysis, complete metabolic profiles can be obtained from a single experimental animal. This approach dramatically decreases the number of experimental animals needed to assess the metabolism of a new drug. [C.E.Lu.]

**Microwave spectroscopy** The study of the interaction of matter and electromagnetic radiation in the microwave region of the spectrum. *See* SPECTROSCOPY.

The interaction of microwaves with matter can be detected by observing the attenuation or phase shift of a microwave field as it passes through matter. These are determined by the imaginary or real parts of the microwave susceptibility (the index of refraction). The absorption of microwaves may also trigger a much more easily observed event like the emission of an optical photon in an optical double-resonance experiment or the deflection of a radioactive atom in an atomic beam.

At room temperature, the relative population difference between the states involved in a microwave transition is a few percent or less. The population difference can be close to 100% at liquid helium temperatures, and microwave spectroscopic experiments are often performed at low temperatures to enhance population differences and to eliminate certain line-broadening mechanisms. The population differences between the states involved in a microwave transition can also be enhanced by artificial means. When the molecules or atoms with inverted populations are placed in an appropriate microwave cavity, the cavity will oscillate spontaneously as a maser (microwave amplification by stimulated emission of radiation).

The magnetic dipole and electric quadrupole interactions between the nuclei and electrons in atoms and molecules can lead to energy splittings in the microwave region of the spectrum. Thus, microwave spectroscopy has been used extensively for precision determinations of spins and moments of nuclei.

The rotational frequencies of molecules often fall within the microwave range, and microwave spectroscopy has contributed a great deal of information about the moments of inertia, the spin-rotation coupling mechanisms, and other physical properties of rotating molecules. *See* MOLECULAR STRUCTURE AND SPECTRA.

The magnetic resonance frequencies of electrons in fields of a few thousand gauss (a few tenths of a tesla) lie in the microwave region. Thus, microwave spectroscopy is used in the study of electron-spin resonance or paramagnetic resonance. *See* ELECTRON PARAMAGNETIC RESONANCE (EPR) SPECTROSCOPY.

The cyclotron resonance frequencies of electrons in solids at magnetic fields of a few thousand gauss (a few tenths of a tesla) lie within the microwave region of the spectrum. Microwave spectroscopy has been used to map out the dependence of the effective mass on the electron momentum. [W.Hap.]

**Mixing** A common operation to effect distribution, intermingling, and homogeneity of matter. Actually the operation is called agitation, with the term mixing being

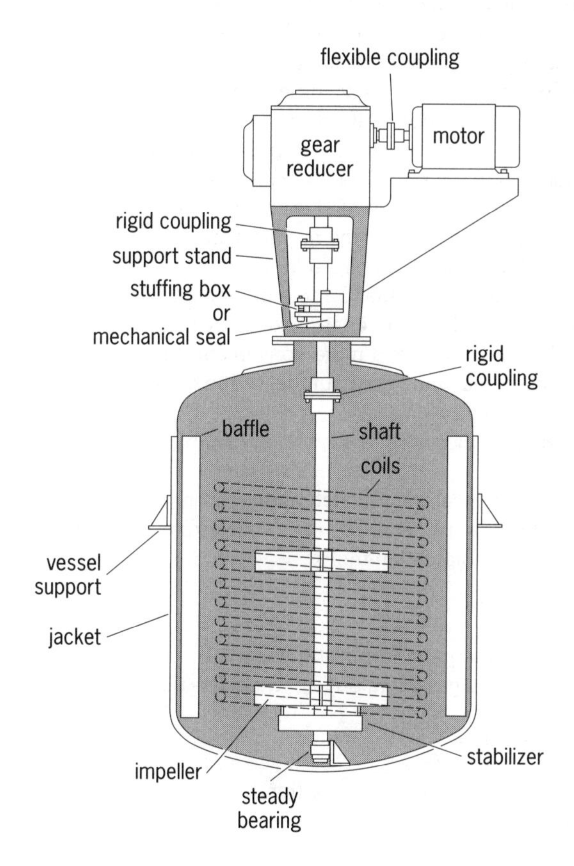


**Typical impeller-type liquid mixer. (*After V. W. Uhl and J. B. Gray, Mixing: Theory and Practice, vol. 2, Academic Press, 1967*)**

applicable when the goal is blending, that is, homogeneity. Other processes, such as reaction, mass transfer (includes solubility and crystallization), heat transfer, and dispersion, are also promoted by agitation. The type, extent, and intensity of agitation determine both the rates and adequacy of a particular process result. The agitation is accomplished by a variety of equipment.

Most liquid mixing is done by rotating impellers in vertical cylindrical vessels. A typical impeller-type liquid mixer with a variety of features is shown in the illustration. The internal features, including the vessel itself, are considered as a whole, that is, as the agitated system. The forces applied by the impeller develop overall circulation or bulk flow. Superimposed on this flow pattern, there is molecular diffusion, and if turbulence is present, also turbulent eddies. These provide micromixing. Solids, granular to powder, are mixed in a variety of contrivances.

Solids of different density and size are mixed in tumblers (a double cone turning end on end) or with agitators (a helical ribbon rotating in a horizontal trough). The duration of mixing is an important additional variable because classification and separation often occur after attainment of the desired distribution if the operation is carried on too long. [V.W.U.]

**Mole (chemistry)** A unit (symbolized mol) used to measure the amount of material in a chemical sample. The mole is defined by international agreement as the amount of substance (chemical amount) of a chemical system that contains as many molecules or entities as there are atoms in 12 g of carbon-12 ($^{12}C$). When the mole is used, the elementary entities need not be molecules, but they must always be specified. They may be atoms, molecules, ions, electrons, or specified groups of such particles.

Three obvious ways of measuring the amount of material in a given sample are to measure the mass of the sample, to measure the volume, or to count the number of molecules in the sample. Although it is more difficult to devise an experiment to count molecules, this third way of measuring amount is of special interest to chemists because molecules react in simple rational proportions (for example, one molecule of A may react with one, or two, or three molecules of B, and so forth). However, to count molecules is inconvenient in practice because the numbers are so large. For any chemical, a mass of 1 kilogram of the sample contains a large number of molecules, of the order $10^{23}$–$10^{24}$. The mole is defined so that 1 mole of any substance always contains the same number of molecules. This number approximately $6.02 \times 10^{23}$, and is known as the Avogadro number. The mole is a more convenient unit in which to measure the amount of a chemical than counting the number of molecules, and it has the same advantages. *See* AVOGADRO NUMBER.

The amount of substance (chemical amount) of a sample, $n$, may be determined in practice by one of three methods.

The value of $n$ (the amount of substance) may be determined from the mass $m$ by dividing by the molar mass $M$ of the sample, as in Eq. (1). If $m$ is expressed in g and $M$ in g/mol, then the value of $n$ will be obtained in mol.

$$n = \frac{m}{M} \qquad (1)$$

For a gas, the value of $n$ may be determined from the volume $V$, pressure $p$, and absolute temperature $T$ by using the ideal gas equation (2), where $R$ is the gas constant

$$n = \frac{pV}{RT} \qquad (2)$$

($R = 8.3145 \text{ J K}^{-1} \text{ mol}^{-1}$). If $pV$ is expressed in $(\text{N m}^{-2}) \times (\text{m}^3) = \text{J}$, and $RT$ in $\text{J mol}^{-1}$, then $pV/RT$ gives the value of $n$ in mol.

For a solution, the amount of solute (or the amount concentration of solution) is frequently determined by titration: if $\nu_A$ molecules of A react with $\nu_B$ molecules of B in the titration, then at the end point the amount of A used ($n_A$) is related to the amount of B ($n_B$) by Eq. (3), so that if one is known the other may be determined.

$$n_A = \frac{\nu_A}{\nu_B} n_B \tag{3}$$

*See* TITRATION.

The concentration of a solution may be recorded as (mass of solute)/(volume of solution), in units gram/liter; or as (chemical amount of solute)/(volume of solution) in units mol/liter. Because of the proportionality of chemical amount to number of molecules, the latter is the more useful measure of concentration and is generally used in chemistry and biochemistry. *See* CONCENTRATION SCALES. [I.M.M.]

**Molecular isomerism** The property of compounds (isomers) which have the same molecular formula but different physical and chemical properties. The difference in properties is caused by a difference in molecular structure (that is, molecular architecture). A typical example is dimethyl ether, $CH_3OCH_3$, a chemically quite inert gas which condenses at −24°C, and ethyl alcohol, $CH_3CH_2OH$, a liquid of substantial chemical reactivity which boils at 78°C; both compounds have the molecular formula $C_2H_6O$.

Isomers may be classified as constitutional isomers or stereoisomers. Constitutional isomers differ in constitution or connectedness, relating to the question as to which atoms are linked to which others and how. Dimethyl ether and ethanol (Fig. 1) are

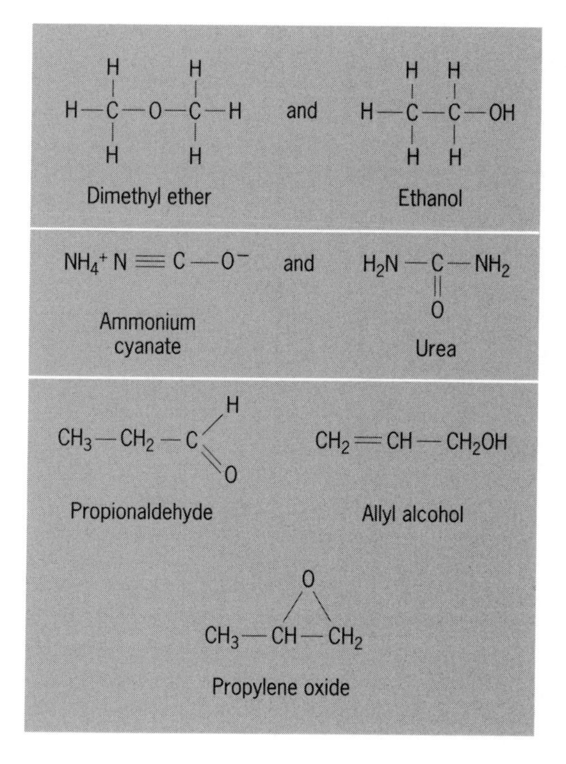


**Fig. 1. Functional isomers.**

constitutional isomers. In dimethyl ether each carbon is connected to three hydrogen atoms and the one oxygen atom; the two carbon atoms are thus equivalent. In ethyl alcohol (ethanol) one carbon is linked to three hydrogen atoms and the other carbon; the second carbon is linked to the first carbon, two hydrogens, and the oxygen atom which, in turn, is linked to the sixth hydrogen atom; the two carbon atoms are not equivalent. Stereoisomers, in contrast, have the same constitution but differ in the three-dimensional array of the atoms in space, called configuration.

**Constitutional isomers.** Constitutional isomers have been subdivided into functional isomers, positional isomers, and chain isomers.

Functional isomers (Fig. 1) differ in functional group, that is, the group (or groups) most material in determining chemical behavior. In the third example shown in Fig. 1 (propionaldehyde) the three compounds all correspond to the molecular formula $C_3H_6O$, but the first one has an aldehyde function, the second combines a double bond with an alcohol function, and the third one has an epoxide function.

Positional isomers (Fig. 2) have the same functional group but differ in its position along a chain or in a ring. Closely related are chain isomers which also have the same functional group or groups but differ in the shape of the carbon chain (Fig. 3*a*); quite similar are ring isomers (Fig. 3*b*) which differ in the size of one or more rings. Ring and chain isomers together are sometimes called skeletal isomers.

**Stereoisomers.** Compounds which have not only the same molecular formula but also the same constitution (connectivity of atoms) but which differ in the disposition of the atoms in space are called stereoisomers. Stereoisomers, in turn, are subdivided into two types: those that are mirror images of each other, called enantiomers, and those which are not mirror images, called diastereomers or diastereoisomers.

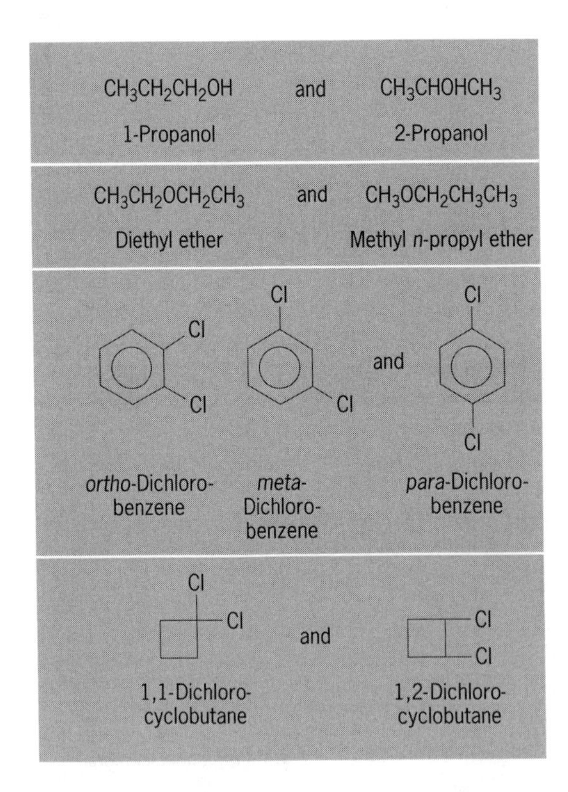


**Fig. 2. Positional isomers.**

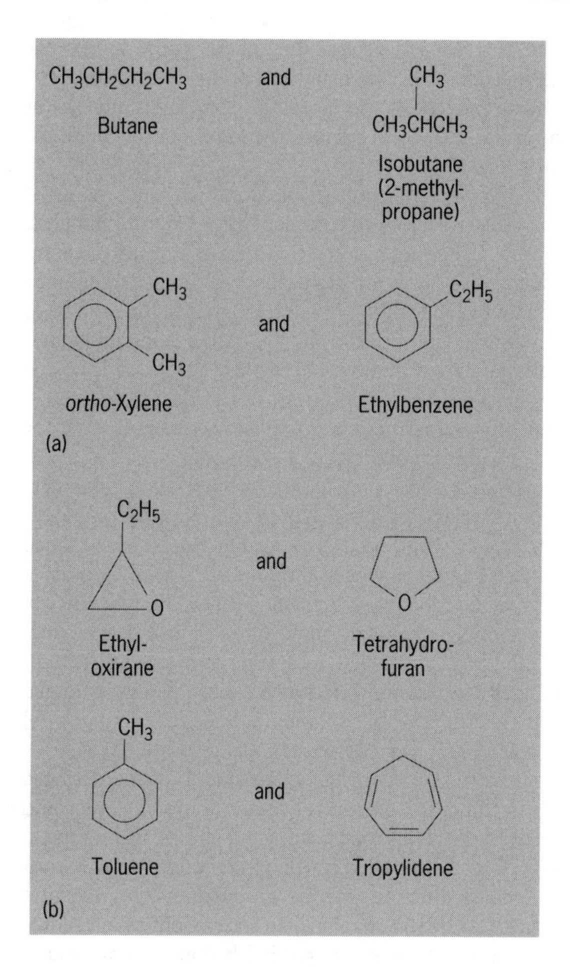


**Fig. 3. Skeletal isomers. (*a*) Chain isomers. (*b*) Ring isomers.**

Enantiomers are unique in that they always come in pairs. Either a molecule is superposable with its mirror image, in which case it does not have an enantiomer, or it is not superposable with its mirror image, in which case it has one and only one enantiomer (since an object can have only one mirror image). Molecules which are not superposable with their mirror images are called chiral; those which are so superposable are called achiral. Enantiomers are much more alike than are other sets of isomers (constitutional isomers or diastereomers); thus they have the same melting point, boiling point, free energy, spectral properties, x-ray diffraction pattern, and so on.

Diastereomers have the same constitution but different spatial arrangement and are not mirror images. They resemble constitutional isomers in that there may be more than two isomers in a set and that their physical, energetic, and spectral properties are generally quite distinct. *See* OPTICAL ACTIVITY; STEREOCHEMISTRY; TAUTOMERISM. [E.L.E.]

**Molecular machine** A molecular device is an assemblage of a discrete number of molecular components (that is, a supramolecular structure) designed to achieve a specific function. Each molecular component performs a single act, while the entire supramolecular structure performs a more complex function, which results from the cooperation of the various molecular components. Molecular devices operate via electronic or nuclear rearrangements. Like any device, they need energy to operate and signals to communicate with the operator. The extension of the concept of a device, so common on a macromolecular level, to the molecular level is of interest not only for basic research but also for the growth of nanoscience and nanotechnology. *See* NANOTECHNOLOGY; SUPRAMOLECULAR CHEMISTRY.

A molecular machine is a particular type of molecular device in which the component parts can display changes in their relative positions as a result of some external stimulus. Such molecular motions usually result in changes of some chemical or physical property of the supramolecular system, resulting in a "readout" signal that can be used to monitor the operation of the machine. The reversibility of the movement, that is, the possibility to restore the initial situation by means of an opposite stimulus, is an essential feature of a molecular machine. Although there are a number of chemical compounds whose structure or shape can be modified by an external stimulus (for example, photoinduced cis-trans isomerization processes), the term "molecular machines" is used only for systems showing large-amplitude movements of molecular components.

The human body can be viewed as a very complex ensemble of molecular-level machines that power motions, repair damage, and orchestrate an inner world of sense, emotion, and thought. Among the most studied natural molecular machines are those based on proteins such as myosin and kynesin, whose motions are driven by adenosine triphosphate (ATP) hydrolysis. One of the most interesting molecular machines of the human body is ATP synthase, a molecular-level rotatory motor. In this machine, a proton flow through a membrane spins a wheellike molecular structure and the attached rodlike species. This changes the structure of catalytic sites, allowing uptake of adenosine diphosphate (ADP) and inorganic phosphate, their reaction to give ATP, and then the release of the synthesized ATP. *See* ADENOSINE TRIPHOSPHATE (ATP).

An artificial molecular machine performs mechanical movements analogous to those observed in artificial macroscopic machines (for example, tweezers, piston/cylinder, and rotating rings). Analogously to what happens for macroscopic machines, the energy to make molecular machines work (that is, the stimulus causing the motion of the molecular components of the supramolecular structure) can be supplied as light, electrical energy, or chemical energy. In most cases, the machinelike movement involves two different, well-defined and stable states, and is accompanied by on/off switching of some chemical or physical signal [absorption and emission spectra, nuclear magnetic resonance (NMR), redox potential, or hydronium ion ($H_3O^+$) concentration]. For this reason, molecular machines can also be regarded as bistable devices for information processing (Fig. 1). *See* ACID AND BASE.

The interest in molecular machines arises not only from their mechanical movements but also from their switching aspects. Computers are based on sets of components constructed by the top-down approach. This approach, however, is now close to its intrinsic limitations. A necessary condition for further miniaturization to increase the power of information processing and computation is the bottom-up construction of molecular-level components capable of performing the functions needed (chemical computer). The molecular machines described above operate according to a binary logic and therefore can be used for switching processes at the molecular level. It has already been shown that suitable designed machinelike systems can be employed to perform complex functions such as multipole switching, plug/socket connection of molecular wires, and XOR logic operation. *See* LOGIC. [V.B.]

**Molecular mechanics** A non-quantum-mechanical way of computing structures, energies, and some properties of molecules. While quantum mechanics treats electrons explicitly, molecular mechanics treats electrons implicitly. For this reason, molecular mechanics calculations are much faster than quantum calculations, and this method is heavily used by scientists who need to quickly determine shapes of molecules. The goal of molecular mechanics is to build a computer model of reality. This is done with potential energy functions. Parameters are then selected for those potential functions so that molecules whose structures, energies, and properties are known can be reproduced with a specified degree of precision. Once that has been accomplished, the molecular mechanics model may be used to compute structures, energies, and properties of unknown molecules.

Molecular mechanics considers molecules as a collection of atomic masses held together by "sticky" forces. These forces represent the electrons holding the atoms together to form molecules. A simple conception is to consider the atoms as balls connected by springs. In this model, as in all real molecules, the atoms migrate toward their lowest-energy, most stable positions. Anything that moves the atoms from their equilibrium positions increases the internal energy of the molecule. This energy, like that of a mechanical spring model, is the potential energy. If one of the bonds were elongated (or compressed) by pulling (or pushing) it from its resting length, the potential energy would increase, and a restoring force would return it to equilibrium position. The magnitude of this restoring force depends on the strength of the spring connecting the masses. This, in turn, is proportional to the strength of the bond connecting the atoms.

Molecular mechanics has several limitations. Most important, it is necessary to have access to a dataset of known, related molecules to use for parametrization. Thus the discovery of completely new molecular species with this computational method is not possible. Also, molecular mechanics cannot account for bond making or bond breaking (the essence of chemical reactions) because of the way it implicitly treats electrons. Finally, the motionless structure computed is not very realistic, so many scientists implement molecular dynamics to achieve better model molecules. *See* CHEMICAL BONDING; COMPUTATIONAL CHEMISTRY. [K.B.L.]

**Molecular orbital theory** A quantum-mechanical model concerned with the description of the discrete energy levels associated with electrons in molecules. One useful way to generate such levels is to assume that the molecular orbital wave function ($\psi_f$) may be written as a simple weighted sum of the constituent atomic orbitals ($\chi_i$) [Eq. (1)]; this is called the linear combination of atomic orbitals approximation.

$$\psi_j = \Sigma c_{ij}\chi_i \tag{1}$$

The $c_{ij}$ coefficients may be determined numerically by substitution of Eq. (1) into the Schrödinger equation and application of the variational theorem. The theorem states that an approximate wave function will always be an upper bound to the true energy; thus minimization of the energy of the system given by the wave function of Eq. (1) will provide the best values of $c_{ij}$. Once the wave function is known, its associated energy may be calculated. The energies of the occupied orbitals in molecules may be probed by using photoelectron spectroscopy, which gives a good check on the accuracy of the theory. There are some simple concepts that contribute to a qualitative understanding of these molecular orbital energy levels and hence an insight into chemical bonding in molecules. They may be illustrated with reference to the hydrogen molecule. *See* CHEMICAL BONDING; ELECTRON SPECTROSCOPY; QUANTUM MECHANICS.

First, the basis orbitals ($\chi_i$) used in the expansion of Eq. (1) can usefully be restricted to include the valence orbitals only. For molecular hydrogen ($H_2$) the 1*s* orbitals on the two hydrogen atoms are then the only two orbitals to be included. Second, since hydrogen atoms are chemically identical, any observable characteristic whose value might be computed with Eq. (1) must be the same for both atoms. This leads to the requirement that $c^2_{1j} = c^2_{2j}$, where the labels 1,2 refer to hydrogen atoms 1 and 2. As a consequence $c_{1j} = \pm c_{2j}$.

When the signs of the two coefficients are the same, the two hydrogen orbitals are mixed in phase; when they are different, the two hydrogen orbitals are mixed out of phase. When the atomic orbitals are mixed in phase, electron density is built up between the two hydrogen nuclei and the potential energy of the nuclei and electrons is lowered. In fact, a reduction of kinetic energy also occurs. An electron lying in the molecular orbital is then of lower energy than an electron associated with an isolated hydrogen 1s orbital. It is called a bonding orbital. The increase in electron density between the two nuclei is the electronic "glue" holding the nuclei together. When the atomic orbitals are mixed out of phase, the opposite behavior occurs. Electron density is removed from the region between the two nuclei, resulting in an increase of both potential and kinetic energy of the electrons. An electron lying in such a molecular orbital would experience an energetic destabilization relative to an electron associated with an isolated hydrogen 1s orbital.

Such a molecular orbital is called an antibonding orbital. Figure 1 shows this information as a molecular orbital diagram. The shading convention of the orbitals has been adopted to indicate the in-phase and out-of-phase mixing of the basis orbitals. Just as the energy levels of atoms are filled in an Aufbau process, so the orbitals of the molecule may be analogously filled up with electrons, each level accommodating two electrons of opposite spin. In $H_2$ there are two electrons to be accounted for. They lie in the bonding orbital, and the stabilization energy relative to two isolated hydrogen atoms (the bond energy) is $2x$. Antibonding orbitals are invariably destabilized more than their bonding counterparts are stabilized. This is shown in Fig. 1 by making $y > x$. With four electrons to be accommodated in this collection of orbitals (this would correspond to the hypothetical case of the $He_2$ molecule), one electron pair resides in the bonding orbital and one pair in the antibonding orbital. Since $y > x$, this molecule is less stable relative to two isolated helium atoms, and as a result the molecule does not exist as a stable entity. $He_2^+$, however, with only three electrons is known.

The size of the interaction energy associated with two atomic orbitals ($x$ in Fig. 1) is controlled by the extent of their spatial overlap. This overlap integral clearly depends upon the internuclear separation. The equilibrium bond length in the hydrogen

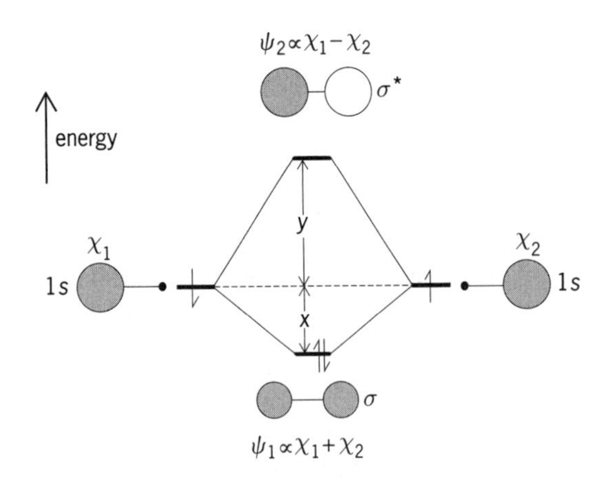


**Fig. 1. Molecular orbital diagram of $H_2$.**

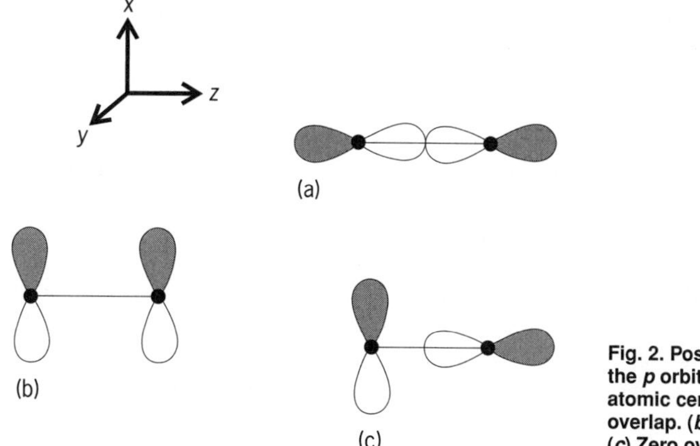


**Fig. 2. Possible orientations of the *p* orbitals on adjacent atomic centers. (*a*) End-on overlap. (*b*) Sideways overlap. (*c*) Zero overlap.**

molecule (and indeed in all molecules) is then a balance between the attractive forces associated with bonding orbital formation and the electrostatic repulsion between the nuclei. Such a molecular parameter is amenable to numerical calculation.

The description of the bonding in the $H_2$ molecule using this model is one where two electrons occupy a bonding orbital and give rise to a simple two-center–two-electron bond traditionally written as H—H. Since the electron density associated with the bonding orbital is cylindrically and symmetrically located about the H—H axis, this bond is called a $\sigma$ bond. In Fig. 1 the bonding orbital is labeled with a $\sigma$ and the corresponding antibonding orbital with a $\sigma^*$.

Ideas similar to those above are readily extended to diatomic molecules from the first row of the periodic table, such as $N_2$ and $O_2$, where the valence orbitals to be considered are the one 2*s* and the three 2*p* orbitals of the atoms. The 2*s* orbitals lie deeper in energy than the triply degenerate 2*p* orbitals. The atomic 2*s* orbitals form bonding and antibonding orbitals ($s\sigma$ and $s\sigma^*$) just as in the case of elemental hydrogen described above, but the behavior of the 2*p* orbitals is a little different. Here there are three possible types of interaction between the *p* orbitals on one center and those on the other. The end-on overlap of two *p* orbitals gives rise to a $\sigma$ interaction (Fig. 2*a*), and the sideways overlap of two *p* orbitals gives rise to a $\pi$ interaction (Fig. 2*b*). The interaction in Fig. 2*c* can be ignored since the overlap between the two orbitals in this orientation can be seen to be identically zero. The result is a $\sigma$-bonding orbital and a $\sigma$-antibonding orbital ($p\sigma$ and $p\sigma^*$), and a pair of $\pi$-bonding and a pair of $\pi$-antibonding orbitals ($\pi$ and $\pi^*$). A larger interaction energy is associated with $p\sigma$ compared to $p\pi$, due to the larger $\sigma$ overlap compared to $\pi$ overlap in Fig. 2.

Filling these orbitals with electrons allows comment on the stability of the resulting diatomics. The molecule $Li_2$ $(s\sigma)^2$ is known and, like $H_2$, may be written as Li—Li to emphasize the single, two-center, two-electron bond between the nuclei. The molecule $Be_2$ which would have the configuration $(s\sigma)^2(s\sigma^*)^2$ is unknown since, just as in $He_2$, $s\sigma^*$ is destabilized more than $s\sigma$ is stabilized relative to an atomic 2*s* level. If the molecular orbital bond order is written as expression (2), then the bond order in $Li_2$ is one but

$$\begin{matrix}\text{Molecular} \\ \text{bond order}\end{matrix} = \begin{pmatrix}\text{number of bonding} \\ \text{electron pairs}\end{pmatrix} - \begin{pmatrix}\text{number of antibonding} \\ \text{electron pairs}\end{pmatrix} \quad (2)$$

the bond order in $Be_2$ is zero.

By filling up the molecular orbital levels derived from the $2p$ orbitals, the bond order associated with the other diatomics may be generated: $B_2$(1), $C_2$(2), $N_2$(3), $O_2$(2), $F_2$(1), and $Ne_2$(0). All of these species are known except $Ne_2$, which is predicted, like $He_2$ and $Be_2$, to have a zero bond order and therefore not to exist as a stable molecule. The molecular orbital bond orders for the three best-known diatomics are consistent with their traditional formulation as N≡N, O═O, and F—F. $N_2$, for example, would be described as having one $\sigma$ and two $\pi$ bonds. With the configuration $(s\sigma)^2(s\sigma^*)^2(p\sigma)^2 - (\pi)^4(\pi^*)^2$, there are four bonding pairs of electrons and two antibonding pairs giving rise to a net bond order of two. The pair of $\pi^*$ orbitals is only doubly occupied, whereas there is space for four electrons. Hund's rules (which for the electronic ground state maximize the number of electrons with parallel spins) identify the lowest-energy arrangement as the one where each of the degenerate $\pi^*$ components is singly occupied, the spins of the two electrons being parallel. Unpaired electrons give rise to paramagnetic behavior, and gaseous oxygen is indeed paramagnetic. [J.K.Bu.]

**Molecular recognition** The ability of biological and chemical systems to distinguish between molecules and regulate behavior accordingly. How molecules fit together is fundamental in disciplines such as biochemistry, medicinal chemistry, materials science, and separation science. A good deal of effort has been expended in trying to evaluate the underlying intermolecular forces. The weak forces that act over short distances (hydrogen bonds, van der Waals interactions, and aryl stacking) provide most of the selectivity observed in biological chemistry and permit molecular recognition. The recognition event initiates behavior such as replication in nucleic acids, immune response in antibodies, signal transduction in receptors, and regulation in enzymes. Most studies of recognition in organic chemistry have been inspired by these biological phenomena. It has been the task of bioorganic chemistry to develop systems capable of such complex behavior with molecules that are comprehensible and manageable in size, that is, with model systems. *See* ENZYME; HYDROGEN BOND; INTERMOLECULAR FORCES; NUCLEIC ACID.

The advantage of cyclic structures lies in their ability to restrict conformation or flexibility. A rigid matrix of binding sites, that is, preorganized sites, is usually associated with high selectivity in binding. A flexible matrix tends to accept several binding partners. Although sacrificing selectivity, this has the advantage of transmitting conformational information and is relevant to biological signaling events. *See* CONFORMATIONAL ANALYSIS.

Macrocyclic (crown) ethers can bind and transport ions and imitate biological processes involving macrolides. Large ring structures that are lined with oxygen present an inner surface which is complementary to the spherical outer surface of positively charged ions.

Cyclophane-type structures offer considerable rigidity because of the aromatic nuclei. Binding forces between host and guest are largely hydrophobic. A typical system is a cyclophane-naphthalene complex (**1**), in which a naphthalene guest is bound by a water-soluble cyclophane derivative. Other macrocyclic structures include the cyclodextrins and hybrid structures assembled from macrocyclic subunits. *See* AROMATIC HYDROCARBON; COORDINATION COMPLEXES.

Because the encircling of larger, more complex molecules with macrocycles poses structural problems, other molecular shapes have been explored. Cleft molecules offer advantages in this regard. The principle underlying these systems involves the shape of the small organic target molecules: convex in surface and bearing functional groups that diverge from their centers. Accordingly, designing a trap for such targets requires molecules of a concave surface in which functional groups converge. This

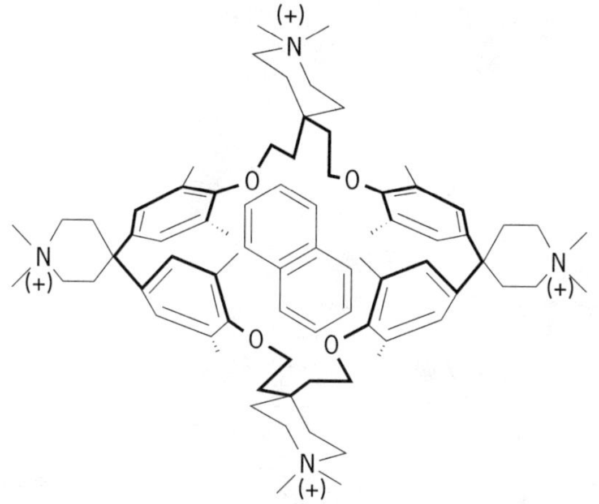

**(1)**

complementarity is also a feature of the immune system: the "hot spots" of an antigen tend to be convex, whereas the binding sites of the antibody are concave.

Systems featuring a cleft have been developed to bind adenine derivatives and other heterocyclic systems through chelation, as shown in (**2**). *See* CHELATION.

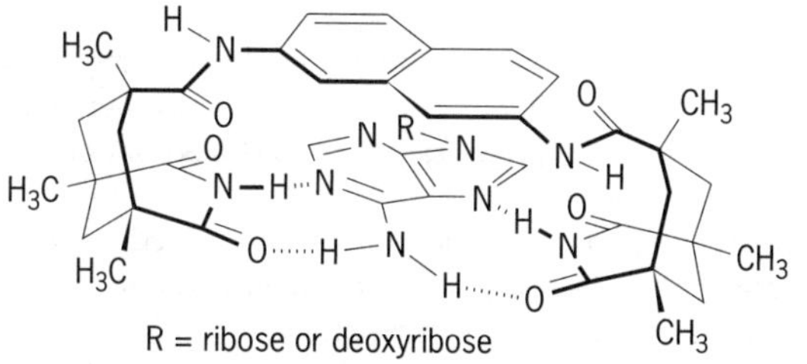

R = ribose or deoxyribose

**(2)**

Apart from the abstract questions concerning articulation of molecules, some practical applications in the pharmaceutical industry may be envisioned. Many of the target structures are biologically active, and the use of synthetic sequestering agents for metabolic substrates can represent a novel approach to biochemical methods and drug delivery. [J.Reb.]

**Molecular sieve** Any one of the crystalline metal aluminosilicates belonging to a class of minerals known as zeolites. An important characteristic of the zeolites is their ability to undergo dehydration with little or no change in crystal structure. The dehydrated crystals are honeycombed with regularly spaced cavities interlaced by channels of molecular dimensions which offer a very high surface area for the absorption of foreign molecules.

The basic formula for all crystalline zeolites can be represented as

$$M_{2/n}O{:}Al_2O_3{:}xSiO_2{:}yH_2O$$

where M represents a metal ion and *n* its valence. The crystal structure consists basically of a three-dimensional framework of $SiO_4$ and $AlO_4$ tetrahedrons (*see* illustration). The tetrahedrons are cross-linked by the sharing of oxygen atoms, so that the ratio of oxygen atoms to the total of silicon and aluminum atoms is equal to 2. The electrovalence of the

**Molecular sieve type-A crystal model. Dark spheres represent the included cations, and light spheres the $SiO_4$ or $AlO_4$ tetrahedrons.**

tetrahedrons containing aluminum is balanced by the inclusion of cations in the crystal. One cation may be exchanged for another by the usual ion-exchange techniques. The size of the cation and its position in the lattice determine the effective diameter of the pore in a given crystal species.

The properties of molecular sieves as adsorbents which distinguish them from nonzeolitic adsorbents are (1) the relatively strong coulomb fields generated by the adsorption surface and (2) the uniform pore size; the pore size is controlled, in a given crystal species, by the associated cation.

The basic characteristics of molecular sieves are utilized commercially in several production and research applications. Their absorption properties make them useful for drying, purification, and separations of gases and liquids. Conversely, molecular sieves can be preloaded with chemical agents, which are thereby isolated from the reactive system in which they are dispersed until released from the adsorbent either thermally or by displacement by a more strongly adsorbed compound. They are also used as cation exchange media and as novel catalysts and catalyst supports. *See* ADSORPTION; GAS CHROMATOGRAPHY; ION EXCHANGE. [R.L.Ma.]

**Molecular simulation** A tool for predicting entirely computationally many useful functional properties of systems of interest in the chemical, pharmaceutical, materials, and related industries. Included are thermodynamic, thermochemical, spectroscopic, mechanical, and transport properties, and morphological information (such as location and shape of binding sites on a biomolecule and crystal structure).

The two main molecular simulation techniques are molecular dynamics and Monte Carlo simulation, both of which are rooted in classical statistical mechanics. Given mathematical models for the internal structure of each molecule (the intramolecular potential which describes the energy of each conformation of the molecule) and the interaction between molecules (the intermolecular potential which describes the energy associated with molecules being in a particular conformation relative to each other), classical statistical mechanics provides a formalism for predicting properties of a macroscopic collection of such molecules based on statistically averaging over the possible microscopic states of the system as it evolves under the rules of classical mechanics. Thus, the building blocks are molecules, the dynamics are described by classical mechanics, and the key concept is statistical averaging. In molecular dynamics, the microscopic states of the system are generated by solving the classical equations of motion as a function of time (typically over a period limited to tens of nanoseconds). Thus, one can observe the relaxation of a system to equilibrium (provided the time for the relaxation falls within the time accessible to molecular dynamics simulation), and so molecular dynamics permits the calculation of transport properties which at the macroscopic scale describe the relaxation of a system in response to inhomogeneities. In Monte Carlo simulation, equilibrium configurations of systems are generated stochastically according to the probabilities rigorously known from classical statistical mechanics. Thus, Monte Carlo simulation generates equilibrium states directly (which has many advantages, including bypassing configurations which are not characteristic of equilibrium but which may be difficult to escape dynamically) and so can be used to study equilibrium configurations of systems which may be expensive or impossible to access via molecular dynamics. The drawback of Monte Carlo simulation is that it cannot yield the kind of dynamical response information that leads directly to transport properties. *See* CHEMICAL DYNAMICS; COMPUTATIONAL CHEMISTRY.

Computational quantum chemistry and molecular simulation methods can be used to predict properties that once were only accessible experimentally, resulting in several significant applications in basic and industrial research. These applications include providing estimates of properties for systems for which little or no experimental data are available, which is especially useful in the early stages of chemical process design; yielding insight into the molecular basis for the behavior of particular systems, which is very useful in developing engineering correlations, design rules, or quantitative structure-property relations; and providing guidance for experimental studies by identifying the interesting systems or properties to be measured. [P.T.C.]

**Molecular structure and spectra** Until the advent of quantum theory, ideas about the structure of molecules evolved gradually from analysis and interpretation of the facts of chemistry. Chemists developed the concept of molecules as built from atoms in definite proportions, and identified and constructed (synthesized) a great variety of molecules. Later, when the structure of atoms as built from nuclei and electrons began to be understood with the help of quantum theory, a beginning was made in seeing why atoms can combine in definite ways to form molecules; also, infrared spectra began to be used to obtain information about the dimensions and the nuclear motions (vibrations) in molecules. However, a fundamental understanding of chemical bonding and molecular structure became possible only by application of the present form of quantum theory, called quantum mechanics. This theory makes it possible to obtain from the spectra of molecules a great deal of information about the nature of molecules in their normal as well as excited states, and about dissociation energies and other characteristics of molecules. *See* CHEMICAL BONDING.

**Molecular sizes.** The size of a molecule varies approximately in proportion to the numbers and sizes of the atoms in the molecule. Simplest are diatomic molecules. These may be thought of as built of two spherical atoms of radii $r$ and $r'$, flattened where they are joined. The equilibrium value $R_e$ of the distance $R$ between their nuclei is then smaller than the sum of the atomic radii. However, the nuclei of atoms in two different molecules cannot normally approach more closely than a distance $r + r'$; $r$ and $r'$ are called the van der Waals radii of the atoms.

To describe a polyatomic molecule, one must specify not merely its size but also its shape or configuration. For example, carbon dioxide ($CO_2$) is a linear symmetrical molecule, the O—C—O angle being 180°. The H—O—H angle in the nonlinear water ($H_2O$) molecule is 105°. Many molecules which are essential for life contain thousands or even millions of atoms. Proteins are often coiled or twisted and cross-linked in ways which are important for their biological functioning.

**Dipole moments.** Most molecules have an electric dipole moment. In atoms, the electron cloud surrounds the nucleus so symmetrically that its electrical center coincides with the nucleus, giving zero dipole moment; in a molecule, however, these coincidences are disturbed, and a dipole moment usually results.

Thus, when the atoms of HCl come together, there is some shifting of the H-atom electron toward the Cl. A complete shift would give $H^+Cl^-$, which would constitute an electric dipole of magnitude $eR_e$, where $e$ is the electronic charge. But in fact the dipole moment is only 0. 17 $eR_e$. This is because the actual electronic shift is only fractional. *See* ELECTRONEGATIVITY.

**Molecular polarizability.** In the preceding consideration of dipole moments, the discussion has been in terms of atoms and molecules free from external forces. An electric field pulls the electrons of an atom or molecule toward it and pushes the nuclei away, or vice versa. This action creates a small induced dipole moment, whose magnitude per unit strength of the field is called the polarizability.

**Molecular energy levels.** The states of motion of nuclei and electrons in a molecule, or of electrons in an atom, are restricted by quantum mechanics to special forms with definite energies. The state of lowest energy is called the ground state; all others are excited states. In analogy to water levels, one speaks of energy levels. Excited states exist only momentarily, following an electrical or other stimulus. *See* QUANTUM CHEMISTRY.

Excitation of an atom consists of a change in the state of motion of its electrons. Electronic excitation of molecules can also occur, but alternatively or additionally, molecules can be excited to discrete states of vibration and rotation.

The total energy of any molecule can be written as Eq. (1). Both the electronic energy

$$E = E_{el} + E_{\upsilon} + (E_r + E_{fs} + E_{hfs} + E_{ext}) \tag{1}$$

$E_{el}$ and vibration energy $E_{\upsilon}$ can be discrete or continuous. The quantities $E_r$, $E_{fs}$, and $E_{hfs}$ denote rotational, fine-structure, and hyperfine-structure energies. The last two appear as small or minute splittings of the rotation levels. The spacings $\Delta E$ of adjacent discrete levels of each type are usually in the order given in notation (2). The $E_{ext}$ term in

$$\Delta E_{el} \gg \Delta E_{\upsilon} \gg \Delta E_r \gg \Delta E_{fs} \gg \Delta E_{hfs} \tag{2}$$

Eq. (1) refers to additional fine structure which appears on subjecting molecules to external magnetic fields (Zeeman effect) or electric fields (Stark effect).

Polyatomic molecules have much more complicated patterns of vibrational and (usually) rotational energy levels than diatomic molecules.

**Molecular spectra.** The frequencies $c\upsilon$ ($c$ = speed of light) of electromagnetic spectra obey the Einstein-Bohr equation (3), where $h$ is Planck's constant. Molecular

$$hc\upsilon = E' - E'' \tag{3}$$

emission spectra accompany jumps in energy from higher to lower levels; absorption spectra accompany jumps from lower to higher levels.

Molecular spectra can be classified as fine-structure or low-frequency spectra, rotation spectra, vibration-rotation spectra, and electronic spectra. *See* ELECTRON PARAMAGNETIC RESONANCE (EPR) SPECTROSCOPY ; MICROWAVE SPECTROSCOPY ; SPECTROSCOPY.

Transitions between energy levels differing only in rotational state give rise to pure rotation spectra. These typically consist of a sequence of lines spaced almost equidistantly, and lying in the far infrared or the microwave region.

Spectra involving only vibrational and rotational state changes consist of bands which lie mainly in the infrared. Each band consists of two sets of closely spaced rotational lines, one on each side of a central frequency. Vibration-rotation absorption bands of liquids and solutions are widely used in chemical analysis. Here the rotational structure is blurred out, and only an "envelope" is seen.

Electronic band spectra are the most general type of molecular spectra. For any one electronic transition, the spectrum consists typically of many bands. *See* ATOMIC STRUCTURE AND SPECTRA; INTERMOLECULAR FORCES; MOLECULAR WEIGHT; RESONANCE (MOLECULAR STRUCTURE); VALENCE. [R.S.M.]

**Molecular weight** The sum of the atomic weights of all atoms making up a molecule. Actually, what is meant by molecular weight is molecular mass. The use of this expression is historical, however, and will be maintained. The atomic weight is the mass, in atomic mass units, of an atom. It is approximately equal to the total number of nucleons, protons and neutrons composing the nucleus. Since 1961 the official definition of the atomic mass unit (amu) has been that it is 1/12 the mass of the carbon-12 isotope, which is assigned the value 12.000 exactly. *See* ATOMIC MASS; ATOMIC MASS UNIT; RELATIVE ATOMIC MASS; RELATIVE MOLECULAR MASS.

A mole is an amount of substance containing the Avogadro number, $N_A$, approximately $6.022 \times 10^{23}$, of molecules or atoms. Molecule, in this definition, is understood to be the smallest unit making up the characteristic compound. Originally, the mole was interpreted as that number of particles whose total mass in grams was numerically equivalent to the atomic or molecular weight in atomic mass units, referred to as gram-atomic or gram-molecular weight. This is how the above value for $N_A$ was calculated. As the ability to make measurements of the absolute masses of single atoms and molecules has improved, however, modern metrology is tending to alter its approach and define the Avogadro number as an exact quantity, thereby changing slightly the definition of the atomic mass unit and removing the need to define atomic weight with respect to a particular isotopic species. The latest and most accurate value for the Avogadro number is $6.0221415(10) \times 10^{23}$ mol$^{-1}$. *See* AVOGADRO NUMBER; MOLE (CHEMISTRY).

As the masses of all the atomic species are now well known, masses of molecules can be determined once the composition of the molecule has been ascertained. Alternatively, if the molecular weight of the molecule is known and enough additional information about composition is available, such as the basic atomic constituents, it is possible to begin to assemble structural information about the molecule. Thus, the determination of the molecular weight is one of the first steps in the analysis of an unknown species. Given the increasing emphasis on the study of biologically important molecules, particular attention has been focused on the determination of molecular weights of larger and larger units. There are a number of methods available, and the one chosen will depend on the size and physical state of the molecule. All processes are physical macroscopic measurements and determine the molecular weight directly. Connection to the absolute mass scale is straightforward by using the Avogadro number, although, for extremely large molecules, this connection is often unnecessary or impossible, as the accuracy of the measurements is not that good. The main function of molecular weight determination of large molecules is elucidation of structure.

Molecular weight determination of materials which are solid or liquid at room temperature is best achieved by taking advantage of one of the colligative properties of solutions, boiling-point elevation, freezing-point lowering, or osmotic pressure, which depend on the number of particles in solution, not on the nature of the particle. The choice of which to use will depend on a number of properties of the substance, the most important of which will be the size. All require that the molecule be small enough to dissolve in the solution but large enough not to participate in the phase change or pass through a semipermeable membrane. Freezing-point lowering is an excellent method for determining molecular weights of smaller organic molecules, and osmometry, as the osmotic pressure determination is called, for determining molecular weights of larger organic molecules, particularly polymeric species. Boiling-point elevation is used less frequently. *See* POLYMER.

The basis of all the methods involving colligative properties of solutions is that the chemical potentials of all phases must be the same. (Chemical potential is the partial change in energy of a system as matter is transferred into or out of it. For two systems in contact at equilibrium, the chemical potentials for each must be equal.) *See* CHEMICAL EQUILIBRIUM; CHEMICAL THERMODYNAMICS.

Another measurement from which molecular weights can be obtained is based on the scattering of light from the molecule. A beam of light falling on a molecule will induce in the molecule a dipole moment which in its turn will radiate. The interference between the radiated beam and the incoming beam produces an angular dependence of the scattered radiation which depends on the molecular weight of the molecule. This occurs whether the molecule is free or in solution. While the theory for this effect is complicated and varies according to the size of the molecule, the general result for molecules whose size is considerably less than that of the wavelength $\lambda$ of the radiation (less than $\lambda/50$) is given by the equation below; $I(\theta)$ is the intensity of radiation at

$$\frac{I(\theta)}{I_0} = \text{constant } (1 + \cos^2\theta)\ Mc$$

angle $\theta$, $I_0$ the intensity of the incoming beam, $M$ the molecular weight, and $c$ the concentration in grams per cubic centimeter of the molecule. If the molecules are much larger than $\lambda/50$ (about 9 nanometers for visible light), this relationship in this simple form is no longer valid, but the method is still viable with appropriate adjustments to the theory. In fact, it can be used in its extended version even for large aggregates. [C.D.C.]

**Molecule** A molecule may be thought of either as a structure built of atoms bound together by chemical forces or as a structure in which two or more nuclei are maintained in some definite geometrical configuration by attractive forces from a surrounding swarm of negative electrons. Besides chemically stable molecules, short-lived molecular fragments called free radicals can be observed under special circumstances. *See* CHEMICAL BONDING; FREE RADICAL; MOLECULAR STRUCTURE AND SPECTRA. [R.S.M.]

**Molybdenum** A chemical element, Mo, atomic number 42, and atomic weight 95.94, in the periodic table in the triad of transition elements that includes chromium (atomic number 24) and tungsten (atomic number 74). Research has revealed it to be one of the most versatile chemical elements, finding applications not only in metallurgy but also in paints, pigments, and dyes; ceramics; electroplating; industrial catalysts; industrial lubricants; and organometallic chemistry. Molybdenum is an essential trace element in soils and in agricultural fertilizers. Molybdenum atoms have been found to perform key functions in enzymes (oxidases and reductases), with particular interest being directed toward its role in nitrogenase, which is employed by bacteria in legumes to convert inert nitrogen ($N_2$) of the air into biologically useful ammonia ($NH_3$). *See* PERIODIC TABLE.

Molybdenum is widely distributed in the Earth's crust at a concentration of 1.5 parts per million by weight in the lithosphere and about 10 parts per billion in the sea. It is found in at least 13 minerals, mainly as a sulfide [molybdenite ($MoS_2$)] or in the form of molybdates [for example, wulfenite ($PbMoO_4$) and magnesium molybdate ($MgMoO_4$)].

Although molybdenum is closer to chromium in atomic weight and atomic number, its chemical behavior is usually very similar to that of tungsten, which has nearly the same atomic radius. (This is due to the so-called lanthanide contraction in which atomic radii decrease for elements 57 to 71 found in the period between molybdenum and tungsten.) *See* CHROMIUM; LANTHANIDE CONTRACTION; TUNGSTEN.

Molybdenum atoms contain six valence electrons ($4d^5 5s^1$), which are employed with great versatility in forming compounds and complexes in which electronic configurations vary from $d^0$ (no *d* electrons in oxidation state + 6) to $d^8$ (8 *d* electrons in oxidation state −2). The +6 state is preferred, but all states from −2 to +6 are known. States usually exhibit a variety of coordination numbers (4 to 9), and include polynuclear complexes and metal-metal bonds in metallic clusters with two to six metal atoms in their metallic cores. Molybdenum forms a very large number of compounds with oxygen. Low-valent molybdenum [for example, $Mo(CO)_6$ and $Mo_2$, $Mo_3$, and $Mo_6$ clusters] has a very rich organometallic chemistry, including clusters that are being studied as models for molybdenum metal surfaces that catalyze organic reactions employed in industrial syntheses and oil refining. The ability of molybdenum atoms to vary oxidation state, coordination number, and coordination geometry and to form metal-metal bonds in clusters accounts in part for the large number of industrial catalysts and biological enzymes in which Mo atoms are found at the active site for catalysis. *See* CHEMICAL BONDING; COORDINATION CHEMISTRY; ELECTRON CONFIGURATION.

Molybdenum is a high-melting silver-gray metal, strong even at high temperatures, hard, and resistant to corrosion (see table). It also exhibits high conductivity, a high modulus of elasticity, high thermal conductivity, and a low coefficient of expansion. Its major use is in alloy steels, for example, as tool steels ($\leq$10% molybdenum), stainless steel, and armor plate. Up to 3% molybdenum is added to cast iron to increase strength. Up to 30% molybdenum may be added to iron-, cobalt-, and nickel-based alloys designed for severe heat- and corrosion-resistant applications. It may be used in filaments for light bulbs, and it has many applications in electronic circuitry. *See* ALLOY.

Molybdenum trioxide, molybdates, sulfo-molybdates, and metallic molybdenum are found in thousands of industrial catalysts used in oil refining, ammonia synthesis, and industrial syntheses of organic chemicals. Monomeric molybdenum(IV) in aqueous

**Physical properties of molybdenum metal**

| Property | Value |
|---|---|
| Density | 10.22 g/cm$^3$ (5.911 oz/in.$^3$) |
| Heat of vaporization | 491 kJ/mol |
| Heat of fusion | 28 kJ/mol |
| Specific heat | 0.267 J/g°C |
| Thermal conductivity | 1.246 J/s/cm$^2$/cm°C (200°C) |
| | 0.923 J/s/cm$^2$/cm°C (2200°C) |
| Electrical conductivity | 34% International Copper Standard |
| Electrical resistivity | 5.2 microhm-cm, 20°C |
| | 78.2 microhm-cm, 2525°C |
| Magnetic susceptibility | 0.93 × 10$^{-6}$ emu, 25°C |
| | 1.11 × 10$^{-6}$ emu, 1825°C |
| Mean linear expansion coefficient | 6.65 × 10$^{-6}$/°C, 20–1600°C |
| Modulus of elasticity | 0.324 N/m$^2$ |
| Lattice parameter | 0.314767 nm (body-centered cube) |

solution is a powerful catalyst for the reduction of inert oxo-anions such as perchlorate ($ClO_4^-$) or nitrate ($NO_3^-$) as well as other oxidized nonmetals such as azide ion ($N_3^-$) and dinitrogen ($N_2$). The trinuclear cation $MO_3O_4^{4+}$ is inert, unreactive, and noncatalytic. *See* CATALYSIS; HOMOGENEOUS CATALYSIS.

The molybdenum enzymes comprise two major categories. The first category contains the single, highly important enzyme nitrogenase, which is responsible for biological nitrogen fixation. The second category contains all other known molybdenum enzymes, which are crucial for the metabolism of bacteria, plants, and animals, including humans. [E.I.S.]

**Monomolecular film** A film one molecule thick; often referred to as a monolayer. Films that form at surfaces or interfaces are of special importance. Such films may reduce friction, wear, and rust, or may stabilize emulsions, foams, and solid dispersions. The broad field of catalysis, which is basic to petroleum refining and many chemical industries, involves chemical reactions that are accelerated in the thin films of reactants at interfaces. Moreover, thin films containing proteins, cholesterol, and related compounds constitute biological membranes, the internal interfaces that control the complex processes of life. *See* CATALYSIS.

In all of these areas, a single monomolecular layer at the interface is the most important. It is held to the adsorbing surface by forces stronger than those that hold any succeeding layer. On solid surfaces, it is the only layer that can be chemisorbed. It may be the site of enhanced chemical reactivity, or the last line of defense.

Monolayers on solids, or at liquid interfaces, may be formed by adsorption from the adjacent bulk phases; the process may show high specificity for particular chemical species. Measurements of the extent of adsorption have historically provided information on the composition and structure of monolayers formed in this way. A variety of surface-sensitive instrumental techniques, such as diffraction and scattering of low-energy electrons, neutrons, and ions, and spectroscopy of adsorbed species, have been brought to bear to obtain information about the structure of the surface layer and chemical perturbations in it. *See* ADSORPTION; SPECTROSCOPY.

In addition, monolayers of a wide variety of substantially insoluble substances can be formed at a liquid-gas interface by allowing them to spread over the surface. The properties of such films at the water-air interface can be manipulated, controled, and measured in simple and elegant ways. A variety of specialized experimental techniques have been developed to study these insoluble monolayers.

In order to form spread monolayers which are sufficiently stable to study, a substance must combine low solubility and volatility with some moiety which attracts it to the liquid surface; for films on water, this generally means one or more polar functional groups. Totally nonpolar substances, such as the higher-molecular-weight paraffin hydrocarbons, will not spread on water (although they can spread on liquids of very high surface tension, such as mercury). Typical among the large group of substances which do form insoluble monolayers on water are the long-chain fatty acids and their derivatives such as glycerides, sterols, and many lipid substances of biological origin, including the fat-soluble vitamins and natural pigments such as chlorophyll. Many polar synthetic polymers, including polyvinyl acetate and polymethyl methacrylate, can be made to spread as monolayers on water; so can many proteins, because their tertiary structure unfolds at the air-water interface. *See* POLAR MOLECULE. [G.L.G.]

**Mordant** A substance or combination of substances that facilitates the fixing of a dye to a fiber. A mordant enables the production of a more permanent and often deeper color. Metallic salts or hydroxides are most frequently used as mordants.

Certain mordants act directly on the fiber, making it more susceptible to the dye. Fabrics are then pretreated with the mordant before exposure to the dye. Other mordants function through the formation of a complex with the dye. The complex acts as the dyeing agent. Mordant and dye in this case are exposed simultaneously to the fabric. *See* DYE.

[F.J.J.]

**Mössbauer effect** Recoil-free gamma-ray resonance absorption. The Mössbauer effect, also called nuclear gamma resonance fluorescence, has become the basis for a type of spectroscopy which has found wide application in nuclear physics, structural and inorganic chemistry, biological sciences, the study of the solid state, and many related areas of science.

The fundamental physics of this effect involves the transition (decay) of a nucleus from an excited state of energy $E_e$ to a ground state of energy $E_g$ with the emission of a gamma ray of energy $E_\gamma$. If the emitting nucleus is free to recoil, so as to conserve momentum, the emitted gamma ray energy is $E_\gamma = (E_e - E_g) - E_r$, where $E_r$ is the recoil energy of the nucleus. The magnitude of $E_r$ is given classically by the relationship $E_r = E_\gamma^2/2mc^2$ where $m$ is the mass of the recoiling atom and $c$ is the speed of light. Since $E_r$ is a positive number, the $E_\gamma$ will always be less than the difference $E_e - E_g$, and if the gamma ray is now absorbed by another nucleus, its energy is insufficient to promote the transition from $E_g$ to $E_e$.

In 1957 R. L. Mössbauer discovered that if the emitting nucleus is held by strong bonding forces in the lattice of a solid, the whole lattice takes up the recoil energy, and the mass in the recoil energy equation given above becomes the mass of the whole lattice. Since this mass typically corresponds to that of $10^{10}$ to $10^{20}$ atoms, the recoil energy is reduced by a factor of $10^{-10}$ to $10^{-20}$, with the important result that $E_r \approx 0$ so that $E_\gamma = E_e - E_g$; that is, the emitted gamma-ray energy is exactly equal to the difference between the nuclear ground-state energy and the excited-state energy. Consequently, absorption of this gamma ray by a nucleus which is also firmly bound to a solid lattice can result in the "pumping" of the absorber nucleus from the ground state to the excited state. *See* EXCITED STATE .

In a typical Mössbauer experiment the radioactive source is mounted on a velocity transducer which imparts a smoothly varying motion (relative to the absorber, which is held stationary), up to a maximum of several centimeters per second, to the source of the gamma rays. These gamma rays are incident on the material to be examined (the absorber). Some of the gamma rays are absorbed and reemitted in all directions, while the remainder of the gamma rays traverse the absorber and are registered in an appropriate detector.

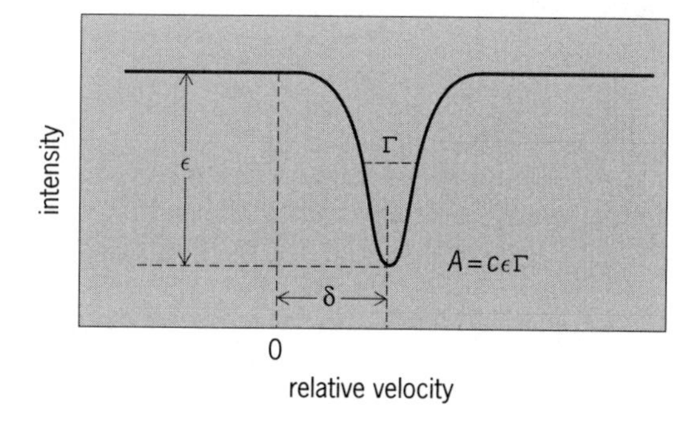


**Mössbauer spectrum of an absorber which gives an unsplit resonance line. The spectrum is characterized by a position $\delta$, a line width $\Gamma$, and an area $A$ related to the effect magnitude $\epsilon$.**

A typical display of a Mössbauer spectrum, which is the result of many repetitive scans through the velocity range of the transducer, is shown in the illustration. In certain nuclides the Mössbauer resonance line displays splitting that arises from the coupling of the nuclear electric quadrupole moment with the electric field gradient or of the nuclear magnetic dipole moment with the magnetic field at the nucleus, providing information on the magnitude of these interactions.

Mössbauer-effect experiments have been used to elucidate problems in a very wide range of scientific disciplines. Applications include the measurement of nuclear magnetic and quadrupole moments and of excited-state lifetimes involved in the nuclear decay process; study of the chemical consequences of nuclear decay; study of the nature of magnetic interactions in iron-containing alloys and of the dependence of the magnetic field in these alloys on various parameters; study of the effects of high pressure on chemical properties of materials; investigation of the relationship between chemical composition and structure on the one hand and the superconductive transition on the other; investigation of the structure of compounds; and study of the structure and bonding properties of metal atoms in complex biological molecules. [R.H.He.]

**Multiple proportions, law of** This law states that, when two elements combine together to form more than one compound, the weights of one element that unite with a given weight of the other are in the ratio of small whole numbers. The law can be illustrated by the composition of the five oxides of nitrogen. One gram of nitrogen is combined with 2.85 g of oxygen in nitrogen pentoxide, $N_2O_5$; with 2.28 g in nitrogen dioxide, $NO_2$; with 1.71 g in nitrogen trioxide, $N_2O_3$; with 1.14 g in nitric oxide, NO; and with 0.57 g in nitrous oxide, $N_2O$. These numbers are in the simple ratio of 5:4:3:2:1. *See* DEFINITE COMPOSITION, LAW OF. [T.C.W.]

# N

**Nanochemistry** The study of the synthesis and characterization of materials in the nanoscale size range (1 to 10 nanometers). These materials include large organic molecules, inorganic cluster compounds, and metallic or semiconductor particles. The synthesis of nanoscale inorganic materials is important because the small size endows these particles with unusual structural and optical properties that may find application in catalysis and electrooptical devices. Approaches to the synthesis of these materials have focused on constraining the reaction environment through the use of surface-bound organic additives, porous glasses, zeolites, clays, or polymers. The use of synthetic approaches that are inspired by the biological processes result in the deposition of inorganic materials such as bones, shells, and teeth (biomineralization). This biomimetic approach involves the use of assemblies of biological molecules that provide nanoscale reaction environments in which inorganic materials can be prepared in an organized and controlled manner. Examples of biological assemblies include phospholipid vesicles and the polypeptide micelle of the iron storage protein, ferritin. *See* MICELLE.

Vesicles are bounded by an organic membrane that provides a spatial limit on the size of the reaction volume. If a chemical reaction is undertaken in this confined space that leads to the formation of an inorganic material, the size of the product will also be constrained to the dimensions of the organic host structure. Provided that the chemical and physical conditions are not too severe to disrupt the organic membrane, these supramolecular assemblies may have advantages over inorganic hosts such as clays and zeolites because the chemical nature of the organic surface can be systematically modified so that controlled reactions can be accomplished. *See* SUPRAMOLECULAR CHEMISTRY.

One problem encountered with the use of phospholipid vesicles is their sensitivity to changes in temperature and ionic strength. Procedures have been developed in which the biomolecular cage of the iron storage protein, ferritin, has been used as a nanoscale reaction environment for the synthesis of inorganic materials. In the simplest approach the native iron oxide core is transformed into another material by chemical reaction within the protein shell. [S.Ma.]

**Nanostructure** A material structure assembled from a layer or cluster of atoms with size of the order of nanometers. Interest in the physics of condensed matter at size scales larger than that of atoms and smaller than that of bulk solids (mesoscopic physics) has grown rapidly since the 1970s, owing to the increasing realization that the properties of these mesoscopic atomic ensembles are different from those of conventional solids. As a consequence, interest in artificially assembling materials from nanometer-sized building blocks arose from discoveries that by controlling the sizes in the range of 1–100 nm and the assembly of such constituents it was possible to begin to alter and prescribe the properties of the assembled nanostructures.

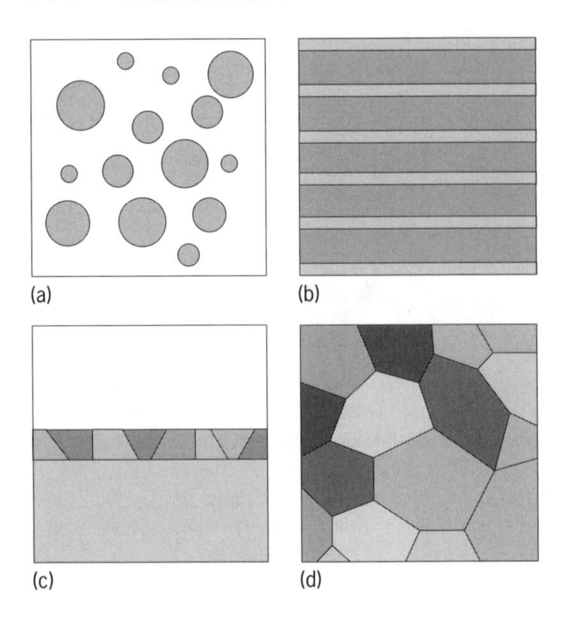


**Schematic of four basic types of nanostructured materials, classified according to integral modulation dimensionality. (*a*) Dimensionality 0: clusters of any aspect ratio from 1 to infinity. (*b*) Dimensionality 1: multilayers. (*c*) Dimensionality 2: ultrafine-grained overlayers (coatings) or buried layers. (*d*) Dimensionality 3: nanophase materials. (*After R. W. Siegel, Nanostructured materials: Mind over matter, Nanostruct. Mat., 3:1–18, 1993*)**

Nanostructured materials are modulated over nanometer length scales in zero to three dimensions. They can be assembled with modulation dimensionalities of zero (atom clusters or filaments), one (multilayers), two (ultrafine-grained overlayers or coatings or buried layers), and three (nanophase materials), or with intermediate dimensionalities (see illustration).

**Multilayers and clusters.** Multilayered materials have had the longest history among the various artificially synthesized nanostructures, with applications to semiconductor devices, strained-layer superlattices, and magnetic multilayers. Recognizing the technological potential of multilayered quantum heterostructure semiconductor devices helped to drive the rapid advances in the electronics and computer industries. A variety of electronic and photonic devices could be engineered by utilizing the low-dimensional quantum states in these multilayers for applications in high-speed field-effect transistors and high-efficiency lasers, for example. Subsequently, a variety of nonlinear optoelectronic devices, such as lasers and light-emitting diodes, have been created by nanostructuring multilayers.

The advent of beams of atom clusters with selected sizes allowed the physics and chemistry of these confined ensembles to be critically explored, leading to increased understanding of their potential, particularly as the constituents of new materials, including metals, ceramics, and composites of these materials. A variety of carbon-based clusters (fullerenes) have also been assembled into materials of much interest. In addition to effects of confinement, interfaces play an important and sometimes dominant role in cluster-assembled nanophase materials, as well as in nanostructured multilayers. *See* ATOM CLUSTER; FULLERENE.

**Synthesis and properties.** A number of methods exist for the synthesis of nanostructured materials. They include synthesis from atomic or molecular precursors (chemical or physical vapor deposition, gas condensation, chemical precipitation, aerosol reactions, biological templating), from processing of bulk precursors (mechanical attrition, crystallization from the amorphous state, phase separation), and from nature (biological systems). Generally, it is preferable to synthesize nanostructured materials from atomic or molecular precursors, in order to gain the most control over a variety of microscopic aspects of the condensed ensemble; however, other methodologies can often yield very useful results. [R.W.Si]

**Nanotechnology** Systems for transforming matter, energy, and information, based on nanometer-scale components with precisely defined molecular features. The term nanotechnology has also been used more broadly to refer to techniques that produce or measure features less than 100 nanometers in size; this meaning embraces advanced microfabrication and metrology. Although complex systems with precise molecular features cannot be made with existing techniques, they can be designed and analyzed. Studies of nanotechnology in this sense remain theoretical, but are intended to guide the development of practical technological systems.

Nanotechnology based on molecular manufacturing requires a combination of familiar chemical and mechanical principles in unfamiliar applications. Molecular manufacturing can exploit mechanosynthesis, that is, using mechanical devices to guide the motions of reactive molecules. By applying the conventional mechanical principle of grasping and positioning to conventional chemical reactions, mechanosynthesis can provide an unconventional ability to cause molecular changes to occur at precise locations in a precise sequence. Reliable positioning is required in order for mechanosynthetic processes to construct objects with millions to billions of precisely arranged atoms.

Mechanosynthetic systems are intended to perform several basic functions. Their first task is to acquire raw materials from an externally provided source, typically a liquid solution containing a variety of useful molecular species. The second task is to process these raw materials through steps that separate molecules of different kinds, bind them reliably to specific sites, and then (often) transform them into highly active chemical species, such as radicals, carbenes, and strained alkenes and alkynes. Finally, mechanical devices can apply these bound, active species to a workpiece in a controlled position and orientation and can deposite or remove a precise number of atoms of specific kinds at specific locations.

Several technologies converge with nanotechnologies, the most important being miniaturization of semiconductor structures, driven by progress in microelectronics. More directly relevant are efforts to extend chemical synthesis to the construction of larger and more complex molecular objects. Protein engineering and supramolecular chemistry are active fields that exploit weak intermolecular forces to organize small parts into larger structures. Scanning probe microscopes are used to move individual atoms and molecules. *See* MOLECULAR RECOGNITION; MONOMOLECULAR FILM; NANOSTRUCTURE; SUPRAMOLECULAR CHEMISTRY. [E.Dr.]

**Naphtha** Any one of a wide variety of volatile hydrocarbon mixtures. They are sometimes obtained from coal tar but are more often derived from petroleum. Physical properties vary widely. The initial boiling point may be as low as 27°C (80°F), and end points may reach 260°C (500°F). Boiling ranges are sometimes as narrow as 11°C (20°F) or as wide as 110°C (200°F). Products sold as naphthas find their greatest use as solvents, thinners, or carriers.

There is a fairly sharp differentiation between aliphatic and aromatic naphthas. Aliphatic naphthas are relatively low in odor and toxicity and tend, also, to be low in solvent power. The aromatic naphthas are highly solvent. Their main components are toluene and xylenes; benzene is less desirable because of the extreme toxicity of its vapors. *See* PETROLEUM PRODUCTS. [J.K.R.]

**Negative ion** An atomic or molecular system with an excess of negative charge. Negative ions, also called anions, are formed in attachment processes in which an additional electron is captured by an atom or molecule. Negative ions were first reported in the early days of mass spectrometry. It was soon learned that even a small concentration of such weakly bound, negatively charged systems had an appreciable effect on the electrical conductivity of gaseous discharges. Negative ions now play a major role in a number of areas of physics and chemistry involving weakly ionized gases and plasmas. Applications include accelerator technology, injection heating of thermonuclear plasmas, material processing, and the development of tailor-made gaseous dielectrics. In nature, negative ions are known to be present in tenuous plasmas such as those found in astrophysical and aeronomical environments. The absorption of radiation by negative hydrogen ions in the solar photosphere, for example, determines the Sun's spectral distribution. *See* ION. [D.J.Pe.]

**Neodymium** A metallic chemical element, Nd, atomic number 60, atomic weight 144.24. Neodymium belongs to the rare-earth group of elements. The naturally occurring element includes the six isotopes. The oxide, $Nd_2O_3$, is a light-blue powder. It dissolves in mineral acids to give reddish-violet solutions. For properties of the metal. *See* PERIODIC TABLE; RARE-EARTH ELEMENTS.

The salts have found application in the ceramic industry for coloring glass and for glazes. The glass is particularly useful in goggles used by glass blowers, since it absorbs the intense yellow D line of sodium present in the flame. The element has found commercial application in the manufacture of lasers. [F.H.Sp.]

**Neon** A gaseous chemical element, Ne, with atomic number 10 and atomic weight 20.183. Neon is a member of the family of noble gases. The only commercial source of neon is the Earth's atmosphere, although traces of neon are found in natural gas, minerals, and meteorites. *See* INERT GASES; PERIODIC TABLE.

Considerable quantities of neon are used in high-energy physics research. Neon fills spark chambers used to detect the passage of nuclear particles. Liquid neon can be utilized as a refrigerant in the temperature range about 25 to 40 K (−416 to −387°F). Neon is also used in some kinds of electron tubes, in Geiger-Müller counters, in spark-plug test lamps, and in warning indicators on high-voltage electric lines. A very small wattage produces visible light in neon-filled glow lamps; such lamps are used as economical night and safety lights.

**Physical properties of neon**

| Property | Value |
|---|---|
| Atomic number | 10 |
| Atomic weight (atmospheric neon only) | 2.183 |
| Melting point, °C | −248.6 |
| Boiling point at 1 atm pressure, °C | −246.1 |
| Gas density at 0°C and 1 atm pressure, g/liter | 0.8999 |
| Liquid density at its boiling point, g/ml | 1.207 |
| Solubility in water at 20°C, ml neon (STP)/1000 g water at 1 atm partial pressure neon | 10.5 |

Neon is colorless, odorless, and tasteless; it is a gas under ordinary conditions. Some of the other properties of neon are given in the table. Neon does not form any chemical compounds in the ordinary sense of the word; there is only one atom in each molecule of gaseous neon. [A.W.F.]

**Neptunium** A chemical element, symbol Np, atomic number 93. Neptunium is a member of the actinide or 5*f* series of elements. It was synthesized as the first transuranium element in 1940 by bombardment of uranium with neutrons to produce neptunium-239. The lighter isotope $^{237}Np$, a long-lived alpha emitter with half-life $2.14 \times 10^6$ years, is particularly important chemically. *See* PERIODIC TABLE.

Neptunium metal is ductile, low-melting (637°C or 1179°F), and in its alpha form is of high density, 20.45 g/cm$^3$ (11.82 oz/in.$^3$). The chemistry of neptunium may be said to be intermediate between that of uranium and plutonium. Neptunium metal is reactive and forms many binary compounds, for example, with hydrogen, carbon, nitrogen, phosphorus, oxygen, sulfur, and the halogens. *See* ACTINIDE ELEMENTS; NUCLEAR CHEMISTRY; TRANSURANIUM ELEMENTS. [R.A.Pe.]

**Neutron** An elementary particle having approximately the same mass as the proton but lacking a net electric charge. It is indispensable in the structure of the elements, and in the free state it is an important reactant in nuclear research and the propagating agent of fission chain reactions.

Neutrons and protons are the constituents of atomic nuclei. The number of protons in the nucleus determines the chemical nature of an atom, but without neutrons it would be impossible for two or more protons to exist stably together within nuclear dimensions. The protons, being positively charged, repel one another by virtue of their electrostatic interactions. The presence of neutrons weakens the electrostatic repulsion, without weakening the nuclear forces of cohesion. In light nuclei the resulting balanced, stable configurations contain protons and neutrons in almost equal numbers, but in heavier elements the neutrons outnumber the protons; in $^{238}U$, for example, 146 neutrons are joined with 92 protons. Only one nucleus, $^1H$, contains no neutrons. For a given number of protons, neutrons in several different numbers within a restricted range often yield nuclear stability—and hence the isotopes of an element.

Free neutrons have to be generated from nuclei, and since they are bound therein by cohesive forces, an amount of energy equal to the binding energy must be expended to get them out. Nuclear machines, such as cyclotrons and electrostatic generators, induce many nuclear reactions when their ion beams strike target material. Some of these reactions release neutrons, and these machines are sources of high neutron flux. Neutrons are released in the act of fission, and nuclear reactors are unexcelled as intense neutron sources.

Having no electric charge, neutrons interact so slightly with atomic electrons in matter that energy loss by ionization and atomic excitation is essentially absent. Consequently they are vastly more penetrating than charged particles of the same energy. The main energy-loss mechanism occurs when they strike nuclei. As with billiard balls, the most efficient slowing-down occurs when the bodies that are struck in an elastic collision have the same mass as the moving bodies; hence the most efficient neutron moderator is hydrogen, followed by other light elements: deuterium, beryllium, and carbon. The great penetrating power of neutrons imposes severe shielding problems for reactors and other nuclear machines, and it is necessary to provide walls, usually of concrete, several feet in thickness to protect personnel.

Free neutrons are themselves radioactive, each transforming spontaneously into a proton, an electron (beta $\beta^-$ particle), and an antineutrino. This instability is a reflection of the fact that neutrons are slightly heavier than hydrogen atoms. Neutrons are, individually, small magnets. This property permits the production of beams of polarized neutrons, that is, beams of neutrons whose magnetic dipoles are aligned predominantly parallel to one direction in space. The magnetic moment is −1.913042. Neutrons spin with an angular momentum of $^1/_2$ in units of $h/2\pi$, where $h$ is Planck's constant. The negative sign which is attached to the magnetic moment indicates that the magnetic moment vector and the angular momentum vector are oppositely directed.

When neutrons are completely slowed down in matter, they have a maxwellian distribution in energy that corresponds to the temperature of the moderator with which they are in equilibrium. The maxwellian distribution has a tail extending to very low energies, and a few neutrons (about $10^{-11}$ of the main neutron flux) at this extreme have energies less than $5 \times 10^{-8}$ attojoule ($3 \times 10^{-7}$ eV), and hence velocities of less than about 7 m/s. The de Broglie wavelength of these ultracold neutrons is greater than 50 nanometers, which is so much larger than interatomic distances in solids that they interact with regions of a surface rather than with individual atoms, and as a result they are reflected from polished surfaces at all angles of incidence. Ultracold neutrons can be expected to be important in basic physics and to have applications in studies of surfaces and of the structure of inhomogeneities and magnetic domains in solids. [A.H.Sn.]

**Nickel** A chemical element, Ni, atomic number 28, a silver-white, ductile, malleable, tough metal. The atomic mass of naturally occurring nickel is 58.71. *See* PERIODIC TABLE.

Nickel consists of five natural isotopes having atomic masses of 58, 60, 61, 62, 64. Seven radioactive isotopes have also been identified, having mass numbers of 56, 57, 59, 63, 65, 66, and 67.

Most commercial nickel goes into stainless steel and other corrosion-resistant alloys. Nickel is also important in coins as a replacement for silver. Finely divided nickel is used as a hydrogenation catalyst.

Nickel is a fairly plentiful element, making up about 0.008% of the Earth's crust and 0.01% of the igneous rocks. Appreciable quantities of nickel are present in some kinds of meteorite, and large quantities are thought to exist in the Earth's core. Two important ores are the iron-nickel sulfides, pentlandite and pyrrhotite $(Ni,Fe)_xS_y$; the ore garnierite, $(Ni,Mg)SiO_3 \cdot nH_2O$, is also commercially important. Nickel occurs in small quantities in plants and animals. It is present in trace amounts in sea water, petroleum, and most coal.

Nickel metal is of moderate strength and hardness (3.8 on Mohs scale). When viewed as very small particles, nickel appears black. The density of nickel is 8.90 times that of water at 20°C (68°F). Nickel melts at 1455°C (2651°F) and boils at 2840°C (5144°F). Nickel is only moderately reactive. It resists alkaline corrosion and does not burn in the massive state, although fine nickel wires can be ignited. Nickel is above hydrogen in the electrochemical series, and it dissolves slowly in dilute acids, releasing hydrogen. In metallic form nickel is a moderately strong reducing agent.

Nickel is usually dipositive in its compounds, but it can also exist in the oxidation states 0, 1+, 3+, and 4+. Besides the simple nickel compounds, or salts, nickel forms a variety of coordination compounds or complexes. Most compounds of nickel are green or blue because of hydration or other ligand bonding to the metal. The nickel ion present in water solutions of simple nickel compounds is itself a complex, $[Ni(H_2O)_6]^{2+}$. [W.E.C.]

**Nicotinamide adenine dinucleotide phosphate (NADP)** A coenzyme and an important component of the enzymatic systems concerned with biological oxidation-reduction systems. It is also known as NAD, triphosphopyridine nucleotide (TPN), coenzyme II, and codehydrogenase II. The compound is similar in structure and function to nicotinamide adenine dinucleotide (NAD). It differs structurally from NAD in having an additional phosphoric acid group esterified at the 2′ position

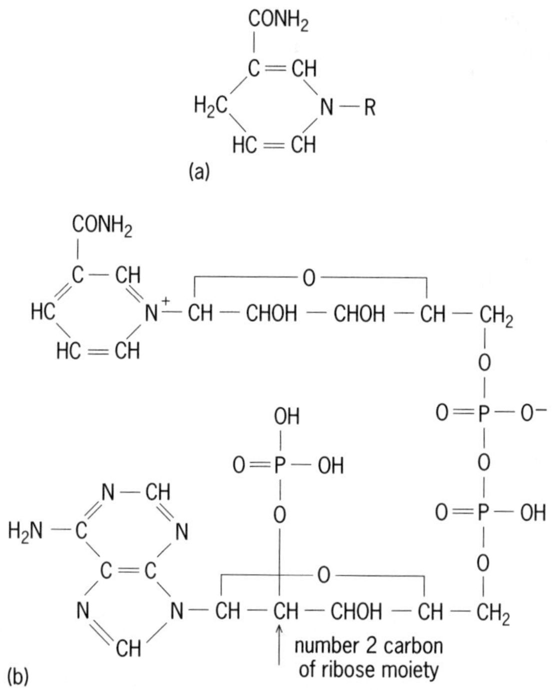

**Triphosphopyridine nucleotide. (*a*) Reduced form of the nicotinamide portion of TPNH. (*b*) Oxidized form of the molecule (TPN).**

of the ribose moiety of the adenylic acid portion. In biological oxidation-reduction reactions the NADP molecule becomes alternately reduced to its hydrogenated form (NADPH) and reoxidized to its initial state (see illustration). *See* COENZYME; ENZYME.

[M.D.]

**Niobium** A chemical element, Nb, atomic number 41 and atomic weight 92.906. In the United States this element was originally called columbium. The metallurgists and metals industry still use this older name. *See* PERIODIC TABLE.

Most niobium is used in special stainless steels, high-temperature alloys, and superconducting alloys such as $Nb_3Sn$. Niobium is also used in nuclear piles.

Niobium metal has a density of 8.6 g/cm$^3$ (5.0 oz/in.$^3$) at 20°C (68°F), a melting point of 2468°C (4474°F), and a boiling point of 4927°C (8900°F). Metallic niobium is quite inert to all acids except hydrofluoric, presumably owing to an oxide film on the surface. Niobium metal is slowly oxidized in alkaline solution. It reacts with oxygen and the halogens upon heating to form the oxidation state V oxide and halides, with nitrogen to form NbN, and with carbon to form NbC, as well as other elements such as arsenic, antimony, tellurium, and selenium.

The oxide $Nb_2O_5$, melting point 1520°C (2768°F), dissolves in fused alkali to yield a soluble complex niobate, $Nb_6O_{19}^{8-}$. Normal niobates such as $NbO_4^{3-}$ are insoluble. The oxide dissolves in hydrofluoric acid to give ionic species such as $NbOF_5^{2-}$ and $NbOF_6^{3-}$, depending on the fluoride and hydrogen-ion concentration. The highest fluoro complex which can exist in solution is $NbF_6^{-}$.

[E.M.L.]

**Nitration** A process in which a nitro group (—$NO_2$) becomes chemically attached to a carbon, oxygen, or nitrogen atom in an organic compound. A hydrogen or halogen atom is often replaced by the nitro group. Three general reactions summarize nitration chemistry:

1. C nitration, in which the nitro group attaches itself to a carbon atom [reaction (1)].

$$\gt C-H + HNO_3 \longrightarrow \gt C-NO_2 + H_2O \qquad (1)$$

2. O nitration (an esterification reaction), in which an O-N bond is formed to produce a nitrate [reaction (2)].

$$\gt C-OH + HNO_3 \longrightarrow \gt C-O+NO_2 + H_2O \qquad (2)$$

3. N nitration, in which a N-N bond is formed [reaction (3)].

$$\gt NH + HNO_3 \longrightarrow \gt N-NO_2 + H_2O \qquad (3)$$

Aromatics, alcohols, glycols, and amines are generally nitrated with mixed acids via an ionic reactions. The mixture includes nitric acid, a strong acid such as sulfuric acid which acts as a catalyst, and a small amount of water. With sulfuric acid, nitric acid is ionized to the nitronium ion, $NO_2^+$, which is the nitrating agent.

Propane is commercially nitrated in relatively large amounts using nitric acid in gas-phase free-radical reactions at temperatures of about 380–420°C. Nitric acid decomposes at these temperatures to produce nitrogen dioxide radicals (actually a mixture of $\cdot NO_2$ and ·ONO) and a hydroxy radical (·OH). In the free-radical reaction, about 35–40% of the nitric acid reacts to form four $C_1$-$C_3$ nitroparaffins; C-C bonds are broken during the nitration. The remaining nitric acid acts mainly as an oxidizing agent to form aldehydes, alcohols, carbon monoxide, carbon dioxide, water, and small amounts of other oxidized materials. Commercially, an adiabatic reactor is used, and the heat of reaction is employed to preheat and vaporize the nitric acid feed (containing water).

The product stream from free-radical nitrations is a condensation of mixed nitroparaffins. This liquid mixture is washed to remove the aldehydes, and then is distilled to recover each of the four nitroparafins—nitromethane, nitroethane, 1-nitropropane, and 2-nitropropane. The unreacted propane is recovered, combined with the feed propane, and returned to the reactor. The oxides of nitrogen are converted back to nitric acid; carbon monoxide, carbon dioxide, and water are discarded.

In the classical Victor Meyer process, an organic halide (often a bromide) is reacted with silver nitrite to produce a nitrohydrocarbon and silver halide. In a modified process, sodium nitrite, dissolved in a suitable solvent, is substituted for the more expensive silver nitrite. The desired nitroalkanes are produced in high yields by these processes, whereas they are produced in rather low yields in free-radical nitrations.

Nitrations can also often be performed by addition reactions using unsaturated hydrocarbons with nitric acid or nitrogen dioxide. [L.F.A.]

**Nitric acid** A strong mineral acid having the formula $HNO_3$. Pure nitric acid is a colorless liquid with a specific gravity of 1.52 at 25°C (77°F); it freezes at −47°C (−53°F). Nitric acid is used in the manufacture of ammonium nitrate and phosphate fertilizers, nitro explosives, plastics, dyes, and lacquers. The principal commercial process for the manufacture of nitric acid is the Ostwald process, in which ammonia, $NH_3$, is catalytically oxidized with air to form nitrogen dioxide, $NO_2$. When the dioxide is dissolved in water, 60% nitric acid is formed. Production of 90–100% nitric acid is based on processes such as the reaction of sulfuric acid with sodium nitrate (an older method of nitric acid manufacture), dehydration of 60% acid, and oxidation of nitrogen dioxide in a solution of dilute nitric acid. *See* AMMONIA; NITROGEN. [F.J.J.]

**Nitric oxide** An important messenger molecule in mammals and other animals. It can be toxic or beneficial, depending on the amount and where in the body it is released. Initial research into the chemistry of nitric oxide (NO) was motivated by its production in car engines, which results in photochemical smog and acid rain. In the late 1980s, researchers in immunology, cardiovascular pharmacology, neurobiology, and toxicology discovered that nitric oxide is a crucial physiological messenger molecule. Nitric oxide is now thought to play a role in blood pressure regulation, control of blood clotting, immune defense, digestion, the senses of sight and smell, and possibly learning and memory. Nitric oxide may also participate in disease processes such as diabetes, stroke, hypertension, impotence, septic shock, and long-term depression. *See* IMMUNOLOGY.

Most cellular messengers are large, unreactive biomolecules that make specific contacts with their targets. In contrast, nitric oxide is a small molecule that contains a free radical—that is, an unpaired electron—making it very reactive. Nitric oxide can freely diffuse through aqueous solutions or membranes, reacting rapidly with metal centers in cellular proteins and with reactive groups in other cellular molecules.

Nitric oxide is produced in the body by an enzyme called nitric oxide synthase, which converts the amino acid L-arginine to nitric oxide and L-citrulline. There are three types of nitric oxide synthase: brain, endothelial, and inducible. Both brain and endothelial enzymes are constitutive, that is, they are always present in cells, while the production of inducible nitric oxide synthase can be turned on or off when a system needs nitric oxide. After nitric oxide is produced in specific areas of the body by nitric oxide synthase, it diffuses to nearby cells. Nitric oxide then reacts preferentially in the interior of these cells with the metal centers of proteins. Nitric oxide binds specifically to the iron (Fe) atom of the heme group in proteins; it can also interact with other metal sites in proteins as well as with the thiol group (SH) of the amino acid cysteine. The interaction of nitric oxide with these proteins causes a cascade of intracellular events that leads to specific physiological changes within cells. For example, nitric oxide causes the smooth muscle cells surrounding blood vessels to relax, decreasing blood pressure. Nitric oxide plays an important role in the central and peripheral nervous systems; the overproduction of nitric oxide in brain tissues has been implicated in stroke and other neurological problems.

Nitric oxide also functions as an important agent in the immune system by killing invading bacterial cells. Nitric oxide released by macrophages can inhibit important cellular processes in the bacteria, including deoxyribonucleic acid (DNA) synthesis and respiration, by binding to and destroying iron-sulfur centers in key enzymes in these pathways.

Although nitric oxide production in the immune system serves a crucial biological function, there can be adverse effects when too much nitric oxide is produced. During a massive bacterial infection, excess nitric oxide can go into the vascular system, causing a dramatic decrease in blood pressure, which may lead to possibly fatal septic shock. Thus, scientists are working on drugs that can selectively inhibit the inducible form of nitric oxide synthase in order to avoid the harmful effects produced by excess nitric oxide without interfering with useful nitric oxide pathways. [J.N.Bu.; M.F.R.]

**Nitrile** One of a group of organic chemical compounds of general formula $RC{\equiv}N$. A nitrile is named from the acid to which it can be hydrolyzed by adding the suffix -onitrile to the acid stem, for example, acetonitrile from acetic acid. An alternative system names the group attached to CN, thus $CH_3CN$ is also named methyl cyanide. In more complex structures the CN group is named as a substituent, cyano.

Industrially, nitriles are formed by heating carboxylic acids with ammonia and a dehydration catalyst under pressure. For the preparation of acrylonitrile, which is used on a large scale in the plastics industry, a vapor-phase catalytic ammoxidation of propylene has been developed. *See* ACRYLONITRILE; AMINE. [P.E.F.]

**Nitro and nitroso compounds** Nitro compounds are derivatives of organic hydrocarbons having one or more —$NO_2$ groups with nitrogen-to-carbon bonding. They differ from the oxygen-linked nitrites, which are esters. The group lacks enough electrons to form double bonds with both oxygens. However, both oxygens react alike; hence the bond is regarded as a resonance hybrid of single and double bonds.

Aromatic nitro compounds have been used chiefly as dye intermediates, explosives, and Pharmaceuticals. They are formed readily by the reaction of aromatic compounds with nitric acid; H is replaced by the —$NO_2$ group, for example,

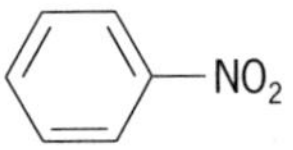

—$NO_2$

Aliphatic nitro compounds are prepared with difficulty and have grown in importance only since the development of vapor-phase nitration of hydrocarbons with nitric acid vapors at 420°C (788°F).

Nitroso compounds contain the —NO group attached to carbon or nitrogen. Many are unstable intermediates, for example, nitrosobenzene formed during the reduction of nitrobenzene. *See* NITRATION. [A.L.H.]

**Nitroaromatic compound** A member of the class of organic compounds in which the nitro group (—$NO_2$) is attached directly to the cyclic, aromatic nucleus. The prototypal compound is nitrobenzene. It is prepared by the reaction of benzene with nitric acid in the presence of sulfuric acid, as shown in the reaction below. The

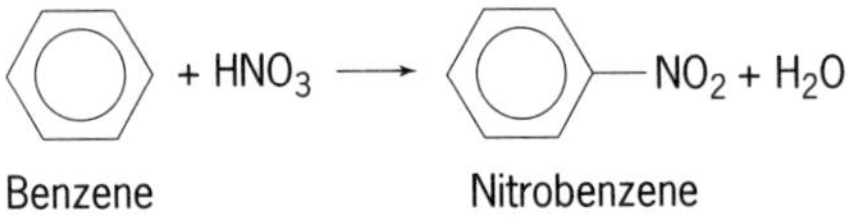

$$+ HNO_3 \longrightarrow \text{—}NO_2 + H_2O$$

Benzene Nitrobenzene

most significant use of nitrobenzene is in the manufacture of aniline (**1**). About 97% of

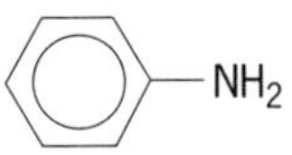

—$NH_2$

(1)

the nitrobenzene produced in the United States is converted to aniline, which is used in the manufacture of plastics, rubber additives, dyes, drugs, and other products. *See* BENZENE.

Second to nitrobenzene in commercial importance are the mononitrotoluenes, particularly the ortho and para isomers, (**2**) and (**3**), respectively. Reaction of toluene with a mixture of nitric and sulfuric acid at about 40°C (104°F) gives a high yield of a mixture of the isomers, which are separated by a combination of fractional distillation

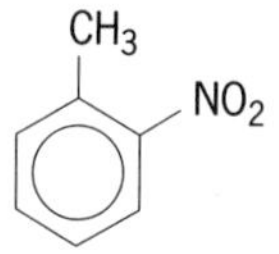

$CH_3$ $NO_2$

(2)

$CH_3$ $NO_2$

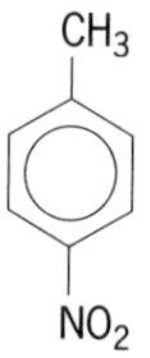

(3)

and crystallization. The nitrotoluenes are important intermediates in the preparation of dyes, rubber chemicals, and agricultural chemicals.

2,4,6-Trinitrotoluene (TNT; **4**) is a military explosive that is stable, nonhygroscopic,

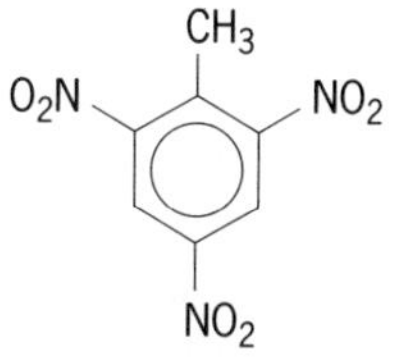

$CH_3$ $O_2N$ $NO_2$ $NO_2$

(4)

and relatively insensitive to impact, friction, shock, and electric spark. It is produced by nitration of toluene in successive stages at progressively higher temperatures and concentrations of acid.

Although literally thousands of other aromatic ring compounds, including the heterocyclics, have been converted to their nitro derivatives, few such compounds have achieved any significant industrial importance. *See* AROMATIC HYDROCARBON; NITRATION.

[P.E.F.]

**Nitrogen** A chemical element, N, atomic number 7, atomic weight 14.0067. Nitrogen, a gas under normal conditions, is the lightest element of periodic group 5 (nitrogen family). *See* PERIODIC TABLE.

At standard temperature and pressure, elemental nitrogen exists as a gas with a density of 1.25046 g/liter. This value indicates that the molecular formula is $N_2$. Some physical properties of elemental nitrogen are listed in Table 1.

Elemental nitrogen has a low reactivity toward most common substances at ordinary temperatures. At high temperatures, molecular nitrogen, $N_2$, reacts with chromium, silicon, titanium, aluminum, boron, beryllium, magnesium, barium, strontium, calcium, and lithium (but not the other alkali metals) to form nitrides; with $O_2$ to form NO; and at moderately high temperatures and pressures in the presence of a catalyst, with hydrogen

**Table 1. Properties of nitrogen**

| Property | Value |
|---|---|
| Heat of transformation ($\alpha$–$\beta$) | 54.71 cal/mole |
| Heat of fusion | 172.3 cal/mole |
| Heat of vaporization | 1332.9 cal/mole |
| Critical temperature | 126.26 ± 0.04 K |
| Critical pressure | 33.54 ± 0.02 atm |
| Density: $\alpha$ form | 1.0265 g/ml at −252.6°C |
| $\beta$ form | 0.8792 g/ml at −210.0°C |
| Liquid | 1.1607–0.0045$T$($T$ = abs temp) |

**Table 2. Compounds of nitrogen**

| Oxidation state | Examples |
|---|---|
| +5 | $N_2O_5$, $HNO_3$, nitrates, $NO_2X$ |
| +4 | $N_2O_4 \rightleftharpoons 2NO_2$ |
| +3 | $N_2O_3$, $HNO_2$, nitrites, NOX, $NX_3$ |
| +2 | NO, $Na_2NO_2$, nitrohydroxylamates |
| +1 | $N_2O$, $H_2N_2O_2$, hyponitrites |
| 0 | $N_2$ |
| −1/3 | $HN_3$, acids |
| −1 | $NH_2OH$, hydroxylammonium salts |
| −2 | $NH_2NH_2$, hydrazinium salts, hydrazides |
| −3 | $NH_3$, ammonium salts, amides, imides, nitrides |

to form ammonia. Above 1800°C (3300°F), nitrogen, carbon, and hydrogen combine to form hydrogen cyanide.

Table 2 lists the principal classes of inorganic nitrogen compounds. Thus, in addition to the typical oxidation states of the family (−3, +3, and +5), nitrogen forms compounds with a variety of additional oxidation states. *See* AMINE; AMMONIA; HYDRAZINE; NITRIC ACID; NITROGEN COMPLEXES; NITROGEN OXIDES.

Molecular nitrogen is the principal constituent of the atmosphere (78% by volume of dry air), in which its concentration is a result of the balance between the fixation of atmospheric nitrogen by bacterial, electrical (lightning), and chemical (industrial) action, and its liberation through the decomposition of organic materials by bacteria or combustion. In the combined state, nitrogen occurs in a variety of forms. It is a constituent of all proteins (both plant and animal) as well as of many other organic materials. Its chief mineral source is sodium nitrate.

The methods for the preparation of elementary nitrogen may be grouped into two classes, separation from the atmosphere and decomposition of nitrogen compounds. The industrial method for the production of nitrogen is the fractional distillation of liquid air. Nitrogen containing about 1% argon and traces of other inert gases may be obtained by the chemical removal of oxygen, carbon dioxide, and water vapor from the atmosphere by appropriate chemical reagents.

Because the importance of nitrogen compounds in agriculture and chemical industry, much of the industrial interest in elementary nitrogen has been in processes for converting elemental nitrogen into nitrogen compounds. The principal methods for doing this are the Haber process for the direct synthesis of ammonia from nitrogen and hydrogen, the electric are process, which involves the direct combination of $N_2$ and $O_2$ to nitric oxide, and the cyanamide process. Nitrogen is also used for filling bulbs of incandescent lamps and, in general, wherever a relatively insert atmosphere is required. [H.H.S.]

**Nitrogen complexes** Compounds containing the dinitrogen molecule, $N_2$, bound to a metal (also called dinitrogen complexes). Outstanding in their ability to form coordination compounds with nitrogen are a number of metals which belong to the group 7 transition metal family. For each metal of this group, several nitrogen complexes have been identified. Nitrogen complexes of these metals occur in low oxidation states, such as Co(I) or Ni(O). The other ligands present in these complexes besides $N_2$ are usually of a type known to stabilize low oxidation states; phosphines appear to be particularly prominent bonding partners in this respect. The illustration shows the structure of a typical $N_2$ complex, elucidated by a crystal structure determination. The N-N bond

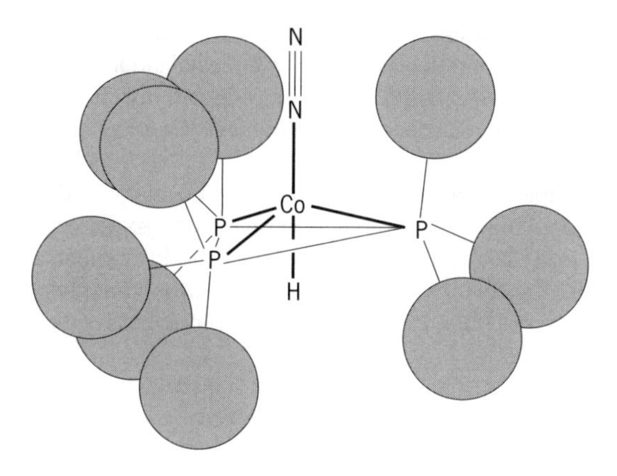


**Structure of a coordination compound with $N_2$ (circles represent phenyl groups).**

axis in this complex is aimed, within the limits of experimental error, directly toward the position of the metal atom. The Co-$N_2$ bond length, 0.18 nanometer, is within the normal range of comparable metal-ligand bonds. *See* COORDINATION CHEMISTRY.

Even in most favorable cases the binding of the dinitrogen molecule to the metal is fairly labile; all the compounds lose their nitrogen on mild heating. Some of the nitrogen complexes are only metastable to loss of dinitrogen even at room temperature; accordingly, they cannot be obtained by direct uptake of gaseous nitrogen. In the synthesis of these metastable complexes, hydrazine or azide compounds serve as a source of nitrogen molecules within the coordination sphere of the metal. Addition of other coordinating agents to the nitrogen complexes usually results in a displacement of $N_2$ from the metal. The cobalt compound in the illustration exchanges its $N_2$ ligand quite reversibly for other ligand molecules, such as $NH_3$ and $H_2C{=}CH_2$. Whereas these ligands are easily displaced again by an excess of $N_2$, an irreversible exchange occurs with carbon monoxide. The bulky organic groups on the phosphine ligands are likely to interfere with the approach to the metal of all but the slimmest ligands and thereby help the "thin" dinitrogen molecule to maintain or regain its position on the metal in competition with most other ligands. [H.B.]

**Nitrogen oxides** Chemical compounds of nitrogen and oxygen. Nitrogen and oxygen do not combine when mixed directly (as in air), but they do combine during chemical reactions of compounds containing them. A number of nitrogen oxides can be isolated which differ from one another in the numbers of nitrogen and oxygen atoms present in each molecule. The table gives data for the five nitrogen oxides which are well established.

**Oxides of nitrogen and their properties**

| Name | Stoichiometric formula | Melting point, °C (°F) | Boiling point, °C (°F) |
|---|---|---|---|
| Nitrous oxide (dinitrogen monoxide) | $N_2O$ | −90.8 (−131) | −88.5 (−127.3) |
| Nitric oxide (nitrogen monoxide) | NO | −163.6 (−262.5) | −151.7 (241.0) |
| Dinitrogen trioxide | $N_2O_3$ | −103 (−155) | 3.5 (38.3) |
| Dinitrogen tetroxide ($\rightleftharpoons$ nitrogen dioxide) | $N_2O_4$ ($\rightleftharpoons NO_2$) | −11.2 (11.8) | 21.2 (70.2) |
| Dinitrogen pentoxide | $N_2O_5$ | 41 (106) | |

**Nitrous oxide and nitric oxide.** When inhaled, nitrous oxide has anesthetic effects; in small amounts it produces mild hysteria and hence is sometimes called laughing gas. It is colorless, is the least reactive of the oxides, and dissolves in water without chemical reaction. Some nitric oxide is formed in an electric arc, as in the technical production of nitric acid.

With oxygen or air, nitric oxide is rapidly converted to nitrogen dioxide. Nitric oxide is colorless and is soluble in water without reaction. It is an important messenger molecule in animals. It is one of the few "odd" molecules which contain an odd number of electrons. As an odd molecule, it has the ability to lose or gain one electron, thus giving the electrically charged ions $NO^+$ and $NO^-$. The important nitrosyl compounds contain these ions.

**Trioxide.** Dinitrogen trioxide exists pure only in the solid state. It is the anhydride of nitrous acid; when the oxide is dissolved in an alkaline solution, nitrite ion is produced.

**Dioxide and tetroxide.** The position of the equilibrium between nitrogen dioxide and dinitrogen tetroxide depends upon temperature and physical state. Dinitrogen tetroxide reacts readily with water to give an equimolecular mixture of nitrous and nitric acids. As temperature is raised, the nitrous acid decomposes to nitric acid and nitric oxide. These reactions are important in the technical production of nitric acid by catalytic oxidation of ammonia. Dinitrogen tetroxide is an oxidizing agent comparable in strength to bromine, and is employed as such in the lead-chamber process for sulfuric acid. In organic chemistry the tetroxide finds use as a special oxidizing agent (for example, in the production of sulfoxides and phosphine oxides) and as a nitrating agent.

**Pentoxide.** Solid dinitrogen pentoxide readily volatilizes, and the molecular type of structure found in the gaseous state is observed also in solutions of the oxide in low dielectric solvents such as carbon tetrachloride and chloroform. Sodium metal reacts with the liquid oxide, liberating nitrogen dioxide and forming sodium nitrate. Gaseous dinitrogen pentoxide decomposes readily, and is a strong oxidizing agent. With water it is converted to nitric acid. *See* NITRIC OXIDE; NITROGEN; OXYGEN. [C.C.A.]

**Nitroparaffin** Any derivative of an aliphatic hydrocarbon that contains one or more $—NO_2$ groups bonded via nitrogen to the carbon framework. Nitroparaffins are also known as nitroalkanes.

Low-molecular-weight nitroparaffins are prepared via the vapor-phase nitration of alkanes at >400°C (750°F). However, the process is not generally satisfactory for higher-molecular-weight nitroparaffins because of polynitration and chain cleavage. The direct nitration of propane is used commercially to prepare nitromethane (boiling point 101°C or 214°F), nitroethane (bp 114°C or 237°F), 1-nitropropane (bp 131°C or 268°F), and 2-nitropropane (bp 120°C or 248°F). Nitroparaffins are prepared in the laboratory by the reaction of nitrite salts with alkyl bromides or iodides, from the oxidation of amines or oximes by using peroxycarboxylic acids, and by the chain homologation of simple nitroparaffins. *See* NITRATION.

Nitromethane, nitroethane, and the nitropropanes are useful solvents with high dielectric constants that readily dissolve many polymers. In addition, these simple nitroparaffins are versatile intermediates for the synthesis of specialty chemicals. [A.G.M.B.]

**Nobelium** A chemical element, No, atomic number 102. Nobelium is a synthetic element produced in the laboratory. It decays by emitting an alpha particle, that is, a doubly charged helium ion. Only atomic quantities of the element have been produced to date. Nobelium is the tenth element heavier than uranium to be produced synthetically. It is the thirteenth member of the actinide series, a rare-earth-like series of elements. *See* ACTINIDE ELEMENTS; PERIODIC TABLE; RARE-EARTH ELEMENTS; TRANSURANIUM ELEMENTS.

[P.R.F.]

**Nonmetal** The elements are conveniently, but arbitrarily, divided into metals and nonmetals. The nonmetals do not conduct electricity readily, are not ductile, do not have a complex refractive index, and in general have high ionization potentials.

If the periodic table is divided diagonally from upper left to lower right, all the nonmetals are on the right-hand side of the diagonal. Examples of elements which do not fit neatly into this useful but arbitrary classification are tin, which exists in two allotropic modifications, one definitely metallic and the other with many properties of a nonmetal, and tellurium and antimony. Such elements are called metalloids. *See* METALLOID; PERIODIC TABLE.

[T.C.W.]

**Nonstoichiometric compounds** Chemical compounds in which the relative number of atoms is not expressible as the ratio of small whole numbers, hence compounds for which the subscripts in the chemical formula are not rational (for example, $Cu_{1.987}S$). Sometimes they are called berthollide compounds to distinguish them from daltonides, in which the ratio of atoms is generally simple. Nonstoichiometry is a property of the solid state and arises because a fraction of the atoms of a given kind may be (1) missing from the regular structure (for example, $Fe_{1-\delta}O$), (2) present in excess over the requirements of the structure (for example, $Zn_{1+\delta}O$), or (3) substituted by atoms of another kind (for example, $Bi_2Te_{3\pm\delta}$). The resulting materials are generally of variable composition, intensely colored, metallic or semiconducting, and different in chemical reactivity from the parent stoichiometric compounds from which they are derived.

Nonstoichiometry is best known in the binary compounds of the transition elements, particularly the hydrides, oxides, chalcogenides, pnictides, carbides, and borides. It is also well represented in the so-called insertion or intercalation compounds, in which a metallic element or neutral molecule has been inserted in a stoichiometric host. Nonstoichiometric compounds are important in some solid-state devices (such as rectifiers, thermoelectric generators, and photodetectors) and are probably formed as chemical intermediates in many reactions involving solids (for example, heterogeneous catalysis and metal corrosion).

The simplest way to classify nonstoichiometric compounds is to consider which element is in excess and how this excess is brought about. A classification scheme largely based on this distinction but which also includes some examples of ternary systems is as follows.

*Binary compounds*:

I. Metal nonmetal ratio greater than stoichiometric
   (*a*) Metal in excess, for example, $Zn_{1+\delta}O$
   (*b*) Missing nonmetal, for example $UH_{3-\delta}$, $WO_{3-\delta}$
II. Metal: nonmetal ratio less than stoichiometric
   (*a*) Metal-deficient, for example, $Co_{1-\delta}O$
   (*b*) Nonmetal in excess, for example, $UO_{2+\delta}$
III. Deviations on both sides of stoichiometry, for example, $TiO_{1\pm 7\delta}$

*Ternary compounds* (insertion compounds):

IV. Oxide "bronzes," for example, $M_\delta WO_3$, $M_\delta V_2O_5$
V. Intercalation compounds, for example, $K_{1.5+\delta}MoO_3$, $Li_\delta TiS_2$

Excluded from consideration are the recognized impurity materials, such as $Na_{1-2x}Ca_xCl$, which are best considered as conventional solid solutions wherein ions of one kind and perhaps vacancies have replaced an equivalent number of ions of another kind. [M.J.Si.]

**Nuclear chemical engineering** The branch of chemical engineering that deals with the production and use of radioisotopes, nuclear power generation, and the nuclear fuel cycle. A nuclear chemical engineer requires training in both nuclear and chemical engineering. As a nuclear engineer, he or she should be familiar with the nuclear reactions that take place in nuclear fission reactors and radioisotope production, with the properties of nuclear species important in nuclear fuels, with the properties of neutrons, gamma rays, and beta rays produced in nuclear reactors, and with the reaction, absorption, and attenuation of these radiations in the materials of reactors. *See* NEUTRON.

As a chemical engineer, he or she should know the properties of materials important in nuclear reactors and the processes used to extract and purify these materials and convert them into the chemical compounds and physical forms used in nuclear systems. *See* CHEMICAL ENGINEERING.

Aspects of nuclear reactors of concern to nuclear chemical engineers include production and purification of the uranium dioxide fuel, production of the hafnium-free zirconium tubing used for fuel cladding, and control of corrosion and radioactive corrosion products by chemical treatment of coolant. A chemical engineering aspect of heavy-water reactor operation is control of the radioactive tritium produced by neutron activation of deuterium. Aspects of liquid-metal fast-breeder reactors of concern to nuclear chemical engineers include fabrication of the mixed uranium dioxide-plutonium dioxide fuel, purity control of sodium coolant to prevent fouling and corrosion, and reprocessing of irradiated fuel to recover plutonium and uranium for recycle. *See* PLUTONIUM; URANIUM. [M.Be.]

**Nuclear chemistry** An interdisciplinary field that, in general, encompasses the application of chemical techniques to the solution of problems in nuclear physics. The discovery of the naturally occurring radioactive elements and of nuclear fission are classical examples of the work of nuclear chemists.

Although chemical techniques that are employed in nuclear chemistry are essentially the same as those in radiochemistry, these fields may be distinguished on the basis of the aims of the investigation. Thus, a nuclear chemist utilizes chemical techniques as a tool for the study of nuclear reactions and properties, whereas a radiochemist utilizes the radioactive properties of certain substances as a tool for the study of chemical reactions and properties. For the application of radioactive tracers to chemical problems *see* RADIOCHEMISTRY. For the chemical effects of radiation on various systems *see* RADIATION CHEMISTRY.

The chemical identification of radioactive nuclides and the determination of their nuclear properties has been one of the major activities of nuclear chemists. Such studies have produced an extensive array of radioisotopes, and present studies are concerned mainly with the more difficult identification of nuclides of very short half-life. Nuclear chemical investigations led to the discovery of the synthetic radioactive elements which do not have any stable isotopes and are not formed in the natural radioactive series (technetium, promethium, astatine, and the transuranium elements). Other major areas of nuclear chemistry include studies of nuclear structure and spectroscopy and of the probability and mechanisms of various nuclear reactions. [E.P.S.]

**Nuclear magnetic resonance (NMR)** A phenomenon exhibited when atomic nuclei in a static magnetic field absorb energy from a radio-frequency field of certain characteristic frequencies. Nuclear magnetic resonance is a powerful analytical tool for the characterization of molecular structure, quantitative analysis, and the examination of dynamic processes. It is based on quantized spectral transitions between nuclear Zeeman levels of stable isotopes, and is unrelated to radioactivity.

The format of nuclear magnetic resonance data is a spectrum that contains peaks referred to as resonances. The resonance of an isotope is distinguished by the transition frequency of the nucleus. The intensity of the resonance is directly proportional to the number of nuclei that produce the signal. Although the majority of nuclear magnetic resonance spectra are measured for samples in solution or as neat liquids, it is possible to measure nuclear magnetic resonance spectra of solid samples. Nuclear magnetic resonance is a nondestructive technique that can be used to measure spectra of cells and living organisms.

**Nuclear magnetic properties.** The nuclei of most atoms possess an intrinsic nuclear angular momentum. The classical picture of nuclear angular momentum is a spherical nucleus rotating about an axis in a manner analogous to the rotation of the Earth. Nuclear angular momentum, like most other atomic quantities, can be expressed as a series of quantized levels.

The transitions that give rise to the nuclear magnetic resonance spectrum are produced when the nuclei are placed in a static magnetic field. The external or applied magnetic field defines a geometric axis, denoted as the $z$ axis. The external magnetic field orients the $z$ components of nuclear angular momentum and the magnetic moment with respect to the $z$ axis. The nuclear magnetic resonance spectrum is produced by spectral transitions between different spin states and is therefore dependent on the value of the nuclear spin. Nuclei for which the nuclear spin value is 0 have a magnetic spin-state value of 0. Therefore, these nuclei do not give rise to a nuclear magnetic resonance spectrum under any circumstances. Many important elements in chemistry have zero values of nuclear spin and are inherently nuclear magnetic resonance inactive, including the most abundant isotopes of carbon-12, oxygen-16, and sulfur-32. Nuclear magnetic resonance spectra are measured most often for nuclei that have nuclear spin values of $\frac{1}{2}$. Chemically important spin-$\frac{1}{2}$ nuclei include the isotopes hydrogen-1, phosphorus-31, flourine-19, carbon-13, nitrogen-15, silicon-29, iron-57, selenium-77, cadmium-113, silver-107, platinum-195, and mercury-199. Nuclear magnetic resonance spectra of nuclei that have nuclear spin values of greater than $\frac{1}{2}$, known as quadrupolar nuclei, can be measured; but the measurements are complicated by faster rates of nuclear relaxation. Nuclei that fall into this class include the isotopes hydrogen-2, lithium-6, nitrogen-14, oxygen-17, boron-11, chlorine-35, sodium-23, aluminum-27, and sulfur-33.

The transition frequency in nuclear magnetic resonance depends on the energy difference of the spin states. The intensity of the resonance is dependent on the population difference of the two spin states, which in turn depends directly on the magnitude of the energy difference. For spin-$\frac{1}{2}$ nuclei, $\Delta E$ is the difference in energy between the $+$ and $-$ values of the magnetic spin state. In contrast to other spectroscopic methods, the nuclear magnetic resonance frequency is variable, depending on the strength of the magnet used to measure the spectrum. Larger magnetic fields produce a greater difference in energy between the spin states. This translates into a larger difference in the population of the spin states and therefore a more intense resonance in the resulting nuclear magnetic resonance spectrum.

Individual nuclei of the same isotope in a molecule have transition frequencies that differ depending on their chemical environment. This phenomenon, called chemical

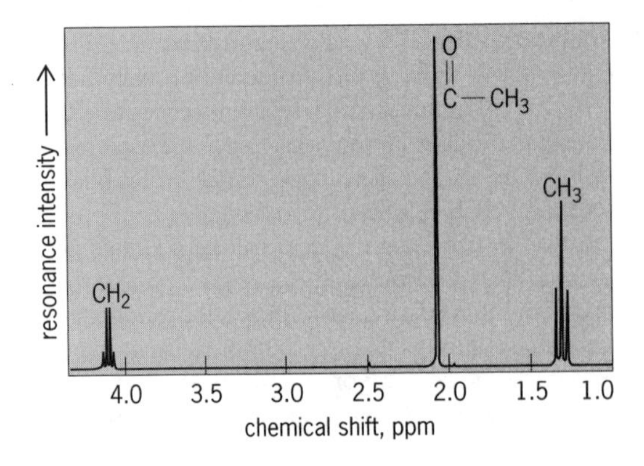


**Proton spectrum of ethyl acetate ($CH_3COOH_2CH_3$). Chemical shift is relative to the protons of tetramethylsilane. The integrated intensity of the resonances corresponds to the relative number of protons which give rise to each resonance.**

shift, occurs because the effective magnetic field at a particular nucleus in a molecule is less than the applied magnetic field due to shielding by electrons.

An example of resonance chemical shift is observed in the $^1$H nuclear magnetic resonance spectrum of ethyl acetate ($CH_3COOCH_2CH_3$; see illustration). Resonances at several frequencies are observed in this spectrum. The methylene ($CH_2$) protons are affected by the electron-withdrawing oxygen atoms of the neighboring ester group, and as a result the chemical shift of the methylene proton resonance is significantly different from the chemical shift of the resonances of the protons of the methyl ($CH_3$) groups of ethyl acetate. The two methyl groups are in different chemical environments and therefore give rise to resonances that have different chemical shifts. Because of the dependence of the transition frequency of a nucleus on its chemical environment, chemical shift is diagnostic of the functional group containing the nucleus of interest. Nuclear magnetic resonance spectroscopy is a frequently employed tool in chemical synthesis studies, because the nuclear magnetic resonance spectrum can confirm the chemical structure of a synthetic product.

In most nuclear magnetic resonance spectra, rather than frequency units, the spectra are plotted in units of chemical shift expressed as part per million (ppm). The ppm scale calculates the ratio of the resonance frequency in hertz (Hz) to the Larmor frequency of the nucleus at the magnetic field strength of the measurement. The ppm scale allows direct comparison of nuclear magnetic resonance spectra acquired by using magnets of differing field strength. This unit of nuclear magnetic resonance chemical shift should not be confused with the concentration units of ppm (mg/kg), often referred to by analytical chemists in the field of trace analysis.

The chemical shift (either in hertz or ppm) of a resonance is assigned relative to the chemical shift of a standard reference material. The nuclear magnetic resonance community has agreed to arbitrarily set the chemical shift of certain standard compounds to 0 ppm. For $^1$H and $^{13}$C nuclear magnetic resonance, the accepted standard is tetramethylsilane, which is defined to have a chemical shift of 0 ppm. However, any molecule with a resonance frequency in the appropriate chemical shift region of the spectrum that does not overlap with the resonances of the sample can be employed as a chemical shift reference. The use of a chemical shift reference compound other than the accepted standard is particularly common for nuclei that have a very large chemical shift range such as fluorine-19 or selenium-77, since it may be impossible to measure the spectrum of the accepted chemical shift reference compound and the analyte of interest simultaneously because of their very different frequencies.

In addition to the differences in chemical shift, resonances in a given spectrum, for example for ethyl acetate (see illustration), also differ in the number of signals composing the resonance detected for each group of protons. Instead of one resonance resulting from a single transition between two spin states, the methylene group ($CH_2$) resonance is actually a quartet composed of four lines. Similarly, although the acetate $CH_3$ resonance is a single resonance, the ethylene group ($CH_3$) resonance consists of three lines. This splitting of some resonances, called spin-spin or scalar coupling, arises from interactions between nuclei through their bonding electrons rather than through space. Scalar coupling is a short-range interaction that usually occurs for nuclei separated by one to three bonds.

The effects of spin-spin coupling can be removed from the spectrum by application of a low-strength magnetic field to one of the coupled spins. The effect of this secondary field is to equalize the population of the coupled transitions and to remove the effects of spin-spin coupling. Decoupling can be homonuclear or heteronuclear. Homonuclear decoupling is useful for assigning the resonances of coupled spin systems in the spectrum of a compound or complex mixture. For example, in ethyl acetate, decoupling at the frequency of the methylene group ($CH_2$) protons would remove the effects of spin-spin coupling, collapsing the ethylene group ($CH_3$) resonance into a singlet and confirming the resonance assignments. [C.K.L.]

**Nucleation** The formation within an unstable, supersaturated solution of the first particles of precipitate capable of spontaneous growth into large crystals of a more stable solid phase. These first viable particles, called nuclei, may either be formed from solid particles already present in the system (heterogeneous nucleation) or be generated spontaneously by the supersaturated solution itself (homogeneous nucleation). *See* SUPERSATURATION.

Nucleation is significant in analytical chemistry because of its influence on the physical characteristics of precipitates. Processes occurring during the nucleation period establish the rate of precipitation, and the number and size of the final crystalline particles. *See* COLLOID; PRECIPITATION (CHEMISTRY). [D.H.K.; L.Go.]

**Nucleic acid** An acidic, chainlike biological macromolecule consisting of multiply repeated units of phosphoric acid, sugar, and purine and pyrimidine bases. Nucleic acids as a class are involved in the preservation, replication, and expression of hereditary information in every living cell. There are two types of nucleic acid: deoxyribonucleic acid (DNA) and ribonucleic acid (RNA). [E.Jo.]

**Nucleoprotein** A generic term for any member of a large class of proteins associated with nucleic acid molecules. Nucleoprotein complexes occur in all living cells and in viruses, where they play vital roles in reproduction and protein synthesis.

Classification of the nucleoproteins depends primarily upon the type of nucleic acid involved—deoxyribonucleic acid (DNA) or ribonucleic acid (RNA)—and on the biological function of the complex. Deoxyribonucleoproteins (complexes of DNA and proteins) constitute the genetic material of all organisms and of many viruses. They function as the chemical basis of heredity and are the primary means of its expression and control. Most of the mass of chromosomes is made up of DNA and proteins whose structural and enzymatic activities are required for the proper assembly and expression of the genetic information encoded in the molecular structure of the nucleic acid.

Ribonucleoproteins (complexes of RNA and proteins) occur in all cells as part of the machinery for protein synthesis. This complex operation requires the participation of messenger RNAs (mRNAs), amino acid transfer RNAs (tRNAs), and ribosomal RNAs (rRNAs), each of which interacts with specific proteins to form functional complexes called polysomes, on which the synthesis of new proteins occurs.

In simpler life forms, such as viruses which infect animal and plant cells and bacteriophages which infect bacteria, most of the mass of the viral particle is due to its nucleoprotein content. The material responsible for the hereditary continuity of the virus may be DNA or RNA, depending on the type of virus, and it is usually enveloped by one or more proteins which protect the nucleic acid and facilitate infection. [V.G.A.]

**Nucleotide** A cellular constituent that is one of the building blocks of ribonucleic acids (RNA) and deoxyribonucleic acid (DNA). In biological systems, nucleotides are linked by enzymes in order to make long, chainlike polynucleotides of defined sequence. The order or sequence of the nucleotide units along a polynucleotide chain plays an important role in the storage and transfer of genetic information. Many nucleotides also perform other important functions in biological systems. *See* COENZYME; CYCLIC NUCLEOTIDES; NUCLEIC ACID.

Nucleotides are generally classified as either ribonucleotides or deoxyribonucleotides. Both classes consist of a phosphorylated pentose sugar that is linked via an *N*-glycosidic bond to a purine or pyrimidine base. The combination of the pentose sugar and the purine or pyrimidine base without the phosphate moiety is called a nucleoside. *See* PURINE; PYRIMIDINE. [E.P.G.]

**Oligonucleotide** A deoxyribonucleic acid (DNA) or ribonucleic acid (RNA) sequence composed of two or more covalently linked nucleotides. Oligonucleotides are classified as deoxyribooligonucleotides or ribooligonucleotides. Fragments containing up to 50 nucleotides are generally termed oligonucleotides, and longer fragments are called polynucleotides.

A deoxyribooligonucleotide consists of a 5-carbon sugar called deoxyribose joined covalently to phosphate at the 5′ and 3′ carbons of this sugar to form an alternating, unbranched polymer. A ribooligonucleotide consists of a similar repeating structure where the 5-carbon sugar is ribose. Chemically synthesized oligonucleotides of predetermined sequence have proven to be very useful for studying a large number of biochemical processes. In the 1960s, these compounds were used to decipher the genetic code. Later, chemically prepared deoxyoligonucleotides were joined to form genes for transfer RNAs. Gene synthesis from synthetic deoxyoligonucleotides is now routinely used to prepare genes and modified genes for proteins having potential clinical applications. Oligonucleotides have also been used to diagnose genetic disorders and bacterial or viral infections. *See* NUCLEIC ACID. [M.H.C.]

**Opalescence** The milky iridescent appearance of a dense transparent medium when the system (or medium) is illuminated by polychromatic radiation in the visible range, such as sunlight. Slight changes in the rainbowlike color of the system can occur, depending on the scattering angle, that is, the angle between the directions of incident radiation and of observation.

Opalescence is a general term which applies to the optical phenomenon of intense scattering in the visible range of the electromagnetic radiation by a system with strong local optical inhomogeneities. The iridescence, or rainbowlike display of interference of colors, arises because the intensity of scattered light is approximately proportional to the reciprocal fourth power of the wavelength of incident light (Rayleigh's law). [B.C.]

**Optical activity** The effect of asymmetric compounds on polarized light. To exhibit this effect, a molecule must be non-superimposable on its mirror image, that is, must be related to its mirror image as the right hand is to the left hand. An optically active compound and its mirror image are called enan-tiomers or optical isomers (see illustration). Enantiomers differ only in their geometric arrangements; they have identical chemical and physical properties. The right-handed and left-handed forms of a molecule can be distinguished only by their optical activity or by their interactions with other asymmetric molecules. Optical activity can be used to probe other aspects of molecular geometry, as well as to identify which enantiomer is present and its purity.

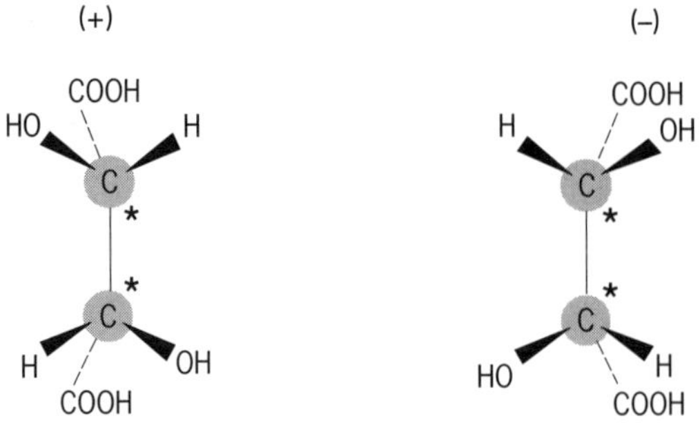

Enantiomers of tartaric acid.

The physical basis of optical activity is the differential interaction of asymmetric substances with left versus right circularly polarized light. If solids and substances in strong magnetic fields are excluded, optical activity is an intrinsic property of the molecular structure and is one of the best methods of obtaining structural information from a sample in which the molecules are randomly oriented. The relationship between optical activity and molecular structure results from the interaction of polarized light with electrons in the molecule. Thus the molecular groups that contribute most directly to optical activity are those that have mobile electrons which can interact with light. Such groups are called chromophores, since their absorption of light is responsible for the color of objects. For example, the chlorophyll chromophore makes plants green. *See* STEREOCHEMISTRY.

Optical activity is measured by two methods, optical rotation and circular dichroism. The optical rotation method depends on the different velocities of left and right circularly polarized light beams in the sample. The velocities are not measured directly, but both beams are passed through the sample simultaneously. This is equivalent to using plane-polarized light. The differing velocities of the left and right circularly polarized components yield a rotation of the plane of polarization. Circular dichroism is the difference in absorption of left and right circularly polarized light. Since this difference is about a millionth of the absorption of either polarization, special techniques are needed to determine it accurately. Circular dichroism is reported as a difference in absorption, or as an ellipticity (a measure of the elliptical polarization of the emergent beam). [V.M.]

**Optical rotatory dispersion** A term used to describe the change in rotation as a function of wavelength experienced by linearly polarized light as it passes through an optically active substance. *See* OPTICAL ACTIVITY.

In all materials the rotation varies with wavelength. The variation is caused by two quite different phenomena. The first accounts in most cases for the majority of the variation in rotation and should not strictly be termed rotatory dispersion. It depends on the fact that optical activity is actually circular birefringence. In other words, a substance which is optically active transmits right circularly polarized light with a different velocity from left circularly polarized light.

In addition to this pseudodispersion which depends on the material thickness, there is a true rotatory dispersion which depends on the variation with wavelength of the indices of refraction for right and left circularly polarized light. [B.H.Bi.]

For wavelengths that are absorbed by the optically active sample, the two circularly polarized components will be absorbed to differing extents. This unequal absorption is known as circular dichroism. Circular dichroism causes incident linearly polarized light to become elliptically polarized. *See* ABSORPTION.

Optical rotatory dispersion and circular dichroism are closely related, just as are ordinary absorption and dispersion. If the entire optical rotatory dispersion spectrum is known, the circular dichroism spectrum can be calculated, and vice versa.

In order for a molecule (or crystal) to exhibit circular birefringence and circular dichroism, it must be distinguishable from its mirror image. An object that cannot be superimposed on its mirror image is said to be chiral, and optical rotatory dispersion and circular dichroism are known as chiroptical properties.

Most biological molecules have one or more chiral centers and undergo enzyme-catalyzed transformations that either maintain or reverse the chirality at one or more of these centers. Still other enzymes produce new chiral centers, always with a high specificity. These properties account for the fact that optical rotatory dispersion and circular dichroism are widely used in organic and inorganic chemistry and in biochemistry. *See* ENZYME; STEREOCHEMISTRY.

In the absence of magnetic fields, only chiral substances exhibit optical rotatory dispersion and circular dichroism. In a magnetic field, even substances that lack chirality rotate the plane of polarized light, as shown by M. Faraday. Magnetic optical rotation is known as the Faraday effect, and its wavelength dependence is known as magnetic optical rotatory dispersion. In regions of absorption, magnetic circular dichroism is observable.

[R.W.Wo.]

**Organic chemistry** The study of the structure, preparation, properties, and reactions of carbon compounds. The term organic was early applied to compounds derived from plant and animal sources. These substances from living systems were usually distillable liquids or low-melting solids and were flammable, in contrast to metals, salts, and oxides from mineral sources. Until about 1830 it was held by some that organic compounds contained some special quality, or vital force. This notion was dispelled, but the term organic remained and became broadened to include carbon compounds in general. *See* CARBON.

**Structure.** The structures of organic compounds are described by a molecular framework of carbon atoms on which substituents may be located at various points.

Structures can be represented in several ways, as illustrated for the three-carbon alcohol 2-propanol (**1**) and the cyclic ketone 2-methyl-3-cyclohexenone (**2**). The

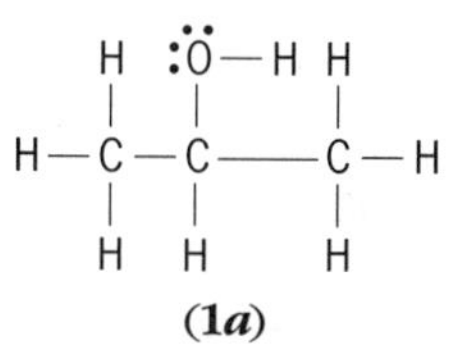

```
    H  :Ö—H  H
    |   |    |
 H— C — C —— C —H
    |   |    |
    H   H    H
```

(**1*a***)

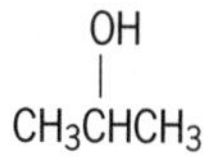

```
   OH
   |
CH3CHCH3
```

(**1*b***)

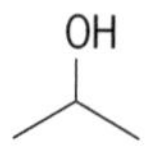

OH

(**1*c***)

O

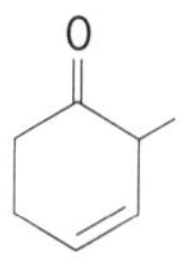

(**2**)

expanded structure of 2-propanol (**1*a***) shows all bonds and electron pairs, including unshared electrons on oxygen. More compact and convenient is the condensed structure (**1*b***) in which the C—C and C—H bonds are implied. In the bond-line convention (**1c**), all C—C bonds are indicated by a line, as shown for 2-propanol. Carbon atoms are not shown explicitly, but rather are implied at the ends of each line segment, together with enough hydrogen atoms to complete the tetravalency at each carbon. The bond-line convention is particularly convenient for cyclic structures such as 2-methyl-3-cyclohexenone; each vertex and the end of each line segment

**Principal organic functional groups**

| Composed class | Group | Structure |
|---|---|---|
| Alkene | Double bond | $>C{=}C<$ |
| Alkyne | Triple bond | $-C{\equiv}C-$ |
| Alcohol | Hydroxyl | $-OH$ |
| Amine | Amino | $-NH_2(-NR_2)^*$ |
| Aldehyde | Carbonyl | $-\overset{O}{\overset{\|\|}{C}}H$ |
| Ketone | Carbonyl | $-\overset{O}{\overset{\|\|}{C}}R$ |
| Acid | Carboxyl | $-\overset{O}{\overset{\|\|}{C}}OH$ |
| Ester | Alkoxycarbonyl | $-\overset{O}{\overset{\|\|}{C}}OR$ |
| Amide | Carbamoyl | $\overset{O}{\overset{\|\|}{C}}N<$ |
| Nitrile | Cyano | $-C{\equiv}N$ |
| Azide | Azido | $-N{=}N{=}N$ |
| Nitro | | $-NO_2$ |
| Sulfide | | $-S-$ |
| Sulfoxide | | $-\overset{O}{\overset{\|\|}{S}}-$ |
| Sulfonic acid | | $-SO_3H$ |

*R = any carbon group, for example, $CH_3$.

represents a carbon and appropriate number of hydrogens. *See* STRUCTURAL CHEMISTRY; VALENCE.

A functional group is an atom other than carbon or a multiple bond, such as the hydroxyl group (OH) of 2-propanol, the double bond (C=C), or the carbonyl group (C=O) of 2-methyl-3-cyclohexenone. The group defines a class of compounds and is the point at which characteristic reactions occur, for example, oxidation, reduction, or addition of an electrophilic or nucleophilic reagent. Some of the principal functional groups are shown in the table. *See* ELECTROPHILIC AND NUCLEOPHILIC REAGENTS.

The fact that there can be two or more compounds, known as isomers, with the same molecular composition was one of the key points in development of a structural theory. One type of isomerism, structural or constitutional, is illustrated by the two isomers that have the formula $C_4H_{10}$, butane (**3*a***) and isobutane (2-methylpropane;

**3*b***). The number of possible structural isomers becomes enormous in larger molecules.

$CH_3—CH_2—CH_2—CH_3$ (3*a*)

$CH_3—CH(CH_3)—CH_3$ (3*b*)

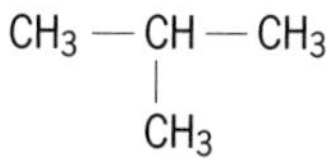

*See* MOLECULAR ISOMERISM.

Several three-dimensional representations of butane, showing the tetrahedral geometry of the carbon atoms, are given in structures (**4**). As indicated in these structures,

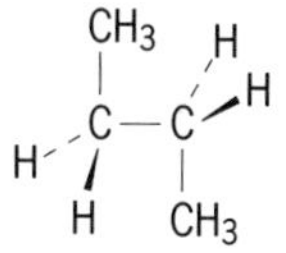

(4*a*) 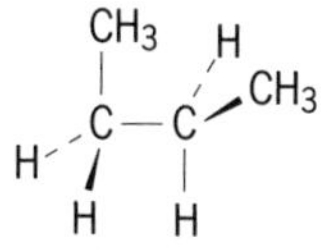 (4*b*) 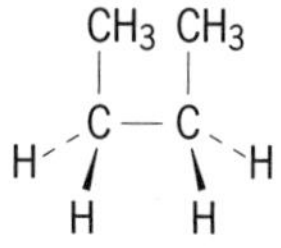 (4*c*)

butane can exist in several forms, called conformations, which differ in the relative positions of the carbon atoms, and thus the overall shape of the molecule. However, the barrier to rotation around the central C—C bond is so low that these individual conformational isomers are not separable, and butane is thus a single compound. *See* CONFORMATIONAL ANALYSIS.

In an alkene, rotation around the C═C bond does not occur, and 2-butene, for example, exists as two isomeric compounds, cis (Z) and trans (E) (**5*a*** and **5*b***, respectively).

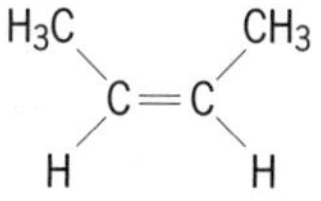

(5*a*) 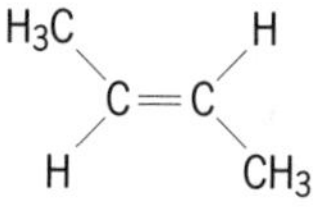 (5*b*)

Stereoisomers are compounds that have the same bond sequence but differ in the spatial array of the bonds. When a carbon atom is bonded to four unlike atoms or groups, the tetrahedral geometry of carbon causes the atom to be dissymmetric or chiral. A compound with a chiral atom can exist in two isomeric forms, known as enantiomers. The relative positions of all atoms is identical in the two enantiomers, but they differ in handedness, a characteristic of an asymmetric object and its nonsuperposable mirror image, as in structures (**6**) of 1-chlorobutane.

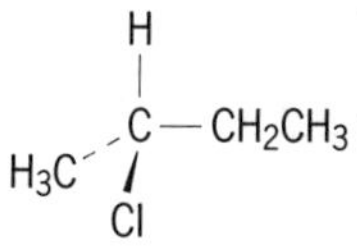

(6*a*) 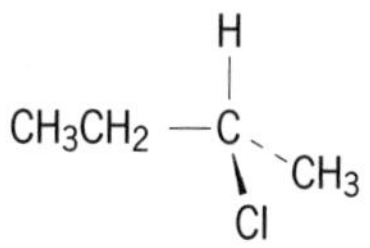 (6*b*)

*See* STEREOCHEMISTRY.

When two chiral centers are present, two stereoisomers can arise from each enantiomer. Thus enantiomer (**7*a***) of chlorobutane can lead to isomeric structures (**7*b***) and (**7*c***), in which the relative positions of the atoms is not identical. In this case,

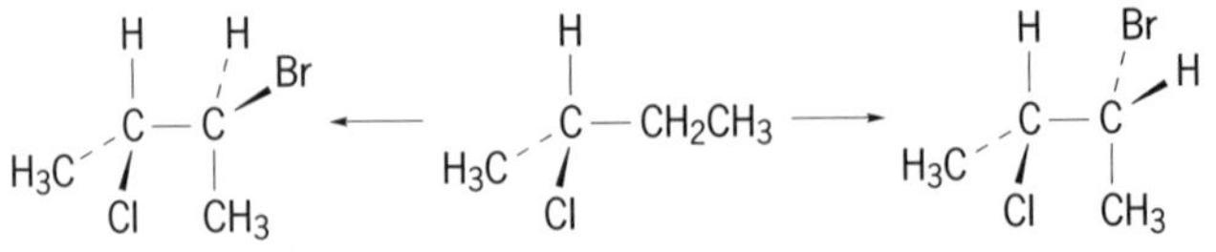

(7b) (7a) (7c)

the isomers are known as diastereoisomers. With $n$ chiral centers, there can be $2^n$ stereoisomers.

**Acyclic compounds.** The simplest organic compound is methane ($CH_4$). It is the first member of the homologous series of alkanes, in which successive compounds differ by an additional —$CH_2$— group ($CH_3CH_3$, $CH_3CH_2CH_3$, and so forth). *See* ALKANE; METHANE.

Higher alkanes, $CH_3(CH_2)_nCH_3$ ($n = 3$–$20$), and also branched isomers and cyclic hydrocarbons are the principal components of petroleum. These compounds have no reactive functional groups.

Both acyclic and cyclic carbon frameworks can contain multiple bonds; oxygen, nitrogen, and sulfur atoms; and other functional groups listed in the table.

**Carbocyclic compounds.** The two large groups of compounds with rings containing only carbon are alicyclic and aromatic. The parent hydrocarbons in the former series are cycloalkanes and in the latter, benzene. The structure of benzene is a planar six-numbered ring with six electrons in a delocalized array. *See* AROMATIC HYDROCARBON; BENZENE.

**Heterocyclic compounds.** A nitrogen, oxygen, or sulfur atom can take the place of carbon in either alicyclic or aromatic rings. The most numerous and important hetrocyclic compounds are those with nitrogen in a five- or six-membered aromatic system.

**Synthesis reactions.** The preparation of compounds occupies much of the effort of organic chemistry, and is the principal business of the chemical industry. The manufacture of drugs, pigments, and polymers entails the preparation of organic compounds on a scale of thousands to billions of kilograms per year, and there is constant research to develop new products and processes. Synthesis of new substances is carried out for many purposes beyond the goal of a commercial product. A compound of a specified structure may be needed to test a mechanistic proposal or to evaluate a biochemical response such as inhibition of an enzyme. Synthesis may provide a more dependable and less expensive source of a naturally occurring compound; moreover, a synthetic approach permits variations in the structure that may lead to enhanced biological activity.

The term synthesis usually implies a planned sequence of steps leading from simple starting compounds to a desired end product. Each of these steps involves a reaction that may lead to formation of a C—C bond or to the introduction, alteration, or removal of a functional group. Progress in synthesis depends on the availability of a wide range of reactions that bring about these changes in good yield, with a minimum of interfering by-products. An integral part of synthesis is the development of new methods and reagents that are selective for a desired transformation, and, very importantly, proceed with control of the stereochemistry. *See* ASYMMETRIC SYNTHESIS; ORGANIC SYNTHESIS.

[J.A.Mo.]

**Organic conductor** An organic substance with low electrical resistance. Two major classes of organic conductors are charge-transfer compounds and conducting polymers.

**Charge-transfer compounds.** The search for organic conductors in the early 1970s led to the observation of metallic-like electrical conduction in well-ordered

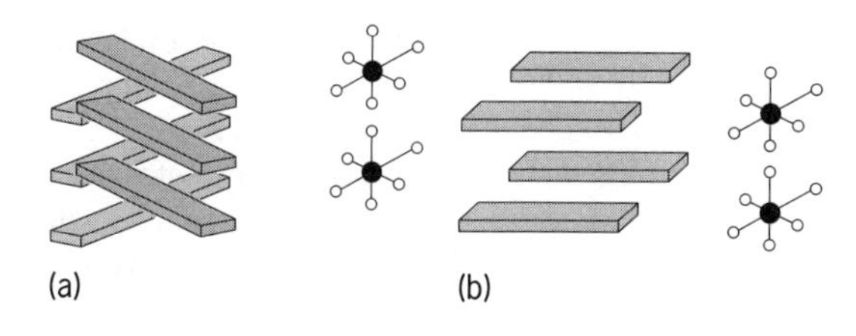


**Stacking of charge-transfer compounds, (*a*) Segregated stacking of TTF-TCNQ-like materials, (*b*) Typical zigzag stacking of the $(TMTSF)_2X$ series.**

molecular crystals and the discovery of many new phenomena such as the stabilization of charge-density waves and spin-density waves, new mechanisms for electronic transport, organic superconductivity, and new states of matter produced under strong magnetic fields. Most of these phenomena are due to the low dimensionality (one or two dimensions) of the electron gas in the charge-transfer compounds where they were observed. Among these properties, superconductivity has created much interest since the zero-resistance state is now observed at temperatures as high as 10 K (–442°F) in certain organic superconductors.

Charge-transfer compounds are two-component materials containing anionic and cationic species originating by charge transfer between donor and acceptor entities; these may be two organic molecules or an organic molecule with an inorganic ion. Tetrathiafulvalene-tetracyanoquinodimethane (TTF-TCNQ) is the prototype of charge-transfer organic crystals. Its crystal structure exhibits piled-up segregated columns of donor TTF and acceptor TCNQ molecules (illus. *a*). In the solid state, the amount of charge transferred from donor to acceptor is determined by the overall crystal stability.

Another class of organic conductors is exemplified by radical cation salts such as $(TMTSF)_2X$ where the organic molecule is tetramethyltetraselenafulvalene and X is an inorganic anion. In this class of materials (Bechgaard salts) the molecules display a zigzag packing along the stacking axis (illus. *b*), where one positive charge (hole) is shared between two organic molecules.

The strong overlap between electron clouds of neighboring molecules along the stacks spreads the partially filled molecular electronic states into an energy band 0.5–1 eV wide. This bandwidth is large enough to allow electron delocalization among all molecules on a given stack and to promote electrical conduction similar to that in metal crystals. *See* DELOCALIZATION; MOLECULAR ORBITAL THEORY.

**Conducting polymers.** Polymeric materials are typically considered as insulators. However, research since the late 1970s has led to the discovery of polymeric materials with extremely high conductivity, approaching that of copper. The prospect of materials combining the properties of plastics and metals or semiconductors has led to a search for applications, made attractive because improved polymers no longer suffer from such drawbacks as low stability, processing difficulties, and brittleness. Most conducting polymers can be switched reversibly between conductive and nonconductive states, with the result that their conductivities can span an enormous range. This switching is accomplished through oxidation-reduction (redox) chemistry, the conductivity being sensitive to the degree of oxidation of the polymer backbone. This property distinguishes conducting polymers from metals and semiconductors and is the basis of many existing and potential applications. In addition, certain polymers become conducting upon oxidation or reduction and thus can exhibit *p*- or *n*-type conduction. *See* OXIDATION-REDUCTION; POLYMER.

In a conducting polymer an oxidant removes electrons from the $\pi$-electron system of the polymer, creating radical cations that, at high concentrations, dimerize to form cation pairs known as bipolarons. Charge-balancing counterions are concomitantly incorporated between polymer chains. The overall process is referred to as doping, and the counteranion (or countercation in the case of reduction) is the dopant.

Considerable progress has been made on the theory of important parameters such as oxidation potentials, band gaps, and band widths, often with good agreement with experiment. Such work is important for the design of new conductive polymers with specific properties. [G.E.W.]

**Organic nomenclature** A system by which a unique and unambiguous name or other designation is assigned to a given organic molecular structure. A set of rules adopted by the International Union of Pure and Applied Chemistry (IUPAC) is the basis for a standardized name for any organic compound. Common or nonsystematic names are used for many compounds that have been known for a long time. The latter names have the advantage of being short and easily recognized, as in the examples below. However, in contrast to systematic names, common or trivial names do not convey information from which the structure can be written by reference to prescribed rules.

In the IUPAC system, a name is formed by combination of a parent alkyl chain or ring with prefixes and suffixes to denote substituents. For aliphatic compounds, the parent name is a stem that denotes the longest straight chain in the structure with the ending –ane. The first four members of the alkane series are methane, ethane, propane, and butane. From five carbons on, the names follow the Greek numerical roots, for example, pentane and hexane. Changing the ending -ane to -yl gives the name of the corresponding radical or group; for example, $CH_3CH_2$— is the ethyl radical or ethyl group. Branches on an alkyl chain are indicated by the name of a radical or group as a prefix. The location of a branch or other substituent is indicated by number; the chain is numbered from whichever end results in the lowest numbering. *See* ALKANE.

Double or triple bonds are indicated by the endings -ene or -yne, respectively. The configuration of the chain at a double bond is denoted by *E*- when two similar groups are on opposite sides of the plane bisecting the bond, and *Z*-when they are on the same side. *See* ALKENE; ALKYNE.

A compound containing a functional group is named by adding to the parent name a suffix characteristic of the group. If there are two or more groups, a principal group is designated by suffix and the other(s) by prefix.

The same general principles apply to cyclic compounds. For alicyclic rings, the prefix cyclo- is followed by a stem indicating the number of carbon atoms in the ring, as is illustrated in the structure below. In bicyclic compounds, the total number of carbons

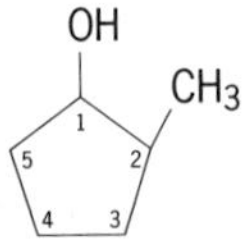

2-Methylcyclopentanol

in the ring system is prefixed by bicyclo- and numbers in brackets which indicate the number of atoms in each connecting chain.

The names of aromatic hydrocarbons have the ending -ene, which denotes a ring system with the maximum number of noncumulative double bonds. Each of the simpler polycyclic hydrocarbons has a different parent name. To number the positions in a polycyclic aromatic ring system, the structure must first be oriented with the maximum number of rings arranged horizontally and to the right. The system is then numbered

**Basis of nomenclature of heterocyclic compounds**

| Heteroatom | Prefix | Ring size | Suffix |
|---|---|---|---|
| O | ox(a)- | 4 | -ete |
| S | thi(a)- | 5 | -ole |
| N | az(a)- | 6 | -ine |
| | | 7 | -epine |

in clockwise sequence starting with the atom in the most counterclockwise position of the upper right-hand ring, omitting atoms that are part of a ring fusion. *See* AROMATIC HYDROCARBON.

Systematic names for rings containing a heteroatom (O, S, N) are based on a combination of a prefix denoting the heteroatom(s) and a suffix denoting the ring size, as indicated in the table. *See* CHEMICAL SYMBOLS AND FORMULAS; HETEROCYCLIC COMPOUNDS; ORGANIC CHEMISTRY. [J.A.Mo.]

**Organic photochemistry** A branch of chemistry that deals with light-induced changes of organic material. Because it studies the interaction of electromagnetic radiation and matter, photochemistry is concerned with both chemistry and physics. In considering photochemical processes, therefore, it is also necessary to consider physical phenomena that do not involve strictly chemical changes, for example, absorption and emission of light, and electronic energy transfer.

Because several natural photochemical processes were known to play important roles (namely, photosynthesis in plants, process of vision, and phototropism), the study of organic photochemistry started very early in the twentieth century. However, the breakthrough occurred only after 1950 with the availability of commercial ultraviolet radiation sources and modern analytical instruments for nuclear magnetic resonance (NMR) spectroscopy and (gas) chromatography.

In general, most organic compounds consist of rapidly interconvertible, nonseparable conformers, because of the free rotation about single bonds. Each conformer has a certain energy associated with it, and its own electronic absorption spectrum. This is one cause of the broad absorption bands produced by organic compounds in solution. The equilibrium of the conformers may be influenced by the solvent and the temperature. According to the Franck-Condon principle, which states that promotion of an electron by the absorption of a photon is much faster than a single vibration, each conformer will have its own excited-state configuration. *See* CONFORMATIONAL ANALYSIS; MOLECULAR STRUCTURE AND SPECTRA.

Because of the change in the pi-bond order upon excitation of the substrate and the short lifetime of the first excited state, the excited conformers are not in equilibrium and each yields its own specific photoproduct. Though different conformers may lead to the same photoproduct and one excited conformer may lead to several photoproducts, a change in solvent, temperature, or wavelength of excitation influences the photoproduct composition. This is especially true with small molecules; larger molecules with aromatic groups are less sensitive for small wavelength differences.

The influence of wavelength is also important when the primary photoproduct also absorbs light and then gives rise to another photoreaction. Excitation with selected wavelengths or monochromatic light by use of light filters or a monochromator, respectively, may then be profitable for the selective production of the primary product. Similarly, irradiation at a low temperature is helpful in detecting a primary photoproduct that is unstable when heated (thermolabile). [W.H.L.]

**Organic reaction mechanism** A complete, step-by-step account of how a reaction of organic compounds takes place. A fully detailed mechanism would correlate the original structure of the reactants with the final structure of the products and account for changes in structure and energy throughout the progress of the reaction. It would also account for the formation of any intermediates and the rates of interconversions of all of the various species. Because it is not possible to detect directly all of these details, evidence for a reaction mechanism is always indirect. Experiments are designed to produce results that provide logical evidence for (but can never unequivocally prove) a mechanism. For most organic reactions, there are mechanisms that are considered to be well established based on bodies of experimental evidence. Nevertheless, new data often become available that provide further insight into new details of a mechanism or that occasionally require a complete revision of an accepted mechanism.

**Classification of organic reactions.** The description of an organic reaction mechanism typically includes designation of the overall reaction (for example, substitution, addition, elimination, oxidation, reduction, or rearrangement), the presence of any reactive intermediates (that is, carbocations, carbanions, free radicals, radical ions, carbenes, or excited states), the nature of the reagent that initiates the reaction (such as electrophilic or nucleophilic), the presence of any catalysis (such as acid or base), and any specific stereochemistry. For example, reaction (1) would be described as a

$$H_3C(H)(D)C{-}Br + I^- \longrightarrow I{-}C(CH_3)(D)(H) + Br^- \qquad (1)$$

concerted nucleophilic substitution of an alkyl halide that proceeds with inversion of stereochemistry. A reaction that proceeds in a single step, without intermediates, is described as concerted or synchronous. Reaction (1) is an example of the $S_N2$ mechanism (substitution, nucleophilic, bimolecular).

**Potential energy diagrams.** A common method for illustrating the progress of a reaction is the potential energy diagram, in which the free energy of the system is plotted as a function of the completion of the reaction (see illustration).

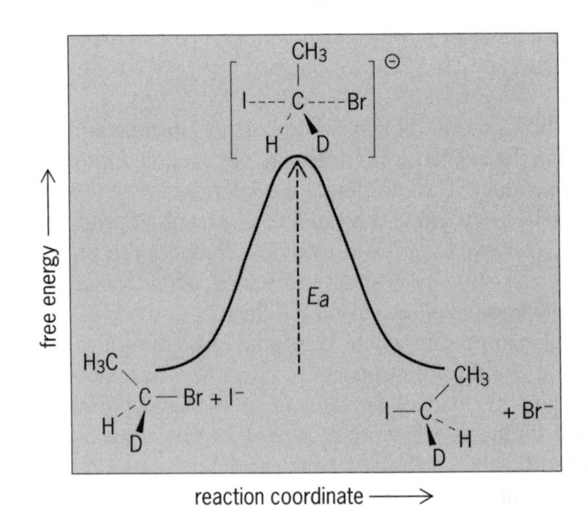


**Potential energy–reaction coordinate diagram for a typical nucleophilic substitution reaction that proceeds by the $S_N2$ mechanism. $E_a$ = activation energy.**

The reaction coordinate is intended to represent the progress of the reaction, and it may or may not correlate with an easily observed or measurable feature. In reaction (1), the reaction coordinate could be considered to be the increasing bond length of the carbon-bromine (C-Br) bond as it is broken, or the decreasing separation of C and iodine (I) as they come together to form a bond. In fact, a complete potential energy diagram should illustrate the variation in energy as a function of both of these (and perhaps several other relevant structural features), but this would require a three-dimensional (or higher) plot.

Besides identifying the energy levels of the original reactants and the final products, the potential energy diagram indicates the energy level of the highest point along the reaction pathway, called the transition state. Because the transition state represents the highest energy that the molecules must attain as they proceed along the reaction pathway, the energy level of the transition state is a key indication of how easily the reaction can occur. Features that tend to make the transition state more stable (lower in energy) make the reaction more favorable. Such stabilizing features could be intramolecular, such as electron donation or withdrawal by substituents, or intermolecular, such as stabilization by solvent. *See* CHEMICAL BONDING; ENERGY.

**Kinetics.** Another way to illustrate the various steps involved in a reaction mechanism is as a kinetic scheme that shows all of the individual steps and their rate constants. The $S_N2$ mechanism is a single step, so the kinetics must represent that step; the rate is observed to depend on the concentrations of both the organic substrate and the nucleophile. However, for multistep mechanisms the kinetics can be a powerful tool for distinguishing the presence of alternative pathways. For example, when more highly substituted alkyl halides undergo nucleophilic substitution, the rate is independent of the concentration of the nucleophile. This evidence suggests a two-step mechanism, called the $S_N1$ mechanism, as shown in reaction scheme (2), where the $k$ terms represent rate constants.

$$(CH_3)_3C-Br \underset{k_{-1}}{\overset{k_1}{\rightleftharpoons}} (CH_3)_3C^{\oplus} + Br^{\ominus} \qquad (2a)$$

$$(CH_3)_3C^{\oplus} + I^{\ominus} \xrightarrow{k_2} (CH_3)_3C-I \qquad (2b)$$

The $S_N1$ mechanism accomplishes the same overall nucleophilic substitution of an alkyl halide, but does so by initial dissociation of the leaving group ($Br^-$) to form a carbocation, step (2*a*). The nucleophile then attaches to the carbocation to form the final product, step (2*b*). Alkyl halides that have bulky groups around the carbon to be substituted are less likely to be substituted by the direct $S_N2$ mechanism, because the nucleophile encounters difficulty in making the bond to the inaccessible site (called steric hindrance). If those alkyl groups have substituents that can support a carbocation structure, generally by electron donation, then the $S_N1$ mechanism becomes preferable.

A crucial feature of a multistep reaction mechanism is the identification of the rate-determining step. The overall rate of reaction can be no faster than its slowest step. In the $S_N1$ mechanism, the bond-breaking reaction (2*a*) is typically much slower than the bond-forming reaction (2*b*). Hence, the observed rate is the rate of the first step only. Thus, kinetics can distinguish the $S_N1$ and $S_N2$ mechanisms, as shown in Eqs. (3) and (4), where R is an alkyl group, X is a halogen or other leaving group, Nu is a nucleophile, and the terms in the brackets represent concentrations.

$$\text{Rate} = k\,[\text{RX}]\,[\text{Nu}] \quad \text{for an } S_N2 \text{ mechanism} \tag{3}$$

$$\text{Rate} = k\,[\text{RX}] \quad \text{for an } S_N1 \text{ mechanism} \tag{4}$$

A more complete description of the $S_N1$ mechanism was recognized when it was observed that the presence of excess leaving group [for example, $Br^-$ in reaction (2)] can affect the rate (called the common ion rate depression). This indicated that the mechanism should include a reverse step [$k_{-1}$ in reaction step (2*a*)] in which the leaving group returns to the cation, regenerating starting material. In this case, the rate depends in a complex manner on the competition of nucleophile and leaving group for reaction with the carbocation. *See* CHEMICAL DYNAMICS; REACTIVE INTERMEDIATES; STERIC EFFECT (CHEMISTRY).

**Activation parameters.** The temperature dependence of the rate constant provides significant information about the transition state of the rate-determining step. The Arrhenius equation (5) expresses that dependence in terms of an exponential function

$$k = Ae^{-E_a/RT} \tag{5}$$

of temperature and an activation energy, $E_a$; $A$ is called the Arrhenius or preexponential factor, and $R$ is the gas constant. *See* GAS.

The activation energy represents the energy difference between the reactants and the transition state, that is, the amount of energy that must be provided in order to proceed along the reaction pathway successfully from reactant to product.

**Stereochemistry.** Careful attention to the stereochemistry of a reaction often provides crucial insight into the specific orientation of the molecules as they proceed through the reaction mechanism. The complete inversion of stereochemistry observed in the $S_N2$ mechanism provides evidence for the backside attack of the nucleophile. Alkyl halides that undergo substitution by the $S_N1$ mechanism do not show specific stereochemistry, since the loss of the leaving group is completely uncorrelated with the bonding of the necleophile.

In addition reactions, possible stereochemical outcomes are addition of the new bonds to the same or opposite sides of the original pi bond, called syn and anti addition, respectively. The anti addition of bromine to double bonds provides evidence for the

$Br_2$ + [cyclohexene] → [bromonium ion, $Br^{\oplus}$] + $Br^{\ominus}$ → [trans-1,2-dibromocyclohexane, Br, Br] + [trans-1,2-dibromocyclohexane, Br, Br] (6)

intermediacy of a bridged bromonium ion, as shown in reaction (6). [C.C.W.]

**Organic synthesis** The making of an organic compound from simpler starting materials. Organic synthesis plays an important role by allowing for the creation of specific molecules for scientific and technological investigations.

The heart of organic synthesis is designing synthetic routes to a molecule. The simplest synthesis of a molecule is one in which the target molecule can be obtained by submitting a readily available starting material to a single reaction that converts it to the desired target molecule. However, in most cases the synthesis is not that straightforward; in order to convert a chosen starting material to the target molecule, numerous

**TABLE 1. Examples of some functional-group interconversions**

| General equation for the reaction* | Net transformation (name) |
|---|---|
| $R_1R_2R_3C{-}X + Nu^- \longrightarrow Nu{-}CR_1R_2R_3 + X^-$ <br> ($X = Cl, Br, I,$ or $OSO_2R$; $Nu = OH, OR, CN, NR_2$, others) | Alkyl halide to various functional groups (alcohols, ethers, nitriles, amines, others) |
| $R_1R_2HC{-}CXR_3R_4$ + base (such as $CH_3O^-$) $\longrightarrow R_1R_2C{=}CR_3R_4$ | Alkyl halide to alkene (elimination) |
| $ROH + HX \longrightarrow RX$ <br> ($X = Cl, Br, I$) | Alcohol to alkyl halide |
| $R_1R_2HC{-}C(OH)R_3R_4 + H_2SO_4 \longrightarrow R_1R_2C{=}CR_3R_4$ | Alcohol to alkene (dehydration) |
| $R_1CH(OH)R_2 \xrightarrow{CrO_3\text{-pyridine}} R_1C({=}O)R_2$ | Oxidation of alcohol to ketone or aldehyde |
| $R_1YH + R_2C({=}O)X \longrightarrow R_2C({=}O)OR_1$ <br> ($Y = O$ or $N$; $X = OH, Cl$, others) | Alcohol and carboxylic acid derivative to ester (esterification); amine and carboxylic acid derivative to amide |
| $R_1R_2C{=}CR_3R_4 + RCO_3H \longrightarrow$ epoxide of $R_1R_2C{-}CR_3R_4$ (O bridging) | Alkene to epoxide (epoxidation) |
| $R_1R_2C{=}CR_3R_4 + H_2 \xrightarrow[\text{(or other catalyst)}]{Pd} R_1R_2HC{-}CHR_3R_4$ | Alkene to alkane (hydrogenation) |
| $R_1C({=}O)R_2 + NaBH_4 \longrightarrow R_1CH(OH)R_2$ | Reduction of ketone or aldehyde to alcohol |
| $R_1COOR_2 \xrightarrow[\text{2. } H_2O \text{ workup}]{\text{1. LiAlH4}} R_1CH_2OH + R_2OH$ | Reduction of ester to two alcohols |
| $R_1COOR_2 \xrightarrow{H_2O\text{, acid or base}} R_1COOH + R_2OH$ | Ester to carboxylic acid and alcohol (ester hydrolysis) |
| $R{-}CN \xrightarrow[\text{2. } H_2O \text{ workup}]{\text{1. LiAlH4}} RCH_2NH_2$ | Reduction of nitrile to amine |
| $C_6H_6 + E^+ \longrightarrow C_6H_5E$ <br> ($E = Br, NO_2, R, RCO$, others) | Benzene to substituted benzene (electrophilic aromatic substitution) |

*R = any organic group (alkyl, aryl, alkenyl) or a hydrogen atom. Nu = nucleophile.

**TABLE 2. Examples of some carbon-carbon bond-forming reactions**

| General equation for the reaction | Name of reaction |
|---|---|
| $R_1X + Mg \longrightarrow R_1MgX \xrightarrow[2.\ H^+]{1.\ R_2COR_3} R_1-C(OH)(R_2)-R_3$ (X = Cl, Br, I) | Grignard |
| $R_1X \xrightarrow[2.\ CuI]{1.\ Li} (R_1)_2CuLi \xrightarrow{R_2X} R_1-R_2$ (X = Cl, Br, I) | Gilman |
| $XC(O)CH_2R_1 \xrightarrow[2.\ R_2CHO,\ 3.\ H^+]{1.\ (C_3H_7)_2NLi} XC(O)CH(R_1)CH(OH)R_2$ (X = R, RO, $NR_2$, others) | Aldol addition |
| $XCH_2CO-Y + RCHO \xrightarrow[2.\ H_2O\ workup]{1.\ Zn} RCH(OH)-CH_2CO-Y$ (X = Cl, Br, or I; Y = R, RO, $NR_2$, others) | Reformatsky |
| $XC(O)C(R_1)=C(R_3)R_2 + H_2C(COOR)_2 \xrightarrow{NaOR} XC(O)CH(R_1)C(R_2)(R_3)CH(COOR)_2$ (X = R, RO, $NR_2$, others) | Michael addition |
| $XC(O)CH_2R_1 + R_2CHO \xrightarrow[2.\ H_3O^+,\ heat]{1.\ NaOCH_3} XC(O)C(R_1)=CHR_2$ (X = R, RO, $NR_2$, others) | Aldol condensation |
| $R_1COOR_2 + R_3CH_2COOR_2 \xrightarrow{NaOR_2} R_1C(O)CH(R_1)C(O)OR_2$ | Claisen condensation |
| $R_1CH_2Br \xrightarrow[2.\ BuLi]{1.\ PPh_3} R_1-CH=PPh_3 \xrightarrow{3.\ R_2CHO} R_1CH=CHR_2$ | Wittig |
| $2RCHO \xrightarrow[2.\ H_2O\ workup]{1.\ Ti(I)} R-CH(OH)-CH(OH)-R$ | Pinacol coupling |
| 6-bromohex-1-ene (Br) $\xrightarrow[Bu_3SnH]{free\text{-}radical\ initiator}$ methylcyclopentane ($CH_3$) | Free-radical cyclization |
| R-substituted diene + Y-substituted alkene $\xrightarrow{heat}$ cyclohexene bearing R and Y (Y = COOR, COR, CN, others) | Diels-Alder |
| $H_2C=C(R_2)CH_2CH_2CH=CHR_1 \xrightarrow{heat} CH_2=C(R_2)CH_2CH(R_1)CH=CH_2$ | Cope rearrangement |

steps that add, change, or remove functional groups, and steps that build up the carbon atom framework of the target molecule may need to be done.

A systematic approach for designing a synthetic route to a molecule is to subject the target molecule to an intellectual exercise called a retrosynthetic analysis. This involves an assessment of each functional group in the target molecule and the overall carbon atom framework in it; a determination of what known reactions form each of those functional groups or that build up the necessary carbon framework as a product; and a determination of what starting materials for each such reaction are required. The resulting starting materials are then subjected to the same retrosynthetic analysis, thus working backward from the target molecule until starting materials are derived.

The retrosynthetic analysis of a target molecule usually results in more than one possible synthetic route. It is therefore necessary to critically assess each derived route in order to chose the single route that is most feasible and most economical. The safety of each possible synthetic route (the toxicity and reactivity hazards associated with the reactions involved) is also considered when assessing alternative synthetic routes to a molecule.

Selectivity is an important consideration in the determination of a synthetic route to a target molecule. Stereoselectivity refers to the selectivity of a reaction for forming one stereoisomer of a product in preference to another. Stereoselectivity cannot be achieved for all organic reactions; the nature of the mechanism of some reactions may not allow for the formation of one particular configuration of a chiral (stereogenic) carbon center or one particular geometry (cis versus trans) for a double bond or ring. When stereoselectivity can be achieved, it requires that the reaction proceed via a geometrically defined transition state and that one or both of the reactants possess a particular geometrical shape during the reaction. For example, if one or both of the reactants is chiral, the absolute configuration of the newly formed stereogenic carbon center can be selected for in many reactions. *See* ASYMMETRIC SYNTHESIS; ORGANIC REACTION MECHANISM; STEREOCHEMISTRY.

Chemoselectivity is the ability of a reagent to react selectively with one functional group in the presence of another similar functional group. An example of a chemoselective reagent is a reducing agent that can reduce an aldehyde and not a ketone. In cases where chemoselectivity cannot be achieved, the functional group that should be prevented from participating in the reaction can be protected by converting it to a derivative that is unreactive to the reagent involved. The usual strategy employed to allow for such selective differentiation of the same or similar groups is to convert each group to a masked (protected) form which is not reactive but which can be unmasked (deprotected) to yield the group when necessary.

A large variety of organic reactions that can be used in syntheses are known. They can be categorized according to whether they feature a functional group interconversion or a carbon-carbon bond formation.

Functional group interconversions (Table 1) are reactions that change one functional group into another functional group. A functional group is a nonhydrogen, non-all-singly-bonded carbon atom or group of atoms. Included in functional group interconversions are nucleophilic substitution reactions, electrophilic additions, oxidations, and reductions. *See* COMPUTATIONAL CHEMISTRY; ELECTROPHILIC AND NUCLEOPHILIC REAGENTS; OXIDATION-REDUCTION; OXIDIZING AGENT; SUBSTITUTION REACTION.

Carbon-carbon bond-forming reactions (Table 2) feature the formation of a single bond or double bond between two carbon atoms. This is a particularly important class of reactions, as the basic strategy of synthesis—to assemble the target molecule from simpler, hence usually smaller, starting materials—implies that most complex molecules must be synthesized by a process that builds up the carbon skeleton of the target by using one or more carbon-carbon bond-forming reactions. [R.D.Wa.]

**Organoactinides** Organometallic compounds of the actinides—elements 90 and beyond in the periodic table. Both the large sizes of actinide ions and the presence of 5*f* valence orbitals are unique features which differ distinctly from most, if not all, other metal ions. *See* PERIODIC TABLE.

Organometallic compounds have been prepared for all actinides through curium (element 96), although most investigations have been conducted with readily available and more easily handled natural isotopes of thorium (Th) and uranium (U). Organic groups (ligands) which bind to actinide ions include both $\pi$- and $\sigma$-bonding functionalities. The importance of this type of compound reflects the ubiquitous character of metal-carbon two-electron sigma bonds in both synthesis and catalysis. *See* CATALYSIS.

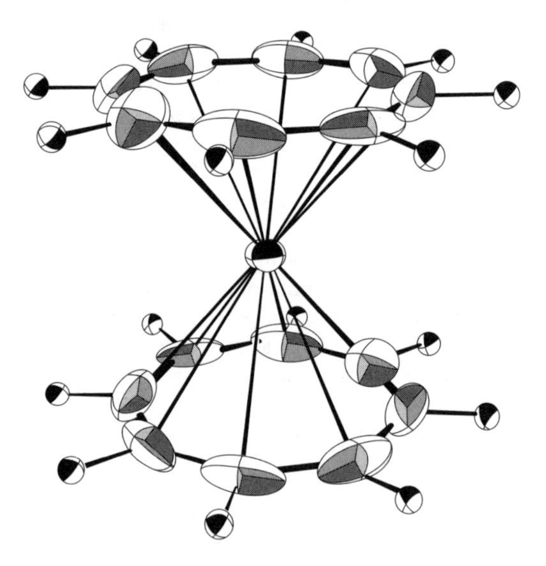

**Molecular structure of $U(C_8H_8)_2$ determined by single-crystal x-ray diffraction. (*After K. O. Hodgson and K. N. Raymond, Inorg. Chem., 12:458, 1973*)**

The molecular structures of a number of organoactinides have been determined by single-crystal x-ray and neutron-diffraction techniques. In almost all cases the large size of the metal ion gives rise to unusually high (as compared to a transition-metal compound) coordination numbers. That is, a greater number of ligands or ligands with greater spatial requirements can be accommodated within the actinide coordination sphere. The sandwich complex bis(cyclooctatetraenyl)-uranium (uranocene), an example of this latter type, is shown in the illustration. *See* ACTINIDE ELEMENTS; COORDINATION CHEMISTRY; METALLOCENES; ORGANOMETALLIC COMPOUND. [T.J.M.]

**Organometallic compound** A member of a broad class of compounds whose structures contain both carbon (C) and a metal (M). Although not a required characteristic of organometallic compounds, the nature of the formal carbon-metal bond can be of the covalent, ionic, or $\pi$-bound type.

The term organometallic chemistry is essentially synonymous with organotransition-metal chemistry; it is associated with a specific portion of the periodic table ranging from groups 3 through 11, and also includes the lanthanides. *See* CHEMICAL BONDING; LIGAND; PERIODIC TABLE; TRANSITION ELEMENTS.

From the perspective of inorganic chemistry, organometallics afford seemingly endless opportunities for structural variations due to changes in the metal coordination number, alterations in ligand-metal attachments, mixed-metal cluster formation, and so forth. From the viewpoint of organic chemistry, organometallics allow for manipulations in the functional groups that in unique ways often result in rapid and efficient elaborations of carbon frameworks for which no comparable direct pathway using nontransition organometallic compounds exists.

In moving across the periodic table, the early transition metals have seen relatively limited use in synthesis, with two exceptions: titanium (Ti) and zirconium (Zr).

Titanium has an important role in a reaction known as the Sharpless asymmetric epoxidation, where an allylic alcohol is converted into a chiral, nonracemic epoxy alcohol with excellent and predictable control of the stereochemistry [reaction (1)]. The

$$R^1R^2C{=}C(R^3)CH_2OH \longrightarrow R^1R^2\overset{*}{C}(O)\overset{*}{C}(R^3)CH_2OH \qquad (1)$$

Allylic alcohol → Optically active epoxy alcohol

significance of the nonracemic product is that the reaction yields a single enantiomer of high purity.

There are many applications utilizing the Sharpless asymmetric synthesis. Examples of synthetic targets that have relied on this chemistry include riboflavin (vitamin $B_2$) and a potent inhibitor of cellular signal transduction known as FK-506. *See* ASYMMETRIC SYNTHESIS; TITANIUM.

Below titanium in group 4 in the periodic table lies zirconium. Most modern organozirconium chemistry concerns zirconium's ready formation of the carbenelike complex [$Cp_2Zr$:], which because of its mode of preparation is more accurately thought of as a $\pi$ complex (2). Also important are reactions of the zirconium chloride hydride, $Cp_2Zr(H)Cl$, commonly referred to as Schwartz's reagent, with alkenes and alkynes, and the subsequent chemistry of the intermediate zirconocenes. *See* METALLOCENES.

When the complex $Cp_2ZrCl_2$ is exposed to two equivalents of ethyl magnesium bromide EtMgBr, the initially formed $Cp_2ZrEt_2$ loses a molecule of ethane ($C_2H_6$) to produce the complexed zirconocene [$Cp_2Zr$: (**1**)], as shown in reaction (2). Upon in-

$$Cp_2ZrCl_2 \xrightarrow[\text{tetrahydrofuran}]{\text{2EtMgBr}} [Cp_2Zr:] \equiv Cp_2Zr{-}\overset{CH_2}{\underset{CH_2}{\|}} \qquad (2)$$

(**1**)

troduction of another alkene or alkyne, a zirconacyclopentane or zirconacyclopentene is formed, respectively [reaction (3)], where the structure above the arrow indicates that

$$Cp_2Zr{-}\overset{CH_2}{\underset{CH_2}{\|}} \xrightarrow[\text{tetrahydrofuran}]{R{-}\equiv{-}R} Cp_2Zr\text{(cyclopentene ring with R, R)} \qquad (3)$$

(**1**)

the chemistry applies to either alkenes (no third bond) or alkynes (with third bond)]. These are reactive species that can be converted to many useful derivatives resulting from reactions such as insertions, halogenations, and transmetalation/quenching. When the preformed complex (**1**) is treated with a substrate containing both an alkene and alkyne, a bicyclic zirconacene results that can ultimately yield polycyclic products (for example, pentalenic acid, a likely intermediate in the biosynthesis of the antibiotic pentalenolactone). *See* REACTIVE INTERMEDIATES.

Among the group 6–8 metals, chromium (Cr), molybdenum (Mo), and tungsten (W) have been extensively utilized in the synthesis of complex organic molecules in the form of their electrophilic Fischer carbene complexes, which are species having, formally, a double bond between carbon and a metal. They are normally generated as heteroatom-stabilized species bearing a "wall" of carbon monoxide ligands (**2**).

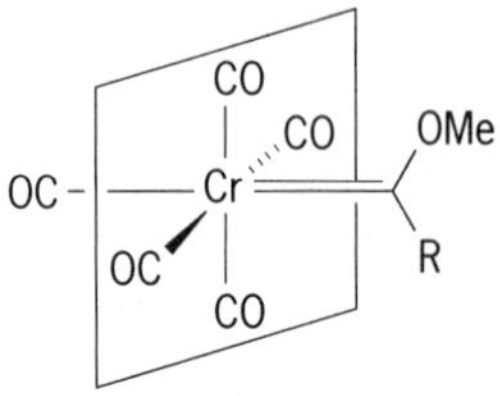

**(2)**

Most of the synthetic chemistry has been performed with chromium derivatives, which are highly electrophilic at the carbene center because of the strongly electron-withdrawing carbonyl (CO) ligands on the metal. Many different types of reactions are characteristic of these complexes, such as $\alpha$ alkylation, Diels-Alder cycloadditions of $\alpha,\beta$-unsaturated systems, cyclopropanation with electron-deficient olefins, and photochemical extrusions/cycloadditions. The most heavily studied and applied in synthesis, however, is the Dötz reaction, which has been applied in the production of antitumor antibiotics in the anthracycline and aureolic acid families. *See* DIELS-ALDER REACTION.

Groups 9–11 contain transition metals that have been the most widely used not only in terms of their abilities to effect C—C bond formations but also organometallic catalysts for some of the most important industrial processes. These include cobalt (Co), rhodium (Rh), palladium (Pd), and copper (Cu). [B.H.L.]

**Organophosphorus compound** One of a series of derivatives of phosphorus that have at least one organic (alkyl or aryl) group attached to the phosphorus atom linked either directly to a carbon atom or indirectly by means of another element (for example, oxygen). The mono-, di-, and trialkylphosphines (and their aryl counterparts) can be regarded formally as the parent compounds of all organophosphorus compounds.

Considering the large number of organic groups that may be joined to phosphorus as well as the incorporation of other elements in these materials, the number of combinations is practically unlimited. A vast family in itself is composed of the heterocyclic phosphorus molecules, in which phosphorus is one of a group of atoms in a ring system.

Some organophosphorus compounds have been used as polymerization catalysts, lubricant additives, flameproofing agents, plant growth regulators, and insecticides. Organophosphorus compounds were made during World War II for use as chemical warfare agents in the form of nerve gases (Sarin, Trilon 46, Soman, and Tabun). *See* PHOSPHORUS. [S.E.Cr.]

**Organoselenium compound** One of a group of compounds that contain both selenium (Se) and carbon (C) and frequently other elements as well, for example, halogen, oxygen (O), sulfur (S), or nitrogen (N). Organoselenium compounds have become common in organic chemistry laboratories, where they have numerous applications, particularly in the area of organic synthesis. Organoselenium compounds formally resemble their sulfur analogs and may be classified similarly. For instance, selenols, selenides, and selenoxides are clearly related to thiols, sulfides, and sulfoxides. Despite structural similarities, however, sulfur and selenium compounds are often strikingly different with respect to their stability, properties, and ease of formation. *See* ORGANOSULFUR COMPOUND; SELENIUM.

Selenols have the general formula RSeH, where R represents either an aryl or alkyl group. They are prepared by the alkylation of hydrogen selenide ($H_2Se$) or selenide salts, as well as by the reaction of Grignard reagents or organolithium compounds with selenium, as in reactions (1).

$$\underset{}{H_2Se\ (\text{or } HSe^-)} \xrightarrow{RX} \underset{\text{Selenol}}{RSeH} \xleftarrow[2.\ H_3O^+]{1.\ Se} \underset{\text{Grignard reagent}}{RMgX} \ \text{or} \ \underset{\text{Organolithium compound}}{RLi} \tag{1}$$

Selenols are stronger acids than thiols. They and their conjugate bases are powerful nucleophiles that react with alkyl halides (R′X; R′ = alkyl group) or similar electrophiles to produce selenides (RSeR′); further alkylation yields selenonium salts, as shown in reactions (2), where R″ represents an alkyl group that may be different from the alkyl

$$\underset{\text{Selenol}}{RSeH} \xrightarrow{R'X} \underset{\text{Selenide}}{RSeR'} \xrightarrow{R''X} \underset{\text{Selenonium salt}}{R{-}\overset{\overset{\displaystyle R'}{|}}{Se^+}{-}R' + X^-} \tag{2}$$

group R′. Both acyclic and cyclic selenides are known. *See* ELECTROPHILIC AND NUCLEOPHILIC REAGENTS; GRIGNARD REACTION; REACTIVE INTERMEDIATES.

Diselenides (RSeSeR) are usually produced by the aerial oxidation of selenols; they react in turn with chlorine or bromine, yielding selenenyl halides or selenium trihalides. Diselenides are easily oxidized to seleninic acids or anhydrides by reagents such as hydrogen peroxide or nitric acid.

Selenoxides are readily obtained from the oxidation of selenides with hydrogen peroxide ($H_2O_2$) or similar oxidants. Selenoxides undergo facile elimination to produce olefins and selenenic acids, a process known as syn-elimination.

Selenocarbonyl compounds tend to be considerably less stable than their thiocarbonyl or carbonyl counterparts because of the weaker double bond between the carbon and selenium atoms. Selenoamides, selenoesters and related compounds can be isolated, but selenoketones (selones) and selenoaldehydes are highly unstable. *See* ALDEHYDE; AMIDE; CHEMICAL BONDING; ESTER; KETONE.

The charge-transfer complexes formed between certain Organoselenium donor molecules and appropriate acceptors such as tetracyanoquinodimethane are capable of conducting electric current. Selenium-containing polymers are also of interest as organic conductors. *See* COORDINATION COMPLEXES; ORGANIC CONDUCTOR.

Selenium is an essential trace element, and its complete absence in the diet is severely detrimental to human and animal health. The element is incorporated into selenoproteins such as glutathione peroxidase, which acts as a natural antioxidant. *See* BIOINORGANIC CHEMISTRY; COORDINATION CHEMISTRY; PROTEIN. [T.G.B.]

**Organosilicon compound** One of a group of compounds in which silicon (Si) is bonded to an organic functional group (R) either directly or indirectly via another atom. Formally, all organosilanes can be viewed as derivatives of silane ($SiH_4$) by the substitution of hydrogen (H) atoms. The most common substituents are methyl ($CH_3$; Me) and phenyl ($C_6H_5$; Ph) groups. However, tremendous diversity results with cyclic structures and the introduction of heteroatoms. *See* SILICON.

Organosilicon compounds are not found in nature and must be prepared in the laboratory. The ultimate starting material is sand (silicon dioxide, $SiO_2$) or other inorganic silicates, which make up over 75% of the Earth's crust. The useful properties of silicone polymers were identified in the 1940s; widespread interest in organosilicon chemistry followed.

The chemistry of organosilanes can be explained in terms of the fundamental electronic structure of silicon and the polar nature of its bonds to other elements. Silicon appears in the third row of the periodic table, immediately below carbon (C) in group IV, and has many similarities with carbon. However, it is the fundamental differences between carbon and silicon that make silicon so useful in organic synthesis and of such great theoretical interest. *See* CARBON.

In the majority of organosilicon compounds, silicon follows the octet rule and is 4-coordinate. This trend can be explained by the electronic configuration of atomic silicon ($3s^2 3p^2 3d^0$) and the formation of $sp^3$-hybrid orbitals for bonding. Unlike carbon ($2s^2 2p^2$), however, silicon is capable of expanding its octet and can form 5- and 6-coordinate species with electronegative substituents, such as the 6-coordinate octahedral dianion, $[Me_2SiF_4]^{2-}$. *See* COORDINATION CHEMISTRY; ELECTRON CONFIGURATION; VALENCE.

Silicon is a relatively electropositive element that forms polar covalent bonds ($Si^{\delta+}—X^{\delta-}$) with carbon and other elements, including the halogens, nitrogen, and oxygen. The strength and reactivity of silicon bonds depend on the relative electronegativities of the two elements. For example, the strongly electronegative elements fluorine (F) and oxygen (O) form bonds that have tremendous thermodynamic stabilities. *See* CHEMICAL THERMODYNAMICS; ELECTRONEGATIVITY.

Before 1981, multiply bonded silicon compounds were identified only as transient species both in solution and in the gas phase. However, when tetramesityldisilene ($Ar_2Si{=}SiAr_2$, where Ar = 2,4,6-trimethylphenyl) was isolated, it was found to be a crystalline, high-melting-point solid having excellent stability in the absence of oxygen and moisture.

Numerous reactive species have been generated and characterized in organosilicon chemistry. Silylenes are divalent silicon species ($SiR_2$, where R = alkyl, aryl, or hydrogen), analogous to carbenes in carbon chemistry. The dimerization of two silylene units gives a disilene. Silicon-based anions, cations, and radicals are important intermediates in the reactions of silicon. *See* CHEMICAL BONDING; FREE RADICAL; REACTIVE INTERMEDIATES.

The direct process for the large-scale preparation of organosilanes has provided a convenient source of raw materials for the development of the silicone industry. The process produces a mixture of chloromethylsilanes from elemental silicon and methyl chloride in the presence of a copper (Cu) catalyst, as in the reaction below.

$$\text{Me—Cl} + \text{Si} \xrightarrow{\text{Cu catalyst}} \text{Cl}_2\text{SiMe}_2 + \text{other Cl}_n\text{SiMe}_{4-n}$$

In spite of concentrated research efforts in this area, the synthetic methods available for the controlled formation of silicon-carbon bonds remain limited to only a few general reaction types. These include the reaction of organometallic reagents with silanes, catalytic hydrosilylation of multiple bonds, and reductive silylation.

The role of silicon in organic synthesis is quite extensive, and chemists exploit the unique reactivity of organosilanes to accomplish a wide variety of transformations. Silicon is usually introduced into a molecule to perform a specific function and is then removed under controlled conditions.

Polysilanes are organosilicon compounds that contain either cyclic arrays or linear chains of silicon atoms. The isolation and characterization of both cyclosilanes (with up to 35 silane units) and high-molecular-weight linear polysilanes (with up to 3000 silane units) have demonstrated that silicon is capable of extended chain formation (catenation). Polysilanes contain only silicon-silicon bonds in their backbone, which differentiates them from polysiloxanes (silicones), which contain alternating silicon and oxygen repeat units. *See* SILICONE RESINS.

Polysilanes have found applications as photoresists in microlithography, charge carriers in electrophotography, and photoinitiators for vinyl polymerization. Polysilanes also function as preceramic polymers for the manufacture of silicon carbide fibers. [H.Y.]

**Organosulfur compound** A member of a class of organic compounds with any of several dozen functional groups containing sulfur (S).

Sulfur is an element of the third row of the periodic table; it is larger and less electronegative than oxygen, which lies above it in the second row. Compounds with an expanded valence shell, that is, compounds bonding to as many as six ligands around sulfur, are therefore possible, and a broad range of compounds can be formed. Moreover, sulfur has a much greater tendency than oxygen to undergo catenation to give chains with several atoms linked together through S—S bonds. *See* CHEMICAL BONDING; PERIODIC TABLE; STRUCTURAL CHEMISTRY; VALENCE.

The structures and names of representative types of organosulfur compounds are shown in the table. Some compounds and groups are named by using the prefix thio to denote replacement of oxygen by sulfur. The prefix thia can be used to indicate that one or more $—CH_2—$ groups have been replaced by sulfur, as in 2,7-dithianonane [$CH_3S(CH_2)_4SCH_2CH_3$].

Thiols and sulfides are sulfur counterparts of alcohols and ethers, respectively, and can be prepared by substitution reactions analogous to those used for the oxygen compounds. Sulfonium salts are obtained by further alkylation of sulfides.

Although thiols and alcohols are structurally analogous, there are significant differences in the properties of these two groups. Hydrogen bonding of the type —S–H—S— is very weak compared to —O—H–O—, and thiols are thus more volatile and have lower boiling points than the corresponding alcohols; for example, methanethiol ($CH_3SH$) has a boiling point of 5.8°C (42.4°F) compared to 65.7°C (150.3°F) for methanol ($CH_3OH$).

Thiols form insoluble precipitates with heavy-metal ions such as lead or mercury. Both thiols and sulfides are extremely malodorous compounds, recalling the stench of rotten eggs (hydrogen sulfide). However, traces of these sulfur compounds are an essential component of the distinctive flavors and aromas of many vegetables, coffee, and roast meat. *See* MAILLARD REACTION; MERCAPTAN.

Thiocarbonyl compounds contain a carbon-sulfur double bond (C═S). Thiocarbonyl compounds (thiones) are much less common than carbonyl compounds (C═O bond). Simple thioaldehydes or thioketones have a strong tendency to form cyclic trimers, polymers, or other products.

Sulfides can be oxidized sequentially to sulfoxides and sulfones, containing the sulfinyl (—SO—) and sulfonyl ($—SO_2—$) groups, respectively, as in the following reaction.

$$RSR' \longrightarrow R\overset{O}{\overset{\|}{S}}R' \longrightarrow R\underset{\underset{O}{\|}}{\overset{\overset{O}{\|}}{S}}R'$$

Sulfoxide   Sulfone

Dimethyl sulfoxide (DMSO) is available in large quantities as a by-product of the Kraft sulfite paper process. It is useful as a polar solvent with a high boiling point and as a selective oxidant and reagent in organic synthesis. *See* DIMETHYL SULFOXIDE.

Compounds containing the sulfonyl group include sulfones, sulfonyl chlorides, sulfonic acids, and sulfonamides. The sulfonyl group resembles a carbonyl in the acidifying effect on an $\alpha$-hydrogen. The diaryl sulfone unit is the central feature of polysulfone resins, used in some high-performance plastics. Sulfonic acids are obtained by oxidation of thiols or by sulfonation. Sulfonamides, prepared from the chlorides, were the mainstay therapeutic agents in infections until the advent of antibiotics; they are still used for some conditions. *See* POLYSULFONE RESINS; SULFONAMIDE; SULFONIC ACID.

**Some types of organosulfur compounds and groups**

| Structure | Name |
|---|---|
| RSH | Thiol (mercaptan) |
| RSR | Sulfide (thioether) |
| RSSR | Disulfide |
| RSSSR | Trisulfide (trisulfane) |
| $R\overset{+}{S}R\ X^-$ with R on S | Sulfonium salt |
| $R_2C{=}S$ | Thioketone |
| $RN{=}C{=}S$ | Isothiocyanate |
| $R\overset{O}{\overset{\|}{C}}SR$ | Thiolate ester (thoic acid S-ester) |
| $R\overset{S}{\overset{\|}{C}}OR$ | Thionoate ester |
| $RCS_2R$ | Dithioate ester |
| RSOH | Sulfenic acid |
| RSCl | Sulfenyl chloride |
| RSOR | Sulfoxide |
| $R\overset{NR}{\overset{\|}{S}}R$ | Sulfimide |
| $RSO_2H$ | Sulfinic acid |
| $RSO_2R$ | Sulfinate ester |
| $R_2S{=}R_2$ | Sulfonium ylide (sulfurane) |
| $RSO_2R$ | Sulfone |
| $RSO_3H$ | Sulfonic acid |
| $RSO_2NH_2$ | Sulfonamide |
| $RSO_2Cl$ | Sulfonyl chloride |
| $ROSO_3R$ | Sulfate ester |

A number of proteins and metabolic pathways in systems of living organisms depend on the amino acid cysteine and other sulfur compounds. In many proteins, for example, in the enzyme insulin, disulfide bonds formed from the —SH groups of cysteine units are an essential part of the structure. The —SH groups of cysteine also play a role in the metal-sulfur proteins that mediate electron-transport reactions in respiration and photosynthesis.

The coenzyme lipoic acid is a cyclic disulfide that functions together with the coenzyme thiamine diphosphate to accept electrons and undergo reduction of the —S—S— bond in the oxidative decarboxylation of pyruvic acid. Two other major pathways in metabolism, the transfer of acetyl groups and of methyl groups, are mediated by organosulfur compounds. Acetyl transfer, a key step in lipid and carbohydrate metabolism, occurs by way of thioesters. *See* COENZYME.

Sulfur is present in numerous other compounds found in natural sources. Petroleum contains variable amounts of sulfur, both as simple thiols and sulfides, and also heterocyclic compounds such as benzothiophene. Removal of these is an important step in petroleum refining. *See* PETROLEUM PROCESSING AND REFINING.

Several sulfur-containing compounds from natural sources have important pharmacological properties. Examples are the $\beta$-lactam antibiotics penicillin, cephalosporin, and thienamycin, and the platelet anticoagulating factor ajoene from garlic, produced by a series of complex enzymatic reactions from allicin. *See* HETEROCYCLIC COMPOUNDS; ORGANIC CHEMISTRY; SULFUR. [J.A.Mo.]

**Oscillatory reaction** A chemical reaction in which some composition variable of a chemical system exhibits regular periodic variations in time or space. It is a basic tenet of chemistry that a closed system moves inexorably toward an unchanging state called chemical equilibrium. That motion can be described by the monotonic increase of entropy if the system is isolated, and by the monotonic decrease of Gibbs free energy if the system is constrained to constant temperature and pressure. *See* CHEMICAL EQUILIBRIUM; ENTROPY.

The species taking part in a chemical reaction can be classified as reactants, products, or intermediates. The concentrations of reactants decrease. Intermediates are formed by some steps and destroyed by others. If there is only one intermediate, and if its concentration is always much less than the initial concentrations of reactants, this intermediate attains a stable steady state in which the rates of formation and destruction are virtually equal. Some oscillations require at least two intermediates which interact in such a way that the steady state of the total system is unstable to the minor fluctuations present in any collection of molecules. The concentrations of the intermediates may then oscillate regularly, although the oscillations must disappear before the inevitable monotonic approach to equilibrium.

The systems whose chemistries are best understood all involve an element that can exist in several different oxidation states. An example is the so-called Belousov-Zhabotinsky reaction. A strong oxidizing agent (bromate) attacks an organic substrate (such as malonic acid), and the reaction is catalyzed by a metal ion (such as cerium) that can exist in two different oxidation states.

As long as bromide ion (Br) is present, it is oxidized by bromate ($BrO_3^-$), as in reaction (1).

$$BrO_3^- + 2Br^- + 3H^+ \rightarrow 3HOBr \quad (1)$$

When bromide ion is almost entirely consumed, the cerous ion ($Ce^{3+}$) is oxidized, as in reaction (2). Reaction (2) is inhibited

$$BrO_3^- + 4Ce^{3+} + 5H^+ \rightarrow HOBr + 4Ce^{4+} + 2H_2O \quad (2)$$

by $Br^-$, but when the concentration of bromide has been reduced to a critical level, reaction (2) accelerates autocatalytically until bromate is being reduced by $Ce^{3+}$ many times as rapidly as it is by $Br^-$ when reaction (3) is dominant.

The hypobromous acid (HOBr) brominates the organic substrate to form bromomalonic acid (BrMA), as in reaction (3). Reaction (3) creates the bromide ion necessary

$$2Ce^{4+} + BrMA \rightarrow 2Ce^{3+} + Br^- + \text{oxidized organic matter} \quad (3)$$

to shut off fast reaction (2) and throw the system back to dominance by slow reaction (1).

As other redox oscillators become understood, they fit the same pattern of a slow reaction destroying a species that inhibits a fast reaction that can be switched on autocatalytically; the fast reaction then generates conditions to produce the inhibitor again. *See* OXIDATION-REDUCTION. [R.M.No.]

**Osmium** A chemical element, Os, atomic number 76, atomic weight 190.2. The element is a hard white metal of rare natural occurrence, usually found in nature alloyed with other platinum metals. *See* PERIODIC TABLE; PLATINUM.

Physical properties of the element, which is found as seven naturally occurring isotopes, are given in the table. The metal is exceeded in density only by iridium. Osmium is a very hard metal and unworkable, and so it must be used in cast form or fabricated by powder metallurgy. Osmium is a third-row transition element and has the electronic configuration $[Xe](4f)^{14}(5d)^6(6s)^2$; in the periodic table it lies below iron (Fe) and ruthenium (Ru). In powder form the metal may be attacked by the oxygen in air at room temperature, and finely divided osmium has a faint odor of the tetraoxide. In bulk form

**Principal properties of osmium**

| Property | Value |
|---|---|
| Density, g/cm$^3$ | 22.6 |
| Naturally occurring isotopes (% abundance) | 184 (0.018) |
| | 186 (1.59) |
| | 187 (1.64) |
| | 188 (13.3) |
| | 189 (16.1) |
| | 190 (26.4) |
| | 192 (41.0) |
| Ionization enthalpy, kJ/mol: 1st 2d | 840 |
| | 1640 |
| Oxidation states | −1 to VIII |
| Most common | IV, VI, VIII |
| Ionic radius, $Os^{4+}$, nm | 0.078 |
| Melting point, °C (°F) | 3050 (5522) |
| Boiling point, °C (°F) | 5500 (9932) |
| Specific heat, cal/g·°C | 0.032 |
| Crystal structure | Hexagonal close-packed |
| Lattice constant *a* at 25°C, nm *c/a* at 25°C | 0.27341 |
| | 0.15799 |
| Thermal neutron capture cross section, barns | 15.3 |
| Thermal conductivity, 0–100°C, (cal · cm)/(cm$^2$ · s · °C) | 0.21 |
| Linear coefficient of thermal expansion at 20–100°C, ($\mu$in./in./°C) | 6.1 |
| Electrical resistivity at 0°C, $\mu\Omega$-cm | 8.12 |
| Temperature coefficient of electrical resistance, 0–100°C/°C | 0.0042 |
| Young's modulus at 20°C, lb/in.$^2$, static | $81 \times 10^6$ |

it does not oxidize in air below 500°C (750°F), but at higher temperatures it yields $OsO_4$. It is attacked by fluorine or chlorine at 100°C (212°F). It dissolves in alkaline oxidizing fluxes to give osmates ($OsO_4^{2-}$). *See* ELECTRON CONFIGURATION; IRIDIUM; IRON; RUTHENIUM.

The chemistry of osmium more closely resembles that of ruthenium than that of iron. The high oxidation states VI and VIII ($OsO_4^{2-}$ and $OsO_4$) are much more accessible than for iron. *See* OXIDATION-REDUCTION.

Osmium forms many complexes. In water, osmium complexes with oxidation states ranging from II to VIII may be obtained. Oxo compounds, which contain Os=O, are very common and occur for oxidation states IV to VIII. Although $OsO_4$ is tetrahedral in the gas phase and in noncomplexing solvents such as dichloromethane ($CH_2Cl_2$), it tends to be six-coordinate when appropriate ligands are available; thus, in sodium hydroxide (NaOH) solution, dark purple $OsO_4(OH)_2^{2-}$ is formed from $OsO_4$. Similarly, $OsO_2(OH)_4^{2-}$ is formed by addition of hydroxide to osmate anion. Analogous reactions with ligands such as halides, cyanide, and amines give osmyl derivatives like the cyanide-deduct $OsO_2(CN)_4^{2-}$, in which the trans dioxo group (O=Os=O) is retained.

Osmium tetraoxide, a commercially available yellow solid (melting point 40°C or 104°F), is used commercially in the important *cis*-hydroxylation of alkenes and as a stain for tissue in microscopy. It is poisonous and attacks the eyes. Osmium metal is catalytically active, but it is not commonly used for this purpose because of its high price. Osmium and its alloys are hard and resistant to corrosion and wear (particularly to rubbing wear). Alloyed with other platinum metals, osmium has been used in needles for record players, fountain-pen tips, and mechanical parts. *See* TRANSITION ELEMENTS.

[C.Cr.]

**Osmosis** The transport of solvent through a semipermeable membrane separating two solutions of different solute concentration. The solvent diffuses from the solution that is dilute in solute to the solution that is concentrated.

The flow of liquid through such a barrier may be stopped by applying pressure to the liquid on the side of higher solute concentration. The applied pressure required to prevent the flow of solvent across a perfectly semipermeable membrane is called the osmotic pressure and is a characteristic of the solution. The walls of cells in living organisms permit the passage of water and certain solutes, while preventing the passage of other solutes, usually of relatively high molecular weight. These walls act as selectively permeable membranes, and allow osmosis to occur between the interior of the cell and the surrounding media. *See* SOLUTION. [F.J.J.]

**Overvoltage** The difference between the electrical potential of an electrode or cell under the passage of current and the thermodynamic value of the electrode or cell potential under identical experimental conditions in the absence of electrolysis; it is also known as overpotential. Overvoltage is expressed in volts, often in absolute value; it is a measure of the rates of the different processes associated with an electrode reaction.

An understanding of the factors that contribute to the overvoltage is important in the operation of practical electrochemical systems. In batteries, the overvoltage plays a significant role in the available voltage and power. In large-scale industrial electrolysis, overvoltage is a major factor in determining the energy efficiency of a process, and hence, the cost of electricity. *See* DECOMPOSITION POTENTIAL; ELECTROCHEMICAL PROCESS; ELECTRODE POTENTIAL; ELECTROLYSIS.

Since the overvoltage is governed by kinetic considerations, all of the experimental conditions that can affect the rate of an electrolytic reaction are of importance.

These include concentration of electrolyzed substance, temperature, composition of solvent and electrolyte, nature of the electrode surface, mode of mass transfer, and the current density (current per unit area of electrode). A rapid reaction occurs with a small overvoltage (a few millivolts). A slow reaction requires a large overvoltage (a few volts).

The rates of electrode reactions are frequently determined from current density-potential curves (see illustration). Since the departure of the electrode or cell potential from the thermodynamic value upon passage of current is sometimes termed polarization, such curves are also known as polarization curves. These curves are obtained by measurements with three-electrode electrolytic cells. The current density that flows through the electrode of interest (the working electrode) is adjusted with an external direct-current power supply, and the potential of this electrode is measured with respect to a reference electrode whose potential is known and fixed. The measured potential contains a contribution, known as the ohmic drop, that results from the flow of current through the solution resistance between the working electrode and the reference electrode. This drop is minimized by placing these electrodes close together and using various experimental approaches. The thermodynamic potential is obtained from available data for the electrode reaction of interest, corrected for the concentration of the reaction species in the solution under the experimental conditions. It is also sometimes given by the working electrode potential in the electrolysis cell when no current flows (the rest potential). The overvoltage can be read directly from the current-potential curve, as shown in the illustration. *See* ELECTRODE.

The total overvoltage can be decomposed into different components, which are assigned to different sources of rate limitations, for example, concentration overvoltage, activation overvoltage, reaction overvoltage, and crystallization overvoltage.

Concentration overvoltage occurs when the concentration of the reactants or products at the electrode surface are different from those in the bulk solution. These differences arise because the electroactive reactant is consumed, and products are produced by the passage of current.

Activation overvoltage arises from slowness in the rate of the electron transfer reaction at the electrode surface. To drive an electrode reaction at a given rate (current density), it is necessary to overcome an energy barrier, the energy of activation for the reaction. The additional energy (that is, beyond the thermodynamic requirements) needed to overcome this barrier is provided by the electrical energy supplied to the cell in the form of an increase in the applied potential. The magnitude of this activation overvoltage often depends upon the nature of the electrode material.

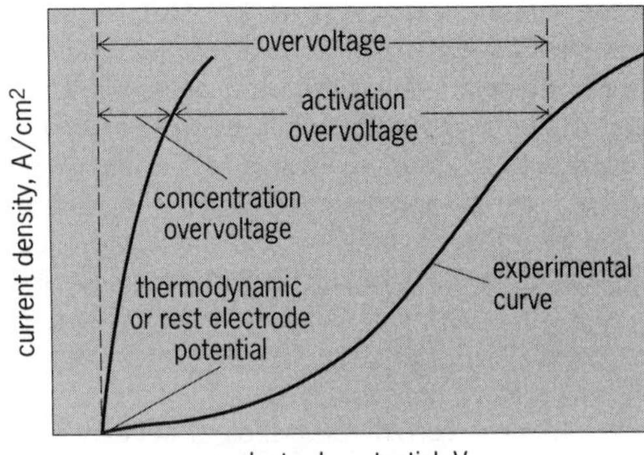


**Current density-potential curve.**

Reaction overvoltage arises when a chemical reaction is associated with the overall electrode reaction. For example, if the electroactive substance is generated from the major reactant by a chemical reaction that precedes the electron transfer, its concentration at the electrode surface will be governed by the rate of this reaction. This preceding reaction will thus affect the potential at which the electrode reaction occurs. Slow steps in the formation of nuclei and the crystal lattice, for example, in the electroplating of a metal, can lead to nucleation and crystallization overvoltages, respectively. [A.J.Ba.]

**Oxidation process** A process in which oxygen is caused to combine with other molecules. The oxygen may be used as elemental oxygen, as in air, or in the form of an oxygen-containing molecule which is capable of giving up all or part of its oxygen. Oxidation in its broadest sense, that is, an increase in positive valence or removal of electrons, is not considered here if oxygen itself is not involved. *See* OXIDATION-REDUCTION.

Most oxidations occur with the liberation of large amounts of energy in the form of either heat, light, or electricity. The stable ultimate products of oxidation are oxides of the elements involved. These oxidations occur in nature as corrosion, decay, and respiration and in the deliberate burning of matter such as wood, petroleum, sulfur, or phosphorus to oxides of the constituent elements.

The principal variables to be considered and controlled in any partial oxidation are temperature, pressure, reaction time (or contact time), nature of catalyst, if any, mole ratio of oxidizing agent, and whether the substance to be oxidized is to be kept in the liquid or vapor phase. Only a narrow range of conditions unique to each substance being oxidized and each product desired will give satisfactory yields. It is also essential to maintain conditions outside the range of spontaneous ignition, to avoid explosive mixtures or the accidental accumulation of unstable peroxides, and to choose materials which not only can resist the environmental conditions but also which do not have adverse catalytic effects or otherwise interfere with the desired reaction. *See* COMBUSTION. [I.E.L.]

**Oxidation-reduction** An important concept of chemical reactions which is useful in systematizing the chemistry of many substances. Oxidation can be represented as involving a loss of electrons by one molecule and reduction as involving an absorption of electrons by another. Both oxidation and reduction occur simultaneously and in equivalent amounts during any reaction involving either process.

**Oxidation number.** The oxidation state is a concept which describes some important aspects of the state of combination of the elements. An element in a given substance is characterized by a number, the oxidation number, which specifies whether the element in question is combined with elements which are more electropositive or more electronegative than it is. It further specifies the combining capacity which the element exhibits in a particular combination. A scale of oxidation numbers is defined by assigning to an oxygen atom in an ion such as $SO_4^{2-}$ the value of 2−. That for sulfur as 6+ then follows from the requirement that the sum of the oxidation numbers of all the atoms add up to the net charge on the species. The value of 2− for oxygen is not chosen arbitrarily. It recognizes that oxygen is more electronegative than sulfur, and that when it reacts with other elements it seeks to acquire two more electrons, by sharing or outright transfer from the electropositive partner, so as to complete a stable valence shell of eight electrons.

Although oxidation number is in some respects similar to valence, the two concepts have distinct meanings. In the substance $H_2$, the valence of hydrogen is 1 because each H makes a single bond to another H, but the oxidation number is 0, because the hydrogen is not combined with a different element. *See* VALENCE.

When the oxidation number of an atom in a species is increased, the process is described as oxidation, no matter what reagent produces it; when a decrease in oxidation number takes place, the process is described as reduction, again without regard to the identity of the reducing agent. The term oxidation has been generalized to imply combination of an element with an element more electronegative than itself.

**Reactions.** In an oxidation-reduction reaction, some element decreases in oxidation state and some element increases in oxidation state. The substances containing these elements are defined as the oxidizing agents and reducing agents, and they are said to be reduced and oxidized, respectively. The processes in question can always be represented formally as involving electron absorption by the oxidizing agent and electron donation by the reducing agent. For example, reaction (1) can be regarded as the sum of the two partial processes, or half-reactions, (2) and (3). Similarly, reaction (4) consists of the two half-reactions (5) and (6), with half-reaction (5) being taken five times to balance the electron flow from reducing agent to oxidizing agent.

$$2Fe^{3+} + 2I^- \rightarrow 2Fe^{2+} + I_2 \qquad (1)$$

$$2I^- \rightarrow I_2 + 2e^- \qquad (2)$$

$$2Fe^{3+} + 2e^- \rightarrow 2Fe^{2+} \qquad (3)$$

$$16H^+ + 2MnO_4{}^- + 10I^- \rightarrow 8H_2O + 2Mn^{2+} + 5I_2 \qquad (4)$$

$$2I^- \rightarrow I_2 + 2e^- \qquad (5)$$

$$16H^+ + 2MnO_4{}^- + 10e^- \rightarrow 2MN^{2+} + 8H_2O \qquad (6)$$

Each half-reaction consists of an oxidation-reduction couple; thus, in half-reaction (6) the reducing agent and oxidizing agent making up the couple are manganous ion, $Mn^{2+}$, and permanganate ion, $MnO_4{}^-$ , respectively; in half-reaction (5) the reducing agent is $I^-$ and the oxidizing agent is $I_2$. The fact that $MnO_4{}^-$ reacts with $I^-$ to produce $I_2$ means that $MnO_4{}^-$ in acid solution is a stronger oxidizing agent than is $I_2$. Because of the reciprocal relation between the oxidizing agent and reducing agent comprising a couple, this statement is equivalent to saying that $I^-$ is a stronger reducing agent than $Mn^{2+}$ in acid solution. Reducing agents may be ranked in order of tendency to react, and this ranking immediately implies an opposite order of tendency to react for the oxidizing agents which complete the couples. In the list below some common oxidation-reduction couples are ranked in this fashion:

Strong reducing agent (top) ↑ Increasing reducing power — Weak reducing agent (bottom)

Weak oxidizing agent (top) ↓ Increasing oxidizing power — Strong oxidizing agent (bottom)

$$Mg = Mg^{2+} + 2e^-$$
$$Zn = Zn^{2} + 2e^-$$
$$H_2 = 2H^+ + 2e^-$$
$$Cu = Cu^{2} + 2e^-$$
$$I^- = \tfrac{1}{2}I_2 + e^-$$
$$Fe^{2+} = Fe^{3+} + e^-$$
$$Br^- = \tfrac{1}{2}Br_2 + e^-$$
$$Cl^- = \tfrac{1}{2}Cl_2 + e^-$$
$$4H_2O + Mn^{2+} = MnO_4{}^- + 8H^+ + 5e^-$$

*See* ELECTROCHEMICAL SERIES; ELECTRONEGATIVITY; OXIDIZING AGENT. [H.Ta.]

**Oxide** A binary compound of oxygen with another element. Oxides have been prepared for essentially all the elements except the noble gases. Often, several different oxides of a given element can be prepared; a number exist naturally in the Earth's crust and atmosphere: silicon dioxide ($SiO_2$) in quartz; aluminum oxide ($Al_2O_3$) in corundum; iron oxide ($Fe_2O_3$) in hematite; carbon dioxide ($CO_2$) gas; and water ($H_2O$).

Most elements will react with oxygen at appropriate temperature and oxygen pressure conditions, and many oxides may thus be directly prepared. Most metals in massive form react with oxygen only slowly at room temperatures because the first thin oxide coat formed protects the metal. The oxides of the alkali and alkaline-earth metals, except for beryllium and magnesium, are porous when formed on the metal surface, and they provide only limited protection to the continuation of oxidation, even at room temperatures. Gold is exceptional in its resistance to oxygen, and its oxide ($Au_2O_3$) must be prepared by indirect means. The other noble metals, although ordinarily resistant to oxygen, will react at high temperatures to form gaseous oxides.

Oxides may be classified as acidic or basic according to the character of the solution resulting from their reactions with water. The nonmetal oxides generally form acid solutions and the metal oxides generally form alkaline solutions. *See* ACID AND BASE; EQUIVALENT WEIGHT; OXYGEN. [R.K.E.]

**Oxidizing agent** A participant in a chemical reaction that absorbs electrons from another reactant. In the process a component atom of this substance undergoes a decrease in oxidation number. In this action as an oxidizing agent, the substance undergoes reduction.

A measure of the effectiveness of a reagent as an oxidizing agent is its reduction potential. This is, in electrochemical terms, the equivalent of the free-energy change for the reduction process. The element with the highest reduction potential (and, therefore, the strongest oxidizing agent) is fluorine, $F_2$. The practical effectiveness of a given oxidizing (or reducing) agent will depend upon both the thermodynamics and the available kinetic pathway for the reaction process. *See* CHEMICAL THERMODYNAMICS.

Substances that are widely used as oxidizing agents in chemistry include ozone ($O_3$), permanganate ion ($MnO_4^-$), nitric acid ($HNO_3$), as well as oxygen itself. Organic chemists have empirically developed combinations of reagents to carry out specific oxidation steps in synthetic processes. The action of molecular oxygen as an oxidizing agent may be made more specific by photochemical excitation to an excited singlet electronic state. [F.J.J.]

**Oxime** One of a group of chemical substances with the general formula RR′C═N—OH, where R and R′ represent any carbon group or hydrogen. Oximes are derived from aldehydes (RHC═NOH, aldoximes) and ketones (RR′C═NOH, where R and R′ are not hydrogen; ketoximes), and they are used to isolate and characterize these carbonyl compounds. Oximes are also useful as intermediates in organic syntheses. *See* ALDEHYDE; KETONE.

Hydroxylamine ($H_2NOH$) reacts readily with aldehydes or ketones to give oximes. The rate of the reaction of hydroxylamine with acetone is greatest at pH 4.5. Oximes are formed by nucleophilic attack of hydroxylamine at the carbonyl carbon (C═C) of an aldehyde or ketone to give an unstable carbinolamine intermediate, as in the reaction below. Since the breakdown of the carbinolamine intermediate to an oxime is acid-catalyzed, the rate of this step is enhanced at low pH. If the pH is too low, however, most of the hydroxylamine will be in the nonnucleophilic protonated form ($NH_3OH^+$), and the rate of the first step will decrease. Thus, in oxime formation the

$$\overset{O}{\overset{\|}{RCR'}} + H_2NOH \rightleftharpoons \left[ R-\overset{OH}{\overset{|}{C}}-R' \atop {\scriptstyle |} \atop NH \atop {\scriptstyle |} \atop OH \right] \rightleftharpoons \overset{N-OH}{\overset{\|}{RCR'}}$$

Carbinolamine intermediate

pH has to be such that there is sufficient free hydroxylamine for the first step and enough acid so that dehydration of the carbinolamine is facile. *See* REACTIVE INTERMEDIATES.

One of the best-known reactions of oximes is their rearrangement to amides. This reaction, the Beckmann rearrangement, can be carried out with a variety of reagents [such as phosphorus pentachloride ($PCl_5$), concentrated sulfuric acid ($H_2SO_4$), and perchloric acid ($HClO_4$)] that induce the rearrangement by converting the oxime hydroxyl group into a group of atoms that easily departs in a displacement reaction by either protonation or formation of a derivative. The industrial synthesis of $\epsilon$-caprolactam is carried out by a Beckmann rearrangement on cyclohexanone oxime. $\epsilon$-Caprolactam is polymerized to the polyamide known as nylon 6, which is used in tire cords. *See* ORGANIC SYNTHESIS. [J.E.Jo.]

**Oxygen** A gaseous chemical element, O, atomic number 8, and atomic weight 15.9994. Oxygen is of great interest because it is the essential element both in the respiration process in most living cells and in combustion processes. It is the most abundant element in the Earth's crust. About one-fifth (by volume) of the air is oxygen.

Oxygen is separated from air by liquefaction and fractional distillation. The chief uses of oxygen in order of their importance are (1) smelting, refining, and fabrication of steel and other metals; (2) manufacture of chemical products by controlled oxidation; (3) rocket propulsion; (4) biological life support and medicine; and (5) mining, production, and fabrication of stone and glass products. *See* PERIODIC TABLE.

Uncombined gaseous oxygen usually exists in the form of diatomic molecules, $O_2$, but oxygen also exists in a unique triatomic form, $O_3$, called ozone. *See* OZONE.

Under ordinary conditions oxygen is a colorless, odorless, and tasteless gas. It condenses to a pale blue liquid, in contrast to nitrogen, which is colorless in the liquid state. Oxygen is one of a small group of slightly paramagnetic gases, and it is the

**Properties of oxygen**

| Property | Value |
|---|---|
| Atomic number | 8 |
| Atomic weight | 15.9994 |
| Triple point (solid, liquid, and gas in equilibrium) | −218.80°C (−139.33°F) |
| Boiling point at 1 atm pressure | −182.97°C (−119.4°F) |
| Gas density at °C and $10^5$ Pa pressure, g/liter | 1.4290 |
| Liquid density at the boiling point, g/ml | 1.142 |
| Solubility in water at 20°C, oxygen (STP) per 1000 g water at $10^5$ Pa partial pressure of oxygen | 30 |

most paramagnetic of the group. Liquid oxygen is also slightly paramagnetic. Some data on oxygen and some properties of its ordinary form, $O_2$, are listed in the table. *See* PARAMAGNETISM.

Practically all chemical elements except the inert gases form compounds with oxygen. Most elements form oxides when heated in an atmosphere containing oxygen gas. Many elements form more than one oxide; for example, sulfur forms sulfur dioxide ($SO_2$) and sulfur trioxide ($SO_3$). Among the most abundant binary oxygen compounds are water, $H_2O$, and silica, $SiO_2$, the latter being the chief ingredient of sand. Among compounds containing more than two elements, the most abundant are the silicates, which constitute most of the rocks and soil. Other widely occurring compounds are calcium carbonate (limestone and marble), calcium sulfate (gypsum), aluminum oxide (bauxite), and the various oxides of iron which are mined as a source of iron. Several other metals are also mined in the form of their oxides. Hydrogen peroxide, $H_2O_2$, is an interesting compound used extensively for bleaching. *See* HYDROGEN PEROXIDE; OXIDATION-REDUCTION; OXIDE; PEROXIDE; WATER. [A.W.F.; L.M.Sa.]

**Ozone** A powerfully oxidizing allotropic form of the element oxygen. The ozone molecule contains three atoms ($O_3$). Ozone gas is decidedly blue, and both liquid and solid ozone are an opaque blue-black color, similar to that of ink. *See* OXYGEN.

Some properties of ozone are given in the table. Ozone has a characteristic, pungent odor familiar to most persons because ozone is formed when electrical apparatus produces sparks in air. Ozone is irritating to mucous membranes and toxic to human beings and lower animals.

**Some properties of ozone**

| Property | Value |
|---|---|
| Density of the gas at 0°C, 1 atm pressure | 2.154 g/liter |
| Density of the liquid | |
| −111.9°C | 1.354 g/ml |
| −183°C | 1.573 g/ml |
| Boiling point at 1 atm pressure | −111.9°C |
| Melting point of the solid | −192.5°C |

Ozone is a more powerful oxidizing agent than oxygen, and oxidation with ozone takes place with evolution of more heat and usually starts at a lower temperature than when oxygen is used. In the presence of water, ozone is a powerful bleaching agent, acting more rapidly than hydrogen peroxide, chlorine, or sulfur dioxide. *See* OXIDIZING AGENT.

Ozone is utilized in the treatment of drinking-water supplies. Odor- and taste-producing hydrocarbons are effectively eliminated by ozone oxidation. Iron and manganese compounds which discolor water are diminished by ozone treatment. Compared to chlorine, bacterial and viral disinfection with ozone is up to 5000 times more rapid.

Ozone occurs to a variable extent in the Earth's atmosphere. Near the Earth's surface the concentration is usually 0.02–0.03 ppm in country air, and less in cities except when there is smog. At vertical elevations above 13 mi (20 km), ozone is formed by photochemical action on atmospheric oxygen. Maximum concentration of $5 \times 10^{12}$ molecules/cm$^3$ (more than 1000 times the normal peak concentration at Earth's surface) occurs at an elevation of 19 mi (30 km). [A.W.F.]

**Ozonolysis** A process which uses ozone to cleave unsaturated organic bonds. Generally, ozonolysis is conducted by bubbling ozone-rich oxygen or air into a solution of the reactant. The reaction is fast at moderate temperatures. Intermediates are usually not isolated but are subjected to further oxidizing conditions to produce acids or to reducing conditions to form alcohols or aldehydes. An unsymmetrical olefin is capable of yielding two different products whose structures are related to the groups substituted on the olefin and the position of the double bond.

Before World War I, ozonolysis was applied commercially to the preparation of vanillin from isoeugenol. The only modem application of the technique in the United States is in the manufacture of azelaic and pelargonic acids from oleic acid. *See* ALKENE; OZONE. [R.K.Ba.]

# P

**Palladium** A chemical element, Pd, atomic number 46, and atomic weight 106.4. A transition metal, palladium occurs in combination with platinum (Pt) and is the second most abundant platinum-group metal, accounting for 38% of the reserves of these metals. *See* PERIODIC TABLE; PLATINUM.

Palladium is soft and ductile and can be fabricated into wire and sheet. The metal forms ductile alloys with a broad range of elements. Palladium is not tarnished by dry or moist air at ordinary temperatures. At temperatures from 350 to 790°C (660 to 1450°F) a thin protective oxide forms in air, but at temperatures from 790°C this film decomposes by oxygen loss, leaving the bright metal. In the presence of industrial sulfur-containing gases a slight brownish tarnish develops; however, alloying palladium with small amounts of iridium or rhodium prevents this action. Important physical properties of palladium are given in the table.

At room temperature, palladium is resistant to nonoxidizing acids such as sulfuric acid, hydrochloric acid, hydrofluoric acid, and acetic acid. The metal is attacked by nitric acid, and a mixture of nitric acid and hydrochloric acid is a solvent for the metal. Palladium is also attacked by moist chlorine (Cl) and bromine (Br). *See* NONSTOICHIOMETRIC COMPOUNDS.

**Physical properties of palladium**

| Property | Value |
|---|---|
| Atomic weight | 106.4 |
| Naturally occurring isotopes (percent abundance) | 102 (0.96) |
| | 104 (10.97) |
| | 105 (22.23) |
| | 106 (27.33) |
| | 108 (26.71) |
| | 110 (11.81) |
| Crystal structure | Face-centered cubic |
| Thermal neutron capture cross section, barns | 8.0 |
| Density at 25°C (77°F), g/cm$^3$ | 12.01 |
| Melting point, °C (°F) | 1554 (2829) |
| Boiling point, °C (°F) | 2900 (5300) |
| Specific heat at 0°C (32°F), cal/g | 0.0584 |
| Thermal conductivity,(cal·cm)(cm$^2$·s·°C) | 0.18 |
| Linear coefficient of thermal expansion, ($\mu$in./in./)/°C | 11.6 |
| Electrical resistivity at 0°C (32°F), $\mu\Omega$-cm | 9.93 |
| Young's modulus, lb/in.$^2$, static, at 20°C (68°F) | $16.7 \times 10^6$ |
| Atomic radius in metal, nm | 0.1375 |
| Ionization potential, eV | 8.33 |
| Binding energy, eV | 3.91 |
| Pauling electronegativity | 2.2 |
| Oxidation potential, V | −0.92 |

The major applications of palladium are in the electronics industry, where it is used as an alloy with silver for electrical contacts or in pastes in miniature solid-state devices and in integrated circuits. Palladium is widely used in dentistry as a substitute for gold. Other consumer applications are in automobile exhaust catalysts and jewelry.

Palladium supported on carbon or alumina is used as a catalyst for hydrogenation and dehydrogenation in both liquid- and gas-phase reactions. Palladium finds widespread use in catalysis because it is frequently very active under ambient conditions, and it can yield very high selectivities. Palladium catalyzes the reaction of hydrogen with oxygen to give water. Palladium also catalyzes isomerization and fragmentation reactions. *See* CATALYSIS.

Halides of divalent palladium can be used as homogeneous catalysts for the oxidation of olefins (Wacker process). This requires water for the oxygen transfer step, and a copper salt to reoxidize the palladium back to its divalent state to complete the catalytic cycle. *See* HOMOGENEOUS CATALYSIS; TRANSITION ELEMENTS. [D.M.Ro.]

**Paraffin** A term used variously for either a waxlike substance or a group of compounds. The former use pertains to the high-boiling residue obtained from certain petroleum crudes. It is recovered by freezing out on a cold drum and is purified by crystallization from methyl ethyl ketone. Paraffin wax is a mixture of 26- to 30-carbon alkane hydrocarbons; it melts at 52–57°C (126–135°F). Microcrystalline wax contains compounds of higher molecular weight and has a melting point as high as 90°C (190°F). The name paraffin was formerly used to designate a group of hydrocarbons—now known as alkanes. *See* ALKANE. [A.L.H.]

**Peptide** A compound that is made up of two or more amino acids joined by covalent bonds which are formed by the elimination of a molecule of $H_2O$ from the amino group of one amino acid and the carboxyl group of the next amino acid. Peptides larger than about 50 amino acid residues are usually classified as proteins. Glutathione is the most abundant peptide in mammalian tissue. Hormones such as oxytocin (8), vasopressin (8), glucagon (29), and adrenocorticotropic hormone (39) are peptides whose structures have been deduced; in parentheses are the numbers of amino acid residues for each peptide.

For each step in the biological synthesis of a peptide or protein there is a specific enzyme or enzyme complex that catalyzes each reaction in an ordered fashion along the biosynthetic route. However, it is noteworthy that, although the biological synthesis of proteins is directed by messenger RNA on cellular structures called ribosomes, the biological synthesis of peptides does not require either messenger RNA or ribosomes. *See* AMINO ACIDS; PROTEIN. [J.M.M.]

**Pericyclic reaction** Concerted (single-step) processes in which bond making and bond breaking occur simultaneously (but not necessarily synchronously) via a cyclic (closed-curve) transition state. Although a given reaction may appear formally to be pericyclic, it cannot be assumed to be a concerted process. In each case, the detailed mechanism of the reaction must be established experimentally. Pericyclic reactions can be promoted either by heat or by light; the stereochemistry of the reaction is determined by the mode of activation employed and the number of electrons that are delocalized in the transition state. *See* CHEMICAL BONDING; PHYSICAL ORGANIC CHEMISTRY; STEREOCHEMISTRY.

Four types of pericyclic reactions that are frequently encountered in organic chemistry are electrocyclic processes, cycloadditions, sigmatropic shifts, and cheletropic reactions.

Electrocyclic processes are reactions that involve their cyclization across the termini of a conjugated $\pi$-system with concomitant formation of a new $\sigma$-bond or the microscopic reverse. The sequence of steps involved in the forward reaction must be the same, in the reverse order, as that in the reverse direction when the forward and reverse reactions are carried out under identical conditions. This statement is known as the principle of microscopic reversibility.

The effect of the mode of activation upon the stereochemistry of an electrocyclic process is shown in the reaction (1), where Me = methyl, for the hexatrienecyclohexadiene

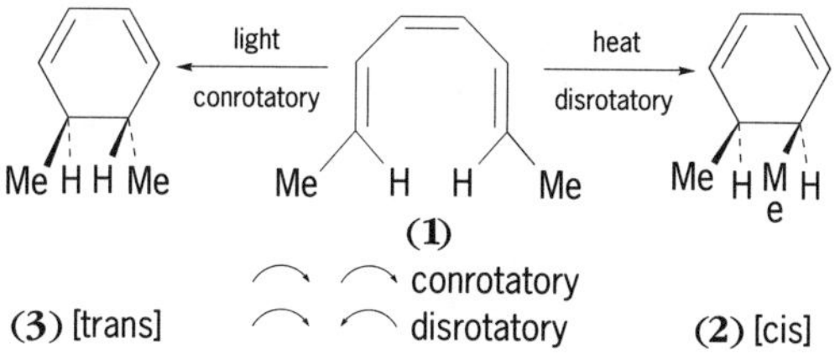

light
conrotatory
heat
disrotatory
Me H H Me
Me H H Me
Me H Me H
(1)
conrotatory
disrotatory
(3) [trans]
(2) [cis]
(1)

interconversion (a six-electron electrocyclic process). Thus, when *trans,cis,trans*-2,4,6-octatriene [structure (**1**)] is heated, disrotatory motion of the two terminal 2*p* orbitals occurs; that is, they rotate in opposite directions thereby resulting in exclusive formation of *cis*-5,6-dimethylcyclohexa-l,3-diene (**2**). The corresponding photochemical process results in conrotatory motion of the termini in structure (**1**); that is, the two terminal 2*p* orbitals rotate in the same direction thereby yielding *trans*-5,6-dimethylcyclohexa-l,3-diene (**3**) exclusively.

Cycloadditions occur when two (or more) $\pi$-electron systems react under the influence of heat or light to form a cyclic compound with concomitant formation of two new $\sigma$-bonds that join the termini of the original $\pi$-systems. The stereochemistry of this reaction is classified with respect to the two molecular planes of the reactants. Thus, if $\sigma$-bond formation occurs from the same face of the molecular plane across the termini of one of the component $\pi$-systems, the reaction is said to be suprafacial on that component. If instead $\sigma$-bond formation occurs from opposite faces of the molecular plane, the reaction is said to be antarafacial on that component. This distinction is illustrated in reaction (2) for two thermal processes, where the symbol $\neq$ indicates the structure of the transition state. Reaction (2*a*) shows the Diels-Alder [4 + 2] cycloaddition of butadiene (**4**) to ethylene (**5**), a six-electron pericyclic reaction in which additions across the termini of the diene (four-electron component) and dienophile (two-electron component) both occur suprafacially. Reaction (2*b*) shows a [14 + 2] cycloaddition in which $\sigma$-bond formation occurs suprafacially on the two-electron component [tetracyanoethylene (**6**)] and antarafacially on the fourteen-electron component [heptafulvalene (**7**)].

Sigmatropic shifts involve migration of a $\sigma$-bond that is flanked at either (or both) ends by conjugated $\pi$-systems. Either one or both ends of the $\sigma$-bond may migrate to a new location within the one or more flanking $\pi$-systems.

Cheletropic reactions involve extrusion of a fragment via concerted cleavage of two $\sigma$-bonds that terminate at a single atom or the reverse process. Cheletropic fragmentations may be either linear or nonlinear [reaction (3)].

R. B. Woodward and R. Hoffmann introduced an application of molecular orbital theory that permits prediction of rates and products of pericyclic reactions. They utilized symmetry properties of molecular orbitals to estimate relative energies of diastereoisomeric transition states for structurally similar pericyclic reactions.

In an alternative theoretical approach to understanding pericyclic reactions, the transition state is examined directly, and attempts to estimate the degree of electronic

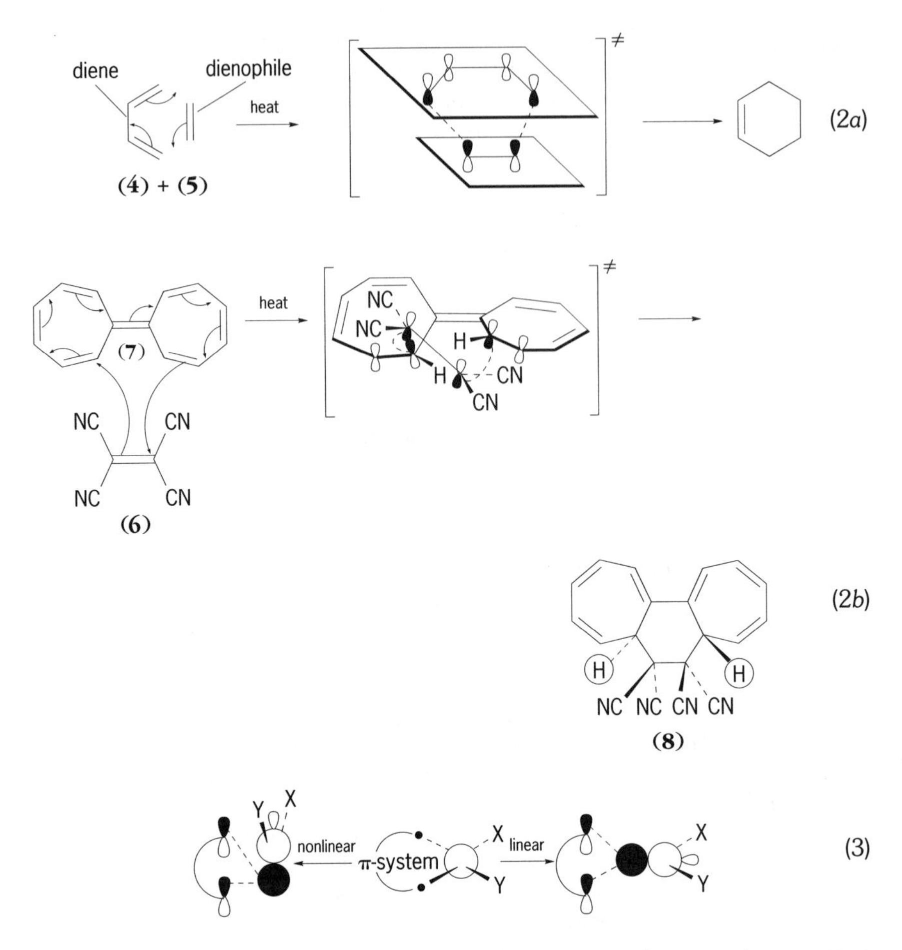


stabilization (allowedness) or destabilization (forbiddenness) inherent in that transition state are made. One such approach emphasizes the importance of frontier orbitals (highest-occupied-lowest-unoccupied molecular orbitals) in determining the course of a pericyclic reaction. *See* DELOCALIZATION; DIELS-ALDER REACTION; ELECTRON CONFIGURATION; MOLECULAR ORBITAL THEORY; ORGANIC REACTION MECHANISM; WOODWARD-HOFFMANN RULE. [A.P.M.]

**Periodic table** A list of elements (atoms) ordered along horizontal rows according to atomic number (the number of electrons in an atom and also the charge on its nucleus). In the periodic table (see illustration), the rows are arranged so that elements with nearly the same chemical properties occur in the same column (group), and each row ends with a noble gas (closed-shell element that is generally inert). For chemists, the position of atoms in the periodic table provides the most powerful guide for classifying the expected properties of molecules and solids made from these particular atoms. *See* INERT GASES.

The origin of the periodic table was explained in the 1920s in terms of the basic physical laws (quantum mechanics) obeyed by the electrons of an atom. Thus, the rows in the periodic table correspond to the shell number, $n$, and groups correspond to a particular electronic configuration designated by the number and type of electrons

| | 1 | 2 | 3 | 4 | 5 | 6 | 7 | 8 | 9 | 10 | 11 | 12 | 13 | 14 | 15 | 16 | 17 | 18 |
|---|---|---|---|---|---|---|---|---|---|---|---|---|---|---|---|---|---|---|
| 1s | | | | | | | | | | | | | | | | | 1 H Hydrogen | 2 He Helium |
| | s | | d | | | | | | | | | | p | | | | | |
| | 3 Li Lithium | 4 Be Beryllium | | | | | | | | | | | 5 B Boron | 6 C Carbon | 7 N Nitrogen | 8 O Oxygen | 9 F Fluorine | 10 Ne Neon |
| | 11 Na Sodium | 12 Mg Magnesium | | | | | | | | | | | 13 Al Aluminum | 14 Si Silicon | 15 P Phosphorus | 16 S Sulfur | 17 Cl Chlorine | 18 Ar Argon |
| | 19 K Potassium | 20 Ca Calcium | 21 Sc Scandium | 22 Ti Titanium | 23 V Vanadium | 24 Cr Chromium | 25 Mn Manganese | 26 Fe Iron | 27 Co Cobalt | 28 Ni Nickel | 29 Cu Copper | 30 Zn Zinc | 31 Ga Gallium | 32 Ge Germanium | 33 As Arsenic | 34 Se Selenium | 35 Br Bromine | 36 Kr Krypton |
| | 37 Rb Rubidium | 38 Sr Strontium | 39 Y Yttrium | 40 Zr Zirconium | 41 Nb Niobium | 42 Mo Molybdenum | 43 Tc Technetium | 44 Ru Ruthenium | 45 Rh Rhodium | 46 Pd Palladium | 47 Ag Silver | 48 Cd Cadmium | 49 In Indium | 50 Sn Tin | 51 Sb Antimony | 52 Te Tellurium | 53 I Iodine | 54 Xe Xenon |
| | 55 Cs Cesium | 56 Ba Barium | 71 Lu Lutetium | 72 Hf Hafnium | 73 Ta Tantalum | 74 W Tungsten | 75 Re Rhenium | 76 Os Osmium | 77 Ir Iridium | 78 Pt Platinum | 79 Au Gold | 80 Hg Mercury | 81 Tl Thallium | 82 Pb Lead | 83 Bi Bismuth | 84 Po Polonium | 85 At Astatine | 86 Rn Radon |
| | 87 Fr Francium | 88 Ra Radium | 103 Lr Lawrencium | 104 Rf Rutherfor-dium | 105 Db Dubnium | 106 Sg Seaborgium | 107 Bh Bohrium | 108 Hs Hassium | 109 Mt Meitnerium | 110 Ds Darmstadtium | 111 | 112 | 113 | 114 | 115 | 116 | 117 | 118 |

| f | | | | | | | | | | | | | |
|---|---|---|---|---|---|---|---|---|---|---|---|---|---|
| 57 La Lanthanum | 58 Ce Cerium | 59 Pr Praseodymium | 60 Nd Neodymium | 61 Pm Promethium | 62 Sm Samarium | 63 Eu Europium | 64 Gd Gadolinium | 65 Tb Terbium | 66 Dy Dysprosium | 67 Ho Holmium | 68 Er Erbium | 69 Tm Thulium | 70 Yb Ytterbium |
| 89 Ac Actinium | 90 Th Thorium | 91 Pa Protactinium | 92 U Uranium | 93 Np Neptunium | 94 Pu Plutonium | 95 Am Americium | 96 Cm Curium | 97 Bk Berkelium | 98 Cf Californium | 99 Es Einsteinium | 100 Fm Fermium | 101 Md Mendelevium | 102 No Nobelium |

**Periodic table. The atomic numbers are listed above the symbols identifying the elements. The heavy line separates metals from nonmetals.**

in the outermost shell. These electrons govern chemical properties and are known as valence electrons. *See* ELECTRON CONFIGURATION; VALENCE.

Additional information from the physical laws of atoms can be incorporated into the periodic table and can greatly enhance its organizing capability. For example, configuration energy adds a third dimension to the periodic table. The configuration energy is defined in terms of the ionization energy ($I$), the energy required to remove an electron from an atom.

Besides enhancing the organizing capability of the periodic table, the concept of configuration energy explains many longstanding puzzles about the table itself. It explains the existence of the metalloid band of elements (configuration energy is nearly constant in this band) and why these elements divide the metals from the nonmetals. Elements possessing configuration energies with magnitudes greater than those of the metalloids are nonmetals; those with lower configuration energies are metals. *See* NONMETAL.

The lack of numerical or analytic connection between the traditional two-dimensional periodic table and methods used to predict the structure and reactivity of molecules and solids has long reduced the table's usefulness. However, configuration energy, introduced as a new dimension of the periodic table, is just the average atomic energy level, and simultaneously the average density of states, for the atoms out of which the molecular-orbit–energy-level diagrams and energy bands in solids are constructed, thereby tying the periodic table directly to present-day research techniques. *See* MOLECULAR ORBITAL THEORY; MOLECULAR STRUCTURE AND SPECTRA. [L.C.A.]

**Peroxide** A chemical compound which contains the peroxy (—O—O—) group, which may be considered to be a derivative of hydrogen peroxide (HOOH). An organic (or inorganic) peroxide is one in which some organic (or inorganic) substituent has replaced one or both hydrogens. Peroxides are used in such diverse reactions as oxidation, synthesis, polymerization, and oxygen generation. Inorganic peroxides include persulfates, hydrogen peroxide ($H_2O_2$), sodium peroxide, bivalent metal peroxides, and $H_2O_2$ addition compounds. Organic peroxides include peroxyacetic acid, dibenzoyl peroxide, and cumene peroxide. *See* HYDROGEN PEROXIDE; OXIDIZING AGENT; OXYGEN. [S.S.N.]

**Peroxynitrite** A nitrogen oxyanion containing an O—O peroxo bond that is a structural isomer of the nitrate ion. These species are generally distinguished as $ONOO^-$ and $NO_3^-$, respectively. Other names for peroxynitrite include pernitrite and peroxonitrite; the systematic name recommended by the International Union of Pure and Applied Chemistry (IUPAC) is oxoperoxonitrate (1−). Energy calculations indicate that there are two stable conformations of $ONOO^-$, for which all of the atoms lie in a plane with the peroxo O—O and N=O bonds forming dihedral angles of approximately 0° (cis isomer) or 180° (trans isomer) [notation (1)].

$$O{=}N{-}O{-}O^- \;(\text{cis}) \rightleftharpoons O{=}N{-}O{-}O^- \;(\text{trans}) \qquad (1)$$

*See* CHEMICAL BONDING.

Peroxynitrite is formed in nitrate salts or nitrate-containing solutions when exposed to ionizing radiation or ultraviolet light. Solutions can also be prepared by a variety of chemical reactions, including the reaction of hydrogen peroxide with nitrous acid (2);

reaction of the hydroperoxide anion with organic and inorganic nitrosating agents (3); reaction of ozone with the azide ion (4); or, apparently, reaction of $O_2$ with compounds capable of generating the nitroxyl anion ($NO^-$) [5]. These preparations invariably

$$HOOH + HNO_2 \rightarrow ONOOH + H_2O \tag{2}$$

$$HOO^- + RONO \rightarrow ROH + ONOO^- \tag{3}$$

$$2O_3 + N_3^- \rightarrow ONOO^- + N_2O + O_2 \tag{4}$$

$$2O_2 + NH_2OH + 2OH^- \rightarrow ONOO^- + HO_2^- + 2H_2O \tag{5}$$

contain unreacted materials or decomposition products, particularly nitrite ion, which can significantly modulate the peroxynitrite chemical reactivity. Peroxynitrite is also formed in radical-radical coupling reactions, notably superoxide ($\cdot O_2^-$) with nitric oxide ($\cdot NO$) [reaction (6)], and hydroxyl radical with nitrogen dioxide [reaction (7)].

$$\cdot NO + \cdot O_2^- \rightarrow ONOO^- \tag{6}$$

$$\cdot OH + \cdot NO_2 \rightarrow ONOOH \tag{7}$$

*See* SUPEROXIDE CHEMISTRY.

Peroxynitrite has been isolated as the tetramethylammonium salt by carrying out reaction (6) in liquid ammonia. Formation of peroxynitrite in both solids and solutions is indicated by the appearance of yellow coloration, which is due to tailing of intense near-ultraviolet absorption bands into the visible region.

Peroxynitrite is a powerful oxidant that has been shown to react with a wide variety of inorganic and organic reductants. Interest in these reactions has been greatly stimulated by recognition that $\cdot NO$ and $\cdot O_2^-$ radicals are generated in the bloodstream, neuronal tissues, and phagocytic cells of animals in sufficient quantities to form peroxynitrite [reaction (6)]. Correspondingly, major roles for this powerful oxidant have been proposed both in diseases and tissue damage associated with oxidative stress, and in natural cellular defense mechanisms against microbial infection. *See* BIOINORGANIC CHEMISTRY.

[J.K.Hu.]

**Petrochemical** Any of the chemicals derived from petroleum or natural gas. The definition of petrochemicals has been broadened to include the whole range of aliphatic, aromatic, and naphthenic organic chemicals, as well as carbon black and such inorganic materials as sulfur and ammonia.

Petrochemicals are made or recovered from the entire range of petroleum fractions, but the bulk of petrochemical products are formed from the lighter ($C_1$–$C_4$) hydrocarbon gases as raw materials. These materials generally occur in natural gas, but they are also recovered from the gas streams produced during refinery operations, especially cracking. Refinery gases are particularly valuable because they contain substantial amounts of olefins that, because of their double bonds, are much more reactive then the saturated (paraffin) hydrocarbons. Also important as raw materials are the aromatic hydrocarbons (benzene, toluene, and xylene) that are obtained from various refinery product streams. For example, catalytic reforming processes convert nonaromatic hydrocarbons to aromatic hydrocarbons by dehydrogenation and cyclization. *See* PETROLEUM.

Thermal cracking processes (such as coking) are focused primarily on increasing the quantity and quality of gasoline and other liquid fuels, but also produce gases, including lower-molecular-weight olefins such as ethylene ($CH_2{=}CH_2$), propylene ($CH_3CH{=}CH_2$), and butylenes (butenes, $CH_3CH{=}CHCH_3$ and $CH_3CH_2CH{=}CH_2$).

Catalytic cracking is a valuable source of propylene and butylene, but it is not a major source of ethylene, the most important of the petrochemical building blocks. *See* CRACKING; ETHYLENE.

The starting materials for the petrochemical industry are obtained from crude petroleum in one of two ways. They may be present in the raw crude oil and are isolated by physical methods, such as distillation or solvent extraction; or they are synthesized during the refining operations. Unsaturated (olefin) hydrocarbons, which are not usually present in natural petroleum, are nearly always manufactured as intermediates during the refining sequences. *See* DISTILLATION; PETROLEUM PROCESSING AND REFINING; SOLVENT EXTRACTION.

The main objective in producing chemicals from petroleum is the formation of a variety of well-defined chemical compounds, including (1) chemicals from aliphatic compounds; (2) chemicals from olefins; (3) chemicals from aromatic compounds; (4) chemicals from natural gas; (5) chemicals from synthesis gas (carbon monoxide and hydrogen); and (6) inorganic petrochemicals.

A significant proportion of the basic petrochemicals are converted into plastics, synthetic rubbers, and synthetic fibers. These materials, known as polymers, are high-molecular-weight compounds made up of repeated structural units. The major polymer products are polyethylene, polyvinyl chloride, and polystyrene, all derived from ethylene, and polypropylene, derived from propylene. Major raw material sources for synthetic rubbers include butadiene, ethylene, benzene, and propylene. Among synthetic fibers the polyesters, which are a combination of ethylene glycol and terephthalic acid (made from xylene), are the most widely used. They account for about one-half of all synthetic fibers. The second major synthetic fiber is nylon, its most important raw material being benzene. Acrylic fibers, in which the major raw material is the propylene derivative acrylonitrile, make up most of the remainder of the synthetic fibers. *See* POLYAMIDE RESINS; POLYESTER RESINS; POLYMER; POLYMERIZATION; POLYOLEFIN RESINS; POLYURETHANE RESINS; POLYVINYL RESINS; RUBBER.

An inorganic petrochemical is one that does not contain carbon atoms; typical examples are sulfur (S), ammonium sulfate [$(NH_4)_2SO_4$], ammonium nitrate ($NH_4NO_3$), and nitric acid ($HNO_3$). Of the inorganic petrochemicals, ammonia is by far the most common. Ammonia is produced by the direct reaction of hydrogen with nitrogen, with air being the source of nitrogen. Refinery gases, steam reforming of natural gas (methane) and naphtha streams, and partial oxidation of hydrocarbons or higher-molecular-weight refinery residual materials (residua, asphalt) are the sources of hydrogen. The ammonia is used predominantly for the production of ammonium nitrate ($NH_4NO_3$) as well as other ammonium salts and urea ($H_2HCONH_2$) that are major constituents of fertilizers. *See* AMMONIA; AMMONIUM SALT; UREA. [J.G.S.]

**Petroleum processing and refining** The separation of petroleum into fractions and the treating of these fractions to yield marketable products. Petroleum is a mixture of gaseous, liquid, and solid hydrocarbon compounds that occurs in sedimentary rock deposits throughout the world. In the crude state, petroleum has little value but, when refined, it provides liquid fuels (gasoline, diesel fuel, aviation fuel), solvents, heating oil, lubricants, and the distillation residuum asphalt, which is used for highway surfaces and roofing materials. *See* PETROLEUM.

Crude petroleum (oil) is a mixture of compounds with different boiling temperatures that can be separated into a variety of fractions (see table). Since there is a wide variation in the composition of crude petroleum, the proportions in which the different fractions occur vary with origin. Some crude oils have higher proportions of

**Petroleum fractions and their uses***

| Fraction | Boiling range °C | Boiling range °F | Uses |
|---|---|---|---|
| Fuel gas | −160 to −40 | −260 to −40 | Refinery fuel |
| Propane | −40 | −40 | Liquefied petroleum gas (LPG) |
| Butane(s) | −12 to −1 | 11–30 | Increases volatility of gasoline, advantageous in cold climates |
| Light naphtha | −1 to 150 | 30–300 | Gasoline components, may be (with heavy naphtha) reformer feedstock |
| Heavy naphtha | 150–205 | 300–400 | Reformer feedstock, with light gas oil, jet fuels |
| Gasoline | −1 to 180 | 30–355 | Motor fuel |
| Kerosine | 205–260 | 400–500 | Fuel oil |
| Stove oil | 205–290 | 400–550 | Fuel oil |
| Light gas oil | 260–315 | 500–600 | Furnace and diesel fuel components |
| Heavy gas oil | 315–425 | 600–800 | Feedstock for catalytic cracker |
| Lubricating oil | >400 | >750 | Lubrication |
| Vacuum gas oil | 425–600 | 800–1100 | Feedstock for catalytic cracker |
| Residuum | >600 | >1100 | Heavy fuel oil, asphalts |

*From J. G. Speight (ed.), *The Chemistry and Technology of Petroleum*, 3d ed., Marcel Dekker, New York, 1999.

lower-boiling components, while others have higher proportions of residuum (asphaltic components).

Petroleum processing and refining involves a series of steps by which the original crude oil is converted into products with desired qualities in the amounts dictated by the market. In fact, a refinery is essentially a group of manufacturing plants that vary in number with the variety of products in the mix. Refinery processes must be selected and products manufactured to give a balanced operation; that is, crude oil must be converted into products according to the demand for each. For example, the manufacture of products from the lower-boiling portion of petroleum automatically produces a certain amount of higher-boiling components. If the latter cannot be sold as, say, heavy fuel oil, these products will accumulate until refinery storage facilities are full. To prevent such a situation, the refinery must be flexible and able to change operations as needed. This usually means more processes, such as thermal processes to change excess heavy fuel oil into gasoline with coke as the residual product, or vacuum distillation processes to separate heavy oil into lubricating oil stocks and asphalt.

**Distillation.** In a petroleum distillation unit, a tower is used for fractionation. The feedstock of crude oil flows through one or more pipes arranged within a large furnace where it is heated to a temperature at which a predetermined portion of the feed changes into vapor. The heated feed is introduced into a fractional distillation tower where the nonvolatiles or liquid portions pass downward to the bottom of the tower and are pumped away, while the vapors pass upward through the tower and are fractionated into gas oils, kerosine, and naphthas.

Vacuum distillation is used in petroleum refining to separate the less volatile products, such as lubricating oils, from petroleum without subjecting the high-boiling products to cracking conditions. Operating pressure for vacuum distillation is usually 50–100 mm of mercury (6.7–13.3 kilopascals) [atmospheric pressure = 760 mm of mercury]. By this means, a heavy gas oil that has a boiling range in excess of 315°C (600°F) at atmospheric pressure may be obtained at temperatures of around 150°C (300°F); and lubricating oil, having a boiling range in excess of 370°C (700°F) at atmospheric pressure may be obtained at temperatures of 250–350°C (480–660°F). Atmospheric and vacuum distillation are major parts of refinery operations, and no doubt will continue to be used as the primary refining operation.

**Thermal processes.** One of the earliest conversion processes used in the petroleum industry was the thermal decomposition of higher-boiling materials into lower-boiling products. This process is known as thermal cracking. The majority of the thermal cracking processes use temperatures of 455–540°C (850–1005°F) and pressures of 100–1000 psi (690–6895 kPa). For example, the feedstock (reduced crude) is preheated by direct exchange with the cracking products in the fractionating columns. Cracked gasoline and heating oil are removed from the upper section of the column. Light and heavy distillate fractions are removed from the lower section and are pumped to separate heaters. Higher temperatures are used to crack the more stable light distillate fraction. The streams from the heaters are combined and sent to a soaking chamber where additional time is provided to complete the cracking reactions. The cracked products are then separated in a low-pressure flash chamber where a heavy fuel oil is removed as bottoms. The remaining cracked products are sent to fractionating columns. The thermal cracking of higher-boiling petroleum fractions to produce gasoline is now virtually obsolete. The antiknock requirements of modern automobile engines together with the different nature of crude oils (compared to those of 50 years ago) has reduced the ability of the thermal cracking process to produce gasoline on an economic basis. *See* DISTILLATION COLUMN.

Visbreaking (viscosity breaking) is a mild thermal cracking operation that can be used to reduce the viscosity of residua to allow the products to meet fuel oil specifications. Alternatively, the visbroken residua can be blended with lighter product oils to produce fuel oils of acceptable viscosity. By reducing the viscosity of the residuum, visbreaking reduces the amount of light heating oil that is required for blending to meet fuel oil specifications.

Delayed coking is a thermal process for converting residua into lower-boiling products, such as gases, naphtha, fuel oil, gas oil, and coke. It is a semicontinuous process in which the heated charge is transferred to large soaking (or coking) drums, which provide the long residence time needed to allow the cracking reactions to proceed to completion. The feedstock is introduced into a product fractionator where it is heated and the lighter fractions are removed as a side streams. Gas oil, often the major product of a coking operation, serves primarily as a feedstock for catalytic cracking units. The coke obtained is typically used as fuel; but specialty uses, such as electrode manufacture, and production of chemicals and metallurgical coke are also possible, increasing the value of the coke. For these uses, the coke may require treatment to remove sulfur and metal impurities. *See* COKE; CRACKING; NAPHTHA.

Catalytic cracking is basically the same as thermal cracking, but differs by the use of a catalyst, which directs the course of the cracking reactions to produce more of the desired higher-octane hydrocarbon products. Catalytic cracking is regarded as the modern method for converting high-boiling petroleum fractions, such as gas oil, into gasoline and other low-boiling fractions. The usual commercial process involves contacting a gas oil faction with an active catalyst at a suitable temperature, pressure, and residence time so that a substantial part (>50%) of the gas oil is converted into gasoline and lower-boiling products, usually in a single-pass operation.

**Hydroprocesses.** The use of hydrogen in thermal processes was perhaps the single most significant advance in refining technology during the twentieth century. The process uses the principle that the presence of hydrogen during a thermal reaction of a petroleum feedstock will terminate many of the coke-forming reactions and enhance the yields of the lower-boiling components, such as gasoline, kerosine, and jet fuel. *See* HYDROGENATION.

Destructive hydrogenation (hydrogenolysis or hydrocracking) is characterized by the conversion of the higher-molecular-weight constituents in a feedstock to lower-boiling products. Such treatment requires severe processing conditions and the use of high hydrogen pressures to minimize the polymerization and condensation reactions that lead to coke formation. *See* HYDROCRACKING.

Nondestructive hydrogenation is used for improving product quality without appreciable alteration of the boiling range. Nitrogen, sulfur, and oxygen compounds undergo reaction with the hydrogen, forming ammonia, hydrogen sulfide, and water, respectively. Unstable compounds that might lead to the formation of gums or insoluble materials are converted to more stable compounds. [J.G.S.]

**pH** An expression for the effective concentration of hydrogen ions in solution. The activity of hydrogen ions or, more correctly, hydronium ions, which are hydrated hydrogen ions $H(H_2O)_n^+$, affects the equilibria and kinetics of a wide variety of chemical and biochemical reactions. Because these effects are activity-dependent, it is extremely important to distinguish between the hydrogen-ion concentration and activity. The concentration, or total acidity, is obtained by titration and corresponds to the total concentration of hydrogen ions available in a solution, that is, free, unbound hydrogen ions as well as hydrogen ions associated with weak acids. The hydrogen-ion activity refers to the effective concentration of unassociated hydrogen ions, the form that directly affects physicochemical reaction rates and equilibria. This activity is therefore of fundamental importance in many areas of science and technology. The relationship between hydrogen-ion activity ($a_{H^+}$) and concentration ($C$) is given by Eq. (1), where

$$a_{H^+} = \gamma C \qquad (1)$$

the activity coefficient $\gamma$ is a function of the total ionic strength (concentration) of the solution and approaches unity as the ionic strength approaches zero; that is, the difference between the activity and the concentration of hydrogen ion diminishes as the solution becomes more dilute. *See* ACTIVITY (THERMODYNAMICS); CHEMICAL EQUILIBRIUM; HYDROGEN ION.

The effective concentration of hydrogen ions in solution is expressed in terms of pH, which is the negative logarithm of the hydrogen-ion activity [Eq. (2)]. Because

$$\text{pH} = -\log_{10} a_{H^+} \qquad (2)$$

of the negative logarithmic (exponential) relationship, the more acidic a solution, the smaller the pH value. The pH of a solution may have little relationship to the titratable acidity of a solution that contains weak acids or buffering substances; the pH of a solution indicates only the free hydrogen-ion activity. If total acid concentration is to be determined, an acid-base titration must be performed. *See* ACID AND BASE; BUFFERS (CHEMISTRY); TITRATION.

Two methods, electrometric and chemical indicator (optical), are used for measuring pH. The more commonly used electrometric method is based on measurement of the difference between the pH of a test solution and that of a standard solution. The pH scale is defined by a series of reference buffer solutions that are used to calibrate the pH measurement system. The instrument measures the potential difference developed between the pH electrode and a reference electrode of constant potential. The difference in potential obtained when the electrode pair is removed from the standard solution and placed in the test solution is converted to the pH value. In the indicator method, the pH value is obtained by simple visual comparison of the color of pH-sensitive dyes to standards (for example, color charts) or by use of calibrated optical readout devices (photometers), often in combination with fiber-optic sensors. *See* ELECTRODE; REFERENCE ELECTRODE. [R.A.D.]

**Pharmaceutical chemistry** The chemistry of drugs and of medicinal and pharmaceutical products. The important aspects of pharmaceutical chemistry are as follows:

1. Isolation, purification, and characterization of medicinally active agents and materials from natural sources used in treatment of disease and in compounding prescriptions.
2. Synthesis of medicinal agents not known from natural sources, or the synthetic duplication, for reasons of economy, purity, or adequate supply, of substances first known from natural sources.
3. Semisynthesis of drugs, whereby natural substances are transformed by means of comparatively simple steps into products which possess more favorable therapeutic or pharmaceutical properties.
4. Determination of the derivative or form of a medicinal agent which exhibits optimum medicinal activity and at the same time lends itself to stable formulation and elegant dispensing.
5. Determination of incompatibilities, chemical and biological, between the various ingredients of a prescription.
6. Establishment of safe and practical standards, with respect to both dosage and quality, to assure uniform and therapeutically reliable forms for all medication.
7. Improvement and promotion of the use of chemical agents for prevention of illness, alleviation of pain, cure of disease, and search for new therapeutic agents, particularly where no satisfactory remedy now exists.

[W.H.H.]

**Phase equilibrium** A general field of physical chemistry dealing with the various situations in which two or more phases (or states of aggregation) can coexist in thermodynamic equilibrium with each other, with the nature of the transitions between phases, and with the effects of temperature and pressure upon these equilibria. Many superficial aspects of the subject are largely qualitative, for example, the empirical classification of types of phase diagrams; but the basic problems always are susceptible to quantitative thermodynamic treatment, and in many cases, statistical thermodynamic methods can be applied to simple molecular models.

Thermodynamics requires that when two phases, $\alpha$ and $\beta$, are free to exchange heat, mechanical work, and matter (chemical species), the temperature $T$, the pressure $P$, and the chemical potential (partial molar free energy) $\mu_i$ of each particular component $i$ must be equal in both phases at equilibrium. Algebraically, equilibrium exists when $T_\alpha = T_\beta$, $P_\alpha = P_\beta$, $\mu_{i,\alpha} = \mu_{i,\beta}$, and $\mu_{j,\alpha} = \mu_{j,\beta}$.

These conditions of thermal, mechanical, and material equilibrium need not all be present if the equilibrium between phases is subject to inhibiting restrictions. Thus, for a solution of a nonvolatile solute in equilibrium with the solvent vapor, the condition of equality of solute chemical potentials $\mu_{2,\alpha} = \mu_{2,\beta}$ need not apply, since there can be no solute molecules in the vapor phase. Similarly, in osmotic equilibria, in which solvent molecules can pass through a semipermeable membrane, whereas solute molecules cannot, $\mu_{1,\alpha} = \mu_{1,\beta}$ and $T_{1,\alpha} = T_{2,\beta}$, but the solute chemical potentials $\mu_2$ are unequal, as are the pressures on opposite sides of the membrane. *See* OSMOSIS; SOLUTION.

If a system consists of $P$ phases and $C$ distinguishable components, there are $C + 2$ thermodynamic variables ($C$ chemical potentials $\mu_i$, plus the temperature and pressure) which are interrelated by an equation for each phase. Since there are $P$ independent equations relating the $C + 2$ variables, only $F = C + 2 - P$ variables need be fixed to define completely the state of the system at equilibrium; the other variables are then beyond control. This relation for the number of degrees of freedom $F$, or variance, is called the phase rule. It has proved to be a powerful tool in interpreting and classifying types of phase equilibria.

When chemical changes may occur in the system, the number of components $C$ is the number of independent components whose amounts can be varied by the experimenter; this is equal to the total number of chemical species present less the number of independent chemical equilibria between them.

An invariant system has no degrees of freedom ($F = 0$), for which the number of phases $P = C + 2$. For a one-component system, such an invariant point is a triple point at which three phases coexist at a single temperature and pressure only; for a two-component system, a quadruple point (four phases) would be invariant. *See* TRIPLE POINT. [R.L.S.]

**Phase rule** A relationship used to determine the number of state variables $F$, usually chosen from among temperature, pressure, and species compositions in each phase, which must be specified to fix the thermodynamic state of a system in equilibrium. It was derived by J. Willard Gibbs. The phase rule (in the absence of electric, magnetic, and gravitational phenomena) is given by the equation below, where $C$ is the

$$F = C - P - M + 2$$

number of chemical species present at equilibrium, $P$ is the number of phases, and $M$ is the number of independent chemical reactions. Here phase is used to indicate a homogeneous, mechanically separable portion of the system, and the term independent reactions refers to the smallest number of chemical reactions which, upon forming various linear combinations, includes all reactions which occur among the species present. The number of independent state variables $F$ is referred to as the degrees of freedom or variance of the system. *See* CHEMICAL EQUILIBRIUM; CHEMICAL THERMODYNAMICS; PHASE EQUILIBRIUM; THERMODYNAMIC PROCESSES. [S.I.S.]

**Phase-transfer catalysis** A process in which the rate of a reaction occurring in a two-phase organic-water system is enhanced by addition of a compound that helps transfer the water soluble reactant across the interface to the organic phase.

An important factor, which contributes to the slowness of many organic reactions, is the lack of homogeneity of the reaction mixture. This is particularly the case with nucleophilic substitution reactions such as the reaction below, where RX is an organic

$$RX + Nu^- \rightarrow RNu + X^-$$

reagent and $Nu^-$ is the nucleophilic reagent. The nucleophilic reagent is frequently an inorganic anion, which is soluble in water in which the organic substrate is insoluble, but is insoluble in the organic phase. The encounter rate between $Nu^-$ and RX is consequently low, as they can only meet at the interface of the heterogeneous system. The water-soluble anion is also frequently highly solvated by water molecules, which stabilize the anion and thus reduce its nucleophilic reactivity. These problems have been overcome in the past by the use of polar aprotic solvents, which will dissolve both the organic and inorganic reagents, or by the use of homogeneous mixed-solvent systems, such as water:ethanol or water:dioxan. *See* ELECTROPHILIC AND NUCLEOPHILIC REAGENTS.

Phase-transfer catalysis involves the transportation of the inorganic anion, $Nu^-$, from the aqueous phase into the organic phase by the formation of a nonsolvated ion-pair with a cationic phase-transfer catalyst, $Q^+$. With highly lipophilic catalysts, the reactive ion pair $[Q^+Nu^-]$ is formed at the interface between the aqueous and organic phases, followed by rapid transportation into the bulk of the organic phase. The rate of the reaction is enhanced, as the encounter rate of the nucleophile, $Nu^-$, with the organic reagent, RX, in the single phase will be significantly higher than at the interface, Moreover, as the anion is transferred without water of solvation, its nucleophilic reactivity can be considerably higher in the organic phase than in the aqueous phase. Rate enhancements of greater than $10^7$ have thus been observed. *See* CATALYSIS; HETEROGENEOUS CATALYSIS; HOMOGENEOUS CATALYSIS; QUATERNARY AMMONIUM SALTS; STEREOCHEMISTRY.

[R.A.J.]

**Phase transitions** Changes of state brought about by a change in an intensive variable (for example, temperature or pressure) of a system. Some familiar examples of phase transitions are the gas-liquid transition (condensation), the liquid-solid transition (freezing), the normal-to-superconducting transition in electrical conductors, the paramagnet-to-ferromagnet transition in magnetic materials, and the superfluid transition in liquid helium. Further examples include transitions involving amorphous or glassy structures, spin glasses, charge-density waves, and spin-density waves.

Typically the phase transition is brought about by a change in the temperature of the system. The temperature at which the change of state occurs is called transition temperature (usually denoted by $T_c$). For example, the liquid-solid transition occurs at the freezing point.

The two phases above and below the phase transition can be distinguished from each other in terms of some ordering that takes place in the phase below the transition temperature. For example, in the liquid-solid transition, the molecules of the liquid get "ordered" in space when they form the solid phase. In a paramagnet, the magnetic moments on the individual atoms can point in any direction (in the absence of an internal magnetic field), but in the ferromagnetic phase the moments are lined up along a particular direction, which is then the direction of ordering. Thus in the phase above the transition, the degree of ordering is smaller than in the phase below the transition. One measure of the amount of disorder in a system is its entropy, which is the negative of the first derivative of the thermodynamic free energy with respect to temperature. When a system possesses more order, the entropy is lower. Thus at the transition temperature the entropy of the system changes from a higher value above the transition to some lower value below the transition. *See* ENTROPY.

This change in entropy can be continuous or discontinuous at the transition temperature. In other words, the development of order in the system at the transition temperature can be gradual or abrupt. This leads to a convenient classification of phase transitions into two types, namely, discontinuous and continuous.

Discontinuous transitions involve a discontinuous change in the entropy at the transition temperature. A familiar example of this type of transition is the freezing of water into ice. As water reaches the freezing point, order develops without any change in temperature. Thus there is a discontinuous decrease in the entropy at the freezing point. This is characterized by the amount of latent heat that must be extracted from the water for it to be "ordered" into the solid phase (ice). Discontinuous transitions are also called first-order transitions.

In a continuous transition, entropy changes continuously, and hence the growth of order below $T_c$ is also continuous. There is no latent heat involved in a continuous transition. Continuous transitions are also called second-order transitions. The paramagnet-to-ferromagnet transition in magnetic materials is an example of such a transition.

The degree of ordering in a system undergoing a phase transition can be made quantitative in terms of an order parameter. At temperatures above the transition temperature the order parameter has a value zero, and below the transition it acquires some nonzero value. For example, in a ferromagnet the order parameter is the magnetic moment per unit volume (in the absence of an externally applied magnetic field). It is zero in the paramagnetic state since the individual magnetic moments in the solid may point in any random direction. Below the transition temperature, however, there exists a preferred direction of ordering, and as the temperature is lowered below $T_c$, more and more individual magnetic moments start to align along the preferred direction of ordering, leading to a continuous growth of the magnetization or the macroscopic magnetic moment per unit volume in the ferromagnetic state. Thus the order parameter changes continuously from zero above to some nonzero value below the transition temperature. In a first-order transition, the order parameter would change discontinuously at the transition temperature. [D.J.S.; S.J.]

**Phenol** The simplest member of a class of organic compounds possessing a hydroxyl group attached to a benzene ring or to a more complex aromatic ring system. Phenol itself, $C_6H_5OH$, may also be called hydroxybenzene or carbolic acid. Pure phenol is a colorless solid melting at 42°C (108°F), moderately soluble in water, and weakly acidic (p*K* 9.9).

Phenol has broad biocidal properties, and dilute aqueous solutions have long been used as an antiseptic. At higher concentrations phenol causes severe skin burns; it is a violent systemic poison.

Phenol has the structure shown. Simple substituted phenols, such as the three isomeric chlorophenols, are named as indicated, using the ortho (*o*), meta (*m*), and para (*p*) prefixes. In more highly substituted phenols the positions of substitution are indicated by numbers (as in 2,4-dichlorophenol). Compounds with more than one hydroxyl group per aromatic ring are known as polyhydric phenols, and include catechol, resorcinol, hydroquinone, phloroglucinol, and pyrogallol.

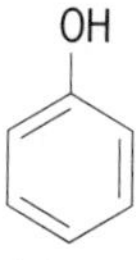

OH OH Cl OH Cl

Phenol 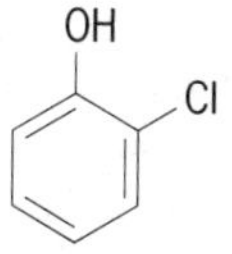 *o*-Chlorophenol 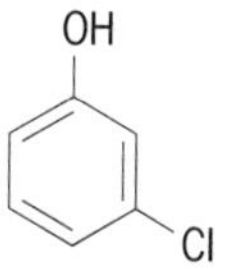 *m*-Chlorophenol

OH Cl OH Cl Cl

*p*-Chlorophenol 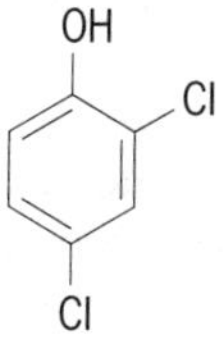 2,4-Dichlorophenol

Until World War I phenol was essentially a natural coal tar product. However, synthetic methods have replaced extraction from natural sources. There are many possible syntheses.

Phenol is one of the most versatile and important industrial organic chemicals. It is the starting point for many diverse products used in the home and industry. A partial list includes: nylon, epoxy resins, surface active agents, synthetic detergents, plasticizers, antioxidants, lube oil additives, phenolic resins (with formaldehyde, furfural, and so on), polyurethanes, aspirin, dyes, wood preservatives, herbicides, drugs, fungicides, gasoline additives, inhibitors, explosives, and pesticides. *See* PHENOLIC RESIN.
[R.I.S.; M.St.]

**Phenolic resin** One of the condensation products of phenols or phenolic derivatives with aldehydes such as formaldehyde and furfural. The phenol-formaldehyde resins, developed commercially between 1905 and 1910, were the first truly synthetic polymers and have found wide usage. They are characterized by low cost, dimensional stability, high strength, and resistance to aging.

Phenolic resins can be cast from syrupy intermediates or molded from B-stage solid resins. Laminated products can be produced by impregnating fiber, cloth, wood, and other materials with the resin. An important type of phenolic resin product is rigid foam. Cured phenolic plastics are rigid, hard, and resistant to chemicals (except strong alkali) and to heat.

Some of the uses for phenolic resins are for making precisely molded articles, such as telephone parts, for manufacturing strong and durable laminated boards, or for impregnating fabrics, wood, or paper. Phenolic resins are also widely used as adhesives, as the binder for grinding wheels, as thermal insulation panels, as ion-exchange resins, and in paints and varnishes. *See* ADHESIVE; ION EXCHANGE; PHENOL; PLASTICS PROCESSING; POLYMERIZATION.
[J.A.M.]

**Phosphorus** A chemical element, P, atomic number 15, atomic weight 30.9738. Phosphorus forms the basis of a very large number of compounds, the most important class of which are the phosphates. For every form of life, phosphates play an essential role in all energy-transfer processes such as metabolism, photosynthesis, nerve function, and muscle action. The nucleic acids which among other things make up the hereditary material (the chromosomes) are phosphates, as are a number of coenzymes. Animal skeletons consist of a calcium phosphate. *See* PERIODIC TABLE.

About three-quarters of the total phosphorus (in all of its chemical forms) used in the United States goes into fertilizers. Other important uses are as builders for detergents, nutrient supplements for animal feeds, water softeners, additives for foods and pharmaceuticals, coating agents for metal-surface treatment, additives in metallurgy, plasticizers, insecticides, and additives for petroleum products.

Of the nearly 200 different phosphate minerals, only one, fluorapatite, $Ca_5F(PO_4)_3$, is mined chiefly from large secondary deposits originating from the bones of dead creatures deposited on the bottom of prehistoric seas and from bird droppings on ancient rookeries.

Research in phosphorus chemistry indicates that there may be as many compounds based on phosphorus as on carbon. In organic chemistry it has been customary to group the various chemical compounds based on carbon into families which are called homologous series. This can also be done in the chemistry of phosphorus compounds, even though many phosphorus-based families are incomplete. The best known of the families of compounds based on phosphorus is the group of chain phosphates.

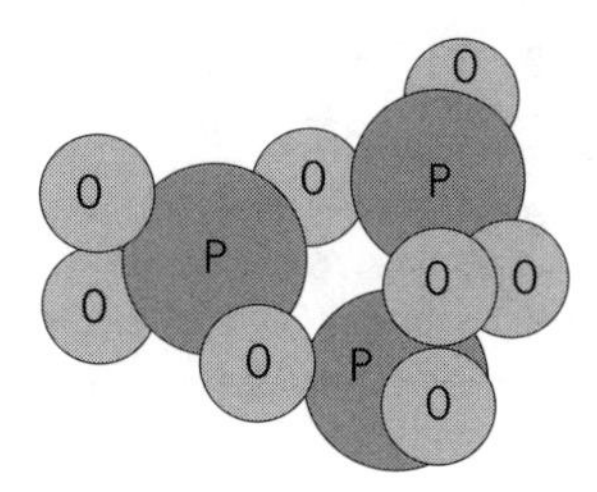


**Fig. 1. Ring phosphate anion, $(P_3O_9)^{3-}$.**

Phosphate salts consist of cations, such as sodium, along with chain anions, such as $(P_nO_{3n+1})^{(n-2)-}$, which may have 1–1,000,000 phosphorus atoms per anion.

The phosphates are based on phosphorus atoms tetrahedrally surrounded by oxygen atoms, with the lowest member of the series being the simple $PO_4^{3-}$ anion (the orthophosphate ion). The family of chain phosphates is based on a row of alternating phosphorus and oxygen atoms in which each phosphorus atom remains in the center of a tetrahedron of four oxygen atoms. There is also a closely related family of ring phosphates, a member of which, the trimetaphosphate, is shown in Fig. 1.

An interesting structural characteristic of many known phosphorus compounds is the formation of cagelike structures. Such cagelike molecules are exemplified by white phosphorus, $P_4$, and one of the phosphorus pentoxides, $P_4O_{10}$ (Fig. 2). Network structures are also common; for example, black phosphorus crystals in which the atoms are bonded together in the form of vast, corrugated planes (Fig. 3).

In the majority of its compounds, phosphorus is chemically bonded to four neighboring atoms. There is a large number of compounds in which one of the four neighboring atoms is absent, and in which its place is taken by an unshared pair of electrons. There are also a few compounds in which there are five or six neighboring atoms bonded to the phosphorus. These compounds are very reactive and tend to be unstable.

During the 1960s and 1970s a large number of organic-phosphorus compounds were prepared. Most of these chemical structures involve three or four neighboring atoms bonded to the phosphorus, but stable structures having two, five, or six neighboring atoms per phosphorus are also known. *See* ORGANOPHOSPHORUS COMPOUND.

Essentially all of the phosphorus used in commerce is in the form of phosphates. The majority of phosphatic fertilizers consist of highly impure monocalcium or dicalcium orthophosphate, $Ca(H_2PO_4)_2$ and $CaHPO_4$. These phosphates are salts of orthophosphoric acid.

The phosphorus compound of major biological importance is adenosine triphosphate (ATP), which is an ester of sodium tripolyphosphate, widely employed in

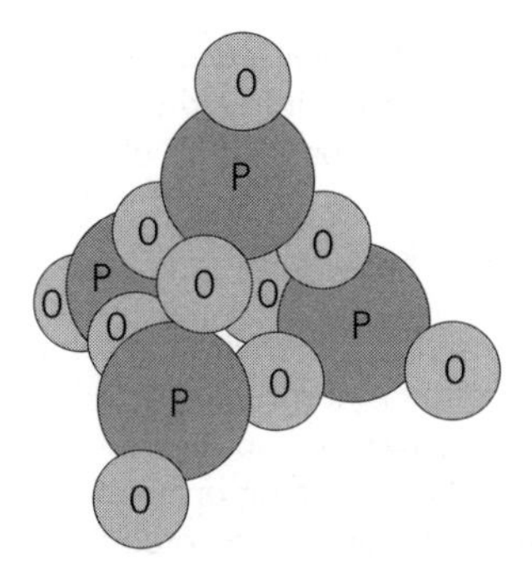


**Fig. 2. Phosphorus pentoxide, $P_4O_{10}$, in vapor state.**

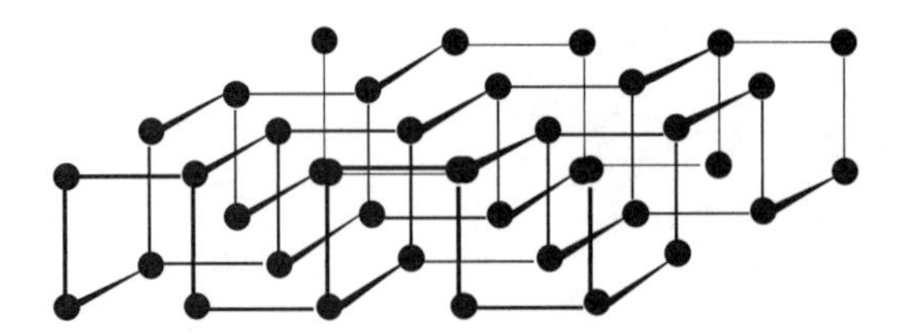

Fig. 3. Black phosphorus, $P_n$.

detergents and water-softening compounds. Practically every reaction in metabolism and photosynthesis involves the hydrolysis of this tripolyphosphate to its pyrophosphate derivative, called adenosine diphosphate (ADP). *See* ADENOSINE TRIPHOSPHATE (ATP). [J.R.V.W.]

**Photoaffinity labeling** A process by which a macromolecule can be labeled at or near its binding or active site. The method can be applied to proteins, nucleic acids, and lipids by chemists who need to identify the parts of these molecules that are significant for particular biological functions. X-ray crystallography can frequently determine the complete structure of proteins and sometimes of nucleic acids, although the method does not necessarily highlight the reactive groups in a biomolecule. Nuclear magnetic resonance spectroscopy is increasingly useful in illuminating such structures.

Many important biological processes involve the formation of a complex between a biopolymer and a chemical reagent; for example, enzymes form complexes with their substrates on the way to catalysis, and antibodies form tight complexes with their antigens. Many biochemical receptors (such as hormone receptors) are complexed with lipid membranes. These complexes cannot usually be crystallized, but information about them can often be obtained by photoaffinity labeling. The idea behind photoaffinity labeling is to place in the binding site a compound that is essentially inert but can be photoactivated at will to yield a highly reactive intermediate. Preferably, this intermediate will react with almost any chemical structure.

The successful photoaffinity labeling reagents include diazirines that yield carbenes on photolysis, arylazides that are photochemically decomposed to nitrenes (highly reactive compounds that contain univalent nitrogen), and ketones that yield free radicals on irradiation. Special structural features (such as trifluromethyl substituents) can enhance the activity of the reagents.

In order to identify the products of reactions, many of which occur only in low yield, the reagent must be made radioactive. This is usually accomplished with carbon-14 ($^{14}C$) or tritium ($^{3}H$), radioactive isotopes of relatively low energy that nevertheless can easily be traced. [F.H.W.]

**Photochemistry** The study of chemical reactions of molecules in electronically excited states produced by the absorption of infrared (700–1000 nanometers), visible (400–700 nm), ultraviolet (200–400 nm), or vacuum ultraviolet (100–200 nm) light. Bond making and bond breaking as well as electron transfer and ionization are often observed in both organic and inorganic compounds as a consequence of such excitation.

**Electronic absorption.** An important generalization sometimes called the first law of photochemistry is that only light that is absorbed can induce chemical change. The absorption of a photon induces an electronic transition in which an electron originally present in a molecular orbital, usually a bonding or nonbonding molecular orbital of the ground state of the absorbing molecule, is promoted to a higher-lying orbital. The excited state produced by absorption of light has a different electronic structure than its ground-state precursor and can reasonably be regarded as an isomeric species with distinct and characteristic chemical and physical properties.

Most organic molecules exist as ground-state singlets in which all electrons are paired. Because photoexcitation causes the promotion of only a single electron, two singly occupied orbitals are produced upon excitation. If the electronic transition takes place without a spin inversion, these two electrons have opposite spins, and a singlet excited state is produced. The number of unpaired electrons in a molecule determines its multiplicity: a molecule with no unpaired spins is a singlet; one with one unpaired spin is a doublet; one with two unpaired spins is a triplet; one with three unpaired spins is a quartet, and so forth. If an electronic transition were to take place with a spin inversion, the two singly occupied orbitals would be populated by electrons with parallel spins, producing a triplet excited state. Spin restrictions forbid spin inversion during excitation, and only singlet-singlet electronic transitions are easily observed spectroscopically. After excitation, however, a change in state multiplicity can take place by a process called intersystem crossing. The facility of intersystem crossing is influenced by the magnitude of spin-orbital coupling, which can be enhanced by the presence of a heavy atom (an atom in the third row or below of the periodic table), either bound to the absorbing molecule or present externally as solvent. *See* PERIODIC TABLE; TRIPLET STATE.

**Transitions.** A chromophore is that part of the molecule that accounts for its absorption of light and its photochemical activity. The absorption corresponding to a particular chromophore depends on the type of transition involved in that particular excitation. The promotion of an electron from a $\pi$-bonding molecular orbital to a $\pi$-antibonding orbital is referred to as a $\pi,\pi^*$ (read pi to pi star) transition. Such transitions are frequently encountered in alkenes, alkynes, aromatic molecules, and other unsaturated compounds. Because the spatial overlap of $\pi$ and $\pi^*$ orbitals is substantial, such a transition typically has high oscillator strength and a large extinction coefficient (absorptivity). Promotion of an electron from a nonbonding molecular orbital to a $\pi$-antibonding orbital, referred to as an $n,\pi^*$ transition, involves orbitals that are nearly orthogonal; and it takes place only inefficiently; that is, it has a low oscillator strength and a small extinction coefficient. Such transitions are often encountered in compounds containing carbon-heteroatom or heteroatom-heteroatom double bonds. Because nonbonding molecular orbitals lie at higher energy than bonding ones, $n,\pi^*$ transitions are of lower energy than the corresponding $\pi,\pi^*$ transitions. Both $n,\pi^*$ and $\pi,\pi^*$ transitions are usually found in the ultraviolet region of the electromagnetic spectrum. Transitions involving sigma ($\sigma$) bonds (for example $n\sigma^*$ transitions in amines, alcohols, ethers, and alkyl halides and $\sigma,\sigma^*$ transitions in alkanes) are usually encountered at the high-energy end of the ultraviolet spectrum or in the vacuum ultraviolet region. *See* CHEMICAL BONDING.

Each allowed transition of a compound registers as a band in the absorption spectrum, with the intensity of the transition (measured by its extinction coefficient) being governed by the operative selection rules. The transition intensity of a given absorption is measured by integrating over the whole absorption band. The resulting integrated absorption coefficient is directly proportional to the oscillator strength of the transition. The oscillator strength, a measure of the allowedness of an electric dipole transition compared to that of a free electron oscillating in the three dimensions, is directly related to an experimentally measured value, the extinction coefficient ($\varepsilon$). Beer's law is given by Eq. (1), where $A$ is the observed absorbance, $\varepsilon$ is the extinction coefficient, $b$ is the

$$A = \varepsilon bc \qquad (1)$$

path length (in centimeters) of the cell used for the measurement, and $c$ is the molar concentration of the absorbing species. This law is used to correlate the observed absorbance with the extinction coefficient and concentration of the absorbing species.

**Photophysics.** The excited state produced by absorption of a photon is not generally a stable species. After a characteristic lifetime that can vary from femtoseconds ($10^{-15}$ s) to hours, the excited molecule will either relax to its ground-state precursor or undergo a chemical transformation. The term photophysics is used to describe nonreactive relaxation processes, which include radiative (taking place with the emission of light) and nonradiative (taking place without the emission of light) pathways.

The energies of the lowest singlet and triplet excited states (relative to the ground state) can be obtained from the longest wavelength band of the fluorescence and phosphorescence spectra, respectively. This band is called a 0,0 band to indicate a transition between the lowest vibrational levels of the lowest-lying states. Singlet and triplet energies can also be determined indirectly by measuring quenching efficiencies. The shift between the 0,0 bands for absorption and emission in a single molecule is called its Stokes shift. A small Stokes shift is usually observed when the excited state has a geometry similar to the ground state. A Jablonski diagram (see illustration) is often used to graphically depict the relationship between competing photophysical processes.

Quantum yield, or quantum efficiency, is defined as the number of molecules participating in a given photophysical process or reaction divided by the number of photons absorbed. The quantum yield ranges between zero and one for photoreactions induced by a single photon; values larger than one are indicative of a chain process in which product is formed in a repeating, dark cycle initiated by the photoexcitation. For a photochemical reaction, the number of molecules participating in the reaction is determined spectroscopically or chromatographically as a chemical yield per volume unit per time. The number of photons absorbed is obtained by measuring with a radiometer the light flux per volume unit per time or by employing a chemical actinometer, a known chemical reaction for which the quantum yield is known and accepted as a standard. *See* QUANTUM CHEMISTRY.

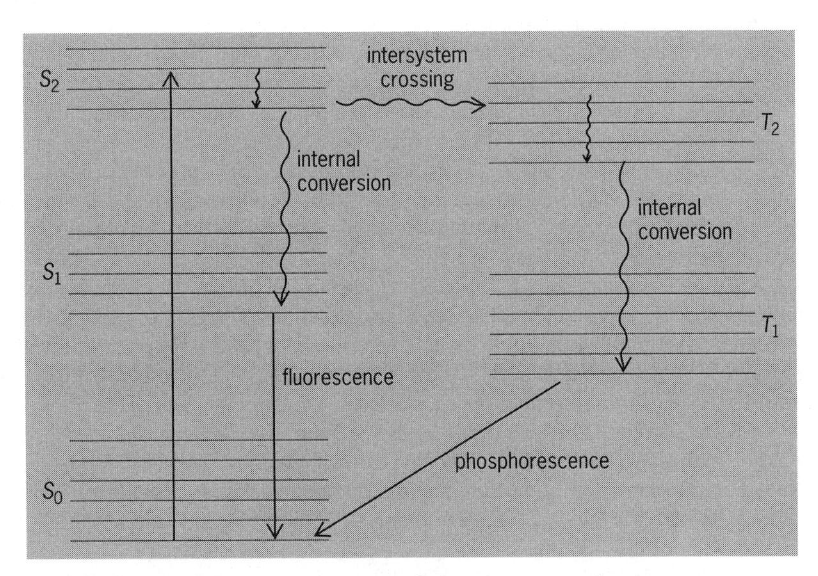


**Jablonski diagram. Solid arrows represent radiative processes; and wavy arrows nonradiative processes. S terms = singlet states; T terms = triplet states.**

**Energy transfer.** The process by which an excited state molecule, M*, in an excited singlet or triplet state transfers all or part of its excitation energy to a reaction partner or quencher, Q, is called energy transfer or quenching when the molecule of interest is M [reaction (2)]. This same process is called sensitization when the molecule of interest

$$M^* + Q \rightarrow M + Q^* \qquad (2)$$

is Q. In the latter case, M is called the sensitizer. Energy transfer permits an exception to the first law of photochemistry in that Q* is produced without having absorbed the incident light.

For energy transfer to take place, an incident wavelength must be chosen so that M is primarily excited, producing an excited state M* whose energy lies above that of Q*. Symmetry selection rules require that all energy transfer events preserve spin multiplicity. Thus, if M* is an excited singlet and Q is a ground-state singlet, M will be produced as a ground-state singlet and Q* as an excited singlet. If M* is an excited triplet and Q is a ground-state singlet, M will be produced as a ground-state singlet and Q* as an excited triplet.

**Photochemical mechanisms.** As in all studies of mechanisms of chemical reactions, determining the structure of all products is the first step in the specification of a photochemical reaction. Spectroscopic (nuclear magnetic resonance spectroscopy, electron spin resonance spectroscopy, infrared spectroscopy, mass spectroscopy, x-ray analysis, absorption spectroscopy) and chromatographic (gas, liquid, or thin-layer chromatography) techniques are used to establish product structure and to determine product yields. Monitoring the effect of solvent polarity on reaction rate, the retention or loss of optical activity during the reaction, the positions of isotopic labels, and the success of intermediate trapping experiments can distinguish step-wise chemical reactions (those that proceed through one or more intermediates) from concerted reactions (those that proceed without intermediates). In addition to these mechanistic approaches, the identity and lifetime of the reactive excited state (singlet, triplet, and so forth) and the quantum yields for both product formation and for other competing photophysical processes are required for a full photochemical mechanistic characterization. Time-resolved flash photolysis and pulse radiolysis measurements can, in addition, be used sometimes for direct spectroscopic detection of absorptive or emissive intermediates encountered in a photochemical mechanism, as well as for their kinetic characterization. The addition of specific reactive quenchers or traps, conducting a photoreaction in a low-temperature matrix in which diffusion processes are stopped, and sensitization experiments are effective means for assigning the observed transient absorptions or emissions. The energetics of a well-defined photochemical reaction can be obtained by photoacoustic calorimetry measurements. *See* CHROMATOGRAPHY; MATRIX ISOLATION; PHOTOLYSIS; SPECTROSCOPY. [M.A.F.]

**Photodegradation** Reduction in the useful properties of materials because of chemical changes resulting from the absorption of light. The chemical changes can include bond scission (especially of the molecular backbone), color formation, cross-linking, and chemical rearrangements. All organic materials can photodegrade, but the process has greatest practical relevance for polymers where scission of the polymer backbone is particularly important. Photodegradations of polymers in the absence of oxygen (photolysis) or using wavelengths shorter (more energetic) than those at the Earth's surface (<280 nanometers) have been studied extensively, but only the more practical situation of polymers exposed to terrestrial sunlight (or its equivalent) in air is discussed in this article.

Although all organic polymers can be degraded by light, the rate of degradation varies enormously from polymer to polymer, and is also dependent on the incident wavelengths. Light containing ultraviolet (uv; shorter-wavelength) components is much more destructive than visible light, so that polymers exposed indoors, behind window glass (transmitting > 330 nm), will degrade much more slowly than samples exposed outdoors.

For many aromatic polymers, such as polyester and the aramids, in which the polymer itself is the chromophore (light-absorbing group), backbone scission results predominantly from this direct absorption of light energy. For many other polymers, including polyolefins, and polyvinyl chloride where only impurities absorb energy from sunlight, scission of a chemical bond by light to give free radicals is followed by reaction of these highly reactive free radicals with atmospheric oxygen. *See* FREE RADICAL.

Although numerous organic materials will undergo photodegradation, hydrocarbon polymers are particularly vulnerable because their useful properties depend entirely on their high molecular weights, in the tens or hundreds of thousands. Anything that reduces the molecular weight of polymeric systems will alter the characteristics of these systems and limit their service life. In fact, the scission of as few as one carbon-carbon bond in a thousand in a polymer molecule can completely destroy its useful physical properties. This sensitivity is not observed in lower-molecular-weight substances such as liquid hydrocarbons.

A general approach to reducing the rates of photodegradation for all types of polymers is the use of low levels of additives. These additives, known as photostabilizers or uv stabilizers, are effective at fractions of a weight percent. *See* PHOTOCHEMISTRY; POLYMER; STABILIZER (CHEMISTRY). [D.M.W.; D.J.Ca.]

**Photolysis** The chemical decomposition of matter due to absorption of incident light. For example, illumination of microcrystals of silver bromide embedded in gelatin results in formation of metallic silver, and is the basis of the photographic process.

Numerous metal complexes, azides, nitrides, and sulfides, and most organometallic compounds undergo decomposition upon illumination, often with concomitant evolution of a gaseous product. In the presence of water and oxygen, illumination of many semiconductors results in their corrosion; for example, an aqueous suspension of cadmium sulfide (CdS) undergoes rapid decomposition upon irradiation with sunlight. The outcome of these photochemical processes can often be controlled by addition of adventitious materials. For example, illumination of cadmium sulfide in aqueous solution containing hydrogen sulfide ($H_2S$) and colloidal platinum results in evolution of hydrogen gas. Here, the semiconductor photosensitizes (or photocatalyzes) decomposition of hydrogen sulfide into its elements, and the reaction can be used to remove sulfides from industrial waste. *See* COORDINATION COMPLEXES.

Many ketones, for example, acetone, abstract hydrogen atoms from adjacent organic matter under illumination. According to the circumstances, the resultant free radicals may be used to initiate polymerization of a monomer or cause decomposition of a plastic film. Both reactions have important commercial applications, and photoinitiators are commonly used for emulsion paints, inks, polymers, explosives, fillings for teeth and for development of photodegradable plastics. Photolysis of carbonyl compounds, released into the atmosphere by combustion of fossil fuels, is responsible for the onset of photochemical smog. Many other types of photochemical transformation of organic molecules are known, including isomerization of unsaturated bonds, cleavage of carbon-halogen bonds, olefin addition reactions, halogenation of aromatic species, hydroxylation, and oxygenation processes. Indeed, photochemistry is often used to produce novel pharmaceutical products that are difficult to synthesize by conventional methods. *See* FREE RADICAL; PHOTODEGRADATION.

The most important photochemical reaction is green plant photosynthesis. Here, chlorophyll that is present in the leaves absorbs incident sunlight and catalyzes reduction of carbon dioxide to carbohydrate.

Excitation of an organic molecule results in spontaneous generation of the singlet excited state. In most molecules, this highly unstable excited singlet state may undergo an intersystem-crossing process that results in population of the corresponding (less energetic) excited triplet state, in competition to fluorescence. The excited triplet state, because of spin restriction rules, retains a significantly longer lifetime than is found for the corresponding excited singlet state, and may be formed in high yield.

Almost without exception, these triplet states react quantitatively with molecular oxygen ($O_2$) present in the system via a triplet energy-transfer process. The resulting product is singlet molecular oxygen. This species is a potent and promiscuous reactant, and it is responsible for widespread damage to both synthetic and natural environments. Indeed, plants and photosynthetic bacteria contain carotenoids to protect the organism against attack by singlet oxygen. The same species is known to be responsible, at least in part, for photodegradation of paint, plastic, fabric, colored paper, and dyed wool. Secondary reactions follow from attack on a substrate by singlet oxygen, resulting in initiation of chain reactions involving free radicals. However, modern technological processes have evolved in which singlet molecular oxygen is used to destroy unwanted organic matter, such as tumors, viruses, and bacteria, in a controlled and specific manner. In photodynamic therapy a dye is injected into a tumor and selectively illuminated with laser light. The resultant singlet oxygen destroys the tumor. Similar methodology can be used to produce photoactive soap powders, bleaches, bactericides, and pest-control reagents. *See* CHAIN REACTION (CHEMISTRY); TRIPLET STATE. [A.Ha.]

**Physical chemistry** The branch of chemistry that deals with the interpretation of chemical phenomena and properties in terms of the underlying physical processes, and with the development of techniques for their investigation. The term chemical physics is often employed to denote a branch of physical chemistry where the emphasis is on the interpretation and analysis of the physical properties of individual molecules and bulk systems, instead of their reactions. Theoretical chemistry is another major branch, where the emphasis is on the calculation of the properties of molecules and systems, and which used the techniques of quantum mechanics and statistical thermodynamics. It is convenient to regard physical chemistry as dealing with three aspects of matter: its equilibrium properties, structure, and ability to change.

**Equilibrium properties.** The study of matter in a state of equilibrium constitutes the field of chemical thermodynamics. In particular, chemical thermodynamics provides a technique for discussing the response of a system to a change in the external conditions (such as the shift in the boiling and freezing point of either a pure substance or a mixture when the applied pressure is changed, or when the composition of the mixture is modified), and for rationalizing the energy changes that occur in the course of a chemical reaction. The branch of thermodynamics dealing with the latter is called thermochemistry. Chemical thermodynamics also provides a framework for the determination of the maximum amount of work that may be generated by a system undergoing a specified change, and it therefore provides a way of establishing bounds for the efficiencies of a variety of devices, including engines, refrigerators, and electrochemical cells. Thermodynamics is used in chemistry to assess the position of equilibrium of a chemical reaction (that is, how far it will proceed), and to determine what conditions are necessary in order to optimize the yield of a particular product. The branch of chemical thermodynamics dealing with ionic reactions occurring in the presence of electrodes constitutes the field of equilibrium electrochemistry. *See* CHEMICAL EQUILIBRIUM; CHEMICAL THERMODYNAMICS; ELECTROCHEMISTRY; ENTHALPY; ENTROPY; FREE ENERGY; THERMOCHEMISTRY.

**Structure.** The principal role of quantum mechanics in chemistry is in the discussion of atomic and molecular structure, and in the interpretation of spectroscopic data. In the branch of physical chemistry known as computational quantum chemistry, interest centers on the numerical solution of the Schrödinger equation in order to obtain wave functions and geometries of molecules. Computational quantum chemistry is so developed that it is capable of being used to map the changes in the structures of molecules while they are in the course of reaction, when atoms and groups of atoms are being transferred from one molecule to another. *See* QUANTUM CHEMISTRY.

Spectroscopic techniques are used not only to identify molecules present in a sample, but also to determine their shape, size, and electron distribution. The techniques fall into four categories: absorption spectroscopy, emission spectroscopy, Raman spectroscopy, and resonance techniques. *See* ELECTRON PARAMAGNETIC RESONANCE (EPR) SPECTROSCOPY; ELECTRON SPECTROSCOPY; MOLECULAR STRUCTURE AND SPECTRA; MÖSSBAUER EFFECT; NUCLEAR MAGNETIC RESONANCE (NMR); PHOTOCHEMISTRY; SPECTROSCOPY.

Techniques for the investigation of molecular structure based on diffraction depend on the observation of the direction through which radiation and particles are scattered when they impinge on a sample. Other techniques for investigating structure include the electric and magnetic properties of molecules, in particular, the determination of electric polarizabilities and dipole moments, magnetic properties, and the properties based on optical birefringence, such as optical activity and the Faraday effect.

Structural properties and thermodynamic properties are brought together by statistical thermodynamics. This major theoretical procedure gives a way of predicting the thermodynamic properties of assemblies of molecules in terms of their individual energy levels.

**Physical and chemical change.** The third major branch of physical chemistry is concerned with change: physical change and chemical change. In particular, it is concerned with the rate of change. Physical change includes the diffusion of one substance into another, or the migration of ions in an electrode solution. The application of thermodynamics to change in general constitutes the field of nonequilibrium thermodynamics. *See* GAS; TRANSPORT PROCESSES.

Chemical change may be studied at a variety of levels. Empirical chemical kinetics is the study of reactions in order to determine how their rates depend on the concentrations of the participants in the reaction and on the conditions, mainly the temperature. Investigation of the time dependence of reactions yields a detailed picture of the sequence of molecular transformations involved in a complex chemical reaction. *See* CHEMICAL DYNAMICS; SHOCK TUBE; ULTRAFAST MOLECULAR PROCESSES.

An important extension of chemical kinetics is to the reactions that occur on surfaces; these are the processes involved in heterogeneous catalysis. A special application of surface chemistry is to the stability of colloidal suspensions of species in fluids, and another is to the processes that occur at the interface between an electrode and the solution in which it is immersed. *See* ADSORPTION; COLLOID; HETEROGENEOUS CATALYSIS.

[P.W.A.]

**Physical organic chemistry** A branch of science concerned with the scope and limitations of the various rules, effects, and generalizations in use in organic chemistry by means of physical and mathematical methods. It includes, but is not limited to, the dynamics and energetics of organic chemical transformations, transient intermediates in these reactions, rate comparisons between families of reactions, dynamic stereochemistry, conservation of orbital symmetry, the least-motion principle, the isomer number for a given elemental composition, conformational analysis, nonexistent compounds, aromaticity, tautomerism, strain and steric hindrance, and the double-bond rule. Spectroscopy is the main tool employed, with nuclear magnetic resonance being the most widely used spectroscopic technique. With the advent of modern fast computers, computational chemistry has also become an important tool. *See* NUCLEAR MAGNETIC RESONANCE (NMR); SPECTROSCOPY.

Physical organic chemistry is traditionally distinguished from, yet totally intertwined with, synthetic organic chemistry, which deals with the question of how to obtain desired products from available compounds. This distinction can be illustrated with a diagram (see illustration) showing how the energy might vary during a chemical reaction in which the reactant R yields a product P (R → P). Whereas the synthetic organic chemist will be interested primarily in the practical problem of how to convert R into P, the physical organic chemist studies the curve or curves connecting R and P as well as the structure and physical properties at all extrema, including R and P. However, the demarcation between synthetic organic chemistry and physical organic chemistry is not sharp. Physical organic chemists have contributed greatly to the understanding of the chemistry of hydrocarbons and their derivatives and have enhanced the repertoire of the synthetic organic chemists. In turn, synthetic organic chemists have made possible the construction of the custom-made, often intricate molecules that physical organic chemists use for their studies. The efforts of both groups, moreover, have made possible the birth of such new fields as molecular biochemistry and computational chemistry. *See* COMPUTATIONAL CHEMISTRY.

**Chemical reaction mechanisms.** The diagram shown in the illustration is also useful in discussions of the dynamics of chemical reactions. It is an attempt to portray how the atoms in the reactant molecule R may move in space to their final positions in the product molecule P, and how the potential energy of the system would vary as a function of these positions. A complete correlation would be multidimensional; what is normally shown is a cross section in which the maximum potential energy is in fact a minimum (saddle point). While an essentially infinite number of pathways between R and P can be imagined and followed, the vast majority of the molecules will in practice use the one that makes the least demand on energy to reach the next maximum; this pathway is known as the reaction mechanism. The maxima (T terms in the illustration) are known as transition states, and the minima (I terms in the illustration) as intermediates.

If a reaction has a single transition state (and, hence, no intermediate), it is known as concerted; alternatively, it is step-wise. A stepwise reaction is simply a succession of concerted steps in which the intermediates are not isolated. *See* FREE RADICAL; ORGANIC REACTION MECHANISM; REACTIVE INTERMEDIATES; PHOTOCHEMISTRY.

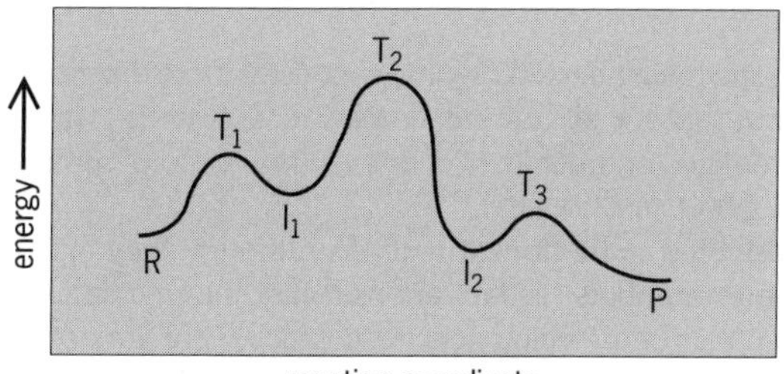


**Energy profile of an organic reaction R → P. T terms indicate transition states and I terms indicate intermediates.**

**Chemical kinetics.** The most important route to quantitative information about a reaction is the study of its kinetics. This must begin with an experimental determination of the rate law: the expression that shows how the rate of formation of product $d[P]/dt$ (or loss of reactant: $-d[R]/dt$) depends on the concentration of all species involved in the reaction other than the solvent. The differential might be found to equal $k[R_1]$, $k[R_1]\ [R_2]$, $k[R_1]^2$, and so on; $k$ is known as the rate constant, and the reaction is described as first-order, second-order, and so forth, depending on the total number of concentration terms. One important feature is that the reaction order equals the sum of all molecules that have participated in the formation of the transition state ($T_2$ in the illustration). Thus, for a reaction to be concerted, it is necessary that the order equal the sum of all reactant molecules involved in the stoichiometry. *See* CATALYSIS; CHEMICAL DYNAMICS.

**Stereochemistry.** This is also a powerful tool in physical organic chemistry. Experimental work has demonstrated that when a chiral compound such as $(-)HCR_1R_2X$ is converted into optically active $HCR_1R_2Y$ by direct displacement with Y, the product obtained has an optical rotation that is opposite to that exhibited by the same material if it is produced in two steps, via initial displacement by A to give intermediate $HCR_1R_2A$ as shown in the reaction scheme below. This result demonstrates that displacement reactions occur with inversion; the reagent approaches in front, and the leaving group departs in the back. *See* OPTICAL ACTIVITY; STEREOCHEMISTRY.

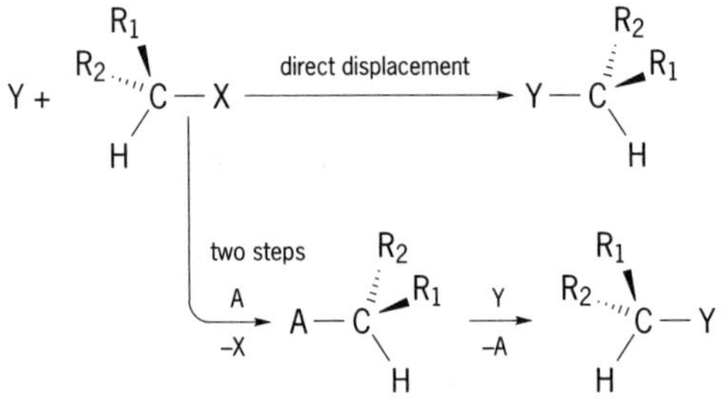

**Isomers.** The isomer number has perhaps been the most important organizing principle in organic chemistry since its inception. Simply put, it means that for every elemental composition and molecular mass the isomer number can be predicted by writing all possible sequences of the atoms present, obeying the valence numbers of the atoms: four for carbon, three for nitrogen, two for oxygen, one for hydrogen and the halogens, and so forth. Once reliable atomic weights became available so that elemental compositions could be reported with confidence, this simple rule proved remarkably successful. *See* CHEMICAL BONDING; ORGANIC CHEMISTRY; VALENCE.

The need to specify molecular mass has proved more troublesome: it requires a definition of the concept of a molecule. Such definitions usually refer to covalent bonds as the entities that hold the atoms together, to rule out ionic species such as sodium chloride as candidates.

Among the extra compounds, none have affected organic chemistry more drastically than the stereoisomers. It turns out that for all but the simplest compounds a given sequence of the atoms may represent two, more than two, or even many more isomers. *See* CONFORMATIONAL ANALYSIS; MOLECULAR ISOMERISM.

Many compounds that are considered to be nonexistent, even though they are allowed by the simple rules of isomer numbers, in fact are transient intermediates in various reactions. Sometimes they can be detected spectroscopically, but cannot be isolated. There are instances in which neither of two isomers can be isolated, but mixtures of the two can. In other words, the barrier between the two is low, and the equilibrium constant is close to unity. An example is acetoacetic ester, which normally contains about 15% of the enol isomer. Such isomers are known as tautomers. *See* TAUTOMERISM. [W.J.LeN.]

**Pine terpene** A major component of the essential oils obtained from various *Pinus* species. The principal terpenes of the oil of southern pines [longleaf pine (*P. palustris*) and slash pine (*P. caribaea*)] are $\alpha$- and $\beta$-pinene, whose structures are shown below.

α-Pinene

β-Pinene

Gum turpentine (gum spirits) is the volatile fraction of the oleoresin that exudes from cuts made in the trunks of live trees. The resin is collected and distilled by a process that yields about 20% turpentine, mainly $\alpha$- and $\beta$-pinene, and 70% rosin; it was the basis of the original naval stores industry.

Wood turpentine is obtained by steam distillation from stumps and other logging residues. The volatile material in this case consists of about 50% turpentine and 30–40% of higher-boiling-point alcohols; the latter fraction is known as pine oil. The bulk of the wood turpentine and pine oil produced by modern industrial processes is a by-product of the sulfate wood-pulping process (sulfate turpentine).

Important uses of turpentine or the purified pinenes derived from turpentine are in terpene resins, as a thinner in paints and varnishes, and as a starting material in the synthesis of other commercially valuable terpenes.

Pine oil is a mixture of monoterpene alcohols, mainly $\alpha$-terpineeol, obtained in large amounts mixed with wood turpentine or sulfate turpentine. The term pine oil is also used to designate the essential oil of various species of pine.

Much of the pine oil of commerce is prepared synthetically by acid-catalyzed hydration of $\alpha$-pinene. This process involves a complex series of reaction that occur via cationic intermediates.

The composition of industrial-grade pine oil is approximately 65% $\alpha$-terpineol, 20–25% of other monoterpene alcohols, and 10–15% hydrocarbons. Pine oil has surfactant and emulsifying properties and is also a disinfectant. Most of the pine oil manufactured is used in the manufacture of cleanser and textile penetrants. *See* SURFACTANT; TERPENE; WOOD CHEMICALS. [J.A.Mo.]

**pK** The logarithm (to the base 10) of the reciprocal of the equilibrium constant for a specified reaction under specified conditions (for example, solvent and temperature). The p$K$ values are often more convenient to tabulate and use than the equilibrium constants themselves. The value of $K$ for the dissociation of the $HSO_4^-$ ion in aqueous solution at 25°C (77°F) is 0.0102 mole/liter. The logarithm is $0.008_6 - 2 = -1.991_4$ The p$K$ is therefore $+1.991_4$. The choice of algebraic sign, although arbitrary, results in positive values for most dissociation constants applicable to aqueous solutions. The concept of p$K$ is especially valuable in the study of solutions. *See* CHEMICAL EQUILIBRIUM; IONIC EQUILIBRIUM; PH. [T.F.Y.]

**Plastics processing** Those methods used to convert plastics materials in the form of pellets, granules, powders, sheets, fluids, or preforms into formed shapes or parts. The plastics materials may contain a variety of additives which influence the properties as well as the processability of the plastics. After forming, the part may be subjected to a variety of ancillary operations such as welding, adhesive bonding, machining, and surface decorating (painting, metallizing).

**Injection molding.** This process consists of heating and homogenizing plastics granules in a cylinder until they are sufficiently fluid to allow for pressure injection into a relatively cold mold where they solidify and take the shape of the mold cavity. Solid particles, in the form of pellets or granules, constitute the main feed for injection moldable plastics. The major advantages of the injection-molding process are the speed of production, minimal requirements for postmolding operations, and simultaneous multipart molding. The development of reaction injection molding (RIM) allowed the rapid molding of liquid materials. This process has proven particularly effective for high-speed molding of such materials as polyurethanes, epoxies, polyesters, and nylons.

**Extrusion.** In this process, plastic pellets or granules are fluidized, homogenized, and continuously formed. Products made this way include tubing, pipe, sheet, wire and substrate coatings, and profile shapes. The process is used to form very long shapes or a large number of small shapes which can be cut from the long shapes. Extrusion can result in the highest output rate of any plastics processes; for example, pipe has been formed at rates of 2000 lb/h (900 kg/h). The extrusion process produces pipe and tubing by forcing the melt through a cylindrical die. *See* EXTRUSION.

**Blow molding.** This process consists of forming a tube (called a parison) and introducing air or other gas to cause the tube to expand into a free-blown hollow object or against a mold for forming into a hollow object with a definite size and shape. The parison is traditionally made by extrusion, although injection-molded tubes have increased in use.

**Thermoforming.** Thermoforming is the forming of plastics sheets into parts through the application of heat and pressure. Tooling for this process is the most inexpensive compared to other plastics processes, accounting for the method's popularity. It can also accommodate very large parts as well as small parts.

**Rotational molding.** In this process, finely ground powders are heated in a rotating mold until melting or fusion occurs. If liquid materials are used, the process is often called slush molding. The melted or fused resin uniformly coats the inner surface of the mold. When cooled, a hollow finished part is removed.

**Compression and transfer molding.** Compression molding consists of charging a plastics powder or preformed plug into a mold cavity, closing a mating mold half, and applying pressure to compress, heat, and cause flow of the plastic to conform to the cavity shape. The process is primarily used for thermosets, and consequently the mold is heated to accelerate the chemical cross-linking. Transfer molding is an adaptation of compression molding in that the molding powder or preform is charged to a separate preheating chamber and, when appropriately fluidized, injected into a closed mold. It is most used for thermosets, and is somewhat faster than compression molding.

**Foam processes.** Foamed plastics materials have achieved a high degree of importance in the plastics industry. Foams can be made in a range from soft and flexible to hard and rigid. There are three types of cellular plastics: blown (expanded matrix, such as a natural sponge), syntactic (the encapsulation of hollow organic or inorganic microspheres in the matrix), and structural (dense outer skin surrounding a foamed core). There are seven basic processes used to generate plastics foams. They include the incorporation of a chemical blowing agent that generates gas (through thermal decomposition) in the polymer liquid or melt; gas injection into the melt which expands during pressure relief; generation of gas as a by-product of a chemical condensation reaction during cross-linking; volatilization of a low-boiling liquid (for example, Freon) through the exothermic heat of reaction; mechanical dispersion of air by mechanical means (whipped cream); incorporation of nonchemical gas-liberating agents (adsorbed gas on finely divided carbon) into the resin mix which is released by heating; and expansion of small beads of thermoplastic resin containing a blowing agent through the external application of heat. *See* FOAM.

**Reinforced plastics/composites.** These are plastics whose mechanical properties are significantly improved because of the inclusion of fibrous reinforcements. The wide variety of resins and reinforcements that constitute this group of materials led to the more generalized description "composites." Composites consist of two main components, the fibrous material in various physical forms and the fluidized resin which will convert to a solid. There are fiber-reinforced thermoplastic materials, and these are typically processed in standard thermoplastic processing equipment. The first step in any composite fabrication procedure is the impregnation of the reinforcement with the resin. The impregnated reinforcement can be subjected to heat to remove impregnating solvents or advance the resin cure to a slightly tacky or dry state. The composite in this form is called a prepreg. Premixes, often called bulk molding compounds, are mixtures of resin, inert fillers, reinforcements, and other formulation additives which form a puttylike rope, sheet, or preformed shape. *See* POLYMERIC COMPOSITE.

**Casting and encapsulation.** Casting is a low-pressure process requiring nothing more than a container in the shape of the desired part. For thermoplastics, liquid monomer is poured into the mold and, with heat, allowed to polymerize in place to a solid mass. For vinyl plastisols, the liquid is fused with heat. Thermosets are poured into a heated mold wherein the cross-linking reaction completes the conversion to a solid. Encapsulation and potting are terms for casting processes in which a unit or assembly is encased or impregnated, respectively, with a liquid plastic which is subsequently hardened by fusion or chemical reaction. These processes are predominant in the electrical and electronic industries for the insulation and protection of components.

**Calendering.** In the calendering process, a plastic is masticated between two rolls that squeeze it out into a film which then passes around one or more additional rolls before being stripped off as a continuous film. Fabric or paper may be fed through the latter rolls, so that they become impregnated with the plastic. *See* POLYMER. [S.H.G.]

**Platinum** A chemical element, Pt, atomic number 78, and atomic weight 195.09. Platinum is a soft, ductile, white noble metal. The platinum-group metals—platinum, palladium, iridium, rhodium, osmium, and ruthenium—are found widely distributed over the Earth. Their extreme dilution, however, precludes their recovery, except in special circumstances. For example, small amounts of the platinum metals, palladium in particular, are recovered during the electrolytic refining of copper. *See* IRIDIUM; OSMIUM; PALLADIUM; PERIODIC TABLE; RHODIUM; RUTHENIUM.

The platinum-group metals have wide chemical use because of their catalytic activity and chemical inertness. As a catalyst, platinum is used in hydrogenation, dehydrogenation, isomerization, cyclization, dehydration, dehalogenation, and oxidation reactions. *See* CATALYSIS; ELECTROCHEMICAL PROCESS.

Platinum is not affected by atmospheric exposure, even in sulfur-bearing industrial atmospheres. Platinum remains bright and does not visually exhibit an oxide film when heated, although a thin, adherent film forms below 450°C (840°F). Platinum may be worked to fine wire and thin sheet, and by special processes, to extremely fine wire. Important physical properties are given in the table.

Platinum can be made into a spongy form by thermally decomposing ammonium chloroplatinate or by reducing it from an aqueous solution. In this form it exhibits a high absorptive power for gases, especially oxygen, hydrogen, and carbon monoxide.

**Physical properties of platinum**

| Properties | Value |
|---|---|
| Atomic weight ($^{12}C = 12.00000$) | 195.09 |
| Naturally occurring isotopes and % abundance | 190, 0.0127% |
| | 192, 0.78% |
| | 194,32.9% |
| | 195, 33.8% |
| | 196, 25.3% |
| | 198, 7.21% |
| Crystal structure | Face-centered cubic |
| Lattice constant $a$ at 25°C, nm | 0.39231 |
| Thermal neutron capture cross section, barns | 8.8 |
| Common chemical valence | 2, 4 |
| Density at 25°C, g/cm$^3$ | 21.46 |
| Melting point | 1772°C (3222°F) |
| Boiling point | 3800°C (6900°F) |
| Specific heat at 0°C, cal/g | 0.0314 |
| Thermal conductivity, 0–100°C, cal cm/cm$^2$ s°C | 0.17 |
| Linear coefficient of thermal expansion, 20–100°C, $\mu$in./in./°C | 9.1 |
| Electrical resistivity at 0°C, microhm-cm | 9.85 |
| Temperature coefficient of electrical resistance, 0–100°C/°C | 0.003927 |
| Tensile strength, 1000 lb/in.$^2$ | |
| Soft | 18–24 |
| Hard | 30–35 |
| Young's modulus at 20°C | |
| lb/in.$^2$, static | $24.8 \times 10^6$ |
| lb/in.$^2$, dynamic | $24.5 \times 10^6$ |
| Hardness, Diamond Pyramid Number (DPN) | |
| Soft | 37–42 |
| Hard | 90–95 |

The high catalytic activity of platinum is related directly to this property. *See* CRACKING; HYDROGENATION.

Platinum strongly tends to form coordination compounds. Platinum dioxide, $PtO_2$, is a dark-brown insoluble compound, commonly known as Adams catalyst. Platinum(II) chloride, $PtCl_2$, is an olive-green water-insoluble solid. Chloroplatinic acid, $H_2PtCl_6$, is the most important platinum compound. *See* COORDINATION CHEMISTRY.

In the glass industry, platinum is used at high temperatures to contain, stir, and convey molten glass. In the electrical industry, platinum is used in contacts and resistance wires because of its low contact resistance and high reliability in contaminated atmospheres. Platinum is clad over tungsten for use in electron tube grid wires. In the medical field, the simple coordination compounds cisplatin and carboplatin are two of the most active clinical anticancer agents. In combination with other agents, cisplatin is potentially curative for all stages of testicular cancer. Both agents are used for advanced gynecologic malignancies, especially ovarian tumors, and for head and neck and lung cancers. Carboplatin was developed in attempts to alleviate the severe toxic side effects of the parent cisplatin, with which is shares a very similar spectrum of anticancer efficacy. *See* CHEMOTHERAPY. [H.J.A.; N.F.]

**Plutonium** A chemical element, Pu, atomic number 94. Plutonium is a reactive, silvery metal in the actinide series of elements. The principal isotope of chemical interest is $^{239}Pu$, with a half-life of 24,131 years. It is formed in nuclear reactors by the process shown in the reaction below. Plutonium-239 is fissionable, but may also capture neutrons to form higher plutonium isotopes. *See* PERIODIC TABLE.

$$^{238}\mathrm{U} + n \longrightarrow {}^{239}\mathrm{U} \xrightarrow[2.35\ \mathrm{min}]{\beta^-} {}^{239}\mathrm{NP} \xrightarrow[2.33\ \mathrm{days}]{\beta^-} {}^{239}\mathrm{Pu}$$

Plutonium-238, with a half-life of 87.7 years, is utilized in heat sources for space application, and has been used for heart pacemakers. Plutonium-239 is used as a nuclear fuel, in the production of radioactive isotopes for research, and as the fissile agent in nuclear weapons.

Plutonium exhibits a variety of valence states in solution and in the solid state. Plutonium metal is highly electropositive. Numerous alloys of plutonium have been prepared, and a large number of intermetallic compounds have been characterized.

Reaction of the metal with hydrogen yields two hydrides. The hydrides are formed at temperatures as low as 150°C (300°F). Their decomposition above 750°C (1400°F) may be used to prepare reactive plutonium powder. The most common oxide is $PuO_2$, which is formed by ignition of hydroxides, oxalates, peroxides, and nitrates of any oxidation state in air of 870–1200°C (1600–2200°F). A very important class of plutonium compounds are the halides and oxyhalides. Plutonium hexafluoride, the most volatile plutonium compound known, is a strong fluorinating agent. A number of other binary compounds are known. Among these are the carbides, silicides, sulfides, and selenides, which are of particular interest because of their refractory nature.

Because of its radiotoxicity, plutonium and its compounds require special handling techniques to prevent ingestion or inhalation. Therefore, all work with plutonium and its compounds must be carried out inside glove boxes. For work with plutonium and its alloys, which are attacked by moisture and by atmospheric gases, these boxes may be filled with helium or argon. *See* ACTINIDE ELEMENTS; NEPTUNIUM; NUCLEAR CHEMISTRY; TRANSURANIUM ELEMENTS; URANIUM. [F.We.]

**Polar molecule** A molecule possessing a permanent electric dipole moment. Molecules containing atoms of more than one element are polar except where forbidden by symmetry; molecules formed from atoms of a single element are nonpolar (except ozone). The dipole moments of polar molecules result in stronger intermolecular attraction, increased viscosities, higher melting and boiling points, and greater solubility in polar solvents than in nonpolar molecules. [R.D.W.]

**Polarimetric analysis** A method of chemical analysis based on the optical activity of the substance being determined. Optically active materials are asymmetric; that is, their molecules or crystals have no plane or center of symmetry. These asymmetric molecules can occur in either of two forms, *d*- and *l*-, called optical isomers. Asymmetric substances possess the power of rotating the plane of polarization of plane-polarized light. Measurement of the extent of this rotation, polarimetry, is performed by an instrument known as a polarimeter. Polarimetry is applied to both organic and inorganic materials. *See* OPTICAL ACTIVITY.

The extent of the rotation depends on the character of the substance, the length of the light path, the temperature of the solution, the wavelength of the light which is being used, the solvent (if there is one), and the concentration of the substance. In most work, the yellow light of the D line of the sodium spectrum (589.3 nanometers) is used to determine the specific rotation, according to the equation below. Here $\alpha$ is

$$\text{Specific rotation} = [\alpha]_{\mathrm{D}}^{20} = \frac{\alpha}{l\rho}$$

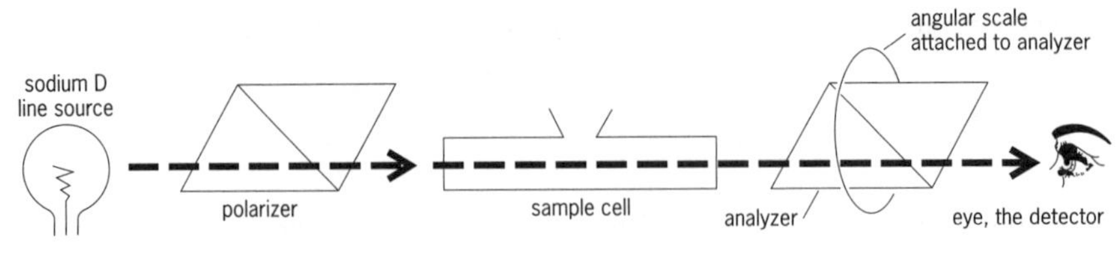


**Simplified diagram of a polarimeter.**

the measured angle of rotation, $l$ is the length of the column of liquid in decimeters, and $\rho$ the density of the solution. In other words, the specific rotation is the rotation in degrees which this plane-polarized light of the sodium D line undergoes in passing through a 10-cm-long (4-in.) sample tube containing a solution of 1 g/ml concentration at 20°C (68°F).

In the illustration, light from the sodium lamp is polarized by the polarizer (prism). It then passes through the cell containing the material being analyzed. After that, it passes through the analyzer (another prism) and then is detected (by eye or photocell). A comparison of the angular orientation of the analyzer as measured on the scale with the cell empty and with the cell filled with solution serves to measure the rotation of the polarized light by the sample. This rotation may be either clockwise (+) or counterclockwise (−).

Polarimetry may be used for either qualitative or quantitative analytical work. In qualitative applications, the presence of an optically active material is shown, and then a calculation of specific rotation often leads to the identification of the unknown. In quantitative work, the concentration of a given optically active material is determined.

Polarimetry is used in carbohydrate chemistry, especially in the analysis of sugar solutions. Since there is great difference between the biological activities of the different optical forms of organic compounds, polarimetry is used in biochemical research to identify the molecular configurations.

Optical rotatory dispersion is the measurement of the specific rotation as a function of wavelength. The information obtained by this method has shown that minor changes in configuration of a molecule have a marked effect on its dispersion properties. By using the properties of compounds of known configuration, it has been possible to determine the absolute configurations of many other molecules and to identify various isomers. Most of the applications have been to steroids, sugars, and other natural products, including amino acids, proteins, and polypeptides. *See* COTTON EFFECT; OPTICAL ROTATORY DISPERSION. [R.F.G.; J.N.L.]

**Polarographic analysis** An electrochemical technique used in analytical chemistry. Polarography involves measurements of current-voltage curves obtained when voltage is applied to electrodes (usually two) immersed in the solution being investigated. One of these electrodes is a reference electrode: its potential remains constant during the measurement. The second electrode is an indicator electrode. Its potential varies in the course of measurement of the current-voltage curve, because of the change of the applied voltage. In the simplest version, so-called dc polarography, the indicator electrode is a dropping-mercury electrode, consisting of a mercury drop hanging at the orifice of a fine-bore glass capillary. The capillary is connected to a mercury reservoir so that mercury flows through it at the rate of a few milligrams per second. The outflowing mercury forms a drop at the orifice, which grows until it falls off. The lifetime of each drop is several seconds (usually 2 to 5). Each drop forms a new electrode; its surface is practically unaffected by processes taking place on the previous drop. Hence each drop represents a well-reproducible electrode with a fresh, clean surface. *See* ELECTRODE; ELECTROCHEMICAL TECHNIQUES.

The dropping-mercury electrode is immersed in the solution to be investigated and placed in a cell containing the reference electrode. Polarographic current-voltage curves can be recorded with a simple instrument consisting of a potentiometer or another source of voltage and a current-measuring device. The voltage can be varied by manually changing the applied voltage in finite increments, measuring current at each, and plotting current as a function of the voltage. Alternatively, commercial instruments are available in which voltage is increased linearly with time (a voltage ramp), and current variations are recorded automatically.

Another polarographic technique is called pulse or differential pulse polarography. This technique is more sensitive by two orders of magnitude than dc polarography, and in inorganic trace analysis competes with atomic absorption and neutron activation analysis. The sensitivity of differential pulse polarography has found application in drug analysis. There is also a polarographic technique called ac polarography that is particularly useful for obtaining information on adsorption-desorption processes at the surface of the dropping-mercury electrode.

Polarographic studies can be applied to investigation of electrochemical problems, to elucidation of some fundamental problems of inorganic and organic chemistry, and to solution of practical problems. In electrochemistry, polarography allows measurement of potentials, and yields information about the rate of the electrode process, adsorption-desorption phenomena, and fast chemical reactions accompanying the electron transfer. In fundamental applications, polarography makes it possible to distinguish the form and charge of the species (for example, inorganic complex or organic ion) in the solution. Polarography also permits the study of equilibria (complex formation, acid-base, tautomeric), rates, and mechanisms.

Polarography can be used for investigation of the relationship between electrochemical data and structure. In inorganic analysis, polarography is used predominantly for trace-metal analysis (with increased sensitivity of differential pulse polarography and stripping analysis). In organic analysis, it is possible in principle to use polarography in elemental analysis and functional group analysis. The most important fields of application of inorganic determinations are in metallurgy, environmental analysis (air, water, and seawater contaminants), food analysis, toxicology, and clinical analysis. The possibility of being able to determine vitamins, alkaloids, hormones, terpenoid substances, and natural coloring substances has made polarography useful in analysis of biological systems, analysis of drugs and pharmaceutical preparations, and determination of pesticide or herbicide residues in foods. [P.Z.]

**Polonium** A chemical element, Po, atomic number 84. Marie Curie discovered the radioisotope $^{210}Po$ in pitchblende. This isotope is the penultimate member of the radium decay series. All polonium isotopes are radioactive, and all are shortlived except the three $\alpha$-emitters, artificially produced $^{208}Po$ (2.9 years) and $^{209}Po$ (100 years), and natural $^{210}Po$ (138.4 days). *See* PERIODIC TABLE.

Polonium ($^{210}Po$) is used mainly for the production of neutron sources. It can also be used in static eliminators and, when incorporated in the electrode alloy of spark plugs, is said to improve the cold-starting properties of internal combustion engines.

Most of the chemistry of polonium has been determined using $^{210}Po$, 1 curie of which weighs 222.2 micrograms; work with weighable amounts is hazardous, requiring special techniques. Polonium is more metallic than its lower homolog, tellurium. The metal is chemically similar to tellurium, forming the bright red compounds $SPoO_3$ and $SePoO_3$. The metal is soft, and its physical properties resemble those of thallium, lead, and bismuth. Valences of 2 and 4 are well established; there is some evidence of hexavalency. Polonium is positioned between silver and tellurium in the electrochemical series.

Two forms of the dioxide are known: low-temperature, yellow, face-centered cubic ($UO_2$ type), and high-temperature, red, tetragonal. The halides are covalent, volatile compounds, resembling their tellurium analogs. *See* TELLURIUM. [K.W.B.]

**Polyacetal** A polyether derived from aldehydes (RCHO) or ketones (RR′CO) and containing —O—R—O— groups in the main chain. Of the many possible polyacetals, the most common is a polymer or copolymer of formaldehyde, polyoxymethylene $(\text{—O—CH}_2\text{—})_n$. While the substance paraformaldehyde contains oligomers or low-molecular-weight polyoxymethylenes ($n$ very small), high-molecular-weight, crystalline polyoxymethylenes constitute an important class of engineering plastics that, in commerce, is often simply referred to as polyacetal. Cellulose and its derivatives also have a polyacetal structure. *See* ACETAL; FORMALDEHYDE; POLYETHER RESINS; POLYMER.

As shown below, formaldehyde can be readily polymerized by using anionic initiators

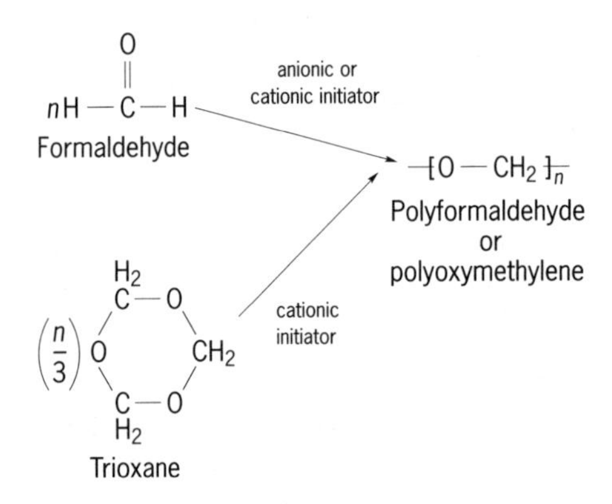


such as triphenylphosphine and, somewhat less readily, by using cationic initiators such as protonic acids. Alternatively, a similar polymer can be obtained by the ring-opening polymerization of trioxane using, for example, a boron trifluoride complex as initiator. *See* POLYMERIZATION.

At temperatures above ~110°C (230°F; the ceiling temperature, above which depolymerization becomes favored over polymerization), the polymers degrade by an unzipping reaction to monomer. To prevent this, one of two approaches is commonly used: esterification of the hydroxyl end groups, or copolymerization with a small amount of a monomer such as ethylene oxide or 1,3-dioxolane.

Polyacetals are typically strong and tough, resistant to fatigue, creep, organic chemicals (but not strong acids or bases), and have low coefficients of friction. Electrical properties are also good. Improved properties for particular applications may be attained by reinforcement with fibers of glass or polytetrafluoroethylene, and by incorporation of an elastomeric toughening phase. The combination of properties has led to many uses such as plumbing fittings, pump and valve components, bearings and gears, computer hardware, automobile body parts, and appliance housings. [J.A.M.]

**Polyacrylate resins** Polymers obtained from a variety of acrylic monomers, such as acrylic and methacrylic acids, their salts, esters, and amides, and the corresponding nitriles. The most important monomers with corresponding repeat units are shown here.

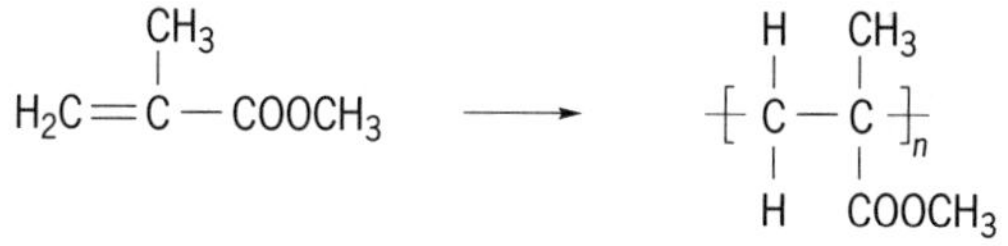

$$H_2C{=}C(CH_3){-}COOCH_3 \longrightarrow \left[ CH_2{-}C(CH_3)(COOCH_3) \right]_n$$

Methyl methacrylate — Poly(methyl methacrylate)

$$H_2C{=}CH{-}COOC_2H_5 \longrightarrow \left[ CH_2{-}CH(COOC_2H_5) \right]_n$$

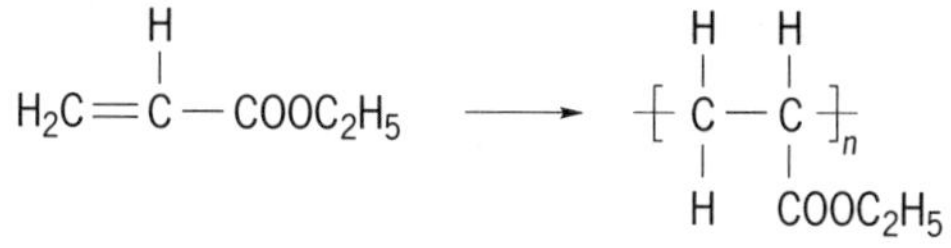

Ethyl acrylate — Poly(ethyl acrylate)

Poly(methyl methacrylate) is a hard, transparent polymer with high optical clarity, high refractive index, and good resistance to the effects of light and aging. It and its copolymers are useful for lenses, signs, indirect lighting fixtures, transparent domes and skylights, dentures, and protective coatings.

Solutions of poly(methyl methacrylate) and its copolymers are useful as lacquers. Aqueous latexes formed by the emulsion polymerization of methyl methacrylate with other monomers are useful as water-based paints and in the treating of textiles and leather.

Poly(ethyl acrylate) is a tough, somewhat rubbery product. The monomer is used mainly as a plasticizing or softening component of copolymers.

Methyl methacrylate is of interest as a polymerizable binder for sand or other aggregates, and as a polymerizable impregnant for concrete; usually a cross-linking acrylic monomer is also incorporated. The binder systems (polymer concrete) are used as overlays for bridge decks as well as for castings, while impregnation is used to restore concrete structures and protect bridge decks against corrosion by deicing salts. *See* PLASTICS PROCESSING; POLYMERIZATION. [J.A.M.]

**Polyacrylonitrile resins** Hard, relatively insoluble, and high-melting materials produced by the polymerization of acrylonitrile, as shown in the reaction below.

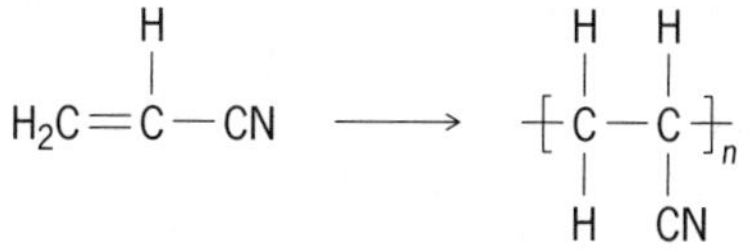

$$H_2C{=}CH{-}CN \longrightarrow \left[ CH_2{-}CH(CN) \right]_n$$

(Acrylonitrile) — (Polyacrylonitrile)

Polyacrylonitrile is used almost entirely in copolymers. The copolymers fall into three groups: fibers, plastics, and rubbers. The presence of acrylonitrile in a polymeric composition tends to increase its resistance to temperature, chemicals, impact, and flexing. The polymerization of acrylonitrile can be readily initiated by means of the conventional free-radical catalysts such as peroxides, by irradiation, or by the use of alkali metal catalysts. Although polymerization in bulk proceeds too rapidly to be commercially feasible, satisfactory control of a polymerization or copolymerization may be achieved in suspension and in emulsion, and in aqueous solutions from which the polymer precipitates. Copolymers containing acrylonitrile may be fabricated in the manner of thermoplastic resins.

The major use of acrylonitrile is in the form of fibers. By definition, an acrylic fiber must contain at least 85% acrylonitrile. The high strength; high softening temperature; resistance to aging, chemicals, water, and cleaning solvents; and the soft woollike feel or fabrics have made the product popular for many uses such as sails, cordage, blankets, and various types of clothing.

Copolymers of vinylidene chloride with small proportions of acrylonitrile are useful as tough, impermeable, and heat-sealable packaging films. Extensive use is made of copolymers of acrylonitrile with butadiene, often called NBR (formerly Buna N) rubbers, which contain 15–40% acrylonitrile. The NBR rubbers resist hydrocarbon solvents such as gasoline, abrasion, and in some cases show high flexibility at low temperatures. *See* RUBBER.

The development of blends and interpolymers of acrylonitrile-containing resins and rubbers represented a significant advance in polymer technology. The products, usually called ABS resins, typically are made by blending acrylonitrile-styrene copolymers with a butadiene-acrylonitrile rubber, or by interpolymerizing polybutadiene with styrene and acrylonitrile. The combination of low cost, good mechanical properties, and ease of fabrication by a variety of methods led to the rapid development of new uses for ABS resins. Applications include products requiring high impact strength, such as pipe, and sheets for structural uses, such as industrial duct work and components of automobile bodies. *See* ACRYLONITRILE; PLASTICS PROCESSING; STYRENE. [J.A.M.]

**Polyamide resins** Products of polymerization of an amino acid or the condensation of a diamine with a dicarboxylic acid. They are used for fibers, bristles, bearings, gears, molded objects, coatings, and adhesives. The term nylon formerly referred specifically to synthetic polyamides as a class. Because of many applications in mechanical engineering, nylons are considered engineering plastics.

The most common commercial aliphatic polamides are nylons-6,6; -6; -6,10; -11; and -12. Nylon-6,6, nylon-6,10, nylon-6,12, and nylon-6 are the most commonly used polyamides for general applications as molded or extruded parts; nylon-6,6 and nylon-6 find general application as fibers.

As a group, nylons are strong and tough. Mechanical properties depend in detail on the degree and distribution of crystallinity, and may be varied by appropriate thermal treatment or by nucleation techniques. Because of their generally good mechanical properties and adaptability to both molding and extrusion, certain nylons are often used for gears, bearings, and electrical mountings. Nylon bearings and gears perform quietly and need little or no lubrication. Nylon resins are also used extensively as filaments, bristles, wire insulation, appliance parts, and film. Properties can also be modified by copolymerization. Reinforcement of nylons with glass fibers results in increased stiffness, lower creep and improved resistance to elevated temperatures. *See* HETEROCYCLIC POLYMER; PLASTICS PROCESSING; POLYETHER RESINS; POLYMERIZATION. [J.A.M.]

**Polychlorinated biphenyls** A generic term for a family of 209 chlorinated isomers of biphenyl. The biphenyl molecule is composed of two six-sided carbon rings connected at one carbon site on each ring. Ten sites remain for chlorine atoms to join the biphenyl molecule. The term polychlorinated biphenyl (PCB) has been used to refer to the biphenyl molecule with one to ten chlorine substitutions, as shown in the following structure.

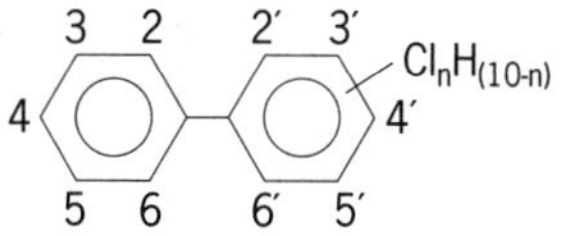

Polychlorinated biphenyl (PCB)

PCBs were introduced into United States industry on a large scale in 1929. The qualities that made PCBs attractive were chemical stability, resistance to heat, low flammability, and high dielectric constant. The PCB mixture is a colorless, viscous fluid, is relatively insoluble in water, and can withstand high temperatures without degradation (higher-chlorinated isomers are not readily degraded in the environment).

The major use of PCBs has been as dielectric fluid in electrical equipment, particularly transformers capacitors, electromagnets, circuit breakers, voltage regulators, and switches. PCBs have also been used in heat transfer systems and hydraulic systems, and as plasticizers and additives in lubricating and cutting oils.

PCBs have been reported in animals, plants, soil, and water all over the world, even in animals living under 11,000 ft (3400 m) of water. These phenomena are the result of bioaccumulation and biomagnification in the food chain. In a few instances, poultry products, cattle, and hogs have been found to contain high concentrations of PCBs after the animals have eaten feed contaminated with PCBs. It is not known what quantities of PCBs have been released to the environment, but major sources are industrial and municipal waste disposal, spills and leaks from PCB-containing equipment, and manufacture and handling of PCB mixtures.

PCBs can enter the body through the lungs, gastrointestinal tract, and skin, circulate throughout the body, and be stored in adipose tissue. PCBs have been detected in human adipose tissues and in the milk of cows and humans. Some PCBs have the ability to alter reproductive processes in mammals. There is concern that PCBs may be carcinogenic in humans. [G.Ku.]

**Polyester resins** Synthetic polymers made by esterification of dicarboxylic acids with diols. The aliphatic polyesters tend to be relatively soft, and the aromatic derivatives are usually hard and brittle, or tough. The properties of either group may be modified by cross linking, crystallization, plasticizers, or fillers. *See* ESTER.

The commercial products are alkyds which are used in paints, enamels, and molding compounds; unsaturated polyesters or unsaturated alkyds which are used extensively with fiber glass for boat hulls and panels; aliphatic saturated polyesters; aromatic polyesters, such as polyethylene terephthalate which is used in the form of fibers and films; and the aromatic polycarbonates. The polydiallyl esters, while frequently listed with the polyesters, are not true polyesters as defined above. *See* POLYVINYL RESINS.

The alkyds are commonly used as coatings. Combinations of conventional vegetable drying oils and alkyd resins represent the basis of most of the oil-soluble paints. The drying oil–alkyd may be further modified by the inclusion of a vinyl monomer, such as styrene. Some of the styrene polymerizes, probably as a graft polymer, and the remainder polymerizes and copolymerizes in the final drying or curing of the pain. *See* POLYMERIZATION.

The unsaturated polyesters, in combination with glass fiber, have found applications as panels, roofing, radar domes, boat hulls, and protective armor for soldiers. The compositions are distinguished by ease of fabrication and high impact resistance. *See* POLYMERIC COMPOSITE.

Saturated aliphatic polyesters have long been frequently used as intermediates in the preparation of prepolymers for making segmented polyurethanes. Lactone rings can also be opened to yield linear polyesters.

The aromatic polyesters which have achieved general importance are the polyethylene terephthalates, which yield very strong and chemically resistant fibers and films. Polyethylene terephthalate is the principal ingredient of polyester fibers. Polyethylene terephthalate may be molded or extruded to yield materials that can replace metals or thermoset resins in some automotive, electrical, and specialty applications, especially when reinforced with glass fibers or mineral fillers.

Aromatic polycarbonates are a strong, tough group of thermoplastic polymers formed most frequently from bisphenol A and phosgene. The products, polycarbonates, are noted for high softening temperatures, and high impact resistance, clarity, and resistance to creep. Polycarbonate is usually available as a molding compound. Because of its high strength, toughness, and softening point, the resin, both by itself and as a glass-reinforced material, has found many electrical domestic and engineering applications. It is often used to replace glass and metals. Examples include bottles, unbreakable windows, appliance parts, electrical housings, marine propellers, and shotgun shells. Flame-retardant grades are of interest because of low toxicity and smoke emission on burning.

Polydiallyl esters are polymers of diallyl esters. Thermosetting molding compounds may be produced by careful limitation of the initial polymerization to yield a product which is fusible. Major applications are in electronic components, sealants, coatings, and glass-fiber composites. *See* PLASTICS PROCESSING. [J.A.M.]

**Polyether resins** Thermoplastic or thermosetting materials which contain ether-oxygen linkages, —C—O—C—, in the polymer chain. Depending upon the nature of the reactants and reaction conditions, a large number of polyethers with a wide range of properties may be prepared. The main groups of polyethers in use are epoxy resins, phenoxy resins, polyethylene oxide and polypropylene oxide resins, polyoxymethylene, and polyphenylene oxides.

The epoxy resins form an important and versatile class of cross-linked polyethers characterized by excellent chemical resistance, adhesion to glass and metals, electrical insulating properties, and ease and precision of fabrication. Various fillers such as calcium carbonate, metal fibers and powders, and glass fibers are commonly used in epoxy formulations in order to improve such properties as the strength and resistance to abrasion and high temperatures. Some reactive plasticizers act as curing agents, become permanently bound to the epoxy groups, and are usually called flexibilizers. Rubbery polymers are added to improve toughness and impact strength. Epoxies are commonly used in protective coatings. They are used as potting or encapsulating compositions for the protection of delicate electronic assemblies from the thermal and mechanical shock of rocket flight, and as dies for stamping metal forms.

Polyethylene oxide and polypropylene oxide are thermoplastic products whose properties are greatly influenced by molecular weight. Low-to-moderate-molecular-weight polyethylene oxides vary in form from oils to waxlike solids. They are relatively nonvolatile, are soluble in a variety of solvents, and have found many uses as thickening agents, plasticizers, lubricants for textile fibers, and components of various sizing, coating, and cosmetic preparations. The polypropylene oxides of similar molecular weight have somewhat similar properties, but tend to be more oil-soluble (hydrophobic) and less water-soluble (hydrophilic). While polyalkylene oxides are not of interest as such in structural materials, polypropylene oxides are used extensively in the preparation of polyurethane foams.

Phenoxy resins are transparent, strong, ductile, and resistant to creep, and, in general, resemble polycarbonates in their behavior. The major application is as a component in protective coatings, especially in metal primers. *See* POLYESTER RESINS.

Polyphenylene oxide (PPO) is the basis for an engineering plastic characterized by chemical, thermal, and dimensional stability. Polyphenylene oxide is outstanding in its resistance to water. Uses include medical instruments, pump parts, and insulation.

Polyoxymethylene, or polyacetal, resins are polymers of formaldehyde. Having high molecular weights and high degrees of crystallinity, they are strong and tough and are established in the general class of engineering thermoplastics. Polyacetals are typically resistant to fatigue, creep, organic chemicals (but not strong acids or bases), and have low coefficients of friction. Electrical properties are also good. The combination of properties has led to many uses such as plumbing fittings, pump and valve components, bearings and gears, computer hardware, automobile body parts, and appliance housings. *See* PLASTICS PROCESSING; POLYMERIZATION. [J.A.M.]

**Poly(ethylene glycol)** Any of a series of water-soluble polymers with the general formula $HO—(CH_2—CH_2—O)_n—H$. These colorless, odorless compounds range in appearance from viscous liquids to waxy solids. The low-molecular-weight members, diethylene glycol ($n = 2$) through tetraethylene glycol ($n = 4$), are produced as pure compounds and find use as humectants, dehydrating solvents for natural gas, textile lubricants, heat-transfer fluids, solvents for aromatic hydrocarbon extractions, and intermediates for polyester resins and plasticizers.

The intermediate members of the series with average molecular weights of 200 to 20,000 are used commercially in ceramic, metal-forming, and rubber-processing operations; as drug suppository bases and in cosmetic creams, lotions, and deodorants; as lubricants; as dispersants for casein, gelatins, and inks; and as antistatic agents. The highest members of the series have molecular weights from 100,000 to 10,000,000. They are of interest because of their ability at very low concentrations to reduce friction of flowing water. *See* ETHYLENE OXIDE; POLYMERIZATION. [R.K.B.]

Poly(ethylene glycol) has a range of properties making it suitable for medical and biotechnical applications. Since poly(ethylene glycol) is soluble both in water and in most organic solvents, many applications are derived from this amphiphilicity. Other properties include lack of toxicity and immunogenicity, and a tendency to avoid other polymers and particles also present in aqueous solution.

Poly(ethylene glycol) is attached to drugs to enhance water and blood solubility. Similarly, poly(ethylene glycol) is attached to enzymes to impart solubility in organic solvents. These poly(ethylene glycol) enzymes are used as catalysts for industrial reactions in organic solvents. *See* CATALYSIS.

The tendency of poly(ethylene glycol) to avoid interaction with cellular and molecular components of the immune system results in the material being nonimmunogenic. This property leads to a greatly enhanced blood circulation lifetime of poly(ethylene glycol) proteins and to application as pharmaceuticals. Similarly, adsorption of proteins and cells to surfaces is greatly reduced by attaching poly(ethylene glycol) to the surface, and such coated materials find wide application as biomaterials. *See* BIOMEDICAL CHEMICAL ENGINEERING. [J.M.Ha.]

**Polyfluoroolefin resins** Resins distinguished by their resistance to heat and chemicals and by the ability to crystallize to a high degree. Several main products are based on tetrafluoroethylene, $F_2C{=}CF_2$ (TFE); hexafluoropropylene, $F_2C{=}CFCF_3$ (HFP); and monochlorotrifluoroethylene, $FClC{=}CF_2$ (CTFE).

Poly(tetrafluoroethylene) is the polymer of tetrafluoroethylene and is commonly known by the trade name Teflon®. It is insoluble, resistant to heat (up to 275°C or 527°F) and chemical attack, and has the lowest coefficient of friction of any solid. However, special surface treatments are required to ensure adhesion because poly(tetrafluoroethylene) does not adhere well to anything. Poly(tetrafluoroethylene) (TFE resin) is used for bearings, valve seats, packings, gaskets, coatings and tubing, and can withstand relatively severe conditions. Because of its excellent electrical properties, poly(tetrafluoroethylene) is useful when a dielectric material is required for service at a high temperature. The nonadhesive quality is often used to coat articles such as rolls and cookware to which materials might otherwise adhere.

The properties of poly(chlorotrifluoroethylene) (CTFE resin) are generally similar to those of poly(tetrafluoroethylene); however, the presence of the chlorine atoms in the former causes the polymer to be a little less resistant to heat and to chemicals. The applications of poly(chlorotrifluoroethylene) are in general similar to those for poly(tetrafluoroethylene). Because of its stability and inertness, the polymer is useful in the manufacture of gaskets, linings, and valve seats that must withstand hot and corrosive conditions. It is also used as a dielectric material, as a vapor and liquid barrier, and for microporous filters.

Poly(vinylidene fluoride) properties has generally similar to those of the other fluorinated resins: relative inertness, low dielectric constant, and thermal stability (up to about 150 or 300°F). The resins ($PVF_2$ resins) are, however, stronger and less susceptible to creep and abrasion than TFE and CTFE resins. Applications of poly(vinylidene fluoride) are mainly as electrical insulation, piping, process equipment, and as a protective coating in the form of a liquid dispersion.

Several types of fluorinated, noncrystallizing elastomers were developed in order to meet needs (usually military) for rubbers which possess good low-temperature behavior with a high degree of resistance to oils and to heat, radiation, and weathering. *See* HALOGENATED HYDROCARBON; PLASTICS PROCESSING; POLYMERIZATION. [J.A.M.]

**Polymer** Polymers, macromolecules, high polymers, and giant molecules are high-molecular-weight materials composed of repeating subunits. These materials may be organic, inorganic, or organometallic, and synthetic or natural in origin. Polymers are essential materials for almost every industry as adhesives, building materials, paper, cloths, fibers, coatings, plastics, ceramics, concretes, liquid crystals, photoresists, and coatings. They are also major components in soils and plant and animal life. They are important in nutrition, engineering, biology, medicine, computers, space exploration, health, and the environment.

Natural inorganic polymers include diamonds, graphite, sand, asbestos, agates, chert, feldspars, mica, quartz, and talc. Natural organic polymers include polysaccharides (or polycarbohydrates) such as starch and cellulose, nucleic acids, and proteins. Synthetic inorganic polymers include boron nitride, concrete, many high-temperature superconductors, and a number of glasses. Siloxanes or polysiloxanes represent synthetic organometallic polymers. *See* SILICONE RESINS.

Synthetic polymers used for structural components weigh considerably less than metals, helping to reduce the consumption of fuel in vehicles and aircraft. They even outperform most metals when measured on a strength-per-weight basis. Polymers have been developed which can also be used for engineering purposes such as gears, bearings, and structural members.

**Nomenclature.** Many polymers have both a common name and a structure-based name specified by the International Union of Pure and Applied Chemistry (IUPAC). Some polymers are commonly known by their acronyms. Some companies use trade names to identify the specific polymeric products they manufacture. For example, Fortrel® polyester is a poly(ethylene terephthalate) (PET) fiber. Polymers are often generically named, such as rayon, polyester, and nylon. *See* ORGANIC NOMENCLATURE; POLYAMIDE RESINS; POLYESTER RESINS.

**Composition.** Polymer structures can be represented by similar or identical repeat units. These are derived from smaller molecules, called monomers, which react to form the polymer. Propylene monomer and the repeat unit it forms in polypropylene are shown below. With the exception of its end groups, polypropylene is composed entirely

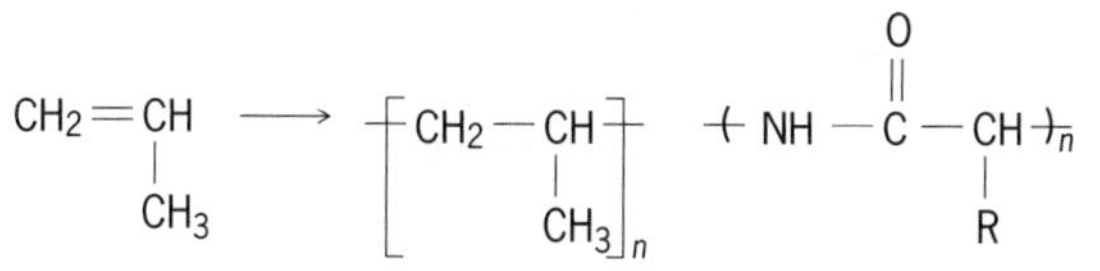

$$CH_2{=}CH(CH_3) \longrightarrow \left[ CH_2 - CH(CH_3) \right]_n \qquad \left( NH - C({=}O) - CH(R) \right)_n$$

Monomer — Polymer repeat unit — Protein repeat unit

of this repeat unit. The number of units ($n$) in a polymer chain is called the degree of polymerization (DP). Other polymers, such as proteins, can be described in terms of the approximate repeat unit where the nature of R (a substituted atom or group of atoms) varies. *See* POLYVINYL RESINS; PROTEIN.

**Primary structure.** The sequence of repeat units within a polymer is called its primary structure. Unsymmetrical reactants, such as substituted vinyl monomers, react almost exclusively to give a "head-to-tail" product, in which the R substituents occur on alternate carbon atoms. A variety of head-to-head structures are also possible.

Each R-substituted carbon atom is a chiral center (an atom in a molecule attached to four different groups) with different geometries possible. Arrangements where the substitutes on the chiral carbon are random are referred to as atactic structures. Arrangements where the geometry about the chiral carbon alternates are said to be syndiotactic. Structures where the geometry about the chiral atom has the same geometry are said to be isotactic or stereoregular.

Stereoregular polymers are produced using special stereoregulating catalyst systems. A series of soluble catalysts have been developed that yield products with high stereoregularity and low chain-size disparity. As expected, polymers with regular structures—that is, isotactic and syndiotactic structures—tend to be more crystalline and stronger.

Polymers can be linear or branched with varying amounts and lengths of branching. Most polymers contain some branching.

Copolymers are derived from two different monomers, which may be represented as A and B. There exists a large variety of possible structures and, with each structure, specific properties. These varieties include alternating, random, block, and graft (see illustration). *See* COPOLYMER.

**Secondary structure.** This refers to the localized shape of the polymer, which is often the consequence of hydrogen bonding. Most flexible to semiflexible linear polymer chains tend toward two structures—helical and pleated sheet/skirtlike. The pleated skirt arrangement is most prevalent for polar materials where hydrogen bonding can occur. In nature, protein tissue is often of a pleated skirt arrangement. For both polar and nonpolar polymer chains, there is a tendency toward helical formation with the inner core having "like" secondary bonding forces. *See* HYDROGEN BOND.

**Tertiary structure.** This refers to the overall shape of a polymer, such as in polypeptide folding. Globular proteins approximate rough spheres because of a complex combination of environmental and molecular constraints, and bonding opportunities. Many natural and synthetic polymers have "superstructures," such as the globular proteins and aggregates of polymer chains, forming bundles and groupings.

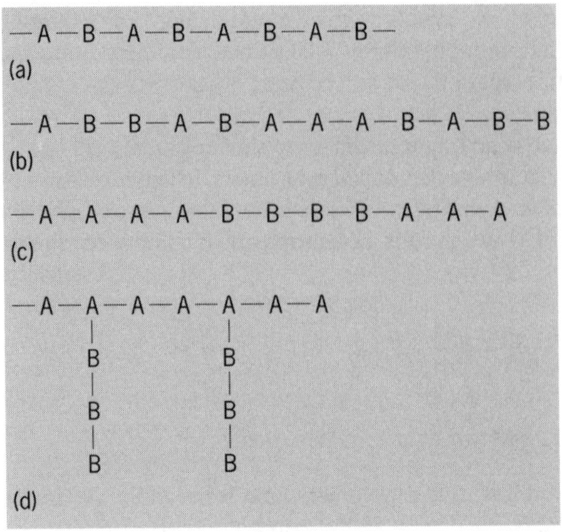

```
—A—B—A—B—A—B—A—B—
(a)

—A—B—B—A—B—A—A—A—B—A—B—B
(b)

—A—A—A—A—B—B—B—B—A—A—A
(c)

—A—A—A—A—A—A—A
   |       |
   B       B
   |       |
   B       B
   |       |
   B       B
(d)
```

**Copolymer structures: (a) alternating, (b) random, (c) block, (d) graft.**

**Quaternary structure.** This refers to the arrangement in space of two or more polymer subunits, often a grouping of tertiary structures. For example, hemoglobin (quaternary structure) is essentially the combination of four myoglobin (tertiary structure) units. Many crystalline synthetic polymers form spherulites.

**Synthesis.** For polymerization to occur, monomers must have at least two reaction points or functional groups. There are two main reaction routes to synthetic polymer formation—addition and condensation. In chain-type kinetics, initiation starts a series of monomer additions that result in the reaction mixture consisting mostly of unreacted monomer and polymer. Vinyl polymers, derived from vinyl monomers and containing only carbon in their backbone, are formed in this way. Examples of vinyl polymers include polystyrene, polyethylene, polybutadiene, polypropylene (see structure), and poly(vinyl chloride).

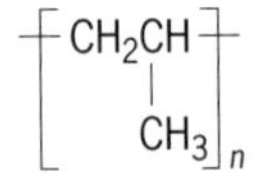

$$\left[ CH_2\underset{\displaystyle CH_3}{\underset{|}{CH}} \right]_n$$

Polypropylene, PP

The second main route is a step-wise polymerization. Polymerization occurs in a step-wise fashion so that the average chain size within the reaction mixture may have an overall degree of polymerization of 2, then 5, then 10, and so on, until the entire mixture contains largely polymer with little or no monomer left. Polymers typically produced using the step-wise process are called condensation polymers, and include polyamides, polycarbonates, polyesters, and polyurethanes (see structures). Condensation polymer

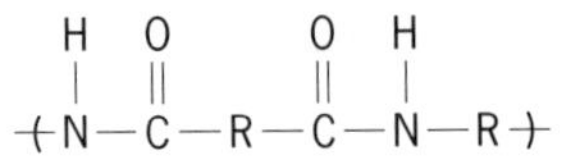

$$\left( \overset{H}{\overset{|}{N}}-\overset{O}{\overset{\|}{C}}-R-\overset{O}{\overset{\|}{C}}-\overset{H}{\overset{|}{N}}-R \right)$$

Polyamide, nylon

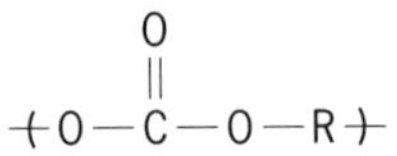

$$\left( O-\overset{O}{\overset{\|}{C}}-O-R \right)$$

Polycarbonate

$$\left( O-R-O-\overset{O}{\overset{\|}{C}}-\overset{H}{\overset{|}{N}}-R'-\overset{H}{\overset{|}{N}}-\overset{O}{\overset{\|}{C}} \right)$$

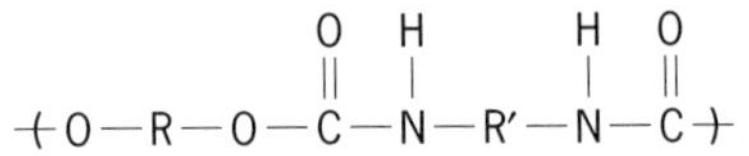

Polyurethane, PU

chains are characterized as having a noncarbon atom in their backbone. For polyamides the noncarbon is nitrogen (N), while for polycarbonates it is oxygen (O). Condensation polymers are synthesized using melt (the reactants are heated causing them to melt), solution (the reactants are dissolved), and interfacial (the reactants are dissolved in immiscible solvents) techniques. *See* POLYMERIZATION; POLYOLEFIN RESINS; POLYURETHANE RESINS.

**Molecular properties.** These are used to help determine the structure and behavior of the polymer. The molecular weight of a particular polymer chain is the product of the number of units times the molecular weight of the repeating unit. Two statistical averages describe polymers, the number-average molecular weight and the weight-average molecular weight. *See* MOLECULAR WEIGHT.

Size is the most important property of polymers allowing for storage of information (nucleic acids and proteins). Polymeric materials remember any action that distorts or moves polymer chains or segments (such as bending, stretching, and melting). Size also accounts for an accumulation of the interchain and intrachain secondary attractive forces called van der Waals forces. For nonpolar polymers, such as polyethylene, the attractive forces for each repeating unit are less than that for polar polymers. Polyvinyl chloride, a polar polymer, has attractive forces that include both dispersion and dipole-dipole forces so that the total attractive forces are proportionally larger than those for polyethylene. Polymers with hydrogen bonding (such as proteins, polysaccharides, nucleic acids, and nylons) have attractive forces that are even greater. Hydrogen bonding is so strong in cellulose that cellulose is not soluble in water until the inter- and intrachain hydrogen bonds are broken.

Polymers often have a combination of ordered regions, called crystalline regions, and disordered or amorphous regions. Crystalline regions are more rigid, contributing to strength and resistance to external forces. The amorphous regions contribute to polymers' flexibility. Most commercial polymers have a balance between amorphous and crystalline regions, allowing a balance between flexibility and strength.

Polymers are viscoelastic materials. Ductile polymers, such as polyethylene and polypropylene, "give" or "yield," and at high elongations some strengthening and orientation occur. A brittle polymer, such as polystryene, does not give much and breaks at a low elongation. A fiber, a polymer material that is much longer than it is wide, exhibits high strength, high stiffness, and little elongation.

**Materials.** Fibers are polymer materials that are strong in one direction, and they are much longer ($>100$ times) than they are wide. Elastomers (or rubbers) are polymeric materials that can be distorted through the application of force, and when the force is removed, the material returns to its original shape. Plastics are materials that have properties between fibers and elastomers—they are hard and flexible. Coatings and adhesives are generally derived from polymers that are members of other groupings (for example, polysiloxanes are elastomers, but also are used as adhesives). Industrially important adhesives and coatings include laminates, sealants and caulks, composites, films, polyblends, liquid crystals, ceramics, cements, and smart materials. *See* ADHESIVE; LIQUID CRYSTALS; POLYMERIC COMPOSITE; RUBBER.

**Additives.** Processed polymeric materials are generally a combination of the polymer and the materials that are added to modify its properties, assist in processing, and introduce new properties. Additives can be solids, liquids, or gases. Typical additives are plasticizers, antioxidants, colorants, fillers, and reinforcements. *See* ANTIOXIDANT; INHIBITOR (CHEMISTRY).

**Recycling.** Many polymers are thermoplastics, that is, they can be reshaped through application of heat and pressure and used in the production of other thermoplastic materials. The recycling of thermosets, polymers that do not melt but degrade prior to softening, is more difficult. These materials are often ground into a fine powder, are blended with additives (often adhesives or binders), and then are reformed. [C.E.Ca.]

**Polymer-supported reaction** An organic chemical reaction where one of the species, such as the substrate, the reagent, or a catalyst, is bound to a cross-linked, and therefore insoluble, polymer support. A major attraction of polymer-supported reactions is that at the end of the reaction period the polymer-supported species can be separated cleanly and easily, usually by filtration, from the soluble species. This easy separation can greatly simplify product isolation procedures, and it may even allow the polymer-supported reactions to be automated. Because it is possible to reuse or recycle polymer-supported reactants and because they are insoluble, involatile, easily handled, and easily recovered, polymer-supported reactants are also attractive from an environmental point of view. *See* ASYMMETRIC SYNTHESIS; CATALYSIS.

Polymer-supported reactants are usually prepared in the form of beads of about 50–100 micrometers' diameter. Such beads can have a practically useful loading only when their interiors are functionalized, that is, carry functional groups. With a typical polymer-supported reactant, more than 99% of the reactive groups are inside the beads. An important consequence is that for soluble species to react with polymer-supported species the former must be able to diffuse freely into the polymer beads. *See* POLYMER.

In a typical application of reactions involving polymer-supported substrates, a substrate is first attached to an appropriately functionalized polymer support. The synthetic reactions of interest are then carried out on the supported species. Finally, the product is detached from the polymer and recovered. Often the polymer-supported species has served as a protecting group that is also a physical "handle" to facilitate separation. Combinatorial synthesis is carried out on polymer-supported substrates. This approach is used in the pharmaceutical industry to identify lead compounds, which can thus be identified in weeks rather than years.

Reactions involving polymer-supported reagents are generally more useful than those involving polymer-supported substrates, because no attachment or detachment reactions are needed, and it is not necessary for all the polymer-supported species to react in high yield. Indeed, polymer-supported reagents are often used in excess to drive reactions to high conversions. Polymer-supported catalysts are the most attractive type of polymer-supported reactants; in such reactions, the loadings of catalytic sites need not be high, not all sites need be active and, in most cases, the polymer-supported catalyst is recovered in a suitable form for reuse. [P.H.]

**Polymeric composite** Any of the combinations or compositions that comprise two or more materials as separate phases, at least one of which is a polymer. By combining a polymer with another material, such as glass, carbon, or another polymer, it is often possible to obtain unique combinations or levels of properties. Typical examples of synthetic polymeric composites include glass-, carbon-, or polymer-fiber-reinforced thermoplastic or thermosetting resins, carbon-reinforced rubber, polymer blends, silica- or mica-reinforced resins, and polymer-bonded or -impregnated concrete or wood. It is also often useful to consider as composites such materials as coatings (pigment-binder combinations) and crystalline polymers (crystallites in a polymer matrix). Typical naturally occurring composites include wood (cellulosic fibers bonded with lignin) and bone (minerals bonded with collagen). On the other hand, polymeric compositions compounded with a plasticizer or very low proportions of pigments or processing aids are not ordinarily considered as composites.

Typically, the goal is to improve strength, stiffness, or toughness, or dimensional stability by embedding particles or fibers in a matrix or binding phase. A second goal is to use inexpensive, readily available fillers to extend a more expensive or scarce resin; this goal is increasingly important as petroleum supplies become costlier and less reliable. Still other applications include the use of some fillers such as glass spheres to improve processability, the incorporation of dry-lubricant particles such as molybdenum sulfide to make a self-lubricating bearing, and the use of fillers to reduce permeability.

The most common fiber-reinforced polymer composites are based on glass fibers, cloth, mat, or roving embedded in a matrix of an epoxy or polyester resin. Reinforced thermosetting resins containing boron, polyaramids, and especially carbon fibers confer especially high levels of strength and stiffness. Carbon-fiber composites have a relative stiffness five times that of steel. Because of these excellent properties, many applications are uniquely suited for epoxy and polyester composites, such as components in new jet aircraft, parts for automobiles, boat hulls, rocket motor cases, and chemical reaction vessels.

Although the most dramatic properties are found with reinforced thermosetting resins such as epoxy and polyester resins, significant improvements can be obtained with many reinforced thermoplastic resins as well. Polycarbonates, polyethylene, and polyesters are among the resins available as glass-reinforced composition. The combination of inexpensive, one-step fabrication by injection molding, with improved properties has made it possible for reinforced thermoplastics to replace metals in many applications in appliances, instruments, automobiles, and tools.

In the development of other composite systems, various matrices are possible; for example, polyimide resins are excellent matrices for glass fibers, and give a high-performance composite. Different fibers are of potential interest, including polymers [such as poly(vinyl alcohol)], single-crystal ceramic whiskers (such as sapphire), and various metallic fibers. *See* GRAPHITE; POLYAMIDE RESINS; POLYESTER RESINS; POLYETHER RESINS; POLYMER; POLYSTYRENE RESIN; POLYVINYL RESINS; RUBBER. [J.A.M.]

**Polymerization** The linking of small molecules (monomers) to make larger molecules. Polymerization requires that each small molecule have at least two reaction points or functional groups. There are two distinct major types of polymerization processes, condensation polymerization, in which the chain growth is accompanied by elimination of small molecules such as $H_2O$ or $CH_3OH$, and addition polymerization, in which the polymer is formed without the loss of other materials. There are many variants and subclasses of polymerization reactions.

An example of the condensation process is the reaction (1) of ε-aminocaproic acid

$$n\,H_2N-(CH_2)_5-\overset{\overset{\displaystyle O}{\|}}{C}-OH \xrightarrow{\text{catalyst}} H\left[-\overset{\overset{\displaystyle H}{|}}{N}-(CH_2)_5-\overset{\overset{\displaystyle O}{\|}}{C}-\right]_n OH + (n-1)H_2O \qquad (1)$$

ε-Aminocaproic acid　　　Polyamide, nylon-6

in the presence of a catalyst to form the polyamide, nylon-6. The repeating structural unit is equivalent to the starting material minus H and OH, the elements of water. The molecules formed are linear because the total functionality of the reaction system (functional groups per molecule) is always two. However, if a trifunctional material, such as a tricarboxylic acid, were added to the nylon-6,6 polymerizing mixture, a branched polymeric structure would result, because two of the carboxylic groups would participate in one polymer chain, and the third carboxylic group would start the growth of another. Under appropriate conditions, these chains can become bridges between linear chains and the polymer becomes cross-linked. The arrangements of the chains are shown in Fig. 1.

**Fig. 1. Polymer chains. (*a*) Linear polymer chain. (*b*) Branched polymer chain. (*c*) Cross-linked polymer chain.**

An example of addition polymerization is reaction (2). The structure of the repeat-

$$nH_2C{=}O \xrightarrow[\text{initiator (I)}]{\text{catalyst or}} [-CH_2-O-]_n \qquad (2)$$

Formaldehyde — Polyoxymethylene

ing unit is the difunctional monomeric unit, or "mer." In the presence of catalysts or initiators, the monomer yields a polymer by the joining together of *n* mers. If *n* is a small number, 2–10, the products are dimers, trimers, tetramers, or oligomers, and the materials are usually gases, liquids, oils, or brittle solids. In most solid polymers, *n* has values ranging from a few score to several hundred thousand, and the corresponding molecular weights range from a few thousand to several million. The end groups of this example of addition polymers are shown to be fragments of the initiator.

If only one monomer is polymerized, the product is called a homopolymer. The polymerization of a mixture of two monomers of suitable reactivity leads to the formation of a copolymer, a polymer in which the two types of mer units have entered the chain in a more or less random fashion. If chains of one homopolymer are chemically joined to chains of another, the product is called a block or graft copolymer:

A — B — A — A — B — B — B

Random copolymer

A — A — A — A — B — B — B — B — B

Block copolymer

A — A — A — A — A
with a side chain B — B — B attached to the third A

Graft polymer

Isotactic and syndiotactic (stereoregular) polymers are formed in the presence of complex catalysts, or by changing polymerization conditions, for example, by lowering the temperature. The groups attached to the chain in a stereoregular polymer are in a spatially ordered arrangement. The configuration of these ordered polymers and the disordered, atactic form is shown in Fig. 2. The regular structures of the isotactic and syndiotactic forms make them often capable of crystallization. The crystalline melting points of isotactic polymers are often substantially higher than the softening points of the atactic product.

In Fig. 2 each carbon atom to which a phenyl group is attached is asymmetrically substituted. For illustration, the heavily marked bonds are assumed to project up from the paper, and the dotted bonds down. Thus in a fully syndiotactic polymer, asymmetric carbons alternate in their left- or right-handedness (alternating *d*, *l* configurations), while in an isotactic polymer, successive carbons have the same steric configuration (*d* or *l*).

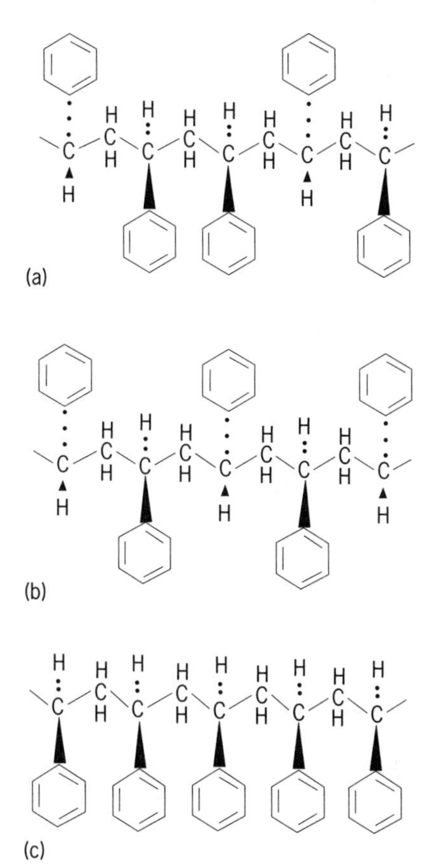


**Fig. 2. Spatially oriented polymers. (*a*) Atactic (random; *dlldl* or *lddld*, and so on). (*b*) Syndiotactic (alternating; *dldl*, and so on). (*c*) Isotactic (right- or left-handed; *dddd*, or *llll*, and so on).**

Among the several kinds of polymerization catalysis, free-radical initiation has been most thoroughly studied and is most widely employed. Atactic polymers are readily formed by free-radical polymerization, at moderate temperatures, of vinyl and diene monomers and some of their derivatives. *See* CATALYSIS; FREE RADICAL.

Some polymerizations can be initiated by materials, often called ionic catalysts, that contain highly polar reactive sites or complexes. The term heterogeneous catalyst is often applicable to these materials because many of the catalyst systems are insoluble in monomers and other solvents. These polymerizations are usually carried out in solution from which the polymer can be obtained by evaporation of the solvent or by precipitation on the addition of a nonsolvent. A distinguishing feature of complex catalysts is the ability of some representatives of each type to initiate stereoregular polymerization at ordinary temperatures or to cause the formation of polymers which can be crystallized. *See* CHEMICAL DYNAMICS; HETEROGENEOUS CATALYSIS; INHIBITOR (CHEMISTRY); INORGANIC POLYMER; ORGANIC REACTION MECHANISM; PLASTICS PROCESSING; POLYMER. [J.A.M.]

**Polynuclear hydrocarbon** One of a class of hydrocarbons possessing more than one ring. The aromatic polynuclear hydrocarbons may be divided into two groups. In the first, the rings are fused, which means that at least two carbon atoms are shared between adjacent rings. Examples are naphthalene (**1**), which has two six-membered rings, and acenaphthene (**2**), which has two six-membered rings and one five-membered ring.

In the second group of polynuclear hydrocarbons, the aromatic rings are joined either directly, as in the case of biphenyl (**3**), or through a chain of one or more carbon atoms, as in 1,2-diphenylethane (**4**).

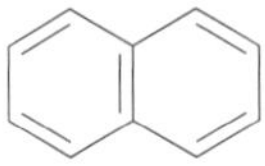

$CH_2-CH_2$

(1) 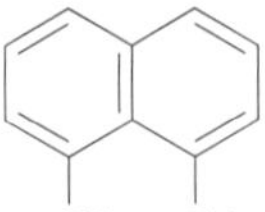 (2) 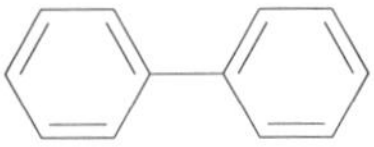 (3)

$C_6H_5CH_2CH_2C_6H_5$

(4)

The higher-boiling polynuclear hydrocarbons found in coal tar or in tars produced by the pyrolysis or incomplete combustion of carbon compounds are frequently fused-ring hydrocarbons, some of which may be carcinogenic. *See* AROMATIC HYDROCARBON; STEROID. [C.K.B.]

**Polyol** A compound containing more than one hydroxyl group (—OH). Each hydroxyl is attached to separate carbon atoms of an aliphatic skeleton. This group includes glycols, glycerol, and pentaerythritol and also such products as trimethylolethane, trimethylolpropane, 1,2,6-hexanetriol, sorbitol, inositol, and poly(vinyl alcohol). Polyols are obtained from many plant and animal sources and are synthesized by a variety of methods.

Polyols such as glycerol, pentaerythritol, trimethylolethane, and trimethylolpropane are used in making alkyd resins for decorative and protective coatings. Glycols, glycerol, 1,2,6-hexanetriol, and sorbitol find application as humectants and plasticizers for gelatin, glue, and cork.

The polymeric polyols used in manufacture of the urethane foams represent a series of synthetic polyols. These polyols are generally poly(oxyethylene) or poly(oxypropylene) adducts of di-to octahydric alcohols. *See* GLYCEROL. [P.C.J.]

**Polyolefin resins** Polymers derived from hydrocarbon molecules that possess one or more alkenyl (or olefinic) groups. The term polyolefin typically is applied to polymers derived from ethylene, propylene, and other alpha-olefins, isobutylene, cyclic olefins, and butadiene, and other diolefins. *See* ALKENE.

**Polyethylene.** Polyethylene is any homopolymer or copolymer in which ethylene is the major component monomer. It is a semicrystalline polymer of low to moderate strength and high toughness; its stiffness, yield strength, and thermal and mechanical properties increase with crystallinity. Toughness and ultimate tensile strength increase with molecular weight. Polyethylene shows excellent toughness at low temperatures. Polyethylene is relatively inexpensive, extremely versatile, and adaptable to a large array of fabrication techniques. It is chemically inert; resistant to solvents, acids, and alkalis; and has good dielectric and barrier properties. It is used in many housewares, films, molded articles, and coatings.

Low-density polyethylene has densities ranging from 0.905 to 0.936 g/cm$^3$. High-pressure low-density polyethylene is referred to simply as LDPE; linear low-density polyethylene (LLDPE) is a copolymer of polyethylene that is produced by a low-pressure polymerization process. High-density polyethylene (HDPE) covers the density range from 0.941 to 0.967 g/cm$^3$. HDPE generally consists of a polymethylene $(CH_2)_n$

chain with no, or very few, side chains to disrupt crystallization, while LLDPE contains side chains whose length depends on the comonomer used. *See* COPOLYMER.

LLDPE finds wide application in plastic films such as garbage bags and stretch cling films. Sheathing and flexible pipe are applications that take advantage of the flexibility and low-temperature toughness of LLDPE. HDPE is used in food packaging, grocery bags, pickup truck bedliners, and large containers. Fibers have been produced that approach the strength of spider silk, and is used in fishing lines and in medical applications.

Rubbery ethylene copolymers are used in compounded mixtures and range in comonomer content from 25 to 60% by weight, with propylene being the most widely used comonomer to form ethylene-propylene rubber. In addition to propylene, small amounts of a diene are sometimes included, forming a terepolymer. Products containing ethylene-propylene rubber and terepolymer have many automotive uses such as in bumpers, facia, dashboard panels, steering wheels, and assorted interior trim. *See* RUBBER.

Ethylene copolymers are used to produce the polymer poly(ethylene-co-vinyl acetate) [EVA]. Applications include specialty film for heat sealing, adhesives, flexible hose and tubing, footwear components, bumper components, and gaskets. Foamed and cross-linked poly(ethylene-co-vinyl acetate) is used in energy-absorbing applications. Ionomers are ethylene copolymers that are produced from the copolymerization of ethylene with a comonomer containing a carboxylic group (COOH) such as methyl acrylic acid. Because of their toughness, ionomers are widely used in the covers for golf balls. [B.Be.]

**Polypropylene.** Commercial polypropylene (PP) homopolymers are isotactic, high-molecular-weight, semicrystalline solids having melting points around 160–165°C (320–329°F), low density (0.90–0.91 g/cm$^3$), and excellent stiffness and tensile strength. They have moderate impact strength (toughness), low density over a wide temperature range, excellent mechanical properties, and low electrical conductivity. Propylene, like ethylene, is produced in large quantities at low cost from the cracking of oil and other hydrocarbon feedstocks. Low-molecular-weight resins are used for melt spun and melt blown fibers and for injection-molding applications. Polypropylene resins are used in extrusion and blow-molding processes and to make cast, slit, and oriented films. Stabilizers are added to polypropylene to protect it from attack by oxygen, ultraviolet light, and thermal degradation; other additives improve resin clarity, flame retardancy, or radiation resistance.

Polypropylene homopolymers, random copolymers, and impact copolymers are used in such products as automotive parts, appliances, battery cases, carpeting, electrical insulation, fiber and fabrics, food packaging, and medical equipment.

**Other poly(alpha-olefins).** Poly(1-butene) is a tough and flexible resin that has been used in the manufacture of film and pipe. Poly(4-methyl-1-pentene) is used in the manufacture of chemical and medical equipment. High-molecular-weight polyisobutylenes are rubbery solids that are used as sealants, inner tubes, and tubeless tire liners. Low-molecular-weight polyisobutylenes are used in formulations for caulking, sealants, and lubricants. Butadiene and isoprene can be polymerized to give a number of polymer structures. The commercially important forms of polybutadiene and polyisoprene are similar in structure to natural rubber. *See* POLYACRYLONITRILE RESINS; POLYMER; POLYSTYRENE RESIN. [S.A.Co.]

**Poly(p-xylylene) resins** Linear, crystallizable resins based on an unusual polymerization of *p*-xylene and derivatives. The polymers are tough and chemically resistant, and may be deposited as adherent coatings by a vacuum process. The vapor deposition process makes it possible to coat small microelectronic parts with a thin layer of the polymer. *See* XYLENE. [J.A.M.]

**Polysaccharide** A class of high-molecular-weight carbohydrates, colloidal complexes, which break down on hydrolysis to monosaccharides containing five or six carbon atoms. The polysaccharides are considered to be polymers in which monosaccharides have been glycosidically joined with the elimination of water. A polysaccharide consisting of hexose monosaccharide units may be represented by the reaction below.

$$nC_6H_{12}O_6 \rightarrow (C_6H_{10}O_5)_n + (n-1)H_2O$$

The term polysaccharide is limited to those polymers which contain 10 or more monosaccharide residues. Polysaccharides such as starch, glycogen, and dextran consist of several thousand D-glucose units. Polymers of relatively low molecular weight, consisting of two to nine monosaccharide residues, are referred to as oligosaccharides. *See* DEXTRAN; GLUCOSE.

Polysaccharides are often classified on the basis of the number of monosaccharide types present in the molecule. Polysaccharides, such as cellulose or starch, that produce only one monosaccharide type (D-glucose) on complete hydrolysis are termed homopolysaccharides. On the other hand, polysaccharides, such as hyaluronic acid, which produce on hydrolysis more than one monosaccharide type (*N*-acetylglucosamine and D-glucuronic acid) are named heteropolysaccharides. *See* CARBOHYDRATE. [W.Z.H.]

**Polystyrene resin** A hard, transparent, glasslike thermoplastic resin (see structure). Polystyrene is characterized by excellent electrical insulation properties, relatively high resistance to water, high refractive index, clarity, and low softening temperature.

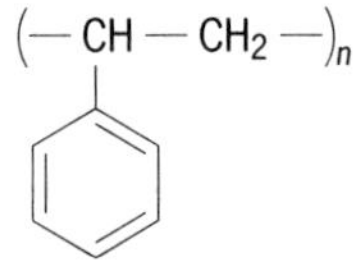

$$(-CH(C_6H_5)-CH_2-)_n$$

Polystyrene

High-molecular-weight homopolymers, copolymers, and polyblends are used as extrusion and molding compounds for packaging, appliance and furniture components, toys, and insulating panels. Styrene-butadiene copolymers are still used for automobile tires and in various rubber articles. The effects of blending small amounts of a rubbery polymer, such as butadiene-styrene rubber, with a hard, brittle polymer are most dramatic when the latter is polystyrene. The polyblend may have impact strength greater than ten times that of polystyrene. Various combinations of complex polyblends and interpolymers of acrylonitrile, styrene, and butadiene (ABS) resins are important as molding resins. ABS resins are also used as toughening agents for polymers such as polyvinyl chloride. Polystyrene is also used in combination with paints. The homopolymer and polyblends are used for panels or liners for refrigerator doors. Polystyrene may also be fabricated in the form of a rigid foam, which is used in packaging, food-service articles, and insulating panels. *See* ACRYLONITRILE; COPOLYMER; PLASTICS PROCESSING; POLYACRYLONITRILE RESINS; POLYMER; POLYMERIZATION; RUBBER; STYRENE. [J.A.M.]

**Polysulfide resins** Resins that vary in properties from viscous liquids to rubberlike solids. Organic polysulfide resins are prepared by the condensation of organic dihalides with a polysulfide.

Compounding and fabrication of the rubbery polymers can be handled on conventional rubber machinery. The polysulfide rubbers are distinguished by their resistance to solvents such as gasoline, and to oxygen and ozone. The polymers are relatively impermeable to gases. The products are used to form coatings which are chemically resistant and special rubber articles, such as gasoline bags. The polysulfide rubbers were among the first polymers to be used in solid-fuel compositions for rockets. *See* ORGANOSULFUR COMPOUND; POLYMERIZATION; RUBBER. [J.A.M.]

**Polysulfone resins** Polymers containing sulfone groups (—$SO_2$—) in the main chain, along with a variety of aromatic or aliphatic constituents. Polysulfones based on aromatic backbones constitute a useful class of engineering plastics, owing to their high strength, stiffness, and toughness together with high thermal and oxidative stability, low creep, transparency, and the ability to be processed by standard techniques for thermoplastics. The aromatic structural elements and the presence of sulfone groups are responsible for the resistance to heat and oxidation; ether and isopropylidene groups contribute some chain flexibility. Aromatic polysulfones can be used over wide temperature ranges. The high-temperature performance of poly(ethersulfones) to 200°C (390°F) is surpassed by few other polymers.

Because of the combination of properties discussed, aromatic polysulfone resins find many applications in electronic and automotive parts, medical instrumentation subject to sterilization, chemical and food processing equipment, and various plumbing and home appliance items. Coating formulations are also available, as well as grades reinforced with glass beads or fibers. Aliphatic polysulfones are less stable, for example, to hydrolysis. However, they have potential use in biomedical applications such as artificial membranes to remove carbon dioxide and perfuse with oxygen. *See* COPOLYMER; HETEROCYCLIC POLYMER; ORGANOSULFUR COMPOUND. [J.A.M.]

**Polyurethane resins** Polyurethane (or polyisocyanate) resins are produced by the reaction of a diisocyanate with a compound containing at least two active hydrogen atoms, such as a diol or diamine. Linear, fiber-forming polymers are formed by the addition of diisocyanates to diols, while cross-linking is made possible by the use of polyols or isocyanates having more than two functional groups. Unique elastomeric or stretch fibers can be made from polyester prepolymers, in which rubbery polyester blocks alternate with rigid urethane units and terminal isocyanate groups provide sites for further chain extension and cross-linking.

There are three major types of polyurethane elastomers. One type is based on ether- or ester-type prepolymers that are chain-extended and cross-linked using polyhydroxyl compounds or amines; alternately, unsaturated groups may be introduced to permit vulcanization with common curing agents such as peroxides. A second type is obtained by first casting a mixture of prepolymer with chain-extending and cross-linking agents, and then cross-linking further by heating. The third type is prepared by reacting a dihydroxy ester- or ether-type prepolymer, or a diacid, with a diisocyanate such as diphenylmethane dissocyanate and a diol.

Polyurethane resins can be produced in forms varying from hard, glossy, solvent-resistant coatings, to abrasion- and solvent-resistant rubbers, fibers, and flexible-to-rigid foams. The foams have found the widest use. The more flexible foams are employed as upholstery material for furniture, as rug backing, insulation, and crash pads. The more rigid foams are employed as the core in structural and insulating laminates and as insulation in refrigerated appliances and vehicles. Polyurethanes are also used as adhesives, for example, in the bonding of rubber and of nylon. The flexible polyurethanes may be used for coating rubber articles to give them additional resistance to abrasion and solvents. Wire insulated with polyurethane resin can be soldered directly without previously removing the coating because the polymer decomposes at the soldering temperature to yield a clean wire surface. *See* PLASTICS PROCESSING; POLYESTER RESINS; POLYETHER RESINS; POLYMERIZATION. [J.A.M.]

**Polyvinyl resins** Polymeric materials generally considered to include polymers derived from monomers having the structure

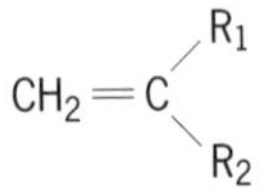

$$CH_2{=}C\begin{matrix} \diagup R_1 \\ \diagdown R_2 \end{matrix}$$

in which $R_1$ and $R_2$ represent hydrogen, alkyl, halogen, or other groups. This article refers to polymers whose names include the term vinyl. For discussions of other vinyl-type polymers *see* POLYACRYLATE RESIN; POLYACRYLONITRILE RESINS; POLYFLUOROOLEFIN RESINS; POLYOLEFIN RESINS; POLYSTYRENE RESIN.

Many of the monomers can be prepared by addition of the appropriate compound to acetylene. For example, vinyl chloride, vinyl fluoride, vinyl acetate, and vinyl methyl ether may be formed by the reactions of acetylene with HCl, HF, $CH_3OOH$, and $CH_3OH$, respectively. Processes based on ethylene as a raw material have also become common for the preparation of vinyl chloride and vinyl acetate.

The polyvinyl resins may be characterized as a group of thermoplastics which, in many cases, are inexpensive and capable of being handled by solution, dispersion, injection molding, and extrusion techniques. The properties vary with chemical structure, crystallinity, and molecular weight.

Poly(vinyl acetals) are relatively soft, water-insoluble thermoplastic products obtained by the reaction of poly(vinyl alcohol) with aldehydes. Properties depend on the extent to which alcohol groups are reacted. Poly(vinyl butyral) is rubbery and tough and is used primarily in plasticized form as the inner layer and binder for safety glass. Poly(vinyl formal) is the hardest of the group; it is used mainly in adhesive, primer, and wire-coating formulations, especially when blended with a phenolic resin.

Poly(vinyl acetate) is a leathery, colorless thermoplastic material which softens at relatively low temperatures and which is relatively stable to light and oxygen. The polymers are clear and noncrystalline. The chief applications are as adhesives and binders for water-based or emulsion paints.

Poly(vinyl alcohol) is a tough, whitish polymer which can be formed into strong films, tubes and fibers that are highly resistant to hydrocarbon solvents. Although poly(vinyl alcohol) is one of the few water-soluble polymers, it can be rendered insoluble in water by drawing or by the use of cross-linking agents. Two groups of products are available, those formed by the essentially complete hydrolysis of poly(vinyl acetate), and those formed by incomplete hydrolysis.

The former may be plasticized with water or glycols and molded or extruded into films, tubes, and filaments which are resistant to hydrocarbons. These products are used for liners in gasoline hoses, for grease-resistant coating and paper adhesives, for treating paper and textiles, and as emulsifiers and thickeners.

Poly(vinyl carbazole) is a tough, glassy thermoplastic with excellent electrical properties and a relatively high softening temperature. Uses of the product has been limited to small-scale electrical applications requiring resistance to high temperatures.

Poly(vinyl chloride) [PVC] is a tough, strong thermoplastic material which has an excellent combination of physical and electrical properties. The products are usually characterized as plasticized or rigid types. Poly(vinyl chloride)[and copolymers] is the second most commonly used polyvinyl resin and one of the most versatile plastics. The plasticized types are somewhat elastic materials which are familiar in the form of shower curtains, floor coverings, raincoats, dishpans, dolls, bottle-top sealers, prosthetic forms, wire insulation, and films. Rigid products, which may consist of the homopolymer, copolymer, or polyblends, are commonly used in the manufacture of phonograph records, pipe, chemically resistant liners for chemical reaction vessels, and siding and window sashes.

Poly(vinylidene chloride) is a tough, hornlike thermoplastic with properties generally similar to those of poly(vinyl chloride). Because of its relatively low solubility and decomposition temperature, the material is most widely used in the form of copolymers with other vinyl monomers, such as vinyl chloride. The copolymers are employed as packaging film, rigid pipe, and as filaments for upholstery and window screens.

Poly(vinyl ethers) exist in several forms varying from soft, balsamlike semisolids to tough, rubbery masses, all of which are readily soluble in organic solvents. Polymers of the alkyl vinyl ethers are used in adhesive formulations and as softening or flexibilizing agents for other polymers.

Poly(vinyl fluoride) is a tough, partially crystalline thermoplastic material which has a higher softening temperature than poly(vinyl chloride). Films and sheets are characterized by high resistance to impact and cracking caused by flexing and temperature and by resistance to weathering.

Poly(vinyl pyrrolidone) is a water-soluble polymer of basic nature which has film-forming properties, strong absorptive or complexing qualities for various reagents, and the ability to form water-solubles salts which are polyelectrolytes. The main uses are as a water-solubilizing agent for medicinal agents such as iodine, and as a semipermanent setting agent in hair sprays. Certain synthetic textile fibers containing small amounts of vinylpyrrolidone as a copolymer have improved affinity for dyes. *See* PLASTICS PROCESSING; POLYMER; POLYMERIZATION. [J.A.M.]

**Porphyrin** One of a class of cyclic compounds in which the parent macrocycle consists of four pyrrole-type units linked together by single carbon bridges. Several porphyrins with selected peripheral substitution and metal coordination carry out vital biochemical processes in living organisms. Chlorins, bacteriochlorins, and corrins (see structures) are related tetrapyrrolic macrocycles that are also observed in biologically important compounds.

The complexity of porphyrin nomenclature parallels the complex structures of the naturally occurring derivatives. Hans Fischer used a simple numbering system for the porphyrin nucleus (see structures) and a set of common names to identify the different porphyrins and their isomers. A systematic naming based on the 1–24 numbering system for the porphyrin nucleus was later developed by the International Union of Pure and Applied Chemistry (IUPAC) and the International Union of Biochemistry (IUB), and this system has gained general acceptance. The need for common names is clear after examination of the systematic names; for example, protoporphyrin IX has the systematic name 2,7,12,18-tetramethyl-3,8-divinyl-13,17-dipropanoic acid.

The aromatic character (hence stability) of porphyrins has been confirmed by measurements of their heats of combustion. In addition, x-ray crystallographic studies have established planarity of the porphyrin macrocycle which is a basic requirement for aromatic character. *See* DELOCALIZATION.

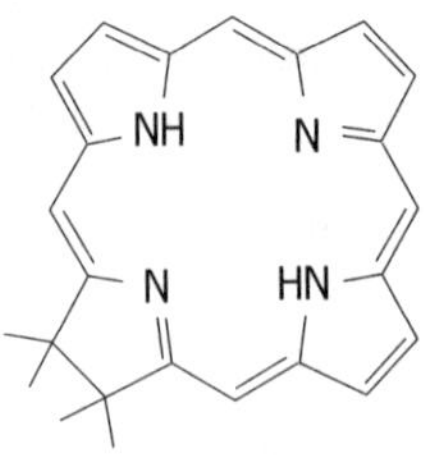

Chlorin

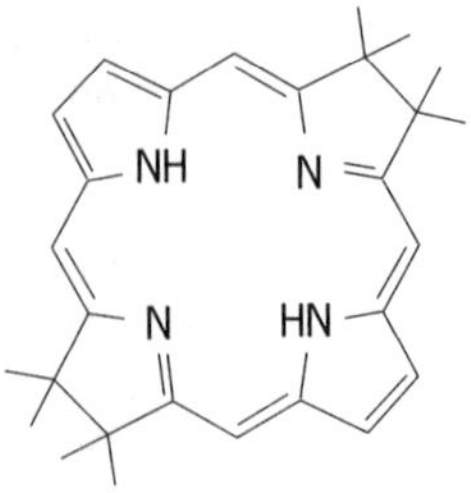

Bacteriochlorin

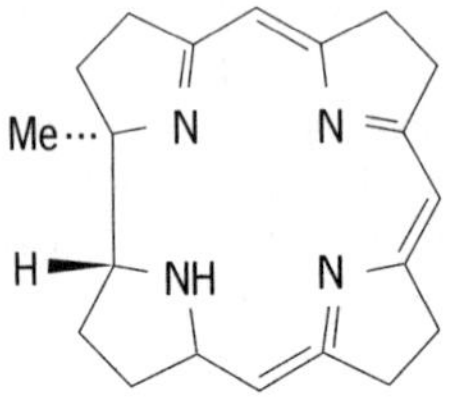

Corrin
Me = $CH_3$

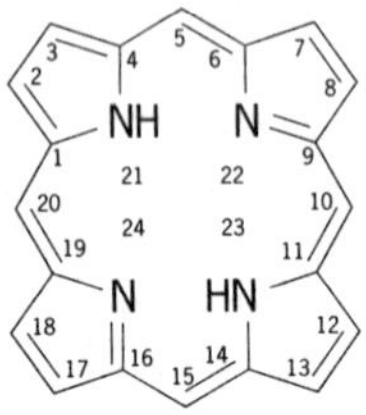

Porphyrin nucleus

Most metals and metalloids have been inserted into the central hole of the porphyrin macrocycle. The resulting metalloporphyrins are usually very stable and can bind a variety of small molecules (known as ligands) to the central metal atom. Heme, the iron complex of protoporphyrin IX, is the prosthetic group of a number of major proteins and enzymes that carry out diverse biological functions. These include binding, transport, and storage of oxygen (hemoglobin and myoglobin), electron-transfer processes (cytochromes), activation and transfer of oxygen to substrates (cytochromes P450), and managing and using hydrogen peroxide (peroxidases and catalases). *See* COORDINATION COMPLEXES; CYTOCHROME.

Chlorophylls and bacteriochlorophylls are magnesium complexes of porphyrin derivatives known as chlorins and bacteriochlorins, respectively. They are the pigments responsible for photosynthesis. Several chlorophylls have been identified, the most common being chlorophyll *a*, which is found in all oxygen-evolving photosynthetic plants. Bacteriochlorophyll *a* is found in many photosynthetic bacteria.

Porphyrins and metalloporphyrins exhibit many potentially important medicinal and industrial properties. Metalloporphyrins are being examined as potential catalysts for a variety of processes, including catalytic oxidations. They are also being examined as possible blood substitutes and as electrocatalysts for fuel cells and for the electrochemical generation of hydrogen peroxide. The unique optical properties of porphyrins make them likely candidates for photovoltaic devices and in photocopying and other optical devices. A major area where porphyrins are showing significant potential is in the treatment of a wide range of diseases, including cancer, using photodynamic therapy. *See* CATALYSIS. [T.W.D.]

**Potassium** A chemical element, K, atomic number 19, and atomic weight 39.102. It stands in the middle of the alkali metal family, below sodium and above rubidium. This lightweight, soft, low-melting, reactive metal (see table) is very similar to sodium in its behavior in metallic forms. *See* ALKALI METALS; PERIODIC TABLE; RUBIDIUM; SODIUM.

**Physical properties of potassium metal**

| Property | Temperature °C | Temperature °F | SI units | Customary (engineering) units |
|---|---|---|---|---|
| Density | 100 | 212 | 0.819 g/cm$^3$ | 51.1 lb/ft$^3$ |
| | 400 | 752 | 0.747 g/cm$^3$ | 46.7 lb/ft$^3$ |
| | 700 | 1292 | 0.676 g/cm$^3$ | 42.2 lb/ft$^3$ |
| Melting point | 63.7 | 147 | | |
| Boiling point | 760 | 1400 | | |
| Heat of fusion | 63.7 | 147 | 14.6 cal/g | 26.3 Btu/lb |
| Heat of vaporization | 760 | 1400 | 496 cal/g | 893 Btu/lb |
| Viscosity | 70 | 158 | 5.15 millipoises | 6.5 kinetic units |
| | 400 | 752 | 2.58 millipoises | 3.5 kinetic units |
| | 800 | 1472 | 1.36 millipoises | 2 kinetic units |
| Vapor pressure | 342 | 648 | 1 mm | 0.019 lb/in.$^2$ |
| | 696 | 1285 | 400 mm | 7.75 lb/in.$^2$ |
| Thermal conductivity | 200 | 392 | 0.017 cal/(s)(cm$^2$)(cm)(°C) | 26.0 Btu(h)(ft$^2$)(°F) |
| | 400 | 752 | 0.09 cal/(s)(cm$^2$)(cm)(°C) | 21.7 Btu/(h)(ft$^2$)(°F) |
| Heat capacity | 200 | 392 | 0.19 cal/(g)(°C) | 0.19 Btu/(lb)(°F) |
| | 800 | 1472 | 0.19 cal/(g)(°C) | 0.19 Btu/(lb)(°F) |
| Electrical resistivity | 150 | 302 | 18.7 microhm-cm | |
| | 300 | 572 | 28.2 microhm-cm | |
| Surface tension | 100–150 | 212–302 | About 80 dynes/cm | |

Potassium chloride, KCl, finds its main use in fertilizer mixtures. It also serves as the raw material for the manufacture of other potassium compounds. Potassium hydroxide, KOH, is used in the manufacture of liquid soaps, and potassium carbonate in making soft soaps. Potassium carbonate, $K_2CO_3$, is also an important raw material for the glass industry. Potassium nitrate, $KNO_3$, is used in matches, in pyrotechnics, and in similar items which require an oxidizing agent. [M.Si.]

Potassium is a very abundant element, ranking seventh among all the elements in the Earth's crust, 2.59% of which is potassium in combined form. Seawater contains 380 parts per million, making potassium the sixth most plentiful element in solution.

Potassium is even more reactive than sodium. It reacts vigorously with the oxygen in air to form the monoxide, $K_2O$, and the peroxide, $K_2O_2$. In the presence of excess oxygen, it readily forms the superoxide, $KO_2$.

Potassium does not react with nitrogen to form a nitride, even at elevated temperatures. With hydrogen, potassium reacts slowly at 200°C (392°F) and rapidly at 350–400°C (662–752°F). It forms the least stable hydride of all the alkali metals.

The reaction between potassium and water or ice is violent, even at temperatures as low as −100°C (−148°F). The hydrogen evolved is usually ignited in reaction at room temperature. Reactions with aqueous acids are even more violent and verge on being explosive.

The potassium ion ($K^+$) is the most common intracellular cation and is essential for maintaining osmotic pressure and electrodynamic cellular properties in organisms. The intracellular potassium ion concentrations are typically high for most cells, whereas the potassium ion concentrations present in extracellular fluids are significantly lower. The hydrolysis of the coenzyme adenosine triphosphate (ATP) is mediated by the membrane-bound enzyme $Na^+$, $K^+$-ATPase. This enzyme is called the sodium pump and it is activated by both potassium and sodium ions; however, many enzymes are activated by potassium ions alone (for example, pyruvate kinase, aldehyde dehydrogenase, and phosphofructokinase). *See* ADENOSINE TRIPHOSPHATE (ATP).

Potassium deficiency may occur in several conditions, including malnutrition and excessive vomiting or diarrhea, and in patients undergoing dialysis; supplementation with potassium salts is sometimes required. [D.M.DeF.]

**Praseodymium** A chemical element, Pr, atomic number 59, and atomic weight 140.91. Praseodymium is a metallic element of the rare-earth group. The stable isotope 140.907 makes up 100% of the naturally occurring element. The oxide is a black powder, the composition of which varies according to the method of preparation. If oxidized under a high pressure of oxygen it can approach the composition $PrO_2$. The black oxide dissolves in acid with the liberation of oxygen to give green solutions or green salts which have found application in the ceramic industry for coloring glass and for glazes. *See* PERIODIC TABLE; RARE-EARTH ELEMENTS. [F.H.Sp.]

**Precipitation (chemistry)** The process of producing a separable solid phase within a liquid medium. In analytical chemistry, precipitation is widely used to effect the separation of a solid phase in an aqueous solution. For example, the addition of a water solution of silver nitrate to a water solution of sodium chloride results in the formation of insoluble silver chloride. Quite often, one of the components in the solution is thus virtually completely separated in a relatively pure form. It can then be isolated from the solution phase by filtration or centrifugation, and the substance determined by weighing. This procedure is known as gravimetric analysis. Precipitation may also be used merely to effect partial or complete separation of a substance for purposes other than that of gravimetric analysis. Such purposes might involve either the isolation of a relatively pure substance or the removal of undesirable components of the solution. *See* GRAVIMETRIC ANALYSIS.

The extent to which a component can be separated from solution can be determined from the solubility-product constant obtained by determining the quantity of dissolved substance present in a known amount of saturated solution. This value is known as the solubility. The solubility can be drastically altered merely by adding to the solution any of the ions that make up the precipitate, for example, by adding varying quantities of either silver nitrate or sodium chloride to a saturated solution of silver chloride. Although solubility can be altered over a wide range, the solubility product itself remains practically constant over this same range. *See* SOLUBILITY PRODUCT CONSTANT.

Various techniques may be employed in order to reduce contamination by foreign ions. Precipitation from dilute solution is often effective. Heating the reaction mixture speeds recrystallization processes by which incorporated foreign ions may be returned to the solution phase. Precipitation from homogeneous solution results in the slow formation of large crystals of small surface area and hence lessens coprecipitation. If all these methods fail to reduce adequately the quantity of foreign ions incorporated in the solid phase, the precipitate is dissolved and reprecipitated by the previous procedure. *See* CHEMICAL SEPARATION TECHNIQUES; CRYSTALLIZATION; NUCLEATION. [L.Go./R.W.Mu.]

**Prochirality** The property displayed by a prochiral molecule or a prochiral atom (prostereoisomerism). A molecule or atom is prochiral if it contains, or is bonded to, two constitutionally identical ligands (atoms or groups), replacement of one of which by a different ligand makes the molecule or atom chiral. Examples are shown in structures (**1**)–(**5**).

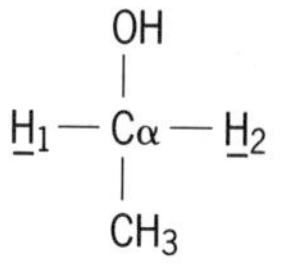

OH
$H_1$ — Cα — $H_2$
$CH_3$
(1)

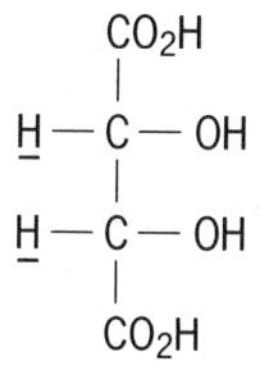

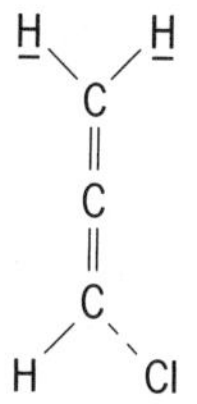

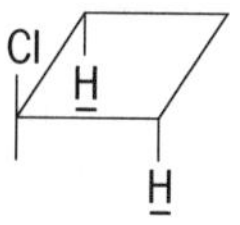

$CO_2H$
H — C — OH
H — C — OH
$CO_2H$
(2)

H H
C
C
C
H Cl
(3)

Cl
H
H
(4)

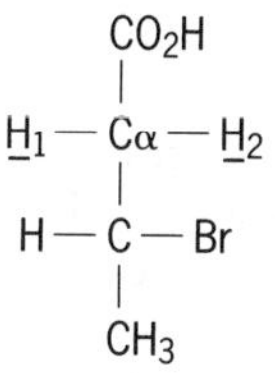

$CO_2H$
$H_1$ — Cα — $H_2$
H — C — Br
$CH_3$
(5)

None of molecules **1–4** is chiral, but if one of the underlined pair of hydrogens is replaced, say, by deuterium, chirality results in all four cases. In compound **1**, ethanol, a prochiral atom or center can be discerned ($C\alpha$:$CH_2$); upon replacement of H by D, a chiral atom or center is generated, whose configuration depends on which of the two pertinent atoms ($H_1$ or $H_2$) is replaced. Molecule **5** is chiral to begin with, but separate replacement of $H_1$ and $H_2$ (say, by bromine) creates a new chiral atom at $C\alpha$ and thus gives rise to a pair of chiral diastereomers. No specific prochiral atom can be discerned in molecules **2–4**, which are nevertheless prochiral (**3** has a prochiral axis).

Faces of double bonds may also be prochiral (and give rise to prochiral molecules), namely, when addition to one or other of the two faces of a double bond gives chiral products.

Although the term prochirality is widely used, especially by biochemists, a preferred term is prostereoisomerism. This is because replacement of one or other of the two corresponding ligands (called heterotopic ligands) or addition to the two heterotopic faces often gives rise to achiral diastereomers without generation of chirality. Thus not all compounds which display prostereoisomerism also display prochirality. *See* MOLECULAR ISOMERISM; STEREOCHEMISTRY. [E.L.E.]

**Promethium** A chemical element, Pm, atomic number 61. Promethium is the "missing" element of the lanthanide rare-earth series. The atomic weight of the most abundant separated radioisotope is 147. *See* PERIODIC TABLE.

Although a number of scientists have claimed to have discovered this element in nature as a result of observing certain spectral lines, no one has succeeded in isolating element 61 from naturally occurring materials. It is produced artificially in nuclear reactors, since it is one of the products that results from the fission of uranium, thorium, and plutonium.

All the known isotopes are radioactive. Its principal uses are for research involving tracers. Its main application is in the phosphor industry. It has also been used to manufacture thickness gages and as a nuclear-powered battery in space applications. *See* RARE-EARTH ELEMENTS. [F.H.Sp.]

**Protactinium** A chemical element, Pa, atomic number 91. Isotopes of mass numbers 216, 217, and 222–238 are known, all of them radioactive. Only $^{231}Pa$, the parent of actinium, $^{234}Pa$, and $^{233}Pa$ occur in nature. The most important of these is $^{231}Pa$, an $\alpha$-emitter with a half-life of 32,500 years. The artificial isotope, $^{233}Pa$, is important as an intermediary in the production of fissile $^{233}U$. Both $^{231}Pa$ and $^{233}Pa$ can be synthesized by neutron irradiation of thorium. *See* ACTINIUM; PERIODIC TABLE; URANIUM.

Protactinium is, formally, the third member of the actinide series of elements and the first in which a 5*f* electron appears, but its chemical behavior in aqueous solution resembles that of tantalum and niobium more closely than that of the other actinides. *See* NIOBIUM; TANTALUM.

Metallic protactinium is silver in color, malleable, and ductile. The crystal structure is body-centered tetragonal. Samples exposed to air at room temperature show little or no tarnishing over a period of several months. The numerous compounds of protactinium that have been prepared and characterized include binary and polynary oxides, halides, oxyhalides, sulfates, oxysulfates, double sulfates, oxynitrates, selenates, carbides, organometallic compounds, and noble metal alloys. *See* ACTINIDE ELEMENTS.

[H.W.Ki.]

**Protein** A biological macromolecule made up of various $\alpha$-amino acids that are joined by peptide bonds. A peptide bond is an amide bond formed by the reaction of an $\alpha$-amino group (—$NH_2$) of one amino acid with the carboxyl group (—COOH) of another, as shown below. Proteins generally contain from 50 to 1000 amino acid residues per polypeptide chain.

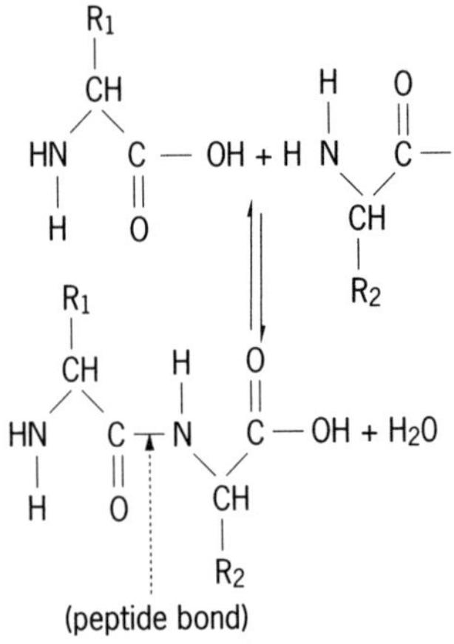

```
          R1
          |
          CH                    H    O
         /  \                   |    ||
      HN      C — OH + H N      C — OH
      |       ||              \   /
      H       O                CH
                                |
                                R2
                  ⇅
          R1
          |
          CH       H    O
         /  \      |    ||
      HN      C ─ N     C — OH + H2O
      |       ||  :  \  /
      H       O   :   CH
                  :    |
                  :    R2
           (peptide bond)
```

*See* PEPTIDE.

**Occurrence.** Proteins are of importance in all biological systems, playing a wide variety of structural and functional roles. They form the primary organic basis of structures such as hair, tendons, muscle, skin, and cartilage. All of the enzymes, the catalysts in biochemical transformations, are protein in nature. Many hormones, such as insulin and growth hormone, are proteins. The substances responsible for oxygen and electron transport (hemoglobin and the cytochromes, respectively) are conjugated proteins that contain a metalloporphyrin as the prosthetic group. Chromosomes are highly complex nucleoproteins, that is, proteins conjugated with nucleic acid. Viruses are also nucleoprotein in nature. Of the more than 200 amino acids that have been discovered either in the free state or in small peptides, only 20 amino acids are present in mammalian proteins. Thus, proteins play a fundamental role in the processes of life. *See* AMINO ACIDS.

**Specificity.** The linear arrangement of the amino acid residues in a protein is termed its sequence (primary structure). The sequence in which the different amino acids are linked in any given protein is highly specific and characteristic for that particular protein.

This specificity of sequence is one of the most remarkable aspects of protein chemistry. The number of possible permutations of sequence in even so small a protein as insulin, of molecular weight 5732 and with 51 amino acid residues, is astronomic: 1051 permutations. Yet it has been established that the pancreatic cell of a given species has only one of these possible sequences. The elucidation of the mechanism conferring such a high degree of specificity on the biosynthetic reactions by which proteins are built up from free amino acids has been one of the key problems of modern biochemistry.

Proteins are not stretched polymers; rather, the polypeptide backbone of the molecule can fold in several ways by means of hydrogen bonds between the carbonyl oxygen and the amide nitrogen. The folding of each protein is determined by its particular sequence of amino acids. The long polypeptide chains of proteins, particularly those of the fibrous proteins, are held together in a rather well-defined configuration. The backbone is coiled in a regular fashion, forming an extended helix. As a result of this coiling, peptide bonds separated from one another by several amino acid residues are brought into close spatial approximation. The stability of the helical configuration can be attributed to hydrogen bonds between these peptide bonds.

In addition to hydrogen bonds, there are electrostatic interactions, such as those between $COO^-$ and $NH_3^+$ groups of the side chains, and van der Waals forces, that is, hydrophobic interactions, which help to determine the configuration of the polypeptide chain. The term secondary structure is used to refer to all those structural features of the polypeptide chain determined by noncovalent bonding interactions.

In addition to the $\alpha$-helical sections of proteins, there are segments that contain $\beta$-structures in which there are hydrogen bonds between two polypeptide chains that run in parallel or antiparallel fashion.

The tertiary structure (third level of folding) of a protein comes about through various interactions between different parts of the molecule. Disulfide bridges formed between cysteine residues at different locations in the molecule can stabilize parts of a three-dimensional structure by introducing a primary valence bond as a cross-link. Hydrogen bonds between different segments of the protein, hydrophobic bonds between nonpolar side chains of amino acids such a phenylalanine and leucine, and salt bridges such as those between positively charged lysyl side chains and negatively charged aspartyl side chains all contribute to the individual tertiary structure of a protein.

Finally, for those proteins that contain more than one polypeptide chain per molecule, there is usually a high degree of interaction between each subunit, for example, between the $\alpha$- and $\beta$-polypeptide chains of hemoglobin. This feature of the protein structure is termed its quarternary structure.

**Properties.** The properties of proteins are determined in part by their amino acid composition. As macromolecules that contain many side chains that can be protonated and unprotonated depending upon the pH of the medium, proteins are excellent buffers. The fact that the pH of blood varies only very slightly in spite of the numerous metabolic processes in which it participates is due to the very large buffering capacity of the blood proteins.

**Biosynthesis.** The processes by which proteins are synthesized biologically have become one of the central themes of molecular biology. The sequence of amino acid residues in a protein is controlled by the sequence of the DNA as expressed in messenger RNA at ribosomes.

[G.E.Pe.; J.M.Ma.]

**Degradation.** As with many other macromolecular components of the organism, most body proteins are in a dynamic state of synthesis and degradation (proteolysis). During proteolysis, the peptide bond that links the amino acids to each other is hydrolyzed, and free amino acids are released. The process is carried out by a diverse group of enzymes called proteases. During proteolysis, the energy invested in generation of the proteins is released. *See* ENZYME.

Distinct proteolytic mechanisms serve different physiological requirements. Proteins can be divided into extracellular and intracellular, and the two groups are degraded by two distinct mechanisms. Extracellular proteins such as the plasma immunoglobulins and albumin are degraded in a process known as receptor-mediated endocytosis. Ubiquitin-mediated proteolysis of a variety of cellular proteins plays an important role in many basic cellular processes such as the regulation of cell cycle and division, differentiation, and development; DNA repair; regulation of the immune and inflammatory responses; and biogenesis of organelles. [A.J.Ci.]

**Molecular chaperones.** Molecular chaperones are specialized cellular proteins that bind nonnative forms of other proteins and assist them to reach a functional conformation. The role of chaperone proteins under conditions of stress, such as heat shock, is to protect proteins by binding to misfolded conformations when they are just starting to form, preventing aggregation; then, following return of normal conditions, they allow refolding to occur. Chaperones also play essential roles in folding under normal conditions, providing kinetic assistance to the folding process, and thus improving the overall rate and extent of productive folding. [A.Ho.]

**Protein engineering.** The amino acid sequences, sizes, and three-dimensional conformations of protein molecules can be manipulated by protein engineering, in which the basic techniques of genetic engineering are used to alter the genes that encode proteins. These manipulations are used to generate proteins with novel activities or properties for specific applications, to discover structure-function relationships, and to generate biologically active minimalist proteins (containing only those sequences necessary for biological activity) that are smaller than their naturally occurring counterparts.

Many subtle variations in a particular protein can be generated by making amino acid replacements at specific positions in the polypeptide sequence. For example, at any specific position an amino acid can be replaced by another to generate a mutant protein that may have different characteristics by virtue of the single replaced amino acid. Amino acids can also be deleted from a protein sequence, either individually or in groups. These proteins are referred to as deletion mutants. Deletion mutants may or may not be missing one or more functions or properties of the full, naturally occurring protein. Moreover, part or all of a protein sequence can be joined or fused to that of another protein. The resulting protein is called a hybrid or fusion protein, which generally has characteristics that combine those of each of the joined partners. [P.Sc.]

**Proton-induced x-ray emission (PIXE)** A highly sensitive analytic technique for determining the composition of elements in small samples. Proton-induced x-ray emission (PIXE) is a nondestructive method capable of analyzing many elements simultaneously at concentrations of parts per million in samples as small as nanograms. PIXE is the preferred technique for surveying the environment for trace quantities of such toxic elements as lead and arsenic. There has also been a rapid development in the use of focused proton beams for PIXE studies in order to produce two-dimensional maps of the elements at spatial resolutions of micrometers.

The typical PIXE apparatus uses a small Van de Graaff machine to accelerate the protons which are then guided to the sample. Nominal proton energies are between

1 and 4 MeV; too low an energy gives too little signal while too high an energy produces too high a background. The energetic protons ionize some of the atoms in the sample, and the subsequent filling of empty inner orbits results in the characteristic x-rays. These monoenergetic x-rays emitted by the sample are then efficiently counted in a high-resolution silicon (lithium) detector which is sensitive to the x-rays of all elements heavier than sodium.

The advantages of PIXE over electron-induced x-ray techniques derive from the heaviness of the proton which permits it to move through matter with little deflection. The absence of scattering results in negligible continuous radiation (bremsstrahlung). As a result, proton-induced x-ray techniques are two to three orders of magnitude more sensitive to trace elements than are techniques based on electron beams. [L.Gro.]

**Purine** A heterocyclic organic compound (**1**) containing fused pyrimidine and imidazole rings. A number of substituted purine derivatives occur in nature; some, as

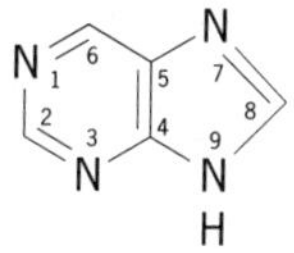

(1)

components of nucleic acids and coenzymes, play vital roles in the genetic and metabolic processes of all living organisms. *See* COENZYME; NUCLEIC ACID.

Purines are generally white solids of amphoteric character. They can form salts with both acids and bases. Conjugated double bonds in purines results in aromatic chemical properties, that confers considerable stability, and accounts for their strong ultraviolet absorption spectra. With the exception of the parent compound, most substituted purines have low solubilities in water and organic solvents.

The purine bases, adenine (**2**) and guanine (**3**), together with pyrimidines, are fundamental components of all nucleic acids. Certain methylated derivatives of adenine and guanine are also present in some nucleic acids in low amounts. In biological systems, hypoxanthine (**4**), adenine, and guanine occur mainly as their 9-glycosides, the sugar

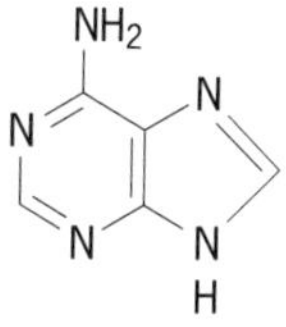

(2) 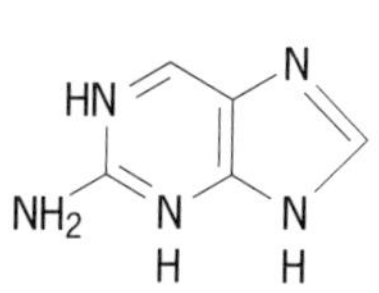 (3) 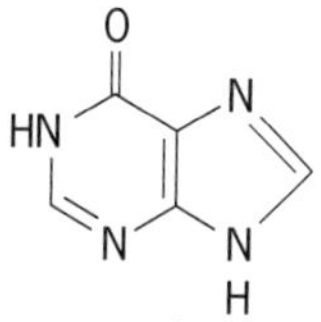 (4)

being either ribose or 2-deoxyribose. Such compounds are termed nucleosides generically, and inosine (hypoxanthine nucleoside), adenosine, or guanosine specifically. The principal nucleotides contain 5′-phosphate groups, as in guanosine 5′-phosphate (GTP) and adenosine 5′-triphosphate (ATP). *See* ADENOSINE TRIPHOSPHATE (ATP).

Most living organisms are capable of synthesizing purine compounds. The sequence of enzymatic reactions by which the initial purine product, inosine 5′-phosphate, is formed utilizes glycine, carbon dioxide, formic acid, and amino groups derived from glutamine and aspartic acid. Adenosine 5′-phosphate and guanosine 5′-phosphate are formed from inosine 5′-phosphate.

Metabolic degradation of purine derivatives may also occur by hydrolysis of nucleotides and nucleosides to the related free bases. Deamination of adenine and guanine produces hypoxanthine and xanthine (**5**), both of which may be oxidized to uric acid (**6**).

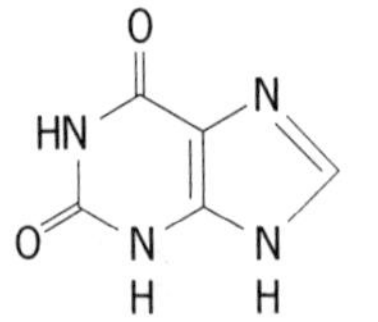

(5) 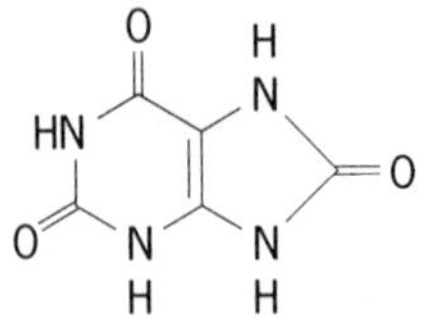 (6)

Purine-related compounds have been investigated as potential chemotherapeutic agents. In particular, 6-mercaptopurine, in the form of its nucleoside phosphate, inhibits several enzymes required for synthesis of adenosine and guanosine nucleotides, and thus proves useful in selectively arresting the growth of tumors. The pyrazolopyrimidine has been used in gout therapy. As a purine analog, this agent serves to block the biosynthesis of inosine phosphate, as well as the oxidation of hypoxanthine and xanthine to uric acid. As a result of its use, overproduction of uric acid is prevented and the primary cause of gout is removed. *See* PYRIMIDINE. [S.C.H.]

**Pyridine** An organic heterocyclic compound containing a triunsaturated six-membered ring of five carbon atoms and one nitrogen atom. Pyridine (**1**) and pyridine

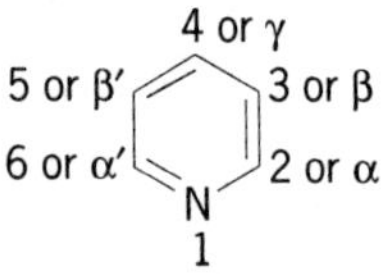

(1)

homologs are obtained by extraction of coal tar or by synthesis. The pyridine system is found in natural products, for example, in nicotine (**2**) from tobacco, in ricinine (**3**) from castor bean, in pyridoxine or vitamin $B_6$ (**4**), in nicotinamide or niacinamide or vitamin P (**5**), and in several groups of alkaloids. *See* HETEROCYCLIC COMPOUNDS.

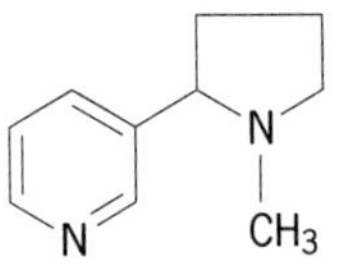

(2) 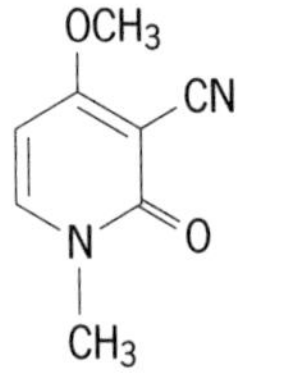 (3)

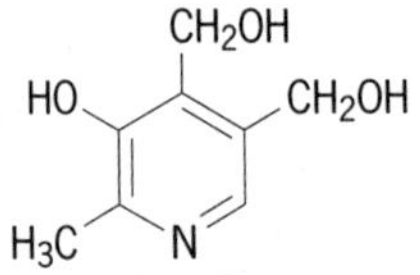

(4) 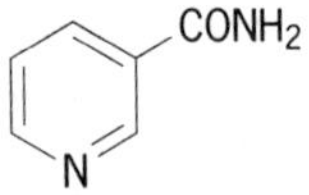 (5)

Pyridine (**1**) is a colorless, hygroscopic liquid with a pungent, unpleasant odor. When anhydrous it boils at 115.2–115.3°C (239.4–239.5°F). Pyridine is miscible with organic solvents as well as with water. The pyridine system is aromatic. It is stable to heat, to acid, and to alkali. Pyridine is used as a solvent for organic and inorganic compounds, as an acid binder, as a basic catalyst, and as a reaction intermediate.

Pyridine is an irritant to skin (eczema) and other tissues (conjunctivitis), and chronic exposure has been known to cause liver and kidney damage. Repeated exposure to atmospheric levels greater than 5 parts per million is considered hazardous. [W.J.Ge.]

**Pyrimidine** A heterocyclic organic compound (**1**) containing nitrogen atoms at positions 1 and 3. Naturally occurring derivatives of the parent compound are of considerable biological importance as components of nucleic acids and coenzymes and, in addition, synthetic members of this group have found use as pharmaceuticals. *See* COENZYME; NUCLEIC ACID.

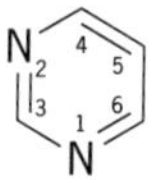

(1)

Pyrimidine compounds which are found universally in living organisms include uracil (**2**), cytosine (**3**), and thymine (**4**). Together with purines these substances make

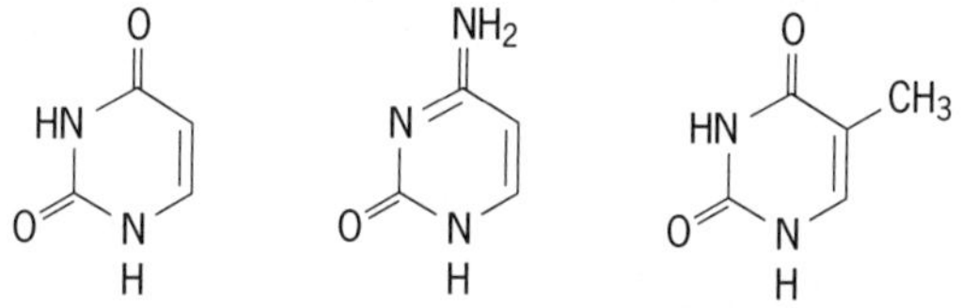

(2) (3) (4)

up the "bases" of nucleic acids, uracil and cytosine being found characteristically in ribonucleic acids, with thymine replacing uracil in deoxyribonucleic acids. A number of related pyrimidines also occur in lesser amounts in certain nucleic acids. Other pyrimidines of general natural occurrence are orotic acid and thiamine (vitamin $B_1$). *See* DEOXYRIBONUCLEIC ACID (DNA); PURINE.

Among the sulfa drugs, the pyrimidine derivatives, sulfadiazine, sulfamerazine, and sulfamethazine, have general formula (**5**). These agents are inhibitors of folic acid

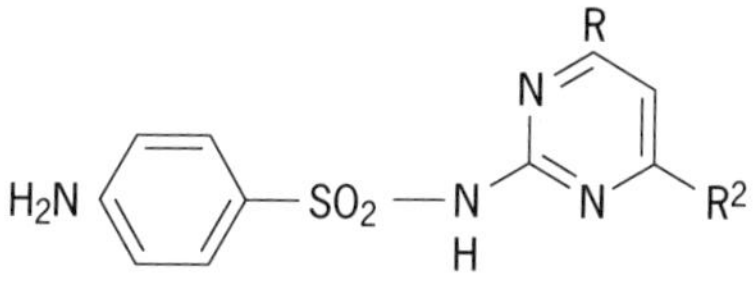

(5)

biosynthesis in microorganisms. The barbiturates are pyrimidine derivatives which possess potent depressant action on the central nervous system. *See* SULFONAMIDE. [S.C.H.]

**Pyrolysis** A chemical process in which a compound is converted to one or more products by heat. By this definition, reactions that occur by heating in the presence of a catalyst, or in the presence of air when oxidation is usually a simultaneous reaction, are excluded. The terms thermolysis or thermal reaction have been used in essentially the same sense as pyrolysis. A simple example of pyrolysis is the classic experiment in which oxygen was first prepared by heating mercuric oxide [reaction (1)]. Similar

$$HgO \xrightarrow[(570°F)]{300°C} Hg + {}^1/_2O_2 \quad (1)$$

reactions occur with numerous other metallic oxides and salts. Thermal decomposition or calcining of limestone (calcium carbonate) is the basic step in the manufacture of lime [reaction (2)].

$$CaCo_3 \xrightarrow[(2200^\circ F)]{1200^\circ C} CaO + CO_2 \quad (2)$$

The term pyrolysis is most commonly associated with thermal reactions of organic compounds. Pyrolysis of material from plant and animal sources provided some of the first clues about constitution, as in the formation of isoprene from the thermal breakdown of rubber. A range of substances, including benzene, naphthalene, pyridine, and many other aromatic compounds, was obtained from coal tar, a pyrolysis product of coal. All of these pyrolysis processes lead to formation of volatile products characteristic of the source and also residues of char with high carbon content.

Pyrolysis reactions have been used as preparative methods and as means of generating transient intermediates that can be trapped or observed spectroscopically, or quenched by a further reaction. For preparative purposes, pyrolysis can generally be carried out by a flow process in which the reactant is vaporized with a stream of inert gas through a heated tube, sometimes at reduced pressure. In flash vacuum pyrolysis, the apparatus is placed under very low pressure, and the material to be pyrolyzed is vaporized by molecular distillation. *See* CHEMICAL DYNAMICS.

**Types of reactions.** At temperatures of 600–800°C (1100–1500°F), most organic compounds acquire sufficient vibrational energy to cause breaking of bonds with formation of free radicals. Alkanes undergo rupture of carbon-hydrogen (C-H) and carbon-carbon (C-C) bonds to two radicals that then react to give lower alkanes, alkenes, hydrogen, and also higher-molecular-weight compounds resulting from their recombination [reactions (3)]. These reactions are the basis of the thermal

$$CH_3CH_2CH_2CH_3 \xrightarrow[(1300^\circ F)]{700^\circ C} 2CH_3CH_2\cdot \rightarrow CH_3CH_3 + H_2C{=}CH_2 \quad (3a)$$

$$CH_3CH_2CH_2CH_3 \xrightarrow[(1300^\circ F)]{700^\circ C} CH_3CH_2\cdot CHCH_3 + H\cdot \rightarrow CH_3CH_2HC{=}CH_2 + H_2 \quad (3b)$$

cracking processes used in petroleum refining. Pyrolysis of simple aromatic hydrocarbons such as benzene or naphthalene produces aryl radicals, which can attack other hydrocarbon molecules to give bi- and polyaryls, as shown in reaction (4) for the formation of biphenyl.

Benzene → Phenyl radical + H• —benzene→ Biphenyl (4)

*See* CRACKING; FREE RADICAL.

Pyrolytic eliminations can result in formation of a multiple bond by loss of HX from a compound H—C—C—X, where X = any leaving group. A typical example is the pyrolysis of an ester, which is one of the general methods for preparing alkenes. Pyrolytic elimination is particularly useful when acid-catalyzed dehydration of the parent alcohol leads to cationic rearrangement. Another useful application of this process is the production of ketenes from acid anhydrides [reaction (5)].

$$\underset{\text{Acid anhydride}}{\mathrm{RCH_2\overset{O}{\overset{\|}{C}}O\overset{O}{\overset{\|}{C}}CH_2R}} \xrightarrow[(1300^\circ F)]{700^\circ C} \underset{\text{Ketene}}{\mathrm{RCH{=}C{=}O}} + \underset{\text{Acid}}{\mathrm{RCH_2CO_2H}} \quad (5)$$

*See* ACID ANHYDRIDE; ALKENE; ESTER.

Another type of thermal elimination occurs by loss of a small molecule such as nitrogen ($N_2$), carbon monoxide (CO), carbon dioxide ($CO_2$), or sulfur dioxide ($SO_2$), leading to reactive intermediates such as arynes, diradicals, carbenes, or nitrenes. The nitrene generated from aminobenzotriazole breaks down to benzyne at 0°C (32°F). Benzyne can be trapped by addition reaction or can dimerize to biphenylene. *See* REACTIVE INTERMEDIATES.

A number of pyrolytic reactions involve cleavage of specific C-C bonds in a carbon chain or ring. Fragmentation accompanied by transfer of hydrogen is a general reaction that occurs by a cyclic process. An example is decarboxylation of acids that contain a carbonyl group, which lose $CO_2$ on relatively mild heating. Acids with a double or triple carbon-to-carbon bond undergo decarboxylation at 300–400°C (570°–750°F). This type of reaction also occurs at higher temperatures with unsaturated alcohols and, by transfer of hydrogen from a C-H bond, with unsaturated ethers. Cleavage of a ring frequently occurs on pyrolysis. With alicyclic or heterocyclic four-membered rings, cleavage into two fragments is the reverse of 2 + 2 cycloaddition, as illustrated by the cracking of diketene [reaction (6)]. Pyrolysis is an important reaction in the chemistry

$$\text{Diketene} \xrightarrow[(840^\circ F)]{450^\circ C} \underset{\text{Ketene}}{\mathrm{CH_2{=}C{=}O}} \quad (6)$$

of the pine terpenes, as in the conversion of $\beta$-pinene to myrcene. Benzocyclobutenes undergo ring opening to *o*-quinone dimethides. By combining this reaction in tandem with formation of the benzocyclobutene and a final Diels-Alder reaction, a versatile one-step synthetic method for the steroid ring system has been developed. *See* DIELS-ALDER REACTION; PINE TERPENE.

Many thermal reactions involve isomerization without elimination or fragmentation. These processes can occur by way of intermediates such as diradicals, as in the pyrolysis of pinene, or they may be concerted pericyclic reactions. An example of the latter is the Claisen-Cope rearrangement of phenyl or vinyl ethers and other 1,5-diene systems. These reactions can be carried out by relatively mild heating, and they are very useful in synthesis. *See* ORGANIC SYNTHESIS; PERICYCLIC REACTION.

**Analytical applications.** Thermal breakdown of complex structures leads to very complex mixtures of products arising from concurrent dissociation, elimination, and bond fission. Separation of these mixtures provides a characteristic pyrogram that is valuable as an analytical method, particularly for polymeric materials of both biological and synthetic origin. In this application, a small sample is heated on a hot filament or by laser. The pyrolysis products are then analyzed by gas chromatography, mass spectrometry, or a combination of both techniques. *See* GAS CHROMATOGRAPHY; MASS SPECTROMETRY; THERMOANALYSIS.

Instrumentation with appropriate interfaces and data-handling systems has been developed to permit rapid and sensitive detection of pyrolysis products for a number of applications. One example is the optimization of conditions in petroleum cracking to produce a desired product from varied crude oils. The profile of pyrolysis fragments from a polymer can also be used to detect impurities. [J.A.Mo.]

**Pyrrole** One of a group of organic compounds containing a doubly unsaturated five-membered ring in which nitrogen occupies one of the ring positions. Pyrrole (**1**) is a representative compound. The pyrrole system is found in the green leaf pigment, chlorophyll, in the red blood pigment, hemoglobin, and in the blue dye, indigo. Interest in these colored bodies has been largely responsible for the intensive study of pyrroles. Tetrahydropyrrole, or pyrrolidine (**2**), is part of the structures of two protein amino

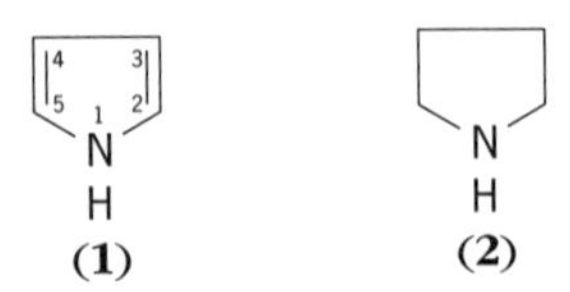


acids, proline and hydroxyproline. *See* HETEROCYCLIC COMPOUNDS; INDOLE; PORPHYRIN.

Pyrrole is a liquid that darkens and resinifies on standing in air, and that polymerizes quickly when treated with mineral acid. Familiar substitution processes, such as halogenation, nitration, sulfonation, and acylation, can be realized. Pyrrole, by virtue of its heterocyclic nitrogen, is very weakly basic. The hydrogen at the 1 position is removable as a proton, and accordingly, pyrrole is also an acid, although a weak one.

Pyrrolidine can be prepared by catalytic hydrogenation of pyrrole or by ring-closure reactions. Another derivative of pyrrole, 2-ketopyrrolidine, or pyrrolidone, is of considerable interest in connection with the preparation of polyvinylpyrrolidone. Pyrrolidone is combined with acetylene to form vinylpyrrolidone; polymerization of this material furnishes polyvinylpyrrolidone, which is suitable for maintaining osmotic pressure in blood and so acting as an extender for plasma or whole blood. *See* PYRIDINE. [W.J.Ge.]

# Q

**Qualitative chemical analysis** The branch of chemistry concerned with identifying the elements and compounds present in a sample of matter. Inorganic qualitative analysis traditionally used classical "wet" methods to detect elements or groups of chemically similar elements, but instrumental methods have largely superseded the test-tube methods. Methods for the detection of organic compounds or classes of compounds have become increasingly available and important in organic, forensic, and clinical chemistry. Once it is known which elements and compounds are present, the role of quantitative analysis is to determine the composition of the sample. *See* ANALYTICAL CHEMISTRY; QUANTITATIVE CHEMICAL ANALYSIS.

**Inorganic analysis.** The operating principles of all systematic inorganic qualitative analysis schemes for the elements are similar: separation into groups by reagents producing a phase change; isolation of individual elements within a group by selective reactions; and confirmation of the presence of individual elements by specific tests.

Through usage and tradition, descriptive terms for sample sizes have been:

| | |
|---|---|
| macro | 0.1 gram or more |
| semimicro | 0.01 to 0.1 gram |
| micro | 1 milligram (1 mg or $10^{-3}$ g) |
| ultramicro | 1 microgram (1 $\mu$g or $10^{-6}$ g) |
| submicrogram | less than 1 microgram |

For defining the smallest amount of a substance that can be detected by a given method, the term "limit of identification" is used. Under favorable conditions, an extremely sensitive method can detect as little as $10^{-15}$ g.

Spot tests are selective or specific single qualitative chemical tests that are carried out on a spot plate (a glass or porcelain plate with small depressions in which drop-size reactions can be carried out), on paper, or on a microscope slide. On paper or specially prepared adsorbent surfaces, spot tests become one- or two-dimensional through the use of solvent migration and differential adsorption (thin-layer chromatography). Containing selected indicator dyes, pH indicator paper strips are widely used for pH estimation. Solid reagent monitoring devices and indicator tubes are used for the detection and estimation of pollutant gases in air. *See* CHROMATOGRAPHY; PH.

By use of the microscope, crystal size and habit can be used for qualitative identification. With the addition of polarized light, chemical microscopy becomes a versatile method. *See* CHEMICAL MICROSCOPY; FORENSIC CHEMISTRY.

Any instrumental method of quantitative analysis can be adapted to qualitative analysis. Some, such as electrochemical methods, are not often used in qualitative analysis (an exception would be the ubiquitous pH meter). Others, such as column chromatography (gas and liquid) and mass spectrometry, are costly and uncomplicated but capable of providing unique results. Emission spectroscopy is important in establishing the presence or absence of a suspected element in forensic analysis. The simple qualitative flame test developed into flame photometry and subsequently into atomic absorption spectrometry. *See* ATOMIC SPECTROMETRY.

Bombardment of surfaces by x-rays, electrons, and positive ions has given rise to a number of methods useful in analytical chemistry. X-ray diffraction is used to determine crystal structure and to identify crystalline substances by means of their diffraction patterns. X-ray fluorescence analysis employs x-rays to excite emission of characteristic x-rays by elements. The electron microprobe uses electron bombardment in a similar manner to excited x-ray emission. *See* SECONDARY ION MASS SPECTROMETRY (SIMS).

Neutron activation analysis, wherein neutron bombardment induces radioactivity in isotopes of elements not naturally radioactive, involves measurement of the characteristic gamma radiation and modes of decay produced in the target atoms. *See* ACTIVATION ANALYSIS. [J.L.L.]

**Organic analysis.** Qualitative analysis of an organic compound is the process by which the characterization of its class and structure is determined. Due to the numerous classes of organic compounds and the complexity of their molecular structures, a systematic analytical procedure is often required. A typical procedure entails an initial assignment of compound classification, followed by a complete identification of the molecular structure.

The initial step is an examination of the physical characteristics. Color, odor, and physical state can be valuable clues. The physical constants of an unknown compound provide pertinent data for the analyst. Constants such as melting point, boiling point, specific gravity, and refractive index are commonly measured.

The preliminary chemical tests which are applied are elemental analysis procedures. The elements generally associated with carbon, hydrogen, and oxygen are sulfur, nitrogen, and the halogens. The analyst is usually interested in the latter group. These elements are converted to water-soluble ionic compounds via sodium fusion. The resulting products are then detected with wet chemical tests. The solubility of a compound in various liquids provides information concerning the molecular weight and functional groups present in the compound. In order to indicate the presence or absence of a functional group, specific classification reactions are tested. The reactions are simple, are rapid, and require a small quantity of sample. No single test is conclusive evidence. A judicious choice of reactions can confirm or negate the presence of a functional group.

**Chemical and physical structure.** Instrumental methods are commonly applied for functional group determination and structure identification. Absorption spectroscopy methods are among the most important techniques. Whenever a molecule is exposed to electromagnetic radiation, certain wavelengths cause vibrational, rotational, or electronic effects within the molecule. The radiation required to cause these effects is absorbed. The nature and configuration of the atoms determine which specific wavelengths are absorbed. Radiation in the ultraviolet and visible regions is associated with electronic effects, and infrared radiation is associated with rotational and vibrational effects.

Infrared absorption spectroscopy is perhaps the most valuable instrument for functional group determination based on the wavelengths at which radiation is absorbed. Ultraviolet absorption spectroscopy is similar to that of infrared spectroscopy, except that ultraviolet scans normally exhibit one or two broad bands of limited utility for qualitative purposes. Visible absorption spectroscopy suffers the same limitations. Raman spectroscopy is slightly different from the other spectroscopic techniques. A sample is irradiated by a monochromatic source. Depending upon the vibrational and rotational energies of the functional groups in the sample, light is scattered from the sample in a way which is characteristic of the functional groups. Raman spectroscopy is a fine complement to infrared spectroscopy because the two are sensitive for different groups. *See* SPECTROSCOPY.

Instrumental techniques such as infrared spectroscopy, nuclear magnetic resonance, and mass spectrometry are commonly used for structure identification. Less common techniques are electron spin resonance, x-ray diffraction, and nuclear quadrupole spectroscopy. An infrared spectrum can identify an unknown compound by comparison to known spectra. In mass spectroscopy a spectrum is obtained by bombarding a sample with an electron beam. The resulting fragmentation pattern and the mass-to-charge ratio of the fragments enables structural information to be obtained. Nuclear magnetic resonance is another powerful structural tool in which a sample is irradiated and the absorbed radiation is monitored. When a sample is placed in a magnetic field, the nuclei are aligned in a given direction. In order to reverse the direction, energy is required. The changing of spin states results in absorption of the applied radio-frequency radiation. Either the radio frequency or the magnetic field can be held constant while the other is changed and the radiation absorption is monitored. Based on the number and intensity of absorptions and the field strength required, structural information can be obtained. *See* MASS SPECTROMETRY; NUCLEAR MAGNETIC RESONANCE (NMR).

In addition to chemical structure, the physical structure of a sample can be important. Thermal analysis is a useful procedure for examining structural characteristics. Among the most popular methods are thermogravimetry, differential thermal analysis, differential scanning calorimetry, and thermal mechanical analysis. *See* CALORIMETRY.

[S.S.; K.Lo.]

**Quantitative chemical analysis** The determination of the amount of an element on compound in a sample. Selection of a technique is based in part on the size of sample available, the quantity of analyte expected to be in the sample, the precision and accuracy of the technique, and the speed of analysis required. All techniques require calibration with respect to some standard of known composition. Caution is necessary to prevent other substances from giving signals falsely attributable to the sought-for substance, called the analyte.

Direct measurement of a signal related to concentration or activity of the chemical species of interest is the most intuitive approach. Generally, a linear relationship between signal (or its logarithm) and concentration (or its logarithm) is sought. The relationship between signal and concentration is called a working curve. The slope of the line describing the relationship is known as sensitivity. The smallest quantity which is measurably different from the absence of analyte is the detection limit. Although a linear working curve is the simplest form to use, nonlinear curves may still be employed (either graphically or with the use of computers).

Titration is the process by which an unknown quantity of analyte (generally in solution) is determined by adding to it a standard reagent with which it reacts in a definite and known proportion. A chemical or instrumental means is provided to indicate when the standard reagent has consumed exactly the amount of analyte initially present. Each determination is performed with a reagent whose concentration is directly traceable to a primary standard, so that accuracy is frequently superior to that of other methods. *See* STOICHIOMETRY; TITRATION.

The sensitivity of a method is related to the method and analyte of interest, as well as to the presence of other species in the sample. The materials other than the analyte constitute a sample matrix. For example, seawater is a matrix quite different from distilled water because of the large amount of dissolved electrolyte present. If a signal can be derived from the analyte which is proportional to the amount of the analyte, but the sensitivity of the signal to concentration varies as a function of the sample matrix, the method of standard additions may be useful. The signal from the analyte is measured, after which small, known quantities of analyte are successively added to the sample, and the signal remeasured. The sensitivity of the method can thus be obtained, and the initial, preaddition signal interpreted to give the amount of analyte ($C_1$) initially present by using the relationship shown in the equation below, where $S_1$ is the initial

$$C_1 = \frac{\Delta C}{\Delta S} \cdot S_1$$

signal, $C_1$ the initial quantity of analyte, $\Delta C$ the standard addition of analyte, and $\Delta S$ the change in the signal caused by the addition. The sensitivity is $\Delta C/\Delta S$.

This approach is applicable only in the absence of reagent blanks, that is, signals caused by the presence of analyte in the reagent used for the determination.

In many methods, aliquots of samples are introduced into the measurement instrument. The signal from the analyte may vary with sample uptake rate or volume. To compensate for such effects, an internal standard, or species other than the analyte, may be added to the analyte in a known concentration prior to determination. The signal due to the internal standard is measured simultaneously with the analyte signal. Variation of the signal of the internal standard is interpreted to indicate the variation in sample uptake, which should be the same for both analyte and internal standard. The ratio of analyte signal to internal standard signal is independent of sample uptake. Thus the ratio of analyte to internal standard signal is used to establish the working curve, rather than using analyte signal alone.

If the relationship between signal and analyte concentration is nonlinear, quantitation may require the use of null comparison. A signal is observed from the analyte and from a standard whose concentration can be adjusted in a known way. When the signal from the adjustable standard equals the signal from the analyte, the two have identical concentrations. This condition is called a null, as there is no detectable difference between the sample and reference signals.

It is good laboratory practice to check every quantitative measurement for the influence of species other than the one being sought. For example, a glass electrode designed to sense hydrogen ion will also respond to high concentrations of sodium ion. The degree to which a given sensor responds to one species in preference to another is called the selectivity coefficient. If a general detector is desired, this coefficient ideally should be 1.0. If a species-specific detector is desired, the coefficient should be infinite. A signal of unknown or general origin which appears to underlie the analyte signal is known as background. This quantity, together with that for the reagent blank, may be subtracted from the raw signal and thus be compensated. However, variations in their level may prevent reliable compensation, particularly when small quantities of analyte are to be determined.

For further information on common quantitative techniques *see* ACTIVATION ANALYSIS; CALORIMETRY; CHEMICAL MICROSCOPY; CHROMATOGRAPHY; COMBUSTION; ELECTRON SPECTROSCOPY; ELECTROPHORESIS; GAS CHROMATOGRAPHY; GEL PERMEATION CHROMATOGRAPHY; ION EXCHANGE; ISOTOPE DILUTION TECHNIQUES; KINETIC METHODS OF ANALYSIS; MASS SPECTROMETRY; NUCLEAR MAGNETIC RESONANCE (NMR); POLARIMETRIC ANALYSIS; SPECTROSCOPY; X-RAY SPECTROMETRY. [A.Sche.]

**Quantum chemistry** A branch of chemistry concerned with the application of quantum mechanics to chemical problems. More specifically, it is concerned with the electronic structure of molecules. Methods developed since 1960 permit the quantum chemist to obtain reliable approximate solutions to the nonrelativistic Schrödinger equation. The method which dominates the field of quantum chemistry is the Hartree-Fock or self-consistent-field approximation.

For closed-shell molecules, the form of the Hartree-Fock wave function is given by Eq. (1), in which $A(n)$, the antisymmetrizer for $n$ electrons, has the effect of making

$$\Psi_{HF} = A(n)\phi_1(1)\phi_2(2)\ldots\phi_n(n) \tag{1}$$

a Slater determinant out of the orbital product on which it operates. The $\phi$'s are spin orbitals, products of a spatial orbital $\chi$ and a one-electron spin function $\alpha$ or $\beta$. For any given molecular system, there are an infinite number of wave functions of form (1), but the Hartree-Fock wave function is the one for which the orbitals $\phi$ have been varied to yield the lowest possible energy [Eq. (2)].

$$E = \int \Psi \times {}_{HF} H \Psi_{HF} d\tau \tag{2}$$

The resulting Hartree-Fock equations are relatively tractable due to the simple form of the energy $E$ for single determinant wave functions [Eq. (3)]

$$E_{HF} = \sum_i I(i \mid i) + \sum_i \sum_{j>i} [(ij \mid ij) - (ij \mid ji)] \tag{3}$$

To solve the Hartree-Fock equations exactly, either the orbitals $\phi$ must be expanded in a complete set of analytic basis functions or strictly numerical (that is, tabulated) orbitals must be obtained. The former approach is impossible from a practical point of view for systems with more than two electrons, and the latter has been accomplished only for atoms and for a few diatomic molecules. Therefore, the exact solution of the Hartree-Fock equations is abandoned for polyatomic molecules. Instead an incomplete (but reasonable) set of analytic basis functions is adopted and solved for the best variational [that is, lowest energy given by Eq. (2)] wave function of form (1). Such a wave function is referred to as being of self-consistent-field (SCF) quality. For very-large-basis sets, then, it is reasonable to refer to the resulting SCF wave function as near-Hartree-Fock.

For large chemical systems, only minimum basis sets (MBS) can be used in ab initio theoretical studies. The term "large" includes molecular systems, with 100 or more electrons.

Ab initio theoretical methods have had the greatest impact on chemistry in the area of structural predictions. The most encouraging aspect of ab initio geometry predictions is their reliability. Essentially all molecular structures appear to be reliably predicted at the Hartree-Fock level of theory. Even more encouraging, many structures are accurately reproduced by using only minimum-basis-set self-consistent-field methods. A fairly typical example is methylenecyclopropane (see structure), with its minimum-basis-set self-consistent-field structure compared with experiment in the table. Carbon-carbon bond distances differ typically by 0.002 nm from experiment, and angles are rarely in error by more than a few degrees. Thus, for many purposes, theory may be considered complementary to experiment in the area of structure prediction.

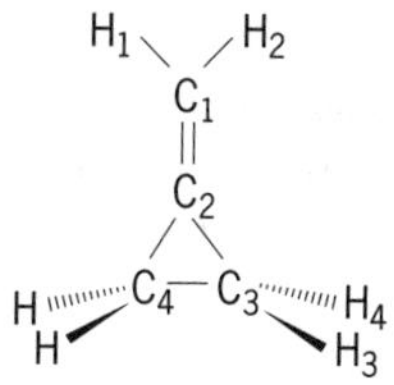

**Minimum-basis-set self-consistent-field geometry prediction compared with experiment for methylenecyclopropane**

| Parameter* | Theory | Experiment |
|---|---|---|
| $r\,(C_1{=}C_2)$ | 0.1298 nm | 0.1332 nm |
| $r\,(C_2{-}C_3)$ | 0.1474 nm | 0.1457 nm |
| $r\,(C_3{-}C_4)$ | 0.1522 nm | 0.1542 nm |
| $r\,(C_1{-}H_1)$ | 0.1083 nm | 0.1088 nm |
| $r\,(C_3H_3)$ | 0.1083 nm | 0.109 |
| $\theta(H_1C_1H_2)$ | 116.0° | 114.3° |
| $\theta(H_3C_3H_4)$ | 113.6° | 113.5° |
| $\theta(H_{34}C_3C_4)$ | 149.4° | 150.8° |

*Here $r$ represents the carbon-carbon bond distance; $\theta$ represents the bond angle in degrees of H-C-H bonds; the numbers on C and H correspond to the numbered atoms in the displayed structure.

The most important post-Hartree-Fock methods for quantum chemistry are perturbation theory and the configuration interaction (CI) and coupled cluster (CC) methods. These three rigorous approaches may be labeled "convergent" quantum-mechanical methods, as each is ultimately capable of yielding exact solutions to Schrödinger's equation. The coupled cluster method treats excitations based on the number of electrons by which they differ from the Hartree-Fock reference function. Thus the CCSD method incorporates amplitudes differing by single (S) and double (D) excitations from Hartree-Fock. The CCSDT method adds all triple excitations to the CCSD treatment. As one goes to higher and higher excitations (for example, CCSDTQ includes all configurations differing by one, two, three, or four electrons from the Hartree-Fock reference configuration), one apporaches the exact quantum-mechanical result.

In fact, coupled cluster theory beyond CCSD becomes impractical for large molecular systems. Thus, although the coupled cluster path to exact results is clear, it becomes a difficult road to follow. Triple excitations are sufficiently important that effective coupled cluster methods have been developed in which the effects of triples are approximated. The best of these methods, CCSD(T), is the closest thing to a panacea that exists today in quantum chemistry for very difficult problems involving smaller molecules. However, the range of applicability of the theoretically superior CCSD(T) method is much narrower than that of the popular hybrid Hartree-Fock/density functional methods. Thus, for most chemists the Hartree-Fock and density functional methods are likely to play central roles in molecular electronic structure theory for many years to come. *See* CHEMICAL BONDING; MOLECULAR ORBITAL THEORY; MOLECULAR STRUCTURE AND SPECTRA; RESONANCE (MOLECULAR STRUCTURE). [H.F.S.]

**Quasielastic light scattering** Small frequency shifts or broadening from the frequency of the incident radiation in the light scattered from a liquid, gas, or solid. The term quasielastic arises from the fact that the frequency changes are usually so small that, without instrumentation specifically designed for their detection, they would not be observed and the scattering process would appear to occur with no frequency changes at all, that is, elastically. The technique is used by chemists, biologists, and physicists to study the dynamics of molecules in fluids, mainly liquids and liquid solutions. It is often identified by a variety of other names, the most common of which is dynamic light scattering (DLS).

Several distinct experimental techniques are grouped under the heading of quasielastic light scattering (QELS). Photon correlation spectroscopy (PCS) is the technique most often used to study such systems as macromolecules in solution, colloids, and critical phenomena where the molecular motions to be studied are rather slow. This technique, also known as intensity fluctuation spectroscopy and, less frequently, optical mixing spectroscopy, is used to measure the dynamical constants of processes with relaxation time scales slower than about $10^{-6}$ s. For faster processes, dynamical constants are obtained by utilizing techniques known as filter methods, which obtain direct measurements of the frequency changes of the scattered light by utilizing a monochromator or filter much as in Raman spectroscopy. [R.P.]

**Quaternary ammonium salts** Analogs of ammonium salts in which organic radicals have been substituted for all four hydrogens of the original ammonium cation. Substituents may be alkyl, aryl, or aralkyl, or the nitrogen may be part of a ring system. Such compounds are usually prepared by treatment of an amine with an alkylating reagent under suitable conditions. They are typically crystalline solids which are soluble in water and are strong electrolytes. Treatment of the salts with silver oxide, potassium hydroxide, or an ion-exchange resin converts them to quaternary ammonium hydroxides, which are very strong bases, as shown in the reaction below.

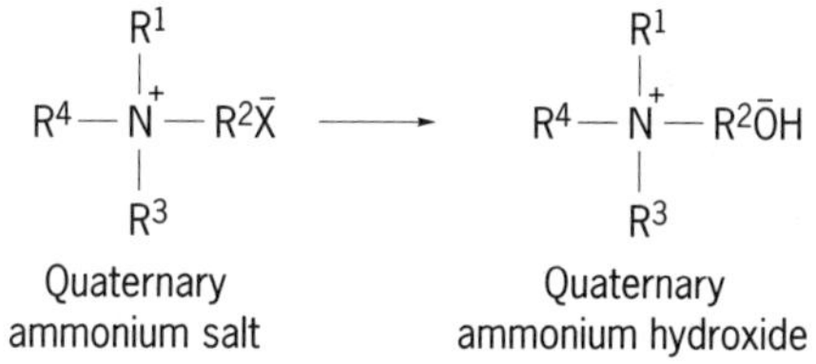

$$R^4-\overset{\overset{R^1}{|}}{\underset{\underset{R^3}{|}}{N^+}}-R^2\bar{X} \longrightarrow R^4-\overset{\overset{R^1}{|}}{\underset{\underset{R^3}{|}}{N^+}}-R^2\bar{O}H$$

Quaternary ammonium salt — Quaternary ammonium hydroxide

Some quaternary ammonium salts have found use as water repellents, fungicides, emulsifiers, paper softeners, antistatic agents, and corrosion inhibitors. *See* AMINE AMMONIUM SALT; SURFACTANT. [P.E.F.]

**Quinine** The chief alkaloid of the bark of the cinchona tree, which is indigenous to certain regions of South America. The structure of quinine is shown below.

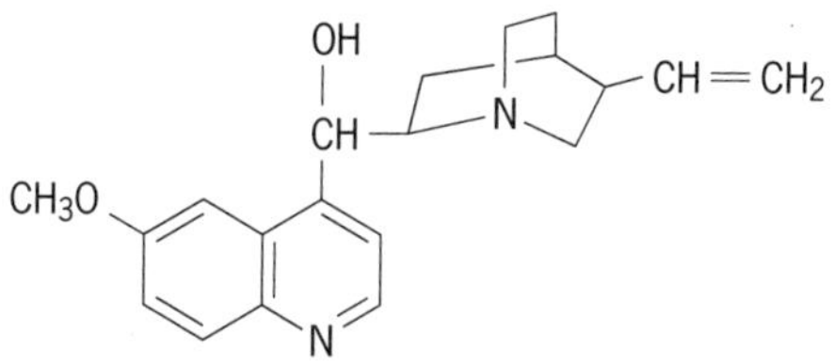

OH

CH

N

$CH=CH_2$

$CH_3O$

N

Until the 1920s quinine was the best chemotherapeutic agent for the treatment of malaria. However, clinical studies definitely established the superiority of the newer synthetic antimalarials such as primaquine, chloroquine, and chloroguanide. *See* ALKALOID. [S.M.K.]

**Quinone** One of a class of aromatic diketones in which the carbon atoms of the carbonyl groups are part of the ring structure. The name quinone is applied to the whole group, but it is often used specifically to refer to *p*-benzoquinone (**1**). *o*-Benzoquinone (**2**) is also known but the meta isomer does not exist.

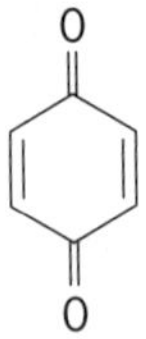

(1) 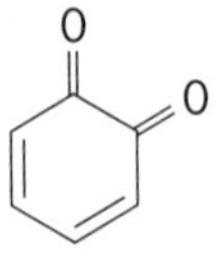 (2)

Quinones are prepared by oxidation of the corresponding aromatic ring systems containing amino (—$NH_2$) or hydroxyl (—OH) groups on one or both of the carbon atoms being converted to the carbonyl group.

Three of the several possible quinones derived from naphthalene are known: 1,4-naphthoquinone (**3**), 1,2-naphthoquinone, and 2,6-naphthoquinone (**4**).

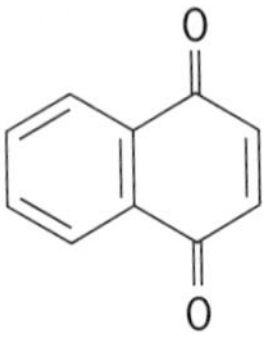

(3) 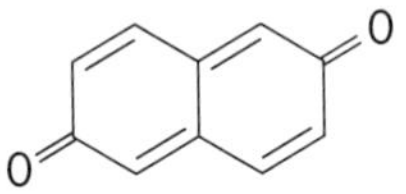 (4)

Important naturally occurring naphthoquinones are vitamins $K_1$ and $K_2$ which are found in blood and are responsible for proper blood clotting reaction. A number of quinone pigments have been isolated from plants and animals. Illustrative of these are juglone found in unripe walnut shells and spinulosin from the mold *Penicillium spinuhsum*. 9,10-Anthraquinone derivatives form an important class of dyes of which alizarin is the parent type. *p*-Benzoquinone is manufactured for use as a photographic developer. *See* AROMATIC HYDROCARBON; DYE; KETONE. [D.A.S.]

# R

**Racemization** The formation of a racemate from a pure enantiomer. Alternatively stated, racemization is the conversion of one enantiomer in a 50:50 mixture of the two enantiomers (+ and −, or R and S) of a substance. Racemization is normally associated with the loss of optical activity over a period of time since 50:50 mixtures of enantiomers are optically inactive. *See* OPTICAL ACTIVITY.

Racemization is an energetically favored process since it reflects a change from a more ordered to a more random state. But the rate at which enantiomers racemize is typically quite slow unless a suitable mechanistic pathway is available, since racemization usually, but not always, requires that a chemical bond at the chiral center of an enantiomer be broken. Racemization of enantiomers possessing more than one chiral center requires that all chiral centers of half of the molecules invert their configurations. *See* ENTROPY.

The observation and study of racemization have important implications for the understanding of the mechanisms of chemical reactions and for the synthesis and analysis of chiral natural products such as peptides. Moreover, racemization is of economic importance since it provides a way of converting an unwanted enantiomer into a useful one. Synthetic medicinal agents are often produced industrially as racemates. After resolution and isolation of the desired enantiomer, half of the product would have to be discarded were it not for the possibility of racemizing the unwanted isomer and of recycling the resultant racemate. [S.H.W.]

**Radiation chemistry** The study of chemical changes resulting from the absorption of high-energy, ionizing radiation, including alpha particles, electrons, gamma rays, fission fragments, protons, deuterons, helium nuclei, and heavier charged projectiles. In absorbing materials of low and intermediate atomic weight such as aqueous systems and most biological systems, such radiation deposits energy in a largely indiscriminate manner, leaving behind a complex mixture of short-lived ions, free radicals, and electronically excited molecules. Radiation-induced chemical changes result from reaction with these intermediates. *See* PHOTOCHEMISTRY.

Sources of high-energy radiation include radioactive nuclides [for example, cobalt-60 ($^{60}Co$), strontium-90 ($^{90}Sr$), and hydrogen-3 ($^{3}H$)] and instruments such as x-ray tubes, Van de Graaff generators, the betatron, the cyclotron, and the synchrotron. An electron accelerator known as the Linac (linear electron accelerator) has proved particularly valuable for the study of transient species that have lifetimes as short as 16 picoseconds; and another electron accelerator, known as the Febetron, has been used for the study of the effects of single pulses of electrons with widths of several nanoseconds at very high currents.

The primary absorption processes for high-energy radiation are ionization and molecular excitation. The distribution of the absorbed energy, however, depends significantly upon the nature of the radiation and absorbing medium.

Evaluation of the yields of radiation-induced reactions requires knowledge of the energy imparted to the reacting system. The energy deposited in the system is termed the dose, and the measurement process is called dosimetry. Absorbed energy from ionizing radiation is described in terms of grays (Gy; joule/kg), in rads (100 ergs/g), or in electronvolts per gram or per cubic centimeter.

Because of its importance in both chemical and biological systems, the radiation chemistry of water has been extensively studied and serves as an example of radiation-induced chemical change. A primary radiation interaction process may be represented by the reaction below, where $H_2O^*$ represents an electronically excited water molecule.

$$H_2O \rightarrow H_2O^+,\ e^-,\ H_2O^*$$

The secondary electron ($e^-$), if formed with sufficient energy, will form its own trail of ionization and excitation. Within $10^{-10}$ to $10^{-8}$ s, reactions within spurs form hydrogen (H) atoms, hydroxyl (OH) radicals, hydrated electrons and molecular products, molecular hydrogen ($H_2$), and hydrogen peroxide ($H_2O_2$). In pure water, radicals escaping the spurs undergo further radical-radical reactions and reactions with molecular products. Upon continuous irradiation, steady-state concentrations of $H_2$, $H_2O_2$, and smaller amounts of dioxygen ($O_2$) result and no further decomposition occurs.

In addition to basic kinetics and mechanistic studies, the principles of radiation chemistry find application in any process in which ionizing radiation is used to study, treat, or modify a biological or chemical system.

In radiation therapy, tumors are destroyed by the application of ionizing radiation from external or internally administered sources. Gamma rays are used for treatment of internal tumors; electron or charged-particle beams are applied to external or invasively accessible lesions.

The physiological concentration of iodine in the thyroid is the basis for the treatment of hyperthyroidism with $^{131}I$. The beta radiation from this isotope is effective in localized tissue destruction.

A goal of any radiation therapy is maximum tumor cell destruction with minimum damage to healthy cells.

Radiation chemistry is used in food preservation by using ionizing radiation in doses that are lethal to microorganisms. The use of ionizing radiation for pathogen control has been approved by most governments for a wide range of foods. In general, limitations on dose have been specified for all products. Radiation is used to control pathogens in meat and meat products.

Processing of commercial quantities of food supplies requires a source of stable intensity and a radiation of sufficient penetrating power to deposit energy throughout the product in an economic, brief time period. [F.J.J.]

**Radiochemical laboratory** A laboratory or facility used for investigation and handling of radioactive chemicals that provides a safe environment for the worker and the public. Features vary depending on the type of radioactive emissions to be handled, the quantity, the half-life, and the physical form (solid, liquid, gas, or powder). Special measures to minimize spread of contaminated material and to dispose of radioactive waste are required. Working surfaces should be smooth and easily washable to permit effective decontamination if necessary. Good ventilation and detectors for monitoring radiation and contamination on surfaces or people are also typical features.

Investigations utilizing only very small amounts (a few microcuries) of beta or gamma emitters which are not readily dispersed (no powders or volatile liquids) may sometimes be performed without special facilities on the bench top. In this case, precautions such as working on plastic-backed absorbent paper and wearing protective gloves and lab coat may be sufficient. A special bag or can for disposing of the paper and gloves as radioactive waste is required. If the radioactive isotopes are solely alpha-particle emitters, containment and isolation from direct contact are more serious concerns. Due to the limited penetration but high biological toxicity of alpha particles, it is essential to avoid ingestion or inhalation. For very small quantities, work may take place with double rubber gloves in a fume hood with appropriate filter. Generally, an enclosed glove box is used, situated inside a hood and maintained at negative pressure with respect to the face of the hood and the room. Sensors to monitor proper differential pressure and adequate airflow are usually used to assure containment and to generate an alarm if conditions degrade.

For work with pure beta-emitting isotopes, long-handled tongs or other tools are used for higher levels of radioactivity in order to shield the hands. Generally, since most beta emission is also accompanied by penetrating gamma emission, the entire work area must be enclosed in heavily shielded enclosures. *See* RADIATION CHEMISTRY; RADIOCHEMISTRY.

Hot laboratories contain walled enclosures for remotely handling larger quantities of gamma-emitting isotopes. A small enclosure is usually referred to as a cave, while large ones are called hot cells. Hot cells are usually equipped with remote manipulators and thick windows made from high-density lead glass. [L.F.M.]

**Radiochemistry** A subject which embraces all applications of radioactive isotopes to chemistry. It is not precisely defined and is closely linked to nuclear chemistry. The widespread use of isotopes in chemistry is based on two fundamental properties exhibited by all radioactive substances. The first property is that the disintegration rate of an isotopic sample is directly proportional to the number of radioactive atoms in the sample. Thus, measurement of its disintegration rate (with a Geiger counter, for example) serves to analyze a radioactive compound. With nearly all chemical elements (notable exceptions being nitrogen and oxygen, which have no suitable radioactive isotopes), an isotope may be incorporated in a chemical compound, and thereafter, masses of this compound as small as $10^{-6}$ to $10^{-10}$ g may be measured with a high precision. The second property is that the disintegration rate is completely unaffected by the chemical form of the isotope, and conversely, the property of radioactivity does not affect the chemical properties of the isotope. By substituting or labeling a particular atom within a molecule, isotopes can be used to trace the fate of that atom during a chemical reaction. Radiochemistry has been used to study the efficiency of chemical separations, rates of chemical reaction and diffusion, isotopic exchange reactions, and chemical reaction mechanisms. *See* NUCLEAR CHEMISTRY; RADIATION CHEMISTRY; RADIOCHEMICAL LABORATORY. [D.R.S.]

**Radium** A chemical element, Ra, with atomic number 88. The atomic weight of the most abundant naturally occurring isotope is 226. Radium is a rare radioactive element found in uranium minerals to the extent of 1 part for about every $3 \times 10^6$ parts of uranium. Chemically, radium is an alkaline-earth metal having properties quite similar to those of barium. Radium is important because of its radioactive properties and is used primarily in medicine for the treatment of cancer, in atomic energy technology for the preparation of standard sources of radiation, as a source for actinium

**Physical properties of radium**

| Property | Value |
|---|---|
| Atomic number | 88 |
| Atomic weight | 226.05 |
| Valence states | 0, 2+ |
| Specific gravity | 6.0 at 20°C |
| Melting point | 700°C (1290°F) |
| Boiling point | ~1140°C (2080°F) |
| Ionic radius, $Ra^{2+}$ | 0.245 nm (estimated) |
| Atomic parachor | ~140 |
| Decomposition potential | 1.718 volt |
| Heat of formation of oxide | 130 kcal/mole |
| Magnetic susceptibility | Feebly paramagnetic |

and protactinium by neutron bombardment, and in certain metallurgical and mining industries for preparing gamma-ray radiographs. *See* PERIODIC TABLE.

Thirteen isotopes of radium are known; all are radioactive; four occur naturally; the rest are produced synthetically. Only $^{226}Ra$ is technologically important. It is distributed widely in nature, usually in exceedingly small quantities. The most concentrated source is pitchblende, a uranium mineral containing about 0.014 oz (0.4 g) of radium per ton of uranium.

Biologically, radium behaves as a typical alkaline-earth element, concentrates in bones by replacing calcium and, as a result of prolonged irradiation, causes anemia and cancerous growths. The tolerance dose for the average human being has been estimated at a total of 1 $\mu$g of radium fixed within the body. However, because radiations from radium and its decay products preferentially destroy malignant tissue, radium and radon, the gaseous decay product of radium, have been used to check the growth of cancer.

When first prepared, nearly all radium compounds are white, but they discolor on standing because of intense radiation. Radiation causes a purple or brown coloration in glass on long contact with radium compounds. Eventually the glass crystallizes and becomes crazed. Radium salts ionize the surrounding atmosphere, thereby appearing to emit a blue glow, the spectrum of which consists of the band spectrum of nitrogen. Radium compounds will discharge an electroscope, fog a light-shielded photographic plate, and produce phosphorescence and fluorescence in certain inorganic compounds such as zinc sulfide. The emission spectrum of radium compounds is similar to those of the other alkaline earths; radium halide imparts a carmine color to a flame.

When freshly prepared, radium metal has a brilliant white metallic luster. Some of its physical properties are shown in the table. Chemically, the metal is highly reactive. It blackens rapidly on exposure to air because of the formation of a nitride. Radium reacts readily with water, evolving hydrogen and forming a soluble hydroxide. *See* ALKALINE-EARTH METALS; NUCLEAR REACTION; RADIOACTIVITY; RADON. [M.L.S.]

**Radon** A chemical element, Rn, atomic number 86. Radon is produced as a gaseous emanation from the radioactive decay of radium. The element is highly radioactive and decays by the emission of energetic alpha particles. Radon is the heaviest of the noble, or inert, gas group and thus is characterized by chemical inertness. More than 25 isotopes of radon have been identified. All isotopes are radioactive with short half-lives. *See* PERIODIC TABLE.

Radon is found in natural sources only because of its continuous replenishment from the radioactive decay of longer-lived precursors in minerals containing uranium, thorium, or actinium. $^{222}$Rn (half-life 3.82 days), $^{220}$Rn (thoron; half-life 55 s), and $^{219}$Rn (actinon; half-life 4.0 s), occur in nature as members of the uranium (U), thorium (Th), and actinium (Ac) series, respectively. All three decay by the emission of energetic alpha particles. *See* ACTINIUM; RADIUM; THORIUM; URANIUM.

Any surface exposed to $^{222}$Rn becomes coated with an active deposit which consists of a group of short-lived daughter products. The radiations of this active deposit include energetic alpha particles, beta particles, and gamma rays. The ultimate decay products of radon following the rapid decay of the active deposit to lead-210 include bismuth-210, polonium-210, and finally, stable lead-206. Radon possesses a particularly stable electronic configuration, which gives it the chemical properties characteristic of noble gas elements. It has a boiling point of −62°C (−80°F) and a melting point of −71°C (−96°F). The spectrum of radon has been extensively studied, and resembles that of the other inert gases. Radon is readily adsorbed on charcoal, silica gel, and other adsorbents, and this property can be used to separate the element from gaseous impurities. [E.K.H.]

The rocks and soils of the Earth's crust contain approximately 3 parts per million of $^{238}$U, the long-lived head of the uranium series; 11 ppm of $^{232}$Th, the head of the thorium series; but only about 0.02 ppm of $^{235}$U, the long-lived member of the actinium group. The radon isotopes $^{222}$Rn and $^{220}$Rn are produced in proportion to the amount of the parent present. Some of the newly formed radon atoms which originate in or on the surface of mineral grains escape into the soil gas, where they are free to diffuse within the soil capillaries. Some of the radon atoms eventually find their way to the surface, where they become a part of the atmosphere. Even though thorium ($^{232}$Th) is generally more abundant than uranium in the Earth's crust, the probability for decay is smaller; hence, the production rate of $^{222}$Rn and $^{220}$Rn in the soil is roughly the same. Much of the $^{220}$Rn decays before reaching the Earth's surface due to its short half-life.

When radon ($^{222}$Rn or $^{220}$Rn) passes from soil to air, it is mixed throughout the lower atmosphere by eddy diffusion and the prevailing winds. Mean radon levels are found to be higher during those times of year when atmospheric stability is the greatest such as may occur during the fall months. Radon and its daughters play an important role in atmospheric electricity. Near the Earth's surface almost half of the ionization of the air is due to $^{220}$Rn and $^{222}$Rn and their daughter products. The alpha emitters from these chains typically produce about $10^7$ ion pairs per second per cubic meter.

Radon is readily soluble in water. Since ground and surface waters are in close contact with soil and rocks containing small quantities of radium, it is not surprising to find radon in public water supplies.

The radon isotopes $^{220}$Rn and $^{222}$Rn are used widely in the study of gaseous transport processes both in the underground environment and in the atmosphere. Radon accumulates to high levels of the order of 4000 becquerels/m$^3$ or more in caves unless natural or artificial ventilation occurs. Changes in $^{222}$Rn concentrations in spring and well water and in soil and rocks have been suggested as a means of predicting earthquakes.

The tendency of the decay products of radon to attach to aerosols means that these nuclides will be inhaled and deposited in the bronchial epithelium and lungs. The daughter products, therefore, make up the major part of the internal radiation dose from radon. Ways of reducing radon levels within homes or workplaces include increased ventilation and sealing of major sources of entry from soil and building materials. Workers in uranium mines may encounter radon and decay product levels of the order of 50,000 Bq/m$^3$ or more. Ventilation procedures and special filters for the miners must be used. [M.Wi.]

**Rare-earth elements** The group of 17 chemical elements with atomic numbers 21, 39, and 57–71; the name lanthanides is reserved for the elements 58–71. The name rare earths is a misnomer, because they are neither rare nor earths. *See* ACTINIDE ELEMENTS; LANTHANIDE CONTRACTION; PERIODIC TABLE.

Most of the early uses of the rare earths took advantage of their common properties and were centered principally in the glass, ceramic, lighting, and metallurgical industries. Today these applications use a very substantial amount of the mixed rare earths just as they are obtained from the minerals, although sometimes these mixtures are supplemented by the addition of extra cerium or have some of their lanthanum and cerium fractions removed.

The elements exhibit very complex spectra, and the mixed oxides, when heated, give off an intense white light which resembles sunlight, a property finding application in cored carbon arcs, such as those employed in the movie industry.

The rare-earth metals have a great affinity for the nonmetallic elements, as, for example, hydrogen, carbon, nitrogen, oxygen, sulfur, phosphorus, and the halides. Considerable amounts of the mixed rare earths are reduced to metals, such as misch metal, and these alloys are used in the metallurgical industry. Alloys made of cerium and the mixed rare earths are used in the manufacture of lighter flints. Rare earths are also used in the petroleum industry as catalysts. Yttrium aluminum garnets (YAG) are used in the jewelry trade as artificial diamonds.

Although the rare earths are widely distributed in nature, they generally occur in low concentrations. They are found in high concentrations as mixtures in a number of minerals. The relative abundance of the different rare earths in various rocks, geological formations, and the stars is of great interest to the geophysicist, astrophysicist, and cosmologist.

The rare-earth elements are metals possessing distinct individual properties. Many of the properties of the rare-earth metals and alloys are quite sensitive to temperature and pressure. They are also different when measured along different crystal axes of the metal; for example, electrical conductivity, elastic constants, and so on. The rare earths form organic salts with certain organic chelate compounds. These chelates, which have replaced some of the water around the ions, enhance the differences in properties among the individual rare earths. Advantage is taken of this technique in the modern ion-exchange methods of separation. *See* CHELATION; ION EXCHANGE; TRANSITION ELEMENTS. [F.H.Sp.]

**Reactive intermediates** Unstable compounds which are formed as necessary intermediate stages during a chemical reaction. Thus, if a reaction in which A is converted to B requires that A first be converted to C, then C is an intermediate in the reaction (A $\rightarrow$ C $\rightarrow$ B ). The term reactive further implies a certain degree of instability of the intermediate; reactive intermediates are typically isolable only under special conditions, and most of the information regarding the structure and properties of reactive intermediates comes from indirect experimental evidence.

In organic reactions the most common types of reactive intermediates are those arising from dissociative reactions, in which carbon has a decreased valence. Associative reactions can also give rise to some of the same intermediates, and to others in which carbon has an increased valence. *See* VALENCE.

Carbocations are compounds in which carbon bears a positive charge. Classical carbocations (also called carbenium ions) are trivalent, and have only six valence electrons. Nonclassical carbocations (also called carbonium ions ) are tetra- or pentavalent, and have eight valence electrons. Examples are the methyl cation (classical) $CH_3^+$, and the methonium ion (nonclassical) $CH_5^+$.

Carbanions are compounds in which carbon bears a negative charge. A carbanion will always have a positive counterion in association with it; depending upon the particular cation and the stability of the carbanion, the association may be ionic, covalent, or some intermediate combination of ionic and covalent bonding, as shown below (M = metal). Carbanions are trivalent, with eight valence electrons.

$$-\overset{|}{\underset{|}{C}}-M \longleftrightarrow -\overset{|}{\underset{|}{C}}\!:^{-}\ {}^{+}M$$

Covalent    Ionic

Free radicals are neutral compounds having an odd number of electrons and therefore one unpaired electron. Carbon free radicals are trivalent, with seven valence electrons, and typically assume a planar structure. Free radicals are primarily electron-deficient species and are stabilized by structural features which donate electron density or delocalize the odd electron by resonance. *See* FREE RADICAL; RESONANCE (MOLECULAR STRUCTURE).

Radical ions are charged compounds with an unpaired electron, and are either radical cations (positively charged) or radical anions (negatively charged). In many cases a radical ion is derived from a stable neutral molecule by addition of one electron (radical anion) or removal of one electron (radical cation).

Carbenes are compounds which have a divalent carbon. The divalent carbon also has two nonbonded electrons, for a total of six valence electrons. The two nonbonded electrons may have either the same spin quantum number, which is a triplet state, or an opposite spin quantum number, which is a singlet state. Generation of carbenes is most commonly by photolysis or thermolysis of diazo compounds or ketenes, or by alpha-elimination reactions.

There are many other kinds of reactive intermediates which do not fit into the previous classifications. Some are simply compounds which are unstable for a variety of possible reasons, such as structural strain or an unusual oxidation state. *See* CHEMICAL DYNAMICS; MOLECULAR ORBITAL THEORY. [C.C.W.]

**Reagent chemicals** High-purity chemicals used for analytical reactions, for the testing of new reactions where the effects of impurities are unknown, and in general for chemical work where impurities must either be absent or at known concentrations.

Chemicals are purified by a variety of methods, the most common being recrystallization from solution. If the desired chemical is volatile and the impurities are not volatile, sublimation is an effective method of purification. For liquid chemicals, distillation is an effective procedure. Finally, the simplest procedure may be to synthesize the desired reagent from pure materials. *See* CHEMICAL SEPARATION TECHNIQUES; CRYSTALLIZATION; DISTILLATION.

Commercial chemicals are available at several levels of purity. Chemicals labeled "technical" or "commercial" are usually quite impure. The grade "USP" indicates only that the chemical meets the requirements of the United States Pharmacopeia. The term "CP" means only that the chemical is purer than "technical." Chemicals designated "reagent grade" or "analyzed reagent" are specially purified materials which usually have been analyzed to establish the levels of impurities. The American Chemical

Society has established specifications and tests for purity for some chemicals. Materials which meet these specifications are labeled "Meets ACS Specifications." [K.G.S./C.Ru.]

**Rearrangement reaction** A reaction in which an atom or bond moves or migrates, having been initially located at one site in a reactant molecule and ultimately located at a different site in a product molecule. A rearrangement reaction may involve several steps, but the key feature defining it as a rearrangement is that a bond shifts from one site of attachment to another. The simplest examples of rearrangement reactions are intramolecular, that is, reactions in which the product is simply a structural isomer of the reactant [reaction (1)].

O → O (1)

*See* MOLECULAR ISOMERISM.

More complex rearrangement reactions occur when the rearrangement is accompanied by another reaction, for example, a substitution reaction (2).

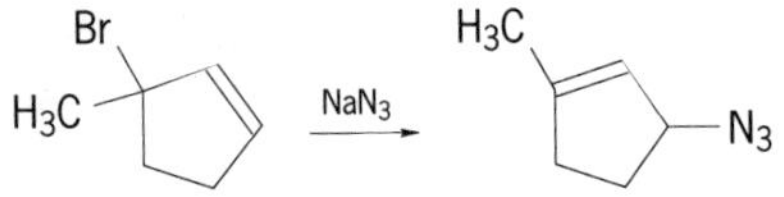

Br, $H_3C$ —($NaN_3$)→ $H_3C$, $N_3$ (2)

Rearrangement reactions are classified and named on the basis of the group that migrates and the initial and final location of the migrating bond. The initial bond location is designated as position 1, and the final location as position i, where the number of atoms is simply counted along the connection from 1 to i. Such a migration is called a [1,i] rearrangement or [1,i] shift. If the migrating group also reattaches itself at a different site from the one to which it had originally been attached, then both shifts are indicated, as in [i,j] shift. Reaction (1) is an example of a [3,3] rearrangement, because the initial carbon-oxygen (C-O) bond that breaks designates the 1 position for each component, and both components then rearrange and form a new C-C bond by reattaching at position 3 for each component. Reaction (2) is an example of a [1,3] rearrangement with substitution, commonly called an $S_N2'$ reaction.

In addition, classification can be based on how many electrons move with the migrating group. Of the two electrons in the initial bond that breaks, the migrating group may bring with it both electrons (nucleophilic or anionotropic), one electron (radical), or no electrons (electrophilic or cationotropic). If the rearrangement is a concerted reaction in which there is a cyclic delocalized transition state that results in shifts of pi bonds as well as sigma bonds, the reaction is called a sigmatropic rearrangement [for example, reaction (1)]. *See* DELOCALIZATION; PERICYCLIC REACTION.

The great majority of rearrangement reactions occur when a molecule develops a severely electron-deficient site. Shift of a nearby atom or group, with its pair of electrons, can serve to satisfy the electron deficiency at the original site, although it typically leaves behind another site with electron deficiency. As long as the final site can bear the electron deficiency better than the original site, the rearrangement will be favorable. *See* CHEMICAL BONDING. [C.C.Wa.]

**Reference electrode** An electrode with an invariant potential. In electrochemical methods, where it is necessary to observe, measure, or control the potential of another electrode (denoted indicator, test, or working electrode), it is necessary to use a reference electrode, which maintains a potential that remains practically unchanged during the course of an electrochemical measurement. Potentials of indicator or working electrodes are measured or expressed relative to reference electrodes.

One such electrode, the normal hydrogen electrode, has been chosen as a reference standard, relative to which potentials of other electrodes and those of oxidation-reduction couples are often expressed. By maintaining a constant pressure of hydrogen gas the potential of a hydrogen electrode can be used for determination of the activity of hydrogen ions in the tested solution. However, in practice the determination of the hydrogen-ion activity (pH) is performed by using a glass electrode. The hydrogen electrode itself is used only in fundamental studies and some nonaqueous solutions. The hydrogen electrode, however, remains important for providing a reference standard. *See* ACTIVITY (THERMODYNAMICS); PH.

In practice, potentials are measured against reference electrodes that are easier to work with than the normal hydrogen electrode. Such electrodes are known as secondary reference electrodes; the most common are the calomel and silver–silver chloride electrodes. *See* ELECTRODE; SOLVENT. [P.Z.]

**Reforming processes** Those processes used to convert, with limited cracking, petroleum liquids into higher-octane gasoline. Due to the demand for higher-octane gasoline, thermal reforming was developed (from thermal cracking processes) to improve the octane number of fractions within the boiling range of gasoline. *See* CRACKING; OCTANE NUMBER.

Upgrading by reforming may be accomplished, in part, by an increase in volatility (reduction of molecular size) or by the conversion of *n*-paraffins to isoparaffins, olefins, and aromatics, and of naphthenes (cycloalkanes) to aromatics. The nature of the final product is influenced by the structure and composition of the straight-run (virgin) naphtha (hydrocarbon mixture) feedstock. In thermal reforming, the reactions resemble those in the cracking of gas oils. The molecular size is reduced, while olefins and some aromatics are synthesized. For example, hydrocracking of high-molecular-weight paraffins yields lower-molecular-weight paraffins and an olefin; dehydrocyclization of paraffin compounds yields aromatic compounds; isomerization of *n*-paraffins yields isoparaffins; and isomerization of methylcyclopentane yields cyclohexane.

In the presence of catalysts and in the presence of the hydrogen available from dehydrogenation reactions, hydrocracking of paraffins to yield two lower-molecular-weight paraffins takes place, and olefins that do not undergo dehydrocyclization are dehydrogenated so that the end product contains only traces of olefins. *See* DEHYDROGENATION; HYDROCRACKING; ISOMERIZATION; PARAFFIN.

Thermal reforming was a natural development from thermal cracking, since reforming is also a thermal decomposition reaction. Cracking converts heavier oils into gasoline constituents, whereas reforming converts these gasoline constituents into higher-octane molecules. The equipment for thermal reforming is essentially the same as for thermal cracking, but higher temperatures are used. The higher octane number of the product (reformate) is due primarily to the cracking of longer-chain paraffins into higher-octane olefins. *See* DISTILLATION.

The products of thermal reforming are gases, gasoline, and residual oil. The amount and quality of the reformate are very dependent on the temperature. As a rule, the higher the reforming temperature, the higher the octane number of the product but the lower the reformate yield. Adding catalysts increases the yield for higher-octane gasolines at a given temperature.

Thermal reforming is less effective and less economical than catalytic processes and has been largely supplanted. The octane number was changed by the severity of the cracking, and the product had increased volatility, compared to the volatility of the feedstock.

Modifications of the thermal reforming process due to the inclusion of hydrocarbon gases with the feedstock are known as gas reversion and polyforming. These are essentially the same but differ in the manner in which the gases and naphtha are passed through the heating furnace. In gas reversion, the naphtha and gases flow through separate lines in the furnace and are heated independently of one another. In naphtha reforming, the $C_3$ and $C_4$ gases are premixed with the naphtha and pass together through the furnace.

Like thermal reforming, catalytic reforming converts low-octane gasoline into high-octane gasoline (reformate). Although thermal reforming can produce reformate with a research octane number of 65–80 depending on the yield, catalytic reforming produces reformate with octane numbers on the order of 90–95. Catalytic reforming is conducted in the presence of hydrogen over hydrogenation-dehydrogenation catalysts. Depending on the catalyst, a definite sequence of reactions takes place, involving structural changes in the feedstock. *See* CATALYSIS; PETROLEUM PROCESSING AND REFINING. [J.G.S.]

**Refractometric analysis** A method of chemical analysis based on the measurement of the index of refraction of a substance. The most common type of refractometer is the Abbe refractometer. It is simple to use, requiring but a drop or two of sample and allowing a measurement of refractive index to be made in 1–2 min, with a precision of 0.0001. More precise measurements of refractive indices may be made by using a dipping or immersion refractometer, the prism of which is completely immersed in the sample. The most precise measurements of the refractive indices of gases or solutions containing small traces of impurities are made with an interferometer. For measurements in flowing systems, differential refractometers are used.

The measurement of refractive index is used to identify compounds whose other physical constants are quite similar. Because minute amounts of impurities often cause a measurable change in the refractive index of a pure material, refractive index is often used as a criterion for purity. A measurement of refractive index gives information as to the gross amount of impurity; it does not serve to identify the impurity. [R.F.G.; J.N.L.]

**Relative atomic mass** The ratio of the average mass per atom of the natural nuclidic composition of an element to $^1/_{12}$ of the mass of an atom of nuclide $^{12}C$. For example, $\mu(Cl) = 35.453$. Relative atomic mass replaces the concept of atomic weight. It is also known as relative nuclidic mass. [T.C.W.]

**Relative molecular mass** The ratio of the average mass per formula unit of the natural nuclidic composition of a substance to $^1/_{12}$ of the mass of an atom of nuclide $^{12}C$. For example, $\mu(KCl) = 74.555$. Relative molecular mass replaces the concept of molecular weight. [T.C.W.]

**Resin** Originally a category of vegetable substances soluble in ethanol but insoluble in water, but generally in modern technology an organic polymer of indeterminate molecular weight. The class of flammable, amorphous secretions of conifers or legumes are considered true resins. Water-swellable secretions of various plants, especially the Burseraceae, are called gum resins. The natural vegetable resins are largely polyterpenes and their acid derivatives, which find application in the manufacture of lacquers, adhesives, varnishes, and inks.

The synthetic resins, originally viewed as substitutes for certain natural resins, have a large place of their own in industry and commerce. Phenol-formaldehyde, phenol-urea, and phenol-melamine resins are important commercially. Any unplasticized organic polymer is considered a resin, thus nearly any one of the common plastics may be viewed as a synthetic resin. Water-soluble resins are marketed chiefly as substitutes for vegetable gums and in their own right for highly specialized applications. Carboxymethyl cellulose, hydroxyalkylated cellulose derivatives, modified starches, polyvinyl alcohol, polyvinylpyrrolindone, and polyacrylamides are widely used as thickening agents for foods, paints, and drilling muds, as fiber sizings, in various kinds of protective coatings, and as encapsulating substances. *See* POLYMER. [F.W.I]

**Resonance (molecular structure)** A feature of the valence-bond method, which is a mathematical procedure to obtain approximate solutions to the Schrödinger equation for molecules. The valence-bond method is based on the theorem that if two or more solutions to the Schrödinger equation are available, certain linear combinations of them will also be solutions. It has this basis in common with its rival, the molecular orbital method. The valence-bond and molecular orbital approaches are both approximations and, if carried out to their logical and exact extremes, must yield identical results; nevertheless, both are often described as theories. In the valence-bond theory, combinations of solutions represent hypothetical structures of the molecule in question. These structures are said to be resonance (or contributing) structures, and the real molecule is said to be the resonance hybrid (or just simply the hybrid) of these structures. *See* MOLECULAR ORBITAL THEORY.

The resonance theory provided a solution for a molecule which had baffled and preoccupied chemists for a century—benzene. The principal use of resonance still lies in the qualitative description of molecules whose properties would otherwise be difficult to understand. *See* BENZENE.

Until the beginning of the 20th century, benzene posed a baffling challenge to organic chemists. In spite of its relatively simple formula, $C_6H_6$, they were unable to conceive of a suitable structure for it. While a great many structures were proposed, the properties of benzene corresponded to none of them.

In the early 1870s F. A. Kekulé proposed a revolutionary idea; benzene must be represented by two structures, (**1**) and (**2**), rather than one, and all compounds containing

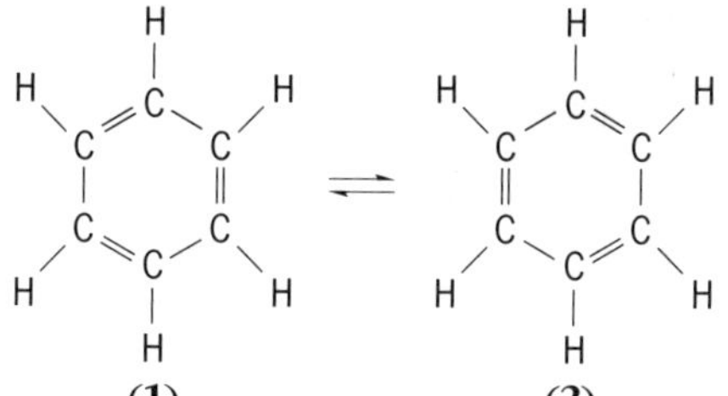

(1) ⇌ (2)

the benzene skeleton must be subject to a rapid equilibration (oscillation) between the two. Kekulé's description of benzene was not completely satisfactory. While it accounted for the number of substituted benzene isomers, it did not explain why the compound failed to exhibit reactivity indicating the presence of multiple bonds. The problem was resolved with the advent of quantum mechanics in the early part of this century. In a sense, this solution is an expansion of Kekulé's oscillating pair; the so-called activation energy (the energy which must be imparted to a molecule in order to make it overcome the barrier that keeps it from being converted into another molecule) is negative in the case of benzene with respect to the oscillation, and this molecule therefore exists neither as (**1**) nor as (**2**) at any time, but it is an intermediate form (**3**) all the time. This

(3)

intermediate structure of benzene is described in terms of Kekulé's structures with the symbol ↔ between them; this is intended to signify that benzene has neither structure, but in fact is a hybrid of the two. The properties of benzene are thereby indicated to be those of neither (**1**) nor (**2**), but to be intermediate between the two.

The only property of the hybrid which is not intermediate between those of the hypothetical contributing structures is the energy: the energy of a resonance hybrid is by definition always at a minimum. This fact is responsible for the abnormal reluctance of benzene to undergo addition reactions; such reactions would lead to products that no longer have the resonance energy.

Although benzene is the classical example of resonance, the phenomenon is certainly not limited to it. Furthermore, the properties of all compounds are affected by resonance to some degree.

Although the molecular orbital approach has largely supplanted the valence-bond method, the resonance language remains so convenient that it is still used. *See* CHEMICAL BONDING; MOLECULAR STRUCTURE AND SPECTRA; QUANTUM CHEMISTRY. [W.J.LeN.]

**Resonance ionization spectroscopy** A form of atomic and molecular spectroscopy in which wavelength-tunable lasers are used to remove electrons from (that is, ionize) a given kind of atom or molecule. Laser-based resonance ionization spectroscopy (RIS) methods have been developed and used with ionization detectors, such as proportional counters, to detect single atoms. Resonance ionization spectroscopy is combined with mass spectrometers to provide analytical systems for a wide range of applications, including physics, chemistry, materials sciences, medicine, and the environmental sciences.

When an atom or molecule is irradiated with a light source of frequency $\nu$, photons at this selected frequency are absorbed only when the energy $h\nu$ ($h$ is Planck's constant) is almost exactly the same as the difference in energy between some excited state and the ground state of the atom or molecule. If a laser source is tuned to a very narrow bandwidth at a frequency that excites a given kind of atom (see illustration), it is highly unlikely that any other kind of atom will be excited. An atom in an excited state can be ionized by photons of the specified frequency $\nu$, provided that $2h\nu$ is greater than the ionization potential of the atom. While the final ionization step can occur

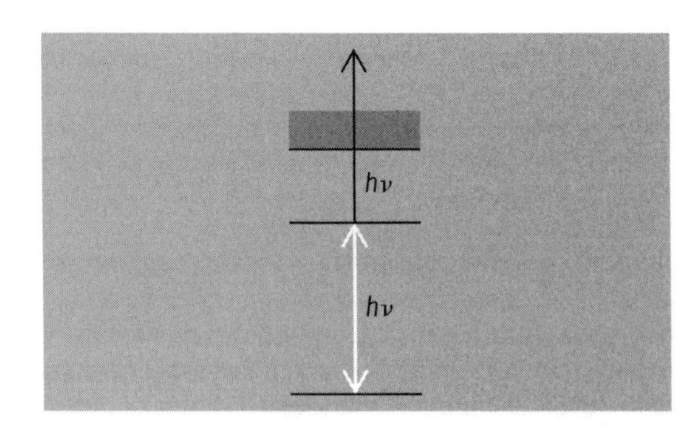


**Basic laser scheme for resonance ionization spectroscopy. The atom or molecule is irradiated by a light source with frequency $\nu$ and photons of energy $h\nu$, where $h$ is Planck's constant.**

with any energy above a threshold, the entire process of ionization is a resonance one. Resonance ionization spectroscopy is a selective process in which only those atoms that are in resonance with the light source are ionized. Modern pulsed lasers have made resonance ionization spectroscopy a practical method for the sensitive (and highly selective) detection of nearly every type of atom in the periodic table. *See* ATOMIC STRUCTURE AND SPECTRA; IONIZATION POTENTIAL; LASER; LASER SPECTROSCOPY; PHOTOIONIZATION.

Resonance ionization spectroscopy is used to analyze very low levels of trace elements in extremely pure materials, for example, semiconductors in the electronics industry. A sputter-initiated resonance ionization spectroscopy (SIRIS) apparatus uses an argon ion beam to sputter a tiny cloud of atoms from a sample placed in a high-vacuum system and a pulsed laser tuned to detect the specified impurity atom.

The sputter-initiated resonance ionization spectroscopy method is also used for chemical and materials research, geophysical research and explorations, medical diagnostics, biological research, and environment analysis. Thermal-atomization resonance ionization spectroscopy (TARIS) may be used for the bulk analysis of materials. By simply using resonance ionization spectroscopy with ionization chambers or proportional counters, gas-phase work can be done to study the diffusion of atoms, measure chemical reaction rates, and investigate the statistical behavior of atoms and molecules. *See* CHEMICAL DYNAMICS; DIFFUSION.

Resonance ionization spectroscopy is used in sophisticated nuclear physics studies involving high-energy accelerators. It is used as an on-line detector to record the hyperfine structure of nuclei with short lifetimes and hence to determine several nuclear properties such as nuclear spin and the shape of nuclei. *See* NUCLEAR STRUCTURE.

Resonance ionization spectroscopy is used for measurements of krypton-81 in the natural environment to determine the ages of polar ice caps and old ground-water deposits. Oceanic circulation and the mixing of oceans could also be studied by measuring the concentrations of noble-gas isotopes by resonance ionization spectroscopy. [G.S.H.]

**Rhenium** A chemical element, Re, with atomic number 75 and atomic weight 186.2. Rhenium is a transition element. It is a dense metal (21.04) with the very high melting point of 3440°C (6220°F). *See* PERIODIC TABLE.

Rhenium is similar to its homolog technetium in that it may be oxidized at elevated temperatures by oxygen to form the volatile heptoxide, $Re_2O_7$; this in turn may be reduced to a lower oxide, $ReO_2$. The compounds $ReO_3$, $Re_2O_3$, and $Re_2O$, are well known. Perrhenic acid, $HReO_4$, is a strong monobasic acid and is only a very weak oxidizing agent. Complex perrhenates, such as cobalt hexammine perrhenate, $[Co(NH_3)_6(ReO_4)_3]$, are also known.

The halogen compounds of rhenium are very complicated, and a large series of halides and oxyhalides have been reported. Rhenium forms two well-characterized sulfides, $Re_2S_7$ and $ReS_2$, as well as two selenides, $Re_2Se_7$ and $ReSe_2$. The sulfides have their counterparts in the technetium compounds, $Tc_2S_7$ and $TcS_2$. *See* TECHNETIUM; TRANSITION ELEMENTS. [S.F.]

**Rhodium** A chemical element, Rh, atomic number 45, relative atomic weight 102.905. Rhodium is a transition metal and one of the group of platinum metals (ruthenium, osmium, rhodium, iridium, palladium, and platinum) that share similar chemical and physical properties. *See* PERIODIC TABLE; PLATINUM.

The terrestrial abundance of rhodium is exceedingly low; it is estimated to be 0.0004 part per million in the Earth's crust. It is found as a single isotope, $^{103}Rh$, with a nuclear spin of 12. Since the platinum metals share common reactivities and are mined from

**Physical properties of rhodium metal**

| Property | Value |
|---|---|
| Crystal structure | Face-centered cubic |
| Lattice constant $a$, at 25°C (77°F), nm | 0.38031 |
| Thermal neutron capture cross section, barns ($10^{-28}$ $m^2$) | 149 |
| Density at 25°C (77°F), g/cm$^3$ | 12.43 |
| Melting point | 1963°C (3565°F) |
| Boiling point | 3700°C (6700°F) |
| Specific heat at 0°C, cal/g (J/kg) | 0.0589 (246) |
| Thermal conductivity, 0–100°C, cal cm/cm$^2$s °C (J · m/m$^2$ · s · °C) | 0.36 (151) |
| Linear coefficient of thermal expansion, 20–100°C, $\mu$in./(in./°C) or m/(m · °C) | 8.3 |
| Electrical resistivity at 0°C, microhm-cm | 4.33 |
| Temperature coefficient of electrical resistance, 0–100°C/°C | 0.00463 |
| Tensile strength, $10^3$ lb/in.$^2$ (6.895 MPa) | |
| Soft | 120–130 |
| Hard | 200–230 |
| Young's modulus at 20°C (68°F), lb/in.$^2$ (Gpa) | |
| Static | 46.2 × $10^6$ (319) |
| Dynamic | 54.8 × $10^6$ (378) |
| Hardness, diamond pyramid number | |
| Soft | 120–140 |
| Hard | 300 |
| $\Delta H_{fusion}$, kJ/mol | 21.6 |
| $\Delta H_{vaporization}$, kJ/mol | 494 |
| $\Delta H_f$ monoatomic gas, kJ/mol | 556 |
| Electronegativity | 2.2 |

a common source, there is an involved chemical process that is used to separate the individual elements, including rhodium.

Metallic rhodium is the whitest of the platinum metals and does not tarnish under atmospheric conditions. Its surface is normally covered by a thin, firmly bound layer of rhodium(IV) oxide ($RhO_2$). Rhodium is insoluble in all acids, including aqua regia. It dissolves in molten potassium bisulfate ($KHSO_4$), a useful property for its extraction from platinum ores, since iridium, ruthenium, and osmium are insoluble in this melt. Important physical properties of metallic rhodium are given in the table. *See* ACID AND BASE; AQUA REGIA; HALOGEN ELEMENTS.

Metallic rhodium is available as powder, sponge, wire, and sheets. It is ductile when hot and retains its ductility when cold. However, it work-hardens rapidly. Molten rhodium dissolves oxygen. Upon cooling, the oxygen gas is liberated, and this can lead to ruptures in the external surface of the crust of the metal. As a result, molten rhodium is best handled under an inert atmosphere of argon, which does not dissolve in rhodium.

Complexes of Rh(III), including $RhCl_3(pyridine)_3$, $Rh(CO)Cl_3[P(C_6H_5)_3]_2$, and $RhCl_6^{3-}$, are diamagnetic six-coordinate with octahedral geometry. The most common chemical form of rhodium is $RhCl_3 \cdot 3H_2O$, a red-brown, deliquescent material that is a useful starting material for the preparation of other rhodium compounds. In contrast to the hydrated material, red anhydrous rhodium(III) chloride ($RhCl_3$) is a polymeric, paramagnetic compound that does not dissolve in water.

The low natural abundance and high cost of rhodium limit its uses to specialty applications. The major use is in catalysis, which accounts for over 60% of its production.

Rhodium is a component of catalytic converters used in the control of exhaust emissions from automobiles.

Rhodium is also used in the hydrogenation of olefins to alkanes. For hydrogenation, both heterogeneous catalysis and homogeneous catalysis are used. Heterogeneous conditions are achieved with rhodium metal finely dispersed on an inert support (activated carbon, charcoal, or alumina).

Rhodium complexes have been developed as catalysts for the synthesis of one optical isomer of L-dopa (used in treatment of Parkinson's disease). Greater selectivity makes rhodium catalysts more useful in hydroformylation or oxo reactions than the less expensive cobalt catalysts. A platinum-rhodium alloy is an efficient commercial catalyst for the formation of nitric acid through ammonia oxidation. *See* CATALYSIS; HETEROGENEOUS CATALYSIS; HOMOGENEOUS CATALYSIS; HYDROFORMYLATION; HYDROGENATION.

Rhodium-platinum alloys are favored for high-temperature applications. The International Temperature Scale over the range 630.5–1063°C (1134.9–1945.4°F) is defined by a thermocouple using a 10% rhodium-platinum alloy. Electroplated rhodium retains its bright surface under atmospheric conditions and finds use as electrical contacts and reflective surfaces. The reflectivity of rhodium surfaces is high (80%) and does not tarnish. About 6% of the rhodium production goes into jewelry manufacturing. *See* TRANSITION ELEMENTS. [A.L.Ba.]

**Rosin** A brittle resin ranging in color from dark brown to pale lemon yellow and derived from the oleoresin of pine trees. Rosin is insoluble in water, but soluble in most organic solvents. It softens at about 180–190°F (80–90°C). Rosin consists of about 90% resin acids and about 10% neutral materials such as anhydrides, sterols, and diterpene aldehydes and alcohols.

Rosin is obtained by wounding living trees and collecting the exudate (gum rosin), by extraction of pine stumps (wood rosin), and as a by-product from the kraft pulping process (sulfate or tall oil rosin).

The largest single use of rosin is in sizing paper to control water absorption, an application in which fortified rosin is important. Rosin soaps are used as emulsifying and tackifying agents in synthetic rubber manufacture. Other rosin uses include adhesives, printing inks, and chewing gum. [I.S.G.]

**Rubber** Originally, a natural or tree rubber, which is a hydrocarbon polymer of isoprene units. With the development of synthetic rubbers having some rubbery characteristics but differing in chemical structure as well as properties, a more general designation was needed to cover both natural and synthetic rubbers. The term elastomer, a contraction of the words elastic and polymer, was introduced, and defined as a substance that can be stretched at room temperature to at least twice its original length and, after having been stretched and the stress removed, returns with force to approximately its original length in a short time.

Three requirements must be met for rubbery properties to be present in both natural and synthetic rubbers: long thread like molecules, flexibility in the molecular chain to allow flexing and coiling, and some mechanical or chemical bonds between molecules.

Natural rubber and most synthetic rubbers are also commercially available in the form of latex, a colloidal suspension of polymers in an aqueous medium. Natural rubber comes from trees in this form; many synthetic rubbers are polymerized in this form; some other solid polymers can be dispersed in water. *See* POLYMER.

Latexes are the basis for a technology and production methods completely different from the conventional methods used with solid rubbers.

In the crude state, natural and synthetic rubbers possess certain physical properties which must be modified to obtain useful end products. The raw or unmodified forms are weak and adhesive. They lose their elasticity with use, change markedly in physical properties with temperature, and are degraded by air and sunlight. Consequently, it is necessary to transform the crude rubbers by compounding and vulcanization procedures into products which can better fulfill a specific function.

Although natural rubber may be obtained from hundreds of different plant species, the most important source is the rubber tree (*Hevea brasiliensis*). Natural rubber is *cis*-1,4-polyisoprene, containing approximately 5000 isoprene units in the average polymer chain.

Styrene-butadiene rubber (SBR) is the most important synthetic rubber and the most widely used rubber in the entire world. Formerly designated GR-S, SBRs are obtained by the emulsion polymerization of butadiene and styrene in varying ratios. However, in the most commonly used type, the ratio of butadiene to styrene is approximately¥break 78:22. Unlike natural rubber, SBR does not crystallize on stretching and thus has low tensile strength unless reinforced. The major use for SBR is in tires and tire products. Other uses include belting, hose, wire and cable coatings, flooring, shoe products, sponge, insulation, and molded goods.

Butyl rubber is essentially a polyisobutylene except for the presence of diolefin, usually isoprene, to provide the unsaturation necessary for vulcanization. Butyl rubbers have excellent resistance to oxygen, ozone, and weathering. In addition, these rubbers exhibit good electrical properties and high impermeability to gases. The high impermeability to gases results in use of butyl as an inner liner in tubeless tires. Other widespread uses are for wire and cable products, injection-molded and extruded products, hose, gaskets, and sealants, and where good damping characteristics are needed.

Ethylene-propylene polymers are produced by the copolymerization of ethylene and propylene. These copolymers exhibit outstanding resistance to heat, oxygen, ozone, and other aging and degrading agents. Abrasion resistance in tire treads is excellent. The mechanical properties of their vulcanizates are generally approximately equivalent to those of SBR.

One of the first synthetic rubbers used commercially in the rubber industry is neoprene, a polymer of chloroprene, 2-chlorobutadiene-1,3. The neoprenes have exceptional resistance to weather, sun, ozone, and abrasion. They are good in resilience, gas impermeability, and resistance to heat, oil, and flame. They are fairly good in low temperature and electrical properties. This versatility makes them useful in many applications requiring oil, weather, abrasion, or electrical resistance or combinations of these properties, such as wire and cable, hose, belts, molded and extruded goods, soles and heels, and adhesives.

Nitrile-type rubbers are copolymers of acrylonitrile and a diene, usually butadiene. The nitrile rubbers can be blended with natural rubber, polysulfide rubbers, and various resins to provide characteristics such as increased tensile strength, better solvent resistance, and improved weathering resistance.

Fluoroelastomers are basically copolymers of vinylidene fluoride and hexafluoropropylene. Because of their fluorine content, they are the most chemically resistant of the elastomers and also have good properties under extremes of temperature conditions. They are useful in the aircraft, automotive, and industrial areas.

Polyurethane elastomers are of interest because of their versatility and variety of properties and uses. They can be used as liquids or solids in a number of manufacturing methods. The largest use has been for making foam for upholstery and bedding. *See* POLYURETHANE RESINS.

Polysulfide rubbers have a large amount of surlfur in the main polymer chain and are therefore very chemically resistant, particularly to oils and solvents. They are used in such applications as putties, caulks, and hose for paint spray, gasoline, and fuel. Polyacrylate rubbers are useful because of their resistance to oils at high temperatures, including sulfur-bearing extreme-pressure lubricants. *See* POLYSULFIDE RESINS.

Proper choice of catalyst and order of procedure in polymerization have led to development of thermoplastic elastomers. The leading commercial types are styrene block copolymers having a structure which consists of polystyrene segments or blocks connected by rubbery polymers such as polybutadiene, polyisoprene, or ethylene-butylene polymer. Thermoplastic elastomers are very useful in providing a fast and economical method of producing a variety of products. One of the disadvantages for many applications is the low softening point of the thermoplastic elastomers. [E.G.P.]

**Rubidium** A chemical element, Rb, atomic number 37, and atomic weight 85.47. Rubidium is an alkali metal. It is a light, low-melting, reactive metal. *See* PERIODIC TABLE.

Most uses of rubidium metal and rubidium compounds are the same as those of cesium and its compounds. The metal is used in the manufacture of electron tubes, and the salts in glass and ceramic production.

Rubidium is a fairly abundant element in the Earth's crust, being present to the extent of 310 parts per million (ppm). This places it just below carbon and chlorine and just above fluorine and strontium in abundance. Sea water contains 0.2 ppm of rubidium, which (although low) is twice the concentration of lithium. Rubidium is like lithium and cesium in that it is tied up in complex minerals; it is not available in nature as simple halide salts as are sodium and potassium.

Rubidium has a density of 1.53 $g/cm^3$ (95.5 $lb/ft^3$, a melting point of 39°C (102°F), and a boiling point of 688°C (1270°F).

Rubidium is so reactive with oxygen that it will ignite spontaneously in pure oxygen. The metal tarnishes very rapidly in air to form an oxide coating, and it may ignite. The oxides formed are a mixture of $Rb_2O$, $Rb_2O_2$, and $RbO_2$. The molten metal is spontaneously flammable in air.

Rubidium reacts violently with water or ice at temperatures down to −100°C (−148°F). It reacts with hydrogen to form a hydride which is one of the least stable of the alkali hydrides. Rubidium does not react with nitrogen. With bromine or chlorine, rubidium reacts vigorously with flame formation. Organorubidium compounds can be prepared by techniques similar to those used for sodium and potassium. *See* ALKALI METALS; CESIUM. [M.Si.]

**Ruthenium** A chemical element, Ru, atomic number 44. The element is a brittle gray-white metal of low natural abundance, usually found alloyed with other platinum metals in nature. Ruthenium occurs as seven stable isotopes, and more than ten radioactive (unstable) isotopes are known. Four allotropes of the metal are known. Ruthenium is a hard white metal, workable only at elevated temperatures. It can be melted with an electric arc or an electron beam. *See* PERIODIC TABLE; TRANSITION ELEMENTS.

The metal is not oxidized by air at room temperature, but it does oxidize to give a surface layer of ruthenium dioxide ($RuO_2$) at about 900ÉC (1650°F); at about 1000°C (1830°F) the volatile compounds ruthenium tetraoxide ($RuO_4$) and ruthenium monoxide (RuO) form, which can result in loss of the metal. Metallic ruthenium is insoluble in common acids and aqua regia up to 100°C (212°F). The principal properties of ruthenium ore given in the table. *See* AQUA REGIA.

**Principal properties of ruthenium**

| Property | Value |
|---|---|
| Atomic number | 44 |
| Atomic weight | 101.07 |
| Crystal structure | Hexagonal close-packed |
| Lattice constant *a* at 25°C (77°F), nm *c/a* at 25°C (77°F) | 0.27056<br>1.5820 |
| Density at 25°C (77°F), g/cm$^3$ | 12.37 |
| Thermal neutron capture cross section, barns ($10^{-28}$ m$^2$) | 2.50 |
| Melting point | 2310°C (4190°F) |
| Boiling point | 4080°C (7380°F) |
| Specific heat at 0°C, cal/g (J/kg) | 0.0551 (231) |
| Thermal conductivity, 0–100°C, cal cm/cm$^2$ °C | 0.25 |
| Linear coefficient of thermal expansion, 20–100°C, $\mu$in./(in.)(°C) or $\mu$m/(m)(°C) | 9.1 |
| Electrical resistivity at 0°C, microhm-cm | 6.80 |
| Temperature coefficient of electrical resistance, 0–100°C/°C | 0.0042 |
| Tensile strength (annealed), kN · m$^{-2}$ | $4.96 \times 10^5$ |
| Young's modulus at 20°C (68°F), lb/in.$^2$ (Pa) | |
| Static | $60 \times 10^6$ ($4.1 \times 10^{11}$) |
| Dynamic | $69 \times 10^6$ ($4.75 \times 10^{11}$) |
| Vickers hardness number (diamond pyramid hardness) | 200–350 |

Ruthenium is relatively rare, having a natural abundance in the Earth's crust of about 0.0004 part per million. It is always found in the presence of other platinum metals. The major commercial sources of the element are the native alloys osmiridium and iridiosmium and the sulfide ore laurite. The element is also separated from other platinum metals by an intricate process, involving treatment with aqua regia (in which ruthenium, osmium, rhodium, and iridium are insoluble), to yield the pure metal.

Ruthenium is used commercially to harden alloys of palladium and platinum. The alloys are used in electrical contacts, jewelry, and fountain-pen tips. Application of ruthenium to industrial catalysis (hydrogenation of alkenes and ketones) and to automobile emission control (catalytic reduction of nitric oxide) and detection have been active areas of research. In medicine, ruthenium complexes have attracted some attention as potential antitumor reagents and imaging reagents. Ruthenium tetraoxide is finding increasing use as an oxidant for organic compounds. *See* CATALYSIS; HYDROGENATION; OSMIUM; PALLADIUM; PLATINUM; TECHNETIUM. [C.Cr.]

**Rutherfordium** A chemical element, symbol Rf, atomic number 104. Rutherfordium is the first element beyond the actinide series. In 1964 G. N. Flerov and coworkers at the Dubna Laboratories in Russia claimed the first identification of rutherfordium. A. Ghiorso and coworkers made a definitive identification at the Lawrence Radiation Laboratory, Berkeley, University of California, in 1969. *See* PERIODIC TABLE.

The Dubna group claimed the preparation of rutherfordium, mass number 260, by irradiating plutonium-242 with neon-22 ions in the heavy-ion cyclotron. The postulated nuclear reaction was $^{242}Pu + ^{22}Ne \rightarrow ^{260}Rf + 4$ neutrons. By 1969 the Berkeley group had succeeded in discovering two alpha-emitting isotopes of rutherfordium with mass numbers 257 and 259 by bombarding $^{249}Cf$ with $^{12}C$ and $^{13}C$ projectiles from the Berkeley heavy-ion linear accelerator (HILAC).

A number of years after the discovery at Berkeley, a team at Oak Ridge National Laboratory confirmed discovery of the isotope $^{257}Rf$ by detecting the characteristic nobelium x-rays following alpha decay. *See* ACTINIDE ELEMENTS; NOBELIUM; TRANSURANIUM ELEMENTS. [A.Gh.]

# S

**Saccharin** The sodium salt of *o*-sulfobenzimide, manufactured by processes that start with toluene or phthalic anhydride. The free imide, called insoluble saccharin because it is insoluble in water, has limited use as a flavoring agent in pharmaceuticals. The sodium and calcium salts are very soluble in water and are widely used as sweetening agents.

Sodium saccharin is 300–500 times sweeter than cane sugar (sucrose). The saccharin salts are used to improve the taste of pharmaceuticals and toothpaste and other toiletries, and as nonnutritive sweeteners in special dietary foods and beverages. Using noncaloric saccharin in place of sugar permits the formulation of low-calorie products for people on calorie-restricted diets and of low-sugar products for diabetics. *See* ASPARTAME. [K.M.B.]

**Salicylate** A salt or ester of salicylic acid having the general formula shown below and formed by replacing the carboxylic hydrogen of the acid by a metal (M) to give a salt or by an organic radical (R) to give an ester. Alkali-metal salts are water-soluble;

$$(o)\text{-}C_6H_4(OH)C(=O)—O—M \quad \text{or} \quad —R$$

the others, insoluble. Sodium salicylate is used in medicines as an antirheumatic and antiseptic, in the manufacture of dyes, and as a preservative (illegal in foods). Salicylic acid is used in the preparation of aspirin. The methyl ester is the chief component of oil of wintergreen. This ester is used in pharmaceuticals as a component of rubbing liniment. It is also used as a flavoring agent and an odorant. *See* ASPIRIN. [E.H.H.]

**Salt (chemistry)** A compound formed when one or more of the hydrogen atoms of an acid are replaced by one or more cations of the base. The common example is sodium chloride in which the hydrogen ions of hydrochloric acid are replaced by the sodium ions (cations) of sodium hydroxide. There is a great variety of salts because of the large number of acids and bases which has become known.

Salts are classified in several ways. One method—normal, acid, and basic salts—depends upon whether all the hydrogen ions of the acid or all the hydroxide ions of the base have been replaced:

| *Class* | *Examples* |
|---|---|
| Normal salts | $NaCl$, $NH_4C1$, $Na_2SO_4$, $Na_2CO_3$, $Na_3PO_4$, $Ca_3(PO_4)_2$ |
| Acid salts | $NaHCO_3$, $NaH_2PO_4$, $Na_2HPO_4$, $NaHSO_4$ |
| Basic salts | $Pb(OH)Cl$, $Sn(OH)Cl$ |

The other method—simple salts, double salts (including alums), and complex salts—depends upon the character of completeness of the ionization:

| *Class* | *Examples* |
|---|---|
| Simple salts | $NaCl$, $NaHCO_3$, $Pb(OH)Cl$ |
| Double salts | $KCl \cdot MgCl_2$ |
| Alums | $KAl(SO_4)_2$, $NaFe(SO_4)_2$ $NH_4Cr(SO_4)_2$ |
| Complex salts | $K_3Fe(CN)_6$, $Cu(NH_3)_4Cl_2$, $K_2Cr_2O_7$ |

*See* ACID AND BASE; CHEMICAL BONDING. [A.B.G.]

**Salt-effect distillation** A process of extractive distillation in which a salt that is soluble in the liquid phase of the system being separated is used in place of the normal liquid additive introduced to the extractive distillation column in order to effect the separation.

Extractive distillation is a process used to separate azeotrope-containing systems or systems in which relative volatility is excessively low. An additive, or separating agent, that is capable of raising relative volatility and eliminating azeotropes in the system being distilled is supplied to the column, where it mixes with the feed components and exerts its effect. The agent is subsequently recovered from one or both product streams by a separate process and recycled for reuse. *See* AZEOTROPIC MIXTURE.

In salt-effect distillation, the process is essentially the same as for a liquid agent, although the subsequent process used to recover the agent for recycling is different; that is, evaporation is used rather than distillation. The salt is added to the system by being dissolved in the reentering reflux stream at the top of the column. Being nonvolatile, it will reside in the liquid phase, flowing down the column and out in the bottom product stream.

The major commercial use of salt-effect distillation is in the concentration of aqueous nitric acid, using the salt magnesium nitrate as the separating agent. Other commercial applications include acetone-methanol separation using calcium chloride and isopropanol-water separation using the same salt. *See* AZEOTROPIC DISTILLATION; DISTILLATION. [W.F.F.]

**Samarium** A chemical element, Sm, atomic number 62, belonging to the rare-earth group. Its atomic weight is 150.35, and there are 7 naturally occurring isotopes; $^{147}Sm$, $^{148}Sm$, and $^{149}Sm$ are radioactive and emit $\alpha$ particles. *See* PERIODIC TABLE.

Samarium oxide is pale yellow, is readily soluble in most acids, and gives topaz-yellow salts in solutions. Samarium has found rather limited use in the ceramic industry, and it is used as a catalyst for certain organic reactions. One of its isotopes has a very high cross section for the capture of neutrons, and therefore there has been some interest in samarium in the atomic industry for use as control rods and nuclear poisons. *See* LANTHANUM; RARE-EARTH ELEMENTS. [F.H.Sp.]

**Scandium** A chemical element, Sc, atomic number 21, atomic weight 44.956. The only naturally occurring isotope is $^{45}Sc$. The electronic configuration of the ground-state, gaseous atom consists of the argon rare-gas core plus three more electrons in the $3d14s2$ levels. It has an unfilled inner shell (only one $3d$ electron) and is the first transition metal. It is one of the elements of the rare-earth group. *See* PERIODIC TABLE; RARE-EARTH ELEMENTS; TRANSITION ELEMENTS.

**Room-temperature properties of scandium metal (unless otherwise specified)**

| Property | Value |
|---|---|
| Atomic number | 21 |
| Atomic weight | 44.9559 ($^{12}C = 12$) |
| Lattice constant (hcp, $\alpha$–Sc), $a_0$ | 0.33088 nm |
| $c_0$ | 0.52680 nm |
| Density | 2.989 g/cm$^3$ |
| Metallic radius | 0.16406 nm |
| Atomic volume | 15.041 cm$^3$/mol |
| Transformation point | 1337°C (2439°F) |
| Melting point | 1541°C (2806°F) |
| Boiling point | 2836°C (5137°F) |
| Heat capacity | 25.51 J/mol K |
| Standard entropy, $S^{\circ}_{298.15}$ | 34.78 J/mol K |
| Heat of transformation | 4.01 kJ/mol |
| Heat of fusion | 14.10 kJ/mol |
| Heat of sublimation (at 298 K) | 377.8 kJ/mol |
| Debye temperature (at 0 K) | 345.3 K |
| Electronic specific heat constant | 10.334 mJ/mol K$^2$ |
| Magnetic susceptibility, $\chi_A^{298}$ ($a$) | 297.6 × 10$^{-6}$ emu/mol |
| $\chi_A^{298}$ ($c$) | 288.6 × 10$^{-6}$ emu/mol |
| Electrical resistivity, $\rho_a^{300}$ | 70.90 $\mu$ohm-cm |
| $\rho_c^{300}$ | 26.88 $\mu$ohm-cm |
| Thermal expansion, $\alpha_{a,i}$ | 7.55 × 10$^{-6}$ |
| $\alpha_{c,i}$ | 15.68 × 10$^{-6}$ |
| Isothermal compressibility | 17.8 × 10$^{-12}$ m$^2$/N |
| Bulk modulus | 5.67 × 10$^{10}$ N/m$^2$ |
| Young's modulus | 7.52 × 10$^{10}$ N/m$^2$ |
| Shear modulus | 2.94 × 10$^{10}$ N/m$^2$ |
| Poisson's ratio | 0.279 |

The principal raw materials for the commercial production of scandium are uranium and tungsten tailings and slags from tin smelters or blast furnaces used in cast iron production. Wolframite ($WO_3$) concentrates contain 500–800 ppm scandium.

Scandium is the least understood of the 3*d* metals. The major reason has been the unavailability of high-purity scandium metal, especially with respect to iron impurities. Many of the physical properties reported in the literature vary considerably, but the availability of electrotransport-purified scandium has allowed measurement of the intrinsic properties of scandium (see table).

Scandium increases the strength of aluminum. It also strengthens magnesium alloys when added to magnesium together with silver, cadmium, or yttrium. Scandium inhibits the oxidation of the light rare earths and, if added along with molybdenum, inhibits the corrosion of zirconium alloys in high-pressure steam. The addition of ScC to TiC has been reported to form the second-hardest material known. $Sc_2O_3$ can be used in many other oxides to improve electrical conductivity, resistance to thermal shock, stability, and density. Scandium is used in the preparation of the laser material $Gd_3ScGa_4O_{12}$, gadolinium scandium gallium garnet (GSGG). This garnet when doped with both $Cr^{3+}$ and $Nd^{3+}$ ions is said to be $3^1/_2$ times as efficient as the widely used $Nd^{3+}$-doped yttrium aluminum garnet (YAG:$Nd^{3+}$) laser. Ferrites and garnets containing scandium are used in switches in computers; in magnetically controlled switches that modulate light passing through the garnet; and in microwave equipment. Scandium is used in high-intensity lights. Scandium iodide is added because of its broad emission spectrum. Bulbs with mercury, NaI, and $ScI_3$ produce a highly efficient light output of a color close to sunlight. This is especially important when televising presentations indoors or at night. When used with night displays, the bulbs give a natural daylight appearance. Scandium metal has been used as a neutron filter. It allows 2-keV neutrons to pass through, but stops other neutrons that have higher or lower energies. *See* YTTRIUM. [J.Cap.; K.A.G.]

**Seaborgium** A chemical element, symbol Sg, atomic number 106. Seaborgium has chemical properties similar to tungsten. It was synthesized and identified in 1974. This discovery of seaborgium took place nearly simultaneously in two nuclear laboratories, the Lawrence Berkeley Laboratory at the University of California and the Joint Institute for Nuclear Research at Dubna in Russia.

The Berkeley group, under the leadership of A. Ghiorso, used as its source of heavy ions the Super-Heavy Ion Linear Accelerator (SuperHILAC). The production and positive identification of the isotope of seaborgium with the mass number 263 decays with a half-life of 0.9 + 0.2 s by the emission of alpha particles of principal energy 9.06 + 0.04 MeV. This isotope is produced in the reaction in which four neutrons are emitted: $^{249}Cf(^{18}O,4n)$.

The Dubna group, under the leadership of G. N. Flerov and Y. T. Oganessian, produced its heavy ions with a heavy-ion cyclotron. They found a product that decays by the spontaneous fission mechanism with the very short half-life of 7 ms. They assigned it to the isotope $^{259}Sg$, suggesting reactions in which two or three neutrons are emitted: $^{207}Pb(^{54}Cr,2n)$ and $^{208}Pb(Cr^{54},3n)$. *See* NOBELIUM; NUCLEAR CHEMISTRY; PERIODIC TABLE; TRANSURANIUM ELEMENTS. [G.T.S.]

**Secondary ion mass spectrometry (SIMS)** An instrumental technique that measures the elemental and molecular composition of solid materials. SIMS provides methods of visualizing the two- and three-dimensional composition of solids at lateral resolutions approaching several hundred nanometers and depth resolutions of 1–10 nm. This technique employs an energetic ion beam to remove or sputter the atomic and molecular constituents from a surface in a very controlled manner. The sputtered products include atoms, molecules, and molecular fragments that are characteristic of the surface composition within each volume element sputtered by the ion beam. A small fraction of the sputtered atoms and molecules are ionized as either positive or negative ions, and a measurement by SIMS determines the mass and intensity of these secondary ions by using various mass analysis or mass spectrometry techniques. In this technique, the sputtering ions are referred to as the primary ions or the primary ion beam, while the ions produced in sputtering the solid are the secondary ions. Most elements in the periodic table produce secondary ions, and SIMS can quantitatively detect elemental concentrations in the part-per-million to part-per-billion range.

Secondary ions are formed by kinetic and chemical ionization processes in which sputtering is achieved by energy transfer from the primary ions to the solid surface. Typically, SIMS primary ions impact the surface with kinetic energies of 5–20 keV, and this energy is ultimately transferred to the sample atoms and molecules. The energy transfer initiates a collision cascade within the solid that ejects atoms and molecules at the solid surface-vacuum interface. *See* IONS.

SIMS analyses are divided into two broad categories known as dynamic and static. In the dynamic type, the most common SIMS method, a relatively intense primary ion beam sputters the sample surface at high sputter rates, providing a very useful way to determine the in-depth concentration of different elements in a solid. The most common secondary ions detected in a dynamic secondary ion mass spectrometry analysis are elemental ions or clusters of elemental ions.

Static or molecular SIMS utilizes a very low intensity primary ion beam, and static analyses are typically completed before a single monolayer has been removed from the surface. Most static analyses are stopped before 1% of the top surface layer has been chemically damaged or eroded; under these conditions, molecular and molecular fragment ions characteristic of the chemical structure of the surface are often detected. Thus, static SIMS is best suited for near-surface analysis of molecular composition or chemical structure information, while dynamic SIMS provides the best technique for bulk and in-depth elemental analysis.

Instrument designs for SIMS require a primary ion gun or column to generate and transport the primary ion beam to the sample surface, a sample chamber with sample mounting facilities, a mass spectrometer which performs mass-to-charge separation of the different secondary ions, and an ion-detection system. The complete instrument is typically housed in an ultrahigh vacuum chamber. *See* MASS SPECTROSCOPE.

Dynamic SIMS has been successfully utilized in diverse applications. The analytical issues are the detection and localization of specific elements on the surface or in the bulk of the materials. The most important and common application area of dynamic SIMS is the bulk and in-depth analysis of semiconductor materials. Another important contribution of SIMS to materials science is the identification and localization of trace elements in metal grain boundaries, providing detailed insight into the chemistry of welds and alloys. Dynamic SIMS also has the unique ability to detect specific catalyst poisons on the surface and in the bulk of used or spent catalysts. In geological sciences, SIMS has been used to detect isotopic anomalies in the composition of various geological and meteoritic samples to help determine the age of the universe. Dynamic SIMS has found extensive applications to the characterization of both hard and soft biological tissue.

Applications of static secondry ion mass spectrometry to analyses of the near-surface region of solids has become increasingly important as ever more sophisticated materials are developed. The near-surface region is generally defined as the top five monolayers (2–3 nm) of a solid. Examples of technology areas in which the chemistry of the near-surface region plays a critical role include high-performance glass coatings, liquid-crystal displays, manufacturing, semiconductor processing, biopolymer and biocompatible materials development, and polymer adhesives and coatings. *See* MASS SPECTROMETRY . [R.W.O.]

**Sedimentation (industry)** The separation of a dilute suspension of solid particles into a supernatant liquid and a concentrated slurry. If the purpose of the process is to concentrate the solids, it is termed thickening; and if the goal is the removal of the solid particles to produce clear liquid, it is called clarification. Thickening is the common operation for separating fine solids from slurries. Examples are magnesia, alumina red mud, copper middlings and concentrates, china clay (kaolin), coal tailings, phosphate slimes, and pulp-mill and other industrial wastes. Clarification is prominent in the treatment of municipal water supplies.

The driving force for separation is the difference in density between the solid and the liquid. Ordinarily, sedimentation is effected by the force of gravity, and the liquid is water or an aqueous solution. For a given density difference, the solid settling process proceeds more rapidly for larger-sized particles. For fine particles or small density differences, gravity settling may be too slow to be practical; then centrifugal force rather than gravity can be used. Further, when centrifugal force is inadequate, the more positive method of filtration may be employed. All those methods of separating solids and liquids belong to the generic group of mechanical separations. *See* CENTRIFUGATION; CLARIFICATION; FILTRATION.

Particles too minute to settle at practical rates may form flocs by the addition of agents such as sodium silicate, alum, lime, and alumina. Because the agglomerated particles act like a single large particle, they settle at a feasible rate and leave a clear liquid behind. [V.W.U.]

**Selenium** A chemical element, Se, atomic number 34, atomic weight 78.96. The properties of this element are similar to those of tellurium. *See* PERIODIC TABLE; TELLURIUM.

Selenium burns in air with a blue flame to give selenium dioxide, $SeO_2$. The element also reacts directly with a variety of metals and nonmetals, including hydrogen and the halogens. Nonoxidizing acids fail to react with selenium, but nitric acid, concentrated sulfuric acid, and strong alkali hydroxides dissolve the element. The only important compound of selenium with hydrogen is hydrogen selenide, $H_2Se$, a colorless flammable gas possessing a distinctly unpleasant odor, and a toxicity greater and a thermal stability less than that of hydrogen sulfide. Selenium oxyhalide, $SeOCl_2$, is a colorless liquid widely used as a nonaqueous solvent. The oxybromide, $SeOBr_2$, is an orange solid having chemical properties similar to those of $SeOCl_2$. The oxyflouride, $SeOF_2$, a colorless liquid with a pungent smell, reacts with water, glass, and silicon, and also forms additional compounds. Compounds in which C-Se bonds appear are numerous and vary from the simple selenols, RSeH, to molecules exhibiting biological activity such as selenoamino acids and selenopeptides. *See* ORGANOSELENIUM COMPOUND.

The abundance of this widely distributed element in the Earth's crust is estimated to be about $7 \times 10^{-5}$ % by weight, occurring as the selenides of heavy elements and to a limited extent as the free element in association with elementary sulfur. Examples of the variety of selenide minerals are berzelianite ($Cu_2Se$), eucairite (AgCuSe), and jermoite [$As(S,Se)_2$]. Selenium minerals do not occur in sufficient quantity to be useful as commercial sources of the element.

Major uses of selenium include the photocopying process of xerography, which depends on the light sensitivity of thin films of amorphous selenium, the decolorization of glasses tinted by the presence of iron compounds, and use as a pigment in plastics, paints, enamels, glass, ceramics, and inks. Selenium is also employed in photographic exposure meters and as a metallurgical additive to improve the machinability of certain steels. Minor uses include application as a nutritional additive for numerous animal species, use in photographic toning, metal-finishing operations, metal plating, high-temperature lubricants, and as catalytic agents, particularly in the isomerization of certain petroleum products. [J.W.Ge.]

The biological importance of selenium is well established, as all classes of organisms metabolize selenium. In humans and other mammals, serious diseases arise from either excessive or insufficient dietary selenium. The toxic effects of selenium have long been known, particularly for grazing animals. In soils with high selenium content, some plants accumulate large amounts of selenium. Animals that ingest these selenium-accumulating plants develop severe toxic reactions.

Although toxic at high levels, selenium is an essential micronutrient for mammalian species. The accepted minimum daily requirement of selenium for adult humans is 70 micrograms. Many types of food provide selenium, particularly seafood, meats, grains, and the onion family. Mammals and birds require selenium for production of the enzyme glutathione peroxidase, which protects against oxidation-induced cancers. Other seleno-proteins of unknown function are found in mammalian blood, various tissues, and spermatozoa. *See* AMINO ACIDS. [M.J.Ax.; T.C.St.]

**Shock tube** A laboratory device for rapidly raising confined samples of fluids (primarily gases) to preselected high temperatures and densities. This is accomplished by a shock wave, generated when a partition (diaphragm) that separates a low-pressure (driven) section from a high-pressure (driver) section is rapidly removed. Shock tubes can be circular, square, or rectangular in cross section. In the driven section, gaseous samples can be heated to temperatures as high as 27,500°F (15,000 K) under strictly homogeneous conditions. At the shock front the transition from the unshocked to the high-temperature condition is of short but of finite duration. The incident shock wave generates a slower-moving wave that additionally heats the compressed gas.

With the diaphragm in place the driven section is filled to a modest pressure with an inert gas plus the gas of interest. The driver section is filled to a high pressure with a low-molecular-weight gas: helium or hydrogen. When the diaphragm is rapidly ruptured, expansion of the high-pressure gas acts as a low-mass piston that generates a steepening pressure front, which moves ahead of the boundary between the driver and the test gases.

Shock tubes are used to investigate the gas dynamics of shocks and for preparing test samples for equilibrium or kinetic studies. Many gas dynamic problems can be investigated in shock tubes, such as thermal boundary-layer growth, shock bifurcation, and shock-wave focusing and reflection. Shock tubes are used to prepare gases for study at very low or very high temperatures without thermal contact with extraneous surfaces. Significant applications of shock tubes to chemical kinetics includes the determination of diatom dissociation rates and the study of polyatomic molecules. [S.H.B.]

**Silicon** A chemical element, Si, atomic number 14, and atomic weight 28.086. Silicon is the most abundant electropositive element in the Earth's crust. The element is a metalloid with a decided metallic luster; it is quite brittle. It has a specific gravity of 2.42 at 20°C (68°F), melts at 1420°C (2588°F), and boils at 3280°C (5936°F). The element is usually tetravalent in its compounds, although sometimes divalent, and is decidedly electropositive in its chemical behavior. In addition, pentacoordinate and hexacoordinate compounds of silicon are known. *See* METALLOID; PERIODIC TABLE.

Crude elementary silicon and its intermetallic compounds are used in alloying constituents to strengthen aluminum, magnesium, copper, and other metals. Metallurgical silicon of 98–99% purity is used as the starting material for manufacturing organosilicon compounds and silicone resins, elastomers, and oils. Silicon chips are used in integrated circuits. Photovoltaic cells for direct conversion of solar energy to electricity use wafers sliced from single crystals of electronic-grade silicon. Silicon dioxide is used as the raw material for making elementary silicon and for silicon carbide. Sizable crystals of it are used for piezoelectric crystals. Fused quartz sand becomes silica glass, used in chemical laboratories and plants as well as an electrical insulator. A colloidal dispersion of silica in water is used as a coating agent and as an ingredient in certain polishes.

Naturally occurring silicon contains 92.2% of the isotope of mass number 28, 4.7% of silicon-29, and 3.1% of silicon-30. In addition to these stable, natural isotopes, several artificially radioactive isotopes are known. Elementary silicon has the physical properties of a metalloid, resembling germanium below it in group 14 of the periodic table. In very pure form silicon is an intrinsic semiconductor, although the extent of its semiconduction is greatly increased by the introduction of minute amounts of impurities. Silicon resembles the metals in its chemical behavior. It is about as electropositive as tin, and decidedly more positive than germanium or lead. In keeping with this rather metallic character, silicon forms tetrapositive ions and a variety of covalent compounds; it appears as a negative ion in only a few silicides and as a positive constituent of oxy acid or complex anions.

Several series of hydrides are formed, a variety of halides (some of which contain silicon-to-silicon bonds), and also many series of oxygen-containing compounds which may be either ironic or covalent in their properties.

Silicon occurs in many forms of the dioxide and as almost numberless variations of the natural silicates.

In abundance, silicon exceeds by far every other element except oxygen. It constitutes 27.72% of the solid crust of the Earth, whereas oxygen constitutes 46.6%, and the next element after silicon, aluminum, accounts for 8.13%.

Silicon is reported to form compounds with 64 of the 96 stable elements, and it probably forms silicides with 18 other elements. Besides the metal silicides, used in large quantities in metallurgy, silicon forms useful and important compounds with hydrogen, carbon, the halogen elements, nitrogen, oxygen, and sulfur. In addition, useful organosilicon derivatives have been prepared. [E.P.P.]

**Silicone resins** Polymers composed of alternating atoms of silicon and oxygen with organic substituents attached to the silicon atoms, as shown in the formula below.

$$\left[ \begin{array}{c} R' \\ | \\ -Si-O- \\ | \\ R \end{array} \right]_n$$

Silicones, also called organopolysiloxanes, may exist as liquids, greases, resins, or rubbers. Silicone polymers have good resistance to water and oxidation, stability at high and low temperatures, and lubricity.

Silicones are obtained by the condensation of hydroxy organosilicon compounds formed by the hydrolysis of organosilicon halides. The first products are usually low in molecular weight ($n = 2$ to 7), and usually consist of a mixture of linear and cyclic species, especially the tetramer. Fluids having a wide range of viscosity are prepared by polymerizing further, using a monofunctional trichlorosilane to limit molecular weights to the value desired. Elastomers are made by polymerization of the purified tetramer using an alkaline catalyst at 100–150°C (212–302°F). Properties can be varied by partial replacement of some of the methyl groups by other substituents and by the use of reinforcing fillers.

The wide range of structural variations makes it possible to tailor compositions for many kinds of applications. Low-molecular-weight silanes containing amino or other functional groups are used as treating or coupling agents for glass fiber and other reinforcements in order to cause unsaturated polyesters and other resins to adhere better.

The liquids, generally dimethyl silicones of relatively low molecular weight, have low surface tension, great wetting power and lubricity for metals, and very small change in viscosity with temperature. They are used as hydraulic fluids, as antifoaming agents, as treating and waterproofing agents for leather, textiles, and masonry, and in cosmetic preparations. The greases are particularly desired for applications requiring effective lubrication at very high and at very low temperatures.

Silicone resins are used for coating applications in which thermal stability in the range 300–500°C (570–930°F) is required. The dielectric properties of the polymers make them suitable for many electrical applications, particularly in electrical insulation that is exposed to high temperatures and as encapsulating materials for electronic devices.

Silicone rubbers are compositions containing high-molecular-weight dimethyl silicone linear polymer, finely divided silicon dioxide as the filler, and a peroxidic curing agent. The silicone rubbers have the remarkable ability of remaining flexible at very low temperatures and stable at high temperatures. *See* INORGANIC POLYMER; PLASTICS PROCESSING; RUBBER; SILICON. [J.A.M.]

**Silver** A chemical element, Ag, atomic number 47, atomic mass 107.868. It is a gray-white, lustrous metal. Chemically it is one of the heavy metals and one of the noble metals; commercially it is a precious metal. There are 25 isotopes of silver with atomic masses ranging from 102 to 117. Ordinary silver is made up of the isotopes of masses 107 (52% of natural silver) and 109 (48%). *See* PERIODIC TABLE.

Although silver is the most active chemically of the noble metals, it is not very active in comparison with most other elements. It does not oxidize as iron does when it rusts, but it reacts with sulfur or hydrogen sulfide to form the familiar silver tarnish. Electroplating silver with rhodium prevents this discoloration. Silver itself does not react with dilute nonoxidizing acids (hydrochloric or sulfuric acids) or strong bases (sodium hydroxide). However, oxidizing acids (nitric or concentrated sulfuric acids) dissolve it by reaction to form the unipositive silver ion, $Ag^+$. The $Ag^+$ ion is colorless, but a number of silver compounds are colored because of the influence of their other constituents.

Silver is almost always monovalent in its compounds, but an oxide, fluoride, and sulfide of divalent silver are known. Some coordination compounds of silver, also called silver complexes, contain divalent and trivalent silver. Although silver does not oxidize when heated, it can be oxidized chemically or electrolytically to form silver oxide or peroxide, a strong oxidizing agent. Because of this activity, silver finds considerable use as an oxidation catalyst in the production of certain organic materials.

Soluble silver salts, especially $AgNO_3$, have proved lethal in doses as small as 0.07 oz (2 g). Silver compounds may be slowly absorbed by the body tissues, with a resulting bluish or blackish pigmentation of the skin (argyria).

Silver is a rather rare element, ranking 63rd in order of abundance. Sometimes it occurs in nature as the free element (native silver) or alloyed with other metals. For the most part, however, silver is found in ores containing silver compounds. The principal silver ores are argentite, $Ag_2S$, cerargyrite or horn silver, AgCl, and several minerals in which silver sulfide is combined with sulfides of other metals; stephanite, $5Ag_2S\text{Ö}Sb_2S_5$; polybasite, $9(Cu_2S, Ag_2S)\text{Ö}(Sb_2S_3, As_2S_3)$; proustite, $3Ag_2S\text{Ö}As_2S_3$; and pyragyrite, $3Ag_2S\text{Ö}Sb_2S_3$. About three-fourths of the silver produced is a by-product of the extraction of other metals, copper and lead in particular.

Pure silver is a white, moderately soft metal (2.5–3 on Mohs hardness scale), somewhat harder than gold. When polished, it has a brilliant luster and reflects 95% of the light falling on it. Silver is second to gold in malleability and ductility. Its density is 10.5 times that of water. The quality of silver, its fineness, is expressed as parts of pure silver per 1000 parts of total metal. Commercial silver is usually 999 fine. Silver is available commercially as sterling silver (7.5% copper) and in ingots, plate, moss, sheets, wire, castings, tubes, and powder.

Silver, with the highest thermal and electrical conductivities of all the metals, is used for electrical and electronic contact points and sometimes for special wiring. Silver has well-known uses in jewelry and silverware. Silver compounds are used in many photographic materials. In most of its uses, silver is alloyed with one or more other metals. Alloys in which silver is an ingredient include dental amalgam and metals for engine pistons and bearings. [W.E.C.]

**Soap** A cleansing agent described chemically as an alkali metal salt of a long-carbon-chain monocarboxylic acid, such as sodium myristate ($NaCOOC_{13}H_{27}$), as represented by the following.

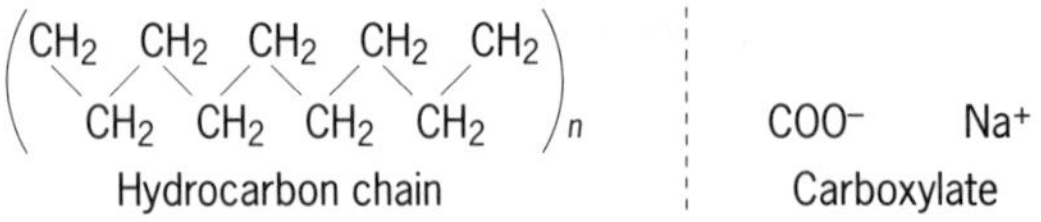

$$\left(\begin{matrix} CH_2 & & CH_2 & & CH_2 & & CH_2 & & CH_2 \\ & CH_2 & & CH_2 & & CH_2 & & CH_2 & \end{matrix}\right)_n \quad \vdots \quad COO^- \quad Na^+$$

Hydrocarbon chain | Carboxylate

The hydrocarbon portion is hydrophobic and the carboxylate portion is hydrophilic. This duality enables soap to physically remove dirt and oils from surfaces and disperse or emulsify them in water. For detergency purposes, the most useful hydrophobic portion contains 12–18 carbon atoms. When the chain length exceeds 18 carbons, they become insoluble in water.

A soap-making process that was formerly used was based on an alkaline hydrolysis reaction, saponification, according to the reaction below, where R represents the hydrocarbon chain.

$$\underset{\text{Fat}}{C_3H_5(OOCR)_3} + \underset{\text{Caustic soda}}{3NaOH} \rightarrow \underset{\text{Soap}}{3NaOOCR} + \underset{\text{Glycerin}}{C_3H_5(OH)_3}$$

Most soap is now made by a continuous process. Fats and oils may be converted directly to soap by the reaction with caustic, and the neat soap separated by a series of centrifuges or by countercurrent washing. The saponification may be carried out with heat, pressure, and recirculation. Other important modern processes hydrolyze the fats directly with water and catalysts at high temperatures. This permits fractionation of the fatty acids, which are neutralized to soap in a continuous process. Advantages for this process are close control of the soap concentration, the preparation of soaps of certain chain lengths for specific purposes, and easy recovery of the by-product glycerin. Tallow and coconut oil are the most common fatty materials used for soapmaking.

Because of the marked imbalance of polarity within the molecules of synthetic detergents and soaps, they have unusual surface and solubility characteristics. The feature of their molecular structure which causes these properties is the location of the hydrophilic function at or near the end of a long hydrocarbon (hydrophobic) chain. One part of the molecule is water-seeking while the other portion is oil-seeking. Thus these molecules concentrate and orient themselves at interfaces, such as an oil-solution interface where the hydrophobic portion enters the oil and the hydrophilic portion stays in the water. Consequently, the interfacial tension is lowered and emulsification can result. At an agitated air-solution interface, the excess detergent concentration leads to sudsing.

Considerable research has also been done on the aqueous solutions of soap. Many of the phase characteristics now known to occur in most surfactant-water systems were first discovered in soap systems. These include micelle formation where, at a critical concentration, soap molecules in solution form clusters (micelles). The hydrocarbon chains associate with each other in the interior, and the polar groups are on the outside. *See* MICELLE; SURFACTANT.

There is one serious disadvantage to the use of soap which has largely caused its replacement by synthetic detergents. The problem is that carboxylate ions of the soap react with the calcium and magnesium ions in natural hard water to form insoluble materials which manifest as a floating curd. Bar soap for personal bathing is the greatest remaining market for soap. Some commercial laundries having soft water continue to use soap powders. Metallic soaps are alkaline-earth or heavy-metal long-chain carboxylates which are insoluble in water but soluble in nonaqueous solvents. They are used as additives to lubricating oils, in greases, as rust inhibitors, and in jellied fuels. [R.C.M.]

**Sodium** A chemical element, Na, atomic number 11, and atomic weight 22.9898. Sodium is between lithium and potassium in the periodic table. The element is a soft, reactive, low-melting metal with a specific gravity of 0.97 at 20°C (68°F). Sodium is commercially the most important alkali metal. The physical properties of metallic sodium are summarized in the table. *See* PERIODIC TABLE.

Sodium ranks sixth in abundance among all the elements in the Earth's crust, which contains 2.83% sodium in combined form. Only oxygen, silicon, aluminum, iron, and calcium are more abundant. Sodium is, after chlorine, the second most abundant element in solution in seawater. The important sodium salts found in nature include sodium chloride (rock salt), sodium carbonate (soda and trona), sodium borate (borax), sodium nitrate (Chile saltpeter), and sodium sulfate. Sodium salts are found in seawater, salt lakes, alkaline lakes, and mineral springs. *See* ALKALI METALS.

Sodium reacts rapidly with water, and even with snow and ice, to give sodium hydroxide and hydrogen. The reaction liberates sufficient heat to melt the sodium and ignite the hydrogen. When exposed to air, freshly cut sodium metal loses its silvery appearance and becomes dull gray because of the formation of a coating of sodium oxide.

Sodium does not react with nitrogen. Sodium and hydrogen react above about 200°C (390°F) to form sodium hydride. Sodium reacts with ammonia, forming sodium amide. Sodium also reacts with ammonia in the presence of coke to form sodium cyanide.

Sodium does not react with paraffin hydrocarbons but does form addition compounds with naphthalene and other polycyclic aromatic compounds and with arylated alkenes. The reaction of sodium with alcohols is similar to, but less rapid than, the reaction of sodium with water. Sodium reacts with organic halides in two general ways. One of these involves condensation of two organic, halogen-bearing compounds by removal of the halogen, allowing the two organic radicals to join directly. The second type of reaction involves replacement of the halogen by sodium, giving an organosodium compound. *See* ORGANOMETALLIC COMPOUND.

**Physical properties of sodium metal**

| Property | Temperature °C | Temperature °F | Metric (scientific) units | British (engineering) units |
|---|---|---|---|---|
| Density | 0 | 32 | 0.972 g/cm$^3$ | 60.8 lb/ft$^3$ |
| | 100 | 212 | 0.928 g/cm$^3$ | 58.0 lb/ft$^3$ |
| | 800 | 1472 | 0.757 g/cm$^3$ | 47.3 lb/ft$^3$ |
| Melting point | 97.5 | 207.5 | | |
| Boiling point | 883 | 1621 | | |
| Heat of fusion | 97.5 | 207.5 | 27.2 cal/g | 48.96 Btu/lb |
| Heat of vaporization | 883 | 1621 | 1005 cal/g | 1809 Btu/lb |
| Viscosity | 250 | 482 | 3.81 millipoises | 4.3 kinetic units |
| | 400 | 752 | 2.69 millipoises | 3.1 kinetic units |
| Vapor pressure | 440 | 824 | 1 mm | 0.019 lb/in.$^2$ |
| | 815 | 1499 | 400 mm | 7.75 lb/in.$^2$ |
| Thermal conductivity | 21.2 | 70.2 | 0.317 cal/(s)(cm)(°C) | 76 Btu/(h)(ft)(°F) |
| | 200 | 392 | 0.193 cal/(s)(cm)(°C) | 46.7 Btu(h)(ft)(°F) |
| Heat capacity | 20 | 68 | 0.30 cal/(g)(°C) | 0.30 Btu/(lb)(°F) |
| | 200 | 392 | 0.32 cal/(g)(°C) | 0.32 Btu/(lb)(°F) |
| Electrical resistivity | 100 | 212 | 965 microhm-cm | |
| Surface tension | 100 | 212 | 206.4 dynes/cm | |
| | 250 | 482 | 199.5 dynes/cm | |

Sodium chloride, or common salt, NaCl, is not only the form in which sodium is found in nature but (in purified form) is the most important sodium compound in commerce as well. Sodium hydroxide, NaOH, is also commonly known as caustic soda. Sodium carbonate, $Na_2CO_3$, is best known under the name soda ash.

The largest single use for sodium metal, accounting for about 60% of total production, is in the synthesis of tetraethyllead, an antiknock agent for automotive gasolines. A second major use is in the reduction of animal and vegetable oils to long-chain fatty alcohols; these alcohols are raw materials for detergent manufacture. Sodium is used to reduce titanium and zirconium halides to their respective metals. Sodium chloride is used in curing fish, meat packing, curing hides, making freezing mixtures, and food preparation (including canning and preserving). Sodium hydroxide is used in the manufacture of chemicals, cellulose film, rayon soap pulp, and paper. Sodium carbonate is used in the glass industry and in the manufacture of soap, detergents, various cleansers, paper and textiles, nonferrous metals, and petroleum products. Sodium sulfate (salt cake) is used in the pulp industry and in the manufacture of flat glass. [M.Si.]

The sodium ion ($Na^+$) is the main positive ion present in extracellular fluids and is essential for maintenance of the osmotic pressure and of the water and electrolyte balances of body fluids. Hydrolysis of adenosine triphosphate (ATP) is mediated by the membrane-bound enzyme $Na^+$, $K^+$-ATPase (this enzyme is also called sodium pump). The potential difference associated with the transmembrane sodium and potassium ion gradients is important for nerve transmission and muscle contraction. Sodium ion gradients are also responsible for various sodium ion–dependent transport processes, including sodium-proton exchange in the heart, sugar transport in the intestine, and sodium-lithium exchange and amino acid transport in red blood cells. *See* ADENOSINE TRIPHOSPHATE (ATP); POTASSIUM. [D.M.DeF.]

**Sol-gel process** A chemical synthesis technique for preparing gels, glasses, and ceramic powders. The sol-gel process generally involves the use of metal alkoxides, which undergo hydrolysis and condensation polymerization reactions to give gels.

The production of glasses by the sol-gel method permits preparation of glasses at far lower temperatures than is possible by using conventional melting. It also makes possible synthesis of compositions that are difficult to obtain by conventional means because of problems associated with volatilization, high melting temperatures, or crystallization. In addition, the sol-gel approach is a high-purity process that leads to excellent homogeneity. Finally, the sol-gel approach is adaptable to producing films and fibers as well as bulk pieces.

The sol-gel process comprises solution, gelation, drying, and densification. The preparation of a silica glass begins with an appropriate alkoxide which is mixed with water and a mutual solvent to form a solution. Hydrolysis leads to the formation of silanol groups (Si—OH). These species are only intermediates. Subsequent condensation reactions produce siloxane bonds (Si—O—Si). The silica gel formed by this process leads to a rigid, interconnected three-dimensional network consisting of submicrometer pores and polymeric chains. During the drying process (at ambient pressure), the solvent liquid is removed and substantial shrinkage occurs. The resulting material is known as a xerogel. When solvent removal occurs under hypercritical (supercritical) conditions, the network does not shrink and a highly porous, low-density material known as an aerogel is produced. Heat treatment of a xerogel at elevated temperature produces viscous sintering (shrinkage of the xerogel due to a small amount of viscous flow) and effectively transforms the porous gel into a dense glass.

Materials used in the sol-gel process include inorganic compositions that possess specific properties such as ferroelectricity, electrochromism, or superconductivity. The most successful applications utilize the composition control, microstructure control, purity, and uniformity of the method combined with the ability to form various shapes at low temperatures. Films and coatings were the first commercial applications of the sol-gel process. The development of sol-gel-based optical materials has also been quite successful, and applications include monoliths (lenses, prisms, lasers), fibers (waveguides), and a wide variety of optical films. Other important applications of sol-gel technology utilize controlled porosity and high surface area for catalyst supports, porous membranes, and thermal insulation. [B.Du.]

**Solid-state chemistry** The science of the elementary, atomic compositions of solids and the transformations that occur in and between solids and between solids and other phases to produce solids. Solid-state chemistry deals primarily with those microscopic features which are uniquely characteristic of solids and which are the causes for the macroscopic chemical properties and the chemical reactions of solids. As with other branches of the physical sciences, solid-state chemistry also includes related areas that furnish concepts and explanations of those phenomena which are more characteristic of the subject itself.

The overlap of solid-state chemistry and solid-state physics is extensive. However, the perspectives of the two are different. In general, solid-state physics treats properties, such as energy and entropy, which are continuously variable in the solid, whereas solid-state chemistry concerns those properties which are discontinuous because of chemical reactions. Also, solid-state chemistry tends to be based on structure in configuration space, whereas solid-state physics tends to be based on momentum space. *See* INORGANIC CHEMISTRY.

Solid-state chemistry has no single unifying theoretical base and tends to be largely an experimental science supported by several theoretical bases. Consequently, its separation into topics is not well established. Aspects of solid-state chemistry include chemical bonding, crystal defects, crystal structures, crystal field theory, diffusion in solids, ionic crystals, lattice vibrations, and nonstoichiometry. *See* NONSTOICHIOMETRIC COMPOUNDS.

Studies of structures provide a basis for understanding the chemical bonding in solids, their properties, reactions among them, and their variabilities of composition. Numerous binary compounds or phases have layered structures. The layers are displaced relative to each other in such a way that the structures can be classified in the space groups, but when other elements or compounds are incorporated between the layers, the structure is highly distorted. The most widely recognized material to possess the layered structure and the associated solid-state chemistry is graphite. Structural information is represented by the location of atoms on the lattice network. The structures are determined by diffraction of x-rays, neutrons, or electrons. Two other features of the structure of solids are the local structure about a given atom and the extended structure on a more global scale. *See* GRAPHITE.

The structure of a solid is the result of the operation of interatomic or interionic forces and the size and shape of the atoms or ions. Hence, logically, bonding should be described first and structure second. However, the detailed role of the electrons in interionic forces is so complex, and the quantitative aspects of the problem of minimizing the potential energy with respect to all the possible configurations is so difficult, that structures cannot be derived. Rather, it is necessary to derive some information about bonding from structures, cohesive energies, refractive indices, electron binding energies, polarizabilities, and other properties through the use of models. Simple, classically based models can be classified generally as ionic, covalent, and metallic bonding and combinations of the three. *See* CHEMICAL BONDING; STRUCTURAL CHEMISTRY.

The mechanisms of chemical reactions within and between solids are through lattice vibrations, lattice defects, and changes in valence states. These are the structural features through which migration of mass, charge, and energy occur. Consequently, diffusion and conductivity are integral, basic parts of solid-state chemistry. *See* DIFFUSION.

A feature of the electronic structure of solids, which is particularly essential to solid-state chemistry, is the valence state of the ion and the energy required to change the valency in the solid. *See* VALENCE.

Many of the photo-induced processes which occur in solids can be imagined to be solid-state chemical reactions. The classical ones, of course, are those involved in photographic plates and films. *See* PHOTOCHEMISTRY. [R.J.T.]

**Solubility product constant** A special type of simplified equilibrium constant (symbol $K_{sp}$ ) defined for, and useful for, equilibria between solids (s) and their respective ions in solution, for example, reaction (1). For this relatively simple equilibrium, Eqs. (2) and (3) apply.

$$AgCl(s) \rightleftharpoons Ag^{+} + Cl^{-} \quad (1)$$

$$[Ag^{+}][Cl^{-}] \cong K_{sp} \quad (2)$$

$$(Ag^{+})(Cl^{-})/AgCl) = K_{sp} \quad (3)$$

It can be demonstrated experimentally that a small increase in the molar concentration of chloride ion $[Cl^{-}]$ causes a reduction in the concentration of silver present as $Ag^{+}$. Similarly, an increase in $[Ag^{+}]$ reduces $[Cl^{-}]$. The product of the two concentrations is approximately constant as indicated by Eq. (2) and equal to the $K_{sp}$ of Eq. (3). Equation (3) is exact since the variables are activities instead of concentrations. In accordance with the choice of standard state usually made for a solid, the activity of solid AgCl is unity, hence Eq. (4) holds.

$$(Ag^{+})(Cl^{-}) = K_{sp} = 1.8 \times 10^{-}\ \text{mole}^{2}\ \text{liter}^{-2} \quad (4)$$

In practice, various complications arise: addition of too much of either ion produces more complicated ions and hence actually increases the apparent concentration of the other ion. Addition of a salt without a common ion (that is, a salt supplying neither $Ag^{+}$ nor $Cl^{-}$) either may react with $Ag^{+}$ or $Cl^{-}$ or may merely increase the concentration of both ions by a lowering of the mean ionic activity coefficient. *See* IONIC EQUILIBRIUM; PRECIPITATION (CHEMISTRY). [T.F.Y.]

**Solubilizing of samples** The process by which samples that do not dissolve easily are converted into different chemical compounds that are soluble. The sample may be heated in air to evolve volatile components or to oxidize a component to a volatile higher oxidation state with the formation of an acid-soluble form, as in the roasting of a sulfide to form the oxide and sulfur dioxide. Most frequently, the sample is treated with a solvent which reacts with one or more constituents of the sample. The choice of solvent is determined by the chemical reactions that are required. Reactions used include solvation, neutralization, complex formation, metathesis, displacement, oxidation-reduction, or combinations of these.

Most water-soluble salts dissolve by solvation. Basic oxides such as ferric oxide dissolve in aqueous hydrochloric acid. Metals above hydrogen in the electrochemical series will dissolve in a nonoxidizing acid by reduction of hydrogen ion. Metals such as copper and lead require an oxidizing acid, usually nitric acid. All of the components of brass and bronze are usually dissolved by nitric acid except tin, arsenic, and antimony, which precipitate as hydrated oxides. Alloy steels are usually dissolved by combinations of hydrochloric, nitric, phosphoric, and hydrofluoric acids. Aluminum-base alloys are treated with sodium hydroxide solution and any residues are dissolved in acid.

Many substances do not dissolve at temperatures obtainable in the presence of liquid water. However, fused salt reactions employing temperatures of 400–1100°C (750–2000°F) are necessary for the attack and decomposition of many types of samples. The material used as the solvent is called a flux, and the process of melting the mixture of dry, solid flux with the sample is called a fusion. [C.L.R.]

**Solution** A homogeneous mixture of two or more components whose properties vary continuously with varying proportions of the components. A liquid solution can be distinguished experimentally from a pure liquid by the fact that during transfers into other single phases at equilibrium (freezing and vaporizing at constant pressure) the temperature and other properties vary continuously, whereas those of a pure liquid remain constant. For an apparent exception *see* AZEOTROPIC MIXTURE; SOLVENT.

Gases, unless highly compressed, are mutually soluble in all proportions.

A solid solution is, similarly, a single phase whose composition and other properties vary continuously with changing composition of the liquid phase with which it is in equilibrium.

The extent to which substances can form solutions depends upon the kind and strength of the attractive forces between the several molecular species involved. It is necessary to consider the attractive forces exerted by molecules of the following types: (1) nonpolar molecules; (2) polar molecules, that is, those containing electric dipoles; (3) ions; and (4) metallic atoms.

Actual solutions may be considered in terms of their departure from a simple idealized model—a mixture of components having the same attractive fields, which mix without change in volume or heat content. This is analogous to an ideal gas mixture, which is formed with no heat of mixing and in which the total pressure is the sum of the partial pressures. In such a solution the escaping tendency of the individual molecules is the same, whether they are surrounded by similar or by different molecules. [J.H.Hi.]

**Solvation** The association or combination of a solute unit (ionic, molecular, or particulate) with solvent molecules. This association may involve chemical or physical forces, or both, and may vary in degree from a loose, indefinite complex to the formation of a distinct chemical compound. Such a compound contains a definite number of solvent molecules per solute molecule. Solvation occurring in aqueous solutions is referred to as hydration. In certain colloidal suspensions, solvation is, to a large extent, responsible for the stability of the sol. *See* COLLOID; HYDRATION; SOLUTION; SOLVENT. [F.J.J.]

**Solvent** By convention, the component present in the greatest proportion in a homogeneous mixture of pure substances (solutions). Components of mixtures present in minor proportions are called solutes. Thus, technically, homogeneous mixtures are possible with liquids, solids, or gases dissolved in liquids; solids in solids; and gases in gases. In common practice this terminology is applied mostly to liquid mixtures for which the solvent is a liquid and the solute can be a liquid, solid, or gas. *See* SOLUTION.

Three broad classes of solvents are recognized—aqueous, nonaqueous, and organic. Formalistically, the nonaqueous and organic classifications are both not aqueous, but the term organic solvents is generally applied to a large body of carbon-based compounds that find use industrially and as media for chemical synthesis. Organic solvents are generally classified by the functional groups that are present in the molecule, for example, alcohols, halogenated hydrocarbons, or hydrocarbons; such groups give an indication of the types of physical or chemical interactions that can occur between solute and solvent. Nonaqueous solvents are generally taken to be inorganic substances and a few of the lower-molecular-weight, carbon-containing substances such as acetic acid, methanol, and dimethylsulfoxide. Nonaqueous solvents can be solids (for example, fused LiI), liquids ($H_2SO_4$), or gases ($NH_3$) at ambient conditions; the solvent properties of fused sodium iodide (NaI) are manifested in the molten state, whereas hydrogen sulfate ($H_2SO_4$) and ammonia ($NH_3$) must be liquefied to act as solvents. [J.J.L.]

**Solvent extraction** A technique, also called liquid extraction, for separating the components of a liquid solution. This technique depends upon the selective dissolving of one or more constituents of the solution into a suitable immiscible liquid solvent. It is particularly useful industrially for separation of the constituents of a mixture according to chemical type, especially when methods that depend upon different physical properties, such as the separation by distillation of substances of different vapor pressures, either fail entirely or become too expensive.

Industrial plants using solvent extraction require equipment for carrying out the extraction itself (extractor) and for essentially complete recovery of the solvent for reuse, usually by distillation. *See* DISTILLATION.

The petroleum refining industry is the largest user of extraction. In refining virtually all automobile lubricating oil, the undesirable constituents such as aromatic hydrocarbons are extracted from the more desirable paraffinic and naphthenic hydrocarbons. By suitable catalytic treatment of lower boiling distillates, naphthas rich in aromatic hydrocarbons such as benzene, toluene, and the xylenes may be produced. The latter are separated from paraffinic hydrocarbons with suitable solvents to produce high-purity aromatic hydrocarbons and high-octane gasoline. Other industrial applications include so-called sweetening of gasoline by extraction of sulfur-containing compounds; separation of vegetable oils into relatively saturated and unsaturated glyceride esters; recovery of valuable chemicals in by-product coke oven plants; pharmaceutical refining processes; and purifying of uranium.

Solvent extraction is carried out regularly in the laboratory by the chemist as a commonplace purification procedure in organic synthesis, and in analytical separations in which the extraordinary ability of certain solvents preferentially to remove one or more constituents from a solution quantitatively is exploited. Batch extractions of this sort, on a small scale, are usually done in separatory funnels, where the mechanical agitation is supplied by handshaking of the funnel. *See* EXTRACTION. [R.E.Tr.]

**Sonochemistry** The study of the chemical changes that occur in the presence of sound or ultrasound. Industrial applications of ultrasound include many physical and chemical effects, for example, cleaning, soldering, welding, dispersion, emulsification, disinfection, pasteurization, extraction, flotation of minerals, degassing of liquids, defoaming, and production of gas-liquid sols.

When liquids are exposed to intense ultrasound, high-energy chemical reactions occur, often accompanied by the emission of light. There are three classes of such reactions: homogeneous sonochemistry of liquids, heterogeneous sonochemistry of liquid-liquid or liquid-solid systems, and sonocatalysis (which overlaps the first two).

In some cases, ultrasonic irradiation can increase reactivity by nearly a millionfold. Especially for liquid-solid reactions, the rate enhancements via ultrasound have proved extremely useful for the synthesis of organic and organometallic compounds. Because cavitation occurs only in liquids, chemical reactions are not generally seen in the ultrasonic irradiation of solids or solid-gas systems.

Ultrasound spans the frequencies of roughly 20 kHz to 10 MHz (human hearing has an upper limit of less than 18 kHz). Ultrasound has acoustic wavelengths of roughly 7.5–0.015 cm which are much larger than molecular dimensions. As a result, the chemical effects of ultrasound are not from direct interaction, but are derived from several different physical mechanisms, depending on the nature of the system. For both sonochemistry and sonoluminescence, the most important of these mechanisms is acoustic cavitation: the formation, growth, and implosive collapse of bubbles in liquids irradiated with high-intensity sound. During the final stages of cavitation, compression of the gas inside the bubbles produces enormous local heating and high pressures. *See* CAVITATION.

When a liquid-solid interface is subjected to ultrasound, cavitation occurs, but a markedly asymmetric bubble collapse occurs, which generates a jet of liquid directed at the surface with velocities greater than 330 ft/s (100 m/s). The impingement of this jet can create a localized erosion (and even melting), responsible for surface pitting and ultrasonic cleaning. Enhanced chemical reactivity of solid surfaces is associated with these processes.

Ultrasonic irradiation of liquid-powder suspensions produces another effect: high-velocity interparticle collisions. Cavitation and the shock waves that it creates in a slurry can accelerate solid particles to high velocities. The resultant collisions are capable of inducing dramatic changes in surface morphology, composition, and reactivity.

The predominant reactions of homogeneous sonochemistry are bond breaking and radical formation. In addition to the initiation or enhancement of chemical reactions, irradiation of liquids with high-intensity ultrasound generates the emission of visible light. The production of such luminescence is a consequence of the localized hot spot created by the implosive collapse of gas- and vapor-filled bubbles during acoustic cavitation. In general, sonoluminescence may be considered a special case of homogeneous sonochemistry. Under conditions where an isolated, single bubble undergoes cavitation, recent studies on the duration of the sonoluminescence flash suggest that a shock wave may be created within the collapsing bubble. *See* CHEMILUMINESCENCE; HOMOGENEOUS CATALYSIS.

A major industrial application of ultrasound is emulsification. The first reported and most studied liquid-liquid heterogeneous systems have involved ultrasonically dispersed mercury. The effect of the ultrasound in this system appears to be due to the large surface area of mercury generated in the emulsion. *See* EMULSION.

The effects of ultrasound on liquid-solid heterogeneous organometallic reactions have been a matter of intense investigation. Various research groups have dealt with extremely reactive metals, such as lithium (Li), magnesium (Mg), or zinc (Zn), as stoichiometric reagents for a variety of common transformations.

Sonochemistry can be used as a synthetic tool for the creation of unusual inorganic materials. Ultrasound has proved extremely useful in the synthesis of a wide range of nanostructured materials, including high-surface-area transition metals, alloys, carbides, oxides, and colloids. Sonochemistry is also proving to have important applications with polymeric materials. Substantial work has been accomplished in the sonochemical initiation of polymerization and in the modification of polymers after synthesis. Sonochemistry has found another recent application in the preparation of unusual biomaterials, notably protein microspheres. The mechanism responsible for microsphere formation is a combination of two acoustic phenomena: emulsification and cavitation. These protein microspheres have a wide range of biomedical applications, including their use as echo contrast agents for sonography, magnetic resonance imaging contrast enhancement, and drug delivery. *See* NANOCHEMISTRY; NANOSTRUCTURE; POLYMER; PROTEIN. [K.S.S.]

**Specific gravity** The specific gravity of a material is defined as the ratio of its density to the density of some standard material, such as water at a specified temperature, for example, 60°F (15°C), or (for gases) air at standard conditions of temperature and pressure. Specific gravity is a convenient concept because it is usually easier to measure than density, and its value is the same in all systems of units. *See* DENSITY. [L.N.]

**Specific heat** A measure of the heat required to raise the temperature of a substance. When the heat $\Delta Q$ is added to a body of mass $m$, raising its temperature by $\Delta T$, the ratio $C$ given in Eq. (1) is defined as the heat capacity of the body. The quantity $c$ defined in Eq. (2) is called the specific heat capacity or specific heat. A commonly

$$C = \frac{\Delta Q}{\Delta T} \tag{1}$$

$$c = \frac{C}{m} = \frac{1}{m}\frac{\Delta Q}{\Delta T} \tag{2}$$

used unit for heat capacity is joule · kelvin$^{-1}$ (J · K$^{-1}$); for specific heat capacity, the unit joule · gram$^{-1}$ · K$^{-1}$ (J · g$^{-1}$ · K$^{-1}$) is often used. Joule should be preferred over the unit calorie = 4.18 J. As a unit of specific heat capacity, Btu · lb$^{-1}$ · °F$^{-1}$ = 4.21 J · g$^{-1}$ · K$^{-1}$ is also still in use in English-language engineering literature. If the heat capacity is referred to the amount of substance in the body, the molar heat capacity $c_m$ results, with the unit J · mol$^{-1}$ · K$^{-1}$.

If the volume of the body is kept constant as the energy $\Delta Q$ is added, the entire energy will go into raising its temperature. If, however, the body is kept at a constant pressure, it will change its volume, usually expanding as it is heated, thus converting some of the heat $\Delta Q$ into mechanical energy. Consequently, its temperature increase will be less than if the volume is kept constant. It is therefore necessary to distinguish between these two processes, which are identified with the subscripts $V$ (constant volume) and $p$ (constant pressure): $C_V$, $c_V$, and $C_p$, $c_p$. For gases at low pressures, which obey the ideal gas law, the molar heat capacities differ by $R$, the molar gas constant, as given in Eq. (3), where $R = 8.31$ J · mol$^{-1}$ · K$^{-1}$; that is, the expanding gas heats up less.

$$c_p - c_V = R \tag{3}$$

For solids, the difference between $c_p$ and $c_V$ is of the order of 1% of the specific heat capacities at room temperature. This small difference can often be ignored. *See* CALORIMETRY; CHEMICAL THERMODYNAMICS; HEAT CAPACITY ; THERMODYNAMIC PROCESSES. [R.O.P.]

**Spectroscopy** An analytic technique concerned with the measurement of the interaction (usually the absorption or the emission) of radiant energy with matter, with the instruments necessary to make such measurements, and with the interpretation of the interaction both at the fundamental level and for practical analysis.

A display of such data is called a spectrum, that is, a plot of the intensity of emitted or transmitted radiant energy (or some function of the intensity) versus the energy of that light. Spectra due to the emission of radiant energy are produced as energy is emitted from matter, after some form of excitation, then collimated by passage through a slit, then separated into components of different energy by transmission through a prism (refraction) or by reflection from a ruled grating or a crystalline solid (diffraction), and finally detected. Spectra due to the absorption of radiant energy are produced when radiant energy from a stable source, collimated and separated into its components in a monochromator, passes through the sample whose absorption spectrum is to be measured, and is detected. Instruments which produce spectra are variously called spectroscopes, spectrometers, spectrographs, and spectrophotometers. *See* SPECTRUM.

Interpretation of spectra provides fundamental information on atomic and molecular energy levels, the distribution of species within those levels, the nature of processes involving change from one level to another, molecular geometries, chemical bonding, and interaction of molecules in solution. At the practical level, comparisons of spectra provide a basis for the determination of qualitative chemical composition and chemical structure, and for quantitative chemical analysis.

**Origin of spectra.** Atoms, ions, and molecules emit or absorb characteristically; only certain energies of these species are possible; the energy of the photon (quantum of radiant energy) emitted or absorbed corresponds to the difference between two permitted values of the energy of the species, or energy levels. (If the flux of photons incident upon the species is great enough, simultaneous absorption of two or more photons may occur.) Thus the energy levels may be studied by observing the differences between them. The absorption of radiant energy is accompanied by the promotion of the species from a lower to a higher energy level; the emission of radiant energy is accompanied by falling from a higher to a lower state; and if both processes occur together, the condition is called resonance.

**Instruments.** Spectroscopic methods involve a number of instruments designed for specialized applications.

An optical instrument consisting of a slit, collimator lens, prism or grating, and a telescope or objective lens which produces a spectrum for visual observation is called a spectroscope.

If a spectroscope is provided with a photographic camera or other device for recording the spectrum, the instrument is called a spectrograph.

A spectroscope that is provided with a calibrated scale either for measurement of wavelength or for measurement of refractive indices of transparent prism materials is called a spectrometer.

A spectrophotometer consists basically of a radiant-energy source, monochromator, sample holder, and detector. It is used for measurement of radiant flux as a function of wavelength and for measurement of absorption spectra.

An interferometer is an optical device that measures differences of geometric path when two beams travel in the same medium, or the difference of refractive index when the geometric paths are equal. Interferometers are employed for high-resolution measurements and for precise determination of relative wavelengths.

**Methods and applications.** Since the early methods of spectroscopy there has been a proliferation of techniques, often incorporating sophisticated technology.

Acoustic spectroscopy uses modulated radiant energy that is absorbed by a sample. The loss of that excess produces a temperature increase that can be monitored around the sample by using a microphone transducer. This is the optoacoustic effect.

In astronomical spectroscopy, the radiant energy emitted by celestial objects is studied by combined spectroscopic and telescopic techniques to obtain information about their chemical composition, temperature, pressure, density, magnetic fields, electric forces, and radial velocity.

Atomic absorption and fluorescence spectroscopy is a branch of electronic spectroscopy that uses line spectra from atomized samples to give quantitative analysis for selected elements at levels down to parts per million, on the average.

Attenuated total reflectance spectroscopy is the study of spectra of substances in thin films or on surfaces obtained by the technique of attenuated total reflectance or by a closely related technique called frustrated multiple internal reflection. In either method the radiant-energy beam penetrates only a few micrometers of the sample. The technique is employed primarily in infrared spectroscopy for qualitative analysis of coatings and of opaque liquids.

Electron spectroscopy includes a number of subdivisions, all of which are associated with electronic energy levels. The outermost or valence levels are studied in photoelectron spectroscopy. Electron impact spectroscopy uses low-energy electrons (0–100 eV).

X-ray photoelectron spectroscopy (XPS), also called electron spectroscopy for chemical analysis (ESCA), and Auger spectroscopy use x-ray photons to remove inner-shell electrons. Ion neutralization spectroscopy uses protons or other charged particles instead of photons. *See* ELECTRON SPECTROSCOPY; SURFACE AND INTERFACIAL CHEMISTRY.

Fourier transform spectroscopy is a technique that has been applied to infrared spectrometry and nuclear magnetic resonance spectrometry to allow the acquisition of spectra from smaller samples in less time, with high resolution and wavelength accuracy.

Gamma-ray spectroscopy employs the techniques of activation analysis and Mössbauer spectroscopy. *See* MÖSSBAUER EFFECT.

Information on processes which occur on a picosecond time scale can be obtained by making use of the coherent properties of laser radiation, as in coherent anti-Stokes-Raman spectroscopy. Laser fluorescence spectroscopy provides the lowest detection limits for many materials of interest in biochemistry and biotechnology. Ultrafast laser spectroscopy may be used to study some aspects of chemical reactions, such as transition states of elementary reactions and orientations in bimolecular reactions. *See* LASER SPECTROSCOPY.

In mass spectrometry, the source of the spectrometer produces ions, often from a gas, but also in some instruments from a liquid, a solid, or a material absorbed on a surface. The dispersive unit provides either temporal or spatial dispersion of ions according to their mass-to-charge ratio. *See* MASS SPECTROMETRY; SECONDARY ION MASS SPECTROMETRY (SIMS); TIME-OF-FLIGHT SPECTROMETERS.

In multiplex or frequency-modulated spectroscopy, each optical wavelength exiting the spectrometer output is encoded or modulated with an audio frequency that contains the optical wavelength information. Use of a wavelength analyzer then allows recovery of the original optical spectrum.

When a beam of light passes through a sample, a small fraction of the light exits the sample at a different angle. If the wavelength of the scattered light is different than the original wavelength, it is called Raman scattering. Raman spectroscopy is used in structural chemistry and is a valuable tool for surface analysis. A related process, resonance Raman spectroscopy, makes use of the fact that Raman probabilities are greatly increased when the exciting radiation has an energy which approaches the energy of an allowed electronic absorption.

In x-ray spectroscopy, the excitation of inner electrons in atoms is manifested as x-ray absorption; emission of a photon as an electron falls from a higher level into the vacancy thus created is x-ray fluorescence. The techniques are used for chemical analysis. *See* X-RAY SPECTROMETRY. [M.M.Bu.]

**Spin label** A molecule which contains an unpaired electron spin which can be detected with electron spin resonance (ESR) spectroscopy. Molecules are labeled when an atom or group of atoms which exhibits some unique physical property is chemically bonded to a molecule of interest. Groups containing unpaired electrons include organic free radicals and a variety of types of transition-metal complexes (such as vanadium, copper, iron, and manganese). Through analysis of ESR spectra, rates of molecular motion whose motion is restrained by surrounding molecules can be determined.

Analysis of the rate and type of motion of a spin label is important for a wide variety of biological problems. The type of label used in these studies is generally a nitroxide free radical. Spin-labeling studies provide a powerful technique for the study of the geometry and dimensions of receptors in enzymes. Spin labels have been used extensively to study the structure of membranes, and can provide important information about the organization and rates of motion in membranes. Spin labels have also been used to study the structure and organization of synthetic polymers and to study phase transitions. *See* ELECTRON PARAMAGNETIC RESONANCE (EPR) SPECTROSCOPY; ENZYME. [R.Kr.]

**Squalene** A $C_{30}$ triterpenoid hydrocarbon. Squalene is made up of six (*trans*-1,4)-isoprene units linked as two farnesyl (head-to-tail) groups that are joined tail to tail in the center (see illustration).

T

T

**Structure of squalene; the tail-to-tail joining is indicated by T.**

Squalene can be isolated in large quantities from the liver oils of the shark and other elasmobranch fishes, and is a relatively inexpensive compound. Complete hydrogenation of the liver oil gives the saturated hydrocarbon squalane, which is used in lotions and skin lubricants.

The major significance of squalene is its role as a central intermediate metabolite in the biogenesis of all steroids and triterpenoids. *See* STEROID; TERPENE. [J.A.Mo.]

**Stabilizer (chemistry)** Any substance that tends to maintain the physical and chemical properties of a material. Degradation, that is, irreversible changes in chemical composition or structure, is responsible for the premature failure of materials. Stabilizers are used to extend the useful life of materials as well as to maintain their critical properties above the design specifications. Oxygen and water are the principal degradants, but ultraviolet radiation also can have a significant effect (photodegradation).

A wide variety of additives has been developed to stabilize polymers against degradation. Stabilizers are available that inhibit thermal oxidation, burning, photodegradation, and ozone deterioration of elastomers. Research in the chemistry of the low-temperature oxidation of natural rubber has revealed that hydrocarbon polymers oxidize by a free-radical chain mechanism. In pure, low-molecular-weight hydrocarbons, an added initiator is required to produce the first radicals. In contrast, initiation of polymer oxidation occurs in the complex molecules of elastomers through impurities already present, for example, hydroperoxides. *See* CHAIN REACTION (CHEMISTRY); FREE RADICAL.

Stabilization of hydrocarbon polymers can be accomplished with preventative or chain-breaking antioxidants. Preventative antioxidants stabilize by reducing the number of radicals formed in the initiation stage. Where hydroperoxides are responsible for initiation, the induced decomposition of these reactive intermediates into nonradical products provides effective stabilization. Suppressing the catalytic effects of metallic impurities that increase the rate of radical formation can also provide stabilization. Chain-breaking antioxidants interrupt the oxidative chain by providing labile hydrogens to compete with the polymer in reaction with the propagating radicals. The by-product of this reaction is a radical which is not capable of continuing the oxidative chain.

Stabilization of polymers against photooxidation, the principal component of outdoor weathering, is accomplished by addition of ultraviolet absorbers, and radical scavengers. Ultraviolet absorbers absorb and harmlessly dissipate damaging radiation. Another class of additives, known as hindered-amine light stabilizers, function by scavenging destructive radicals. *See* ANTIOXIDANT ; INHIBITOR (CHEMISTRY); PHOTODEGRADATION. [W.L.H.]

**Stereochemistry** The study of the three-dimensional arrangement of atoms or groups within molecules and the properties which follow from such arrangement. Molecules that have identical molecular structures but differ in the relative spatial arrangement of component parts are stereoisomers. Inorganic and organic compounds exhibit stereoisomerism. Examples are structures (**1**)–(**8**).

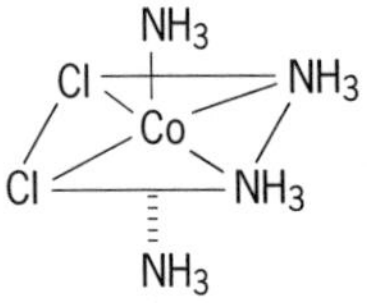

(1) 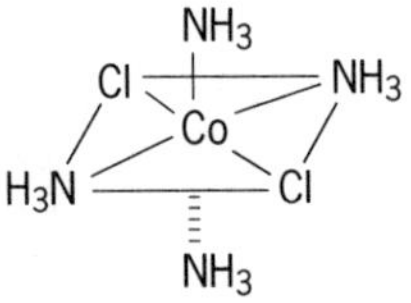 (2) 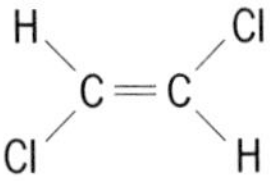 (3)

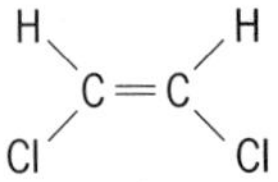

(4) 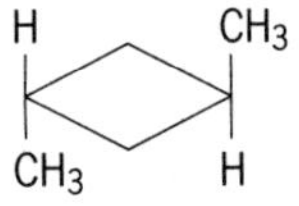 (5) 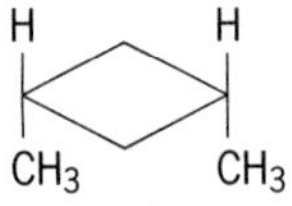 (6)

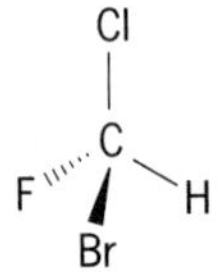

(7) 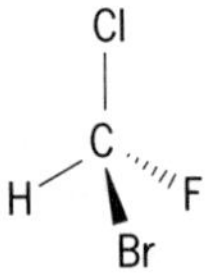 (8)

The nature of the stereochemistry of a molecule is determined by its symmetry. The symmetry elements to be considered are: planes of symmetry, axes of symmetry, centers of symmetry, and reflection or mirror symmetry. Two types of stereoisomers are known. Those such as (**7**) and (**8**), which are devoid of reflection symmetry—which cannot be superimposed on their image in a mirror—are called enantiomers. All other stereoisomers, such as the pairs (**1**)–(**2**), (**3**)–(**4**), and (**5**)–(**6**), are called diastereomers. The configuration of a stereoisomer designates the relative position of the atoms associated with a specific structure. The structures of stereoisomers (**1**) and (**2**) differ only in configuration. The same is true for (**3**) and (**4**), (**5**) and (**6**), and (**7**) and (**8**). [S.H.W.]

**Stereospecific catalyst** Stereospecific polymerization catalysts lead to the formation of stereoregular (tactic) polymers, that is, polymers where the centers of steric isomerism in the main chain are arranged in a regular fashion with respect to their configurations. Three factors determine the tacticity of a polymeric chain during its formation: (1) the kind of monomer approach to the growing chain end, (2) the kind of attack of the growing chain end on the double bond (cis or trans opening), and (3) the configuration in the initiation step. In addition to a regular head-to-tail configuration and absence of branching, reactions have to be assumed. The kind of monomer approach is strongly affected by electrical and stereochemical forces, and therefore changes in monomer structure and environment greatly influence polymer tacticity. The intermediate radical or ion can be assumed to have a planar or near-planar structure, and in an uncomplexed form it should be able to rotate freely around its axis with cis and trans addition being equally possible. Upon addition of the next monomer this carbon changes into a tetrahedral structure, thereby creating two isomeric forms, isotactic or syndiotactic. *See* ASYMMETRIC SYNTHESIS; CATALYSIS; POLYMERIZATION; STEREOCHEMISTRY. [A.Sc.]

**Steric effect (chemistry)** The influence of the spatial configuration of reacting substances upon the rate, nature, and extent of reaction. The sizes and shapes of atoms and molecules, the electrical charge distribution, and the geometry of bond angles influence the courses of chemical reactions. The steric course of organochemical reactions is greatly dependent on the mode of bond cleavage and formation, the environment of the reaction site, and the nature of the reaction conditions (reagents, reaction time, and temperature). The effect of steric factors is best understood in ionic reactions in solution. *See* STEREOCHEMISTRY. [E.W.]

**Steroid** Any of a group of organic compounds belonging to the general class of biochemicals called lipids, which are easily soluble in organic solvents and slightly soluble in water. Additional members of the lipid class include fatty acids, phospholipids, and triacylglycerides. The unique structural characteristic of steroids is a four-fused ring system. Members of the steroid family are ubiquitous, occurring, for example, in plants, yeast, protozoa, and higher forms of life. Steroids exhibit a variety of biological functions, from participation in cell membrane structure to regulation of physiological events. Naturally occurring steroids and their synthetic analogs are used extensively in medical practice.

Each steroid contains three fused cyclohexane (six-carbon) rings plus a fourth cyclopentane ring (see illustration). Naturally occurring steroids have an oxygen-containing group at carbon-3. Shorthand formulas for steroids indicate the presence of double bonds, as well as the structure and position of oxygen-containing or other organic groups.

The most abundant steroid in mammalian cells is cholesterol. The levels and locations of planar cholesterol molecules, embedded in the phospholipid bilayers that form cell and organelle membranes, are known to influence the structure and function of the membranes. A second major function of cholesterol is to serve as a precursor of steroids acting as physiological regulators (such as the steroid hormones). Enzyme systems present in a hormone-secreting gland convert cholesterol to the hormone specific for that gland. For example, the ovary produces estrogens (such as estradiol and progesterone); the testis produces androgens (such as testosterone); the adrenal cortex produces hormones that regulate metabolism (such as cortisol) and sodium ion transport (such as aldosterone). A third major function of cholesterol is to serve as a precursor of the bile acids. These detergentlike molecules are produced in the liver and

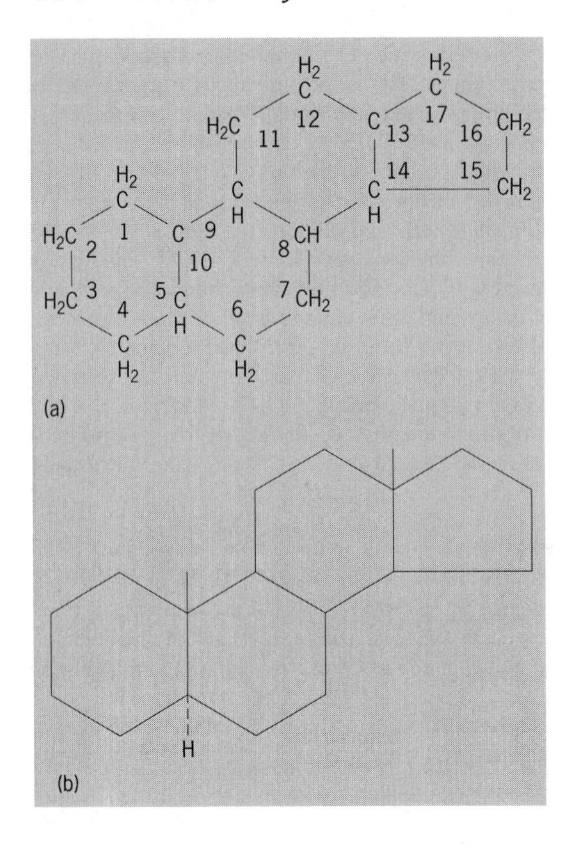


**Steroid skeleton. (*a*) Structure and numbering. (*b*) Shorthand formulation; the lines attached to the rings represent methyl groups.**

stored in the gall bladder until needed to assist in the absorption of dietary fat and fat-soluble vitamins and in the digestion of dietary fat by intestinal enzymes.

Some examples of diseases treated with naturally occurring or synthetic steroids are allergic reactions, arthritis, some malignancies, and diseases resulting from hormone deficiencies or abnormal production. In addition, synthetic steroids that mimic an action of progesterone are widely used oral contraceptive agents. Other synthetic steroids are designed to mimic the stimulation of protein synthesis and muscle-building action of naturally occurring androgens. [M.E.D.]

**Stoichiometry** All chemical measurements, such as the measurements of atomic and molecular weights and sizes, gas volumes, vapor densities, deviation from the gas laws, and the structure of molecules. In determining the relative weights of the atoms, scientists have relied upon combining ratios, specific heats, and measurements of gas volumes. All such measurements, and the calculations that relate them to each other, constitute the field of stoichiometry. Since measurements are expressed in mathematical terms, stoichiometry can be considered to be the mathematics of general chemistry.

In a general usage, the term stoichiometry refers to the relationships between the measured quantitites of substances or of energy involved in a chemical reaction; the calculations of these quantities include the assumption of the validity of the laws of definite proportions and of conservation of matter and energy.

A typical stoichiometric problem involves predicting the weight of reactant needed to produce a desired amount of a product in a chemical reaction. For example, phosphorus can be extracted from calcium phosphate, $Ca_3(PO_4)_2$, by a certain process with a 90% yield (some calcium phosphate fails to react or some phosphorus is lost). In a specific problem, it might be necessary to determine the mass of calcium phosphate required to prepare 16.12 lb of phosphorus by this process. The balanced equation for the preparation is shown in reaction (1). In this reaction, 2 moles of calcium phosphate are

$$2Ca_3(PO_4)_2 + 10C + 6SiO_2 \rightarrow 6CaSiO_3 + 10CO + P_4 \quad (1)$$

required to produce 1 mole of phosphorus. Two moles of calcium phosphate have a mass of 620 lb, and 1 mole of phosphorus as $P_4$ has a mass of 124 lb. Using these relationships, calculation (2) is made. Since the yield of phosphorus is only 90%, extra

$$\frac{2 \text{ moles } Ca_3(PO_4)}{1 \text{ mole } P_4} = \frac{620 \text{ lb } Ca_3(PO_4)_2}{124 \text{ lb } P_4} = \frac{X \text{ lb } Ca_3(PO_4)_2}{16.12 \text{ lb } P_4} \quad (2)$$

$$X = 80.6 \text{ lb}$$

$Ca_3(PO_4)_2$ must be used: 88.1 lb is the mass of calcium phosphate required to yield 16.12 lb of phosphorus by this process.

Calculations of this sort are important in chemical engineering processes, in which amounts and yields of products must be known. The same reasoning is used in calculations of energy generated or required. In this case, the energy involved in the reaction of a known weight of the material in question must be known or determined.

The calculations discussed involve compounds in which the ratio of atoms is generally simple. For a discussion of compounds in which the relative number of atoms cannot be expressed as ratios of small whole numbers, *see* NONSTOICHIOMETRIC COMPOUNDS

[J.C.Ba.]

**Streaming potential** The potential which is produced when a liquid is forced to flow through a capillary or a porous solid. The streaming potential is one of four related electrokinetic phenomena which depend upon the presence of an electrical double layer at a solid-liquid interface. This electrical double layer is made up of ions of one charge type which are fixed to the surface of the solid and an equal number of mobile ions of the opposite charge which are distributed through the neighboring region of the liquid phase. In such a system the movement of liquid over the surface of the solid produces an electric current, because the flow of liquid causes a displacement of the mobile counterions with respect to the fixed charges on the solid surface. The applied potential necessary to reduce the net flow of electricity to zero is the streaming potential.

[Q.V.W.]

**Stripping** The removal of volatile component from a liquid by vaporization. The stripping operation is an important step in many industrial processes which employ absorption to purify gases and to recover valuable components from the vapor phase. In such processes, the rich solution from the absorption step must be stripped in order to permit recovery of the absorbed solute and recycle of the solvent. *See* GAS ABSORPTION OPERATIONS.

Stripping may be accomplished by pressure reduction, the application of heat, or the use of an inert gas (stripping vapor). Many processes employ a combination of all three; that is, after absorption at elevated pressure, the solvent is flashed to atmospheric pressure, heated, and admitted into a stripping column which is provided with a bottom heater (reboiler). Solvent vapor generated in the rebolier or inert gas injected at the bottom of the column serves as stripping vapor which rises countercurrently to the downflowing solvent. When steam is used as stripping vapor for a system not miscible with water, the process is called steam stripping.

In addition to its use in conjunction with gas absorption, the term stripping is also used quite generally in technical fields to denote the removal of one or more components from a mixed system. Such usage covers (1) the distillation operation which takes place in a distilling column in the zone below the feed point, (2) the extraction of one or more components from a liquid by contact with a solvent liquid, (3) the removal of organic or metal coatings from solid surfaces, and (4) the removal of color from dyed fabrics. *See* DISTILLATION; SOLVENT EXTRACTION. [A.L.K.]

**Strontium** A chemical element, Sr, atomic number 38, and atomic weight 87.62. Strontium is the least abundant of the alkaline-earth metals. The crust of the Earth is 0.042% strontium, making this element as abundant as chlorine and sulfur. The main ores are celestite, $SrSO_4$, and strontianite, $SrCO_3$. *See* ALKALINE-EARTH METALS.

**Properties of strontium**

| Property | Value |
|---|---|
| Atomic number | 38 |
| Atomic weight | 87.62 |
| Isotopes (stable) | 84,86,87,88,90 |
| Boiling point, °C | 1638(?) |
| Melting point, °C | 704(?) |
| Density, g/cm$^3$ at 20°C | 2.6 |

Strontium nitrate is used in pyrotechnics, railroad flares, and tracer bullet formulations. Strontium hydroxide forms soaps and greases with a number of organic acids which are structurally stable, resistant to oxidation and breakdown over a wide temperature range.

Strontium is divalent in all its compounds which are, aside from the hydroxide, fluoride, and sulfate, quite soluble. Strontium is a weaker complex former than calcium, giving a few weak oxy complexes with tartrates, citrates, and so on. Some physical properties of the element are given in the table. [R.F.R.]

**Structural chemistry** Much of chemistry is explainable in terms of the structures of chemical compounds. The understanding of these structures hinges very strongly on understanding the electronic configurations of the elements. The union of atoms, and therefore the formation of compounds from the elements, is associated with interactions among the extranuclear electrons of the individual atoms. Electronic interactions among atoms may occur in two ways: Electrons may be transferred from one atom to another, or they may be shared by two (or more) atoms. The first type of interaction is called electrovalence and results in the formation of electrically charged monatomic ions. The second, covalence, leads to the formation of molecules and complex ions. *See* CHEMICAL BONDING. [D.H.B.]

In considering structures more complex than those derived from simple monoatomic ions, the logical step is to consider single polyhedral aggregates of atoms. In its most precise sense, structure is used to denote a knowledge of the bonding distances and angles between atoms in chemical compounds and, in turn, the geometrical arrangements which they form. These atomic arrangements and the associated distances and angles serve uniquely as "fingerprints" of these atom spatial configurations, and depend very much on the electronic configurations around atoms. The chemical combination of neutral atoms to produce uncharged species results in molecule formation, whereas the similar combination of atoms or ions possessing a net charge results in the formation of complex ions. A basic understanding of the species formed involves the concept of the coordination polyhedron, which allows a simple classification of the structures of many polyatomic molecules and ions. This type of classification is particularly useful because it conveniently explains the packing together of simple chemical molecules or ions in terms of highly symmetrical polyhedra. There is an obvious connection between polyhedra and the structures found in crystalline solids formed from them. Crystal formation often involves the linking of convex polyhedra by the sharing of corners, edges, or faces, ultimately forming space-filling assemblies in which all faces of each polyhedron are in contact with faces of other polyhedra. The most important simple polyhedrons are the tetrahedron, the trigonal bipyramid, the octahedron, the pentagonal bipyramid, and the cube. The most commonly observed of these polyhedral configurations are the tetrahedron (four faces) and the octahedron (six faces).

The simplest correlative device which accurately summarizes a very large number of structures and enables the chemist to predict, with a good chance of success, the geometric array of the atoms in a compound of known composition, is based on an extreme electrostatic model. This model, or theory, represents the bonds in a purely formal way. The central atom is considered to be a positive ion having a charge equal to its oxidation state. The groups attached to the central atom (the ligands) are then treated either as negative ions or as neutral dipolar molecules. The principal justification for this approach lies in its successful correlation of a vast amount of information.

A number of significant observations can be made with regard to these formulations. There are series of ions, or ions and molecules, having the same type of composition, differing only in the nature of the central ion and the net charge on the aggregate. Examples are found in the series: $NO_3^-$, $CO_3^{2-}$, $BO_3^{3-}$, $ClO_3^-$, $SO_3^{2-}$, $PO_3^{3-}$; $ClO_4^-$, $SO_4^{2-}$, $PO_4^{3-}$, $SiO_4^{4-}$; $AlF_6^{3-}$, $SiF_6^{2-}$, $PF_6^-$. The numbers of atomic nuclei and of electrons are the same for all the members of each series; consequently, these are called isoelectronic series. Not only are the several chemical entities in such series isoelectronic, but they are usually identical in geometrical structure (isostructural).

It may also be observed that corresponding ions from a given vertical family of the periodic table commonly vary in coordination number. A useful example is found in $N^{5+}$ and $P^{5+}$ which form $NO_3^-$ and $PO_4^{3-}$, respectively. In addition, some neutral molecules expand their coordination numbers to form stable anionic halo complexes, whereas others do not. Thus, $SiF_4$ reacts with fluoride ion to form $SiF_6^{2-}$, whereas $CF_4$ does not form a similar complex ion. The most satisfactory explanation of these and many related observations is conveniently formulated in terms of the electrostatic model chosen here.

The necessary condition for stability of the coordination polyhedron $MA_n$ requires that the anions A are each in contact with the central atom M. As a consequence of this condition, the limit of stability of the structure arises in those cases where the anions are also mutually in contact. Larger ligands, or anions, would not be in contact with the central ion. This relationship is usually summarized in terms of the limiting ratio of the radius of the cation, $r_M$, to that of the anion, $r_A$, below which the anions would no longer be in contact with the cation.

According to the valence-bond theory, the principal requirements for the formation of a covalent bond are a pair of electrons and suitably oriented electron orbitals on each of the atoms being bonded. The geometry of the atoms in the resulting coordination polyhedron is correlated with the orientation of the orbitals on the central atom. The orbitals used depend on the energies of the electrons in them. In general, the order of increasing energy of the electron orbitals is $(n-1)d < ns < np < nd$. It is concluded that a nontransition atom having one valence electron will form a covalent bond utilizing an *s* orbital. In those cases where an unshared pair of electrons may be assigned to the *ns* orbital, as many as three equivalent bonds may be formed by utilizing the three *np* orbitals of the central atom. Because of the orientation of these *p* orbitals with respect to each other, the three resulting bonds should be at 90° to each other. This expectation is nearly realized in $PH_3$. In order to account for four or six equivalent bonds, or for that matter in order to account for all the remaining polyhedral and polygonal structures, except the angular structure for a coordination number of 2 (with two unshared pairs of electrons on the central atom), an additional assumption is necessary. It is assumed that *s* and *p*, *s* and *d*, or *s*, *p*, and *d* orbitals, are replaced by new orbitals, called hybridized orbitals. These hybridized orbitals are derived from the original orbitals (mathematically) in such a way that the required number of equivalent bonds may be formed. In the simplest case, it is shown that *s* and *p* may be combined to form two equivalent *sp* hybridized orbitals directed at 180° to each other. Other sets of hybridized orbitals have been shown to be appropriate to describe the bonding in other structures. *See* LIGAND FIELD THEORY.

Among inert-gas ions of the first row of eight elements in the periodic table, there are four orbitals available for covalent bond formation, one 2*s* and three 2*p*. Consequently, a maximum of four bonds may be formed. This is in general agreement with the existence of the tetrahedron as the limiting coordination polyhedron among these elements, for example, $BeF_4^{2-}$, $BF_4^-$, $CCl_4$, $NH_4^+$. Although only $Li^+$ deviates from this pattern, having a coordination number of 6 in its crystalline halides, these compounds are best treated as simple electrovalent salts. In keeping with the limitation of only four orbitals, the formation of double or triple bonds between atoms of these elements reduces the coordination number of the central atom. Thus, the highest coordination number of a first-row element forming one double bond is 3. This is illustrated by the structures below.

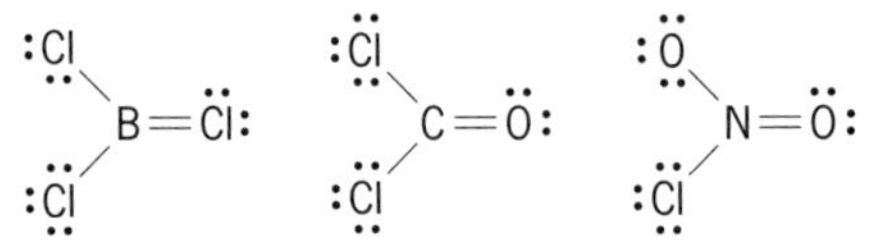

```
:Cl                :Cl                :O
    \                  \                  \
     B=Cl:              C=O:               N=O:
    /                  /                  /
:Cl                :Cl                :Cl
```

In these and similar examples, the geometric array is determined by the formation of three single bonds utilizing $sp^2$ hybridized orbitals on the central atom and *p* orbitals on the ligand. In general, the bonds determining the geometry of a molecule or ion (in this way) are called $\sigma$ bonds. The double bond results from the superposition of a second bond, a $\pi$ bond, between two atoms. In this example the formation of a $\pi$ bond reduces the number of $\sigma$ bonds from four to three, thus changing the geometries of the corresponding molecules or ions from tetrahedral to trigonal planar.

The formation of a second $\pi$ bond (a triple bond or two double bonds) reduces the coordination number of the atom in question still further, resulting in the linear *sp* set of hybridized orbitals being utilized in $\sigma$-bond formation. *See* VALENCE.

With regard to the nature of doubly bonded compounds, another problem arises when such structures are viewed from the standpoint of valence-bond theory. In the species $BCl_3$, $COCl_2$, $NO_2Cl$, and many similar substances, nonequivalent bonds are predicted. The doubly bonded oxygen should be closer to the central carbon atom than the singly bonded ones. This is not found to be true experimentally so long as the similar atoms are otherwise equivalent. There is only one observable C—O distance in carbonate, one N—O distance in nitrate, and so on. To account for such facts as these, the concept of resonance must be introduced. If the $\pi$ bond exists, it must exist equally between the central atom and all the equivalent oxygen atoms. The resonance method of describing this situation is to say that one of the pictorial structures

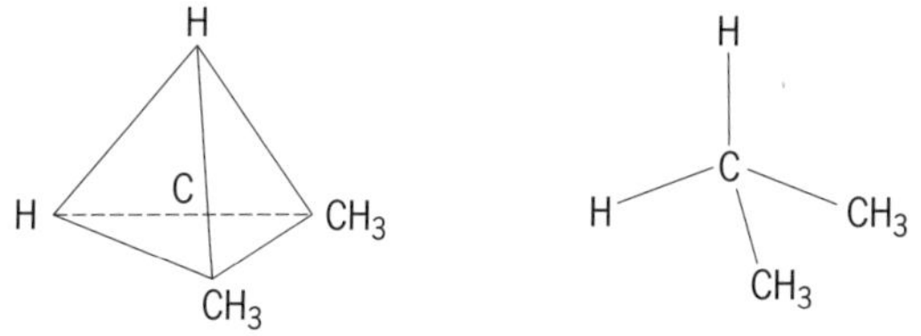

Fig. 1. Geometric structure of propane.

is inadequate to describe the substance properly, but that enough pictorial structures (resonance structures) should be considered to permute the double bond about all the equivalent bonds. The true structure is assumed to be something intermediate to all the resonance structures and more stable than any of them because it exists in preference to any one of them. The resonance structures for $CO_3^{2-}$ are the following:

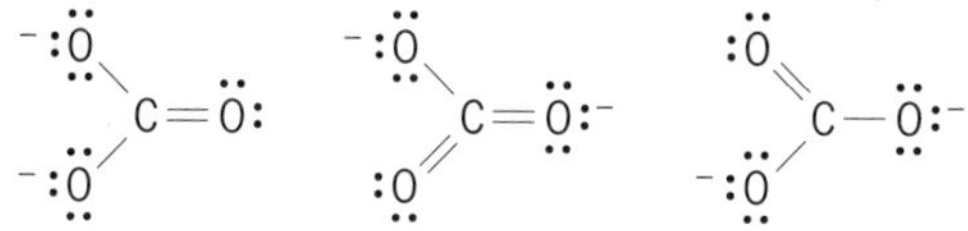

*See* CONJUGATION AND HYPERCONJUGATION; RESONANCE (MOLECULAR STRUCTURE).

The classic homologous series of compounds in organic chemistry provide useful examples involving the condensation of polyhedrons containing the same central element in the individual units. The general formula, $C_nH_{2n} + 2$, represents a large number of compounds extending from the lowest member, methane, $CH_4$, to polyethylene, a plastic of economic importance in which $n$ is a very large number. Two ways exist for the linking together of these tetrahedrons. This gives rise to two molecular forms, both of which are stable, well-known compounds. It is an essential part of these structures that each —C link is linear (because it is merely a $\sigma$-bond); however, when two carbon atoms are linked to a third, the C—C—C angle is essentially determined by the bond angle of the central carbon atom (that is, the other carbons may be treated as ligands to the first (Fig. 1)). The other familiar homologous series of organic chemistry differ from the saturated hydrocarbons in having at least one unique coordination polyhedron of a different type. The olefins contain two doubly bonded, or unsaturated, carbon atoms, whose polyhedral structures are trigonal planar, and the remainder of the carbons are tetrahedral. As in the case of the aliphatic hydrocarbons, the olefins exhibit an isomerism which is associated with the branching of the chain structure. In addition, the presence of two linked trigonal planar carbon atoms and the fact that the polyhedrons cannot rotate about the double bond give rise to a different kind of isomerism, called cis-trans isomerism (Fig. 2). *See* BOND ANGLE AND DISTANCE.

The existence of a predicted isomerism provides one of the most important confirmations of the theories of chemical structure. In general, the polyhedral view of molecular structure, as described here, has been thoroughly verified by the discovery of the many

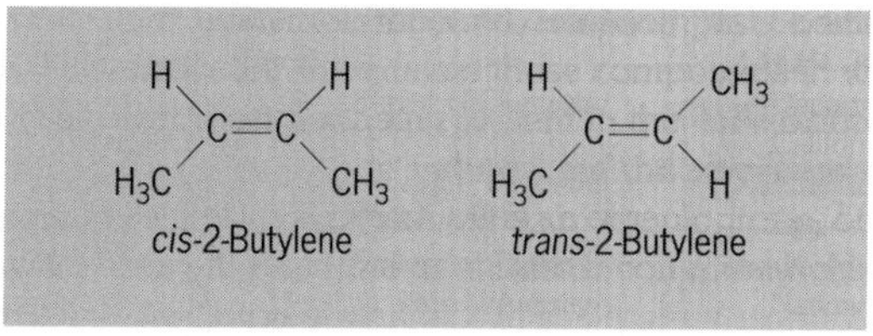

*cis*-2-Butylene *trans*-2-Butylene

Fig. 2. Cis-trans isomerism among olefins.

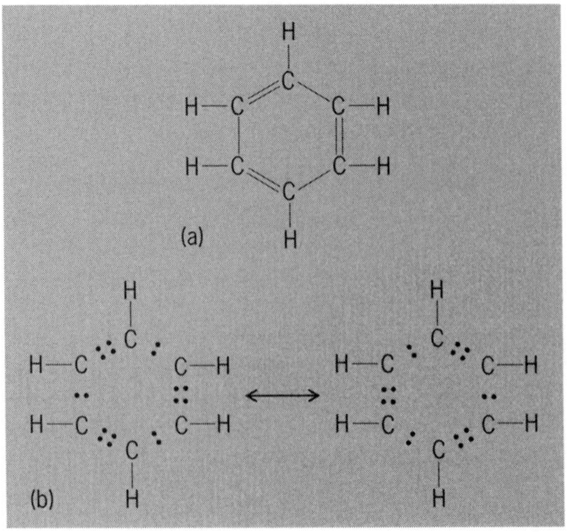

**Fig. 3. Benzene molecule. (*a*) Structural formula. (*b*) Two forms in resonance.**

types of predicted isomerism. The first really convincing proof of the tetrahedral structures of saturated carbon atoms involved optical isomerism. *See* OPTICAL ACTIVITY.

The aromatic hydrocarbons are characterized by cyclic arrangements of trigonal planar carbon atoms (Fig. 3*a*). The highly symmetrical nature of the benzene molecule is not fully represented by such a structure. The figure indicates the presence of three double and three single bonds in the ring. It has been shown that the C—C bonds are all the same and, consequently, the true structure of the substance must be represented by two resonance structures which interchange the single and double bonds (Fig. 3*b*). *See* BENZENE. [J.M.Wi.]

**Strychnine alkaloids** Alkaloid substances derived from the seeds and bark of plants of the genus *Strychnos* (family Loganiaceae). This genus serves as the source of poisonous, nitrogen-containing plant materials, such as strychnine (see structure; R = H). The seeds of the Asian species of *Strychnos* contain 2–3% alkaloids, of which about half is strychnine and the rest is closely related materials; for example, brucine (see structure; R = $OCH_3$) is a more highly oxygenated relative. Strychnine and brucine

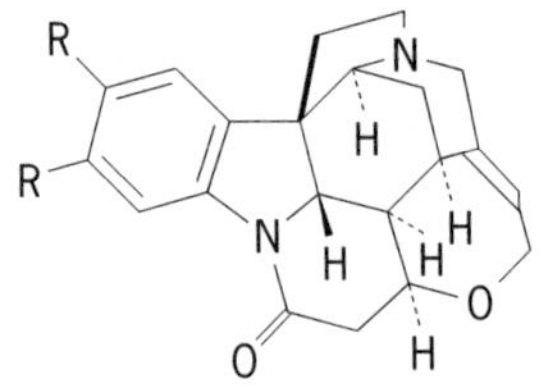

are isolated by extraction of basified plant residue with chloroform and then, from the chloroform solution, by dilute sulfuric acid. Precipitation from the dilute acid is accomplished with ammonium hydroxide. Strychnine is separated from brucine by fractional crystallization from ethanol. *See* CRYSTALLIZATION.

At one time strychnine was used as a tonic and a central nervous system stimulant, but because of its high toxicity (5 mg/kg is a lethal dose in the rat) and the availability of more effective substances, it no longer has a place in human medicine. [D.D.]

**Styrene** A colorless, liquid hydrocarbon with the formula $C_6H_5CH{=}CH_2$. It boils at 145.2°C (293.4°F) and freezes at −30.6°C (−23.1°F). The ethylenic linkage of styrene readily undergoes addition reactions and under the influence of light, heat, or catalysts undergoes self-addition or polymerization to yield polystyrene.

The majority of the styrene used is converted into polystyrene, but other thermoplastic or even thermosetting resins are prepared from styrene by copolymerization with suitable comonomers. A smaller quantity of styrene goes into the manufacture of elastomers or synthetic rubbers.

Styrene is a skin irritant. Prolonged breathing of air containing more than 400 ppm of styrene vapor may be injurious to health. *See* POLYMERIZATION; POLYSTYRENE RESIN. [C.K.B.]

**Sublimation** The process by which solids are transformed directly to the vapor state without passing through the liquid phase. Sublimation is of considerable importance in the purification of certain substances such as iodine, naphthalene, and sulfur.

Sublimation is a universal phenomenon exhibited by all solids at temperatures below their triple points. For example, it is a common experience to observe the disappearance of snow from the ground even though the temperature is below the freezing point and liquid water is never present. The rate of disappearance is low, of course, because the vapor pressure of ice is low below its triple point. Sublimation is a scientifically and technically useful phenomenon, therefore, only when the vapor pressure of the solid phase is high enough for the rate of vaporization to be rapid. *See* PHASE EQUILIBRIUM; TRIPLE POINT; VAPOR PRESSURE. [N.H.N.]

**Substitution reaction** One of a class of chemical reactions in which one atom or group (of atoms) replaces another atom or group in the structure of a molecule or ion. Usually, the new group takes the same structural position that was occupied by the group replaced.

Substitution reactions involve the attack of a reagent, which is the source of the new atom or group, on the substrate, the molecule or ion in which the replacement occurs. They involve the formation of a new bond and the breaking of an old bond. Substitution reactions are classified according to the nature of the reagent (electrophilic, nucleophilic, or radical) and according to the nature of the site of substitution (saturated carbon atom or aromatic carbon atom).

Systematic names for substitution reactions are composed of the parts: name of group introduced + de + name of group replaced + ation, with suitable elision or change of vowels for euphony. Thus, the replacement of bromine by a methoxy group is called methoxydebromination. *See* ORGANIC REACTION MECHANISM. [J.F.B.]

**Sucrose** An oligosaccharide, α-D-glucopyranosyl-β-D-fructofuranoside, also known as saccharose, cane sugar, or beet sugar. The structure is shown below. Sucrose

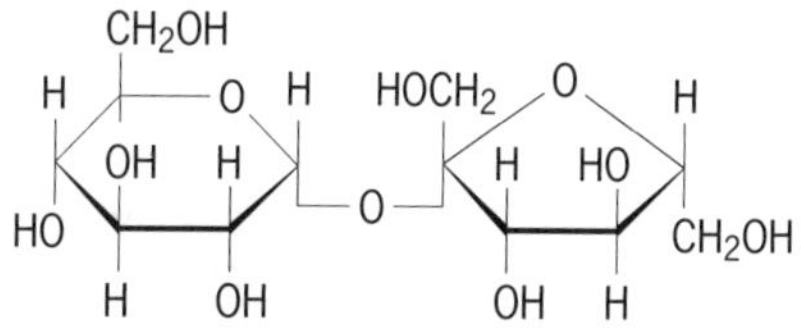

$CH_2OH$
H O H $HOCH_2$ O H
OH H H HO
HO O $CH_2OH$
H OH OH H

is very soluble in water and crystallizes from the medium in the anhydrous form. The sugar occurs universally throughout the plant kingdom in fruits, seeds, flowers, and roots of plants. Honey consists principally of sucrose and its hydrolysis products. Sugarcane and sugarbeets are the chief sources for the preparation of sucrose on a large scale. Another source of commercial interest is the sap of maple trees. [W.Z.H.]

**Sulfonamide** One of the group of organosulfur compounds, $RSO_2NH_2$. Many sulfonamides, which are the amides of sulfonic acids, have the marked ability to halt the growth of bacteria. The therapeutic drugs of this group are known as sulfa drugs. *See* ORGANOSULFUR COMPOUND.

The antibacterial spectrum of sulfonamides comprises a wide variety of gram-positive and gram-negative bacteria, including staphylococci, streptococci, meningococci, and gonococci, as well as the gangrene, tetanus, coli, dysentery, and cholera bacilli. They have only slight activity against *Mycobacterium tuberculosis*, while certain closely related sulfones are quite active against *M. leprae*. The use of sulfonmethoxine has proved to be effective in the treatment of chloroquine-resistant malaria. The relative potency of sulfonamides against the different microorganisms varies, and their action is bacteriostatic rather than bactericidal.

The antibacterial effect, both in patients and in test tube cultures, is antagonized by *p*-aminobenzoic acid (PABA) and PABA-containing natural or synthetic products, such as folic acid and procaine. Accordingly, the mode of action of sulfonamides is considered to be an antimetabolite activity, dependent upon the inhibition of enzyme systems involving the essential PABA.

Bacterial resistance has developed to all known sulfonamides, and many sulfonamide-resistant strains are encountered among the gram-positive and gram-negative bacteria. The emergence of resistance to sulfonamides, however, seems less rapid and less widespread than resistance to most antibiotics.

Sulfonamides are used today mostly as auxiliary drugs or in combination with antibiotics. In certain infectious diseases, however (for instance, in meningococcal infections and most infections of the urinary tract), sulfonamides deserve preference over antibiotics. [R.L.M.; E.Gru.]

**Sulfonation and sulfation** Sulfonation is a chemical reaction in which a sulfonic acid group, $—SO_3H$, is introduced into the structure of a molecule or ion in place of a hydrogen atom. Sulfation involves the attachment of the $—OSO_2OH$ group to carbon, yielding an acid sulfate, $ROSO_2OH$, or of the $—SO_4—$ group between two carbons, forming the sulfate, $ROSO_2OR$.

Sulfonation of aromatic compounds is the most important type of sulfonation. This is accomplished by treating the aromatic compound with sulfuric acid, as in the reaction below. The product of sulfonation is a sulfonic acid.

$$\text{Napthalene} + H_2SO_4 \xrightarrow{320°F \text{ or } 160°C} \beta\text{-Napthalenesulfonic acid } (SO_3H) + H_2O$$

Napthalene β-Napthalenesulfonic acid

Sulfonation may also be defined as any chemical process by which the sulfonic acid group, $—SO_2OH$, or the corresponding salt or sulfonyl halide group, for example, $—SO_2Cl$, is introduced into an organic compound. These groups may be bonded to either a carbon or a nitrogen atom. The latter compounds are designated *N*-sulfonates or sulfamates.

Most sulfonates are employed as such in acid or salt form for applications where the strongly polar hydrophilic $—SO_2OH$ group confers needed properties on a comparatively hydrophobic nonpolar organic molecule. Some sulfonates, such as methanesulfonic and toluenesulfonic acids, are used as catalysts. The major quantity of sulfonates and sulfates is both marketed and used in salt form. This category includes detergents; emulsifying, demulsifying, wetting, and solubilizing agents; lubricant additives; and rust inhibitors. *See* ORGANOSULFUR COMPOUND; SUBSTITUTION REACTION. [P.H.G./R.S.K.]

**Sulfonic acid** A derivative of sulfuric acid ($HOSO_2OH$) in which an OH has been replaced by a carbon group, as shown in the structure below. Sulfonic acids

$$\mathrm{R}-\overset{\overset{\displaystyle\mathrm{O}}{\|}}{\underset{\underset{\displaystyle\mathrm{O}}{\|}}{\mathrm{S}}}-\mathrm{OH}$$

are strongly acidic, water-soluble, nonvolatile, and hygroscopic; they do not act as oxidizing agents and are typically highly stable compounds. Sulfonic acids rarely occur naturally. An exception, taurine, $NH_2CH_2CH_2SO_3H$, occurs in bile.

The aliphatic sulfonic acids are generally made by oxidation of thiols. Several have unique properties. For example, trifluoromethanesulfonic acid, $CF_3SO_3H$, is such a strong acid that it will protonate sulfuric acid. A compound derived from natural camphor, 10-camphorsulfonic acid, is used extensively in the optical resolution of amines.

Aromatic sulfonic acids are much more important than those of the aliphatic series. Aromatic sulfonic acids are produced by sulfonation of aromatic compounds with sulfuric acid or fuming sulfuric acid. Sulfonation of aromatic hydrocarbons is a reversible process; treatment of an aromatic sulfonic acid with superheated steam removes the $—SO_3H$ group. This process can be used in purifying aromatic hydrocarbons. Aromatic sulfonic acids and their derivatives, especially metal salts, are important industrial chemicals. *See* SULFURIC ACID.

The most extensive use of the sulfonation reaction is in the production of detergents. The most widely used synthetic detergents are sodium salts of straight-chain alkylbenzenesulfonic acids.

Sulfonated polymers, particularly sulfonated polystyrenes, act as ion-exchange resins which have important applications in water softening, ion-exchange chromatography, and metal separation technology. Both sulfonated polymers and simple aromatic sulfonic acids, particularly *p*-toluenesulfonic acid, are frequently used as acid catalysts in organic reactions such as esterification and hydrolysis. *See* HOMOGENEOUS CATALYSIS; ION EXCHANGE.

The sulfonic group, in either acid or salt form, is capable of making many substances water soluble, increasing their usefulness. This application is particularly significant in the dying industry, in which a majority of the dyes are complex sodium sulfonates. Many acid-base indicators are soluble due to the presence of a sodium sulfonate moiety. Some pigments used in the paint and ink industry are insoluble metal salts or complexes of sulfonic acid derivatives. Most of the brighteners used in detergents compounded for laundering are sulfonic acid derivatives of heterocyclic compounds. *See* ACID-BASE INDICATOR; DYE; ORGANOSULFUR COMPOUND. [C.R.J.]

**Sulfur** A chemical element, S, atomic number 16, and atomic weight 32.064. The atomic weight reflects the fact that sulfur is composed of the isotopes $^{32}S$ (95.1%),

$^{33}S$ (0.74%), $^{34}S$ (4.2%), and $^{36}S$ (0.016%). The ratios of the various isotopes vary slightly but measurably according to the history of the sample. By virtue of its position in the periodic table, sulfur is classified as a main-group element. *See* PERIODIC TABLE.

The chemistry of sulfur is more complex than that of any other elemental substance, because sulfur itself exists in the largest variety of structural forms. At room temperature, all the stable forms of sulfur are molecular; that is, the individual atoms aggregate into discrete molecules, which in turn pack together to form the solid material. In contrast, other elements near sulfur in the periodic table normally exist as polymers (silicon, phosphorus, arsenic, selenium, tellurium) or as diatomic molecules (oxygen, nitrogen, chlorine). Selenium and phosphorus can exist as molecular solids, but the stable forms of these elements are polymeric.

At room temperature the most stable form of sulfur is the cyclic molecule $S_8$. The molecule adopts a crownlike structure, consisting of two interconnected layers of four sulfur atoms each. The S—S bond distances are 0.206 nanometer and the S—S—S bond angles are 108°. Three allotropes are known for cyclo-$S_8$. The most common form is orthorhombic $\alpha$-sulfur, which has a density of 2.069 g/cm$^3$ (1.200 oz/in.$^3$) and a hardness of 2.5 on the Mohs scale. It is an excellent electrical insulator, with a room temperature conductivity of $10^{18}$ ohm$^{-1}$ cm$^{-1}$. Sublimed sulfur and "flowers" of sulfur are generally composed of $\alpha$-$S_8$. Sulfur is quite soluble in carbon disulfide ($CS_2$; 35.5/100 g or 1.23 oz/3.52 oz at 25°C or 77°F), poorly soluble in alcohols, and practically insoluble in water. At 95.3°C (203°F), sulfur changes into the monoclinic $\beta$ allotrope. This form of sulfur also consists of cyclic $S_8$ molecules, but it has a slightly lower density at 1.94–2.01 g/cm$^3$ (1.12–1.16 oz/in.$^3$). A third allotrope containing $S_8$ is triclinic $\gamma$-sulfur. The $\beta$ and $\gamma$ allotropes of sulfur slowly revert to the $\alpha$ form at room temperature. Crystals of sulfur are yellow and have an absorption maximum in the ultraviolet at 285 nm, which shifts to higher energy as the temperature decreases. At low temperatures, $S_8$ is colorless. Even at room temperature, however, finely powdered sulfur can appear to be nearly white.

The best-studied system is $\alpha$-$S_8$, which converts to the $\beta$ form at 90°C (194°F), which then melts at 120°C (248°F) to give a golden yellow liquid. If this melt is quickly recooled, it refreezes at 120°C (248°F), thus indicating that it consists primarily of $S_8$ molecules. If the melt is maintained longer at 120°C (248°F), then the freezing point is lowered about 5°C (9°F), indicating the formation of about 5% of other rings and some polymer. At 159.4°C (318.9°F), the melt suddenly assumes a red-brown color. Over the range 159.4–195°C (318.9–383°F), the viscosity of the melt increases 10,000-fold before gradually decreasing again. This behavior is very unusual, since the viscosity of most liquids decreases with increasing temperature. The strong temperature dependence of the viscosity is due to the polymerization and eventual depolymerization of sulfur. Polymeric sulfur retains its elastomeric character even after being cooled to room temperature. There are several polymeric forms of sulfur, but all of them revert to $\alpha$-$S_8$ after a few hours.

Sublimination of $S_8$ occurs when it is maintained in a vacuum at a temperature below its melting point. It vaporizes at 444.61°C (832.30°F). Below 600°C (1110°F), the predominant species in the gas are $S_8$ followed by $S_7$ and $S_6$. Above 720°C (1328°F), violet $S_2$ is the major species.

**Principal inorganic compounds.** Hydrogen sulfide ($H_2S$) is the most important compound that contains only sulfur and hydrogen. It is a gas at room temperature with a boiling point of −61.8°C (−79.2°F) and a freezing point of −82.9°C (−117°F). The low boiling point of hydrogen sulfide is attributed to the weakness of intermolecular S···H hydrogen bonding; the O···H hydrogen bond is much stronger, as evidenced by the high boiling point of water. Gaseous hydrogen sulfide is 1.19 times more dense than air, and air-$H_2S$ mixtures are explosive. Hydrogen sulfide has a strong odor similar to that of rotten eggs; its odor is detectable at concentrations below 1 microgram/$m^3$. At high concentrations, $H_2S$ has a paralyzing effect on the olfactory system, which is very hazardous because $H_2S$ is even more toxic than carbon monoxide (CO).

The most common compound that contains only carbon and sulfur is carbon disulfide ($CS_2$). Carbon disulfide molecules are linear, consisting of two sulfur atoms connected to a central carbon atom. Carbon disulfide is a toxic, highly flammable, and volatile liquid that melts at −111°C (−168°F) and boils at 46°C (115°F). Commercial carbon disulfide has a strong unpleasant odor due to impurities. It is manufactured from methane and elemental sulfur and is used for the production of carbon tetrachloride, rayon, and cellophane. Structurally related to carbon disulfide is carbonyl sulfide (SCO), which forms from carbon monoxide and elemental sulfur. The chlorination of $CS_2$ gives $Cl_3CSCl$, which can be reduced by $H_2S$ to thiophosphene, $CSCl_2$. Thiophosgene ($CSCl_2$) [boiling point 73°C or 163°F] is a planar molecule with the carbon at the center of a triangle defined by the sulfur and two chlorine atoms. Thiocyanate, the linear anion $NCS^-$, is prepared by the reaction of cyanide (—CN) with elemental sulfur.

Several sulfur oxides exist, but the dioxide and trioxide are of preeminent importance. Sulfur dioxide ($SO_2$) is a colorless gas that boils at −10.02°C (113.97°F) and freezes at −75.46°C (−103.8°F). The density of liquid sulfur dioxide at −10°C (14°F) is 1.46 g/$cm^3$ (0.84 oz/in.$^3$). Liquid sulfur dioxide is an excellent solvent. The sulfur dioxide molecule is bent, with an O—S—O angle of 119°.

Sulfur trioxide ($SO_3$) is a planar molecule that is a liquid at room temperature that exists in equilibrium with a cyclic trimeric structure known as $\beta$-$SO_3$. When $\beta$-$SO_3$, actually $S_3O_9$, is treated with traces of water, it converts to either of two polymeric forms referred to as $\gamma$- and $\alpha$-sulfur trioxide. These are fibrous materials, proposed to have the formula $(SO_3)_xH_2$, where $x$ is in the thousands. Sulfur trioxide is prepared by the oxidation of sulfur dioxide, although at very high temperatures this reaction reverses. Exposure of sulfur trioxide to water yields sulfuric acid ($H_2SO_4$); exposure of $SO_3$ to sulfuric acid yields disulfuric acid ($H_2S_2O_7$). *See* SULFURIC ACID.

Chlorine and sulfur react to give a family of compounds with the general formula $S_xCl_2$, several members of which have been obtained in pure form. The structures of these compounds are based on an atom or chain of sulfur atoms terminated with Cl. Sulfur monochloride ($S_2Cl_2$), also known as sulfur monochloride, is the most widely available of the series. It is a yellow oil that boils at 138°C (280°F), and reacts with chlorine in the presence of iron(III) chloride ($FeCl_3$) catalyst to give sulfur dichloride ($SC_2$), which is a red volatile liquid with a boiling point of 59°C (138°F). Treatment of sulfur dichloride with sodium fluoride (NaF) gives $SF_4$.

Thionyl chloride ($OSCl_2$) is a colorless reactive compound with a boiling point of 76°C (169°F); it is used to convert hydroxy compounds to chlorides. Important applications include the preparation of anhydrous metal halides and alkyl halides. Sulfuryl chloride ($O_2SCl_2$; boiling point 69°C or 156°F) is used as a source of chlorine.

**Organosulfur compounds.** This family of compounds contains carbon, hydrogen, and sulfur, and it is a particularly vast area of sulfur chemistry. Thiols, also known as mercaptans, feature the linkage C—S—H. Mercaptans are foul-smelling compounds. They are the sulfur analogs of alcohols, but they are more volatile. They can be prepared by the action of hydrogen sulfide ($H_2S$) on olefins. Deprotonation of thiols gives thiolate anions, which form stable compounds with many heavy metals. Thiols and especially thiolates can be oxidized to form disulfides (persulfides), which have the connectivity of C—S—S—C. The organic persulfides are also related to organic polysulfides, which have chains of sulfur atoms terminated with carbon. The introduction of such mono-, di-, and polysulfide linkages is the basis of the vulcanization process, which imparts desirable mechanical properties to natural or synthetic polyolefin rubbers. This is accomplished by heating the polymer with sulfur in the presence of a zinc catalyst. *See* RUBBER.

Thioethers, also known as organic sulfides, feature the connectivity C—S—C and are often prepared from the reaction of thiolates and alkyl halides. Like mercaptans, thioethers often have strong unpleasant odors, but they are also responsible for the pleasant odors of many foods and perfumes. They are intentionally introduced at trace levels in order to impart an odor to gaseous hydrocarbon fuels. The reaction of alkyl dihalides and sodium polysulfides affords organic polysulfide polymers known as thiokols.

There are many organic sulfur oxides; prominent are sulfonic acids ($RSO_3H$), which are the organic derivatives of sulfuric acid. These compounds are prepared by the oxidation of thiols as well as by treatment of benzene derivatives with sulfuric acid, for example, benzene sulfonic acid. Most detergents are salts of sulfonic acids.

**Biochemistry.** Sulfur is required for life. Typical organisms contain 2% sulfur dry weight. Three amino acids contain sulfur, as do many prosthetic groups in enzymes. Some noteworthy sulfur compounds include the disulfide lipoic acid, the thioethers biotin and thiamine (vitamin $B_1$), and the thiol coenzyme A. Sulfide ions, $S^{2-}$, are found incorporated in metalloproteins and metalloenzymes such as the ferredoxins, nitrogenases, and hydrogenases. *See* AMINO ACIDS; ENZYME; PROTEIN.

Many bacterial species obtain energy by the oxidations of sulfides. Bacteria of the genus *Thiobacillus* couple the conversion of carbon dioxide ($CO_2$) to carbohydrates to the aerobic oxidation of mineral sulfides to sulfuric acid. This activity can be turned to good use for leaching low-grade mineral ores. Often, however, the sulfuric acid runoff (such as in mines or sewers) has negative environmental consequences. The purple and green bacteria as well as the blue-green algae are remarkable because they are photosynthetic but anaerobic; they oxidize sulfide, not water (as do most photosynthetic organisms). Depending on the species, the sulfur produced in this energy-producing pathway can accumulate inside or outside the cell wall.

**Minerals.** Sulfide minerals are among the most important ores for several metals. These compounds are two- or three-dimensional polymers containing interconnected metal cations and sulfide $S^{2-}$ or persulfido $S_2{}^{2-}$ anions. In general, metal sulfides are darkly colored, often black, and they are not soluble in water. They can sometimes be decomposed by using strong acids, with liberation of hydrogen sulfide. Certain sulfides will also dissolve in the presence of excess sulfide or polysulfide ions.

Pyrites ($FeS_2$), also known as iron pyrites or fool's gold, are the most common sulfide minerals and can be obtained as very large crystals that have a golden luster. Sphalerite (zinc blende; ZnS) and galena (PbS) are major sources of zinc and lead. Orange cinnabar (HgS) and yellow greenockite (CdS) are the major ores for mercury and cadmium, respectively. Molybdenite ($MoS_2$) is the major ore of molybdenum.

The sulfur content of fossil fuels results from the sulfur in the ancient organisms as well as from subsequent incorporation of mineral sulfur into the hydrocarbon matrix. Gaseous fossil fuels are often contaminated with hydrogen sulfide, which is an increasingly important source of sulfur. Organic derivatives containing the C—S—C linkage are primarily responsible for the sulfur content of petroleum and coal. The so-called organic sulfur in petroleum can be removed by hydrodesulfurization catalysis, involving reaction with hydrogen over a molybdenum catalyst, to give hydrocarbons and hydrogen sulfide. [T.B.R.]

**Sulfuric acid** A strong mineral acid with the chemical formula $H_2SO_4$. It is a colorless, oily liquid, sometimes called oil of vitriol or vitriolic acid. The pure acid has a density of 1.834 at 25°C (77°F) and freezes at 10.5°C (50.90°F). It is an important industrial commodity, used extensively in petroleum refining and in the manufacture of fertilizers, paints, pigments, dyes, and explosives.

Sulfuric acid is produced on a large scale by two commercial processes, the contact process and the lead-chamber process. In the contact process, sulfur dioxide, $SO_2$, is converted to sulfur trioxide, $SO_3$, by reaction with oxygen in the presence of a catalyst. Sulfuric acid is produced by the reaction of the sulfur trioxide with water. The lead-chamber process depends upon the oxidation of sulfur dioxide by nitric acid in the presence of water, the reaction being carried out in large lead rooms.

Sulfuric acid reacts vigorously with water to form several hydrates. The concentrated acid, therefore, acts as an efficient drying agent, taking up moisture from the air and even abstracting the elements of water from such compounds as sugar and starch. The concentrated acid also acts as a strong oxidizing agent. It reacts with most metals upon heating to produce sulfur dioxide. *See* SULFUR. [F.J.J.]

**Superacid** An acid which has an extremely great proton-donating ability. It has proved convenient to define a superacid somewhat arbitrarily as an acid, or more generally, an acidic medium, which has a proton-donating ability equal to or greater than that of anhydrous (100%) sulfuric acid.

Superacids belong to the general class of proton or Brønsted acids. A proton acid is defined as any species which can act as a source of protons and which will therefore protonate a suitable base, as in reaction (1).

$$HA + B \rightleftharpoons BH^+ + A^- \qquad (1)$$

The strengths of acids are often compared by measuring the extent of their ionization in water, that is, the extent to which they can protonate the base water, as in reaction (2).

$$HA + H_2O \rightleftharpoons H_3O^+ + A^- \qquad (2)$$

However, all strong acids are fully ionized in dilute aqueous solution, and they therefore appear to have the same strength. Their strengths are said to be reduced or leveled to that of the hydronium ion ($H_3O^+$), which is the most highly acidic species that can exist in water. In any case, many of the superacids react with and are destroyed by water. For these reasons, the strengths of superacids cannot be measured by the conventional means of utilizing their aqueous solutions. The acidities of superacids can, however, be conveniently measured in terms of the Hammett acidity function. *See* ACID AND BASE.

**Hammett acidity function values for several superacids**

| Superacid | Formula | $-H_0$ |
|---|---|---|
| Sulfuric acid | $H_2SO_4$ | 11.9 |
| Chlorosulfuric acid | $HSO_3Cl$ | 13.8 |
| Trifluoromethane sulfonic acid | $HSO_3CF_3$ | 14.0 |
| Disulfuric acid | $H_2S_2O_7$ | 14.4 |
| Fluorosulfuric acid | $HSO_3F$ | 15.1 |
| Hydrogen fluoride | HF | 15.1 |

The Hammett acidity function is a method of measuring acidity based on the determination of the ionization ratios of suitable weak bases (indicators), usually by means of the change in absorption spectrum that occurs on protonation of the base, although the nuclear magnetic resonance (NMR) spectrum has also been used. The Hammett acidity function ($H_0$) is defined by Eq. (3), where $K_{BH^+}$ is the dissociation constant

$$H_0 = pK_{BH^+} - \log\frac{[BH^+]}{[B]} \tag{3}$$

of the acid form of the indicator and $[BH^+]/[B]$ is the ionization ratio of the indicator. Hammett acidity function ($H_0$) values for a number of superacids are given in the table. In each case the value refers to the 100% (anhydrous) acid. Each of the superacids in the table is a liquid at room temperature, and each forms the basis of a solvent system. *See* IONIC EQUILIBRIUM; SOLUTION. [R.J.Gi.]

**Supercritical fluid** Any fluid at a temperature and a pressure above its critical point; also, a fluid above its critical temperature regardless of pressure. Below the critical point the fluid can coexist in both gas and liquid phases, but above the critical point there can be only one phase. Supercritical fluids are of interest because their properties are intermediate between those of gases and liquids, and are readily adjustable. *See* CRITICAL PHENOMENA; PHASE EQUILIBRIUM.

In a given supercritical fluid the thermodynamic and transport properties are a function of density, which depends strongly on the fluid's pressure and temperature. The density may be adjusted from a gaslike value of 0.1 g/ml to a liquidlike value as high as 1.2 g/ml. Furthermore, as conditions approach the critical point, the effect of temperature and pressure on density becomes much more significant. Increasing the density of supercritical carbon dioxide from 0.2 to 0.5 g/ml, for example, requires raising the pressure from 85 to 140 atm (8.6 to 14.2 megapascals) at 158°F (70°C), but at 95°F (35°C) the required change is only from 65 to 80 atm (6.6 to 8.1 MPa).

For a given fluid, the logarithm of the solubility of a solute is approximately proportional to the solvent density at constant temperature. Therefore, a small increase in pressure, which causes a large increase in the density, can raise the solubility a few orders of magnitude. While almost all of a supercritical fluid's properties vary with density, some of these properties are more like those of a liquid while others are more like those of a gas. *See* SUPERCRITICAL-FLUID CHROMATOGRAPHY.

In most supercritical-fluid applications the fluid's critical temperature is less than 392°F (200°C) and its critical pressure is less than 80 atm (8.1 MPa). High critical temperatures require operating temperatures that can damage the desired product, while high critical pressures result in excessive compression costs. In addition to these pure fluids, mixed solvents can be used to improve the solvent strength. [K.Jo.; R.Len.]

**Supercritical-fluid chromatography** Any separation technique in which a supercritical fluid is used as the mobile phase. For any fluid, a phase diagram can be constructed to show the regions of temperature and pressure at which gases and liquids, gases and solids, and liquids and solids can coexist. For the gas–lliquid equilibrium, there is a certain temperature and pressure, known as the critical temperature and pressure, below which a gas and a liquid can coexist but above which only a single phase (known as a supercritical fluid) can form.

For example, the density of supercritical fluids is usually between 0.25 and 1.2 g/ml and is strongly pressure-dependent. Their solvent strength increases with density, so that molecules that are retained on the column can often be eluted simply by increasing the pressure under which the fluid is compressed.

Diffusion coefficients of solutes in supercritical fluids are tenfold greater than the corresponding values in liquid solvents (although about three orders of magnitude less than the corresponding values in gases). The high diffusivity of solutes in supercritical fluids decreases their resistance to mass transfer in a chromatographic column and hence allows separations to be made either very quickly or at high resolution. Molecules can be separated in a few minutes or less by using packed columns of the type used for high-performance liquid chromatography (HPLC).

Despite the relatively large number of separations made using packed-column supercritical-fluid chromatography that have been described since 1962, supercritical-fluid chromatography became popular only in the late 1980s with the commercial introduction of capillary (open-tubular) supercritical-fluid chromatographs. While supercritical-fluid chromatography may never be as widely used as gas chromatography or high-performance liquid chromatography, it provides a useful complement to these chromatographies. *See* CHEMICAL SEPARATION TECHNIQUES; CHROMATOGRAPHY; GAS CHROMATOGRAPHY; LIQUID CHROMATOGRAPHY. [P.R.G.]

**Superoxide chemistry** A branch of chemistry that deals with the reactivity of the superoxide ion ($O_2^-$), a one-electron ($e^-$) adduct of molecular oxygen (dioxygen; $O_2$) formed by the combination of $O_2$ and $e^-$. Because 1–15% of the $O_2$ that is respired by mammals goes through the $O_2^-$ oxidation state, the biochemistry and reaction chemistry of the species are important to those concerned with oxygen toxicity, carcinogenesis, and aging. Although the name superoxide has prompted many to assume an exceptional degree of reactivity for $O_2^-$, the use of the prefix in fact was chosen to indicate stoichiometry. Superoxide was the name given in 1934 to the newly synthesized potassium salt ($KO_2$) to differentiate its two-oxygens-per-metal stoichiometry from that of most other metal-oxygen compounds ($NA_2O$, $Na_2O_2$, NaOH, $Fe_2O_3$).

Ionic salts of superoxide (yellow-to-orange solids), which form from the reaction of dioxygen with metals such as potassium, rubidium, or cesium, are paramagnetic, with one unpaired electron per two oxygen atoms.

In 1969, by means of electron spin resonance (ESR) spectroscopy, superoxide ion was detected as a respiratory intermediate, and metalloproteins were discovered that catalyze the disproportionation of superoxide, that is, superoxide dismutases (SODs), as shown in the reaction below.

$$2O_2^- + 2H^+ \xrightarrow{\text{SOD}} O_2 + H_2O_2$$

The biological function of superoxide dismutases is believed to be the protection of living cells against the toxic effects of superoxide. The possibility that superoxide might be an important intermediate in aerobic life provided an impetus to the study of superoxide reactivity.

The most general and universal property of $O_2^-$ is its tendency to act as a strong Brønsted base. Its strong proton affinity manifests itself in any media. Another characteristic of $O_2^-$ is its ability to act as a moderate one-electron reducing agent. *See* ACID AND BASE; OXIDATION-REDUCTION.

In general, superoxide ion chemistry does not appear to be sufficiently robust to make superoxide ion a toxin. However, it can interact with protons, halogenated carbons, and carbonyl compounds to yield peroxy radicals that are toxic. *See* REACTIVE INTERMEDIATES.

Superoxide does not appear to have exceptional reactivity. Nevertheless, superoxide will continue to be an interesting species for study because of the multiplicity of its chemical reactions and because of its importance as an intermediate in reactions that involve dioxygen and hydrogen peroxide. *See* BIOINORGANIC CHEMISTRY; OXYGEN. [D.T.S.]

**Supersaturation** A solution is at the saturation point when dissolved solute in its crystallizes from it at the same rate at which it dissolves. Under prescribed experimental conditions of temperature and pressure, a solution can contain at saturation only one fixed amount of dissolved solute. However, it is possible to prepare relatively stable solutions which contain a quantity of a dissolved solute greater than that of the saturation value provided solute phase is absent. Such solutions are said to be supersaturated. They can be prepared by changing the experimental conditions of a system so that greater solubility is obtained, perhaps by heating the solution, and then carefully returning the system to or near its original state. The addition of solute phase will immediately relieve supersaturation. Solutions in which there is no spontaneous formation of solute phase for extended periods of time are said to be metastable. There is no sharp line of demarcation between an unstable and metastable solution. The process whereby initial aggregates within a supersaturated solution develop spontaneously into particles of new stable phase is known as nucleation. The greater the degree of supersaturation, the greater will be the number of nuclei formed. *See* NUCLEATION; PHASE EQUILIBRIUM. [L.Go./R.W.Mu.]

**Supramolecular chemistry** A highly interdisciplinary field covering the chemical, physical, and biological features of complex chemical species held together and organized by means of intermolecular (noncovalent) bonding interactions. *See* CHEMICAL BONDING; INTERMOLECULAR FORCES.

When a substrate binds to an enzyme or a drug to its target, and when signals propagate between cells, highly selective interactions occur between the partners that control the processes. Supramolecular chemistry is concerned with the study of the basic features of these interactions and with their implementation in biological systems as well as in specially designed nonnatural ones. In addition to biochemistry, its roots extend into organic chemistry and the synthetic procedures for receptor construction, into coordination chemistry and metal ion-ligand complexes, and into physical chemistry and the experimental and theoretical studies of interactions. *See* BIOINORGANIC CHEMISTRY; ENZYME; LIGAND; PHYSICAL ORGANIC CHEMISTRY; PROTEIN.

The field started with the selective binding of alkali metal cations by natural as well as synthetic macrocyclic and macropolycyclic ligands, the crown ethers and cryptands. This led to the emergence of molecular recognition as a new domain of chemical research that, by encompassing all types of molecular components and interactions as well as both oligo and polymolecular entities, became supramolecular chemistry. It underwent rapid growth with the development of synthetic receptor molecules of numerous types for the strong and selective binding of cationic, anionic, or neutral complementary substrates of organic, inorganic, or biological nature by means of various interactions (electrostatic, hydrogen bonding, van der Waals, and donor-acceptor). Molecular recognition implies the (molecular) storage and the (supramolecular) retrieval and processing of molecular structural (geometrical and interactional) information. *See* HYDROGEN BOND; MACROCYCLIC COMPOUND; MOLECULAR RECOGNITION.

Many types of receptor molecules have been explored (crown ethers, cryptands, cyclodextrins, calixarenes, cavitands, cyclophanes, cryptophanes, and so on), and many others may be imagined for the binding of complementary substrates of chemical or biological significance. They allow, for instance, the development of substrate-specific sensors or the recognition of structural features in biomolecules (for example, nucleic acid probes, affinity cleavage reagents, and enzyme inhibitors). *See* BIOPOLYMER; CYCLOPHANE.

A major step in the development of supramolecular chemistry over the last 20 years involved the design of systems capable of spontaneously generating well-defined, supramolecular entities by self-assembly under a given set of conditions.

The information necessary for supramolecular self-assembly to take place is stored in the components, and the program that it follows operates via specific interactional algorithms based on binding patterns and molecular recognition events. Thus, rather than being preorganized, constructed entities, these systems may be considered as self-organizing, programmed supramolecular systems.

Self-assembly and self-organization have recently been implemented in numerous types of organic and inorganic systems. By clever use of metal coordination, hydrogen bonding, and donor-acceptor interactions, researchers have achieved the spontaneous formation of a variety of novel and intriguing species such as inorganic double and triple helices termed helicates, catenates, threaded entities (rotaxanes), cage compounds, grids of metal ions, and so on.

Another major development concerns the design of molecular species displaying the ability to perform self-replication, based on components containing suitable recognition groups and reactive functions. Self-recognition processes involve the spontaneous selection of the correct partner(s) in a self-assembly event—for instance, the correct ligand strand in helicate formation.

A major area of interest is the design of supramolecular devices built on photoactive, electroactive, or ionoactive components, operating respectively with photons, electrons, or ions. Thus, a variety of photonic devices based on photoinduced energy and electron transfer may be imagined. Molecular wires, ion carriers, and channels facilitate the flow of electrons and ions through membranes. Such functional entities represent entries into molecular photonics, electronics, and ionics, which deal with the storage, processing, and transfer of materials, signals, and information at the molecular and supramolecular levels. Dynamic and mechanical devices exploit the control of motion within molecular and supramolecular entities. *See* INORGANIC PHOTOCHEMISTRY; ION TRANSPORT; PHOTOCHEMISTRY.

The design of systems that are controlled, programmed, and functionally self-organized by means of molecular information contained in their components represents new horizons in supramolecular chemistry and provides an original approach to nanoscience and nanotechnology. In particular, the spontaneous but controlled generation of well-defined, functional supramolecular architectures of nanometric size through self-organization—supramolecular nanochemistry—represents a means of performing programmed engineering and processing of nanomaterials. It offers a powerful alternative to the demanding procedures of nanofabrication and nanomanipulation, bypassing the need for external intervention. A rich variety of architectures, properties, and processes should result from this blending of supramolecular chemistry with materials science. *See* NANOCHEMISTRY; NANOTECHNOLOGY. [J.M.Le.]

**Surface and interfacial chemistry** Chemical processes that occur at the phase boundary between gas–liquid, liquid–liquid, liquid–solid, or gas–solid interfaces.

The chemistry and physics at surfaces and interfaces govern a wide variety of technologically significant processes. Chemical reactions for the production of low-molecular-weight hydrocarbons for gasoline by the cracking and reforming of the high-molecular-weight hydrocarbons in oil are catalyzed at acidic oxide materials. Surface and interfacial chemistry are also relevant to adhesion, corrosion control, tribology (friction and wear), microelectronics, and biocompatible materials. In the last case, schemes to reduce bacterial adhesion while enhancing tissue integration are critical to the implantation of complex prosthetic devices, such as joint replacements and artificial hearts. *See* CRACKING; HETEROGENEOUS CATALYSIS.

Interactions with the substrate may alter the electronic structure of the adsorbate. Those interactions that lower the activation energy of a chemical reaction result in a catalytic process. Adsorption of reactants on a surface also confines the reaction to two dimensions as opposed to the three dimensions available for a homogeneous process. The two-dimentional confinement of reactants in a bimolecular event seems to drive biochemical processes with higher reaction efficiencies at proteins and lipid membranes. *See* ADSORPTION.

A limitation in the study of surfaces and interfaces rests with the low concentrations of the participants in the chemical process. Concentrations of reactants at surfaces are on the order of $10^{-10}$ to $10^{-8}$ mole/cm$^2$. Such low concentrations pose a sensitivity problem from the perspective of surface analysis. Experimental techniques with high sensitivity are required to examine the low concentrations of a surface species at interfaces.

Electron spectroscopy methods are widely used in the study of surfaces because of the small penetration depth of electrons through solids. This attribute makes electron spectroscopy inherently surface-sensitive, since only a few of the outermost atomic layers are accessible. The methods of electron spectroscopy used in surface studies have several common characteristics (see table). A source provides the incident radiation to the sample, which can be in the form of electrons, x-radiation, or ultraviolet radiation. Electron beams are generated from the thermionic emission of metal filaments or metal oxide pellets. The incident radiation induces an excitation at the surface of the sample, which alters the energy distribution of electrons that leave the surface. This distribution provides a diagnostic of the composition or structure of the interface. *See* ELECTRON SPECTROSCOPY.

**Surface-sensitive experimental techniques**

| Technique | Source* | Detectors | Level of information |
|---|---|---|---|
| Auger electron spectroscopy (AES) | Electrons 2–3 keV | Cylindrical mirror of retarding field | Elemental composition |
| X-ray photoelectron spectroscopy (XPS) | X-rays 1254 eV (Mg) 1487 eV (Al) | Hemispherical or cylindrical mirror | Elemental composition and oxidation state |
| Ultraviolet photoelectron spectroscopy (UPS) | UV radiation 21 eV He (I) 41 ev He (II) | Hemispherical or cylindrical mirror | Electronic properties of adsorbate and/or bulk material |
| Energy loss spectroscopy (ELS) | Electrons 50–1000 eV | Electron energy analyzer | Electronic structure of surface |
| High-resolution electron energy loss spectroscopy (HREELS) | Electrons 1–10 eV | Electron energy analyzer | Vibrational losses |
| Low-energy electron diffraction (LEED) | Electrons 20–500 eV | Retarding fields and phosphorescent screen | Surface structure or periodicity |
| Infrared spectroscopy (IRS) | Photons | Mercury-cadmium-telluride or indium antimony | Molecular identity |
| Optical ellipsometry | Photons | Photomultiplier | Adsorbate layer thickness |
| Scanning tunneling microscopy (STM) | Tunneling current | Ammeter | Substrate roughness and texture |

*Mg = magnesium; Al = aluminum; He = helium.

Optical spectroscopy techniques (visible and infrared) are also useful for probing the chemical composition and molecular arrangement of surface species. Typical application configurations are the transmission and reflection (both external and internal) modes. Transmission spectroscopy relies on the passage of the probe beam through the sample. External and internal reflection spectroscopies involve the reflection of the probe beam from a medium with a lower refractive index to a medium with a higher refractive index, and from a higher to lower refractive index, respectively. The sample support must be optically transparent to the probe beam for the internal reflection mode. In both cases, the substrates are polished to a smooth, mirrorlike finish. *See* SPECTROSCOPY. [M.M.W.; M.D.P.]

**Surface tension** The force acting in the surface of a liquid, tending to minimize the area of the surface. Surface forces, or more generally, interfacial forces, govern such phenomena as the wetting or nonwetting of solids by liquids, the capillary rise of liquids in fine tubes and wicks, and the curvature of free-liquid surfaces. The action of detergents and antifrothing agents and the flotation separation of minerals depend upon the surface tensions of liquids.

In the body of a liquid, the time-averaged force exerted on any given molecule by its neighbors is zero. Even though such a molecule may undergo diffusive displacements because of random collisions with other molecules, there exist no directed forces upon it of long duration. It is equally likely to be momentarily displaced in one direction as in any other. In the surface of a liquid, the situation is quite different; beyond the free surface, there exist no molecules to counteract the forces of attraction exerted by molecules in the interior for molecules in the surface. In consequence, molecules in the surface of a liquid experience a net attraction toward the interior of a drop. These centrally directed forces cause the droplet to assume a spherical shape, thereby minimizing both the free energy and surface area.

Liquids which wet the walls of fine capillary tubes rise to a height which depends upon the tube radius, the surface tension, the liquid density, and the contact angle between the solid and the liquid (measured through the liquid). In the illustration a liquid of a certain density is shown as having risen to a height $h$ in a capillary whose radius is $r$. A balance exists between the force exerted by gravity on the mass of liquid raised in the capillary and the opposing force caused by surface tension.

Detergents, soaps, and flotation agents owe their usefulness to their ability to lower the surface tension of water, thereby stabilizing the formation of small bubbles of air.

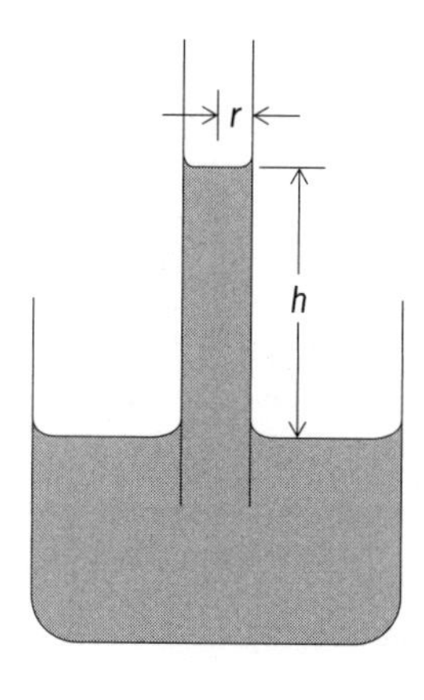


**Rise of liquid in capillary tube.**

At the same time, the interfacial tension between solid particles and the liquid phase is lowered, so that the particles are more readily wetted and floated after attachment to air bubbles. *See* INTERFACE OF PHASES; SURFACTANT. [N.H.N.]

**Surfactant** A member of the class of materials that, in small quantity, markedly affect the surface characteristics of a system; also known as surface-active agent. In a two-phase system, for example, liquid-liquid or solid-liquid, a surfactant tends to locate at the interface of the two phases, where it introduces a degree of continuity between the two different materials. Soaps and detergents are classic examples of surfactants due to their dual (amphipathic) character. These substances consist of a hydrophobic tail portion, usually a long-chain hydrocarbon, and a hydrophilic polar head group, which is often ionic. A material possessing these characteristics is known as an amphiphile. It tends to dissolve in both aqueous and oil phase and to locate at the oil-water interface. *See* INTERFACE OF PHASES; SOAP.

Surfactants are employed to increase the contact of two materials, sometimes known as wettability. Surfactants and surface activity are controlling features in many important systems, including emulsification, detergency, foaming, wetting, lubrication, water repellance, waterproofing, spreading and dispersion, and colloid stability. *See* EMULSION; MICELLE.

In general, surfactants are divided into four classes: amphoteric, with zwitterionic head groups; anionic, with negatively charged head groups; cationic, with positively charged head groups; and nonionic, with uncharged hydrophilic head groups. Those with anionic head groups include long-chain fatty acids, sulfosuccinates, alkyl sulfates, phosphates, and sulfonates. Cationic surfactants may be protonated long-chain amines and long-chain quaternary ammonium compounds. The class of amphoteric surfactants is represented by betaines and certain lecithins, while nonionic surfactants include polyethylene oxide, alcohols, and other polar groups.

Quite different materials, such as polymers and clays, can also exhibit surface activity; many polymeric materials, for example, polyvinyl alcohol and polyacrylamide, are excellent stabilizers for a variety of colloid systems. These entities adsorb at the colloid interface and, by means of steric effects, prevent colloid-colloid adhesion and flocculation. Clays readily adsorb other materials or adsorb onto large particles suspended in solution, so that the particle interface consists of charged clay particles, which increase colloid stability by electrostatic and steric effects. *See* ADSORPTION; COLLOID; ION EXCHANGE; POLYMER; SURFACE AND INTERFACIAL CHEMISTRY. [J.K.T.]

**Synthetic fuel** A gaseous, liquid, or solid fuel that does not occur naturally; also known as synfuel. Synthetic fuels can be made from coal, oil shale, or tar sands. Included in the category are various fuel gases, such as substitute natural gas and synthesis gas.

Syncrude is a synthetic crude oil, a complex mixture of hydrocarbons somewhat similar to petroleum. It is obtained from coal (liquefaction), from synthesis gas (a mixture of carbon monoxide and hydrogen), or from oil shale and tar sands. Syncrudes generally differ in composition from petroleum; for example, syncrude from coal usually contains more aromatic hydrocarbons than petroleum. Gaseous fuels can be produced from sources other than petroleum and natural gas.

The most important source of synthetic crude oil is the tar sand deposit that occurs in northeastern Alberta, Canada. Tar sand is a common term for oil-impregnated sediments that can be found in almost every continent. The routes by which synthetic fuels can be prepared from coal involve either gasification or liquefaction.

Gasification can yield clean gases for combustion or synthesis gas, which has a controlled ratio of hydrogen to carbon monoxide. Catalytic conversion of synthesis gas to liquids (indirect liquefaction) can be carried out in fixed- and fluidized-bed reactors and in dilute-phase systems. Another method for producing synthetic fuels from coal involves gasification of the coal to a fuel gas that may be used as such or as a source of synthetic liquids. *See* COAL GASIFICATION; FLUIDIZATION.

Coal liquefaction is accomplished by four principal methods: direct catalytic hydrogenation, solvent extraction, pyrolysis, and indirect catalytic hydrogenation (of carbon monoxide). *See* COAL LIQUEFACTION; HYDROGENATION; PYROLYSIS; SOLVENT EXTRACTION.

Shale oil is readily produced by the thermal processing of oil shales. The basic technology is available, and commercial plants are operated in many parts of the world. [J.G.S.]

# T

**Tantalum** A chemical element, symbol Ta, atomic number 73, and atomic weight 180.948. It is a member of the vanadium group of the periodic table and is in the 5*d* transitional series. Oxidation states of IV, III, and II are also known. *See* PERIODIC TABLE; TRANSITION ELEMENTS.

Tantalum metal is used in the manufacture of capacitors for electronic equipment, including citizen band radios, smoke detectors, heart pacemakers, and automobiles. It is also used for heat-transfer surfaces in chemical production equipment, especially where extraordinarily corrosive conditions exist. Its chemical inertness has led to dental and surgical applications. Tantalum forms alloys with a large number of metals. Of special importance is ferrotantalum, which is added to austentitic steels to reduce intergranular corrosion.

The metal is quite inert to acid attack except by hydrofluoric acid. It is very slowly oxidized in alkaline solutions. The halogens and oxygen react with it on heating to form the oxidation-state-V halides and oxide. At high temperature it absorbs hydrogen and combines with nitrogen, phosphorus, arsenic, antimony, silicon, carbon, and boron. Tantalum also forms compounds by direct reaction with sulfur, selenium, and tellurium at elevated temperatures. [E.M.L.]

**Tartaric acid** Any of the stereoisomeric forms of 2,3-dihydroxybutanedioic acid: L(+), D(−), and meso [(**1**), (**2**), and (**3**), respectively]. L(+)-Tartaric acid is present in

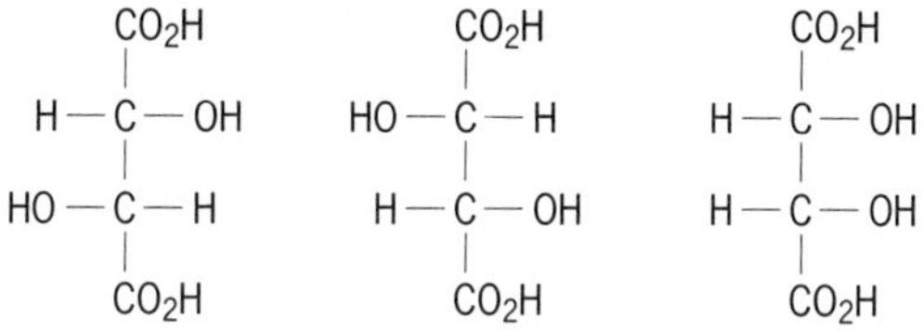

$$
\begin{array}{c}
CO_2H \\
| \\
H-C-OH \\
| \\
HO-C-H \\
| \\
CO_2H \\
\textbf{(1)}
\end{array}
\qquad
\begin{array}{c}
CO_2H \\
| \\
HO-C-H \\
| \\
H-C-OH \\
| \\
CO_2H \\
\textbf{(2)}
\end{array}
\qquad
\begin{array}{c}
CO_2H \\
| \\
H-C-OH \\
| \\
H-C-OH \\
| \\
CO_2H \\
\textbf{(3)}
\end{array}
$$

the juice of various fruits and is produced from grape juice as a by-product of the wine industry. The monopotassium salt precipitates in wine vats, and L(+)-tartaric acid is recovered from this residue. On heating in alkaline solution, the L(+) acid is converted to the racemic mixture of (**1**) and (**2**), plus a small amount of the meso acid (**3**).

Tartaric acid has played a central role in the discovery of several landmark stereochemical phenomena. In 1848, L. Pasteur isolated enantiomers (**1**) and (**2**) by mechanical separation of hemihedral crystals of the racemic mixture. He also used tartaric acid and its salts to demonstrate a distinction between the meso isomer (**3**) and the racemic mixture (**1**) + (**2**), and between enantiomers and diastereoisomers in general. The difference in properties between (**1**) [or (**2**)] and the meso form (**3**) was later a key in establishing the relative configuration of the pentose and hexose sugars.

Both L(+)- and D(−)-tartaric acid and the esters are inexpensive compounds and are used as chiral auxiliary reagents in the oxidation of alkenes to enantiomerically pure epoxides. This method employs a hydroperoxide oxidant, titanium alkoxide catalyst, and L(+)- or D(−)-tartrate, and involves chirality transfer from the tartrate to the product. *See* ASYMMETRIC SYNTHESIS; EPOXIDE.

Tartaric acid has some use as an acidulant in foods and also as a chelating agent. Potassium hydrogen tartrate (cream of tartar) is an ingredient of baking powder. The potassium sodium salt, commonly called Rochelle salt, was the first compound used as a piezoelectric crystal. *See* CHELATION. [J.A.Mo.]

**Tautomerism** The reversible interconversion of structural isomers of organic chemical compounds. Such interconversions usually involve transfer of a proton, but anionotropic rearrangements may be reversible and so be classed as tautomeric interconversions. A cyclic system containing the grouping —CONH— is called a lactam, and the isomeric form, —COH=N—, a lactim. These terms have been extended to include the same structures in open-chain compounds when considering the shift of the hydrogen from nitrogen to oxygen.

Molecular grouping (**1**) may in certain substances exist partly or wholly as (**2**). The former constitutes the keto form and the latter the enol form. The existence of an enol in an acyclic system requires that a second carbonyl group (or its equivalent, for example, (**3**) be attached to the same (**4**) as an aldehyde or ketone carbonyl. Thus,

$$\underset{(1)}{-\mathrm{COCH}<} \qquad \underset{(2)}{-\mathrm{COH{=}C}<} \qquad \underset{(3)}{>\mathrm{C{=}N}-} \qquad \underset{(4)}{-\mathrm{CN}<}$$

ethyl acetoacetate tautomerizes demonstrably, but ethyl malonate does not. Where the enol form includes an aromatic ring such as phenol, the existence of the keto form is often not demonstrable, although in some substances there may be either chemical or spectroscopic evidence for both forms. Closely related to keto-enol tautomerism is the prototropic interconversion of nitro and aci forms of aliphatic nitro compounds such as nitromethane.

In general, tautomeric forms exist in substances possessing functional groups which can interact additively and which are so placed that intramolecular reaction leads to a stable cyclic system. The cyclic form usually predominates (especially if it contains five or six members). *See* MOLECULAR ISOMERISM. [W.R.V.]

**Technetium** A chemical element, Tc, atomic number 43, discovered by Carlo Perrier and Emilio Segre in 1937. They separated and isolated it from molybdenum (Mo; atomic number 42), which had been bombarded with deuterons in a cyclotron. Technetium does not occur naturally. *See* ATOMIC NUMBER; ELEMENT (CHEMISTRY).

In the periodic table, technetium is located in the middle of the second-row transition series, situated between manganese and rhenium. Because of the lanthanide contraction, the chemistry of technetium is much more like that of rhenium, its third-row congener, than it is like that of manganese. Its location in the center of the periodic table gives technetium a rich and diverse chemistry. Oxidation states −1 to +7 are known, and complexes with a wide variety of coordination numbers and geometries have been reported. *See* LANTHANIDE CONTRACTION; PERIODIC TABLE; RHENIUM; TRANSITION ELEMENTS.

The most readily available chemical form of technetium is the ion pertechnetate ($TcO_4^-$), the starting point for all of its chemistry. In its higher oxidation states (+4 to +7), technetium is dominated by oxo chemistry, which is dominated by complexes containing one or two multiply bonded oxygen (oxo) groups.

Since about 1979, inorganic chemists have been very interested in understanding and developing the fundamental chemistry of technetium, utilizing the isotope $^{99g}Tc$. Most of this interest has arisen from the utility of another isotope of technetium, $^{99m}Tc$ (where m designates metastable), to diagnostic nuclear medicine. In fact, $^{99m}Tc$ in some chemical form is used in about 90% of all diagnostic scans performed in hospitals in the United States. The nuclear properties of $^{99m}Tc$ make it an ideal radionuclide for diagnostic imaging. This technetium isotope has a 6-h half-life, emits a 140-keV gamma ray which is ideal for detection by the gamma cameras used in hospitals, and emits no alpha or beta particles. [S.S.J.]

**Tellurium** A chemical element, Te, atomic number 52, and chemical atomic weight 127.60. There are eight stable isotopes of natural tellurium. Tellurium makes up approximately $10^{-9}$% of the Earth's igneous rock. It is found as the free element, sometimes associated with selenium. It is more often found as the telluride sylvanite (graphic tellurium), nagyagite (black tellurium), hessite, tetradymite, altaite, coloradoite, and other silver-gold tellurides, as well as the oxide, tellurium ocher. *See* PERIODIC TABLE.

There are two important allotropic modifications of elemental tellurium, the crystalline and the amorphous forms. The crystalline form has a silver-white color and metallic appearance. This form melts at 841.6°F (449.8°C) and boils at 2534°F (1390°C). It has a specific gravity of 6.25, and a hardness of 2.5 on Mohs scale. The amorphous form (brown) has a specific gravity of 6.015. Tellurium burns in air with a blue flame, forming tellurium dioxide, $TeO_2$. It reacts with halogens, but not sulfur or selenium, and forms, among other products, both the dinegative telluride anion ($Te^{2-}$), which resembles selenide, and the tetrapositive tellurium cation ($Te^{4+}$) which resembles platinum(IV).

Tellurium is used primarily as an additive to steel to increase its ductility, as a brightener in electroplating baths, as an additive to catalysts for the cracking of petroleum, as a coloring material for glasses, and as an additive to lead to increase its strength and corrosion resistance. [S.Ki.]

**Terbium** Element number 65, terbium, Tb, is a very rare metallic element of the rare-earth group. Its atomic weight is 158.924, and the stable isotope $^{159}Tb$ makes up 100% of the naturally occurring element. *See* PERIODIC TABLE.

The common oxide, $Tb_4O_7$, is brown and is obtained when its salts are ignited in air. Its salts are all trivalent and white in color and, when dissolved, give colorless solutions. The higher oxides slowly decompose when treated with dilute acid to give the trivalent ions in solution. Although the metal is attacked readily at high temperatures by air, the attack is extremely slow at room temperatures. The metal has a Néel point at about 229 K and a Curie point at about 220 K. For properties of the metal *see* RARE-EARTH ELEMENTS. [F.H.Sp.]

**Terpene** A class of natural products having a structural relationship to isoprene, as shown below. Over 5000 structurally determined terpenes are known; many of these

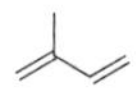

CH3 — C(=CH2 via H2C) — C(H)=CH2

Isoprene unit

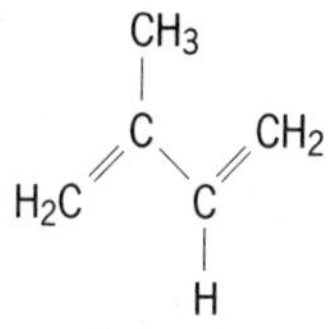

Isoprene

have also been synthesized in the laboratory. Historically terpenes have been isolated from green plants, but new compounds structurally related to isoprene continue to be isolated from other sources as well, so the class is also referred to as terpenoids, reflecting the biochemical origin without specification of the natural source. *See* ISOPRENE.

Terpenes are classified according to the number of isoprene units of which they are composed, as follows:

| | | | |
|---|---|---|---|
| 5 | hemi | 25 | ses- |
| 10 | mono- | 30 | tri- |
| 15 | sesqui- | 40 | tetra- |
| 20 | di- | $(5)_n$ | poly- |

Although they may be named according to the systematic nomenclature and numbering systems set by the International Union of Pure and Applied Chemistry for all organic compounds, it is often easier to refer to terpenes by their common names, which usually reflect the botanical or zoological name of their source. [T.Hu.]

**Thallium** A chemical element, Tl, atomic number 81, relative atomic weight of 204.37. The valence electron notation corresponding to its ground state term is $6s^2 6p^1$, which accounts for the maximum oxidation state of III in its compounds. Compounds of oxidation state I and apparent oxidation state II are also known. *See* PERIODIC TABLE.

Thallium occurs in the Earth's crust to the extent of 0.00006%, mainly as a minor constituent in iron, copper, sulfide, and selenide ores. Minerals of thallium are considered rare. Thallium compounds are extremely toxic to humans and other forms of life.

The insolubility of thallium(I) chloride, bromide, and iodide permits their preparation by direct precipitation from aqueous solution; the fluoride, on the other hand, is water-soluble. Thallium(I) chloride resembles silver chloride in its photosensitivity.

Thallium(I) oxide is a black powder which reacts with water to give a solution from which yellow thallium hydroxide can be crystallized. The hydroxide is a strong base and will take up carbon dioxide from the atmosphere.

Thallium also forms organometallic compounds of the following general classes, $R_3Tl$, $R_2TlX$, and $RTlX_2$, where R may be an alkyl or aryl group and X a halogen. *See* ORGANOMETALLIC COMPOUND. [E.M.L.]

**Thermal analysis** A group of analytical techniques developed to continuously monitor physical or chemical changes of a sample which occur as the temperature of a sample is increased or decreased. Thermogravimetry, differential thermal analysis, and differential scanning calorimetry are the three principal thermoanalytical methods. *See* ANALYTICAL CHEMISTRY; THERMOCHEMISTRY.

The occurrence of physical or chemical changes upon heating a sample may be explained from either a kinetic or thermodynamic viewpoint. Kinetically the rate of a process may be increased by raising the temperature as shown by the Arrhenius equation (1), where $A$, $E_a$, and $R$ represent the preexponential factor, activation energy,

$$\text{Rate} = Ae^{-E_a/RT} \tag{1}$$

and the gas law constant, respectively. At some point the rate becomes significant and readily observable. Similarly an increase in temperature can change the Gibbs free energy [Eq. (2), where $\Delta G^\circ$ is the Gibbs free energy, $\Delta H^\circ$ is the reaction enthalpy, and

$$\Delta G^\circ = \Delta H^\circ - T\Delta S^\circ \tag{2}$$

$\Delta S^\circ$ is the entropy change for the process] to a more favorable (that is, more negative) value. In particular, $\Delta G^\circ$ will become more negative if $\Delta S^\circ$ is positive and the temperature is increased. In many cases a combination of these factors causes the observed physiochemical process. *See* CHEMICAL THERMODYNAMICS; KINETICS (CLASSICAL MECHANICS).

Thermogravimetry involves measuring the changes in mass of a substance, typically a solid, as it is heated. Specially designed thermobalances are required to continuously monitor sample mass during the heating process. Modern balances have a capacity of 1–1500 milligrams and can accurately detect mass changes of 0.1 microgram.

Any type of physiochemical process which involves a change in sample mass may be observed by using thermogravimetry. Mass losses are observed for dehydration, decomposition, desorption, vaporization, sublimation, pyrolysis, and chemical reactions with gaseous products. Mass increases are noted with adsorption, absorption, and chemical reactions of the sample with the atmosphere in the oven, such as the oxidation of metals.

Quantitative gravimetric analyses may be performed due to the precise measure of the mass change obtained. Rates of mass change have been used to evaluate the kinetics of a process and to estimate activation energies. Fine details of these thermograms may also be used to deduce reaction intermediates and reaction mechanisms.

Primary applications of thermogravimetry are to deduce stabilities of compounds and mixtures of elevated temperatures and to determine appropriate drying temperatures for compounds and mixtures. Evaluation of polymers, food products, and pharmaceuticals is a major application of thermogravimetry.

Differential thermal analysis involves the monitoring of the temperature difference $T_D$ between a sample and inert reference material (such as aluminum oxide) as they are simultaneously heated, or cooled, at a predetermined rate. Multijunction thermocouples and thermistors are the most common temperature sensors used for this purpose; they are arranged in an oven. As enthalpic changes occur, $T_D$ will be positive if the process is exothermic and negative if it is endothermic.

More physical and chemical processes may be observed using differential thermal analysis as compared to thermogravimetry. Endothermic physical processes include crystalline transitions, fusion, vaporization, sublimation, desorption, and adsorption. Endothermic physical processes include crystalline transitions, fusion, vaporization, sublimation, desorption, and adsorption. Endothermic chemical processes include dehydration, decomposition, gaseous reduction, redox reactions, and solid-state metathesis. Exothermic processes include adsorption, chemisorption, decomposition, oxidation, redox reactions, and solid-state metathesis reactions. Both solids and liquids can be studied by differential thermal analysis. Hermetically sealed capsules are often used for liquids and some solids. Other samples are studied in open or crimped pans.

Analytical applications of this technique include the identification, characterization, and quantitation of a wide variety of materials, including polymers, pharmaceuticals, metals, clays, minerals, and inorganic and organic compounds. Characteristic thermograms can be used to determine purity, heats of reaction, thermal stability, phase diagrams, catalytic properties, and radiation damage.

In differential scanning calorimetry a sample and a reference are individually heated, by separately controlled resistance heaters, at a predetermined rate. Enthalpic (heat-generating or -absorbing) processes are detected as differences in electrical energy supplied to either the sample or the reference material to maintain this heating rate. This difference in electrical energy, in milliwatts per second, of the heat flow into or out of the sample is due to the occurrence of a physical or chemical process. Modulated differential scanning calorimetry is a new method that superimposes a sine wave on the heating ramp. A significant increase in sensitivity is often observed with modulated differential scanning calorimetry.

Analytical uses of differential scanning calorimetry are very similar to those of differential thermal analysis. Usually one calibration standard is sufficient to calibrate the entire operating range of the instrument. Differential scanning calorimetry instruments are highly sensitive and may measure heat flows as small as 1 nanowatt. Differential scanning calorimetry is very useful in determining heat capacities of substances over large temperature ranges. Such evaluation has become important in polymer and biochemical studies. Small (approximately 1–10 mg) samples are used in most cases, although some instruments have been developed which use up to 1 ml of a liquid sample. *See* CALORIMETRY. [N.D.J.]

**Thermal hysteresis** A phenomenon in which a physical quantity depends not only on the temperature but also on the preceding thermal history. It is usual to compare the behavior of the physical quantity while heating and the behavior while cooling through the same temperature range. The illustration shows the thermal hysteresis which has been observed in the behavior of the dielectric constant of single crystals of barium titanate. On heating, the dielectric constant was observed to follow the path *ABCD*, and on cooling the path *DCEFG*.

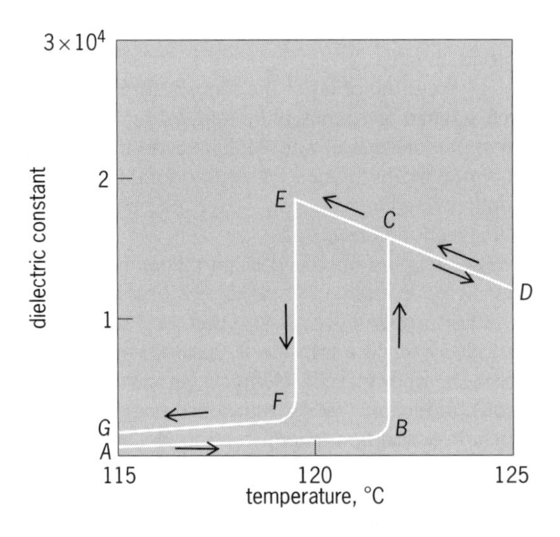


**Plot of dielectric constant versus temperature for a single crystal of barium titanate. (*After M. E. Drougard and D. R. Young, Phys. Rev., 95:1152–1153, 1954*)**

Perhaps the most common example of thermal hysteresis involves a phase change such as solidification from the liquid phase. In many cases these liquids can be dramatically supercooled. Elaborate precautions to eliminate impurities and outside disturbances can be instrumental in supercooling 60 to 80°C. On raising the temperature after freezing, however, the system follows a completely different path, with melting coming at the prescribed temperature for the phase change. *See* NUCLEATION; PHASE TRANSITIONS. [H.B.H.; R.K.Mac.C.]

**Thermochemistry** A branch of physical chemistry concerned with the absorption or evolution of heat that accompanies chemical reactions. Closely related topics are the latent heat associated with a change in phase (crystal, liquid, gas), the chemical composition of reacting systems at equilibrium, and the electrical potentials of galvanic cells. Thermodynamics provides the link among these phenomena.

A knowledge of such heat effects is important to the chemical engineer for the design and operation of chemical reactors, the determination of the heating values of fuels, the design and operation of refrigerators, the selection of heat storage systems, and the assessment of chemical hazards. Thermochemical information is used by the physiologist and biochemist to study the energetics of living organisms and to determine the calorific values of foods. Thermochemical data give the chemist an insight to the energies of, and interactions among, molecules. *See* CHEMICAL THERMODYNAMICS.

A calorimeter is an instrument for measuring the heat added to or removed from a process. There are many designs, but the following parts can generally be identified: the vessel in which the process is confined, the thermometer which measures its temperature, and the surrounding environment called the jacket. The heat associated¥break with the process is calculated by the equation below, where $T$ is the temperature. The

$$q = C[(T(\text{final}) - T(\text{initial})] - q_{\text{ex}} - w$$

quantity $C$, the energy equivalent of the calorimeter, is obtained from a separate calibration experiment. The work transferred to the process, $w$, is generally in the form of an electric current (as supplied to a heater, for example) or as mechanical work (as supplied to a stirrer, for example) and can be calculated from appropriate auxiliary measurements. The quantity $q_{ex}$ is the heat exchanged between the container and its jacket during the experiment. It is calculated from the temperature gradients in the system and the measured thermal conductivities of its parts.

Two principal types of calorimeters are used to measure heats of chemical reactions. In a batch calorimeter, known quantities of reactants are placed in the vessel and the initial temperature is measured. The reaction is allowed to occur and then the final equilibrium temperature is measured. If necessary, the final contents are analyzed to determine the amount of reaction which occurred.

In a flow calorimeter, the reactants are directed to the reaction vessel in two or more steady streams. The reaction takes place quickly and the products emerge in a steady stream. The rate of heat production is calculated from the temperatures, flow velocities, and heat capacities of the incoming and outgoing streams, and the rates of work production and heat transfer to the jacket. Dividing this result by the rate of reaction gives the heat of reaction. *See* CALORIMETRY.

In the past, thermochemical quantities usually have been given in units of calories. A calorie is defined as the amount of heat needed to raise the temperature of 1 gram of water 1°C. However, since this depends on the initial temperature of the water, various calories have been defined, for example, the 15° calorie, the 20° calorie, and the mean calorie (average from 0 to 100°C). In addition, a number of dry calories have been defined. Those still used are the thermochemical calorie (exactly 4.184 joules) and the International Steam Table calorie (exactly 4.1868 J). [R.C.W.]

**Thermodynamic principles** Laws governing the conversion of energy from one form to another. Among the many consequences of these laws are relationships between the properties of matter and the effects of changes in pressure, temperatures, electric field, magnetic field, and composition. The great practicality of the science arises from the foundations of the subject. Thermodynamics is based upon observations of common experience that have been formulated into the thermodynamic laws. From these few laws all of the remaining laws of the science are deducible by purely logical reasoning.

For exposition, it is necessary to define a few terms. A system is that part of the physical world under consideration. The rest of the world is the surroundings. An open system may exchange mass, heat, and work with the surroundings. A closed system may exchange heat and work but not mass with the surroundings. An isolated system has no exchange with the surroundings. Those parts of a system spatially uniform and homogeneous are called phases. For example, a liquid together with its vapor may be considered a two-phase system. It is a fact of experience that an isolated system approaches a particularly simple terminal state called an equilibrium state.

**Temperature.** It is within the scope of thermodynamics to refine the primitive notion of hotness and coldness into an operational and precise concept of temperature. The equilibrium states of a single-component, single-phase fluid provide a starting point. For such a fluid the equilibrium state is defined by fixing two of its properties. For example, one could construct a mercury-in-glass thermometer that has its pressure held constant; then only one other property could be varied independently. If the volume (height of mercury) is observed at any equilibrium state, there is a 1:1 correspondence between it and any other property excepting the pressure. The degree of hotness is one of these properties.

Also note that thermal equilibrium between different systems exists. For example, if the mercury thermometer is placed in contact with a body of quiet water, the mercury will either expand or contract. The volume change of the mercury will eventually stop, and the properties of the mercury will be constant, indicating an equilibrium state; moreover, the water will also have the constant properties of an equilibrium state. Bodies in thermal equilibrium are said to have the same temperature. Thus, one arrives at a method of measuring the temperature of a body of water.

Now suppose there is a large body of water in thermal contact through a wall with a large body of another fluid such as alcohol, both at equilibrium. It is an experimental fact that the mercury-in-glass thermometer will register the same volume when placed in either of the fluids. This fact of experience is designated as the zeroth law of thermodynamics: If two bodies A and B are separately in thermal equilibrium with C, then A and B are in equilibrium with each other. Thus, a useful empirical temperature measurement based upon the volume of mercury under constant pressure is established.

**Internal energy.** Thermodynamics does not define the concepts of energy or work but adopts them from the other macroscopic sciences of mechanics and electromagnetism. Also, the conservation of energy is taken as axiomatic. Therefore, if an isolated system is formed from any part of the world, a definite amount of energy will be trapped in the system. The energy resides in the kinetic and potential energy of the trapped molecules. This trapped energy is called the internal energy $U$.

Because of the conservation of energy, the internal energy of a closed system can be altered only by an exchange of energy with the surroundings. There are only three modes by which the exchange can occur: by mass transfer, heat transfer, or work exchange. So for a closed (no mass transfer), adiabatic (no heat transfer) system the change in internal energy $\Delta U$ is equal to the work done by the surroundings on the system, as defined by Eq. (1). Here a convention has been adopted that work done on

$$\Delta U = W_{AD} \tag{1}$$

the system is positive. There is a great mass of experimental information where work has been done on a closed system enclosed within adiabatic walls. Among these are experiments performed by J. Joule more than a century ago. It may be concluded from Joule's experiments that the expenditure of a given quantity of work always causes the same change of state regardless of how the work is carried out. Both $W_{AD}$ and $\Delta U$ are independent of the path. It is concluded that $U$ is a state function.

It is known from experience that the same change in state of a system can be effected by either supplying work to the system in an adiabatic enclosure or by contacting the system through a conducting wall with a higher temperature system. The latter method is a different means of transferring energy than work and is termed heat and given the symbol $Q$.

The first law of thermodynamics, Eq. (2), states that the algebraic sum of heat and

$$\Delta U = Q + W \tag{2}$$

work during a process is equal to the change in the state function $U$. The term $(Q + W)$ is therefore independent of the path taken between the two states. The differential form of the first law is given by Eq. (3), where $q$ and $w$ represent small quantities.

$$dU = q + w \tag{3}$$

**Reversible and irreversible processes.** Any process that occurs in nature is in agreement with the first law, but many processes permissible by the first law never occur. There is an overwhelming preference for processes to proceed in one direction. In one of Joule's experiments, a falling weight caused a paddle to do work on an adiabatically enclosed body of water. The total effect of the experiment was to increase the internal energy of the water and to lower the weight. The surroundings remained unchanged. There is no way one can reverse this process, that is, restore the water to its original state and raise the weight to its original height without also making some additional change in the surroundings. The process is irreversible.

Thermodynamics makes use of an idealization, called a reversible process, that is a limiting case of the natural or irreversible process. The reversible process may be defined as one which can be completely reversed without leaving more than a vanishingly small change in the surroundings. It is a consequence of the definition that a reversible process proceeds through a succession of equilibrium states and may be reversed by an infinitesimal change in the external conditions. Imagine having a cylinder of gas fitted with a frictionless piston. If the piston is moved so slowly that pressure gradients are absent, the gas will be in an equilibrium state at all times. The difference between the gas pressure and the external pressure needs only to be infinitesimal in order to move the frictionless piston. Under the rather restrictive conditions of a reversible process, Eq. (4) may be quite properly written, where $p$ is the gas pressure and $V$ is the gas volume.

$$dW = -pdV \tag{4}$$

**Entropy.** The discussion on irreversible processes has led to the second law of thermodynamics, which is just a general statement of the idea that there is a preferred direction for a given process. There are many physical statements of the second law, all being equivalent and leading to the same mathematical statement. The statement of R. Clausius is: "It is *not* possible that, at the end of a cycle of changes, heat has been transferred from a colder to a hotter body without producing some other effect." Lord Kelvin's statement is: "It is *not* possible that, at the end of a cycle of changes, heat has been extracted from a reservoir and an equal amount of work has been produced without producing some other effect."

The most efficient way of developing the mathematical consequences of the second law is to proceed from Carathéodory's principle, which can be either taken as another physical expression of the second law or derived from the Clausius or Kelvin statement. Carathéodory's principle is: "In the neighborhood of any equilibrium state of a system there are states which are not accessible by an adiabatic process."

Carathéodory used this principle together with a mathematical theorem that he developed to infer the existence of a state function $S$ and an integrating factor $1/T$, where $T$ is the thermodynamic temperature such that Eq. (5) holds for a reversible change. The state function $S$ is called the entropy. It can also be shown that the entropy

$$dQ_{REV} = TdS \tag{5}$$

in an adiabatic system increases for an irreversible change and remains constant for a reversible change as in Eq. (6).

$$\Delta S_{AD} \geq 0 \tag{6}$$

The first part of the mathematical statement of the second law allows one to write a very important thermodynamic equation (7). Although this equation was derived for

$$dU = TdS - pdV \tag{7}$$

reversible changes, it is valid for all changes. All the quantities are functions of state. Therefore, for a change between two states the integral of the equation will be valid even if the path is not reversible.

The second part of the mathematical statement is a concise summary of physical statements on the direction of processes. Also contained in the second part of the entropy statement is the key idea of equilibrium. The equilibrium state of an adiabatic or isolated system is characterized by entropy being at its maximum value consistent with the physical constraints. *See* ENTROPY. [W.F.J.]

**Thermodynamic processes** Changes of any property of an aggregation of matter and energy, accompanied by thermal effects.

**Systems and processes.** To evaluate the results of a process, it is necessary to know the participants that undergo the process, and their mass and energy. A region, or a system, is selected for study, and its contents determined. This region may have both mass and energy entering or leaving during a particular change of conditions, and these mass and energy transfers may result in changes both within the system and within the surroundings which envelop the system.

To establish the exact path of a process, the initial state of the system must be determined, specifying the values of variables such as temperature, pressure, volume, and quantity of material. The number of properties required to specify the state of a system depends upon the complexity of the system. Whenever a system changes from one state to another, a process occurs.

The path of a change of state is the locus of the whole series of states through which the system passes when going from an initial to a final state. For example, suppose a gas expands to twice its volume and that its initial and final temperatures are the same. An extremely large number of paths connect these initial and final states. The detailed path must be specified if the heat or work is to be a known quantity; however, changes in the thermodynamic properties depend only on the initial and final states and not upon the path. A quantity whose change is fixed by the end states and is independent of the path is a point function or a property.

**Pressure-volume-temperature diagram.** Whereas the state of a system is a point function, the change of state of a system, or a process, is a path function. Various processes or methods of change of a system from one state to another may be depicted graphically as a path on a plot using thermodynamic properties as coordinates.

The variable properties most frequently and conveniently measured are pressure, volume, and temperature. If any two of these are held fixed (independent variables), the third is determined (dependent variable). To depict the relationship among these physical properties of the particular working substance, these three variables may be used as the coordinates of a three-dimensional space. The resulting surface is a graphic presentation of the equation of state for this working substance, and all possible equilibrium states of the substance lie on this *P-V-T* surface.

Because a *P-V-T* surface represents all equilibrium conditions of the working substance, any line on the surface represents a possible reversible process, or a succession of equilibrium states.

The portion of the *P-V-T* surface shown in Fig. 1 typifies most real substances; it is characterized by contraction of the substance on freezing. Going from the liquid surface to the liquid-solid surface onto the solid surface involves a decrease in both temperature and volume. Water is one of the few exceptions to this condition; it expands upon freezing, and its resultant *P-V-T* surface is somewhat modified where the solid and liquid phases abut.

One can project the three-dimensional surface onto the *P-T* plane as in Fig. 2. The triple point is the point where the three phases are in equilibrium. When the temperature exceeds the critical temperature (at the critical point), only the gaseous phase is possible.

**Temperature-entropy diagram.** Energy quantities may be depicted as the product of two factors: an intensive property and an extensive one. Examples of intensive properties are pressure, temperature, and magnetic field; extensive ones are volume, magnetization, and mass. Thus, in differential form, work is the product of a pressure exerted against an area which sweeps through an infinitesimal volume, as in Eq. (1). As

$$dW = P\,dV \qquad (1)$$

a gas expands, it is doing work on its environment. However, a number of different kinds of work are known. For example, one could have work of polarization of a dielectric, of magnetization, of stretching a wire, or of making new surface area. In all cases, the

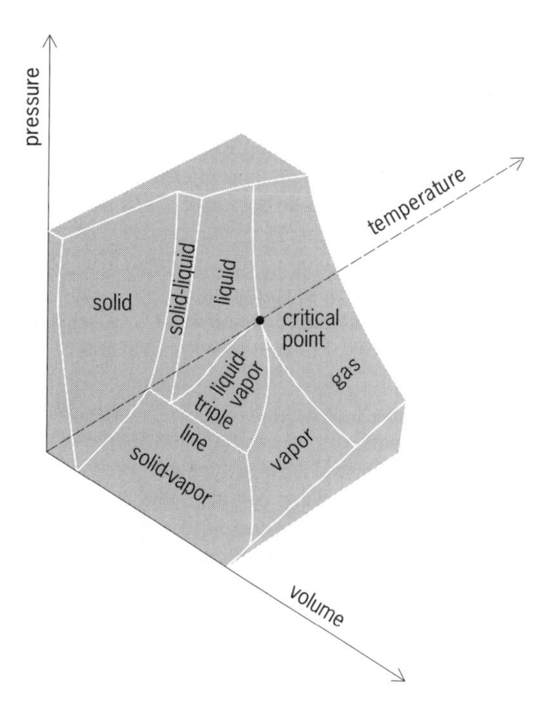


**Fig. 1. Portion of pressure-volume-temperature (*P-V-T*) surface for a typical substance.**

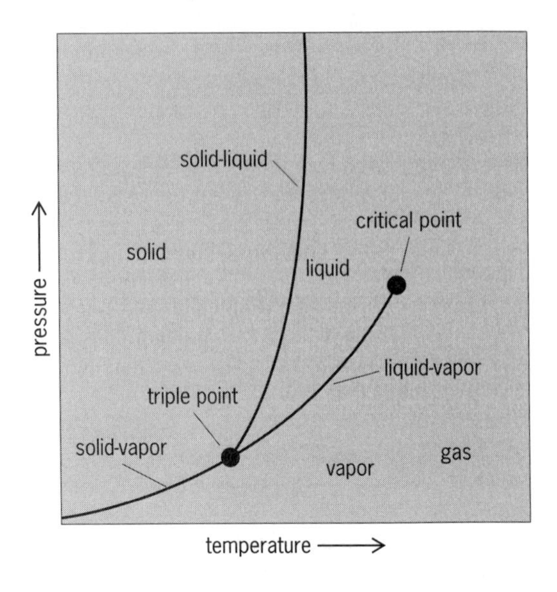


**Fig. 2. Portion of equilibrium surface projected on pressure-temperature (*P-T*) plane.**

infinitesimal work is given by Eq. (2), where X is a generalized applied force which is

$$dW = X\,dx \tag{2}$$

an intensive quantity, and $dx$ is a generalized displacement of the system and is thus extensive.

By extending this approach, one can depict transferred heat as the product of an intensive property, temperature, and a distributed or extensive property defined as entropy, for which the symbol is $S$. *See* ENTROPY.

**Reversible and irreversible processes.** Not all energy contained in or associated with a mass can be converted into useful work. Under ideal conditions only a fraction of the total energy present can be converted into work. The ideal conversions which retain the maximum available useful energy are reversible processes.

Characteristics of a reversible process are that the working substance is always in thermodynamic equilibrium and the process involves no dissipative effects such as viscosity, friction, inelasticity, electrical resistance, or magnetic hysteresis. Thus, reversible processes proceed quasi-statically so that the system passes through a series of states of thermodynamic equilibrium, both internally and with its surroundings. This series of states may be traversed just as well in one direction as in the other.

Actual changes of a system deviate from the idealized situation of a quasi-static process devoid of dissipative effects. The extent of the deviation from ideality is correspondingly the extent of the irreversibility of the process. *See* THERMODYNAMIC PRINCIPLES. [P.E.Bl.; W.A.S.]

**Thiocyanate** A compound containing the —SCN group, typically a salt or ester of thiocyanic acid (HSCN). Thiocyanates are bonded through the sulfur(s) and have the structure R—S—C≡N. They are isomeric with the isothiocyanates, R—N=C=S, which are the sulfur analogs of isocyanates (—NCO). The thiocyanates may be viewed as structural analogs of the cyanates (—OCN), where the oxygen (O) atom is replaced by a sulfur atom.

The principal commercial derivatives of thiocyanic acid are ammonium and sodium thiocyanates. Thiocyanates and isothiocyanates have been used as insecticides and herbicides. Specifically, ammonium thiocyanate is used as an intermediate in the manufacture of herbicides and as a stabilizing agent in photography. Sodium and potassium thiocyanates are used in the manufacture of textiles and the preparation of organic thiocyanates.

In living systems, thiocyanates are the product of the detoxification of cyanide ion (CN—) by the action of 3-mercaptopyruvate sulfur transferase. In addition, thiocyanates can interfere with thyroxine synthesis in the thyroid gland and are part of a class known as goitrogenic compounds. *See* CYANIDE; SULFUR; THYROXINE. [T.J.Me.]

**Thiophene** An organic heterocyclic compound containing a diunsaturated ring of four carbon atoms and one sulfur atom. *See* HETEROCYCLIC COMPOUNDS.

Thiophene (**1**), methylthiophenes, and other alkylthiophenes are found in relatively small amounts in coal tar and petroleum. Thiophene accompanies benzene in the fractional distillation of coal tar. 2,5-Dithienylthiophene (**2**) has been found in the marigold plant. Biotin, a water-soluble vitamin, is a tetrahydrothiophene derivative.

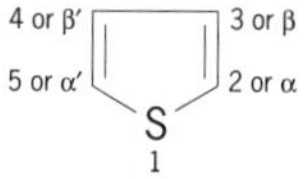

4 or β′ 3 or β 5 or α′ 2 or α S 1 (1)

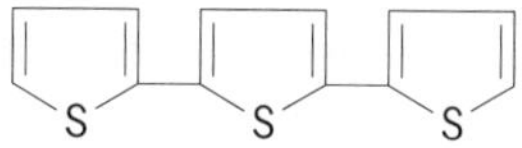

S S S (2)

The parent compound (**1**) is nearly insoluble in water, with mp −38.2°C (−36.8°F), bp 84.2°C (183.6°F), and specific gravity (20/4) 1.0644. Thiophene is considered to be an aromatic compound. Thiophenes are stable to alkali and other nucleophilic agents, and are relatively resistant to disruption by acid. *See* AROMATIC HYDROCARBON; ORGANOSULFUR COMPOUND. [W.J.Ge.; M.St.]

**Thiosulfate** A salt containing the negative ion $S_2O_3^{2-}$. This species is an important reducing agent and may be viewed as a structural analog of the sulfate ion ($SO_4^{2-}$) where one of the oxygen (O) atoms has been replaced by a sulfur (S) atom. The sulfur atoms of the thiosulfate ion are not equivalent. Thiosulfate is tetrahedral, and the central sulfur is in the formal oxidation state 6+ and the terminal sulfur is in the formal oxidation state 2−.

Principal uses of thiosulfates include agricultural, photographic, and analytical applications. Ammonium thiosulfate [$(NH_4)_2S_2O_3$] is exploited for both the nitrogen and sulfur content, and it is combined with other nitrogen fertilizers such as urea. Thiosulfate ion is an excellent complexing agent for silver ions (bound through sulfur). The sodium salt and the ammonium salt are well known as the fixing agent "hypo" used in photography. The aqueous thiosulfate ion functions as a scavenger for unreacted solid silver bromide on exposed film and therefore prevents further reaction with light. In nature, thiosulfate is converted into hydrogen sulfide ($H_2S$) via enzymatic reduction. Hydrogen sulfide, in turn, is converted into the thiol group of cysteine by the reaction with *O*-acetylserine. *See* COORDINATION COMPLEXES; OXIDATION-REDUCTION; SULFUR. [T.J.Me.]

**Thorium** A chemical element, Th, atomic number 90. Thorium is a member of the actinide series of elements. It is radioactive with a half-life of about $1.4 \times 10^{10}$ years. *See* PERIODIC TABLE.

Thorium oxide compounds are used in the production of incandescent gas mantles. Thorium oxide has also been incorporated in tungsten metal, which is used for electric light filaments. It is employed in catalysts for the promotion of certain organic chemical reactions and has special uses as a high-temperature ceramic material. The metal or its oxide is employed in some electronic tubes, photocells, and special welding electrodes. Thorium has important applications as an alloying agent in some structural metals. Perhaps the major use for thorium metal, outside the nuclear field, is in magnesium technology. Thorium can be converted in a nuclear reactor to uranium-233, an atomic fuel. The energy available from the world's supply of thorium has been estimated as greater than the energy available from all of the world's uranium, coal, and oil combined.

Monazite, the most common and commercially most important thorium-bearing mineral, is widely distributed in nature. Monazite is chiefly obtained as a sand, which is separated from other sands by physical or mechanical means.

Thorium has an atomic weight of 232. The temperature at which pure thorium melts is not known with certainty; it is thought to be about 1750°C (3182°F). Good-quality thorium metal is relatively soft and ductile. It can be shaped readily by any of the ordinary metal-forming operations. The massive metal is silvery in color, but it tarnishes on long exposure to the atmosphere; finely divided thorium has a tendency to be pyrophoric in air.

All of the nonmetallic elements, except the rare gases, form binary compounds with thorium. With minor exceptions, thorium exhibits a valence of 4+ in all of its salts. Chemically, it has some resemblance to zirconium and hafnium. The most common soluble compound of thorium is the nitrate which, as generally prepared, appears to have the formula $Th(NO_3)_4 \cdot 4H_2O$. The common oxide of thorium is $ThO_2$, thoria. Thorium combines with halogens to form a variety of salts. Thorium sulfate can be obtained in the anhydrous form or as a number of hydrates. Thorium carbonates, phosphates, iodates, chlorates, chromates, molybdates, and other inorganic salts of thorium are well known. Thorium also forms salts with many organic acids, of which the water-insoluble oxalate, $Th(C_2O_4)_2 \cdot 6H_2O$, is important in preparing pure compounds of thorium. *See* ACTINIDE ELEMENTS. [H.A.W.]

**Thulium** A chemical element, Tm, atomic number 69, atomic weight 168.934. It is a rare metallic element belonging to the rare-earth group. The stable isotope $^{169}Tm$ makes up 100% of the naturally occurring element. *See* PERIODIC TABLE.

The salts of thulium possess a pale green color and the solutions have a slight greenish tint. The metal has a high vapor pressure at the melting point. When $^{169}Tm$ is irradiated in a nuclear reactor, $^{170}Tm$ is formed. The isotope then emits strongly an 84-keV x-ray, and this material is useful in making small portable x-ray units for medical use. *See* RARE-EARTH ELEMENTS. [F.H.Sp.]

**Time-of-flight spectrometer** Any of a general class of instruments in which the speed of a particle is determined directly by measuring the time that it takes to travel a measured distance. By knowing the particle's mass, its energy can be calculated. If the particles are uncharged (for example, neutrons), difficulties arise because standard methods of measurement (such as deflection in electric and magnetic fields) are not possible. The time-of-flight method is a powerful alternative, suitable for both uncharged and charged particles.

The time intervals are best measured by counting the number of oscillations of a stable oscillator that occur between the instants that the particle begins and ends its journey. Oscillators operating at 100 MHz are in common use. *See* MASS SPECTROSCOPE. [F.W.K.F.]

**Tin** A chemical element, symbol Sn, atomic number 50, atomic weight 118.69. Tin forms tin(II) or stannous ($Sn^{2+}$), and tin(IV) or stannic ($Sn^{4+}$) compounds, as well as complex salts of the stannite ($M_2SnX_4$) and stannate ($M_2SnX_6$) types. *See* PERIODIC TABLE.

**Properties of tin**

| Property | Value |
|---|---|
| Melting point, °C | 231.9 |
| Boiling point, °C | 2270 |
| Specific gravity, $\alpha$ form (gray tin) | 5.77 |
| $\beta$ form (white tin) | 7.29 |
| Specific heat, cal/g*, white tin at 25°C | 0.053 |
| Gray tin at 10°C | 0.049 |

*1 cal = 4.184 joules.

Tin melts at a low temperature, is highly fluid when molten, and has a high boiling point. It is soft and pliable and is corrosion-resistant to many media. An important use of tin has been for tin-coated steel containers (tin cans) used for preserving foods and beverages. Other important uses are solder alloys, bearing metals, bronzes, pewter, and miscellaneous industrial alloys. Tin chemicals, both inorganic and organic, find extensive use in the electroplating, ceramic, plastic, and agricultural industries.

The most important tin-bearing mineral is cassiterite, $SnO_2$. No high-grade deposits of this mineral are known. The bulk of the world's tin ore is obtained from low-grade alluvial deposits.

Two allotropic forms of tin exist: white ($\beta$) and gray ($\alpha$) tin. Tin reacts with both strong acids and strong bases, but it is relatively resistant to solutions that are nearly neutral. In a wide variety of corrosive conditions, hydrogen gas is not evolved from tin and the rate of corrosion becomes controlled by the supply of oxygen or other oxidizing agents. In their absence, corrosion is negligible. A thin film of stannic oxide forms on tin upon exposure to air and provides surface protection. Salts that have an acid reaction in solution, such as aluminum chloride and ferric chloride, attack tin in the presence of oxidizers or air. Most nonaqueous liquids, such as oils, alcohols, or chlorinated hydrocarbons, have slight or no obvious effect on tin. Tin metal and the simple inorganic salts of tin are nontoxic. Some forms of organotin compounds, on the other hand, are toxic. Some important physical constants for tin are shown in the table.

Stannous oxide, SnO, is a blue-black, crystalline product which is soluble in common acids and strong alkalies. It is used in making stannous salts for plating and glass manufacture. Stannic oxide, $SnO_2$, is a white powder, insoluble in acids and alkalies. It is an excellent glaze opacifier, a component of pink, yellow, and maroon ceramic stains and of dielectric and refractory bodies. It is an important polishing agent for marble and decorative stones.

Stannous chloride, $SnCl_2$, is the major ingredient in the acid electrotinning electrolyte and is an intermediate for tin chemicals. Stannic chloride, $SnCl_4$, in the pentahydrate form is a white solid. It is used in the preparation of organotin compounds and chemicals to weight silk and to stabilize perfume and colors in soap. Stannous fluoride, $SnF_2$, a white water-soluble compound, is a toothpaste additive.

Organotin compounds are those compounds in which at least one tin-carbon bond exists, the tin usually being present in the + IV oxidation state. Organotin compounds that find applications in industry are the compounds with the general formula $R_4Sn$, $R_3SnX$, $R_2SnX_2$, and $RSnX_3$. R is an organic group, often methyl, butyl, octyl, or phenyl, while X is an inorganic substituent, commonly chloride, fluoride, oxide, hydroxide, carboxylate, or thiolate.

[J.B.Lo.]

**Titanium** A chemical element, Ti, atomic number 22, and atomic weight 47.90. It occurs in the fourth group of the periodic table, and its chemistry shows many similarities to that of silicon and zirconium. On the other hand, as a first-row transition element, titanium has an aqueous solution chemistry, especially of the lower oxidation states, showing some resemblances to that of vanadium and chromium. *See* PERIODIC TABLE; TRANSITION ELEMENTS.

The catalytic activity of titanium complexes forms the basis of the well-known Ziegler process for the polymerization of ethylene. This type of polymerization is of great industrial interest since, with its use, high-molecular-weight polymers can be formed. In some cases, desirable special properties can be obtained by forming isotactic polymers, or polymers in which there is a uniform stereochemical relationship along the chain. *See* POLYOLEFIN RESINS.

The dioxide of titanium, $TiO_2$, occurs most commonly in a black or brown tetragonal form known as rutile. Less prominent naturally occurring forms are anatase and brookite (rhombohedral). Both rutile and anatase are white when pure. The dioxide may be fused with other metal oxides to yield titanates, for example, $K_2TiO_3$, $ZnTiO_3$, $PbTiO_3$, and $BaTiO_3$. The black basic oxide, $FeTiO_3$, occurs naturally as the mineral ilmenite; this is a principal commercial source of titanium.

Titanium dioxide is widely used as a white pigment for exterior paints because of its chemical inertness, superior covering power, opacity to damaging ultraviolet light, and self-cleaning ability. The dioxide has also been used as a whitening or opacifying agent in numerous situations, for example as a filler in paper, a coloring agent for rubber and leather products, a pigment in ink, and a component of ceramics. It has found important use as an opacifying agent in porcelain enamels, giving a finish coat of great brilliance, hardness, and acid resistance. Rutile has also been found as brilliant, diamondlike crystals, and some artificial production of it in this form has been achieved. Because of its high dielectric constant, it has found some use in dielectrics.

The alkaline-earth titanates show some remarkable properties. The dielectric constants range from 13 for $MgTiO_3$ to several thousand for solid solutions of $SrTiO_3$ in $BaTiO_3$. Barium titanate itself has a dielectric constant of 10,000 near 120°C (248°F), its Curie point; it has a low dielectric hysteresis. These properties are associated with a stable polarized state of the material analogous to the magnetic condition of a permanent magnet, and such substances are known as ferroelectrics. In addition to the ability to retain a charged condition, barium titanate is piezoelectric and may be used as a transducer for the interconversion of sound and electrical energy. Ceramic transducers containing barium titanate compare favorably with Rochelle salt and quartz, with respect to thermal stability in the first case, and with respect to the strength of the effect and the ability to form the ceramic in various shapes, in the second case. The compound has been used both as a generator for ultrasonic vibrations and as a sound detector. [A.W.A.]

In addition to important uses in applications such as structural materials, pigments, and industrial catalysis, titanium has a rich coordination chemistry. The formal oxidation of titanium in molecules and ions ranges from −II to +IV. The lower oxidation states of −II and −I occur only in a few complexes containing strongly electron-withdrawing carbon monoxide ligands.

The lower oxidation states of titanium are all strongly reducing. Thus, unless specific precautions are taken, titanium complexes are typically oxidized rapidly to the +IV state. Moreover, many titanium complexes are extremely susceptible to hydrolysis. Consequently, the handling of titanium complexes normally requires oxygen- and water-free conditions. *See* COORDINATION CHEMISTRY. [L.K.W.]

**Titration** A quantitative analytical process that is basically volumetric. However, in high-precision titrimetry the titrant solution is sometimes delivered from a weight buret, so that the volumetric aspect is indirect. Generally, a standard solution, that is, one containing a known concentration of substance X (titrant), is progressively added to a measured volume of a solution of a substance Y (titrand) that will react with the titrant. The addition is continued until the end point is reached. Ideally, this is the same as the equivalence point, at which an excess of neither X nor Y remains. If the stoichiometry or exact ratio in which X and Y react is known, it is possible to calculate the amount of Y in the unknown solution.

The normal requirements for the performance of a titration are: a standard titrant solution; calibrated volumetric apparatus, including burets, pipets, and volumetric flasks; and some means of detecting the end point.

**Classification by chemical reaction.** For the purposes of titrimetry, chemical reactions can be placed in three general categories: acid-base or neutralization, combination, and oxidation-reduction.

Acid-base titrations involve neutralization of an acid by titration with a base, or vice versa. However, the process is often nonspecific; in the titration of a mixture of nitric and hydrochloric acids, only the total acidity can be found without recourse to additional measurements. A salt derived from a strong base and a very weak acid can often be titrated just as if it were a base. *See* ACID AND BASE.

In titrimetry, attention is usually focused upon the combination of an ion in the titrant with one of the opposite sign in the titrand solution. Sometimes the combination may involve more than two species, some of which may be nonionic. The combinations may result in precipitation or formation of a complex. *See* COORDINATION COMPLEXES; PRECIPITATION (CHEMISTRY).

In so-called redox titrations the titrant is usually an oxidizing agent, and is used to determine a substance that can be oxidized and hence can act as a reducing agent. *See* OXIDATION-REDUCTION.

**Coulometric titration.** The passage of a uniform current for a measured period of time can be used to generate a known amount of a product such as a titrant. This fact is the basis of the technique known as coulometric titration. An obvious requirement is that generation shall proceed with a fixed, preferably 100%, current efficiency. The uniform current is then analogous to the concentration of an ordinary titrant solution, while the total time of passage is analogous to the volume of such a solution that would be needed to reach the end point. *See* ELECTROLYSIS.

**Classification by end-point techniques.** The precision and accuracy with which the end point can be detected is a vital factor in all titrations. Because of its simplicity and versatility, chemical indication is quite common, especially in acid-base titrimetry.

*Indicators.* An acid-base indicator is a weak acid or a weak base that changes color when it is transformed from the molecular to the ionized form, or vice versa. The color change is normally intense, so that only a low concentration of indicator is needed. The working range, or visual color change, of a typical acid-base indicator is spread over about a hundredfold (~2 pH units) change in hydrogen ion concentration. Available indicators have individual working ranges that together cover the entire range of hydrogen ion concentration likely to be encountered in general acid-base titration. *See* ACID-BASE INDICATOR; HYDROGEN ION; PH.

Sometimes no suitable chemical indicator can be found for a desired titration. Possibly the concentrations involved may be so low that chemical indication functions poorly. Other situations might be the need for high precision or for the automatic arrest of the titration. Recourse is then made to some physical method of end-point detection.

*Potentiometric titration.* If a pH meter is used, its associated electrodes are first standardized by use of a buffer solution of known pH. By suitable choice of electrodes, potentiometric methods can also be applied to combination titrations and to oxidation-reduction titrations. The advent of modern ion-selective electrodes has greatly extended the scope of potentiometric titration and of other branches of titrimetry. *See* ELECTRODE POTENTIAL; ION-SELECTIVE MEMBRANES AND ELECTRODES.

*Conductometric titration.* Conductometric titration is sometimes successful when chemical indication fails. The underlying principles of conductometric titration are that the solvent and any molecular species in solution exhibit only negligible conductance; that the conductance of a dilute solution rises as the concentration of ions is increased; and that at a given concentration the hydrogen ion and the hydroxyl ion are much better conductors than any of the other ions. *See* ELECTROLYTIC CONDUCTANCE.

*Spectrophotometric titration.* The spectrophotometer is an optical device that responds only to radiation within a selected very narrow band of wavelengths in the visual, ultraviolet, or infrared regions of the spectrum. The response can be made both quantitative and linearly related to the concentration of a species that absorbs radiation within this band. Titrations at wavelengths within the visual region are by far the most common. *See* SPECTROPHOTOMETRIC ANALYSIS.

*Amperometric titration.* By use of a dropping-mercury or other suitable microelectrode, it is possible to find a region of applied electromotive force (emf) in which the current is proportional to the concentration of one or both of the reactants in a titration.

Biamperometric titration is a closely related technique. An emf that is usually small is applied across two identical microelectrodes that dip into the titrand solution. This arrangement, which involves no liquid-liquid junctions, is valuable in nonaqueous titrations, but also finds much use in aqueous titrimetry. *See* POLAROGRAPHIC ANALYSIS.

*Thermometric or enthalpimetric titration.* Many chemical reactions proceed with the evolution of heat. If one of these is used as the basis of a titration, the temperature first rises progressively and then remains unchanged as the titration is continued past the end point. If the reaction is endothermic, the temperature falls instead of rising. Thermometric titration is applicable to all classes of reactions. *See* THERMOCHEMISTRY.

*Nonaqueous titration.* This technique is used to perform titrations that give poor or no end points in water. Although applicable in principle to all classes of reactions, acid-base applications have greatly exceeded all others. Nonaqueous titrations in which the solvent is a molten salt or salt mixture are also possible.

*Automatic titration.* Automation is particularly valuable in routine titrations, which are usually performed repeatedly. One approach is to record the titration curve and to interpret it later. Another method is to stop titrant addition or generation automatically at, or very near to, the end point. Although a constant-delivery device is desirable, an ordinary buret with an electromagnetically controlled valve is often used.

Microcomputer control permits such refinements as the continuous adjustment of the titrant flow rate during the titration. In some cases, it is possible to automate an entire analysis, from the measurement of the sample to the final washout of the titration vessel and the printout of the result of the analysis. *See* ANALYTICAL CHEMISTRY. [J.T.St.]

**Transamination** The transfer of an amino group from one molecule to another without the intermediate formation of ammonia. Enzymatic reactions of this type play a prominent role in the formation and ultimate breakdown of amino acids by living organisms. Enzymes that catalyze such reactions are widely distributed and are termed transaminases, or amino-transferases. Perhaps the most prominent transamination reactions in higher animals are those in which glutamate is formed from $\alpha$-ketoglutarate and other amino acids. [E.E.S.]

**Transition elements** In broad definition, the elements of atomic numbers 21–31, 39–49, and 71–81, inclusive. A more restricted classification of the transition elements, preferred by many chemists, is limited to elements with atomic numbers 22–28, 40–46, and 72–78, inclusive. All of the elements in this classification have one or more electrons present in an unfilled *d* subshell in at least one well-known oxidation state.

All the transition elements are metals and, in general, are characterized by high densities, high melting points, and low vapor pressures. Within a given subgroup, these properties tend to increase with increasing atomic weight. Facility in the formation of metallic bonds is demonstrated also by the existence of a wide variety of alloys between different transition metals.

The transition elements include most metals of major economic importance, such as the relatively abundant iron, nickel, and zinc, on one hand, and the rarer coinage metals, copper, silver, and gold, on the other. Also included are the rare and relatively unfamiliar element, rhenium, and technetium, which is not found naturally in the terrestrial environment, but is available in small amounts as a product of nuclear fission.

In their compounds, the transition elements tend to exhibit multiple valency, the maximum valence increasing from +3 at the beginning of a series (Sc, Y, Lu) to +8 at the fifth member (Mn, Re). One of the most characteristic features of the transition elements is the ease with which most of them form stable complex ions. Features which contribute to this ability are favorably high charge-to-radius ratios and the availability of unfilled *d* orbitals which may be used in bonding. Most of the ions and compounds of the transition metals are colored, and many of them are paramagnetic. Both color and paramagnetism are related to the presence of unpaired electrons in the *d* subshell. Because of their ability to accept electrons in unoccupied *d* orbitals, transition elements and their compounds frequently exhibit catalytic properties.

Broadly speaking, the properties of the transition elements are intermediate between those of the so-called representative elements, in which the subshells are completely occupied by electrons (alkali metals, halogen elements), and those of the inner or *f* transition elements, in which the subshell orbitals play a much less significant role in influencing chemical properties (rare-earth elements, actinide elements). *See* ACTINIDE ELEMENTS; ATOMIC STRUCTURE AND SPECTRA; RARE-EARTH ELEMENTS. [B.B.Cu.]

**Transition point** The point at which a substance changes from one state of aggregation to another. This general definition would include the melting point (transition from solid to liquid), boiling point (liquid to gas), or sublimation point (solid to gas); but in practice the term transition point is usually restricted to the transition from one solid phase to another, that is, to the temperature (for a fixed pressure, usually 1 atm or $10^5$ pascals) at which a substance changes from one crystal structure to another.

Another kind of transition point is the culmination of a gradual change (for example, the loss of ferromagnetism in iron or nickel) at the lambda point, or Curie point. This behavior is typical of second-order transitions. *See* BOILING POINT; MELTING POINT; PHASE EQUILIBRIUM; SUBLIMATION; TRIPLE POINT. [R.L.S.]

**Transmutation** The nuclear change of one element into another, either naturally, in radioactive elements, or artificially, by bombarding an element with electrons, deuterons, or $\alpha$-particles in particle accelerators or with neutrons in atomic piles.

Natural transmutation was first explained by Marie Curie about 1900 as the result of the decay of radioactive elements into others of lower atomic weight. Ernest Rutherford produced the first artificial transmutation (nitrogen into oxygen and hydrogen) in 1919. Artificial transmutation is the method of origin of the heavier, artificial transuranium elements, and also of hundreds of radioactive isotopes of most of the chemical elements in the periodic table. *See* PERIODIC TABLE. [F.H.R.]

**Transport processes** The processes whereby mass, energy, or momentum are transported from one region of a material to another under the influence of composition, temperature, or velocity gradients. If a sample of a material in which the chemical composition, the temperature, or the velocity vary from point to point is isolated from its surroundings, the transport processes act so as to eventually render these quantities uniform throughout the material. The nonuniform state required to generate these transport processes causes them to be known also as nonequilibrium processes. Associated with gradients of composition, temperature, and velocity in a material are the transport processes of diffusion, thermal conduction, and viscosity, respectively. *See* DIFFUSION IN GASES AND LIQUIDS. [W.A.W.]

**Transuranium elements** Those synthetic elements with atomic numbers larger than that of uranium (atomic number 92). They are the members of the actinide series, from neptunium (atomic number 93) through lawrencium (atomic number 103), and the transactinide elements (with higher atomic numbers than 103). Of these elements, plutonium, an explosive ingredient for nuclear weapons and a fuel for nuclear power because it is fissionable, has been prepared on the largest (ton) scale, while some of the others have been produced in kilograms (neptunium, americium, curium) and in much smaller quantities (berkelium, californium, and einsteinium).

The concept of atomic weight in the sense applied to naturally occurring elements is not applicable to the transuranium elements, since the isotopic composition of any given sample depends on its source. In most cases the use of the mass number of the longest-lived isotope in combination with an evaluation of its availability has been adequate. Good choices at present are neptunium, 237; plutonium, 242; americium, 243; curium, 248; berkelium, 249; californium, 249; einsteinium, 254; fermium, 257; mendelevium, 258; nobelium, 259; lawrencium, 260; rutherfordium, 261; dubnium, 262; and seaborgium, 263. The actinide elements are chemically similar and have a strong chemical resemblance to the lanthanide, or rare-earth, elements (atomic numbers 57–71). The transactinide elements, with atomic numbers 104–118, appear in an expanded periodic table under the row of elements beginning with hafnium, number 72, and ending with radon, number 86. This arrangement allows prediction of the chemical properties of these elements and suggests that they will have a chemical analogy with the elements which appear immediately above them in the periodic table.

The transuranium elements up to and including fermium (atomic number 100) are produced in largest quantity through the successive capture of neutrons in nuclear reactors. The yield decreases with increasing atomic number, and the heaviest to be produced in weighable quantity is einsteinium (number 99). Many additional isotopes are produced by bombardment of heavy target isotopes with charged atomic projectiles in accelerators; beyond fermium, all elements are produced by bombardment with heavy ions.

Beyond darmstadtium (atomic number 110), transactinide elements 111–116 have been produced, although their acceptance is pending. *See* ACTINIDE ELEMENTS; AMERICIUM; BERKELIUM; BOHRIUM; CALIFORNIUM; CURIUM; DARMSTADTIUM; DUBNIUM; EINSTEINIUM; ELEMENT 111; ELEMENT 112; FERMIUM; HASSIUM; LAWRENCIUM; MEITNERIUM; MENDELEVIUM; NEPTUNIUM; NOBELIUM; NUCLEAR CHEMISTRY; PERIODIC TABLE; PLUTONIUM; RUTHERFORDIUM; SEABORGIUM. [P.J.R.; G.T.S.]

**Triglyceride** A simple lipid. Triglycerides are fatty acid triesters of the trihydroxy alcohol glycerol which are present in plant and animal tissues, particularly in the food storage depots, either as simple esters in which all the fatty acids are the same or as mixed esters in which the fatty acids are different. The triglycerides constitute the main component of natural fats and oils.

The generic formula of a triglyceride is shown below, where $RCO_2H$, $R'CO_2H$, and

$$\begin{array}{l} CH_2-OOC-R \\ | \\ CH-OOC-R' \\ | \\ CH_2-OOC-R'' \end{array}$$

$R''CO_2H$ represent molecules of either the same or different fatty acids, such as butyric or caproic (short chain), palmitic or stearic (long chain), oleic, linoleic, or linolenic (unsaturated). Saponification with alkali releases glycerol and the alkali metal salts of the fatty acids (soaps). The triglycerides in the food storage depots represent a concentrated energy source, since oxidation provides more energy than an equivalent weight of protein or carbohydrate. *See* SOAP.

The physical and chemical properties of fats and oils depend on the nature of the fatty acids present. Saturated fatty acids give higher-melting fats and represent the main constituents of solid fats, for example, lard and butter. Unsaturation lowers the melting point of fatty acids and fats. Thus, in the oil of plants, unsaturated fatty acids are present in large amounts, for example, oleic acid in olive oil and linoleic and linolenic acids in linseed soil. [R.H.G.; H.E.Ca.]

**Triple point** A particular temperature and pressure at which three different phases of one substance can coexist in equilibrium. In common usage these three phases are normally solid, liquid, and gas, although triple points can also occur with two solid phases and one liquid phase, with two solid phases and one gas phase, or with three solid phases.

According to the Gibbs phase rule, a three-phase situation in a one-component system has no degrees of freedom (that is, it is invariant). Consequently, a triple point occurs at a unique temperature and pressure, because any change in either variable

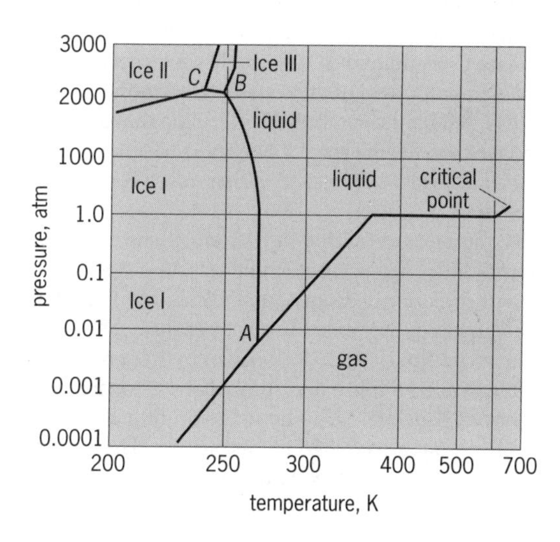


**Phase diagram for water, showing gas, liquid, and several solid (ice) phases; triple points at *A*, *B*, and *C*. The pressure scale changes at 1 atm from logarithmic scale at low pressure to linear at high pressure. 1 atm = $10^2$ kilopascals.**

will result in the disappearance of at least one of the three phases. Triple points are shown in the illustration of part of the phase diagram for water. *See* PHASE EQUILIBRIUM.

For most substances the solid-liquid-vapor triple point has a pressure less than 1 atm ($10^2$ kilopascals); such substances then have a liquid-vapor transition at 1 atm (normal boiling point). However, if this triple point has a pressure above 1 atm, the substance passes directly from solid to vapor at 1 atm. *See* BOILING POINT; MELTING POINT; SUBLIMATION; TRANSITION POINT; VAPOR PRESSURE; WATER. [R.L.S.]

**Triplet state** An electronic state of a molecule that occurs when its total spin angular momentum quantum number $S$ is equal to one. The triplet state is an important intermediate of organic chemistry. In addition to the wide range of triplet molecules available through photochemical excitation techniques, numerous molecules exist in stable triplet ground states, for example, oxygen molecules. *See* ATOMIC STRUCTURE AND SPECTRA; MOLECULAR STRUCTURE AND SPECTRA.

A triplet may result whenever a molecule possesses two electrons which are both orbitally unpaired and spin unpaired. Orbital unpairing of electrons results when a molecule absorbs a photon of visible or ultraviolet light. Direct formation of a triplet as a result of this photon absorption is a very improbable process since both the orbit and spin of the electron would have to change simultaneously. Thus, a singlet state is generally formed by absorption of light. However, quite often the lifetime of this singlet state is sufficiently long to allow the spin of one of the two electrons to invert, thereby producing a triplet. *See* MOLECULAR ORBITAL THEORY. [N.J.T.]

**Triterpene** A hydrocarbon or its oxygenated analog containing 30 carbon atoms and composed of six isoprene units. Triterpenes form the largest group of terpenoids, but are classified into only a few major categories. Resins and saps contain triterpenes in the free state as well as in the form of esters and glycosides.

Biogenetically triterpenes arise by the cyclization of squalene and subsequent skeletal rearrangements. Apart from the linear squalene itself and some bicyclic, highly substituted skeletons, most triterpenes are either tetracyclic or pentacyclic compounds. The various structural classes are designated by the names of representative members. *See* TERPENE. [T.Hu.]

**Tritium** The heaviest isotope of the element hydrogen and the only one which is radioactive. Tritium occurs in very small amounts in nature but is generally prepared artificially by processes known as nuclear transmutations. It is widely used as a tracer in chemical and biological research and is a component of the so-called thermonuclear or hydrogen bomb. It is commonly represented by the symbol $^3_1H$ indicating that it has an atomic number of 1 and an atomic mass of 3, or by the special symbol T. For information about the other hydrogen isotopes *see* DEUTERIUM; HYDROGEN. *See also* TRANSMUTATION.

Both molecular tritium, $T_2$, and its counterpart hydrogen, $H_2$, are gases under ordinary conditions. Because of the great difference in mass, many of the properties of tritium differ substantially from those of ordinary hydrogen, as indicated in the table. Chemically, tritium behaves quite similarly to hydrogen. However, because of its larger mass, many of its reactions take place more slowly than do those of hydrogen.

The nucleus of the tritium atom, often called a triton and symbolized $t$, consists of a proton and two neutrons. It undergoes radioactive decay by emission of a $\beta$-particle to leave a helium nucleus of mass 3. No $\gamma$-rays are emitted in this process. The half-life for the decay is 12.26 years. When tritium is bombarded with deuterons of sufficient energy, a nuclear reaction known as fusion occurs and energy considerably greater

**Properties of hydrogen and tritium**

| Property | $H_2$ | $T_2$ |
|---|---|---|
| Melting point | −259.20°C (−434.56°F) | −252.54°C (−422.57°F) |
| Boiling point at 1 atm ($10^5$ pascals) | −252.77°C (−423.00°F) | −248.12°C (−414.62°F) |
| Heat of vaporization | 216 cal/mol (904 J/mol) | 333 cal/mol (1390 J/mol) |
| Heat of sublimation | 247 cal/mol (1030 J/mol) | 393 cal/mol (1640 J/mol) |

than that of the bombarding particle is released. This reaction is one of those which supply the energy of the thermonuclear bomb. It is also of major importance in the development of controlled thermonuclear reactors. *See* HEAVY WATER . [L.K.]

**Tungsten** A chemical element, W, atomic number 74, and atomic weight 183.85. Naturally occurring tungsten consists of five stable isotopes having the following mass numbers and relative abundances: 180 (0.14%), 182 (26.4%), 183 (14.4%), 184 (30.6%), and 186 (28.4%). Twelve radioactive isotopes ranging from 173 to 189 also have been characterized. *See* PERIODIC TABLE.

Tungsten crystallizes in a body-centered cubic structure in which the shortest interatomic distance is 274.1 picometers at 25°C (77°F). The pure metal has a lustrous, silver-white appearance. It possesses the highest melting point, lowest vapor pressure, and the highest tensile strength at elevated temperature of all metals. Some important physical properties of tungsten are compiled in the table.

**Physical properties of tungsten**

| Property | Value |
|---|---|
| Melting point | 3410 ± 20°C (6170 ± 36°F) |
| Boiling point | 5700 ± 200°C (10,300 ± 360°F) |
| Density, 27°C (81°F) | 19.3 g/cm$^3$ (11.2 oz/in.$^3$) |
| Specific heat, 25°C (77°F) | 0.032 cal/g-°C (0.13 J/g-°C) |
| Heat of fusion | 52.2 ± 8.7 cal/g (218 ± 36 J/g) |
| Vapor pressure, 2027°C (3681°F) | $6.4 \times 10^{-12}$ atm ($6.5 \times 10^{-7}$ Pa) |
| 3382°C (6120°F) | $2.3 \times 10^{-5}$ atm (2.3 Pa) |
| 5470°C (9878°F) | 0.53 atm ($5.4 \times 10^4$ Pa) |
| Electrical resistivity, 27°C (81°F) | 5.65 microhm-cm |
| 1027°C (1881°F) | 34.1 |
| 3027°C (5481°F) | 103.3 |
| Thermal conductivity, 27°C (81°F) | 0.43 cal/cm-s-°C (1.8 J/cm-s-°C) |
| 1027°C (1881°F) | 0.27 (1.1) |
| Absorption cross section, 0.025-eV neutrons | 18.5 ± 0.5 barns ($18.5 \pm 0.5 \times 10^{-24}$ cm$^2$) |

At room temperature tungsten is chemically resistant to water, oxygen, most acids, and aqueous alkaline solutions, but it is attacked by fluorine or a mixture of concentrated nitric and hydrofluoric acids.

Tungsten is used widely as a constituent in the alloys of other metals, since it generally enhances high-temperature strength. Several types of tool steels and some stainless steels contain tungsten. Heat-resistant alloys, also termed superalloys, are nickel-, cobalt-, or iron-based systems containing varying amounts (typically 1.5–25 wt %) of tungsten. Wear-resistant alloys having the trade name Stellites are composed mainly of cobalt, chromium, and tungsten.

The major use of tungsten in the United States is in the production of cutting and wear-resistant materials. Tungsten carbides (representing 60% of total tungsten consumption) are used for cutting tools, mining and drilling tools, dies, bearings, and armor-piercing projectiles.

Unalloyed tungsten (25% of tungsten consumption) in the form of wire is used as filaments in incandescent and fluorescent lamps, and as heating elements for furnaces and heaters. Because of its high electron emissivity, thorium-doped (thoriated) tungsten wire is employed for direct cathode electronic filaments. Tungsten rods find use as lamp filament supports, electrical contacts, and electrodes for arc lamps.

Tungsten compounds (5% of tungsten consumption) have a number of industrial applications. Calcium and magnesium tungstates are used as phosphors in fluorescent lights and television tubes. Sodium tungstate is employed in the fireproofing of fabrics and in the preparation of tungsten-containing dyes and pigments used in paints and printing inks. Compounds such as $WO_3$ and $WS_2$ are catalysts for various chemical processes in the petroleum industry. Both $WS_2$ and $WSe_2$ are dry, high-temperature lubricants. Other applications of tungsten compounds have been made in the glass, ceramics, and tanning industries.

Miscellaneous uses of tungsten account for the remainder (2%) of the metal consumed. [C.Ku.]

# U

**Ultracentrifuge** A centrifuge of high or low speed which provides convection-free conditions and which is used for quantitative measurement of sedimentation velocity or sedimentation equilibrium or for the separation of solutes in liquid solution. *See* CENTRIFUGATION.

The ultracentrifuge is used (1) to measure molecular weights of solutes and to provide data on molecular weight distributions in polydisperse systems; (2) to determine the frictional coefficients, and thereby the sizes and shapes, of solutes; and (3) to characterize and separate macromolecules on the basis of their buoyant densities in density gradients. *See* MOLECULAR WEIGHT.

The ultracentrifuge is most widely used to study high polymers, particularly proteins, nucleic acids, viruses, and other macromolecules of biological origin. However, it is also used to study solution properties of small solutes. In applications to macromolecules, the analytical ultracentrifuge, which is used for accurate determination of sedimentation velocity or equilibrium, is distinguished from the preparative ultracentrifuge, which is used to separate solutes on the basis of their sedimentation velocities or buoyant densities.

The application of a centrifugal field to a solution causes a net motion of the solute. If the solution is denser than the solvent, the motion will be away from the axis of rotation. The nonuniform concentration distribution produced in this way leads to an opposing diffusion flux tending to reestablish uniformity. In sedimentation-velocity experiments, sedimentation prevails over diffusion, and the solute sediments with finite velocity toward the bottom of the cell, although the concentration profile may be markedly influenced by diffusion. In sedimentation-equilibrium experiments, centrifugal and diffusive forces balance out, and an equilibrium concentration distribution results which may be analyzed by thermodynamic methods. [V.A.B.]

**Ultrafast molecular processes** Various types of physical and molecular changes occurring on time scales of $10^{-14}$ to $10^{-9}$ s that are studied in the field of photophysics, photochemistry, and photobiography. The time scale ranges from near the femtosecond regime ($10^{-15}$ s) on the fast side, embodies the entire picosecond regime ($10^{-12}$ s), and borders the nanosecond regime ($10^{-9}$ s) on the slow side.

A typical experiment is initiated with an ultrashort pulse of energy—radiation (light) or particles (electrons). Rapid changes in the system under study are brought about by the absorption of these ultrashort energy pulses. These changes can be measured by ultrafast detection methods. Such methods are usually based on some type of linear or nonlinear spectroscopic monitoring of the system, or on optical delay lines that employ the speed of light itself to create a yardstick of time (a distance of 3 mm equals a time of 10 picoseconds). These studies are important because elementary motion, such as rotation, vibrational exchange, chemical bond breaking, and charge transfer, takes place on these ultrafast time scales. The relationship between such motions and overall physical, chemical, or biological changes can thus be observed directly. *See* CHEMICAL DYNAMICS; SPECTROSCOPY.

Just as a very fast shutter speed is necessary to obtain a sharp photograph of a fast-moving object, energy pulses for the study of ultrafast molecular motions must be extremely narrow in time. The basic technique for producing ultrashort light pulses is mode-locking a laser. One way to do this is to put a nonlinear absorption medium in the laser cavity, which functions somewhat like a shutter. In the time domain, the developing light pulse, bouncing back and forth between the laser cavity reflectors, experiences an intensity-dependent loss each time it passes through the nonlinear absorption medium. Highintensity light penetrates the medium, and its intensity is allowed to build up in the laser cavity; weak-intensity light is blocked. This process shaves off the low-amplitude edges of a pulse, thus shortening it. One of the laser cavity reflectors has less than 100% reflectance at the laser wavelength, so part of the pulse energy is coupled out of the cavity each time the pulse reaches that reflector. In this way, a train of pulses is emitted from the mode-locked laser, each pulse separated from the next by the round-trip time for a pulse traveling at the speed of light in the laser cavity, about 13 nanoseconds for a 2-m (6.6-ft) laser cavity length.

In addition to the extremely narrow energy pulses in the time domain, special ultrafast detection techniques are required in order to preserve the time resolution of an experiment.

A photobiological process, which has received considerable attention by spectroscopists studying the femtosecond regime, is light-driven transmembrane proton pumping in purple bacteria (*Halobacterium halobium*). Each link in the complex chain of molecular processes, such as an ultrafast photoionization event, is "fingerprinted" by a somewhat different absorption spectrum, making it possible to sort out these events by methods of ultrafast laser spectroscopy. The necessity for use of fast primary events in nature concerns overriding unwanted competing chemical and energy loss processes. The faster the wanted process, the less is the likelihood that unwanted, energy-wasting processes will take place. *See* LASER SPECTROSCOPY. [G.W.R.; N.L.]

**Ultrafiltration** A filtration process in which particles of colloidal size are retained by a filter medium while solvent and accompanying low-molecular-weight solutes are allowed to pass through. Ultrafilters are used (1) to separate colloid from suspending medium, (2) to separate particles of one size from particles of another size, and (3) to determine the distribution of particle sizes in colloidal systems by the use of filters of graded pore size.

Ultrafilter membranes have been prepared from various types of gel-forming substances: Unglazed porcelain has been impregnated with gels such as gelatin or silicic acid. Filter paper has been impregnated with glacial acetic acid collodions. Another type of ultrafilter membrane is made up of a thin plastic sheet containing millions of tiny pores evenly distributed over its surface. *See* COLLOID; FILTRATION. [Q.V.W.]

**Unit operations** A structure of logic used for synthesizing and analyzing processing schemes in the chemical and allied industries, in which the basic underlying concept is that all processing schemes can be composed from and decomposed into a series of individual, or unit, steps. If a step involves a chemical change, it is called a unit process; if physical change, a unit operation. These unit operations cut across widely different processing applications, including the manufacture of chemicals, fuels, pharmaceuticals, pulp and paper, processed foods, and primary metals. The unit operations approach serves as a very powerful form of morphological analysis, which systematizes process design, and greatly reduces both the number of concepts that must be taught and the number of possibilities that should be considered in synthesizing a particular process.

Most unit operations are based mechanistically upon the fundamental transport processes of mass transfer, heat transfer, and fluid flow (momentum transfer). Unit operations based on fluid mechanics include fluid transport (such as pumping), mixing/agitation, filtration, clarification, thickening or sedimentation, classification, and centrifugation. Operations based on heat transfer include heat exchange, condensation, evaporation, furnaces or kilns, drying, cooling towers, and freezing or thawing. Operations that are based on mass transfer include distillation, solvent extraction, leaching, absorption or desorption, adsorption, ion exchange, humidification or dehumidification, gaseous diffusion, crystallization, and thermal diffusion. Operations that are based on mechanical principles include screening, solids handling, size reduction, flotation, magnetic separation, and electrostatic precipitation. The study of transport phenomena provides a unifying and powerful basis for an understanding of the different unit operations. *See* ABSORPTION; ADSORPTION; CENTRIFUGATION; CHEMICAL ENGINEERING; CLARIFICATION; CRYSTALLIZATION; DIFFUSION; DISTILLATION; DRYING; ELECTROSTATIC PRECIPITATOR; EVAPORATION; FILTRATION; ION EXCHANGE; MECHANICAL SEPARATION TECHNIQUES; MIXING; SEDIMENTATION (INDUSTRY); SOLVENT EXTRACTION; TRANSPORT PROCESSES; UNIT PROCESSES. [C.J.Ki.]

**Unit processes** Processes that involve making chemical changes to materials, as a result of chemical reaction taking place. For instance, in the combustion of coal, the entering and leaving materials differ from each other chemically: coal and air enter, and flue gases and residues leave the combustion chamber. Combustion is therefore a unit process. Unit processes are also referred to as chemical conversions.

Together with unit operations (physical conversions), unit processes (chemical conversions) form the basic building blocks of a chemical manufacturing process. Most chemical processes consist of a combination of various unit operations and unit processes.

The basic tools of the chemical engineer for the design, study, or improvement of a unit process are the mass balance, the energy balance, kinetic rate of reaction, and position of equilibrium (the last is included only if the reaction does not go to completion). *See* CHEMICAL ENGINEERING; UNIT OPERATIONS. [W.F.F.]

**Uranium** A chemical element, symbol U, atomic number 92, atomic weight 238.03. The melting point is 1132°C (2070°F) and the boiling point is 3818°C (6904°F). Uranium is one of the actinide series. *See* ACTINIDE ELEMENTS; PERIODIC TABLE.

Uranium in nature is a mixture of three isotopes: $^{234}U$, $^{235}U$, and $^{238}U$. Uranium is believed to be concentrated largely in the Earth's crust, where the average concentration is 4 parts per million (ppm). The total uranium content of the Earth's crust to a depth of 15 mi (25 km) is calculated to be $2.2 \times 10^{17}$ lb ($10^{17}$ kg); the oceans may contain $2.2 \times 10^{13}$ lb ($10^{13}$ kg) of uranium. Several hundred uranium-containing minerals have been identified, but only a few are of commercial interest.

Because of the great importance of the fissile isotope $^{235}U$, rather sophisticated industrial methods for its separation from the natural isotope mixture have been devised. The gaseous diffusion process, which in the United States is operated in three large plants (at Oak Ridge, Tennessee; Paducah, Kentucky; and Portsmouth, Ohio) has been the established industrial process. Other processes applied to the separation of uranium include the centrifuge process, in which gaseous uranium hexafluoride is separated in centrifuge cascades, the liquid thermal diffusion process, the separation nozzle, and laser excitation.

Uranium is a very dense, strongly electropositive, reactive metal; it is ductile and malleable, but a poor conductor of electricity. Many uranium alloys are of great interest in nuclear technology because the pure metal is chemically active and anisotropic and has poor mechanical properties. However, cylindrical rods of pure uranium coated with silicon and canned in aluminum tubes (slugs) are used in production reactors. Uranium alloys can also be useful in diluting enriched uranium for reactors and in providing liquid fuels. Uranium depleted of the fissile isotope $^{235}U$ has been used in shielded containers for storage and transport of radioactive materials.

Uranium reacts with nearly all nonmetallic elements and their binary compounds. Uranium dissolves in hydrochloric acid and nitric acid, but nonoxidizing acids, such as sulfuric, phosphoric, or hydrofluoric acid, react very slowly. Uranium metal is inert to alkalies, but addition of peroxide causes formation of water-soluble peruranates.

Uranium reacts reversibly with hydrogen to form $UH_3$ at 250°C (482°F). Correspondingly, the hydrogen isotopes form uranium deuteride, $UD_3$, and uranium tritide, $UT_3$. The uranium-oxygen system is extremely complicated. Uranium monoxide, UO, is a gaseous species which is not stable below 1800°C (3270°F). In the range $UO_2$ to $UO_3$, a large number of phases exist. The uranium halides constitute an important group of compounds. Uranium tetrafluoride is an intermediate in the preparation of the metal and the hexafluoride. Uranium hexafluoride, which is the most volatile uranium compound, is used in the isotope separation of $^{235}U$ and $^{238}U$. The halides react with oxygen at elevated temperatures to form uranyl compounds and ultimately $U_3O_8$.

[F.We.]

**Urea** A colorless crystalline compound, formula $CH_4N_2O$, melting point 132.7°C (270.9°F). Urea is also known as carbamide and carbonyl diamide, and has numerous trade names as well. It is highly soluble in water and is odorless in its purest state, although most samples of even high purity have an ammonia odor. The diamide of carbonic acid, urea has the structure below.

$$H_2N-\overset{\overset{\displaystyle O}{\|}}{C}-NH_2$$

Urea occurs in nature as the major nitrogen-containing end product of protein metabolism by mammals, which excrete urea in the urine. The adult human body discharges almost 50 g (1.8 oz) of urea daily. Urea was first isolated in 1773 by G. F. Rouelle. By preparing urea from potassium cyanate (KCNO) and ammonium sulfate ($NH_4SO_4$) in 1828, F. Wöhler achieved a milestone, the first synthesis of an organic molecule from inorganic starting materials, and thus heralded the modern science of organic chemistry. *See* NITROGEN.

Because of its high nitrogen content (46.65% by weight), urea is a popular fertilizer. About three-fourths of the urea produced commercially is used for this purpose. After application to soil, usually as a solution in water, urea gradually undergoes hydrolysis to ammonia (or ammonium ion) and carbonate (or carbon dioxide). Another major use of urea is as an ingredient for the production of urea-formaldehyde resins, extremely effective adhesives used for laminating plywood and in manufacturing particle board, and the basis for such plastics as malamine. *See* UREA-FORMALDEHYDE RESINS.

Other uses of urea include its utilization in medicine as a diuretic. In the past, it was used to reduce intracranial and intraocular pressure, and as a topical antiseptic. It is still used for these purposes, to some extent, in veterinary medicine and animal husbandry, where it also finds application as a protein feed supplement for cattle and sheep. Urea has been used to brown baked goods such as pretzels. It is a stabilizer for nitrocellulose explosives because of its ability to neutralize the nitric acid that is formed from, and accelerates, the decomposition of the nitrocellulose. Urea was once used for flameproofing fabrics. Mixed with barium hydroxide, urea is applied to limestone monuments to slow erosion by acid rain and acidic pollutants. [R.D.Wa.]

**Urea-formaldehyde-type resins** The condensation products obtained by the reaction of urea or melamine with formaldehyde. Resinous condensation products of formaldehyde with other nitrogen-containing compounds, for example, aniline and amides, also belong to this group of resins but have gained only limited utility.

The urea- and melamine-formaldehyde resins (amino resins) possess an excellent combination of physical properties and can be easily fabricated in a variety of colors. They are widely used as adhesives, laminating resins, molding compounds, paper and textile finishes, and surface coatings.

The aniline- and sulfonamide-formaldehyde resins are produced from other compounds containing $—NH_2$ groups that condense with formaldehyde to form methylol derivatives which are capable of further reaction. These resins have received limited attention except for aniline and *p*-toluenesul-fonamide. Because of their resistance to the absorption of water, the aniline-formaldehyde resins have been used in electrical applications, such as insulation or panels, where their natural brown color is not objectionable. The sulfonamide-formaldehyde polymers are less colored than the aniline-type resins and have been employed in surface coatings. *See* FORMALDEHYDE; POLYMERIZATION; UREA. [J.A.M.]

# V

**Vacuum fusion** A technique of analytical chemistry for determining the oxygen, hydrogen, and sometimes nitrogen content of metals.

The metal sample is either fused or dissolved in a bath, or flux, of a second metal in a heated graphite crucible supported inside an evacuated glass or quartz vessel. Oxygen is released from the metal as carbon monoxide by reaction of oxides or dissolved oxygen with carbon from the graphite crucible at high temperature. Metal nitrides dissociate to form elemental nitrogen. Hydrogen is evolved as elemental hydrogen. The mixture of carbon monoxide, nitrogen, and hydrogen is analyzed to determine individual component concentrations by various techniques.

Inert gas fusion has been developed for determination of gases in metals. The techniques used are similar to vacuum fusion but substitute inert gases for the vacuum environment. [F.C.B.]

**Valence** A term commonly used by chemists to characterize the combining power of an element for other elements, as measured by the number of bonds to other atoms which one atom of the given element forms upon chemical combination. The term also has come to signify the theory of all the physical and chemical properties of molecules that specially depend on molecular electronic structure.

Thus, in water, $H_2O$, the valence of each hydrogen atom is 1; the valence of oxygen, 2. In methane, $CH_4$, the valence of hydrogen again is 1; of carbon, 4. In NaCl and $CCl_4$ the valence of chlorine is 1, and in $CH_2$ the valence of carbon is 2. *See* CHEMICAL BONDING.

Most of the simple facts of valence (though certainly not all) follow from the postulate that atoms combine in such a way as to seek closed-shell or inert-gas structures (rule of eight) by the transfer of electrons between them or the sharing of a pair of electrons between them. Many molecular structures may be obtained by inspection using these rules; letting a dot represent an electron:

$$\begin{array}{c} \mathrm{H} \\ \ddot{} \\ :\!\underset{\cdot\cdot}{\mathrm{O}}\!:\!\mathrm{H} \end{array} \qquad \begin{array}{c} \mathrm{H} \\ \mathrm{H}\!:\!\underset{\cdot\cdot}{\overset{\cdot\cdot}{\mathrm{C}}}\!:\!\mathrm{H} \\ \mathrm{H} \end{array} \qquad [\mathrm{Na}]^{+} \qquad \left[:\!\underset{\cdot\cdot}{\overset{\cdot\cdot}{\mathrm{F}}}\!:\right]^{-}$$

In these electron-dot symbols, the electrons in the *K* shell are not included for atoms after He, nor are the electrons in the *K* and *L* shells for atoms following Ne.

As generally used and here defined, the word valence is ambiguous. Before a value can be assigned to the valence of an atom in a molecule, the electronic structure of the molecule must be exactly known, and this structure must be describable simply in terms of simple bonds. In practice neither of these conditions is ever precisely fulfilled. A term not so ambiguous is oxidation number or valence number. Oxidation numbers are useful for the balancing of oxidation-reduction equations, but they are not related simply to ordinary valences. Thus the valence of carbon in $CH_4$, $CHCl_3$, and $CCl_4$ is 4; oxidation numbers of carbon in these three substances are −4, +2, and +4. *See* ELECTRONEGATIVITY; MOLECULAR ORBITAL THEORY; OXIDATION-REDUCTION. [R.G.P.]

**Vanadium** A chemical element, V, with atomic number 23. Natural deposits contain two isotopes, $^{50}V$ (0.24%), which is weakly radioactive, and $^{51}V$ (99.76%). Commercially important as an oxidation catalyst, vanadium also is used in the production of alloy steel and ceramics and as a colorizing agent. Studies have demonstrated the biological occurrence of vanadium, especially in marine species; in mammals, vanadium has a pronounced effect on heart muscle contraction and renal function. *See* PERIODIC TABLE; TRANSITION ELEMENTS.

Very pure vanadium is difficult to prepare because the metal is highly reactive at temperatures above the melting point of its oxide (663°C or 1225°F) from which it is produced. Vanadium is a bright white metal that is soft and ductile. It has a melting point of 1890°C (3434°F), a boiling point of 3380°C (6116°F), and a density of 6.11 g/cm$^3$ (3.53 oz/in.$^3$) at 18.7°C (65.7°F). The thermal and electrical conductivity of vanadium is superior to that of titanium.

At room temperature, the metal is resistant to corrosion by oxygen, salt water, alkalies, and nonoxidizing acids, the exception being hydrogen fluoride (HF). Vanadium cannot withstand the oxidizing conditions presented by nitric acid or aqua regia. At elevated temperatures it will combine with most nonmetals to form oxides, nitrides, carbides, arsenides, and other such compounds.

The physical properties of vanadium are very sensitive to interstitial impurities. The strength varies from 30,000 lb/in.$^2$ (200 megapascals) in the purest form to 80,000 lb/in.$^2$ (550 MPa) in the commercial grade. The melting point is markedly altered by small impurities; vanadium containing 10% carbon has a melting point of 2700°C (4892°F).

Vanadium has a low fission neutron cross section. This property combined with the metal's excellent retention of strength at elevated temperatures has made its use in atomic energy applications attractive.

Carbon and alloy steels consume more than half the vanadium produced in the United States. Many plate, structural, bar, and pipe steels contain vanadium to enhance strength and toughness. The basis for the unique properties of these carbon alloy steels is the formation of vanadium carbide. These carbides are extremely hard and wear-resistant; they do not coalesce readily, but maintain a state of fine dispersion. Many large steel forgings contain vanadium in the range 0.05–0.15%; here vanadium acts as a grain refiner, and also improves the mechanical properties of the forgings. Tool steels are another large class of vanadium-containing steels; vanadium ensures the retention of hardness and cutting ability at the elevated temperatures generated by the rapid cutting of metals.

The production of ferrovanadium, an iron alloy, is very important since the primary commercial use of vanadium is in steel. Ferrovanadium is produced by aluminum or silicon reduction of $V_2O_5$ in the presence of iron in an electric arc furnace. The commonly practiced aluminum reduction is exothermic, so that little additional heat from the arc is required. Silicon processing requires a two-stage reduction to achieve efficient operation.

Vanadium compounds, especially $V_2O_5$ and $NH_4VO_3$, are excellent oxidation catalysts in the chemical industry. Processes that employ such catalysts include the manufacture of polyamides, such as nylon; sulfuric acid production by the contact process; phthalic and maleic anhydride syntheses; and various oxidations of organic compounds such as the conversion of anthracene to anthraquinone, ethanol to acetaldehyde, and sugar to oxalic acid. Vanadium pentoxide is used as a mordant in dyeing and printing fabrics and in producing aniline black for the dye industry. Vanadium compounds are used in the ceramics industry for glazes and enamels. A wide range of colors can be obtained with combinations of vanadium oxide, zirconia, silica, lead, tin, zinc, cadmium, and selenium. *See* MORDANT.

Vanadium has long been recognized as an essential element in biological systems; however, the role of the metal often is obscure. Tunicates accumulate vanadium to levels 1 million times greater than the surrounding seawater. This vanadium was once thought to act as an oxygen carrier but now is believed to be an oxidation catalyst that repairs damage to the polymeric, protective tunic of these animals. The first vanadium-dependent enzyme, vanadium bromoperoxidase, was isolated from brown, red, and green marine algae (for example, *Ascophyllum nodosum*); this enzyme catalyzes the bromination of a variety of organic molecules by using hydrogen peroxide and bromide. This activity may be the source of many important brominated compounds that potentially may be used as antifungal and antineoplastic agents. *See* ENZYME.

A variety of physiological effects in mammalian systems have been reported, the most significant being in cardiovascular and renal function. Vanadate causes constriction of veins in the kidney and can alter the retention and excretion of sodium and chloride ions. Cardiac effects of vanadium are species-specific, with observed increases (rabbit and rat) and decreases (guinea pig and cat) in heart muscle contractility. Because vanadate is a potent inhibitor of Na,K-ATPase in the laboratory, it has been suggested that this is the site of the metal'saction. However, the physiology of vanadate is probably more complicated, since vanadium behaves as a hormone mimic by elevating intracellular calcium ion ($Ca^{2+}$) levels in a process that is poorly understood. *See* BIOINORGANIC CHEMISTRY. [V.L.P.]

**Van der Waals equation** An equation of state of gases and liquids proposed by J. D. van der Waals in 1873 that takes into account the nonzero size of molecules and the attractive forces between them. He expressed the pressure $p$ as a function of the absolute temperature $T$ and the molar volume $V_m = V/n$, where $n$ is the number of moles of gas molecules in a volume $V$ (see equation below). Here $R = 8.3145$ J K$^{-1}$ mol$^{-1}$

$$p = \frac{RT}{V_m - b} - \frac{a}{V_m^2}$$

is the universal gas constant, and $a$ and $b$ are parameters that depend on the nature of the gas. Parameter $a$ is a measure of the strength of the attractive forces between the molecules, and $b$ is approximately equal to four times the volume of the molecules in one mole, if those molecules can be represented as elastic spheres. The equation has no rigorous theoretical basis for real molecular systems, but is important because it was the first to take reasonable account of molecular attractions and repulsions, and to emphasize the fact that the intermolecular forces acted in the same way in both gases and liquids. It is accurate enough to account for the fact that all gases have a critical temperature $T_c$ above which they cannot be condensed to a liquid. The expression that follows from this equation is $T_c = 8a/27\ Rb$. *See* CRITICAL PHENOMENA; GAS; INTERMOLECULAR FORCES; LIQUID.

In a gas mixture, the parameters $a$ and $b$ are taken to be quadratic functions of the mole fractions of the components since they are supposed to arise from the interaction of the molecules in pairs. The resulting equation for a binary mixture accounts in a qualitative but surprisingly complete way for the many kinds of gas-gas, gas-liquid, and liquid-liquid phase equilibria that have been observed in mixtures. *See* PHASE EQUILIBRIUM.

The equation is too simple to represent quantitatively the behavior of real gases, and so the parameters $a$ and $b$ cannot be determined uniquely; their values depend on the ranges of density and temperature used in their determination. For this reason, the equation now has little practical value, but it remains important for its historical interest and for the concepts that led to its derivation. *See* THERMODYNAMIC PRINCIPLES. [J.S.Ro.]

**Vapor pressure** The saturation pressures exerted by vapors which are in equilibrium with their liquid or solid forms. One of the most important physical properties of a liquid, the vapor pressure, enters into many thermodynamic calculations and underlies several methods for the determination of the molecular weights of substances dissolved in liquids. For a discussion of the vapor pressure relationships of solids *see* SUBLIMATION. *See also* MOLECULAR WEIGHT; SOLUTION.

If a liquid is introduced into an evacuated vessel at a given temperature, some of the liquid will vaporize, and the pressure of the vapor will attain a maximum value which is termed the vapor pressure of the liquid at that temperature. Although the quantity of liquid remaining does not diminish thereafter, the process of evaporation does not cease. A dynamic equilibrium is established, in which molecules escape from the liquid phase and return from the vapor phase at equal rates. *See* EVAPORATION.

It is important to make a distinction between the vapor pressure of a liquid, as described above, and the pressure of a vapor. The vapor pressure of a pure liquid is a unique and characteristic property of the liquid and depends only upon the temperature. A gas or vapor may, on the other hand, exert any pressure within reason, depending upon the volume to which it is confined, provided it is not in contact with its liquid phase. *See* PHASE EQUILIBRIUM. [N.H.N.]

**Volatilization** The process of converting a chemical substance from a liquid or solid state to a gaseous or vapor state. Other terms used to describe the same process are vaporization, distillation, and sublimation. A substance can often be separated from another by volatilization and can then be recovered by condensation of the vapor. The substance can be made to volatilize more rapidly either by heating to increase its vapor pressure or by removal of the vapor using a stream of inert gas or a vacuum pump. Chemical reactions are sometimes utilized to produce volatile products. Volatilization methods are generally characterized by great simplicity and ease of operation, except when high temperatures or highly corrosion-resistant materials are needed. *See* CHEMICAL SEPARATION TECHNIQUES; DISTILLATION; SUBLIMATION; VAPOR PRESSURE. [L.Go./R.W.Mu.]

**Water** The chemical compound with two atoms of hydrogen and one atom of oxygen in each of its molecules. It is formed by the direct reaction (1) of hydro-

$$2H_2 + O_2 \rightarrow 2H_2O \quad (1)$$

gen with oxygen. The other compound of hydrogen and oxygen, hydrogen peroxide, readily decomposes to form water, reaction (2). Water also is formed in the combus-

$$2H_2O_2 \rightarrow 2H_2O + O_2 \quad (2)$$

tion of hydrogen-containing compounds, in the pyrolysis of hydrates, and in animal metabolism. Some properties of water are given in the table.

**Properties of water**

| Property | Value |
|---|---|
| Freezing point | 0°C (32°F) |
| Density of ice, 0°C (32°F) | 0.92 g/cm$^3$ (0.53 oz/in.$^3$) |
| Density of water, 0°C (32°F) | 1.00 g/cm$^3$ (0.578 oz/in.$^3$) |
| Heat of fusion | 80 cal/g (335 J/g) |
| Boiling point | 100°C (212°F) |
| Heat of vaporization | 540 cal/g (2260 J/g) |
| Critical temperature | 347°C (657°F) |
| Critical pressure | 217 atm (22.0 MPa) |
| Specific electrical conductivity at 25°C | $1 \times 10^{-7}$/ohm-cm |
| Dielectric constant, 25°C (77°F) | 78 |

**Gaseous state.** Water vapor consists of water molecules which move nearly independently of each other. The relative positions of the atoms in a water molecule are shown in the illustration. The dotted circles show the effective sizes of the isolated atoms. The atoms are held together in the molecule by chemical bonds which are very polar, the hydrogen end of each bond being electrically positive relative to the oxygen. When two molecules near each other are suitably oriented, the positive hydrogen of one molecule attracts the negative oxygen of the other, and while in this orientation, the repulsion of the like charges is comparatively small. The net attraction is strong enough to hold the molecules together in many circumstances and is called a hydrogen bond.

**Solid state.** Ordinary ice consists of water molecules joined together by hydrogen bonds in a regular arrangement. It appears that there is considerable empty space between the molecules. This unusual feature is a result of the strong and directional hydrogen bonds taking precedence over all other intermolecular forces in determining the structure of the crystal. *See* HYDROGEN BOND.

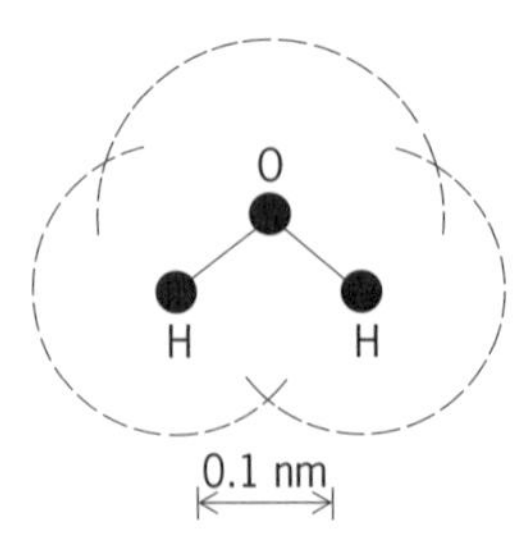


The water molecule.

**Liquid state.** The molecules in liquid water also are held together by hydrogen bonds. When ice melts, many of the hydrogen bonds are broken, and those that remain are not numerous enough to keep the molecules in a regular arrangement. As water is heated from 0°C (32°F), it contracts until 4°C (39°F) is reached and then begins the expansion which is normally associated with increasing temperature. This phenomenon and the increase in density when ice melts both result from a breaking down of the open, hydrogen-bonded structure as the temperature is raised. *See* LIQUID.

**Properties.** The electrical conductivity of water is at least 1,000,000 times larger than that of most other nonmetallic liquids at room temperature. The current in this case is carried by ions produced by the dissociation of water according to reaction (3).

$$H_2O \rightleftharpoons H^+ + OH^- \qquad (3)$$

Water is an excellent solvent for many substances, but particularly for those which dissociate to form ions. Its principal scientific and industrial use as a solvent is to furnish a medium for purifying such substances and for carrying out reactions between them.

Water is not a strong oxidizing agent, although it may enhance the oxidizing action of other oxidizing agents, notably oxygen. Water is an even poorer reducing agent than oxidizing agent. One of the few substances that it reduces rapidly is fluorine. *See* OXIDATION-REDCUTION.

Substances with strong acidic or basic character react with water. For example, calcium oxide, a basic oxide, reacts in a process called the slaking of lime, reaction (4).

$$CaO + H_2O \rightarrow Ca(OH)_2 \qquad (4)$$

Another type of substance with strong acidic character is an acid chloride. An example and its reaction with water is boron trichloride, reaction (5). This is termed hydrolysis,

$$BCl_3 + 3H_2O \rightarrow H_3BO_3 + 3HCL \qquad (5)$$

as is the reaction of an ester with water. *See* HYDROLYSIS.

Water also reacts with a variety of substances to form solid compounds in which the water molecule is intact, but in which it becomes a part of the structure of the solid. Such compounds are called hydrates. *See* HYDRATE.

For various aspects of water, its uses, and occurrence *see* HEAVY WATER; HYDROGEN; OXYGEN; TRIPLE POINT; VAPOR PRESSURE. [H.L.F.]

**Water desalination** The removal of dissolved minerals (including salts) from seawater or brackish water. This may occur naturally as part of the hydrologic cycle, or as an engineered process. Engineered water desalination processes, which produce potable water from seawater or brackish water, have become important because many regions throughout the world suffer from water shortages caused by the uneven distribution of the natural water supply and by human use.

Seawater, brackish water, and fresh water have different levels of salinity, which is often expressed by the total dissolved solids (TDS) concentration. Seawater has a TDS concentration of about 35,000 mg/L, and brackish water has a TDS concentration of 1000–10,000 mg/L. Water is considered fresh when its TDS concentration is below 500 mg/L, which is the secondary (voluntary) drinking water standard for the United States. Salinity is also expressed by the water's chloride concentration, which is about half of its TDS concentration.

Water desalination processes separate feed water into two streams: a fresh-water stream with a TDS concentration much less than that of the feed water, and a brine stream with a TDS concentration higher than that of the feed water.

Distillation is a process that turns seawater into vapor by boiling, and then condenses the vapor to produce fresh water. Boiling water is an energy-intensive operation, requiring about 4.2 kilojoules of energy (or latent heat) to raise the temperature of 1 kg of water by 1°C. After water reaches its boiling point, another 2257 kJ of energy (or the heat of vaporization) is required to convert it to vapor. The boiling point depends on ambient atmospheric pressure—at lower pressure, the boiling point of water is lower. Therefore, keeping water boiling can be accomplished either by providing a constant energy supply or by reducing the ambient atmospheric pressure. *See* DISTILLATION.

Reverse osmosis, the process that causes water in a salt solution to move through a semipermeable membrane to the fresh-water side, is accomplished by applying pressure in excess of the natural osmotic pressure to the salt solution. The operational pressure of reverse osmosis for seawater desalination is much higher than that for brackish water, as the osmotic pressure of seawater at a TDS concentration of 35,000 mg/L is about 2700 kJ while the osmotic pressure of brackish water at a TDS concentration of 3000 mg/L is only about 230 kJ.

Salts dissociate into positively and negatively charged ions in water. The electrodialysis process uses semipermeable and ion-specific membranes, which allow the passage of either positively or negatively charged ions while blocking the passage of the oppositely charged ions. An electrodialysis membrane unit consists of a number of cell pairs bound together with electrodes on the outside. These cells contain an anion exchange membrane and cation exchange membrane. Feed water passes simultaneously in parallel paths through all of the cells, separating the product (water) and ion concentrate. *See* DIALYSIS; ION EXCHANGE. [C.C.K.L.; J.W.P.]

**Wax, petroleum** A substance produced primarily from the dewaxing of lubricating-oil fractions of petroleum. It may be of either the crystalline or microcrystalline type. Petroleum wax has a wide variety of uses. It is used to coat paper products, to blend with other waxes for the manufacture of candles, in the manufacture of electrical equipment and many polishes for home and industry, and as a source material for oxidized products. The softer waxes, such as petroleum jelly, after proper purification, are being used as medicinal products. *See* DEWAXING OF PETROLEUM. [W.E.Ku.]

**Wood chemicals** Substances derived from wood. Woody plants comprise the greatest part of the organic materials produced by photosynthesis on a renewable basis, and were the precursors of the fossil coal deposits. The derivation of chemicals from wood is carried out wherever technical utility and economic conditions have combined to make it feasible.

Wood is a mixture of three natural polymers—cellulose, hemicelluloses, and lignin—in an approximate abundance of 50:25:25. In addition to these polymeric cell wall components which make up the major portion of the wood, different species contain varying amounts and kinds of extraneous materials called extractives. The nature of the chemicals derived from wood depends on the wood component involved.

Chemicals derived from wood include: bark products, cellulose, cellulose esters, cellulose ethers, charcoal, dimethyl sulfoxide, ethyl alcohol, fatty acids, furfural, hemicellulose extracts, kraft lignin, lignin sulfonates, pine oil, rayons, rosin, sugars, tall oil, turpentine, and vanillin. Most of these are either direct products or by-products of wood pulping, in which the lignin that cements the wood fibers together and stiffens them is dissolved away from the cellulose. High-purity chemical cellulose or dissolving pulp is the starting material for such polymeric cellulose derivatives as viscose rayon and cellophane, cellulose esters such as the acetate and butyrate for fiber, film, and molding applications, and cellulose ethers such as carboxymethylcellulose, ethylcellulose, and hydroxyethylcellulose for use as gums. *See* CELLOPHANE; DIMETHYL SULFOXIDE; ETHYL ALCOHOL; ROSIN. [I.S.G.]

**Woodward-Hoffmann rule** A concept which can predict or explain the stereochemistry of certain types of reactions in organic chemistry. It is also described as the conservation of orbital symmetry, and is named for its developers, R. B. Woodward and Roald Hoffmann. The rule applies to a limited group of reactions, called pericyclic, which are characterized by being more or less concerted (that is, one-step, without a distinct intermediate between reactants and products) and having a cyclic arrangement of the reacting atoms of the molecule in the transition state. Most pericyclic reactions fall into one of three major classes: electrocyclic, cycloaddition, or sigmatropic. *See* STEREOCHEMISTRY. [D.L.D.]

**Work function (thermodynamics)** The thermodynamic function better known as the Helmholtz energy, $A = U - TS$, where $U$ is the internal energy, $T$ is the thermodynamic (absolute) temperature, and $S$ is the entropy of the system. At constant temperature, the change in work function is equal to the maximum work that can be done by a system ($\Delta A = w_{max}$). *See* FREE ENERGY. [P.W.A.]

**Xenon** A chemical element, Xe, atomic number 54. It is a member of the family of noble gases, group 18 in the periodic table. Xenon is colorless, odorless, and tasteless; it is a gas under ordinary conditions (see table). *See* INERT GASES; PERIODIC TABLE.

Xenon is the only one of the nonradioactive noble gases which forms chemical compounds that are stable at room temperature. Xenon also forms weakly bonded clathrates with such substances as water, hydroquinone, and phenol. *See* CLATHRATE COMPOUNDS.

**Physical properties of xenon**

| Property | Value |
|---|---|
| Atomic number | 54 |
| Atomic weight (atmospheric xenon only) | 131.30 |
| Melting point (triple point) | −111.8°C (−169.2°F) |
| Boiling point at 1 atm pressure | −108.1°C (−162.6°F) |
| Gas density at 0°C and 1 atm pressure, g/liter | 5.8971 |
| Liquid density at its boiling point, g/ml | 3.057 |
| Solubility in water at 20°C, ml xenon (STP) per 1000 g water at 1 atm partial pressure of xenon | 108.1 |

The three fluorides, $XeF_2$, $XeF_4$, and $XeF_6$, are thermodynamically stable compounds at room temperature, and they may be prepared simply by heating mixtures of xenon and fluorine at 300–400°C (570–750°F).

The reaction of $XeF_6$ with water gives $XeOF_4$; if the reaction is allowed to continue, $XeO_3$ is formed. $XeO_3$ is a colorless, odorless, and dangerously explosive white solid of low volatility. Gaseous xenon tetroxide, $XeO_4$, is formed by the reaction of sodium perxenate, $Na_4XeO_6$, with concentrated $H_2SO_4$. The vapor pressure of $XeO_4$ is about 3.3 kPa at 0°C (32°F). It is unstable and has a tendency to explode.

Xenon is produced commercially in an air-separation plant. The air is liquefied and distilled. The oxygen is redistilled; the least volatile portion contains small amounts of xenon and krypton, which are adsorbed on silica gel directly from the liquid oxygen. The crude xenon and krypton thus obtained are separated and further purified by distillation and selective absorption, at controlled low temperatures, on activated carbon. Remaining impurities are removed by passing the xenon over hot titanium, which reacts with all but the inert gases. *See* AIR SEPARATION.

Xenon is used to fill a type of flashbulb used in photography and called an electronic speed light. These bulbs produce a white light with a good balance of all the colors in the visible spectrum, and can be used 10,000 times or more before burning out.

A xenon-filled arc lamp gives a light intensity approaching that of the carbon arc; it is particularly valuable in projecting motion pictures.

An important development in high-energy physics was the detection of nuclear radiation, such as gamma rays and mesons, by bubble chambers, in which a liquid is kept at a temperature just above its boiling point. Nucleation by the radiation results in bubble formation along the path of the particle. The tracks made by the particles are then photographed. Liquid xenon is one of the liquids used in these bubble chambers.

Xenon is used to fill neutron counters, x-ray counters, gas-filled thyratrons, and ionization chambers for cosmic rays; it is also used in high-pressure arc lamps to produce ultraviolet radiation.

Between 3 and 5% of the fissions in a nuclear reactor using uranium as fuel lead to the formation of xenon-135. [A.W.F.]

**X-ray spectrometry** A rapid and economical technique for quantitative analysis of the elemental composition of specimens. It differs from x-ray diffraction, whose purpose is the identification of crystalline compounds. It differs from spectrometry in the visible region of the spectrum in that the x-ray photons have energies of thousands of electronvolts and come from tightly bound inner-shell electrons in the atoms, whereas visible photons come from the outer electrons and have energies of only a few electronvolts.

In x-ray spectrometry the irradiation of a sample by high-energy electrons, protons, or photons ionizes some of the atoms, which then emit characteristic x-rays whose wavelength depends on the atomic number of the element, and whose intensity is related to the concentration of that element. Generally speaking, the characteristic x-ray lines are independent of the physical state (solid or liquid) and of the type of compound (valence) in which an element is present, because the x-ray emission comes from inner, well-shielded electrons in the atom. The illustration shows the removal of one of the innermost, *K*-shell, electrons by a high-energy photon. The photon energy must be greater than the binding energy of the electron; the difference in energy appears as the kinetic energy of the ejected electron. The *K* ionized atom is unstable, and one of the *L*- or *M*-shell electrons drops into the *K*-shell vacancy. As this transition occurs, a characteristic x-ray photon is emitted with an energy equal to the difference in energy between the *K* and the *L* (or *M*) shell, or an additional electron, called an Auger electron, is ejected from the atom. Either the x-rays or the Auger electrons may be used for analysis. *See* AUGER EFFECT.

X-ray spectrometry generally does not require any separation of elements before measuring, because the x-ray lines are easily resolved. However, preconcentration

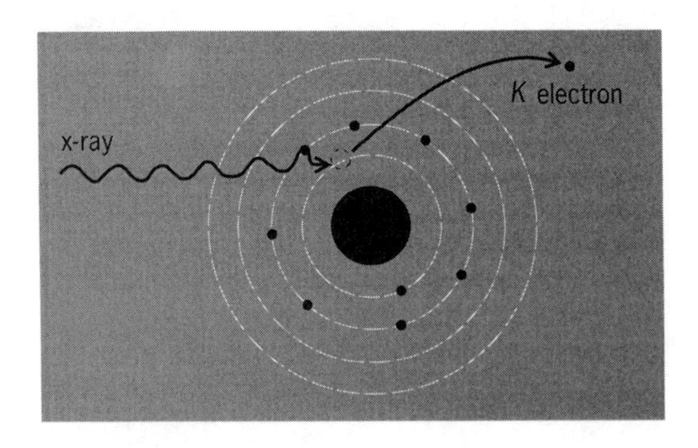


**Removal of a *K* election from an atom by a primary x-ray photon. (*After L. S. Birks, X-Ray Spectrochemical Analysis, 2d ed., Wiley-Interscience, 1969*)**

methods are sometimes useful as a means for improving the limit of detection. One limitation of x-ray spectrometry is the progressive difficulty of measurement below atomic number 11. *See* SPECTROSCOPY. [L.S.B.]

**Xylose** A pentose sugar, referred to in the early literature as L-xylose. It is present in many woody materials. The polysaccharide xylan, which is closely associated with cellulose, consists practically entirely of D-xylose. Corncobs, cottonseed hulls, pecan shells, and straw contain considerable amounts of this sugar. This pentose sugar is also a component of the hemicelluloses and the rare disaccharide, primeverose. *See* CARBOHYDRATE; POLYSACCHARIDE. [W.Z.H.]

# Y

**Ylide** A variety of organic compounds that contain two adjacent atoms bearing formal positive and negative charges, and in which both atoms have full octets of electrons. Heteroatoms most commonly utilized as the positive atom are phosphorus, nitrogen, sulfur, selenium and oxygen, and the negative atom usually involves carbon, nitrogen, oxygen, or sulfur. The ylide may be in an alicyclic or cyclic environment, and in the latter the ylide function may be endocyclic or exo-cyclic. Ylides most useful in organic synthesis are those containing phosphorus or sulfur and an adjacent carbanion.

A phosphorus ylide is known as a phosphorane. Ylides of this type are highly reactive. Sulfur-containing ylides ($\pi$-sulfuranes) are of two types, sulfonium ylides and oxosulfonium ylides, which differ in having an oxygen atom attached to the sulfur in the latter. *See* REACTIVE INTERMEDIATES.

The ylide function may also be incorporated into heterocyclic systems. For example, the meso-ionic sydnone molecule contains an azomethine imine ylide, and the mesomeric betaine derived from 3-hydroxypyridine contains an azomethine ylide. These ylides are often referred to as masked ylides. [K.T.P.]

**Ytterbium** A chemical element, Yb, atomic number 70, and atomic weight 173.04. Ytterbium is a metal element of the rare-earth group. There are 7 naturally occurring stable isotopes. *See* PERIODIC TABLE.

The common oxide, $Yb_2O_3$, is colorless and dissolves readily in acids to form colorless solutions of trivalent salts which are paramagnetic. Ytterbium also forms a series of divalent compounds. The divalent salts are soluble in water but react very slowly with water to liberate hydrogen.

The metal is best prepared by distillation. It is a silvery soft metal which corrodes slowly in air and resembles the calcium-strontium-barium series more than the rare-earth series. For a discussion of the properties of the metal and its salts *see* RARE-EARTH ELEMENTS. [F.H.Sp.]

**Yttrium** A chemical element, Y, atomic number 39, and atomic weight 88.905. Yttrium resembles the rare-earth elements closely. The stable isotope $^{89}Y$ constitutes 100% of the natural element, which is always found associated with the rare earths and is frequently classified as one. *See* PERIODIC TABLE.

Yttrium metal absorbs hydrogen, and in alloys up to a composition of $YH_2$ they resemble metals very closely. In fact, in certain composition ranges, the alloy is a better conductor of electricity than the pure metal.

Yttrium forms the matrix for the europium-activated yttrium phosphors which emit a brilliant, clear-red light when excited by electrons. The television industry uses these phosphors in manufacturing television screens.

Yttrium is used commercially in the metal industry for alloy purposes and as a "getter" to remove oxygen and nonmetallic impurities in other metals. For properties of the metal and its salts *see* RARE-EARTH ELEMENTS. [F.H.Sp.]

# Z

**Zinc** A chemical element, Zn, atomic number 30, and atomic weight 65.38. Zinc is a malleable, ductile, gray metal. Because of chemical similarities among zinc, cadmium, and mercury, these three metals are classed together in a transition-elements subgroup of the periodic table. *See* PERIODIC TABLE; TRANSITION ELEMENTS.

Fifteen isotopes of zinc are known, of which five are stable, having atomic masses of 64, 66, 67, 68, and 70. About half of ordinary zinc occurs as the isotope of atomic mass 64. The half-lives of the radioactive isotopes range from 88 s for $^{61}Zn$ to 244 days for $^{65}Zn$.

Zinc is a fairly active metal chemically. It can be ignited with some difficulty to give a blue-green flame in air and to discharge clouds of zinc oxide smoke. Zinc ranks above hydrogen in the electrochemical series, so that metallic zinc in an acidic solution will react to liberate hydrogen gas as the zinc passes into solution to form dipositively charged zinc ions, $Zn^{2+}$. This reaction is slow with very pure zinc, but the presence of small amounts of impurities, addition of a trace of copper sulfate, or contact between the zinc surface and such metals as nickel or platinum facilitates formation of gaseous hydrogen and speeds the reaction. The combination of zinc and dilute acid is often used to generate small quantities of hydrogen in the laboratory. Zinc also dissolves in strongly alkaline solutions, such as sodium hydroxide, to liberate hydrogen and form dinegatively charged tetrahydroxozincate ions, $Zn(OH)_4{}^{2-}$, sometimes written as $ZnO_2{}^{2-}$ in the formulas of the zincate compounds. Zinc also dissolves in solutions of ammonia or ammonium salts. The common soluble zinc compounds undergo to some extent the process of hydrolysis, which makes their solutions slightly acidic. The ion $Zn^{2+}$ is colorless, so that the relatively few zinc compounds that are not colorless in large crystals, or white as powders, receive their color through the influence of the other constituents. Some of the atomic and ionic properties of zinc are shown in the table. *See* ELECTROCHEMICAL SERIES; HYDROLYSIS.

**Atomic and ionic properties of zinc**

| Property | Value |
|---|---|
| Electronic configuration | $1s^2, 2s^2, 2p^6, 3s^2, 3p^6, 3d^{10}, 4s^2$ |
| Ionization potentials | |
| 1st electron loss | 9.39 eV |
| 2d electron loss | 17.9 eV |
| Ionic radius, $Zn^{2+}$ | 0.072 nm |
| Covalent radius (tetrahedral) | 0.131 nm |
| Oxidation potentials | $Zn \rightleftharpoons Zn^{2+} + 2e^-, E^\circ = 0.76$ V |
| | $Zn + 4OH^- \rightleftharpoons ZnO_2{}^{2-} + 2H_2O + 2e^-, E^\circ = 1.22$ V |

Zinc also forms many coordination compounds. The zincates are actually coordination compounds, or complexes, in which hydroxide ions, $OH^-$, are bound to the zinc ions. Ammonia, $NH_3$, forms complexes with zinc, such as the typical tetrammine zinc ion, $[Zn(NH_3)_4]^{2+}$. Zinc cyanide, usually given the simple formula $Zn(CN)_2$, is a coordination compound in which many alternating zinc and cyanide ions are three-dimensionally bound together in a very large molecule. This compound is still widely used in zinc plating, but concern over environmental pollution has led to increasing use of zinc chloride plating baths. In most coordination compounds of zinc, the fundamental structural unit is a central zinc ion surrounded by four coordinated groups arranged spatially at the corners of a regular tetrahedron. *See* COORDINATION CHEMISTRY; COORDINATION COMPLEXES.

Pure, freshly polished zinc is bluish-white, lustrous, and moderately hard (2.5 on Mohs scale). Moist air brings about a superficial tarnish to give the metal its usual grayish color. Pure zinc is malleable and ductile enough to be rolled or drawn, but small amounts of other metals present as contaminants may render it brittle. Malleability of even pure zinc is improved by heating zinc to 100–150°C (212–300°F). If heated zinc is mechanically worked; it does not embrittle on cooling. Zinc melts at 420°C (788°F) and boils at 907°C (1665°F). Its density is 7.13 times that of water, so that 1 $ft^3$ (0.028 $m^3$) of zinc weighs 445 lb (200 kg).

As a conductor of heat and of electricity, zinc ranks fairly high. However, its electrical resistivity (5.92 microhm-cm at 20°C or 68°F) is almost four times that of silver, the best conductor. As a conductor of heat, zinc is likewise only about one-fourth as efficient as silver. At 0.91 K zinc is an electrical superconductor. Pure zinc is not ferromagnetic, but the alloy compound $ZrZn_2$ displays ferromagnetism below 35 K.

The most important uses of zinc are in its alloys and as a protective coating on other metals. Coating iron or steel with zinc is known as galvanizing, and it may be done by immersing the article in melted zinc (hot-dip process), depositing zinc electrolytically onto the article in a plating bath (electrogalvanizing), exposing the article to powdered zinc near its melting point (sherardizing), or spraying the article with melted zinc (metallizing). The mere physical presence of the zinc coat prevents corrosion of iron, and even if breaks in the coat expose portions of the iron, the greater chemical activity of the zinc causes it to be consumed in preference to the iron. Adding small amounts of other metals to galvanizing baths has been found to improve the adhesion and weathering qualities of the coating.

Even such nonstructural materials as cardboard can be zinc-coated by low-temperature flame spraying. Other important uses of zinc are in brass and zinc die-casting alloys, in zinc sheet and strip, in electrical dry cells, in making certain zinc compounds, and as a reducing agent in chemical preparations.

A so-called tumble-plating process coats small metal parts by applying zinc powder to them with an adhesive, then tumbling them with glass beads to roll out the powder into a continuous coat of zinc. Rechargeable nickel-zinc batteries offer higher energy densities than conventional dry cells. Foamed zinc metal has been suggested for use in lightweight structures such as aircraft and spacecraft. Some other uses of zinc are in dry cells, roofing, lithographic plates, fuses, organ pipes, and wire coatings. Zinc dust, a flammable material when dry, is used in fireworks and as a chemical catalyst and reducing agent. Radioactive $^{65}Zn$ is used medically in the study of metabolism of zinc, and also in determining rates of wear for zinc-containing alloys. [W.E.C.]

Zinc is believed to be needed for normal growth and development of all living species, including humans; actually, life without zinc would be impossible. Zinc is a common element that is present in virtually every type of human food, and zinc deficiency is therefore not considered to be a common problem in humans. Zinc is a trace element;

that is, it is present in biological fluids at a concentration below 1 ppm, and only a small amount (normally <25 mg) is required in the daily diet. (The recommended daily allowance for zinc is 15 mg/day for adults and 10 mg/day for growing children.) It is relatively nontoxic, without noticeable side effects at intake levels of up to 10 times the normal daily requirement. [J.F.Ri.; K.H.F.]

**Zirconium** A chemical element, Zr, atomic number 40, atomic weight 91.22. Its naturally occurring isotopes are 90, 91, 92, 94, and 96. Zirconium is one of the more abundant elements, and is widely distributed in the Earth's crust. Being very reactive chemically, it is found only in the combined state. Under most conditions, it bonds with oxygen in preference to any other element, and it occurs in the Earth's crust only as the oxide, $ZrO_2$, baddeleyite, or as part of a complex of oxides as in zircon, elpidite, and eudialyte. Zircon is commercially the most important ore. Zirconium and hafnium are practically indistinguishable in chemical properties, and occur only together. *See* HAFNIUM; PERIODIC TABLE.

Most of the zirconium used has been as compounds for the ceramic industry: refractories, glazes, enamels, foundry mold and core washes, abrasive grits, and components of electrical ceramics, The incorporation of zirconium oxide in glass significantly increases its resistance to alkali. The use of zirconium metal is almost entirely for cladding uranium fuel elements for nuclear power plants. Another significant use has been in photo flashbulbs.

Zirconium is a lustrous, silvery metal, with a density of 6.5 g/cm$^3$ (3.8 oz/in.$^3$) at 20°C (68°F). It melts at about 1850°C (3362°F). Estimates of the boiling point from appropriate data have commonly been of the order of 3600°C (6500°F), but observations suggest about 8600°C (15,500°F). The free energies of formation of its compounds indicate that zirconium should react with any nonmetal, other than the inert gases, at ordinary temperatures. In practice, the metal is found to be nonreactive near room temperature because of an invisible, impervious oxide film on its surface. The film renders the metal passive, and it remains bright and shiny in ordinary air indefinitely. At elevated temperatures it is very reactive to the non-metallic elements and many of the metallic elements, forming either solid solutions or compounds.

Zirconium generally has normal covalency of 4, and commonly exhibits coordinate covalencies of 5, 6, 7, and 8. Zirconium is at oxidation number 4 in nearly all of its compounds, Halides in which its oxidation numbers are 3 and 2 have been prepared. While zirconium is often part of cationic or anionic complexes, there is no definite evidence for a monatomic zirconium ion in any of its compounds.

Most handling and testing of zirconium compounds have indicated no toxicity. There has generally been no ill consequence of contact of zirconium compounds with the unabraded skin. However, some individuals appear to have allergic sensitivity to zirconium compounds, characteristically manifested by appearance of nonmalignant granulomas. Inhalation of sprays containing some zirconium compounds and of metallic zirconium dusts have had inflammatory effects. [W.B.B.]

# BIBLIOGRAPHIES


## ANALYTICAL CHEMISTRY

Bard, A.J., and I. Rubenstein (eds.), *Electroanalytical Chemistry: A Series of Advances*, vol. 22, 2003.
Bockris, J.O., et al. (eds.), *Modern Aspects of Electrochemistry*, vol. 34, 2001.
Christian, G.D., *Analytical Chemistry*, 5th ed., 1994.
Crow, D.R., *Principles and Applications of Electrochemistry*, 4th ed., 1994.
Kennedy, J.H., *Analytical Chemistry: Principles*, 2d ed., 1990.
Koryta, J., *Ions, Electrodes, and Membranes*, 2d ed., 1991.
Patnaik, P., *Dean's Analytical Chemistry Handbook*, 2d ed., 2004.
Sawyer, D.T., A. Sobkowiak, and J.L. Roberts, *Electrochemistry for Chemists*, 2d ed., 1995.
Settle, F.A. (ed.), *Handbook of Instrumental Techniques for Analytical Chemistry*, 1997.
Skoog, D.A., D.M. West, and F.J. Holler, *Fundamentals of Analytical Chemistry*, 7th ed., 1996.
Stock, J.T., and M.V. Orna (eds.), *Electrochemistry, Past & Present*, 1989.
Winefordner, J.D., et al., *Treatise on Analytical Chemistry*, 2d ed., vol. 13, pt. 1, 1993.

### Journals:

*American Laboratory*, monthly.
*Analytical Chemistry*, American Chemical Society, semimonthly.
*Journal of the Electrochemical Society*, monthly.
*Trends in Analytical Chemistry*, Elsevier, monthly.

## CHEMICAL ENGINEERING

Bailey, J.E., and D.F. Ollis, *Biochemical Engineering Fundamentals*, 2d ed., 1986.
Espenson, J.H., *Chemical Kinetics and Reaction Mechanisms*, 2d ed., 1995.
Furter, W.F. (ed.), *A Century of Chemical Engineering*, 1981.
Himmelblau, D.M., and J.B. Riggs, *Basic Principles and Calculations in Chemical Engineering*, 7th ed., 2003.
Humphrey, J.L., and G.E. Keller II, *Separation Process Technology*, 1997.
*Kirk-Othmer Encyclopedia of Chemical Technology*, 4th ed., 27 vols., 1998.
McGraw-Hill, *Chemical Engineer's Solutions Suite* (CD-ROM), 1996.
Miller, R.W., *Flow Measurement Engineering Handbook*, 3d ed., 1996.
Perry, R.H., et al., *Perry's Chemical Engineers' Handbook*, 7th ed., 1997.
Schweitzer, P.A., *Handbook of Separation Techniques for Chemical Engineers*, 3d ed., 1997.

### Journals:

*Biotechnology Progress*, American Chemical Society and American Institute of Chemical Engineers (copublished), bimonthly.
*Chemical and Engineering News*, American Chemical Society, weekly.
*Chemical Engineering*, monthly.
*Chemical Engineering Progress*, American Institute of Chemical Engineers, monthly.
*Industrial & Engineering Chemistry Research*, American Chemical Society, biweekly.

## GENERAL CHEMISTRY

Bell, J.A., *Chemistry: A Project of the American Chemical Society*, 2004.
Brady, J.E., J.W. Russell, and J.R. Holum, *Chemistry: The Study of Matter and Its Changes*, 3d ed., 2000.
Chang, R., *Chemistry*, 7th ed., 2001.

Dean, J.A. (ed.), *Lange's Handbook of Chemistry*, 15th ed., 1998.
Dickson, T.R., *Introduction to Chemistry*, 8th ed., 1999.
Oxtoby, D.W., et al., *Principles of Modern Chemistry*, 5th ed., 2002.
Timberlake, K.C., *Chemistry: An Introduction to General, Organic, and Biological Chemistry*, 8th ed., 2002.
Zumdahl, S.S., and S. Zumdahl, *Chemistry*, 2000.

# INORGANIC CHEMISTRY

Bowser, J. R., *Inorganic Chemistry*, 1993.
Butler, I.S., and J.F. Harrod, *Inorganic Chemistry: Principles and Applications*, 1989.
Cotton, F.A., and G. Wilkinson, *Advanced Inorganic Chemistry*, 6th ed., 1999.
Douglas, B.E., *Concepts & Models of Inorganic Chemistry*, 2d ed., 1993.
Greenwood, N.N., and A. Earnshaw, *Chemistry of the Elements*, 2d ed., 2002.
Housecroft, C.E., and A.G. Sharp, *Inorganic Chemistry*, 2001.
Hueehy, J.E., E.A. Keiter, and R.L. Keiter, *Inorganic Chemistry: Principles of Structure & Reactivity*, 4th ed., 1993.
Jolly, W.J., *Modern Inorganic Chemistry*, 2d ed., 1991.
Lee, J.D., *Concise Inorganic Chemistry*, 5th ed., 1996.
Mackay, K.M., and R.A. Mackay, *Introduction to Modern Inorganic Chemistry*, 5th ed., 1996.
Miessler, G.L., and D.A. Tarr, *Inorganic Chemistry*, 3d ed., 1991.
Mingos, D.M.P., *Essential Trends in Inorganic Chemistry*, 1998.
Porterfield, W.W., *Inorganic Chemistry: A Unified Approach*, 2d ed., 1984.
Rayner-Canham, G., and T. Overton, *Descriptive Inorganic Chemistry*, 3d ed., 2002.
Shriver, D., and P. Atkins., *Inorganic Chemistry*, 3d ed., 1934.
Wulfsberg, G., *Inorganic Chemistry*, 1944.

### Journals:

*Inorganic Chemistry*, American Chemical Society.
*Inorganica Chimica Acta*, Elsevier.
*Dalton Transactions*, Royal Society of Chemistry.

# ORGANIC CHEMISTRY

Bassindale, A., *The Third Dimension in Organic Chemistry*, 1984.
Carey, F.A., and R.J. Sundberg, *Advanced Organic Chemistry: Structure and Mechanisms*, 4th ed., 2 vols., 2000.
Carey, F.A., *Organic Chemistry*, 5th ed., 2003.
Ege, S.N., *Organic Chemistry: Structure and Reactivity*, 4th ed., 1999.
Eliel, E., and S.H. Wilen, *Stereochemistry of Organic Compounds*, 1994.
Fleming, I., *Frontier Orbitals in Organic Chemical Reactions*, 1976.
Holum, J.R., *Fundamentals of General, Organic and Biological Chemistry*, 6th ed., 1997.
Juaristi, E., *Introduction to Stereochemistry and Conformational Analysis*, 1991.
March, J., and M.B. Smith, *March's Advanced Organic Chemistry: Reactions, Mechanisms, and Structure*, 5th ed., 2001.
McMurry, J.E., *Organic Chemistry*, 6th ed., 2004.
Mehrotra, R.C., and A. Singh, *Organometallic Chemistry: A Unified Approach*, 1991.
Morrison, R.T., and R.N. Boyd, *Organic Chemistry*, 7th ed., 2003.
Nicolaou, K.C., and E.J. Sorenson, *Classics in Total Synthesis: Targets, Strategies, Methods*, 1996.
*Science of Synthesis: Houben-Weyl Methods of Molecular Transformations*, 48 vols., 2001–2008.
Smith, M.B., *Organic Synthesis*, 2d ed., 2001.
Solomons, T.W.G., *Fundamentals of Organic Chemistry*, 5th ed., 1998.
Traynham, J.G., *Organic Nomenclature: A Programmed Introduction*, 5th ed., 1997.

Trost, B.M. (ed.), *Comprehensive Organic Synthesis: Selectivity, Strategy & Efficiency in Modern Organic Chemistry*, 9 vols., 1991.
Von Zelewsky, A., *Stereochemistry of Coordination Compounds*, 1996.

**Journals:**

*European Journal of Organic Chemistry*, Wiley-VCH, twice monthly.
*Journal of Organic Chemistry*, American Chemical Society, biweekly.
*Journal of Organometallic Chemistry*, Elsevier (Switzerland), 48 issues per year.
*Organic Letters*, American Chemical Society.
*Synlett: Accounts and Rapid Communications in Synthetic Organic Chemistry*, Thieme, 15 issues per year.
*Synthesis: Journal of Synthetic Organic Chemistry*, Thieme, 18 issues per year.
*Tetrahedron*, Elsevier, weekly.
*Tetrahedron Letters*, Elsevier, weekly.

## PHYSICAL CHEMISTRY

Adamson, A.W., and P.D. Fleischauer (eds.), *Concepts of Inorganic Photochemistry*, 1975, reprint.
Alberty, R.A., and R.J. Sibley, *Physical Chemistry*, 2d ed., 1996.
Atkins, P.W., *Concepts in Physical Chemistry*, 1995.
Baggott, J., *The Meaning of Quantum Theory: A Guide for Students of Chemistry and Physics*, 1992.
Barrow, G., *Physical Chemistry*, 6th ed., 1996.
Berry, R.S., et al., *Physical Chemistry*, 2d ed., 3 vols., reprinted 2002.
Bruce, P.G. (ed.), *Solid State Electrochemistry*, 1997.
Burdett, J.K., *Chemical Bonding in Solids*, 1995.
Calais, J.-L., *Quantum Chemistry Workbook: Basic Concepts and Procedures in the Theory of the Electronic Structure of Matter*, 1994.
Cheetham, A.K., and P. Day, *Solid-State Chemistry*, vol. 1: *Techniques*, 1990, vol. 2: *Compounds*, 1992.
Denbigh, K.G., *The Principles of Chemical Equilibrium*, 4th ed., 1981.
Evans, T.R., and A. Weissburger (eds.), *Applications of Lasers to Chemical Problems*, 1982.
Ferraudi, G.J., *Elements of Inorganic Photochemistry*, 1988.
Hoffman, R., *Solids and Surfaces: A Chemist's View of Bonding in Extended Structures*, 1989.
Horspool, W.M., and P.-S. Song (eds.), *Handbook of Organic Photochemistry and Photobiology*, 1995.
Houston, P.L., *Chemical Kinetics and Reaction Dynamics*, 2001.
Kopecky, J., *Organic Photochemistry: A Visual Approach*, 1992.
Ladd, M.F.C., *Structure and Bonding in Solid State Chemistry*, 1979.
Laidler, K.L., and J. Keith, *Chemical Kinetics*, 3d ed., 1987.
Levine, I.N., *Physical Chemistry*, 5th ed., 2001.
Levine, R.D., and R.B. Bernstein, *Molecular Reaction Dynamics and Chemical Reactivity*, 1987.
Lowe, J.P., *Quantum Chemistry*, 2d ed., 1993.
McQuarrie, D.A., *Quantum Chemistry*, 1983.
McQuarrie, D.A., and J.D. Simon, *Physical Chemistry: A Molecular Approach*, 1997.
Meites, L., *An Introduction to Chemical Equilibrium and Kinetics*, 1981.
Moore, J.W., and R.G. Pearson, *Kinetics and Mechanism*, 3d ed., 1981.
Rabek, J.F., *Photochemistry and Photophysics*, vol. 4, 1991.
Reid, C.E., *Chemical Thermodynamics*, 1990.
Schmalzried, H., *Solid State Reactions*, 2d ed., 1981.
Servos, J.W., *Physical Chemistry from Ostwald to Pauling: The Making of a Science in America*, 1996.
Smith, E.B., *Basic Chemical Thermodynamics*, 4th ed., 1990.
Smith, V.H., Jr., et al., *Applied Quantum Chemistry*, 1986.

Szabo, A., and N.S. Ostlund, *Modern Quantum Chemistry: Introduction to Advanced Electronic Structure Theory*, 1996.
Turro, N.J., *Modern Molecular Photochemistry*, 1981, reprint 1991.
West, A.R., *Basic Solid State Chemistry*, 2d ed., 1999.

### Journals:

*International Journal of Quantum Chemistry*, semimonthly.
*Journal of Chemical Physics*, American Institute of Physics, weekly.
*Journal of Physical Chemistry*, American Chemical Society, biweekly.
*Photochemistry and Photobiology* (text in English, French, or German), American Society for Photobiology, monthly.

## POLYMER CHEMISTRY

Allcock, H., et al., *Contemporary Polymer Chemistry*, 3d ed., 2004.
Allen, G., and J.C. Bevington (eds.), *Comprehensive Polymer Science: The Synthesis, Characterization, Reactions and Applications of Polymers: Polymer Characterization*, 1990.
Elias, H.G., *An Introduction to Polymer Science*, 1997.
Feldman, D., and A. Barbalata, *Synthetic Polymers: Technology, Properties, and Applications*, 1996.
Fried, J., and J. Hunger, *Polymer Science and Technology*, 2d ed., 2003.
*Kirk-Othmer Encyclopedia of Chemical Technology*, 4th ed., 27 vols., 1998.
Mark, H.F. (ed.), *Encyclopedia of Polymer Science and Technology*, 3d ed., 19 vols., 2004.
Morawetz, H., *Polymers: The Origins and Growth of a Science*, 1985.
Odian, G., *Principles of Polymerization*, 3d ed., 1991.
Seymour, R.B., and C.E. Carraher, Jr., *Seymour/Carraher's Polymer Chemistry*, 5th ed., 2000.
Stevens, M.P., *Polymer Chemistry: An Introduction*, 3d ed., 1998.
Young, R.J., and P. Lovell, *Introduction to Polymers*, 2d rev. ed., 1991.

### Journals:

*Macromolecules*, American Chemical Society.
*Polymer*, Elsevier.
*Progress in Polymer Science*, Elsevier.

**Equivalents of commonly used units for the U.S. Customary System and the metric system**

| | | |
|---|---|---|
| 1 inch = 2.5 centimeters (25 millimeters) | 1 centimeter = 0.4 inch | 1 inch = 0.083 foot |
| 1 foot = 0.3 meter (30 centimeters) | 1 meter = 3.3 feet | 1 foot = 0.33 yard (12 inches) |
| 1 yard = 0.9 meter | 1 meter = 1.1 yards | 1 yard = 3 feet (36 inches) |
| 1 mile = 1.6 kilometers | 1 kilometer = 0.62 mile | 1 mile = 5280 feet (1760 yards) |
| 1 acre = 0.4 hectare | 1 hectare = 2.47 acres | |
| 1 acre = 4047 square meters | 1 square meter = 0.00025 acre | |
| 1 gallon = 3.8 liters | 1 liter = 1.06 quarts = 0.26 gallon | 1 quart = 0.25 gallon (32 ounces; 2 pints) |
| 1 fluid ounce = 29.6 milliliters | 1 milliliter = 0.034 fluid ounce | 1 pint = 0.125 gallon (16 ounces) |
| 32 fluid ounces = 946.4 milliliters | | 1 gallon = 4 quarts (8 pints) |
| 1 quart = 0.95 liter | 1 gram = 0.035 ounce | 1 ounce = 0.0625 pound |
| 1 ounce = 28.35 grams | 1 kilogram = 2.2 pounds | 1 pound = 16 ounces |
| 1 pound = 0.45 kilogram | 1 kilogram = $1.1 \times 10^{-3}$ ton | 1 ton = 2000 pounds |
| 1 ton = 907.18 kilograms | | |
| $°F = (1.8 \times °C) + 32$ | $°C = (°F - 32) \div 1.8$ | |

**Conversion factors for the U.S. Customary System, metric system, and International System**

**A. Units of length**

| *Units* | *cm* | *m* | *in.* | *ft* | *yd* | *mi* |
|---|---|---|---|---|---|---|
| 1 cm | = 1 | 0.01 | 0.3937008 | 0.03280840 | 0.01093613 | $6.213712 \times 10^{-6}$ |
| 1 m | = 100. | 1 | 39.37008 | 3.280840 | 1.093613 | $6.213712 \times 10^{-4}$ |
| 1 in. | = 2.54 | 0.0254 | 1 | 0.08333333 . . . | 0.02777777 . . . | $1.578283 \times 10^{-5}$ |
| 1 ft | = 30.48 | 0.3048 | 12. | 1 | 0.3333333 . . . | $1.893939\ldots \times 10^{-4}$ |
| 1 yd | = 91.44 | 0.9144 | 36. | 3. | 1 | $5.681818\ldots \times 10^{-4}$ |
| 1 mi | $= 1.609344 \times 10^{5}$ | $1.609344 \times 10^{3}$ | $6.336 \times 10^{4}$ | 5280. | 1760. | 1 |

**B. Units of area**

| *Units* | *cm²* | *m²* | *in.²* | *ft²* | *yd²* | *mi²* |
|---|---|---|---|---|---|---|
| 1 $cm^2$ | = 1 | $10^{-4}$ | 0.1550003 | $1.076391 \times 10^{-3}$ | $1.195990 \times 10^{-4}$ | $3.861022 \times 10^{-11}$ |
| 1 $m^2$ | $= 10^{4}$ | 1 | 1550.003 | 10.76391 | 1.195990 | $3.861022 \times 10^{-7}$ |
| 1 $in.^2$ | = 6.4516 | $6.4516 \times 10^{-4}$ | 1 | $6.944444\ldots \times 10^{-3}$ | $7.716049 \times 10^{-4}$ | $2.490977 \times 10^{-10}$ |
| 1 $ft^2$ | = 929.0304 | 0.09290304 | 144. | 1 | 0.1111111 . . . | $3.587007 \times 10^{-8}$ |
| 1 $yd^2$ | = 8361.273 | 0.8361273 | 1296. | 9. | 1 | $3.228306 \times 10^{-7}$ |
| 1 $mi^2$ | $= 2.589988 \times 10^{10}$ | $2.589988 \times 10^{6}$ | $4.014490 \times 10^{9}$ | $2.78784 \times 10^{7}$ | $3.0976 \times 10^{6}$ | 1 |

**C. Units of volume**

| *Units* | *m³* | *cm³* | *liter* | *in.³* | *ft³* | *qt* | *gal* |
|---|---|---|---|---|---|---|---|
| 1 $m^3$ | = 1 | $10^6$ | $10^3$ | $6.102374 \times 10^4$ | $35.31467 \times 10^{-3}$ | 1.056688 | 264.1721 |
| 1 $cm^3$ | $= 10^{-6}$ | 1 | $10^{-3}$ | 0.06102374 | $3.531467 \times 10^{-5}$ | $1.056688 \times 10^{-3}$ | $2.641721 \times 10^{-4}$ |
| 1 liter | $= 10^{-3}$ | 1000. | 1 | 61.02374 | 0.03531467 | 1.056688 | 0.2641721 |
| 1 $in.^3$ | $= 1.638706 \times 10^{-5}$ | 16.38706 | 0.01638706 | 1 | $5.787037 \times 10^{-4}$ | 0.01731602 | $4.329004 \times 10^{-3}$ |
| 1 $ft^3$ | $= 2.831685 \times 10^{-2}$ | 28316.85 | 28.31685 | 1728. | 1 | 2.992208 | 7.480520 |
| 1 qt | $= 9.463529 \times 10^{-4}$ | 946.3529 | 0.9463529 | 57.75 | 0.03342014 | 1 | 0.25 |
| 1 gal (U.S.) | $= 3.785412 \times 10^{-3}$ | 3785.412 | 3.785412 | 231. | 0.1336806 | 4. | 1 |

**D. Units of mass**

| *Units* | *g* | *kg* | *oz* | *lb* | *metric ton* | *ton* |
|---|---|---|---|---|---|---|
| 1 g | = 1 | $10^{-3}$ | 0.03527396 | $2.204623 \times 10^{-3}$ | $10^{-6}$ | $1.102311 \times 10^{-6}$ |
| 1 kg | = 1000. | 1 | 35.27396 | 2.204623 | $10^{-3}$ | $1.102311 \times 10^{-3}$ |
| 1 oz (avdp) | = 28.34952 | 0.02834952 | 1 | 0.0625 | $2.834952 \times 10^{-5}$ | $3.125 \times 10^{-5}$ |
| 1 lb (avdp) | = 453.5924 | 0.4535924 | 16. | 1 | $4.535924 \times 10^{-4}$ | $5. \times 10^{-4}$ |
| 1 metric ton | $= 10^6$ | 1000. | 35273.96 | 2204.623 | 1 | 1.102311 |
| 1 ton | = 907184.7 | 907.1847 | 32000. | 2000. | 0.9071847 | 1 |

**Conversion factors for the U.S. Customary System, metric system, and International System (*cont.*)**

**E. Units of density**

| *Units* | $g \cdot cm^{-3}$ | $g \cdot L^{-1}$, $kg \cdot m^{-3}$ | $oz \cdot in.^{-3}$ | $lb \cdot in.^{-3}$ | $lb \cdot ft^{-3}$ | $lb \cdot gal^{-1}$ |
|---|---|---|---|---|---|---|
| 1 $g \cdot cm^{-3}$ | = 1 | 1000. | 0.5780365 | 0.03612728 | 62.42795 | 8.345403 |
| 1 $g \cdot L^{-1}$, $kg \cdot m^{-3}$ | = $10^{-3}$ | 1 | $5.780365 \times 10^{-4}$ | $3.612728 \times 10^{-5}$ | 0.06242795 | $8.345403 \times 10^{-3}$ |
| 1 $oz \cdot in.^{-3}$ | = 1.729994 | 1729.994 | 1 | 0.0625 | 108. | 14.4375 |
| 1 $lb \cdot in.^{-3}$ | = 27.67991 | 27679.91 | 16. | 1 | 1728. | 231. |
| 1 $lb \cdot ft^{-3}$ | = 0.01601847 | 16.01847 | $9.259259 \times 10^{-3}$ | $5.787037 \times 10^{-4}$ | 1 | 0.1336806 |
| 1 $lb \cdot gal^{-1}$ | = 0.1198264 | 119.8264 | $4.749536 \times 10^{-3}$ | $4.329004 \times 10^{-3}$ | 7.480519 | 1 |

**F. Units of pressure**

| *Units* | Pa, $N \cdot m^{-2}$ | $dyn \cdot cm^{-2}$ | *bar* | *atm* | $kgf \cdot cm^{-2}$ | *mmHg (torr)* | *in. Hg* | $lbf \cdot in.^{-2}$ |
|---|---|---|---|---|---|---|---|---|
| 1 Pa, 1 $N \cdot m^{-2}$ | = 1 | 10 | $10^{-5}$ | $9.869233 \times 10^{-6}$ | $1.019716 \times 10^{-5}$ | $7.500617 \times 10^{-3}$ | $2.952999 \times 10^{-4}$ | $1.450377 \times 10^{-4}$ |
| 1 $dyn \cdot cm^{-2}$ | = 0.1 | 1 | $10^{-6}$ | $9.869233 \times 10^{-7}$ | $1.019716 \times 10^{-6}$ | $7.500617 \times 10^{-4}$ | $2.952999 \times 10^{-5}$ | $1.450377 \times 10^{-5}$ |
| 1 bar | = $10^5$ | $10^6$ | 1. | 0.9869233 | 1.019716 | 750.0617 | 29.52999 | 14.50377 |
| 1 atm | = 101325 | 1013250 | 1.01325 | 1 | 1.033227 | 760. | 29.92126 | 14.69595 |
| 1 $kgf \cdot cm^{-2}$ | = 98066.5 | 980665 | 0.980665 | 0.9678411 | 1 | 735.5592 | 28.95903 | 14.22334 |
| 1 mmHg (torr) | = 133.3224 | 1333.224 | $1.333224 \times 10^{3}$ | $1.315789 \times 10^{-3}$ | $1.359510 \times 10^{-3}$ | 1 | 0.03937008 | 0.01933678 |
| 1 in. Hg | = 3386.388 | 33863.88 | 0.03386388 | 0.03342105 | 0.03453155 | 25.4 | 1 | 0.4911541 |
| 1 $lbf \cdot in.^{-2}$ | = 6894.757 | 68947.57 | 0.06894757 | 0.06804596 | 0.07030696 | 51.71493 | 2.036021 | 1 |

**G. Units of energy**

| Units | g mass (energy equiv) | J | eV | cal | $cal_{IT}$ | $Btu_{IT}$ | kWh | hp-h | ft-lbf | $ft^3 \cdot lbf \cdot in.^{-2}$ | liter-atm |
|---|---|---|---|---|---|---|---|---|---|---|---|
| 1 g mass (energy equiv) | = 1 | $8.987552 \times 10^{13}$ | $5.609589 \times 10^{32}$ | $2.148076 \times 10^{3}$ | $2.146640 \times 10^{13}$ | $8.518555 \times 10^{10}$ | $2.496542 \times 10^{7}$ | $3.347918 \times 10^{7}$ | $6.628878 \times 10^{13}$ | $4.603388 \times 10^{11}$ | $8.870024 \times 10^{11}$ |
| 1 J | $= 1.112650 \times 10^{-14}$ | 1 | $6.241509 \times 10^{18}$ | 0.2390057 | 0.2388459 | $9.478172 \times 10^{-4}$ | $2.777777\ldots \times 10^{-7}$ | 3.725062 | 0.7375622 | $5.121960 \times 10^{-3}$ | $9.869233 \times 10^{-3}$ |
| 1 eV | $= 1.782662 \times 10^{-33}$ | $1.602177 \times 10^{-19}$ | 1 | $3.829294 \times 10^{-20}$ | $3.826733 \times 10^{-20}$ | $1.518570 \times 10^{-22}$ | $4.450490 \times 10^{-26}$ | $5.968206 \times 10^{-26}$ | $1.181705 \times 10^{-19}$ | $8.206283 \times 10^{-22}$ | $1.581225 \times 10^{-21}$ |
| 1 cal | $= 4.655328 \times 10^{-14}$ | 4.184 | $2.611448 \times 10^{19}$ | 1 | 0.9993312 | $3.965667 \times 10^{-3}$ | $1.1622222\ldots \times 10^{-6}$ | $1.558562 \times 10^{-6}$ | 3.085960 | $2.143028 \times 10^{-2}$ | 0.04129287 |
| 1 $cal_{IT}$ | $= 4.658443 \times 10^{-14}$ | 4.1868 | $2.613195 \times 10^{19}$ | 1.000669 | 1 | $3.968321 \times 10^{-3}$ | $1.163 \times 10^{-6}$ | $1.559609 \times 10^{-6}$ | 3.088025 | $2.144462 \times 10^{-2}$ | 0.04132050 |
| 1 $Btu_{IT}$ | $= 1.173908 \times 10^{-11}$ | 1055.056 | $6.585141 \times 10^{21}$ | 252.1644 | 251.9958 | 1 | $2.930711 \times 10^{-4}$ | $3.930148 \times 10^{-4}$ | 778.1693 | 5.403953 | 10.41259 |
| 1 kWh | $= 4.005540 \times 10^{-8}$ | 3600000. | $2.246943 \times 10^{25}$ | 860420.7 | 859845.2 | 3412.142 | 1 | 1.341022 | 2655224. | 18349.06 | 35529.24 |
| 1 hp-h | $= 2.986931 \times 10^{-8}$ | 2384519. | $1.675545 \times 10^{25}$ | 641615.6 | 641186.5 | 2544.33 | 0.7456998 | 1 | 1980000. | 13750. | 26494.15 |
| 1 ft-lbf | $= 1.508551 \times 10^{-14}$ | 1.355818 | $8.462351 \times 10^{18}$ | 0.3240483 | 0.3238315 | $1.285067 \times 10^{-3}$ | $3.766161 \times 10^{-7}$ | $5.050505\ldots \times 10^{-7}$ | 1 | $6.944444\ldots \times 10^{-3}$ | 0.01338088 |
| 1 $ft^3 \cdot lbf \cdot in.^{-2}$ | $= 2.172313 \times 10^{-12}$ | 195.2378 | $1.218578 \times 10^{21}$ | 46.66295. | 46.63174 | 0.1850497 | $5.423272 \times 10^{-5}$ | $7.272727\ldots \times 10^{-5}$ | 144. | 1 | 1.926847 |
| 1 liter-atm | $= 1.127393 \times 10^{-12}$ | 101.325 | $6.324209 \times 10^{20}$ | 24.21726 | 24.20106 | 0.09603757 | $2.814583 \times 10^{-5}$ | $3.774419 \times 10^{-5}$ | 74.73349 | 0.5189825 | 1 |

**Defining fixed points of the International Temperature Scale of 1990 (ITS-90)**

| *Equilibrium state* | *Temperature* K | °C | °F |
|---|---|---|---|
| Vapor pressure equation of helium | 3 to 5 | −270.15 to −268.15 | −454.27 to 450.67 |
| Triple point of equilibrium hydrogen | 13.80 | −259.35 | −434.82 |
| Vapor pressure point of equilibrium hydrogen | ≈17 | ≈−256.15 | −429.07 |
| (or constant volume gas thermometer point of helium) | ≈20.3 | ≈−252.85 | −423.13 |
| Triple point of neon | 24.56 | −248.59 | −415.47 |
| Triple point of oxygen | 54.36 | −218.79 | −361.82 |
| Triple point of argon | 83.81 | −189.34 | −308.82 |
| Triple point of mercury | 234.32 | −38.83 | −37.90 |
| Triple point of water | 273.16 | 0.01 | 32.02 |
| Melting point of gallium | 302.91 | 29.76 | 85.58 |
| Freezing point of indium | 429.75 | 156.60 | 313.88 |
| Freezing point of tin | 505.08 | 231.93 | 449.47 |
| Freezing point of zinc | 692.68 | 419.53 | 787.15 |
| Freezing point of aluminum | 933.47 | 660.32 | 1220.58 |
| Freezing point of silver | 1234.93 | 961.78 | 1763.20 |
| Freezing point of gold | 1337.33 | 1064.18 | 1947.52 |
| Freezing point of copper | 1357.77 | 1084.62 | 1984.32 |

**Primary thermometry methods**

| Method | Approximate useful range of T, K | Principal measured variables | Relation of measured variables to T | Remarks |
|---|---|---|---|---|
| Gas thermometry | 1.3–950 | Pressure P and volume V | Ideal gas law plus correction: $PV \propto k_B T$ plus corrections | Careful determination of corrections necessary, but capable of high accuracy |
| Acoustic interferometry | 1.5–3000 | Speed of sound W | $W^2 \propto k_B T$ plus corrections | |
| Magnetic thermometry | | Magnetic susceptibility | Curie's law plus correction: $\chi \propto 1/k_B T$ plus corrections | |
| 1. Electron paramagnetism | 0.001–35 | | | |
| 2. Nuclear paramagnetism | 0.000001–1 | | | |
| Gamma-ray anisotropy or nuclear orientation thermometry | 0.01–1 | Spatial distribution of gamma-ray emission | Spatial distribution related to Boltzmann factor for nuclear spin states | Useful standard for $T < 1$ K |
| Thermal electric noise thermometry | | Mean square voltage fluctuation $\bar{V}^2$ | Nyquist's law: $\bar{V}^2 \propto k_B T$ | Other sources of noise are a serious problem for $T > 4$ K |
| 1. Josephson junction point contact | 0.001–1 | | | |
| 2. Conventional amplifier | 4–1400 | | | |
| Radiation thermometry (visual, photoelectric, or photodiode) | 500–50,000 | Spectral intensity I at wavelength $\lambda$ | Planck's radiation law, related to Boltzmann factor for radiation quanta | Needs blackbody conditions or well-defined emittance |
| Infrared spectroscopy | 100–1500 | Intensity I of rotational lines of light molecules | Boltzmann factor for rotational levels related to I | Also Doppler line broadening ($\propto \sqrt{k_B T}$) is useful; principal applications are to plasmas and astrophysical observations; proper sampling, lack of equilibrium, atmospheric absorption are often problems |
| Ultraviolet and x-ray spectroscopy | 5000–2,000,000 | Emission spectra from ionized atoms—H, He, Fe, Ca, and so on | Boltzmann factor for electron states related to band structure and line density | |

**Principal spectral regions and fields of spectroscopy**

| *Spectral region* | *Approx. wavelength range* | *Typical source* | *Typical detector* | *Energy transitions studied in matter* |
|---|---|---|---|---|
| Gamma | 1–100 pm | Radioactive nuclei | Geiger counter; scintillation counter | Nuclear transitions and disintegrations |
| X-rays | 6 pm–100 nm | X-ray tube (electron bombardment of metals) | Geiger counter | Ionization by inner electron removal |
| Vacuum ultraviolet | 10–200 nm | High-violet discharge; high-vacuum spark | Photomultiplier | Ionization by outer electron removal |
| Ultraviolet | 200–400 nm | Hydrogen-discharge lamp | Photomultiplier | Excitation of valence electrons |
| Visible | 400–800 nm | Tungsten lamp | Phototubes | Excitation of valence electrons |
| Near-infrared | 0.8–2.5 μm | Tungsten lamp | Photocells | Excitation of valence electrons; molecular vibrational overtones |
| Infrared | 2.5–50 μm | Nernst glower; Globar lamp | Thermocouple; bolometer | Molecular vibrations; stretching, bending, and rocking |
| Far-infrared | 50–1000 μm | Mercury lamp (high-pressure) | Thermocouple; bolometer | Molecular rotations |
| Microwave | 0.1–30 cm | Klystrons; magnetrons | Silicon-tungsten crystal; bolometer | Molecular rotations; electron spin resonance |
| Radio-frequency | $10^{-1}$–$10^{3}$ m | Radio transmitter | Radio receiver | Molecular rotations; nuclear magnetic resonance |

**Recommended values (2002) of selected fundamental physical constants**

| Quantity | Symbol* | Numerical value† | Units† | Relative uncertainty (standard deviation) |
|---|---|---|---|---|
| Speed of light in vacuum | $c$ | 299792458 | m/s | (defined) |
| Permeability of vacuum | $\mu_0$ | $4\pi \times 10^{-7}$ | $N/A^2$ | (defined) |
| Permittivity of vacuum | $\varepsilon_0$ | 8.854187817... | $10^{-12}$ F/m | (defined) |
| Constant of gravitation | G | 6.6742 (10) | $10^{-11}$ $m^3/(kg \cdot s^2)$ | $1.5 \times 10^{-4}$ |
| Planck constant | $h$ | 6.6260693 (11) | $10^{-34}$ J · s | $1.7 \times 10^{-7}$ |
| Elementary charge | $e$ | 1.60217653 (14) | $10^{-19}$ C | $8.5 \times 10^{-8}$ |
| Magnetic flux quantum, $h/(2e)$ | $\Phi_0$ | 2.06783372 (18) | $10^{-15}$ Wb | $8.5 \times 10^{-8}$ |
| Fine-structure constant, | $\alpha$ | 7.297352568 (24) | $10^{-3}$ | $3.3 \times 10^{-9}$ |
| $\mu_0 ce^2/(2h)$ | $\alpha^{-1}$ | 137.03599911 (46) | | $3.3 \times 10^{-9}$ |
| Electron mass | $m_e$ | 9.1093826 (16) | $10^{-31}$ kg | $1.7 \times 10^{-7}$ |
| Proton mass | $m_p$ | 1.67262171 (29) | $10^{-27}$ kg | $1.7 \times 10^{-7}$ |
| Neutron mass | $m_n$ | 1.67492728 (29) | $10^{-27}$ kg | $1.7 \times 10^{-7}$ |
| Proton-electron mass ratio | $m_p/m_e$ | 1836.15267261 (85) | | $4.6 \times 10^{-10}$ |
| Rydberg constant, $m_e c\alpha^2/(2h)$ | $R_\infty$ | 10973731.568525 (73) | $m^{-1}$ | $6.6 \times 10^{-12}$ |
| Bohr radius, $\alpha/(4\pi R_\infty)$ | $\alpha_0$ | 5.291772108 (18) | $10^{-11}$ m | $3.3 \times 10^{-9}$ |
| Compton wavelength of the electron, $h/(m_e c) = \alpha^2/(2R_\infty)$ | $\lambda_c$ | 2.426310238 (16) | $10^{-12}$ m | $6.7 \times 10^{-9}$ |
| Classical electron radius, $\mu_0 e^2/(4\pi m_e) = \alpha^3/(4\pi R_\infty)$ | $r_e$ | 2.817940325 (28) | $10^{-15}$ m | $1.0 \times 10^{-8}$ |
| Bohr magneton, $eh/(4\pi m_e)$ | $\mu_B$ | 9.27400949 (80) | $10^{-24}$ J/T | $8.6 \times 10^{-8}$ |
| Electron magnetic moment $\mu_e$ | | −9.28476412 (80) | $10^{-24}$ J/T | $8.6 \times 10^{-8}$ |
| Electron magnetic moment/Bohr magneton ratio | $\mu_e/\mu_B$ | −1.0011596521859 (38) | | $3.8 \times 10^{-12}$ |
| Nuclear magneton, $eh/(4\pi m_p)$ | $\mu_N$ | 5.05078343 (43) | $10^{-27}$ J/T | $8.6 \times 10^{-8}$ |
| Proton magnetic moment/nuclear magneton ratio | $\mu_p/\mu_N$ | 2.792847351 (28) | | $1.0 \times 10^{-8}$ |
| Avogadro constant | $N_A$ | 6.0221415 (10) | $10^{23}$ | $1.7 \times 10^{-7}$ |
| Faraday constant, $N_A e$ | F | 96485.3383 (83) | C/mol | $8.6 \times 10^{-8}$ |
| Molar gas constant | R | 8.314472 (15) | J/(mol · K) | $1.7 \times 10^{-6}$ |
| Boltzmann constant, $R/N_A$ | $k$ | 1.3806505 (24) | $10^{-23}$ J/K | $1.7 \times 10^{-6}$ |

*A = ampere, C = coulomb, F = farad, J = joule, kg = kilogram, K = kelvin, m = meter, mol = mole, N = newton, s = second, T = tesla, Wb = weber.
†Recommended by CODATA Task Group on Fundamental Constants. Digits in parentheses represent one-standard-deviation uncertainties in final two digits of quoted value.

**Electromagnetic spectrum**

| *Frequency, Hz* | *Wavelength, m* | *Nomenclature* | *Typical source* |
|---|---|---|---|
| $10^{23}$ | $3 \times 10^{-15}$ | Cosmic photons | Astronomical |
| $10^{22}$ | $3 \times 10^{-14}$ | $\gamma$-rays | Radioactive nuclei |
| $10^{21}$ | $3 \times 10^{-13}$ | $\gamma$-rays, x-rays | |
| $10^{20}$ | $3 \times 10^{-12}$ | x-rays | Atomic inner shell, positron-electron annihilation |
| $10^{19}$ | $3 \times 10^{-11}$ | Soft x-rays | Electron impact on a solid |
| $10^{18}$ | $3 \times 10^{-10}$ | Ultraviolet, x-rays | Atoms in sparks |
| $10^{17}$ | $3 \times 10^{-9}$ | Ultraviolet | Atoms in sparks and arcs |
| $10^{16}$ | $3 \times 10^{-8}$ | Ultraviolet | Atoms in sparks and arcs |
| $10^{15}$ | $3 \times 10^{-7}$ | Visible spectrum | Atoms, hot bodies, molecules |
| $10^{14}$ | $3 \times 10^{-6}$ | Infrared | Hot bodies, molecules |
| $10^{13}$ | $3 \times 10^{-5}$ | Infrared | Hot bodies, molecules |
| $10^{12}$ | $3 \times 10^{-4}$ | Far-infrared | Hot bodies, molecules |
| $10^{11}$ | $3 \times 10^{-3}$ | Microwaves | Electronic devices |
| $10^{10}$ | $3 \times 10^{-2}$ | Microwaves, radar | Electronic devices |
| $10^{9}$ | $3 \times 10^{-1}$ | Radar | Electronic devices, interstellar hydrogen |
| $10^{8}$ | 3 | Television, FM radio | Electronic devices |
| $10^{7}$ | 30 | Short-wave radio | Electronic devices |
| $10^{6}$ | 300 | AM radio | Electronic devices |
| $10^{5}$ | 3000 | Long-wave radio | Electronic devices |
| $10^{4}$ | $3 \times 10^{4}$ | Induction heating | Electronic devices |
| $10^{3}$ | $3 \times 10^{5}$ | | Electronic devices |
| 100 | $3 \times 10^{6}$ | Power | Rotating machinery |
| 10 | $3 \times 10^{7}$ | Power | Rotating machinery |
| 1 | $3 \times 10^{8}$ | | Commutated direct current |
| 0 | Infinity | Direct current | Batteries |

**Standard atomic weights**

| *Atomic number* | *Symbol* | *Name* | *Atomic weight** |
|---|---|---|---|
| 89 | Ac | Actinium | [227] |
| 13 | Al | Aluminum | 26.981538(2) |
| 95 | Am | Americium | [243] |
| 51 | Sb | Antimony | 121.760(1) |
| 18 | Ar | Argon | 39.948(1) |
| 33 | As | Arsenic | 74.92160(2) |
| 85 | At | Astatine | [210] |
| 56 | Ba | Barium | 137.327(7) |
| 97 | Bk | Berkelium | [247] |
| 4 | Be | Beryllium | 9.012182(3) |
| 83 | Bi | Bismuth | 208.98038(2) |
| 107 | Bh | Bohrium | [264] |
| 5 | B | Boron | 10.811(7) |
| 35 | Br | Bromine | 79.904(1) |
| 48 | Cd | Cadmium | 112.411(8) |
| 20 | Ca | Calcium | 40.078(4) |
| 98 | Cf | Californium | [251] |
| 6 | C | Carbon | 12.0107(8) |
| 58 | Ce | Cerium | 140.116(1) |
| 55 | Cs | Cesium | 132.90545(2) |
| 17 | Cl | Chlorine | 35.453(2) |
| 24 | Cr | Chromium | 51.9961(6) |
| 27 | Co | Cobalt | 58.933200(9) |
| 29 | Cu | Copper | 63.546(3) |
| 96 | Cm | Curium | [247] |
| 110 | Ds | Darmstadtium | [281] |
| 105 | Db | Dubnium | [262] |
| 66 | Dy | Dysprosium | 162.500(1) |
| 99 | Es | Einsteinium | [252] |
| 68 | Er | Erbium | 167.259(3) |
| 63 | Eu | Europium | 151.964(1) |
| 100 | Fm | Fermium | [257] |
| 9 | F | Fluorine | 18.9984032(5) |
| 87 | Fr | Francium | [223] |
| 64 | Gd | Gadolinium | 157.25(3) |
| 31 | Ga | Gallium | 69.723(1) |
| 32 | Ge | Germanium | 72.64(1) |
| 79 | Au | Gold | 196.96655(2) |
| 72 | Hf | Hafnium | 178.49(2) |
| 108 | Hs | Hassium | [277] |
| 2 | He | Helium | 4.002602(2) |
| 67 | Ho | Holmium | 164.93032(2) |
| 1 | H | Hydrogen | 1.00794(7) |
| 49 | In | Indium | 114.818(3) |
| 53 | I | Iodine | 126.90447(3) |
| 77 | Ir | Iridium | 192.217(3) |
| 26 | Fe | Iron | 55.845(2) |
| 36 | Kr | Krypton | 83.798(2) |
| 57 | La | Lanthanum | 138.9055(2) |
| 103 | Lf | Lawrencium | [262] |
| 82 | Pb | Lead | 207.2(1) |
| 3 | Li | Lithium | [6.941(2)] |
| 71 | Lu | Lutetium | 174.967(1) |
| 12 | Mg | Magnesium | 24.3050(6) |
| 25 | Mn | Manganese | 54.938049(9) |
| 109 | Mt | Meitnerium | [268] |
| 101 | Md | Mendelevium | [258] |
| 80 | Hg | Mercury | 200.59(2) |
| 42 | Mo | Molybdenum | 95.94(2) |

**Standard atomic weights (*cont.*)**

| *Atomic number* | *Symbol* | *Name* | *Atomic weight** |
|---|---|---|---|
| 60 | Nd | Neodymium | 144.24(3) |
| 10 | Ne | Neon | 20.1797(6) |
| 93 | Np | Neptunium | [237] |
| 28 | Ni | Nickel | 58.6934(2) |
| 41 | Nb | Niobium | 92.90638(2) |
| 7 | N | Nitrogen | 14.0067(2) |
| 102 | No | Nobelium | [259] |
| 76 | Os | Osmium | 190.23(3) |
| 8 | O | Oxygen | 15.9994(3) |
| 46 | Pd | Palladium | 106.42(1) |
| 15 | P | Phosphorus | 30.973761(2) |
| 78 | Pt | Platinum | 195.078(2) |
| 94 | Pu | Plutonium | [244] |
| 84 | Po | Polonium | [209] |
| 19 | K | Potassium | 39.0983(1) |
| 59 | Pr | Praseodymium | 140.90765(2) |
| 61 | Pm | Promethium | [145] |
| 91 | Pa | Protactinium | 231.03588(2) |
| 88 | Ra | Radium | [226] |
| 86 | Rn | Radon | [222] |
| 75 | Re | Rhenium | 186.207(1) |
| 45 | Rh | Rhodium | 102.90550(2) |
| 37 | Rb | Rubidium | 85.4678(3) |
| 44 | Ru | Ruthenium | 101.07(2) |
| 104 | Rf | Rutherfordium | [261] |
| 62 | Sm | Samarium | 150.36(3) |
| 21 | Sc | Scandium | 44.955910(8) |
| 106 | Sg | Seaborgium | [266] |
| 34 | Se | Selenium | 78.96(3) |
| 14 | Si | Silicon | 28.0855(3) |
| 47 | Ag | Silver | 107.8682(2) |
| 11 | Na | Sodium | 22.989770(2) |
| 38 | Sr | Strontium | 87.62(1) |
| 16 | S | Sulfur | 32.065(5) |
| 73 | Ta | Tantalum | 180.9479(1) |
| 43 | Tc | Technetium | [98] |
| 52 | Te | Tellurium | 127.60(3) |
| 65 | Tb | Terbium | 158.92534(2) |
| 81 | Tl | Thallium | 204.3833(2) |
| 90 | Th | Thorium | 232.0381(1) |
| 69 | Tm | Thulium | 168.93421(2) |
| 50 | Sn | Tin | 118.710(7) |
| 22 | Ti | Titanium | 47.867(1) |
| 74 | W | Tungsten | 183.84(1) |
| 112 | Uub | Ununbium | [285] |
| 111 | Uuu | Unununium | [272] |
| 92 | U | Uranium | 238.02891(3) |
| 23 | V | Vanadium | 50.9415(1) |
| 54 | Xe | Xenon | 131.293(6) |
| 70 | Yb | Ytterbium | 173.04(3) |
| 39 | Y | Yttrium | 88.90585(2) |
| 30 | Zn | Zinc | 65.409(4) |
| 40 | Zr | Zirconium | 91.224(2) |

*Atomic weights are those of the most commonly available long-lived isotopes on the 1999 IUPAC Atomic Weights of the Elements. A value given in square brackets denotes the mass number at the longest-lived isotope.

# BIOGRAPHICAL LISTING

**Abel, Frederick Augustus** (1827–1902), English chemist. Expert on the chemistry of explosives; originated the Abel test for determination of the flash point of petroleum.

**Abney, William de Wiveleslie** (1843–1920), English photographic chemist and physicist. Photographed the infrared solar spectrum.

**Agre, Peter** (1949– ), American medical doctor and scientist. Discovered water channels in cell membranes; Nobel Prize, 2003.

**Alder, Kurt** (1902–1958), German chemist. Codeveloper of the Diels-Alder reaction for diene synthesis; contributed to stereochemistry; Nobel Prize, 1950.

**Altman, Sidney** (1939– ), American chemist. Discovered an unusual enzyme that contains ribonucleic acid (RNA) in addition to a protein, leading to the discovery that RNA molecules have catalytic properties similar to those of enzymes; Nobel Prize, 1989.

**Amagat, Emile** (1841–1915), French physicist. Investigated relationship of pressure, density, and temperature in gases and liquids, particularly at high pressure.

**Anfinsen, Christian Boehmer** (1916–1995), American biochemist. Discovered how three-dimensional structures of ribonuclease and other proteins are formed; Nobel Prize, 1972.

**Angström, Anders Jonas** (1814–1874), Swedish physicist. Mapped the solar spectrum; discovered hydrogen in the solar atmosphere.

**Arber, Werner** (1929– ), Swiss molecular biologist. Determined the molecular mechanism of host-controlled restriction modification of bacterial viruses and discovered the restriction enzymes; Nobel Prize, 1978.

**Arrhenius, Svante August** (1859–1927), Swedish physicist and chemist. Developed theory of electrolytic dissociation; investigated osmosis and viscosity of solutions; Nobel Prize, 1903.

**Aston, Francis William** (1877–1945), English physicist and chemist. Discovered isotopes in nonradioactive elements by using the mass spectrograph he invented; Nobel Prize, 1922.

**Avogadro, Amedeo** (1776–1856), Italian physicist. Formulated Avogadro's law.

**Axelrod, Julius** (1912– ), American biochemist and pharmacologist. Showed that many drugs act by modifying storage of neurotransmitters at nerve terminals; made discoveries concerning metabolism, and mechanisms for formation and inactivation of norepinephrine; Nobel Prize, 1970.

**Babcock, Stephen Moulton** (1843–1931), American agricultural chemist. Pioneer in nutrition; devised the Babcock test to measure fat content in milk.

**Back, Ernst E. A.** (1881–1959), German physicist. Developed improved spectrographs; made spectroscopic observations leading to Paschen-Back effect.

**Badger, Richard McLean** (1896–1974), American physical chemist and spectroscopist. Studied structures of polyatomic molecules; formulated Badger's rule concerning molecular bonds.

**Baekeland, Leo Hendrik** (1864–1944), Belgian-born, American chemist. Invented the phenol-formaldehyde polymer, Bakelite, the first commercial synthetic polymer.

**Baeyer, Johann Friedrich Wilhelm Adolf von** (1835–1917), German chemist. Synthesized indigo and hydroaromatic compounds; Nobel Prize, 1905.

**Balmer, Johann Jakob** (1825–1898), Swiss physicist. Expressed the mathematical formula for frequencies of hydrogen lines in the visible spectrum.

**Bardeen, John** (1908–1991), American physicist. With L. N. Cooper and J. R. Schrieffer, formulated a theory of superconductivity; invented the transistor; Nobel Prize, 1956 and 1972.

**Barton, Derek Harold Richard** (1918–1998), British chemist. Developed and expanded concept of conformation to include large molecules with complex ring systems; Nobel Prize, 1969.

**Bassham, James Alan** (1922– ), American chemist. Helped to elucidate basic photosynthetic carbon cycle.

**Baumé, Antoine** (1728–1804), French chemist. Invented a graduated hydrometer which utilizes the Baumé scale.

**Beams, Jesse Wakefield** (1898–1977), American physicist. Developed vacuum-type ultracentrifuges, used in purification and molecular-weight determination of large-molecular-weight substances, isotope separation, and determination of the gravitational constant.

**Beattle, James Alexander** (1895–1981), American chemist and physicist. Studied ionic theory and thermodynamics; with P. W. Bridgman, proposed Beattie and Bridgman equation for gases.

**Béchamp, Pierre Jacques Antoine** (1816–1908), French chemist. Discovered a method of preparing aniline.

**Beckmann, Ernst Otto** (1853–1923), German chemist. Discovered Beckmann molecular transformation; invented the Beckmann thermometer.

**Becquerel, Antoine César** (1788–1878), French physicist. Pioneer in electrochemistry; first to extract metals from ore by electrolysis.

**Becquerel, Antoine Henri** (1852–1908), French physicist. A discoverer of radioactivity in uranium.

**Bednorz, Johannes Georg** (1950– ), German physicist. With K. A. Müller, discovered high-temperature superconductivity in copper oxide ceramic materials; Nobel Prize, 1987.

**Beer, August** (1825–1863), German physicist. Discovered Beer's law of light absorption.

**Berg, Paul** (1926– ), American biochemist. Investigated the biochemistry of deoxyribonucleic acid (DNA) and designed a technique for gene splicing; Nobel Prize, 1980.

**Bergius, Friedrich** (1884–1949), Polish-born German chemist. Developed Bergius process for hydrogenation of coal to a petroleumlike oil; Nobel Prize, 1931.

**Bergström, Sune Karl** (1916– ), Swedish biochemist and medical scientist. Studied the metabolism of unsaturated fatty acids and determined the chemical structure of prostaglandins; Nobel Prize, 1982.

**Berthelot, Pierre Eugéne Marcellin** (1827–1907), French chemist. Founder of thermochemistry; first to synthesize organic compounds; demonstrated nitrogen fixation.

**Berzelius, Jons Jakob** (1779–1848), Swedish chemist. Discovered the elements cerium, selenium, thorium, and silicon; developed a system for classification and nomenclature of compounds.

**Binnig, Gerd** (1947– ), German physicist. With H. Rohrer, developed scanning tunneling microscope; Nobel Prize, 1986.

**Bloch, Felix** (1905–1983), Swiss-born American physicist. Discovered a technique for studying magnetism of atomic nuclei in normal matter; Nobel Prize, 1952.

**Bloch, Konrad Emil** (1912–2000), German-born American biochemist. Traced the transformations of fat and carbohydrate metabolites to cholesterol; Nobel Prize, 1964.

**Blodgett, Katharine Burr** (1898–1979), American chemical physicist. Studied surface science, and is best know for her work with Irving Langmuir and the development of the Langmuir-Blodgett film.

**Bohr, Aage** (1922– ), Danish physicist. With B. R. Mottelson, developed theory which unifies shell and liquid-drop models of the atomic nucleus, and which explains nonspherical nuclei; Nobel Prize, 1975.

**Bohr, Niels** (1885–1962), Danish physicist. Devised an atomic model; codeveloped the quantum theory, applying it to atomic structure in Bohr's theory; Nobel Prize, 1922.

**Bosch, Carl** (1874–1940), German chemist. Developed chemical high-pressure methods, and the Haber-Bosch process for ammonia synthesis; Nobel Prize, 1931.

**Boyle, Robert** (1627–1691), British physicist and chemist. Conducted experiments on properties of the air pump; Boyle's law concerning gases is named for him; advanced the atomistic theory of matter.

**Brønsted, Johannes Nicolaus** (1879–1947), Danish chemist. Researched kinetic properties of ions, catalysis, and nitramide; formulated the Brønsted theory of acid-base reactions.

**Brown, Herbert Charles** (1912– ), British-born American chemist. Developed methods for chemical synthesis of diborane and organoboranes; Nobel Prize, 1979.

**Brunauer, Stephen** (1903–1986), Hungarian-born American chemist. Contributed to surface and colloid chemistry; with P. H. Emmett and E. Teller, developed Brunauer-Emmett-Teller equation for surface area determinations.

**Buchner, Eduard** (1860–1917), German chemist. Studied alcoholic fermentation of sucrose; Nobel Prize, 1907.

**Bunsen, Robert Wilhelm** (1811–1899), German chemist. Discovered, with G. R. Kirchhoff, spectrum analysis; invented the Bunsen burner, Bunsen cell, and Bunsen ice calorimeter; formulated law of reciprocity with H. E. Roscoe.

**Butenandt, Adolph Friedrich Johann** (1903–1995), German chemist. Researched sex hormones; Nobel Prize (declined), 1939.

**Cailletet, Louis Paul** (1832–1913), French chemist. Researched liquefaction of gases; first to obtain liquid oxygen, hydrogen, nitrogen, and air.

**Calvin, Melvin** (1911–1997), American chemist. With J. A. Bassham, traced the path of carbon in photosynthesis; Nobel Prize, 1961.

**Cannizzaro, Stanislao** (1826–1910), Italian chemist. Promulgated Avogadro's work as related to atomic weights; discovered Cannizzaro's reaction in organic chemistry.

**Cavendish, Henry** (1731–1810), French physicist and chemist. Determined the density of the Earth and the composition of the atmosphere; studied properties of carbon dioxide and hydrogen.

**Cech, Thomas R.** (1947– ), American chemist. By studying the single-cell organisms *Tetrahymena* and *Thermophila*, discovered that molecules of ribonucleic acid (RNA) have catalytic properties similar to those of enzymes; Nobel Prize, 1989.

**Chain, Ernst Boris** (1906–1979), German-born British biochemist. With H. W. Florey, worked on the chemical structure of penicillin and its first clinical trials; Nobel Prize, 1945.

**Chaptal, Jean Antoine Claude, Comte de Chanteloup** (1756–1832), French chemist. Wrote on technical chemistry; introduced the metric system after the Revolution.

**Charles, Jacques Alexandre César** (1746–1823), French physicist, chemist, and inventor. Formulated Charles' law, relating gas volume to pressure.

**Claisen, Ludwig** (1851–1930), German organic chemist. Developed Claisen condensation; contributed to understanding of tautomerism; worked on rearrangement of allyl aryl ethers into phenols.

**Clapeyron, Benoit Paul Émile** (1799–1864). French engineer. Developed N. L. S. Carnot's concept of a universal function of temperature.

**Clausius, Rudolf Julius Emmanuel** (1822–1888), German physicist. A founder of thermodynamics; worked out the Clausius-Clapeyron equation for the universal temperature function.

**Cole, Robert Hugh** (1914–1990), American chemist and physicist. Research on dielectric properties of matter and intermolecular forces; with brother, K. S. Cole, introduced Cole-Cole plot of dielectric behavior.

**Conant, James Bryant** (1893–1978), American chemist. Researched free radicals, hemoglobin, and chlorophyll; contributed to atomic energy development.

**Corey, Elias James** (1928– ), American chemist. Developed theories and methods of organic chemical synthesis that have made possible the production of a wide variety of complex biologically active substances and useful chemicals; Nobel Prize, 1990.

**Cori, Carl Ferdinand** (1896–1984), **and Cori, Gerty Theresa Radnitz** (1896–1957), Czechoslovakian-born American biochemists. Discovered the enzymatic mechanism of glucose-glycogen interconversion and the effects of hormones on this mechanism; Nobel Prize, 1947.

**Cornforth, John Warcup** (1917– ), Australian-born British chemist. Investigated stereochemistry of enzyme-catalyzed reactions; Nobel Prize, 1975.

**Cottrell, Frederick Gardner** (1877–1948), American chemist. Invented the Cottrell process for precipitation of particles from gas; researched nitrogen fixation, liquefaction of gases, and recovery of helium.

**Crafts, James Mason** (1839–1917), American chemist. With C. Friedel, discovered the Friedel-Crafts reaction, wherein anhydrous aluminum chloride acts as a catalyst.

**Cram, Donald J.** (1919–2001), American chemist. Expanded the field of crown ether chemistry by using crown ethers to synthesize structures that mimic the action of biological molecules; Nobel Prize, 1987.

**Crick, Francis Harry Compton** (1916– ), English molecular biologist. With J. D. Watson, proposed a double-helix structure for the deoxyribonucleic acid molecule; Nobel Prize, 1962.

**Cronstedt, Axel Fredrik, Baron** (1722–1765), Swedish mineralogist. Discovered nickel; developed a chemical classification system for minerals.

**Crookes, William** (1832–1919), English physicist and chemist. Invented Crookes tube to study electrical discharges in high vacuum, and a radiometer; discovered thallium.

**Crutzen, Paul J.** (1933– ), Dutch-born meteorologist. Contributed to the understanding of how atmospheric ozone forms and decomposes; Nobel Prize, 1995.

**Curie, Marie,** born **Marya Sklodowska** (1867–1934), Polish physical chemist in France. Explored nature of radioactivity; codiscoverer of radium, and first to separate polonium; Nobel Prize, 1903 and 1911.

**Curie, Pierre** (1859–1906), French chemist and physicist. Codiscoverer of radium; formulated the Curie point, relating magnetic properties and temperature; discovered the piezoelectric effect; Nobel Prize, 1903.

**Curl, Robert F., Jr.** (1933– ), American chemist. With Harold Kroto and Richard Smalley, discovered fullerenes; Nobel Prize, 1996.

**Dalton, John** (1766–1844), English chemist and physicist. Proposed the atomic theory of chemical reactions; developed the law of partial pressures of gases; studied color-blindness.

**Danckwerts, Peter Victor** (1916–1984), British chemical engineer. Proposed the surface-renewal model of liquids.

**Daniell, John Frederic** (1790–1845), English physicist and chemist. Invented the Daniell cell.

**Davy, Humphry** (1778–1829), English chemist. Discovered potassium and sodium; invented the Davy safety lamp for use in coal mines; proposed theoretical explanations of electrolysis and voltaic action.

**Debye, Peter Joseph William** (1884–1966), American physical chemist born in the Netherlands. Worked on dipole moments and the diffraction of x-rays in gases; formulated Debye-Hückel theory on the behavior of strong electrolytes; Nobel Prize, 1936.

**de Gennes, Pierre-Gilles** (1932– ), French physicist. Applied physical principles to the study of complex systems, including liquid crystals and polymers; Nobel Prize, 1991.

**Deisenhofer, Johann** (1943– ), German chemist. With R. Huber and H. Michel, elucidated the structure of a bacterial protein that performs photosynthesis; Nobel Prize, 1988.

**Dewar, James** (1842–1923), British chemist. Made pioneering studies of matter at low temperatures; first to liquefy hydrogen; invented the Dewar vacuum flask.

**Diels, Otto Paul Hermann** (1876–1954), German chemist. Codiscoverer of the Diels-Alder reaction (diene synthesis); worked on sterol chemistry; discovered carbon suboxide; Nobel Prize, 1950.

**Djerassi, Carl** (1923– ), Austrian-born, American chemist. Synthesized the first oral contraceptive.

**Donnan, Frederick George** (1870–1956), Irish chemist born in Ceylon. Research in chemical kinetics; originated the Donnan theory of membrane equilibrium.

**Dufay, Charles François de Cisternay** (1698–1739), French chemist. Discovered positive and negative types of electricity.

**Dulong, Pierre Louis** (1785–1838), French chemist and physicist. With A. T. Petit, formulated the law of the constancy of atomic heats; developed the Dulong formula for heat value of fuels.

**Dumas, Jean Baptiste André** (1800–1884), French chemist. Research on organic compounds; determined many atomic weights.

**Eigen, Manfred** (1927– ), German chemist. Devised relaxation techniques to study high-speed chemical reactions; Nobel Prize, 1967.

**Emmett, Paul Hugh** (1900–1985), American chemist. Worked on catalysts for ammonia synthesis and the water-gas conversion reaction; with S. Brunauer and E. Teller, formulated Brunauer-Emmett-Teller equation for surface area determinations.

**Erlenmeyer, Richard August Carl Emil** (1825–1909), German organic chemist. Research on synthesis and constitution of aliphatic compounds; introduced modern structural notation; invented Erlenmeyer flask.

**Ernst, Richard R.** (1933– ). Swiss chemist. Developed methods that transformed nuclear magnetic resonance (NMR) spectroscopy from a tool with a narrow application to a key analytical technique in chemistry as well as many other fields: Nobel Prize, 1991.

**Euler-Chelpin, Hans Karl August Simon von** (1873–1964), German-Swedish chemist. Research on enzyme action and fermentation of sugars; Nobel Prize, 1929.

**Eyring, Henry** (1901–1981), Mexican-born American chemist. Pioneered in the application of quantum and statistical mechanics to chemistry; conceived the theory of absolute reaction rates and the significant structures theory of liquids.

**Faraday, Michael** (1791–1867), English chemist and physicist. Discovered electromagnetic induction; formulated two laws of electrolysis; invented the dynamo.

**Fenn, John B.** (1917– ), American chemist. Developed soft desorption ionization method for mass spectrometric analysis of biological macromolecules; Nobel Prize, 2002.

**Fischer, Edmond H.** (1920– ), American biochemist. With E. G. Krebs, discovered phosphorylation processes that play a critical role in cell-protein regulation; they isolated the first protein kinase, a class of enzymes that transfer phosphate from adenosine triphosphate to proteins; Nobel Prize, 1992.

**Fischer, Emil Hermann** (1852–1919), German chemist. Synthesized many natural substances, including purines, D-glucose and other sugars, and the first nucleotide; studied polypeptides and proteins; Nobel Prize, 1902.

**Fischer, Ernst Otto** (1918– ), German chemist. Studied how metals and organic molecules combine to form unique molecules with sandwichlike structures; Nobel Prize, 1973.

**Fischer, Hans** (1881–1945), German organic chemist. Investigated and synthesized pyrrole pigments; studied structure of chlorophylls; Nobel Prize, 1930.

**Flory, Paul John** (1910–1985), American physical chemist. Developed analytic techniques to explore properties and molecular structures of long-chain molecules; Nobel Prize, 1974.

**Friedel, Charles** (1832–1899), French chemist and mineralogist. With J. M. Crafts, described the Friedel-Crafts reaction; work on artificial production of minerals; studied crystals, ketones, and aldehydes.

**Fukui, Kenichi** (1918–1998), Japanese chemist. Developed frontier orbital theory, a quantum-mechanical model useful in prediction of the combinative properties of molecules; Nobel Prize, 1981.

**Fuller, R. Buckminster** (1895–1983), American engineer and architect. Designed geodesic dome; the carbon molecular form $C_{60}$ was named buckminsterfullerene because of its structured resemblance to the geodesic dome, and the name fullerene was given to any closed-cage molecule containing an even number of carbon atoms.

**Gatterman, Friedrich August Ludwig** (1860–1920), German chemist. Originated Gatterman-Koch synthesis of aldehydes; isolated and analyzed nitrogen trichloride; synthesized aromatic carboxylic acids, thionaphthene, and thioanilide.

**Gay-Lussac, Joseph Louis** (1778–1850), French chemist and physicist. Discovered the law of expansion of gases by heat, and the law of combining volumes of gases; studied chemistry of iodine and cyanogen.

**Giauque, William Francis** (1895–1982), Canadian-born American chemist. Developed adiabatic demagnetization technique for production of extremely low temperatures; collaborated in discovery of isotopes of oxygen; Nobel Prize, 1949.

**Glauber, Johann Rudolf** (1604–1670), German chemist. Discovered Glauber's salt (sodium sulfate) and hydrochloric acid; conducted experiments on compounds of mercury, arsenic, and antimony.

**Gmelin, Leopold** (1788–1853), German chemist. Wrote *Handbuch der Chemie*, first systematic treatment of chemical knowledge; devised Gmelin's test for presence of bile pigments; studied cyanides.

**Grignard, Francois Auguste Victor** (1871–1935), French chemist. Discovered organomagnesium compounds, or Grignard reagents, useful in synthesis of organic and organometallic compounds; Nobel Prize, 1912.

**Haber, Fritz** (1868–1934), German chemist. Developed the Haber-Boch process for synthesis of ammonia; made electrochemical studies; Nobel Prize, 1918.

**Hahn, Otto** (1879–1968), German chemist. With L. Meitner and F. Strassman, discovered that fission of heavy nuclei was possible by irradiation with neutrons; discovered protactinium with Meitner; Nobel Prize, 1944.

**Hall, Charles Martin** (1863–1914), American commercial chemist. Discovered Hall process for extracting aluminum.

**Harden, Arthur** (1865–1940), English chemist. Research on enzymes and alcoholic fermentation; Nobel Prize, 1929.

**Hassel, Odd** (1897–1981), Norwegian chemist. Developed concept of conformation by studying three-dimensional structure of cyclohexane molecule, and explaining the orientation of attached atoms or functional groups; Nobel Prize, 1969.

**Hauptman, Herbert Aaron** (1917– ), American chemist. With J. Karle, developed computer-aided mathematical techniques for use in x-ray crystallography to determine three-dimensional structures of molecules; Nobel Prize, 1985.

**Haworth, Walter Norman** (1883–1950), English chemist. Synthesized ascorbic acid; studied carbohydrates, including the structure of sugars; Nobel Prize, 1937.

**Heeger, Alan J.** (1936– ), American physicist. Discovered and developed conductive polymers with Alan MacDiarmid and Hideki Shirakawa; Nobel Prize, 2000.

**Henry, William** (1775–1836), English chemist and physician. Formulated Henry's law of solubility of gases in liquid.

**Herschbach, Dudley Robert** (1932– ), American chemist. With Y. T. Lee, developed crossed molecular-beam technique for tracing chemical reactions; Nobel Prize, 1986.

**Hevesy, George de** (1885–1966), Hungarian chemist. Experimented with radioisotope indication, leading to the technique of isotope tracing of biological and chemical processes; Nobel Prize, 1943.

**Heyrovský, Jaroslav** (1890–1967), Czechoslovakian physical chemist. Developed the technique of polarographic analysis; Nobel Prize, 1959.

**Hinshelwood, Cyril Norman** (1897–1967), British chemist. Elucidated chain reaction and chain branching mechanisms; Nobel Prize, 1956.

**Hoffmann, Roald** (1937– ), Polish-born American chemist. Developed methods for predicting whether a chemical reaction is possible based on molecular orbital models; Nobel Prize, 1981.

**Hofmann, August Wilhelm von** (1818–1892), German chemist. Studied reactions of derivatives; developed the Hofmann reaction for preparing primary amines.

**Huber, Robert** (1937– ), German chemist. With J. Deisenhofer and H. Michel, elucidated the structure of a bacterial protein that performs photosynthesis; Nobel Prize, 1988.

**Hückel, Erich** (1896–1980), German chemist. With P. J. W. Debye, formulated Debye-Hückel theory of strong electrolytes; devised theoretical explanation of electron properties of aromatic hydrocarbons.

**Jaeger, Frans Maurits** (1877–1945), Dutch crystallographer and physical chemist. Measured physical properties of molten salts and silicates at extremely high temperatures.

**Karle, Jerome** (1918– ), American crystallographer. With H. A. Hauptman, developed computer-aided mathematical techniques for use in x-ray crystallography to determine three-dimensional structures of molecules; Nobel Prize, 1985.

**Karrer, Paul** (1889–1971), Swiss chemist. Pioneering research on vitamins A and $B_2$ and on the flavins and carotenoids; Nobel Prize, 1937.

**Kastler, Alfred** (1902–1984), French physicist. Developed a double-resonance method to study energy levels of atoms in excited states; Nobel Prize, 1966.

**Kekulé von Stradonitz, Friedrich August** (1829–1896), German chemist. A founder of structural organic chemistry; made theoretical proposal of the structure of benzene.

**Kendall, Edward Calvin** (1886–1972), American biochemist. Chemical investigation of the adrenal cortex, leading to the isolation of crystalline cortical hormones, especially cortisone; Nobel Prize, 1950.

**Knowles, William S.** (1917– ), American chemist. Developed chirally catalyzed hydrogenation reactions; Nobel Prize, 2001.

**Kolbe, Adolf Wilhelm Hermann** (1818–1884), German chemist. Contributed to the synthesis concept of compound formation, doing much to eliminate the division of chemistry into two branches: organic and inorganic.

**Kornberg, Arthur** (1918– ), American biochemist. Discovered deoxyribonucleic acid polymerase, providing the first rational enzymatic mechanism for the replication of genetic material of the cell; Nobel Prize, 1959.

**Kossel, Albrecht** (1853–1927), German chemist. Investigated the chemistry of cells and of proteins; Nobel Prize, 1910.

**Krebs, Hans Adolf** (1900–1981), German-born British biochemist. Elucidated metabolic pathways, including the tricarboxylic acid cycle; Nobel Prize, 1953.

**Kroto, Harold W.** (1939– ), British chemist. With Richard Smalley and Robert Curl, discovered fullernes; Nobel Prize, 1996.

**Kuhn, Richard** (1900–1967), German chemist. Research on the structures and synthesis of vitamins and carotenoids; Nobel Prize, 1938 (declined).

**Kwolek, Stephanie** (1923– ), American polymer chemist. Developed Kevlar, poly(*p*-phenylene terephthalamide), a lightweight synthetic fiber that is stronger than steel.

**Langmuir, Irving** (1881–1957), American chemist. With G. N. Lewis, proposed the Lewis-Langmuir atomic theory; studied surface chemistry and thermionic emission; Nobel Prize, 1932.

**Lavoisier, Antoine Laurent** (1743–1794), French chemist. The founder of modern chemistry; studied combustion and respiration; published a table of the elements.

**Le Chatelier, Henry Louis** (1850–1936), French chemist and metallurgist. Research on cement chemistry, gas combustion, blast furnace reactions, chemical equilibria, alloy properties, and chemistry and metallurgy of iron and steel; formulated Le Chatelier's principle.

**Leclanché, Georges** (1839–1882), French chemist and electrician. Invented the Leclanché galvanic cell.

**Lee, Yuan Tseh** (1936– ), American chemist. With Dudley R. Herschbach, developed crossed molecular-beam technique for tracing chemical reactions; Nobel Prize, 1986.

**Lehn, Jean-Marie** (1939– ), French chemist. Studied crown ethers and developed the synthesis of related structures known as cryptands; Nobel Prize, 1987.

**Leloir, Luis Federico** (1906–1987), French-born Argentine biochemist. Discovered sugar nucleotides and their role in carbohydrate biosynthesis; Nobel Prize, 1970.

**Lennard-Jones, John Edward** (1894–1954), English physicist and chemist. Proposed Lennard-Jones potential for interatomic forces; contributed to quantum theory of molecular structure and statistical mechanics of liquids, gases, and surfaces.

**Lewis, Gilbert Newton** (1875–1946), American chemist. Collaborated in developing the Lewis-Langmuir atomic theory; worked on the electronic theory of valency and chemical thermodynamics.

**Libby, Willard Frank** (1908–1980), American chemist. Developed the method of radiocarbon dating; Nobel Prize, 1960.

**Liebig, Justus, Baron von** (1803–1873), German chemist. Discovered chloroform and chloral; founded agricultural chemistry; invented the Liebig condenser.

**Lipscomb, William Nunn, Jr.** (1919– ), American physical chemist. Studied structure and bonding of boranes, providing insight into nature of chemical bonding; Nobel Prize, 1976.

**Loschmidt, Johann Joseph** (1821–1895), Austrian physicist and chemist, born in Bohemia. Worked on graphical and structural molecular formulas; attempted to estimate size of air molecules and number of air molecules per unit volume.

**Lowry, Thomas Martin** (1874–1936), British chemist. Simultaneously with Johannes Brønsted, defined acid-base theory in terms of proton transfer—the Brønsted- Lowry theory (of acids and bases).

**MacDiarmid, Alan G.** (1927– ), New Zealand-born American chemist. Discovered and developed conductive polymers with Hideki Shirakawa and Alan Heeger; Nobel Prize, 2000.

**MacKinnon, Roderick** (1956– ), American medical doctor and scientist. Determined structure and mechanism of ion channels in cell membranes; Nobel Prize, 2003.

**Marcus, Rudolph Arthur** (1923– ), Canadian-born American chemist. Developed the mathematical analysis for electron transfer reactions in chemical systems; Nobel Prize, 1992.

**Mark, Herman Francis** (1895–1992), Austrian-born American chemist. Elucidated molecular structures of natural and synthetic polymers; developed theory of polymerization; studied relation between structure and properties of macromolecular systems.

**Martin, Archer John Porter** (1910– ), English chemist. With R. L. M. Synge, developed partition chromatography; Nobel Prize, 1952.

**McMillan, Edwin Mattison** (1907–1991), American physicist. Discovered element 93 (neptunium), which led to the creation of element 94 (plutonium); conceived the theory of phase stability; Nobel Prize, 1951.

**Meitner, Lise** (1878–1968), German physicist. With O. Hahn, discovered protactinium; found evidence of four other radioactive elements; with Hahn and F. Strassmann, accomplished fission of uranium.

**Mercer, John** (1791–1866), English chemist. Invented the mercerizing process for cotton.

**Merrifield, Robert Bruce** (1921– ), American biochemist. Developed methods of protein synthesis, including solid-phase peptide synthesis that produces proteins by assembling amino acids sequentially into peptide chains; Nobel Prize, 1984.

**Michaelis, Leonor** (1875–1949), German-born American biochemist. Developed theory of kinetics of enzyme-catalyzed reactions.

**Michel, Hartmut** (1948– ), German chemist. With J. Deisenhofer and R. Huber, elucidated the structure of a bacterial protein that performs photosynthesis; Nobel Prize, 1988.

**Mitchell, Peter** (1920–1992), British chemist. Explained how plant and animal cells store and transfer energy by creating protonic gradients in the oxidative and photosynthetic phosphorylation processes; Nobel Prize, 1978.

**Mohr, Carl Friedrich** (1806–1879), German chemist. Developed titration procedures, including use of Mohr's salt.

**Moissan, Ferdinand Frédéric Henri** (1852–1907), French chemist. First to isolate fluorine; invented an electric furnace and used it to produce synthetic metal compounds and samples of less common metals; Nobel Prize, 1906.

**Molina, Mario J.** (1943– ), Mexican-born American chemist. Contributed to the understanding of how atmospheric ozone forms and decomposes; Nobel Prize, 1995.

**Morley, Edward Williams** (1838–1923), American chemist and physicist. Associated with A. A. Michelson in an experiment on ether drift; research on variations of atmospheric oxygen content.

**Müller, Paul Hermann** (1899–1965), Swiss chemist. Discovered the insecticidal properties of DDT; Nobel Prize, 1948.

**Mulliken, Robert Sanderson** (1896–1986), American chemist. Applied principles of quantum mechanics to study of chemical bonding; with F. Hund, systematized electronic states of molecules in terms of molecular orbitals; Nobel Prize, 1966.

**Mullis, Kary B.** (1994– ), American biochemist. Invented the polymerase chain reaction method; Nobel Prize, 1993.

**Natta, Giulio** (1903–1979), Italian chemist. Discovered stereospecific polymerization, making possible the production of new classes of macromolecules from inexpensive raw materials; Nobel Prize, 1963.

**Nernst, Hermann Walther** (1864–1941), German chemist. Proposed the heat theorem (third law of thermodynamics); determined the specific heat of solids at low temperatures; proposed the chain reaction theory in photochemistry; Nobel Prize, 1920.

**Nobel, Alfred Bernhard** (1833–1896), Swedish chemist and engineer. Invented dynamite and a blasting gelatin containing nitroglycerin; established the annual Nobel prizes.

**Norrish, Ronald George Wreyford** (1897–1978), British physical chemist. With G. Porter and colleagues, developed methods of flash photolysis and kinetic spectroscopy for the study of very fast reactions; Nobel Prize, 1967.

**Noyori, Ryoji** (1938– ), Japanese chemist. Developed chirally catalyzed hydrogenation reactions; Nobel Prize, 2001.

**Olah, George A.** (1927– ), American chemist. Prepared and studied the first stable carbocations; Nobel Prize, 1994.

**Ostwald, Friedrich Wilhelm** (1853–1932), German chemist born in Latvia. Researches on affinity and mass action; discovered the Ostwald dilution law; worked on the catalytic oxidation of ammonia; Nobel Prize, 1909.

**Pauling, Linus Carl** (1901–1994), American chemist. Applied quantum theory to chemistry; did research on molecular structure and chemical bonds; contributed to electrochemical theory of valency; Nobel Prize, 1954; Nobel Peace Prize, 1963.

**Pedersen, Charles J.** (1904–1989), American chemist. Developed the synthesis of cyclic polyethers known as crown ethers; Nobel Prize, 1987.

**Pelletier, Pierre Joseph** (1788–1842), French chemist. Discovered quinine, strychnine, and other alkaloids.

**Petit, Alexis Thérèse** (1791–1820), French physicist. With P. L. Dulong, formulated the law of constancy of atomic heats; devised methods for determining thermal expansion and specific heats of solids.

**Pitzer, Kenneth Sanborn** (1914–1997), American chemist. Pioneered in the development of useful approximations which made possible the calculation of chemical thermodynamic properties of broad classes of chemical substances.

**Pople, John A.** (1925– ), British-born chemist. Developed computational methods in quantum chemistry; Nobel Prize, 1998.
**Porter, George** (1920– ), British chemist. With R. G. W. Norrish, developed the technique of flash photolysis to initiate and record very fast chemical reactions; Nobel Prize, 1967.
**Pregl, Fritz** (1869–1930), Austrian chemist. Developed microchemical methods of analysis; Nobel Prize, 1923.
**Prelog, Vladimir** (1906–1998), Yugoslavian-born Swiss chemist. Investigated stereochemistry of organic molecules and reactions; Nobel Prize, 1975.
**Priestley, Joseph** (1733–1804), English chemist and physicist. Discovered oxygen, ammonia, oxides of nitrogen, hydrochloric acid gas, nitrogen, carbon monoxide, and sulfur dioxide.
**Prigogine, Ilya** (1917–2003), Soviet-born Belgian chemist. Contributed to nonequilibrium thermodynamics, particularly the theory of dissipative structures; Nobel Prize, 1977.

**Ramsay, William** (1852–1916), British chemist. With J. W. S. Rayleigh, discovered argon; with M. W. Travers, discovered neon, krypton, and xenon; Nobel Prize, 1904.
**Raoult, François Marie** (1830–1901), French chemist. Formulated Raoult's law concerning vapor pressure of a solution.
**Reichstein, Tadeus** (1897–1996), Polish-born Swiss organic chemist. Isolated about 30 of the 40 substances produced by the adrenal cortex; synthesized and described the structure and properties of many of these substances; Nobel Prize, 1950.
**Richards, Theodore William** (1868–1928), American chemist. Worked on atomic weights; experimentally confirmed the existence of isotopes of lead from uranium and thorium; Nobel Prize, 1914.
**Richter, Jeremias Benjamin** (1762–1807), German chemist. Discovered the law of equivalent proportions.
**Robinson, Robert** (1886–1975), English chemist. Worked on plant pigments, alkaloids, and phenanthrene derivatives; Nobel Prize, 1947.
**Rohrer, Heinrich** (1933– ), Swiss physicist. With G. Binnig, developed scanning tunneling microscope; Nobel Prize, 1986.
**Rowland, F. Sherwood** (1927– ), American chemist. Contributed to the understanding of how atmospheric ozone forms and decomposes; Nobel Prize, 1995.
**Rutherford, Ernest, 1st Baron** (1871–1937), British physicist. Discovered alpha, beta, and gamma rays; suggested the divisible nuclear atom; effected the transmutation of an atom: Nobel Prize, 1908.
**Ružička, Leopold** (1887–1976), Swiss chemist born in Croatia. Research on many-membered rings and higher terpenes (including male sex hormones); Nobel Prize, 1939.

**Sabatier, Paul** (1854–1941), French chemist. Discovered, with J. B. Senderens, the process for catalytic hydrogenation of oils to solid fat; Nobel Prize, 1912.
**Sanger, Frederick** (1918– ), English chemist. Determined the exact order of amino acids in insulin; first to establish amino acid sequence for a protein; developed methods for determining nucleotide sequences (independently of W. Gilbert), advancing the technology of DNA recombination; Nobel Prizes, 1958 and 1980.
**Scheele, Karl Wihelm** (1742–1786), Swedish chemist. Made many discoveries, including oxygen (independently of J. Priestley), chlorine, and glycerin; synthesized many organic acids.
**Schiff, Hugo Josef** (1834–1915), German-born Italian organic chemist. Discovered Schiff bases; devised Schiff test; devised an improved nitrometer.
**Seaborg, Glenn Theodore** (1912–1999), American chemist. Synthesized and identified eight transuranium elements and over a hundred isotopes; Nobel Prize, 1951.
**Segrè, Emilio Gino** (1905–1989), Italian-born American physicist. Codiscovered the elements technetium, astatine, and plutonium, slow neutrons, and the antiproton; Nobel Prize, 1959.
**Semenov, Nikolai Nikolaevich** (1896–1986), Soviet chemist. Elucidated the mechanisms of chemical reactions, especially the chain mechanism; Nobel Prize, 1956.
**Senderens, Jean Baptiste** (1856–1937), French chemist. With P. Sabatier, discovered hydrolysis of oils by catalysis.
**Sharpless, K. Barry** (1941– ), American chemist. Developed chirally catalyzed oxidation reactions; Nobel Prize, 2001.
**Shirakawa, Hideki** (1936– ), Japanese polymer scientist. Discovered and developed conductive polymers with Alan Heeger and Alan MacDiarmid; Nobel Prize, 2000.
**Smalley, Richard E.** (1943– ), American chemist. with Robert Curl and Harold Kroto, discovered fullerenes; Nobel Prize, 1996.
**Soddy, Frederick** (1877–1956), English chemist. With E. Rutherford, developed theory of atomic disintegration of radioactive substances; research on isotopes; Nobel Prize, 1921.
**Solvay, Ernest** (1838–1922), Belgian industrial chemist. Developed the Solvay process for production of sodium carbonate.
**Staudinger, Hermann** (1881–1965), German chemist. Conceived and elaborated the explanation of phenomenon of polymerization; Nobel Prize, 1953.
**Strassmann, Fritz** (1902–1980), German chemist. With O. Hahn and L. Meitner, discovered nuclear fission; research on uranium and thorium isotopes.
**Svedberg, Theodor** (1884–1971), Swedish chemist. An authority on colloid chemistry (dispersed phase); developed a centrifuge for colloidal particles and protein molecules; Nobel Prize, 1926.
**Synge, Richard Laurence Millington** (1914–1994), English chemist. Developed partition chromatography with A. J. P. Martin; Nobel Prize, 1952.

**Tanaka, Kiochi** (1959– ), Japanese engineer. Developed soft desorption ionization method for mass spectrometric analyses of biological macromolecules; Nobel Prize, 2002.

**Taube, Henry** (1915– ), American chemist. Elucidated the mechanisms of electron transfer reactions, especially in metal complexes; Nobel Prize, 1983.

**Thiele, F. K. Johannes** (1865–1918), German chemist. Research on nitrogen compounds and the theory of unsaturated organic molecules.

**Todd of Trumpington, Alexander Robertus Todd, Baron** (1907–1997), British chemist. Worked on the structure and synthesis of nucleotides, and nucleotide coenzymes, and the related problem of phosphorylation; Nobel Prize, 1957.

**Travers, Morris William** (1872–1961), English chemist. Discovered, with W. Ramsay, krypton, xenon, and neon; investigated low-temperature phenomena.

**Urey, Harold Clayton** (1893–1981), American chemist. Isolated heavy water and thus discovered the heavy isotope of hydrogen; Nobel Prize, 1934.

**van der Waals, Johannes Diderik** (1837–1923), Dutch physicist. Formulated van der Waals equation; investigated van der Waals forces, concerning intermolecular attraction; Nobel Prize, 1910.

**van't Hoff, Jacobus Hendricus** (1852–1911), Dutch chemist. Pioneered in the study of stereochemistry; studied reaction rates, thermodynamics applied to chemistry, and the theory of dilute solutions; Nobel Prize, 1901.

**Vauquelin, Louis Nicolas** (1763–1829), French chemist. Discovered chromium and its compounds and beryllium compounds.

**Wallach, Otto** (1847–1931), German chemist. Research on essential oils and the terpenes; Nobel Prize, 1910.

**Watson, James Dewey** (1928– ), American biochemist. With F. H. C. Crick, determined the double-helix structure of deoxyribonucleic acid; Nobel Prize, 1962.

**Warner, Alfred** (1866–1999), Swiss chemist. Formulated the coordination theory of valency; Nobel Prize, 1913.

**Wiedemann, Gustave Heinrich** (1826–1899), German physicist and physical chemist. With R. Franz, discovered Wiedemann-Franz law of thermal conductivity of metals; discovered Wiedemann effect.

**Wieland, Heinrich** (1877–1957), German chemist. Studied bile acids, chlorophyll, and hemoglobin; Nobel Prize, 1927.

**Wilkinson, Geoffrey** (1921–1996), British chemist. Did research to determine how metals and organic molecules combine to form unique molecules which have sandwichlike structures; Nobel Prize, 1973.

**Willstätter, Richard** (1872–1942), German chemist. Worked on plant pigments; investigated alkaloids and their derivatives; Nobel Prize, 1915.

**Windaus, Adolf** (1876–1959), German chemist. Worked on sterols; discovered that ultraviolet light activates ergosterol and gives vitamin $D_2$; Nobel Prize, 1928.

**Wittig, Georg** (1897–1987), German chemist. Work on the linking of carbon and phosphorus (Wittig reaction) made it possible to synthesize new types of compounds, including metal-organic complex compounds; Nobel Prize, 1979.

**Wöhler, Friedrich** (1800–1882), German chemist. First to synthesize an organic compound, urea.

**Wollaston, William Hyde** (1766–1828), English chemist and physicist. Discovered the lines in the solar spectrum; discovered palladium and rhodium; invented the Wollaston lens.

**Woodward, Robert Burns** (1917–1979), American chemist. Contributed to the development of total synthesis of complex natural products, and structural determination of several complex natural molecules, later confirmed by total synthesis; Nobel Prize, 1965.

**Wurtz, Charles Adolphe** (1817–1884), French chemist. Discovered methyl and ethyl amines; evolved the Wurtz reaction for synthesis of hydrocarbons.

**Wüthrich, Kurt** (1938– ), Swiss chemist. Developed nuclear magnetic resonance spectroscopy for determining the three-dimentional structure of biological macromolecules in solution; Nobel Prize, 2002.

**Ziegler, Karl** (1898–1973), German organic chemist. Developed a low-pressure process for production of polyethylene; Nobel Prize, 1963.

**Zsigmondy, Richard** (1865–1929), German chemist. Studied colloidal solutions; introduced the ultramicroscope; Nobel Prize, 1925.

# Contributors

## A

**A.A.G.** Albert A. Gunkler, Chief Process Engineer, Midland Division, Dow Chemical Company, Midland, Michigan.

**A.B.G.** Alfred B. Garrett, Department of Chemistry, Ohio State University.

**A.E.L.** Albert E. Litherland, Director, Isotrace Laboratory, Department of Physics, University of Toronto, Ontario, Canada.

**A.E.M.** A. E. Martell, Department of Chemistry, Texas A&M University.

**A.Gh.** Albert Ghiorso, Department of Chemistry, Lawrence Berkeley Laboratory, University of California, Berkeley.

**A.G.M.B.** A.G.M. Barrett, Department of Chemistry, Northwestern University.

**A.Ha.** Anthony Harriman, Laboratorie de Photochimie, Ecole Européenne Chimie Polymeres Materiaux, Strasbourg, France.

**A.Ho.** Arthur Horwich, Yale University School of Medicine, Howard Hughes Medical Institute, New Haven, Connecticut.

**A.H.Sn.** Arthur H. Snell, Associate Director, Oak Ridge National Laboratory, Oak Ridge, Tennessee.

**A.J.A.** Arnold P. Appleby, Professor Emeritus of Crop Science, Oregon State University, Corvallis.

**A.J.Bo.** Allen J. Bard, Department of Chemistry and Biochemistry, University of Texas, Austin.

**A.J.Cl.** Alex Clegler, Southern Regional Research Center, USDA Science and Education Administration, New Orleans, Louisiana.

**A.J.Fr.** Albert J. Fry, Department of Chemistry, Wesleyan University, Middletown, Connecticut.

**A.J.S.** Anthony J. Stone, Department of Chemistry, University of Cambridge, University Chemical Laboratory, Cambridge, England.

**A.J.T.** Aaron J. Teller, Teller Environmental Systems, Worcester, Massachusetts.

**A.L.A.** Albert L. Allred, Department of Chemistry, Northwestern University.

**A.L.Ba.** Alan L. Baich, Department of Chemistry, University of California, Davis.

**A.L.H.** Allen L. Hanson, Department of Chemistry, Saint Olaf College.

**A.L.K.** Arthur L. Kohl, Project Engineer, Advanced Development, Atomics International Division, North American Rockwell, Woodland Hills, California.

**A.L.M.** A. L. Myers, Department of Chemical Engineering, University of Pennsylvania.

**A.M.H.** A. M. Hartley, Department of Chemistry and Chemical Engineering, University of Illinois, Urbana.

**A.Mo.** Albert Moscowitz, Department of Chemistry, University of Minnesota.

**A.O.N.** Alfred O. Nier, School of Physics and Astronomy, University of Minnesota.

**A.P.M.** Alan P. Marchand, College of Arts and Sciences, Department of Chemistry, University of Texas, Denton.

**A.P.S.** Andrew Paul Somlyo, Department of Physiology, School of Medicine, University of Virginia.

**A.Ri.** Arthur Rich, deceased; formerly, Department of Physics, University of Michigan.

**A.R.C.** Anthony R. Cooper, Dynapol, Palo Alto, California.

**A.R.Ra.** A.R. Ravishankara, National Oceanic and Atmospheric Administration, Aeronomy Laboratory, Boulder, Colorado.

**A.Sc.** Anton Schindler, Camille Dreyfus Laboratory, Research Triangle Institute, Research Triangle Park, North Carolina.

**A.Sche.** Alexander Scheeline, School of Chemical Sciences, University of Illinois, Urbana-Champaign.

**A.S.L.H.** A. S. L. Hu, Department of Biochemistry, University of Kentucky.

**A.S.Ru.** Allen S. Russell, Vice President, Alcoa Laboratories, Aluminum Company of America, Pittsburgh, Pennsylvania.

**A.T.Z.** Andrew T. Zander, Varian Associates, Palo Alto, California.

**A.V.E.** Alexey V. Eliseev, Assistant Professor, Department of Medicinal Chemistry, State University of New York at Buffalo.

**A.V.P.** Alphonsus V. Pocius, AC&S Division, 3M Company, St. Paul, Minnesota.

**A.W.A.** Arthur W. Adamson, Department of Chemistry, University of Southern California.

**A.W.Cz.** Anthony W. Czarnik, Senior Director, Chemistry, IRORI Quantum Microchemistry, La Jolla, California.

**A.W.F.** Arthur W. Francis, Union Carbide Corporation, Tarrytown, New York.

## B

**B.Be.** Bruce Bersted, Amoco Polymers, Inc., Alpharetta, Georgia.

**B.B.Cu.** Burris B. Cunningham, deceased; formerly, Radiation Laboratory, University of California, Berkeley.

**B.C.** Benjamin Chu, Department of Chemistry, State University of New York, Stony Brook.

**B.Du.** Bruce Dunn, Department of Materials Science & Engineering, University of California, Los Angeles.

**B.E.Be.** Brian E. Bent, Department of Chemistry, Columbia University, New York.

**B.E.Bo.** Bruce E. Bowler, Division of Chemistry and Chemical Engineering, California Institute of Technology.

**B.H.Bi.** Bruce H. Billings, Special Assistant to the Ambassador of Science and Technology, Embassy of the United States of America, Taipei.

**B.H.L.** Bruce H. Lipshutz, Department of Chemistry, University of California, Santa Barbara.

**B.Ko.** Bruce Kowalski, Department of Chemistry, University of Washington, Seattle.

**B.M.A.** Bernard M. Abraham, Solid State Science Division, Argonne National Laboratory, Argonne, Illinois.

**B.M.S.** Bruce M. Sankey, Petroleum Products Division, Imperial Oil, Ltd., Sarnia, Ontario, Canada.

**B.R.W.** Bennie R. Ware, Department of Chemistry, Syracuse University.

**B.S.G.** Bernard S. Greensfelder, deceased; formerly, Director of Oil Research, Shell Development Company, Emeryville, California.

## C

**C.A.** Chris Adams, Unilever Research, Merseyside, England.

**C.A.Co.** Charles A. Cohen, Research Associate (retired), Exxon Research and Engineering Company, Florham Park, New Jersey.

**C.A.Sw.** Clayton A. Swenson, Department of Physics, Iowa State University.

**C.A.V.** Chris A. Veale, Zeneca, Inc., Wilmington, Delaware.

**C.Cr.** Carol Creutz, Department of Chemistry, Brookhaven National Laboratory, Associated University Inc., Upton, New York.

**C.C.K.L.** Clark C. K. Liu, Department of Civil Engineering, University of Hawaii at Manoa.

**C.C.W.** Carl C. Wamser, Department of Chemistry, California State University.

**C.C.Wa.** Carl C. Wamser, Department of Chemistry, Portland State University, Portland, Oregon.

**C.D.C.** C. Denise Caldwell, Department of Physics, Yale University.

**C.E.Ca.** Charles E. Carraher, Department of Chemistry, Florida Atlantic University, Boca Raton.

**C.E.Lu.** Craig E. Lunte, Department of Chemistry, University of Kansas, Lawrence.

**C.E.V.** Cecil E. Vanderzee, Department of Chemistry, University of Nebraska.

**C.F.B.** Charles F. Beam, Department of Chemistry, College of Charleston, South Carolina.

**C.F.C.** Charles F. Curtis, Department of Chemistry, University of Wisconsin.

**C.F.K.** Carl F. Kayan, deceased; formerly, Department of Mechanical Engineering, School of Engineering, Columbia University.

**C.F.P.** Colin F. Poole, Department of Chemistry, Wayne State University, Detroit, Michigan.

**C.J.Bi.** Christopher Blermann, Department of Forest Products, Forest Research Laboratory, Oregon State University, Corvallis.

**C.J.G.** Charles J. Goebel, Department of Physics, University of Wisconsin.

**C.Ku.** Charles Kutal, Department of Chemistry, University of Georgia.

**C.K.B.** Charles K. Bradsher, Department of Chemistry, Duke University.

**C.K.L.** Cynthia K. Larive, Department of Chemistry, University of Kansas, Lawrence.

**C.L.R.** Charles L. Rulfs, Department of Chemistry, University of Michigan.

**C.O.D.B.** C. O. Dietrich-Buchecker, Laboratoire de Chimie Organo-Minérale, Institut Le Bel, Université Louis Pasteur, Strasbourg, France.

**C.P.B.** Charles P. Brewer, Shell Development Corp., Emeryville, California.

**C.R.J.** Carl R. Johnson, Department of Chemistry, Wayne State University.

## D

**D.A.S.** David A. Shirley, deceased; formerly, Department of Chemistry, University of Tennessee.

**D.B.J.** Deane B. Judd, deceased; formerly, Office of Colorimetry, National Bureau of Standards.

**D.D.** David Dalton, Department of Chemistry, College of Arts and Sciences, Temple University.

**D.E.P.** David E. Pritchard, Department of Physics, Massachusetts Institute of Technology, Cambridge.

**D.F.** Denis Forster, Monsanto Company, St. Louis, Missouri.

**D.Fi.** Daniele Finotello, Liquid Crystal Institute, Kent State University, Kent, Ohio.

**D.F.O.** Donald F. Othmer, Department of Chemical Engineering, Polytechnic Institute of Brooklyn.

**D.G.** D. Gauss, Max-Planck Institut für Experimentelle Medizin, Gottingen, Germany.

**D.H.A.** David H. Abrahams, Dexter Chemical Corporation, Bronx, New York.

**D.H.B.** Daryle H. Busch, Department of Chemistry, Ohio State University.

**D.H.K.** David H. Klein, Department of Chemistry, Hope College.

**D.J.Ca.** D. J. Carlsson, National Research Council of Canada, Ottawa, Ontario.

**D.J.Pe.** David J. Pegg, Department of Physics and Astronomy, University of Tennessee, Knoxville.

**D.J.S.** David J. Sellmyer, Behlen Laboratory of Physics, University of Nebraska.

**D.L.A.** Daniel L. Azarnoff, President, Searle Research and Development, G. D. Searle & Co., Skokie, Illinois.

**D.L.D.** David L. Dalrymple, Department of Chemistry, University of Delaware.

**D.L.H.** Donald L. Holt, Monsanto Company, St. Louis, Missouri.

**D.M.DeF.** Duarte Mota de Freitas, Department of Chemistry, Loyola University, Chicago, Illinois.

**D.M.Gr.** Dieter M. Gruen, Argonne National Laboratory, Argonne, Illinois.
**D.M.Ro.** D. Max Roundhill, Department of Chemistry, Tulane University.
**D.M.W.** D. M. Wiles, National Research Council of Canada, Ottawa, Ontario.
**D.R.B.** Donald R. Baer, Jackson Laboratory, Organic Chemical Department, E. I. du Pont de Nemours and Company, Wilmington, Delaware.
**D.R.D.** Donald R. Dimmel, Research Associate, Wood Sciences, Chemical Sciences Division, Institute of Paper Chemistry, Appleton, Wisconsin.
**D.R.S.** Donald R. Stranks, Department of Physical and Inorganic Chemistry, University of Adelaide, Australia.
**D.Sn.** Deanne Snavely, Center for Photochemical Sciences, Bowling Green State University, Bowling Green, Ohio.
**D.S.Br.** David S. Breslow, David S. Breslow Associates, Wilmington, Delaware.
**D.T.S.** Donald T. Sawyer, Distinguished Professor, Department of Chemistry, Texas A&M University.
**D.W.** Daniel Wellner, Department of Biochemistry, Cornell University Medical College.
**D.W.Ku.** Donald W. Kupke, Department of Biochemistry, School of Medicine, University of Virginia.
**D.W.Sl.** Donald W. Slocum, Argonne National Laboratory, Argonne, Illinois.

## E

**E.B.R.** Evans B. Reid, Professor Emeritus, Department of Chemistry, Colby College.
**E.C.** Ellen Clarke, Columbia Organic Chemicals Company, Inc., Columbia, South Carolina.
**E.Ca.** Eleanor Campbell, Max-Born Institut, Berlin, Germany.
**E.C.L.** Edward C. Luckenbach, Senior Engineering Associate, Exxon Research and Engineering Company, Florham Park, New Jersey.
**E.Dr.** Eric Drexler, Foresight Institute, Palo Alto, California.
**E.E.S.** Esmond E. Snell, Department of Biochemistry, University of California, Berkeley.
**E.E.W.** E. Eugene Weaver, Research Scientist, Product Development Group, Ford Motor Company, Dearborn, Michigan.
**E.Gru.** Emanuel Grunberg, Director, Department of Chemotherapy, Hoffmann-LaRoche, Inc., Nutley, New Jersey.
**E.G.P.** Edward G. Partridge, Redondo Beach, California.
**E.G.Ro.** Eugene G. Rochow (retired), Department of Chemistry, Harvard University.
**E.H.H.** Elbert H. Hadley, Assistant Dean (retired), College of Liberal Arts and Sciences, and Professor of Chemistry, Southern Illinois University.
**E.I.S.** Edward I. Stiefel, Scientific Advisor, Exxon Research and Engineering Co., Annandale, New Jersey.
**E.Jo.** Edward Johnson, Department of Cell Biology, Rockefeller University.
**E.K.H.** Earl K. Hyde, Lawrence Berkeley Laboratory, University of California, Berkeley.
**E.L.E.** Ernest L. Eliel, Department of Chemistry, University of North Carolina.
**E.L.S.** Everett L. Saul, Director of Scientific Liaison, Products Division, Bristol-Myers Company.
**E.M.Ey.** Edward M. Eyring, Department of Chemistry, University of Utah.
**E.M.L.** Edwin M. Larsen, Department of Chemistry, University of Wisconsin.
**E.M.M.** Emil M. Mrak, Office of Chancellor Emeritus, University of California, Davis.
**E.P.G.** E. Patrick Groody, Integrated Genetics, Framingham, Massachusetts.
**E.P.P.** Edwin P. Plueddemann, Organic Laboratories, Dow Corning Corporation, Midland, Michigan.
**E.P.S.** Ellis P. Steinberg, Senior Scientist, Argonne National Laboratory, Argonne, Illinois.
**E.R.Co.** E. Richard Cohen, retired; formerly, Adjunct Professor of Physics, Science Center, Rockwell International, California State University, Encino.
**E.W.** Ernest Wenkert, Department of Chemistry, Indiana University.
**E.W.C.** Edward W. Comings, University of Petroleum and Minerals, Dhahran, Saudi Arabia.

## F

**F.A.J.** Francis A. Jenkins, deceased; formerly, Department of Physics, University of California, Berkeley.
**F.A.L.** Franklin A. Long, formerly, Department of Chemistry, Cornell University.
**F.Ba.** Fred Basolo, Department of Chemistry, Northwestern University.
**F.Cr.** Friedrich Cramer, Max Planck Institut für Experimentelle Medizin, Gottingen, Germany.
**F.C.B.** Frank C. Benner, Assistant Director, Research and Development Department, Norton Company, Worcester, Maine.
**F.C.S.** Ferris C. Standiford, Jr., W. L. Badger Associates, Inc., Consulting Engineers, Ann Arbor, Michigan.
**F.D.** Farrington Daniels, deceased; formerly, Professor Emeritus, Solar Energy Laboratory, University of Wisconsin.
**F.G.** Fabio Garbassi, Instituto Guiido Donegani, EniChem, Polymeric Materials Department, Novara, Italy.
**F.H.M.** F. H. May, Consultant, Kerr-McGee Corporation, Whittier, California.
**F.H.R.** Frank H. Rockett, Engineering Consultant, Chariottesville, Virginia.
**F.H.S.** Frank H. Shelton, Chief Scientist, Kaman Sciences Corporation, Colorado Springs, Colorado.
**F.H.Sp.** Frank H. Spedding, deceased; formerly, Ames Laboratory, Energy Research and Development Administration, Iowa State University.

**F.H.W.** F. H. Westheimer, Chemistry Department, Harvard University, Cambridge, Massachusetts.
**F.J.J.** Francis J. Johnston, Department of Chemistry, University of Georgia.
**F.J.L.** Frank J. Lockhart, Department of Chemical Engineering, University of Southern California.
**F.L.M.** Florence Lansana Margai, Department of Geography, State University of New York at Binghamton.
**F.L.P.** Fred L. Palmer, Lighting Research Center, Osram Sylvania, Beverly, Massachusetts.
**F.Sa.** Fred Sauer, Technical Service Department, Pfizer, Inc., Groton, Connecticut.
**F.V.** Felix Viro, Kind & Knox, Sioux City, Iowa.
**F.Vo.** Fritz Vogtle, Department of Chemistry, Friedrich-Wilhelms University, Bonn, Germany.
**F.W.** Frank Wagner, Technical Center, Celanese Chemical Company, Corpus Cristi, Texas.
**F.We.** Fritz Weigel, Institut für Anorganische Chemie, Universität München, Germany.
**F.W.K.F.** Frank W. K. Firk, Electron Accelerator Laboratory, Yale University.
**F.W.Ko.** Frederick W. Koerker, retired; formerly, Technical Expert, Dow Chemical Company, Midland, Michigan.
**F.W.Sw.** F. W. Sweat, Department of Chemistry, University of Utah.

# G

**G.A.** Gordon Atkinson, Department of Chemistry, University of Oklahoma.
**G.C.** George Cocks, Los Alamos National Laboratory, Los Alamos, New Mexico.
**G.Ch.** George Christou, Department of Chemistry, Indiana University.
**G.E.L.** George E. Liedholm, Department Head (retired), Petroleum Processing, Shell Development Company, Emeryville, California.
**G.E.Pe.** Gertrude E. Perlmann, deceased; formerly, Rockefeller University.
**G.E.W.** Gary E. Wnek, Director, Polymer Science and Engineering Program, Department of Chemistry, School of Science, Rensselaer Polytechnic Institute.
**G.H.B.** Glenn H. Brown, Professor of Chemistry and Director of the Liquid Crystal Institute, Kent State University.
**G.H.M.** Glenn H. Miller, Weapons Effects Division, Sandia Laboratories, Albuquerque, New Mexico.
**G.H.Mo.** George H. Morrison, Professor of Chemistry, and Director, Analytical Facility of Materials Science Center, Cornell University.
**G.L.G.** George L. Gaines, Research and Development Center, General Electric Company, Schenectady, New York.
**G.L.H.** Gary L. Haller, Professor of Engineering and Applied Science, Jonathan Edwards College, Yale University, New Haven, Connecticut.
**G.Ma.** Gleb Mamantov, Chemistry Department, University of Tennessee.
**G.M.H.** Gary M. Hieftje, Department of Chemistry, Indiana University.
**G.S.H.** G. S. Hurst, Department of Physics, Oak Ridge National Laboratory, Oak Ridge, Tennessee.
**G.S.M.** George S. Mill, Research Chemist, Shell Oil Company, New York.
**G.S.W.** George S. Wilson, Department of Chemistry, University of Kansas, Lawrence.
**G.T.S.** Glenn T. Seaborg, Lawrence Berkeley Laboratory, University of California, Berkeley.
**G.W.G.** Gordon W. Groves, Institute de Geofisica, Torre de Ciencias, Ciudad Universitaria, Mexico.
**G.W.R.** G. Wilse Robinson, Department of Chemistry, Texas Tech University.
**G.W.Sc.** George W. Scherer, DuPont Central Research and Development, Wilmington, Delaware.

# H

**H.A.L.** Herbert A. Laitinen, Department of Chemistry, University of Florida.
**H.A.W.** Harley A. Wilhelm, Principal Scientist, EMRR Institute and Ames Laboratory of U.S. Department of Energy, Iowa State University.
**H.B.** Hans Brintzinger, Fachbereich Biologie-Chemie, Universität Konstanz, Germany.
**H.B.A.** H. Burnham Allport, Project Manager, Carbon Products Division, Union Carbide Corporation, Cleveland, Ohio.
**H.B.Gr.** Harry B. Gray, Division of Chemistry and Chemical Engineering, California Institute of Technology.
**H.B.H.** H. B. Huntington, Department of Physics, Rensselaer Polytechnic Institute.
**H.B.Ma.** Harry B. Mark, Jr., Department of Chemistry, University of Cincinnati.
**H.B.Yok.** Howard B. Yokelson, Research and Development Department, Amoco Chemical Company, Naperville, Illinois.
**H.C.B.** Herbert C. Brown, Department of Chemistry, Purdue University.
**H.C.W.** Harold C. Weber, Chemical Engineer, Boston, Massachusetts.
**H.E.Ca.** Herbert E. Carter, Vice Chancellor for Academic Affairs, University of Illinois.
**H.E.D.** Henry E. Duckworth, Department of Physics, University of Manitoba, Canada.
**H.Frei.** Henry Freiser, Department of Chemistry, University of Arizona, Tucson.
**H.F.S.** Henry F. Schaefer, III, Department of Chemistry, University of California, Berkeley.
**H.Ha.** Harold Hart, Professor Emeritus, Department of Chemistry, Michigan State University.
**H.H.S.** Harry H. Sisler, Department of Chemistry, University of Florida.
**H.H.St.** Henry H. Storch, deceased; formerly, Assistant Professor of Chemistry, New York University.
**H.H.T.** H. Holden Thorp, Department of Chemistry, University of North Carolina, Chapel Hill.

**H.J.A.** Henry J. Albert, Manager, Physics and Metallurgy Laboratory, Research and Development Department, Engelhard Industries Division, Engelhard Minerals and Chemicals Corporation, Newark, New Jersey.

**H.J.P.** Herman J. Phaff, Department of Food Science and Technology, College of Agriculture and Environmental Science, University of California, Davis.

**H.K.C.** H. Keith Chenault, Department of Chemistry, University of Delaware, Newark.

**H.L.F.** Harold L. Friedman, Department of Chemistry, State University of New York, Stony Brook.

**H.Me.** Helmut Mehrer, Institut für Metallforschung, Münster, Germany.

**H.M.F.** Harry M. Freeman, Chief, Waste Minimization Branch, Risk Reduction Engineering Laboratory, U.S. Environmental Protection Agency, Cincinnati, Ohio.

**H.M.H.** Howard M. Hickman, Section Manager, Research and Development, Sherex Chemical Company, Dublin, Ohio.

**H.M.Ts.** Henry M. Tsuchiya, Department of Chemical Engineering, University of Minnesota.

**H.S.** Hymin Shapiro, Research and Development Department, Ethyl Corporation, Baton Rouge, Louisiana.

**H.Sc.** Harvey Schugar, Department of Chemistry, Wright and Rieman Laboratories, New Brunswick, Rutgers University.

**H.S.B.** Howard S. Bean, deceased; formerly, Consultant on Fluid Metering, Liquids and Gases, Sedona, Arizona.

**H.Ta.** Henry Taube, Department of Chemistry, Stanford University.

**H.W.Ki.** H.W. Kirby, Mound Facility, Monsanto Research Corporation, Miamisburg, Ohio.

**H.W.Kr.** Harold W. Kroto, School of Chemistry and Molecular Sciences, University of Sussex, United Kingdom.

**H.W.W.** Howard W. Wainwright, Coal Research Center, U.S. Bureau of Mines, Morgantown, West Virginia.

**H.Y.** Howard B. Yokelson, Research and Development Department, Amoco Chemical Company, Naperville, Illinois.

# I

**I.A.S.** Ivan A. Sellin, Department of Physics, University of Tennessee.

**I.B.W.** Irwin B. Wilson, Department of Chemistry and Biochemistry, University of Colorado.

**I.E.L.** I. E. Levine, Chevron Research Company, Richmond, California.

**I.N.L.** Ira N. Levine, Department of Chemistry, Brooklyn College.

**I.S.** Irving Sheft, Chemistry Division, Argonne National Laboratory, Argonne, Illinois.

**I.S.G.** Irving S. Goldstein, College of Forestry Resources, North Carolina State University.

# J

**J.A.M.** John A. Manson, deceased; formerly, Department of Chemistry, Lehigh University.

**J.A.Mo.** James Alexander Moore, Consulting Editor, deceased; formerly, Department of Chemistry, University of Delaware, Newark.

**J.Big.** Jacob Bigeleisen, Department of Chemistry, University of Rochester.

**J.B.Lo.** Joseph B. Long, Consultant, Columbus, Ohio.

**J.Cap.** Jennings Capellen, Rare Earth Information Center, Iowa State University.

**J.C.Ch.** Jean-Claude Chambron, Laboratoire de Chimie Organo-Minérale, Institut le Bel, Université Louis Pasteur, Strasbourg, France.

**J.C.Sm.** Julian C. Smith, Department of Chemical Engineering, Cornell University.

**J.C.St.** James C. Sternberg, Applied Research Department, Beckman Instruments, Brea, California.

**J.C.W.** James C. Warf, Department of Chemistry, University of Southern California.

**J.D.J.** James D. Johnston, Senior Research Advisor, Pioneering Research, Ethyl Corporation, Baton Rouge, Louisiana.

**J.E.B.** James E. Bayfield, Department of Physics, University of Pittsburgh.

**J.E.Fi.** J.E. Fischer, Department of Materials Science and Engineering, University of Pennsylvania, Philadelphia.

**J.E.Jo.** James G. Johnson, deceased; formerly, Department of Geology, Oregon State University.

**J.F.B.** Joseph F. Bunnett, Department of Chemistry, University of California, Santa Cruz.

**J.F.E.** John F. Endicott, Department of Chemistry, Wayne State University.

**J.F.Li.** Joel F. Liebman, Department of Chemistry and Biochemistry, University of Maryland, Baltimore.

**J.F.Ri.** James F. Riordan, Center for Biochemical and Biophysical Sciences and Medicine, Harvard Medical School, Boston, Massachusetts.

**J.F.We.** Jonathan F. Weil, Senior Editor, "McGraw-Hill Encyclopedia of Science and Technology," McGraw-Hill, New York.

**J.G.S.** James G. Speight, Western Research Institute, Laramie, Wyoming.

**J.Hare.** Jonathan Hare, School of Chemistry and Molecular Sciences, University of Sussex, United Kingdom.

**J.H.F.** J. Homer Ferguson, Department of Biological Science, University of Idaho.

**J.H.Fe.** Janos H. Fendler, Department of Chemistry, Center for Research in Membranes and Colloid Science, Syracuse University.

**J.H.Hi.** Joel H. Hildebrand, deceased; formerly, Department of Chemistry, University of California, Berkeley.

**J.J.Ca.** James J. Carberry, Department of Chemical Engineering, University of Notre Dame.

**J.J.K.** Joseph J. Katz, Chemistry Division, Argonne National Laboratory, Argonne, Illinois.

**J.J.L.** J. J. Lagowski, Department of Chemistry, University of Michigan.

**J.K.Bu.** Jeremy K. Burdett, Department of Chemistry, University of Chicago.

**J.K.Hu.** James K. Hurst, Department of Biochemistry and Biophysics, Washington State University, Pullman.

**J.K.Le.** Jonathan K. Leland, IGEN, Inc., Rockville, Maryland.

**J.K.R.** J. K. Roberts, retired; formerly, Research and Development Department, Standard Oil Company, Chicago, Illinois.

**J.K.T.** J. Kerry Thomas, Department of Chemistry, University of Notre Dame.

**J.L.L.** Jack L. Lambert, Department of Chemistry, Kansas State University.

**J.L.M.** John L. Margrave, Dean of Research, Rice University.

**J.L.T.W.** John L.T. Waugh, Department of Chemistry, University of Hawaii.

**J.M.Be.** János M. Beér, Emeritus Professor, Department of Chemical Engineering, Massachusetts Institute of Technology, Cambridge.

**J.M.Bo.** James M. Bobbitt, Department of Chemistry, University of Connecticut.

**J.M.Ha.** J. Milton Harris, Department of Chemistry, University of Alabama, Huntsville.

**J.M.Le.** Jean-Marie Lehn, Collège de France, Paris, and Institute de Science et Ingénierie Supramoléculaires-Université Louis Pasteur, Strasbourg, France.

**J.M.M.** James M. Manning, Department of Biochemistry, Rockefeller University.

**J.M.Ma.** James M. Manning, Department of Biochemistry, Rockefeller University, New York, New York.

**J.M.Wi.** Jack M. Williams, Chemistry Division, Argonne National Laboratory, Argonne, Illinois.

**J.N.Bu.** Judith N. Burstyn, Department of Chemistry, University of Wisconsin, Madison.

**J.N.L.** James N. Little, Research Division, Waters Associates, Milford, Massachusetts.

**J.O.E.** John O. Edwards, Department of Chemistry, Brown University.

**J.O.H.** J. O. Hirschfelder, Department of Chemistry, University of Wisconsin.

**J.P.Fr.** Jeremiah P. Freeman, Department of Chemistry, University of Notre Dame.

**J.P.Ha.** James Penner-Hahn, Department of Chemistry, University of Michigan, Ann Arbor.

**J.P.Sa.** Jean-Pierre Sauvage, Laboratoire de Chimie Organo-Minérale, Institut le Bel, Université Louis Pasteur, Strasbourg, France.

**J.Reb.** Julius Rebek, Jr. Department of Chemistry, Massachusetts Institute of Technology.

**J.R.Bo.** James R. Bolton, Department of Chemistry, University of Western Ontario, London, Canada.

**J.R.V.W.** John R. Van Wazer, Department of Chemistry, Vanderbilt University.

**J.Sch.** Jack Schubert, Radiation Health, Graduate School of Public Health, University of Pittsburgh.

**J.S.L.** J. S. Lindsey, Liquid Carbonic Corporation, Chicago, Illinois.

**J.S.Mo.** Jeffrey S. Moore, Department of Chemistry, University of Illinois, Urbana-Champaign.

**J.S.Ro.** J.S. Rowlinson, Physical Chemistry Laboratory, University of Oxford, England.

**J.T.St.** John T. Stock, Department of Chemistry, University of Connecticut.

**J.Va.** Jacob Vaya, Migal-Galilee Technological Center, Rosh-Pina, Israel.

**J.W.D.** J. William Doane, Liquid Crystal Institute, Kent State University, Kent, Ohio.

**J.W.F.** John W. Faller, Department of Chemistry, Yale University.

**J.W.Fu.** James W. Fulton, Monsanto Company, St. Louis, Missouri.

**J.W.Ge.** James W. Gewartowski

**J.W.P.** Jae-Woo Park, Department of Civil Engineering, University of Hawaii at Manoa.

# K

**K.Ak.** Kin-ya Akiba, Hiroshima University, Department of Chemistry, Faculty of Science, Kagamiyama, Higashi-Hiroshima, Japan.

**K.B.** Klaus Brodersen, Professor of Inorganic and Analytical Chemistry, University of Erlangen-Nurnberg, Germany.

**K.B.L.** Kenneth B. Lipkowitz, Department of Chemistry, Indiana-Purdue University, Indianapolis.

**K.G.S.** Kenneth G. Stone, deceased; formerly, Department of Chemistry, Michigan State University.

**K.H.F.** Kenneth A. Falchuck, Center for Biochemical and Biophysical Sciences and Medicine, Harvard Medical School.

**K.Jo.** Keith Johnston, Department of Chemical Engineering, University of Texas, Austin.

**K.Lo.** Kenneth Longmore, Department of Chemistry, University of Massachusetts.

**K.M.B.** Karl M. Beck, Abbott Laboratories, North Chicago, Illinois.

**K.N.M.** Kenneth N. Marsh, Thermodynamic Research Center, Texas A & M University.

**K.S.** Kai Siegbahn, Institute of Physics, University of Uppsala, Sweden.

**K.S.S.** Kenneth S. Suslick, Professor, School of Chemical Sciences, Noyes Laboratory, University of Illinois, Urbana.

**K.T.P.** Kevin T. Potts, Department of Chemistry, Rensselaer Polytechnic Institute.

**K.W.B.** Kenneth W. Bagnall, Research Chemist, Atomic Energy Research Establishment, Harwell, England.

# L

**L.A.** Lester Andrews, Department of Chemistry, University of Virginia.

**L.C.A.** Leland C. Allen, Department of Chemistry, Princeton University.

**L.C.F.** Leonard C. Feldman, AT&T Bell Laboratories, Murray Hill, New Jersey.
**L.Fr.** Lewis Friedman, Brookhaven National Laboratory, Upton, New York.
**L.F.A.** Lyle F. Albright, Department of Chemical Engineering, Purdue University.
**L.F.M.** Leonard F. Mausner, Radionuclide/Radiopharmaceutical Research, Brookhaven National Laboratory, Upton, New York.
**L.Go.** Louis Gordon, deceased; formerly, Case Institute of Technology.
**L.Gro.** Lee Grodzins, Physics Department, Massachusetts Institute of Technology.
**L.K.** Louis Kaplan, Senior Chemist, Argonne National Laboratory, Argonne, Illinois.
**L.K.W.** L. Keith Woo, Department of Chemistry, Iowa State University of Science and Technology, Ames.
**L.Lo.** Luca Longo, Instituto Guido Donegani, Eni-Chem, Polymeric Materials Department, Novara, Italy.
**L.Ma.** Luigi G. Marzilli, Department of Chemistry, Emory University.
**L.M.Sa.** Lawrence M. Sayre, Department of Chemistry, Case Western Reserve University, Cleveland, Ohio.
**L.N.** Leo Nedelsky, Department of Physical Science, University of Chicago.
**L.Na.** Laurette Nacamulli, IGEN, Inc., Rockville, Maryland.
**L.P.** Lester Packer, University of California, Department of Molecular and Cell Biology, Berkeley, California.
**L.R.M.** Lee R. Mahoney, Chemistry Department, Ford Motor Company, Dearborn, Michigan.
**L.S.** Louis Schmerling, Research Associate, Universal Oil Products Company, Des Plaines, Illinois.
**L.S.B.** L. S. Birks, retired; formerly, Radiation Technology Division, Naval Research Laboratory, Washington, D.C.
**L.S.Be.** Laurence S. Bernson, Environmental Management Department, AT&T Bell Laboratories, Murray Hill, New Jersey.

## M

**M.A.F.** Marye Anne Fox, Department of Chemistry, University of Texas, Austin.
**M.Be.** Manson Benedict, Institute Professor Emeritus, Department of Nuclear Engineering, Massachusetts Institute of Technology.
**M.Boh.** M. Böhme, Department of Chemistry, Friedrich-Wilhelms University, Germany.
**M.B.McC.** Mary B. McCann, Nutrition Program, Center for Disease Control, Health Services and Mental Health Administration, U.S. Department of Health, Education and Welfare, Rockville, Maryland.
**M.D.** Michael Doudoroff, deceased; formerly, Department of Bacteriology, University of California, Berkeley.
**M.D.Ar.** Mary D. Archer, Grantchester, Cambridge, United Kingdom.
**M.D.P.** M. David Potter, School of Business, San Francisco State College.
**M.D.U.** Michael D. Uhler, Department of Biological Chemistry, University of Michigan, Ann Arbor.
**M.E.B.** Milton E. Bailey, Department of Food Science and Nutrition, College of Agriculture, University of Missouri-Columbia.
**M.F.D.** Michael F. Doherty, Head, Department of Chemical Engineering, University of Massachusetts.
**M.F.R.** Mark F. Reynolds, Graduate Research Assistant, Department of Chemistry, University of Wisconsin, Madison.
**M.F.S.** Matthew F. Schlecht, Section Research Chemist, E.I. du Pont de Nemours & Company, Inc., Agricultural Products Department, Newark, Delaware.
**M.H.C.** Marvin H. Carruthers, Department of Chemistry, University of Colorado.
**M.H.Ch.** Malcolm H. Chisholm, Department of Chemistry, Indiana University.
**M.H.H.** McAllister H. Hull, Department of Physics, University of New Mexico.
**M.H.W.C.** Moses H. W. Chan, Department of Physics, Pennsylvania State University.
**M.J.Ax.** Milton J. Axley, Department of Health and Human Services, National Heart, Lung and Blood Institute, Bethesda, Maryland.
**M.J.C.** Michael J. Camp, Director, State Crime Laboratory, Milwaukee, Wisconsin.
**M.J.Si.** Michell J. Sienko, deceased; formerly, Department of Chemistry, Cornell University.
**M.Ka.** Miriam Kastner, Geologic, Research Division, Scripps Institute of Oceanography, La Jolla, CA.
**M.L.K.** M. L. Knotek, Sandia National Laboratories, Albuquerque, New Mexico.
**M.L.S.** Murrell L. Salutsky, Vice President, Research, Dearborn Chemical Division, W. R. Grace and Company, Lake Zurich, Illinois.
**M.M.Bu.** Maurice M. Bursey, Department of Chemistry, University of North Carolina.
**M.M.W.** Mary M. Walczak, Department of Chemistry, Iowa State University of Science and Technology.
**M.P.** Michael Perch, Koppers Company, Inc., Monroeville, Pennsylvania.
**M.S.** Max Samter, Institute of Allergy and Clinical Immunology, Grant Hospital of Chicago.
**M.Si.** Marshall Sittig, Assistant Director, Office of Research Project Administration, Princeton University.
**M.Sou.** Mott Souders, deceased; formerly, Director, Oil Development, Shell Oil Company, Emeryville, California.
**M.St.** Martin Stiles, Department of Chemistry, University of Kentucky.
**M.Wi.** M. Wilkening, Department of Physics, New Mexico Institute of Mining and Technology.

# N–O

**N.A.C.** Noel A. Clark, Department of Physics and Astrophysics, University of Colorado.

**N.F.** Nicholas Farrel, Department of Chemistry, Virginia Commonwealth University, Richmond.

**N.H.N.** Norman H. Nachtrieb, Chairman, Department of Chemistry, University of Chicago.

**N.H.T.** Noel H. Turner, Chemistry Division, Naval Research Laboratory, Washington, D.C.

**N.J.T.** Nicholas J. Turro, Department of Chemistry, Columbia University.

**N.K.** Norman Kharasch, Department of Biomedicinal Chemistry, University of Southern California.

**N.L.** Ningyl Luo, Subpicosecond and Quantum Radiation Laboratory, Texas Tech University, Lubbock.

**N.N.L.** Norman N. Li, Director, Separations Research, UOP Inc., World Headquarters, Des Plaines, Illinois.

**N.N.Li.** Norman N. Li, Director, Separations Research, UOP Inc., World Headquarters, Des Plaines, Illinois.

**N.S.B.** Norbert S. Baer, Conservation Center of the Institute of Fine Arts, New York University.

**N.W.** Nicholas Winograd, Department of Chemistry, Pennsylvania State University, University Park.

**O.W.W.** Owen W. Webster, DuPont Experimental Station, Wilmington, Delaware.

# P–Q

**P.Ar.** Peter Armbruster, Gesellschaft für Schwerionenforschung, MBH, Germany.

**P.A.G.** Paul A. Giguère, Emeritus Professor, St. Petersburg Beach, Florida.

**P.A.Mar.** Patricia A. Marzilli, Department of Chemistry, Emory University.

**P.B.M.** Paul B. Moore, Department of the Geophysical Sciences, University of Chicago.

**P.C.J.** Philip C. Johnson, Research and Development Department, Chemicals and Plastics Operations Division, Union Carbide Corporation, South Charleston, West Virginia.

**P.De.** Paul Delahay, Department of Chemistry, New York University.

**P.E.Bl.** Philip E. Bloomfield, Department of Physics, University of Pennsylvania; City College of the City University of New York, Bloomfield.

**P.E.F.** Paul E. Fanta, Department of Chemistry, Illinois Institute of Technology.

**P.H.** Philip Hodge, Chair of Polymer Chemistry, Department of Chemistry, University of Manchester, England.

**P.H.C.** Philip H. Cook, Plastics Department, Texas Division, Dow Chemical U.S.A., Freeport, Texas.

**P.H.E.M.** Paul H. E. Meijer, Department of Physics, Catholic University of America.

**P.H.G.** P. H. Groggins, deceased; formerly, Chemical Division, Food Machinery and Chemical Corporation.

**P.J.B.** Paul J. Bender, Professor of Physical Chemistry, University of Wisconsin.

**P.M.K.** Peter M. Koch, Department of Physics, State University of New York at Stony Brook.

**P.R.B.** Philip R. Brooks, Department of Chemistry, Wise School of Natural Sciences, Rice University.

**P.R.F.** Paul R. Fields, Senior Chemist, Argonne National Laboratory, Argonne, Illinois.

**P.R.G.** Peter R. Griffiths, Chairman, College of Letters and Science, Department of Chemistry, University of Idaho.

**P.T.C.** Peter T. Cummins, Department of Chemical Engineering, University of Tennessee, Knoxville.

**P.W.A.** P. W. Atkins, Department of Chemistry, Oxford University, England.

**P.Z.** Petr Zuman, Department of Chemistry, Clarkson College of Technology.

**Q.V.W.** Quentin Van Winkle, Department of Chemistry, Ohio State University.

# R

**R.A.A.** Rudolph A. Abramovitch, Department of Chemistry and Geology, Clemson University.

**R.A.D.** Richard A. Davis, Jr., Department of Geology, University of South Florida.

**R.A.D.W.** R. A. D. Wentworth, Department of Chemistry, Indiana University.

**R.A.J.** R. Alan Jones, School of Chemical Sciences, University of East Anglia, England.

**R.A.Pe.** Robert A. Penneman, Los Alamos Scientific Laboratory, Los Alamos, New Mexico.

**R.A.Pi.** Robert A. Pierotti, Department of Chemistry, Georgia Institute of Technology.

**R.A.S.** Robert A. Shaw, Department of Chemistry, Birkbeck College, University of London, England.

**R.A.We.** Robert A. Weller, Department of Physics, Vanderbilt University, Nashville, Tennessee.

**R.C.M.** Roy C. Mast, Miami Valley Laboratories, Proctor and Gamble Company, Cincinnati, Ohio.

**R.C.S.** Randy C. Stauffer, Halogens Research Laboratory, Dow Chemical Company, Midland, Michigan.

**R.C.W.** Randolph C. Wilhoit, Thermodynamics Research Center, Texas A&M University.

**R.DeL.** Robert de Levie, Department of Chemistry, Georgetown University, Washington, DC.

**R.D.E.** Robley D. Evans, retired; Department of Physics, Massachusetts Institute of Technology.

**R.D.W.** Robert D. Waldron, Director, Research Enterprises, Scottsdale, Arizona.

**R.D.Wa.** Robert D. Wlkup, Department of Chemistry and Biochemistry, Texas Tech University.

**R.E.Th.** Roy E. Thoma, Oak Ridge National Laboratory, Oak Ridge, Tennessee.

**R.E.Tr.** Robert E. Treybal, deceased; formerly, Department of Chemical Engineering, New York University.

**R.F.** Roberto Fusco, Instituto Guiido Donegani, EniChem, Polymeric Materials Department, Novara, Italy.

**R.F.G.** Robert F. Goddu, Manager, Fibers and Film Research Division, Hercules, Inc., Wilmington, Delaware.

**R.F.R.** Reed F. Riley, Quantum Electronics Group, General Telephone and Electronics, Bayside, New York.

**R.G.P.** Robert G. Parr, Department of Chemistry, University of North Carolina.

**R.H.Bo.** Richard H. Boyd, Department of Chemical Engineering, University of Utah.

**R.H.Cr.** Robert H. Crabtree, Department of Chemistry, Yale University, New Haven, Connecticut.

**R.H.G.** Roy H. Gigg, Chemistry Division, National Institute for Medical Research, London, England.

**R.H.He.** Rolfe H. Herber, Department of Chemistry, Rutgers University.

**R.H.Ri.** Reinhard H. Richter, Manager, Polyurethane Research, Dow Chemical U.S.A., North Haven Laboratories, North Haven, Connecticut.

**R.H.W.** Robert H. Wentorf, Jr., General Electric Research Laboratory, Schenectady, New York.

**R.I.S.** Robert I. Stirton, Vice President (retired), Roger Williams Technical and Economic Services, Inc., Berkeley, California.

**R.J.C.** Richard J. Campana, Department of Botany and Plant Pathology, University of Maine.

**R.J.Gi.** Ronald J. Gillespie, Department of Chemistry, McMaster University, Hamilton, Ontario, Canada.

**R.J.Mo.** R. J. Motekaitis, Department of Chemistry, Texas A&M University.

**R.J.T.** Robert J. Thorn, Divisions of Chemistry and Materials Science, Argonne National Laboratory, Argonne, Illinois.

**R.Kr.** Robert Kreilik, Department of Chemistry, University of Rochester.

**R.K.Ba.** Robert K. Barnes, Research and Development Department, Chemicals and Plastics Operations Division, Union Carbide Corporation, South Charleston, West Virginia.

**R.K.E.** Russell K. Edwards, Argonne National Laboratory, Argonne, Illinois.

**R.K.MacC.** Robert K. MacCrone, Department of Physics, Rensselaer Polytechnic Institute.

**R.K.Mu.** Roger K. Murray, Jr., Department of Chemistry and Biochemistry, University of Delaware, Newark.

**R.Le.** Roberto Lee, Monsanto Company, St. Louis, Missouri.

**R.Len.** Richard Lennert, Department of Chemical Engineering, University of Texas, Austin.

**R.L.Bu.** Robert L. Burwell, Department of Chemistry, Northwestern University.

**R.L.Ma.** Rolland L. Mays, Materials System Division, Union Carbide Corporation, Tarrytown, New York.

**R.L.S.** Robert L. Scott, Department of Chemistry, University of California, Los Angeles.

**R.L.Wh.** Roy L. Whistler, Department of Biochemistry, Purdue University.

**R.M.Ke.** Robert M. Kelley, Department of Chemical Engineering, North Carolina State University, Raleigh.

**R.M.No.** Richard M. Noyes, Department of Chemistry, College of Arts and Sciences, University of Oregon.

**R.M.Th.** Robert M. Thorogood, Department of Chemical Engineering, North Carolina State University, Raleigh.

**R.O.P.** Robert O. Pohl, Laboratory of Atomic and Solid State Physics, Cornell University.

**R.P.** Robert Pecora, Department of Chemistry, Stanford University.

**R.P.B.** Richard P. Buck, Department of Chemistry, University of North Carolina.

**R.S.J.** Richard S. Juvet, Jr., Department of Chemistry, Arizona State University.

**R.S.K.** Robert S. Kapner, Department of Chemical Engineering, Cooper Union, New York.

**R.S.M.** Robert S. Mulliken, Institute of Molecular Biophysics, Florida State University.

**R.W.B.** Robert W. Belfit, Jr., Manager, Quality Standards, Quality Assurance Department, Dow Chemical Company, Midland, Michigan.

**R.W.Mu.** Royce W. Murray, Department of Chemistry, University of North Carolina.

**R.W.O.** Robert W. Odom, Charies Evans Associates, Redwood City, California.

**R.W.Si.** Richard W. Siegel, Materials Science Division, Argonne National Laboratory, Argonne, Illinois.

**R.W.Wo.** R. W. Woody, Department of Biochemistry, Colorado State University.

## S

**S.A.Co.** Steven A. Cohen, Senior Research Scientist, Polymers Business Group, Amoco Corporation, Alpharetta, Georgia.

**S.A.M.** Shelby A. Miller, Argonne National Laboratory, Argonne, Illinois.

**S.C.E.** Stephen C. Encleson, Technical Editor, Corporate Communications, Dow Chemical Company, Midland, Michigan.

**S.C.H.** Standish C. Hartman, Department of Chemistry, Boston University.

**S.E.Cr.** Sheldon E. Cremer, Department of Chemistry, Marquette University.

**S.E.Wh.** Sidney E. White, Department of Geology and Mineralogy, Ohio State University.

**S.F.** Sherman Fried, Chemistry Division, Argonne National Laboratory, Argonne, Illinois.

**S.F.P.** Stephen F. Perry, Exxon Research, Florham Park, New Jersey.

**S.F.Y.** S.F. Yates, Allied Signal Research Center, Inc., Des Plaines, Illinois.

**S.G.S.** Stephen G. Simpson, Department of Chemistry, Massachusetts Institute of Technology.

**S.G.T.** Stanley G. Thompson, Senior Staff Member, Lawrence Berkeley Laboratory, University of California, Berkeley.

**S.H.B.** Simon H. Bauer, Department of Chemistry, Cornell University.

**S.H.G.** Sidney H. Goodman, Manager, Materials Products Department, Technology Support Division, Hughes Aircraft Co., Culver City, California; Senior Lecturer, Department of Chemical Engineering, University of Southern California.

**S.H.W.** Samuel H. Wilen, Department of Chemistry, City University of New York.

**S.I.S.** Stanley I. Sandier, Department of Chemical Engineering, University of Delaware.

**S.J.** S. Jafarey, Behlen Laboratory of Physics, University of Nebraska.

**S.J.Y.** Samuel J. Yosim, Atomics International Division, North American Rockwell, Canoga Park, California.

**S.Ki.** Stanley Kirschner, Department of Chemistry, Wayne State University.

**S.K.F.** Sheldon K. Friedlander, Department of Chemical Engineering, University of California, Los Angeles.

**S.Ma.** Stephen Mann, School of Chemistry, University of Bath, United Kingdom.

**S.M.K.** S. Morris Kupchan, deceased; formerly, Department of Chemistry, University of Virginia.

**S.N.G.** Stanley N. Gershoff, Department of Nutrition, Harvard School of Public Health, Boston, Massachusetts.

**S.Ro.** Sidney Ross, Department of Chemistry, Rensselaer Polytechnic Institute.

**S.S.** Sidney Siggia, Department of Chemistry, University of Massachusetts.

**S.Sir.** S. Sircar, Senior Research Associate, Air Products and Chemicals, Inc., Allentown, Pennsylvania.

**S.S.B.** Steven S. Brown, National Oceanic and Atmospheric Administration Aeronomy Laboratory, Boulder, Colorado, and Cooperative Institute for Research in the Environmental Sciences (CIRES), University of Colorado, Boulder.

**S.S.J.** Silvia S. Jurisson, Department of Chemistry, University of Missouri, Columbia.

**S.S.K.** Sudhir S. Kulkarni, Corporate Research Center, UOP Inc., Des Plaines, Illinois.

**S.S.N.** Samuel S. Naistat, Department of Chemistry, Stephen F. Austin State College, Nacogdoches Texas.

**S.W.F.** Steven W. Feldberg, Department of Applied Science, Brookhaven National Laboratory, Upton, New York.

**S.W.Pe.** S. William Pelletier, Director, Institute for Natural Product Research, University of Georgia, Athens.

# T

**T.C.St.** Theressa C. Stadtman, Department of Health and Human Services, National Heart, Lung and Blood Institute, National Institutes of Health, Bethesda, Maryland.

**T.C.W.** Thomas C. Waddington, deceased; formerly, Department of Chemistry, University of Durham, England.

**T.D.G.** Thomas D. Getman, Department of Chemistry, Northern Michigan University.

**T.D.Wi.** Todd D. Williams, Department of Chemistry, University of Kansas, Lawrence.

**T.F.Y.** Thomas F. Young, deceased; formerly, Department of Chemistry, University of Chicago.

**T.G.B.** Thomas G. Back, Department of Chemistry, Faculty of Science, University of Calgary, Alberta, Canada.

**T.Hu.** Tomas Hudlicky, Department of Chemistry, Illinois Institute of Technology.

**T.H.D.** Theodore H. Dexter, Research Supervisor, Inorganic Chemistry, Hooker Chemical Corporation Research Center, Niagara Falls, New York.

**T.J.Gr.** Thomas J. Greytak, Department of Physics, Massachusetts Institute of Technology, Cambridge.

**T.J.M.** Tobin J. Marks, Department of Chemistry, Northwestern University.

**T.J.Me.** Thomas J. Meade, Division of Chemistry and Chemical Engineering, California Institute of Technology.

**T.Ki.** Toichiro Kinoshita, Laboratory of Nuclear Studies, Cornell University.

**T.Su.** Terry Surles, Limnetics, Inc., Milwaukee, Wisconsin.

**T.W.** Thomas Wartik, Department of Chemistry, Pennsylvania State University.

**T.Wi.** Therese Wilson, Biological Laboratories, Harvard University.

**T.W.Ha.** Theo W. Hansch, Department of Physics, Stanford University.

# V

**V.A.B.** Victor A. Bloomfield, Department of Chemistry, University of Minnesota.

**V.B.** Vincenzo Balzani, Departimento de Chimica, Università di Bologna, Italy.

**V.C.C.** Valeria Cizewski Culotta, Associate Professor, Johns Hopkins University School of Public Health, Department of Environmental Health Sciences, Baltimore, Maryland.

**V.G.A.** Vincent G. Allfrey, Department of Cell Biology, Rockefeller University.

**V.L.P.** Vincent L. Pecoraro, Department of Chemistry, University of Michigan.

**V.M.** Vincent Madison, Department of Medicinal Chemistry, School of Pharmacy, University of Illinois, Chicago.

**V.V.L.** V. V. Levasheff, American Potash and Chemical Corporation, Whittier, California.

**V.W.U.** Vincent W. Uhl, Department of Chemical Engineering, University of Virginia.

# W

**W.A.Li.** Werner A. Lindenmaier, Hoffmann-LaRoche, Inc., Nutley, New Jersey.

**W.A.S.** William A. Steele, Department of Chemistry, Pennsylvania State University.

**W.A.W.** W. A. Wakeham, Department of Chemical Engineering and Chemical Technology, Imperial College of Science and Technology, London, England.

**W.B.B.** Warren B. Blumenthal, Blumenthal-Zirconium, North Tonawanda, New York.

**W.Ch.** Wallace Chinitz, Department of Mechanical Engineering, Cooper Union.

**W.C.L.** W. C. Lineberger, Joint Institute for Laboratory Astrophysics, National Institute for Standards and Technology, University of Colorado.

**W.E.C.** William E. Cooley, Winton Hill Technical Center, Proctor and Gamble Company, Cincinnati, Ohio.

**W.E.Ku.** Wayne E. Kuhn, Professional Engineer, Portland, Oregon.

**W.E.Kup.** Walter E. Kupper, Manager, Technical Marketing Services, Mettler Instrument Corporation, Hightstown, New Jersey.

**W.F.F.** W. F. Furter, Dean of Graduate Studies and Research, Royal Military College of Canada.

**W.F.J.** William F. Jaep, Central Research Department, Experimental Station, E. I. DuPont de Nemours and Company, Wilmington, Delaware.

**W.H.** Wilbur Hague, Oxy Metal Industries Corporation, Warren, Michigan.

**W.Hap.** William Happer, Department of Physics, Columbia University.

**W.H.Gr.** William H. Gross, Technical Editor, Corporate Communications, Dow Chemical Company, Midland, Michigan.

**W.H.H.** Walter H. Hartung, formerly, Professor of Pharmaceutical Chemistry, Medical College of Virginia.

**W.H.L.** William H. Laarhoven, Department of Organic Chemistry, University of Nijmegen, The Netherlands.

**W.H.M.** William H. Miller, Department of Chemistry, University of California, Berkeley.

**W.H.S.** Wendell H. Slabaugh, deceased; formerly, Department of Chemistry, Oregon State University.

**W.J.Ge.** Walter J. Gensler, deceased; formerly, Department of Chemistry, Boston University.

**W.J.H.** Walter J. Hamer, Institute for Basic Standards, National Bureau of Standards.

**W.J.LeN.** William J. Le Noble, Department of Chemistry, State University of New York, Stony Brook.

**W.L.H.** W. Lincoln Hawkins, Bell Laboratories, Murray Hill, New Jersey.

**W.Mi.** Wesley Minnis, formerly, Assistant to the Director of Research and Development, Allied Chemical Corporation, New York.

**W.O.M.** W. O. Milligan, Robert A. Welch Foundation, Houston, Texas.

**W.R.Bu.** William R. Burg, W. R. Burg and Associates, Inc., Lilburn, Georgia.

**W.R.El.** Walther R. Ellis, Jr., Division of Chemistry and Chemical Engineering, California Institute of Technology.

**W.R.Ep.** W. R. Epperly, Exxon Research and Engineering, Annandale, New Jersey.

**W.R.M.** William R. Marshall, Jr., Associate Dean, College of Engineering, University of Wisconsin.

**W.R.Se.** W. Rudolf Seitz, Department of Chemistry, College of Engineering and Physical Sciences, University of New Hampshire.

**W.R.V.** Wyman R. Vaughan, Head, Department of Chemistry, University of Connecticut.

**W.R.W.** William R. Wilcox, Department of Chemical Engineering, Clarkson College of Technology.

**W.S.** Wolfram Saenger, Max Planck Institut für Experimentelle Medizen, Gottingen, Germany.

**W.S.L.** W. S. Lyon, Analytical Division, Oak Ridge National Laboratory, Oak Ridge, Tennessee.

**W.W.E.** W. W. Epstein, Department of Chemistry, University of Utah.

**W.W.W.** W. W. Watson, Professor Emeritus of Physics, Yale University.

**W.W.We.** Wesley W. Wendlandt, Department of Chemistry, University of Houston.

**W.Z.H.** William Z. Hassid, deceased; formerly, Department of Biochemistry, University of California, Berkeley.

# Y–Z

**Y.U.** Yasuhiro Uozumi, Laboratory of Complex Catalysis, Institute for Molecular Science (IMS), Okazaki, Japan.

**Z.R.W.** Zelda R. Wasserman, Department of Central Research and Development, E. I. du Pont de Nemours and Company, Wilmington, Delaware.

# Index

The asterisk indicates page numbers of an article title.

## A

# B



# C

# D

# F



# G

# H

# I



# J, K



# L

# P

# Q



# R

# S

# T

# U



# V



# W



# X, Y, Z